Guide to English Language Teaching

2007

KEYWAYS
PUBLISHING

Guide to English Language Teaching 2007

This Fourth Edition
Published 2007 by: Keyways Publishing Ltd
PO Box 100, Chichester, West Sussex, PO18 8HD, UK
Tel: +44 (0)1243 576444
Fax: +44 (0)1243 576456
Web: www.CareersInELT.com
Email: info@KeywaysPublishing.com

Fourth edition 2007

Editor Simon Collin
Editorial Director Peter Collin

Design Keyways Publishing

Section Acknowledgments:
Training: Mary Ellen Butler-Pascoe, Susan Matson.
Working Overseas: The British Council, James Beetham, Roger Chrisman, Guillaume Gazembetti, Chung Han, Michael Howard, Alistair Kennedy, Dianne James, Ian McNamara, Martha Oral, Edward Prados, Mark Sigrist, Yuri Tambovtsev, Celia Thompson, Mark Warschauer, Colin Underwood, Paul Vreeland, Joe Ziegler, Brian Garity, Tony Donovan.
Career Development: Tina Carver, Lin Lougheed, Andy Martin, Mark Peterson, Mona Scheraga, Barbara Stipek.

British Library Cataloguing-in-Publication Data

A catalogue record for this book is available from the British Library

ISBN 1-898789-17-7

Printed by Nuffield Press, UK

NOTE: Every effort has been made to ensure that the information contained in this book is accurate and up-to-date at the time of going to press. However, details are bound to change, especially with regard to information that refers to visa requirements, state certification, exchange rates and working conditions. The publisher is not responsible for any problems or disappointments encountered during your training, qualifications or travels.

If your school is mentioned in this guide and you would like to update your information for the next edition of the guide, please visit our website **www.CareersInELT.com** where you can update your information using the forms online.

Guide to

Contents

continued: WORKING AROUND THE WORLD

Foreword

Being an English language teacher can be whatever you want it to be: a passport to world travel; a stable and fulfilling career; a way of helping others or the first step to owning your own business. This guide will help you choose how best to get qualified, find a job, develop professionally and travel the world. World demographics will result in sustained demand for teachers of English for at least the next twenty years – that's about as secure a job market as you will find these days.

Like it or not, English has become the world's language and it is growing in importance every day. Globalization and developments in communications are fuelling demand for the language around the world. About 85% of the Internet's web-pages are in English and everyone has acknowledged that English is essential for efficient use of the web – recently the governments of Japan and Korea announced policies to encourage all of their working populations to acquire a basic level of English.

If you are looking for a way to travel and work overseas, then teaching English is an obvious path. There are teachers from the UK, USA, Canada, Australia and New Zealand working as English teachers in almost every country in the world. Hundreds of thousands of young people have used teaching English as a means to see the world and understand other cultures.

By the year 2050 it is estimated that half the world's population will be competent users of English. This learning frenzy will not take place because of some love for the language, but due to the realization that people will be left behind in work and social situations unless they can communicate with the rest of the world.

And even in native-English-speaking countries, the demand for learning English is high: for example, in the UK, there are hundreds of ELT schools that gain most of their student intake from overseas visitors combining lessons with travel – from young learners visiting with a school for a week to professionals learning business English for their job development. And in the USA, teaching people to speak English accounts for almost half of all adult education!

This book will show you how you can work in English language teaching: if you are looking for a stable, long term career; to give something back to your community or to the rest of the world through voluntary work; or a good way to pay for your round-the-world adventure, this guide will explain how to do it through teaching English.

Using This Guide

This *Guide* is designed for use by people thinking of becoming teachers as well as those who have started training and experienced teachers. Although you can read the *Guide* from cover to cover, you may find it easier to start with the section that is most suited to your own particular needs.

One: Training to Teach (p.12) – is aimed at readers who are thinking of becoming teachers of English as a second or foreign language and who are interested in getting qualified to teach. We provide an overview of the different training courses on offer including short term introductory and certificate courses.

Two: Finding a job (p.24) – explains how best to find a job as a newly qualified teacher. This covers job agencies, applying directly to schools and using the Internet. Country-specific information and a directory of schools to approach for a job are listed in section 4.

Three: Preparation – before you go (p.34) – covers the topics that you need to address before you set off and travel the world! This includes insurance, tax issues and how to handle problems when you reach your school.

Four: Working Around the World (p.42) – is written for readers who want to explore the exciting prospects of combining teaching with travel. The section provides detailed profiles of over 100 countries, together with their work prospects, regulations (including tax and visas), contact details of the local embassy and British Council office and, most importantly, a directory of the main job prospects: the schools in the country together with their contact details.

Five: Career Development (p.176) – is the section intended for language teachers with experience who want to develop their skills and professional lives. We include directories of Diploma/Certificate courses, universities that provide Master's courses and doctorates. We also cover specialist English subjects, including young learners and business English. Lastly, we cover school management and how to get published.

Appendix (p.230) contains a list of useful websites, addresses for teacher-associations, schools, publishers, suppliers, travel-related sites, British Council offices, Canadian embassies and US embassies around the world.

Acronyms

For many new teachers, the first lesson is to learn the acronyms that are used in the profession. Here are some of the more common acronyms.

Basic

CALL: Computer Assisted Language Learning.

DOS: Director of Studies

EAP: English for Academic Purposes–The study or teaching of English with specific reference to an academic (usually a university- or college-based) course.

ECIS: European Council of International Schools

EFL: English as a Foreign Language–English language programmes in countries where English is not the common or official language. It is used in American university programmes where international students study English although the use of the word "foreign" is now avoided in some schools because of its xenophobic connotations.

ELICOS: English Language Intensive Courses for Overseas Students. Acronym used in Australia to refer to EFL.

ELL: English Language Learner–a term that has become popular in California designed to replace the acronym "LEP" (see below) which many teachers felt to be pejorative.

ELT: English Language Teaching or Training– a term coined in the UK and designed to replace EFL. It is in use around the world but has yet to catch on in the USA.

EOP: English for Occupational Purposes

ESL: English as a Second Language–English language programmes in countries where English is the dominant or official language. In the U.S. programmes designed for non-English-speaking students in the U.S. are ESL programmes; in the UK, this term is not used - ESOL is used instead.

ESOL: English to Speakers of Other Languages – in the U.S. this acronym is often used to describe elementary and secondary English language programmes; in the UK it refers to a specific type of English language teaching for newcomers to the country who might have complex requirements (perhaps asylum seekers or refugees) and learn English to integrate and take further education. It is sometimes used to distinguish ESL classes within adult basic education programmes.

ESP: English for Specific Purposes–a term that refers to teaching or studying English for a particular career (like law or medicine) or for business in general.

IEP: Intensive English Program–refers to an intensive course designed to help non-English-speaking students prepare for academic study at a university or college.

L1: First language (**L2**: Second language)

LEP: Limited English Proficient–a term used for many years to designate children in the school systems for whom English was not their first language. Now replaced by terms like ELL.

Teacher Training

ACE: Access Certificate in Education. An entry-level training certificate being piloted by Pitmans/City and Guilds in the UK.

CELTA: Certificate in English Language Teaching to Adults. Widely-recognised certificate course developed in the UK by University of Cambridge ESOL (UCLES) and RSA. DELTA is the advanced Diploma course.

CELTYL: Certificate in English Language Teaching to Young Learners. A certificate course developed in the UK by University of Cambridge ESOL (UCLES); normally taken as an add-on option with CELTA.

Cert.TEB: Certificate in Teaching English for Business. A certificate course developed by LCCI for teachers specialising in business English (see LCCIEB, below)

Cert.TESOL: Certificate in TESOL. Widely-recognised certificate course developed in the UK by Trinity College London. The advanced version for experienced teachers is the Lic.Dip.TESOL.

Cert.TEYL: Certificate in Teaching English to Young Learners. A certificate course extension developed by Trinity College London; normally taken as an add-on option with Cert.TESOL.

COTE: Certificate for Overseas Teachers of English. A certificate-level course developed by University of Cambridge ESOL (UCLES) - now replaced by ICELT

DELTA: Diploma in English Language Teaching to Adults. The advanced (post-experience) qualification from University of Cambridge ESOL (UCLES)

FTBE: (Further Certificate for Teachers of Business English) - awarded by LCCIEB

ICELT: In-Service Certificate in English Language Teaching - certificate for teachers awarded by CambridgeESOL, designed to replace COTE, and used to help improve a working teacher's skills

LTCL.Dip.TESOL: Licentiate Diploma in TESOL. The advanced (post-experience) qualification from Trinity College London

TEFL: Teaching English as a Foreign Language – a term that refers to teacher training programmes in EFL.

TESL: Teaching English as a Second Language–a term that refers to teacher training programmes in ESL.

TESOL: Teaching English to Speakers of Other Languages–a term that is used to distinguish English language teaching as a professional activity that requires specialized training. Also refers to the teacher examinations developed by Trinity College London (Cert.TESOL and LTCL.Dip.TESOL)

TKT: Teaching Knowledge Test - flexible award developed by Cambridge ESOL to provide and test the basic teaching requirements of any teacher.

Associations

AAIEP: American Association of Intensive English Programmes is a group of university and college-based intensive English programmes

ACELS: Advisory Council for English Language Schools in Ireland

ARELS: Association of Recognised English Language Schools in the UK

BASELT: British Association of State English Language Teaching schools in the UK

CRELS: Combined Registered English Language Schools of New Zealand

ELTAs: English Language Teacher Associations groups for teachers in Germany and Austria

IATEFL: International Association of Teachers of English as a Foreign Language is based in the UK with members around the world.

JALT: Japanese Association for Language Teaching

JET: Japanese Exchange and Teaching Programme

NABE: National Association for Bilingual Education is an association that represents the interests of bilingual teachers in the USA.

NEAS: National ELT Accreditation Scheme, for course providers in Australia

RELSA: Recognised English Language Schools Association. The organisation of independent language schools in Ireland

TESL: Teaching English as a Second Language, Canada – national federation of teachers and providers in Canada

TESOL: US-based international association of teachers of English as a second or foreign language. There are regional affiliates and many countries have their own affiliated associations.

Exams/Exam Boards

Assessment and Qualifications Alliance (AQA) (formerly AEB & NEAB) Certificate in English Language Skills (ESOL), JET SET, range of graded exams for ESOL. www.aqa.org.uk

Cambridge *see* University of Cambridge

Chauncey Group (part of ETS, below) US-based group that administers the TOEIC student examination. www.toeic.com

ESB (English-Speaking Board International) Oral assessments in (spoken) English www.esbuk.demon.co.uk

ETS (Educational Testing Service) Based in Princeton, NJ, the world's biggest examination board, administrators of the TOEFL student examination. www.ets.com

City & Guilds Based in the UK, offers a wide range of student exams and the ACE certificate for teachers.

Institute of Linguists Diploma in English for International Communication. www.iol.org.uk

International English Language Testing System (IELTS) Managed by UCLES, the British Council and IDP Australia for academic and vocational English. www.ielts.org

London Chamber of Commerce and Industry Exam Board (LCCIEB) Range of business and specialist English examinations. www.lccieb.com

London Examinations – Edexcel International London Tests of English range of exams graded from basic to proficient. www.edexcel.org.uk

Pitman Qualifications Range of general ESOL exams, including spoken English and business English. www.pitmanqualifications.com

RSA Royal Society of Arts is a body that works with UCLES (*see below*).

Trinity College London Responsible for the Certificate in TESOL and the Licentiate Diploma in TESOL examinations. www.trinitycollege.co.uk

UCLES (University of Cambridge Local Examinations Syndicate) Syndicate of local examination centres around the world that administer the University of Cambridge ESOL examinations. www.ucles.org.uk

University of Cambridge ESOL (administered locally by UCLES) is a British-based organization responsible for developing a number of important English language exams (including PET, FCE, CAE) and teacher training programmes including TKT, CELTA, CELTYL, and DELTA examinations. www.cambridgeESOL.org

Training To Teach

This chapter offers a comprehensive guide to achieving one of the teaching credentials that will be essential for your working career. Although the profession is not governed by a single body or a single qualifying entrance exam, there are many ways to become a qualified teacher – the following sections explain how.

Whether you plan to teach in your own country or travel overseas, you will need to consider the type of training you will require in order to be employed; opinions within the teaching field are varied on this question.

Some teachers argue that you don't need a TEFL Certificate to teach English around the world. Others argue that a Master's degree is an unnecessary waste of time and money and that it is easy to get a job at a school or college without one. Teachers of this mindset are convinced that the teacher training industry is some sort of vast conspiracy set up in order to fleece unsuspecting students of their money by conning them into taking courses for qualifications which are almost useless. The reality is that you can get a job teaching English without a formal qualification, though it is increasingly difficult.

And another consideration: Do you really want to start teaching with no idea of what you are doing? Is this fair on you? Is it fair on your students? There are gifted teachers born with a natural ability, but even these teachers need guidance and feedback from peers and mentors.

This section of the *Guide* is designed to give you an idea of the courses and programmes on offer and which qualification suits your needs best when training to be a teacher. For post-experience courses (i.e. once you have taught for several years and want to develop your career), see page 176.

The English language teaching profession is not governed by a single body. Instead of a single qualifying entrance exam, there are multiple routes to qualification as an English language teacher. You can start with a simple introductory course – often designed to help you decide if this is the career for you – then move on to a basic certificate course. You can then teach around the world, before perhaps working on a diploma, Master's degree or postgraduate qualification.

Training and Qualifications

There are several ways to become a qualified English language teacher and your choice of qualification route mainly depends on where and who you want to teach. The time you have to train and the cost of training are also important considerations.

Can anyone be an English language teacher?

It is still possible to get a job teaching EFL/ESL without any formal qualification, but this is becoming increasingly rare. There is a global shortage of teachers, but unqualified teachers are only recruited locally and the pay is usually poor. In major cities and tourist centres only qualified teachers will be taken on.

Is there a standard TEFL qualification?

No. There are many routes to qualification as a TEFL teacher ranging from intensive 3-weekend Certificate courses to two-year Master's Degrees. In the USA and some of the Far East, a good first degree is adequate; whereas in Europe, the Commonwealth and South America teachers are preferred to have taken a TEFL course. Of these TEFL courses, the two best-known practical teaching certificates are CELTA and Certificate in TESOL, validated by University of Cambridge ESOL (administered around the world by UCLES) and Trinity College London respectively. Once you have experience as a teacher, you can move on to a post-experience course such as the Diploma in ELT (from Cambridge) or the Licentiate Diploma in TESOL (from Trinity College London) – see page 176 for details of these courses, MAs and doctorates in ELT.

Am I too old to teach ?

Many middle-aged people faced with redundancy or early retirement decide to embark on a career in English language teaching. Although age can be a barrier in some parts of the world, most schools welcome the added depth of life experience that a more mature teacher can bring to the classroom.

I'm not sure if I want to commit to long course. What should I do?

There are a number of introductory courses in teaching EFL/ESL offered by various institutions including Oxford Seminars (www.oxfordseminars.com), if you think that you might not want to commit yourself to a career in teaching, but many employers may not consider you qualified unless you have at least a Certificate (see below).

Choosing a Course

How do you choose a Certificate course? There are hundreds of schools, several accredited courses and various levels and ways to study. However, they will all cost you money and take a couple of months (or longer) of study to complete.

Those teachers who want to grow, personally and professionally, are turning to TEFL Certificate courses. But not every course will be right for you – and you can only find a suitable course by asking the right questions. Here are some essential questions to ensure you choose the best school for you:

What are the qualifications of your trainers?

A Diploma or Master's degree (MATESOL) for all or most trainers is a good sign. If those who will be your role models only have a TEFL Cert. themselves, look deeper into the curriculum offered and ask about their years of experience.

While a Master's is not a guarantee of good training, it does suggest deeper understanding about the theory behind the methods and a clear commitment to the field.

Have any complaints ever been filed against your school?

If you are studying in the USA, any complaints must, by State licensing agencies, be reported – justified or not. The nature of the complaint will tell you what the programme may not publicly advertise.

What about the duration of your programme and practicum?

Ask about both classroom time and practicum hours. Intensive programmes can run 120 to 150 hours. The more practicum hours, closely supervised, the better. It is very difficult for a teacher to learn from his or her mistakes if given only three or four opportunities to teach. The ideal programme will allow you chances to work with beginning, intermediate, and advanced speakers of English. Note: part-time programmes lack the immediacy of feedback seen in intensive, full-time programmes.

What is your hire/placement track record?

Proprietary programmes are required to keep close records on how many graduates get jobs. The vast majority, over 80%, should be getting jobs within two months of hire, given the intense market for EFL/ESL teachers.

Tell me about your job network

At minimum, a school should have an extensive list of employers throughout the world and access to Internet postings. Some programmes also offer job counselling, CV advice, and a means of providing feedback from graduates who are now teaching overseas.

In the USA, are you licensed by any state agency?

If you are studying in the USA, this is an important point: licensure for proprietary schools does not mean that you will qualify for a public school job, but rather that the school meets stringent state standards. For example, in California, these include standards for qualifications of trainers and administrators, strict record keeping, and a refund procedure.

How are trainees tested?

If you do poorly at written tests, you won't perform well in a TEFL programme that requires passing a test for graduation. All programmes should have highly specific criteria for evaluating you, particularly in the area of practicum teaching. These should be in written form, to help prevent any bias on the part of an observer. And if you do find there is a chance of failing, you'll want to know it early on, while there is still a chance to get a refund or improve your performance.

What kind of coursework and topics do you have?

A comprehensive programme will at least introduce you to the basics of classroom management, lesson planning, student evaluation, and techniques for teaching grammar, reading, writing, listening, speaking, and pronunciation.

What is your trainee/trainer ratio?

The programme should allow for easy exchanges in class work and for closer counselling as needed. Classes of more than 15 trainees may make it more difficult to get the best from your teacher. Your practical classes should be observed by more than one trainer, as opinions on your strengths and weaknesses will vary.

How do you get your students for the practicals?

The ideal is students who really do not know the material you will be teaching, so that their responses will be genuine. Make sure you will have a guaranteed number of students for your practical sessions in order to do pair and group work.

How are refunds handled?

Many unanticipated events may force a drop-out – family crisis, illness, or a mismatch of trainee and programme goals. The best programmes allow you to leave within the first few days with few, if any, financial penalty. Others will pro-rata the amount coming back to you depending on the time spent in the programme.

What will I receive on completion?

Trainee teachers disperse quickly after a course ends. An efficient programme will make sure that on graduation day, they receive a certificate, a letter of recommendation, and, perhaps, a transcript of courses taken and the results.

How much will it all cost?

Of course you will want to know about course tuition, but don’t forget to add in extras such as books and day trips, and accommodation.

Training: Types of Course

Before you start to teach, you should take an initial teacher training course; when applying for a job, some school do not require any qualifications of their teachers, but with a qualification you will find it easier to get a job, easier to manage classes and organise lessons and essential if you want to offer good training for you and your prospective students.

Introductory Courses

Introductory courses are designed for prospective teachers who want to experience teaching before making the decision to train formally and obtain qualifications. The courses usually last from a week up to four weeks. These are not certificate courses and you will not be regarded as a qualified teacher after completion of your Introductory course. You will, however, be able to say that you have had some experience of English language teaching. Some schools require trainee teachers to take an introductory course before starting a Certificate course.

Certificate courses

The majority of trainee teachers who would like to teach take a certificate course in teaching EFL/ESL. A certificate course provides basic grounding in teaching, lesson development, managing students and, in particular, the different stages of teaching English. Certificate courses are great for trainee teachers who have no experience of teaching – perhaps have a degree or experience in a totally different field – and want to ensure that they are well prepared for their first 'real' class (as a trainee teacher, part of your certificate course will be to experience practical teaching in a classroom). Most certificate courses are intensive and run over a four or six week period. You can take distance-study courses or take a certificate course on a part-time basis over several months.

Once you have a certificate in teaching English, you can start to apply for jobs at schools around the world. Although there is no single qualification, there are two main certificate courses: University of Cambridge CELTA and Trinity College London Cert.TESOL. Either one is well recognised around the world.

Aside from these two main certificate qualifications, many school, colleges and universities have their own certificate course developed in-house. Make sure that you ask the course provider about their qualification and if they will provide assistance finding a job once you have completed your course.

As well as the basic certificate course, you can also study for additional modules that will gain you experience and qualifications in specialist subjects (see also the section on ESP); for example, teaching young learners or teaching business English.

Certificate Qualification Providers

When you look at the different courses on offer you will find they provide a range of different certificate courses. There are a number of certificates awarded by different organisations - they have different names, provide emphasis on different types of teaching but all have the same goal to ensure that you are able to manage your first class teaching English. The main organisations offering certificate awards are:

University of Cambridge ESOL

TKT

The TKT (Teaching Knowledge Test) is a new award from Cambridge. Its focus is on the core teaching skills and requirements needed by teachers of English and provides a foundation for new or prospective teachers who could then progress on to CELTA or ICELT awards.

CELTA

The Cambridge CELTA (Certificate in English Language Teaching to Adults) is the longest established qualification for English language teachers and has over 8,000 student-teacher enrolments per year. It was developed by the University of Cambridge ESOL and managed around the world by the University of Cambridge Local Examination Syndicate (UCLES).

The CELTA is a pre-experience course, usually run on a four-week intensive basis, though some schools do run part-time courses. CELTA is run in over 40 countries around the world and is externally validated by University of Cambridge ESOL; a course normally costs between £700-1000 ($1300-1900).

A CELTA course is normally a 4-week intensive programme, with an examination at the end (the fee for the examination is normally included as part of the course fee). There are a few part-time CELTA courses, but these are very much the exception.

CELTA courses are, like the Trinity College London Certificate course (see next page), based on practical teaching – with observed teaching practice integral to any course. Pass rates are generally high, since applicants are carefully selected: when applying, you will be interviewed and might have to take a language awareness test. Candidates are assessed over the duration of the course (with moderation by external examiners), and there is no final examination. Successful candidates are awarded a Pass, a Pass B or a Pass A.

Many candidates are recent university graduates, but the course attracts a wide range of applicants, including mature professionals looking for a change of career or even an alternative to retirement. Although certain levels of previous academic achievement are recommended by the administrators, there are no specific entry requirements, and admission is at the discretion of the course provider. Most centres will accept non-native speakers who can demonstrate an acceptable level of written and spoken English.

The course is available throughout the year in more than 50 countries, and is taken by more than 10,000 candidates annually. It consists of a minimum of 114 hours, and includes an element of teaching practice to genuine ESOL students.

CELTYL

Cambridge CELTYL (Certificate in English Language Teaching to Young Learners) is administered by the University of Cambridge Local Examination Syndicate (UCLES). This course provides a course in teaching age groups of 5–10, 8–13 or 11–16. Some candidates are new to TESOL, others are practising teachers who wish to develop their skills. Most of the details given above about the CELTA also apply to CELTYL. As an alternative to the full CELTYL course, successful CELTA candidates can take a Young Learners extension course (CELTYL Ext).

COTE

Cambridge COTE (Certificate for Overseas Teachers of English) is a pre-experience course that is taken as an alternative to CELTA by non-native-speakers who already have teaching experience in their own language. You will probably still see this in brochures and advertisments, but it has now been replaced by the similar ICELT examination (below).

ICELT (In-service Certificate in English Language Teaching)

An in-service award for practising teachers who want to improve their skills in the context in which they are currently working. Candidates taking ICELT courses are normally assessed teaching their own classes. A modular option enables courses to focus on language for teaching purposes only or for the full award to be gained in two phases. ICELT replaced the broadly similar COTE (Certificate for Overseas Teachers of English) from the start of 2004.

University of Cambridge
ESOL Examinations
1 Hills Road
Cambridge, CB1 2EU, UK
Tel: + 44 (0)1223 553355
Fax: +44 (0)1223 460278
Email: esolhelpdesk@ucles.org.uk
Website: www.cambridgeESOL.org

Trinity College London

Certificate in TESOL

The Certificate in TESOL (Teaching English to Speakers of Other Languages) – often referred to as *certTESOL* is administered by Trinity College London. It has over 4,000 student-teachers enrolments per year and is, like the CELTA (above) a well-recognised Certificate in practical ELT teaching. Trainees are expected to take courses in a foreign language, to understand the difficulties in teaching a foreign language. No two courses are the same, as course designers can introduce their own ideas and elements, so you will need to verify the details of the specific course. Cert.TESOL courses are usually full-time intensive and last between four to six weeks, although part-time courses are available at a few centres. The costs are very similar to a CELTA, £700-1000 ($1300-1900).

Certificate in TEYL

The Certificate in Teaching English to Young Learners is administered by Trinity College London. This is a pre-experience course that is normally taken as an add-on module to a Cert.TESOL course and is designed for teachers who would like to specialise in teaching to young learners of English (which is a strong market that is growing rapidly, especially in Italy, Korea, Japan, and Greece).

Trinity-ARELS One-to-One Certificate

A one-week/25-hour course for teachers of individual students on a one-to-one basis. Most candidates have an initial certificate (for example CertTESOL), but relevant experience is in some cases acceptable as an alternative. Availability is limited at present, but may increase.

Trinity College London
89 Albert Embankment
London , SE1 7TP, UK
Tel: +44 (0)20 7820 6100
Fax: +44 (0)20 7820 6161
Email: info@trinitycollege.co.uk
Website: www.trinitycollege.co.uk

Pitman / City & Guilds

ACE

The Access Certificate in English language teaching (ACE) is a Pitman/City & Guilds qualification developed with Manchester University in the UK. It offers a initial-level certificate for people who have never taught before or for teachers with some experience who want a formal qualification or for teachers experienced in teaching another subject who want to switch to teaching English. The training is in two parts covering theory and practical teaching activities.

City & Guilds
1 Giltspur StreetLondon, EC1A 9DD, UK
Tel: 020 7294 2800
Fax: 020 7294 2400
Email: enquiry@city-and-guilds.co.uk
Website: www.city-and-guilds.co.uk

School for International Training (SIT)

SIT TESOL Certificate

SIT is a US-based international organisation offering two qualifications for teachers of English (TESOL). The SIT Certificate in Teaching English to Speakers of Other Languages is available from a few dozen test centres around the world; the course of study is normally offered as an intensive four-week course. SIT also offer an advanced diploma (the IDLTM - International Diploma in Language Teaching Management) as a graduate-level diploma.

SIT
Kipling Road, P.O. Box 676,
Brattleboro, Vermont 05302-0676, USA
Tel: (802) 257-7751
Fax: (802) 258-3248
Website: www.sit.edu

Private Language School Certificates

Some chains of private language schools offer their own 'in-house' training courses and certificates. As a general rule, if these are free and likely to lead to employment or promotion within the organisation, they may be well worth considering. If, however, fees are payable, it is not really advisable to accept any, often vague, assurances that the certificate is internationally-known or respected or that it is equivalent to a CELTA or Cert.TESOL.

Ask to see a list of employers who accept the qualification and contact them to ask if they really do consider this to be a full equivalent of an independent, mainstream certificate.

The best private language school TESOL courses leading to in-house certificates tend to be offered as a supplement to mainstream qualifications (for example, a refresher course) rather than as an alternative.

University Certificate Courses

University Certificate courses are usually short courses running from one to six months and are an alternative to the Cambridge or Trinity certificates (above). These are the most common pre-experience courses available, with almost every university and college in the UK offering some form of English language teaching course. There are also 'in-service' certificate courses available for those teachers who have classroom experience but no formal qualifications.

Distance Learning Courses

Some training courses in English language teaching are offered on a distance basis, though these are not generally popular with employers unless they include an observed period of teaching practice and are externally validated. But if you are already teaching English abroad, they can be a viable option.

Certificates for Teaching English for Business

There are a number of different examinations for teachers who would like to train to teach English for business: Chauncey group, LCCIEB, Pitman and University of Cambridge ESOL all offer specialist qualifications. These are covered in detail on page 176 in the section on Professional Development.

Do I need a degree?

In some countries such as Japan and Korea a university degree is a legal requirement (the requirement is for a degree in any subject rather than in TESOL/TEFL). In Spain some schools look for teachers to have a degree because government contracts demand it. We cover degrees, diplomas and post-graduate qualifications in chapter 5.

University Degree Courses

Although qualifications in TEFL/TESL and TESOL are usually offered at the postgraduate level, there are a few programmes which enable undergraduate students to study for a Bachelor's degree in this area. Contact individual universities for information regarding admission requirements. For Diploma and post-experience university courses, see the chapter on Career Development.

NEW ZEALAND (+64)

Victoria University, Wellington
PO Box 600, Wellington 6001, New Zealand
Tel: 4 472 1000
Website: www.vuw.ac.nz
Courses Offered: B.A. Linguistics.

UK (+44)

Coventry University
Priory Street, Coventry CV1 5FB
Tel: 024 7688 7688
Website: www.coventry.ac.uk
Courses Offered: English and TEFL BA (Hons)

De Montfort University
The Gateway, Leicester, LE1 9BH
Tel: (0116) 255 1551
Website: www.dmu.ac.uk
Courses Offered: Education Studies and TEFL

Northumbria University
Ellison Place, Newcastle upon Tyne, NE1 8ST
Tel: 0191 238 6002
Website: www.northumbria.ac.uk
Courses Offered: TESOL BA

Nottingham Trent University
Burton Street, Nottingham NG1 4BU
Tel: 0115 941 8418
Website: www.ntu.ac.uk
Courses offered: English and TESOL BA (Hons)

University of Wales Swansea
Singleton Park, Swansea, SA2 8PP
Tel: 01792 205678
Website: www.swan.ac.uk
Courses Offered: English and TEFL BA

University of Sunderland
Edinburgh Building, City Campus, Chester Road, Sunderland SR1 3SD.
Tel: 0191 515 3154
Website: www.sunderland.ac.uk
Courses Offered: English and TESOL BA

University of Middlesex
North London Business Park, Oakleigh Road South, London N11 1QS
Tel: 0208 411 5000
Website: www.mdx.ac.uk
Courses Offered: English and TEFL BA

University of Manchester
Department of Linguistics, Oxford Road, Manchester M13 9PL, UK

Tel: 0161 275 3187
Website: www.manchester.ac.uk
Courses Offered: B.A. Linguistics; B.A. English Language.

University of Sterling

centre for English Language Teaching, Stirling FK9 4LA, Scotland UK
Website: www.stir.ac.uk
Courses Offered: B.A. ELT; B.A. EFL

University of Sussex

Falmer, Brighton BN1 9QH, UK
Tel: 01273 678195
Fax: 01273 671320
Website: www.sussex.ac.uk
Courses Offered: B.A. Linguistics.

Wolverhampton University

Wulfruna St., Wolverhampton WV1 1SB, UK
Tel: 01902 322222
Website: www.wlv.ac.uk
Course Length: 3 years.
Courses Offered: B.A. TESOL.

USA & CANADA (+1)

Brigham Young University, UT

Dept. of Linguistics, 2129 JKHB, Provo, UT 84602-6278
Tel: (801) 378-4789
Website: www.byu.edu
Courses Offered: B.A. Linguistics.

Brigham Young University, HI

Hawaii Campus, P.O. Box 1940, Laie, HI 96762.
Tel: (808) 293 3614
Website: www.byuh.edu
Courses Offered: B.A. TESOL.

Cal. State, Northridge

Linguistics Program, 18111 Nordhoff St., Northridge, CA 91330-8306
Website: www.csun.edu
Start Dates: Any semester.
Courses Offered: B.A. Linguistics.

Caroll College

161 N. Benton Ave., Helena, MT 59625
Website: www.caroll.edu
Courses Offered: BA TESOL.
Start Dates: Any semester.

Central Washington University

400 E. 8th Ave., Ellensburg, WA 98926-7562.
Website: www.cwu.edu
Start Dates: Any quarter.
Courses Offered: B.Ed. TESOL.

Eastern Kentucky University

Case Annex 270-1340, Richmond, KY 40475
Website: www.eku.edu
Start Dates: Any semester.
Course Offered: B.A. English with TESL.

Eastern Mennonite University

1200 Park Road, Harrisonburg, VA 22802–2462
Tel: (540) 432 4000
Website: www.emu.edu
Start Dates: Any semester.
Courses Offered: Minor in TESL.

Iowa State University

300 Pearson Hall, Ames, IA 50011.
Tel: (515) 294 4555
Website: www.iastate.edu
Courses Offered: Linguistics; K-12 ESL Teaching

US International University

10455 Pomerado Road, San Diego, CA 92131
Tel: (858) 635 4772
Courses Offered: B.A. TESOL.

University of Florida

Gainesville, FL 32611
Tel: (352) 392 3261
Website: www.ufl.edu
Courses Offered: B.A in Linguistics/Minor in TESOL

University of Rochester

Dept. of Linguistics, 503 Lattimore Hall, P.O. Box 270096, Rochester, NY 14627-0096.
Tel: (716) 275 8053
Website: www.rochester.edu
Courses Offered: B.A. Linguistics.

University of Texas, Austin

Dept. of Linguistics, Austin TX 78712-1196.
Tel: (512) 471 1701
Courses Offered: B.A. Linguistics.

University of Calgary

2500 University Drive NW, Calgary, Alberta, T2N 1N4, Canada.
Tel: (403) 220 5110
Website: www.ucalgary.edu
Courses Offered: B.A. Linguistics

Finding a Job

This chapter gives you leads and ideas on how best to find a job as an English language teacher. The main employers are ELT/ESL schools around the world, but there are also jobs in large international companies, and other specialist organisations. You can find jobs on our website, apply directly to a school or through a recruitment agency, use one of the hundreds of websites, or look for adverts in the trade newspapers and journals.

One of the great advantages of being an English language teacher is that you can travel the world and still get a job; almost every country has a requirement for English language teachers – some need teachers specialising in a particular accent (British or American), others have a focus on a particular age-range. However, once you have either a recognised qualification (see previous chapter) or basic experience, you can travel and find work around the world. This means that any ELT job section must cover travel and working practice around the world (see the next chapter) as well as covering agencies that can get you a job in a particular country. Over the next few pages, you will see how to find a job, how to get through an interview, how to get a job in a different country and how to prepare for the culture shock of travelling and teaching in a new environment.

How Do I Find A Job?

Probably the easiest way to find a job is to use the resources we provide! Either contact the schools and teacher recruitment agencies listed in this book or visit the Oxford Seminars teacher placement service (www.oxfordseminars.com) or visit our sister magazine websites (www.eslmag.com for American English teachers and www.etprofessional.com for international English language teachers) and view 100s of job vacancies online.

To apply directly to a school, use the information listed in the country sections of this guide – you'll find this starting on page 42. If you want to teach English abroad as a volunteer, turn to page 31 for more information. When it comes to preparing your CV, be sure to stress any practical teaching you may have as well as any qualifications you have earned.

What Should I Do When I Receive A Job Offer?

Ask your prospective employer if you can contact teachers who are working at the school for references. If the school is reputable, the owner or principal should have no problems with this request. If no other native-speaking teachers are currently employed at the school ask to speak to former employees. If this request does cause a problem, then you should be wary about accepting a job offer from this school. Another point to remember is that some teachers do criticize schools they have worked for even if the problems were mostly of their own making. Lazy and bad teachers do exist – even if they speak the same language as you do! Use common sense in a situation where you are confronted with conflicting stories from schools and teachers. After all, you're going to have to rely on your own judgement when you are living and working in a new environment.

What If A Dispute Arises?

No single aspect of teaching EFL overseas causes more heartache, grief, and anguish than disputes between teachers and employers arising from contractual problems. Historically, certain countries, such as South Korea, were notorious for such disputes although there have been moves to stop these unpleasant episodes from arising. A few words of advice: don't be surprised if you find yourself caught in a dispute over a matter like wages, hours or accommodation. It has happened to thousands of EFL teachers before you and you won't be the last. Try to make sure you have a written contract and insist that it is translated into English. In many countries, the English language contract will not be valid so try to find a third party who knows both the language of your employer and English who can check that both contracts say the same thing.

If a dispute arises, try to be civil and courteous with your employer. Again, if you can find a third party to mediate the dispute this may help diffuse the situation. Some experienced teachers have reported that yelling at an employer has brought about the desired results but this is not advisable unless you are confident you can find a job if you get fired! If you find that your employer is consistently underpaying you or is increasing your hours without giving you money then start looking for another job. Unless you are in a remote region, there will probably be other language schools that will welcome your experience.

Recruitment Agencies and International School Groups

There are plenty of commercial recruitment agencies specialising in the English language market – working for a school to help find teachers or helping a teacher find a job. The agencies are normally paid by the employer (the school), though make sure that you check before you sign up! Agencies normally cover a particular area or country – and will help you put together a good CV, send you new job opportunities and even advice on relocating, accommodation and visas.

Almost all the jobs for ELT or ESL teachers are with English language schools. The majority of all schools are run on an individual basis, with the director of studies also the person who interviews and appoints new teachers. But there are some major chains of schools who have their own internal recruitment agencies who can help you find a job within the schools in their group. Once you have a job in the group, the internal job-agency often works to help find a job in another school within the group or in a different country.

The biggest ELT employer in the world is the British Council, which has its own recruitment service – see page 27 for more details on the British Council.

As a good starting point, try these commercial agencies and major international school groups:

Anglo Pacific
Suite 32, Nevilles Court, Dollis Hill Lane, London NW2 6HG
Tel: +44 (0)20 8452 7836
Agency that specializes in recruitment in SE Asia

Avalon House
8 Denmark Street, London WC2H 8LS
Tel: +44 (0)20 7279 1998
Email: info@avalonschool.co.uk
Web: www.avalonschool.co.uk
Recruitment for its schools in China, France, Spain, Poland, UK

Bell Educational Trust
Overseas Department, Hillscross, Red Cross Lane, Cambridge, CB2 2QX
Tel: +44 (0)1223 246644
Fax: +44 (0)1223 414080
Email: info.overseas@bell-schools.ac.uk
Web: www.bell-schools.ac.uk
Major group that has schools around the world; most schools recruit locally, but this office can provide contact information

Benedict Schools
3 Place Chauderon, PO Box 270, 1000 Lausanne 9, Switzerland
Tel: +21 323 6655
Fax: +21 323 6777
Email: info@benedict-schools.com
Web: www.benedict-schools.com
Major group with over 80 schools (in Europe, Africa and America); most schools recruit locally, but this office can provide contact information

Berlitz
Lincoln House, 296-302 High Holborn, London WC1V 7JH
Tel: +44 (0)20 7915 0909
Fax: +44 (0)20 7915 0222
Web: www.berlitz.com
Major group that has over 400 centres around the world; teachers are trained in the Berlitz way of teaching

Billington Recruitment
1 Mariners Close, St James Court, Victoria Dock, Kingston upon Hull, HU9 1QE
Email: bill1312@hotmail.com
Agency that covers Europe

British Council
Education and Training Group, 10 Spring Gardens, London SW1A 2BN
Tel: +44 (0)20 7389 4596
Fax: +44 (0)20 7389 4594
Email: assistants@britishcouncil.org
Web: www.languageassistant.co.uk
Provides language assistants to schools around the world. (See section on British Council on next page)

CG Associates
83 Clarence Mews, London
Tel: +44 (0)7802 211542
Email: info@christopherg.co.uk
Agency offering worldwide coverage.

CIEE
52 Poland Street, London W1F 7AB
Tel: +44 (0)20 7478 2020
Fax: +44 (0)20 7734 7322
Web: www.councilexchanges.org.uk
Looks after the JET (Japan Exchange Teaching) programme as well as Teach in China and Teach in Thailand programmes

EF - English First
Teacher Recruitment, 36-38 St Aubyns, Hove, East Sussex, BN3 2TD
Tel: +44 (0)1273 201431
Fax: +44 (0)1273 746742
Email: recruitment.uk@englishfirst.com
Web: www.englishfirst.com
Major group that has schools around the world

ELS Language Centers
400 Alexander Park, Princeton, NJ 08540
Tel: +1 609 750 3512
Fax: +1 609 750 3596
Email: info@els.com
Web: www.els.com
Major group that works with Berlitz for its language centers – however its 40-plus schools still recruit through this central address

ERC Recruitment
New Tyning, Stone Allerton, Axbridge, BS26 2NJ

Tel: +44 (0)1934 713892
Email: info@erc-recruitment.co.uk
Web: www.erc-recruitment.co.uk
Agency offering worldwide coverage

GEOS English Academy
Compton Park, Compton Park Road, Eastbourne, BN21 1EH
Email: info@geos.com
Web: www.geos.com
Major group that has over 500 schools around the world

GTCE
52 Loampit Hill, Lewisham, London SE13 7SW
Tel: +44 (0)20 8691 6569
Email: info@gtce.co.uk
Web: www.gtce.co.uk
Agency that concentrates on China

International House
106 Piccadilly, London W1V 7NL
Tel: +44 (0)20 7518 6970
Fax: +44 (0)20 7518 6971
Email: hr@ihlondon.co.uk
Web: www.ihworld.com
Major group that has over 120 schools around the world

Linguarama
Personnel Department, 89 High Street, Alton, Hampshire, GU34, 1LG
Tel: +44 (0)1420 80899
Fax: +44 (0)1420 80856
Email: personnel@linguarama.com
Web: www.linguarama.com
Major group that has schools around the world

Protocol Professional
Tel: +44 (0)115 911 1177
Email: recruitment@protocol-professional.co.uk
Web: www.protocol-professional.co.uk
Agency that specializes on the UK

Saxoncourt
124 New Bond Street, London W1S 1DX
Tel: +44 (0)20 7491 1911
Fax: +44 (0)20 7493 3657
Email: recruit@saxoncourt.com
Web: www.saxoncourt.com
Major group of schools and recruitment centre that places over 500 teachers per year around the world

Sterling Recruitment
49 Sudbury Ave, Wembley, Middx, HA0 3AN
Tel: +44 (0)20 8903 4424
Fax: +44 (0)20 8903 3566
Recruitment Agency for jobs in London

Wall Street Institute
Paca de Catalunya 9, 4th Floor, 08002 Barcelona, Spain
Tel: +93 306 33 00
Fax: +93 302 08 29
Web: www.wallstreetinstitute.com
Major chain of over 400 schools around the world.

The British Council

The British Council has over 60 language teaching centres, offices in over 100 countries and is the by far the biggest ELT employer in the world, with over 7500 staff (most of whom are involved in teaching). The Council is funded by the British Government, but works in a non-political way promoting British English and culture around the world.

For job-hunters, the British Council can help in two ways:

1. Each local office is a mine of information about local schools, job opportunities and conditions of work. Many offices have pre-printed information for prospective job-hunters and sometimes provide contacts to known schools in the country. For a list of local offices, see our directory on page 302 or visit www.britishcouncil.org

2. The British Council has over 60 teaching centres and it recruits around 300 teachers per year. For more information, contact Educational Enterprises, British Council, 10 Spring Gardens, London SW1A 2BN, Tel: +44 (0)20 7389 4931, Fax: +44 (0)20 7389 4140, Email: teacher.vacancies@britishcouncil.org

The British Council also publishes a wide range of leaflets and information packs about training to be an English language teacher and opportunities within the organisation. Either visit the information-packed website (www.britishcouncil.org) or contact its information centre: Bridgewater House, 58 Whitworth Street, Manchester M1 6BB, Tel: +44 (0)161 957 7755, Fax: +44 (0)161 957 7762

UK Summer Schools

The UK has a very strong summer-school industry (with several hundred schools working the season) that always has a need for qualified teachers. The summer schools cater for students who want to learn English over a summer holiday; these students can be either young learners or young adults – but, even with the current downturn in international travel, the summer schools are still very busy – as is their requirement for teachers.

The rates of pay for work in summer schools is often higher than usual. In the past, some summer schools have not offered very good quality teaching, but with the work of industry associations, particularly English UK (www.EnglishUK.com) - the newly-merged organisation previously known as ARELS (Association of Recognised English Language Schools) and BASELT (British Association of State English Language Teaching) - and the British Council (www.britishcouncil.org), the schools are being accredited and standards are now high.

With so many schools on their lists, it's impossible to fit all these into this book: the best way to search for a suitable school to approach for a job is to use a site such as English in Britain (www.EnglishInBritain.com) or the English UK (www.EnglishUK.com) websites – both have search features.

Jobs in Canada

To teach in most private language schools in Canada you will need a TESOL or TESL certificate. To find a list of schools, the simplest route is to use the Internet. The next few pages include ESL job sites in Canada. Alternatively, use the website directory for the individual states to search for schools in the state.

Jobs in the USA

In the USA, each state has its own requirements for teacher qualifications (read the requirements later in this *Guide*). The schools within a state must also be registered with their state education body – and each state, again, has its own requirements for schools operating in their area.

Jobs on the Internet

There are hundreds of websites specialising in the ELT/ESL job sector. Some simply provide advice, others offer help with interviews, others provide lists of jobs available in a particular region or country.

Starting Points

Try these sites for a varied flavour of what the Internet offers job-hunters – these are some of the main sites from each of the following sections.

British Council Vacancies
www.britishcouncil.org/teacherrecruitment.htm

English Teaching Professional
www.etprofessional.com

ESL Magazine
www.eslmag.com

International House Recruitment Services
www.ihworld.com/recruitment/

Oxford Seminars
www.oxfordseminars.com

Peace Corps
www.peacecorps.gov

Saxoncourt
www.saxoncourt.com

The Times Educational Supplement
www.tesjobs.co.uk/homepage.asp

VSO (Voluntary Service Overseas)
www.vso.org.uk/volunteering/education_tefl.htm

Career Advice

Starting a Career in TESOL (tesol.org)
www.tesol.org/careers/counsel/

Teaching English Abroad
www.cal.org/ericcll/digest/snow0001.html

Best Bets for Teaching Abroad
www.uci.edu/~cie/iop/teaching.html

Creating a CV
www.onestopenglish.com/Jobs/

200 FREE Cover Letters For Job Hunters
www.careerlab.com/letters/

Finding a Job – all-in-one sites

BestPeople
www.bestpeople.co.uk

CareerCentral
www.careercentral.com

CareerMosaic UK
www.careermosaic-uk.co.uk

Gradunet-Virtual Careers Office
www.gradunet.co.uk

JobHunter
www.jobhunter.co.uk

Jobs Unlimited
www.jobsunlimited.co.uk

JobSearch
www.jobsearch.co.uk

JobSite UK
www.jobsite.co.uk

Monster
www.monster.co.uk

Reed Online
www.reed.co.uk

ELT-specific Jobs

All English Job
www.allenglishjob.com

Canadian Institute for Teaching Overseas (CITO)
www.nsis.com/~cito/CITO.html

Dave's ESL Café: Job centre
www.eslcafe.com/jobs/

Easter School Agency
www.easterschool.com

Edufind ELT Job Centre
www.jobs.edufind.com

EFLTeachingJobs.com
www.eflteachingjobs.com

English Club ESL Jobs Centre
http://jobs.englishclub.com

English Job Maze
www.englishjobmaze.com

English Teacher Recruitment: Greece
www.teach.english.freeservers.com

English Worldwide – Recruitment & Training
www.englishworldwide.com

English-International.com
www.english-international.com

Erudite Café Job centre
www.eruditecafe.net/job_centre1.php

ESL Career.com
www.eslcareer.com

ESLclassifieds.com
www.eslclassifieds.com/esljobsoffered.html
ESL Employment
www.eslemployment.com
ESL Job Search
www.esljobsearch.com
ESL Magazine
www.eslmag.com
ESL Resume Database
my.globalesl.net/
ESL Teachers Club
www.eslteachersclub.com
ESL WideWord Teach English
www.eslwideworld.com
ESLworldwide.com
www.ESLworldwide.com
ETNI – Advertising Teaching Positions in Israel
www.etni.org.il/anntchadver.htm
Foreignteacher.com
www.foreignteacher.com/currentjobs.htm
Global ESL Network
www.globalesl.net
Jobs Offered
www.englishclub.net/cgi-bin/jobs/jobs_o.pl
Russian School Placement
www.sv-agency.udm.ru/svfiles/school.htm
Saxoncourt & English Worldwide
www.saxoncourt.com
TeachAbroad.com
www.teachabroad.com
Teaching Abroad
www.teaching-abroad.co.uk
Teaching Abroad .net
www.teachingabroad.net
Teaching ESL at Kuwait University
www.iteslj.org/Articles/Martin-Kuwait
TEFL Jobs
www.tefljobs.net
TEFLnet
www.tefl.net/jobs/
TEFL Professional Network
www.tefl.com
University EFL/ESL Positions
www.ultimateenglish.com
Work Abroad ESL.com
www.workabroadesl.com

ELT Jobs in Asia

Collection of Teaching Positions in Taiwan
http://esltaiwan.20M.com
Ding Ding Dang Recruiting
www.dingdingdang.com
EFL in Asia
www.geocities.com/Tokyo/Flats/7947/eflasia.htm
English Teaching programme in Thailand
www.taiteach.com
EnglishWork.com
www.englishwork.com
ESL City
www.eslcityjobs.com
ESL House: China
www.eslhouseonline.com
HBS Consulting: Korea
www.hbscompany.com
International Avenue Consulting
www.iacc.com.tw
Japanese Jobs
www.japanesejobs.com/textonly/index.html
Jobs in Japan
www.eltnews.com/jobsinjapan.shtml
KOAM Academy: Korea
www.koam.org
O-Hayo Sensei: Japan
www.ohayosensei.com
Reena Recruiting Co. Ltd.
www.reena-recruit.com
Russell Recruiting
www.asiangateway.net
Taiwan Teacher
www.geocities.com/Athens/Delphi/1979/
Teach English in South Korea
www.teachenglish.co.uk/jobs.htm
teacheslabroad.com: Korea
www.teacheslabroad.com
Teaching English and Educating in Korea
www.keepro.com
Teaching English in China
www.EnglishTeachersExpress.com
teach-in-thailand.com
www.teach-in-thailand.com
TeachinginJapan
www.teachinginjapan.com
teachkorea.com
www.teachkorea.com/mb.htm

Volunteer Programmes

In some countries the demand for adult English language teachers outstrips the supply; as state-funded programmes struggle to provide needed services with shrinking funds, the role of volunteers in teaching English becomes crucial. And this happens in developing countries as well as western countries such as the USA and Canada.

The following organisations arrange for volunteer teachers to travel and teach in volunteer programmes around the world, particularly into China and Africa. There are also volunteer groups providing support and teachers into local communities, for example, into regions of the USA where there is a great need for volunteer teachers. The volunteer groups should help arrange your travel, training, accommodation and any visas.

AmeriCorps

www.americorps.org

AmeriCorps is a United States-based organization. The domestic Peace Corps has more than 40,000 Americans in intensive, results-driven service each year. AmeriCorps has built its foundation on teaching children to read, attempting to make neighborhoods safer, building affordable homes, and responding to natural disasters. AmeriCorps serves with projects such as Habitat for Humanity, the American Red Cross, and Boys and Girls Clubs. After a term of service, AmeriCorps members receive education awards to help finance college tuition or pay back student loans.

AmeriSpan Unlimited

www.amerispan.com

AmeriSpan has become a major force for anyone wanting to study language abroad in Latin America. AmeriSpan's network offers language programmes in nearly every Spanish-speaking country in the Americas.

Amity Institute

www.amity.org

Amity Institute was founded in 1962. It is a non-profit educational exchange programme that gives young people around the world the opportunity to represent their countries, by sharing their languages and cultures in language classrooms of all levels. Amity tries to encourage and enhance international understanding and friendship through the study of world languages and cultures.

Christians Abroad

www.cabroad.org.uk

Provides advice and information for volunteers of any faith who are considering working abroad as a volunteer. It provides job vacancies through a monthly listing newsletter – the majority of its postings are to Asia and Africa. Also visit www.wse.org.uk for online information packs.

Colorado China Council

www.asiacouncil.org

The primary focus of CCC is to send people to teach English in China, as well as other academic subjects, at universities and secondary schools throughout the country. The Council has placed over 200 teachers, making it one of the U.S.'s largest non-religious providers of teachers to China.

Cross-Cultural Solutions

www.crossculturalsolutions.org

Cross-Cultural Solutions offers short-term and long-term programmes that give volunteers from all over the world the opportunity to come face to face with global issues and become part of productive solutions. CCS has close partnerships with social service pioneers in our host countries. The focus of CCS's work is health care, education and social development

Friends of World Teaching

www.fowt.com

Friends of World Teaching helps teachers find jobs around the world. Friends of World Teaching works with English-speaking schools and colleges throughout the world that offer employment opportunities to American and Canadian teachers. There are some student teaching programmes as well.

Global Routes

www.globalroutes.org

Global Routes is a tax-exempt non-profit, is a non-governmental, non-sectarian organization whose aim is to strengthen the global community. It has designed community-service/cross-cultural-exchange programmes that bring people with different world-views together.

India Literacy Project

www.ilpnet.org

India Literacy Project is a US-based non-profit, volunteer organization that focuses on literacy in India. Its aim is to empower every individual we serve with functional literacy and an understanding of their basic rights and responsibilities. India Literacy Project's goal is 100% literacy in India.

Intl. Foundation for Ed. & Self-help

www.ifesh.org

The International Foundation for Education and Self-Help (IFESH) was established as a non-governmental, nonprofit, charitable organization. IFESH aims to reduce hunger and poverty, empower people through literacy, train and place the unskilled and unemployed in jobs, provide preventive and basic health care to individuals in need, deal with population and environment problems, develop employment through economic development activities, foster cultural, social, and economic relations between Africans and Americans, particularly African Americans, and others.

International Rescue Committee

www.intrescom.org

International Rescue Committee was founded in 1933. The IRC is a nonsectarian, voluntary organization that provides relief, protection and resettlement services for refugees and victims of oppression or violent conflict throughout the world.

Intl. Schools Internship programme

www.tieonline.com

A TIE subscription gives you immediate access to the most prominent schools looking for staff: from 90 to 150 schools are always on the Job Ads web page. And the TIE Resume Bank means over 200 headmasters of International schools can access your file; and from there they can contact you through email, phone, your website, etc. Schools can, and often do, change their vacancies weekly at so you can be the first to know when a vacancy in the country you want opens up!

Laubach Literacy

www.laubach.org/home.html

Laubach Literacy Action (LLA), the U.S. programme Division of Laubach Literacy, is the largest volunteer-based literacy organization in the United States. We provide a full range of literacy services to more than 175,000 students annually through more than 1,000 local member programmes, 45 state organizations. Instruction is provided by a national network of more than 80,000 volunteer trainers and tutors .

Literacy Volunteers

www.literacyvolunteers.org

Literacy Volunteers of America, Inc. has served adults at the lowest levels of literacy and their families since 1962, and has assisted more than half a million people to acquire literacy skills. LVA offers a professionally designed and field-tested workshop that enables volunteers to tutor adults in English for speakers of other languages. After being matched with an adult learner or a small group of learners, tutors receive regular support and opportunities for additional training through the local LVA affiliate.

National Adult Literacy Database

www.nald.ca

The National Adult Literacy Database Inc. (NALD) is a federally incorporated, non-profit service organization which fills the crucial need for a single-source, comprehensive, up-to-date and easily accessible database of adult literacy programmes, resources, services and activities across Canada. It also links with other services and databases in North America and overseas.

PeaceCorps

www.peacecorps.org

As an English teaching volunteer, you will help expand the horizons and opportunities of students and teachers alike in Peace Corps' largest programme. There is no other experience like helping a middle school, secondary school, or university. More than 7,000 Peace Corps volunteers are serving in 76 countries. They are working to bring clean water to communities, teach children, help start new small businesses, and stop the spread of AIDS. Since 1961, more than 161,000 Americans have joined the Peace Corps, serving in 134 nations.

VolunteerMatch

www.volunteermatch.org

VolunteerMatch, ImpactOnline's main service, uses the power of the Internet to help individuals find volunteer opportunities posted by local non-profit and public sector organizations.

Volunteers in Asia

www.volasia.org

VIA is a private, non-profit, non-sectarian organization whose aim it is to increase understanding between the United States and Asia. Its Volunteer Programs in Asia have provided young Americans with an opportunity to work and live within an Asian culture while meeting the needs of Asian host institutions, since 1962.

VSO (Voluntary Service Overseas)

www.vso.org.uk

The VSO is one of the biggest and best-known of the volunteer agencies. It works in over 70 countries around the world, with a current majority of postings to China and Africa. The VSO offers volunteers a complete package, including basic training (volunteers are expected to have a degree or be experienced TEFL teachers), airfares and, importantly, in-country support centres.

VSO Canada

www.vsocan.com

People are what VSO Canada is all about– and you just might be one of them. VSO's people share their skills in developing countries.

WorldTeach

www.worldteach.org

WorldTeach is a non-profit, non-governmental organization based at the centre for International Development at Harvard University which provides opportunities for individuals to make a meaningful contribution to international education by living and working as volunteer teachers in developing countries.

Gap year travel (huge in the UK, still relatively unknown in Canada and the USA) is a way for students to take a year out and travel before settling down to their studies at university, often working as a volunteer (and teaching English is a popular option). The needs and expectations of both sides (agency and volunteer) are very different for gap year compared to a qualified teacher taking a career break or travelling, so there are agencies that specialise in this work. For information about working in your gap year, visit these websites:

The gap year company	www.gapyear.com
GAP Activity projects	www.gap.org.uk
Gap Challenge	www. world-challenge.co.uk

Preparation: Before you go

This section covers the preparation you need to consider before you head off on your travels around the world: in particular, income tax on earnings and insurance.

Before you go

If you are thinking of teaching English in another country, you should have a lot of questions on your mind. Here are some of the most frequent questions asked by prospective teachers and the answers.

When Should I Plan To Go?

This depends on where you want to teach. In many countries, everything shuts down during certain months of the year, usually in the summer. Our country profiles, later in this book, provide general information about the best time to apply for jobs. You can apply for jobs from home or you can turn up on spec. Both approaches have their own merits and drawbacks depending on the country you want to teach in. The country guides offer more detailed answers to this question.

What Should I Take With Me?

Essentials – like obtaining health insurance valid for your stay overseas – are dealt with later in this chapter. Do remember to check if your passport will be valid for the period of time you plan to be away – it's amazing the number of teachers who forget to do this!

Make sure you take clothes appropriate for the climate you will be living in. Remember that a country that is hot and sunny in the summer can be freezing cold in the winter. Clothing may be expensive in your new country so take a good supply of things you like to wear. Check with the school to see if you will be required to dress formally when you teach. If you will be going out to look for a teaching job on spec it would not hurt to take clothes that you would wear to an interview. Appearance really does count when you are job hunting, almost as much your CV in some cases! You might want to invest in some ESL/ EFL teaching books as these may be hard to obtain in your destination country – they may be in short supply. A good dictionary and a grammar guidebook will always come in handy when you are faced with any awkward questions about word definitions or the nature of past participles. Also, make sure you have a few good novels for your leisure time. If you are learning the language of your destination country (it's a good idea), then take along some tapes, books and a Walkman.

Will I Need A Teaching Qualification?

It's a good idea to get some training before you go or why not take a teaching course while you are on your travels? Whether you train in your home country or while you are away is your decision, but if you plan to train in another country make sure that there are appropriate courses on offer.

Insurance

When planning to travel abroad you should you have adequate health insurance. With so many different types of coverage available it can be very confusing deciding which plan is right for you. Here's what to look for.

Don't leave home without it

All European Union citizens are entitled to the same level of medical care in any other country of the EU as they would receive at home. To secure this coverage as a UK citizen, you must obtain form E111 from the Department of Health and Social Security before you leave the UK.

If you plan to stay for more than a few months, you should check what your local social security contributions entitle you to and decide whether you need to take out a private health care scheme. A private scheme gives you more flexibility about where you are treated, but you should check carefully exactly what your coverage is. It is also possible to obtain coverage for repatriation in a real emergency.

Home Country Coverage: your existing health plan cover

This section will be of particular importance to citizens of the US (and other countries that do not have a good, free, state-funded health service) who typically have their own private health and medical insurance polices.

For US citizens, if you already have strong benefits at home, it is wise to double-check what exactly is covered while you are out of the country. Most health plans limit coverage to a maximum of 30 or 60 days outside the States, and HMOs and PPOs will likely impose severe – out of network – penalties for all but the most basic emergency care. More importantly, you will want to insure you have 24 hour access to emergency evacuation if you are sick or hurt in an area where quality care is not available. For those over 65, Medicare will not cover treatment outside the U.S. If you do not have any domestic coverage, travel insurance is a must. Once you have answered these questions, you will need to know what benefits and policy provisions to look for in a travel medical plan.

Short Trips (two weeks or less)

Assuming you have solid coverage at home you may only need a small supplemental plan, with features like emergency evacuation and 24 hour worldwide assistance, along with a limited benefit for medical expenses (£10,000 to £50,000 is typical). These plans are readily available through your travel agent and may include lost luggage and trip cancellation protection too. Be wary of plans that require you settle all bills yourself and then seek reimbursement when you return home.

Intermediate Trips (two weeks to six months)

The longer you are away the less likely it is your own plan will cover you and therefore the more importance you should place on the medical benefits of the plan. In addition to a higher emergency evacuation benefit, your plan should cover between £100,000 to £1 million in medical expenses. Other features to look for include coverage for a family member to come to your assistance if you are taken to hospital in a foreign country and an option to include "hazardous activities" such as scuba diving, skiing and bungee jumping.

Longer Trips (six months+)

The travel insurance you purchase will most likely be your primary or only health insurance. Be sure that your plan includes all of the above, with higher maximum benefits (£1 million or more) as well as some provision for coverage when you return home. If you spend a year abroad but return home for a vacation, make sure you are still covered.

Multiyear or Open Ended Trips

Standard medical plans are usually limited to one year, with the option to renew for another. If you anticipate being abroad for a longer period of time, look into a permanent international major medical plan. This type of coverage contains all the benefits of a travel medical plan, but once you are accepted you can maintain coverage year after year simply by paying the premium. Most importantly, you are covered for medical treatment anywhere in the world, including back at home. Note that these plans are medically underwritten, and people with prior health problems may have trouble obtaining coverage.

All of the plan types discussed share some common features; you will normally have a deductible and co-insurance payment or premium to meet, and pre-existing conditions are usually excluded. Worldwide assistance is sometimes done by the plan administrator, or through an independent assistance company. The market for these plans is growing rapidly and new companies enter the market all the time. Be sure the company you are dealing with has at least five to ten years' experience in the international market. Doing a little research before you go can mean security and peace of mind while you are away, and allows you to concentrate on the exciting and fulfilling adventure that international travel and work can be!

Brian Garity, founder of International Health Insurance.com, has specialized in international health insurance for over ten years.

Tax

If you are planning to work abroad, it is important that you know your financial position, especially what your liability for tax will be. In the directory of countries, page 87, we note any special requirements of local tax authorities – often, you will have to register with the local tax office before starting work. However, there are plenty of potential problems when declaring your earned income in your home country. Regulations differ for citizens of different countries. Here are some guidelines for teachers from some Commonwealth countries and from the USA.

Australian nationals

If you are an Australian national, you have to fill in your own tax return. If you work abroad for less than a full tax year, when you return you will need to declare your earnings and tax, and will either receive a tax credit for the excess tax you have paid or be taxed for the shortfall. If you are abroad for longer periods, there will be a break in your tax requirements in Australia for the period you are away. But check your social security arrangement as payments made overseas will not be credited.

Canadian nationals

If you are a Canadian national, you will need to consider what your residency status is while you are working abroad. If you work abroad for a short term but keep your ties with Canada, you are considered a factual resident and will be liable to pay federal taxes, etc. If you work abroad for a longer period, you can be considered a non-resident. In this case your tax position in Canada will change. You will need to check with the authorities or consult a financial advisor to confirm your position, however the following general advice applies to anyone who works abroad.

If you move abroad (therefore live abroad for a significant period), taxation is complex - especially if you retain assets in Canada. Double taxation is one major concern - however, most countries have taxation treaties with Canada so this is relatively straightforward to explain to the tax department.

If you plan to move abroad, you might consider the radical step of establishing non-resident status with Canada to be able to take advantage of the (often lower) tax rates in other countries. However, this has serious implications on your return to Canada, so you should talk through the implications with the tax office or a tax advisor. For the latest regulations, visit: www.cra-arc.gc.ca/tax/nonresidents

Lastly, if you work abroad for a period of time then move back to Canada, you could still reclaim some taxes (for up to six years' absence) - again, visit the www.cra-arc.gc.ca for more details and forms.

UK nationals

It is most important to check whether or not the country you are going to has a reciprocal tax agreement with your home country.

If you are a UK national, you will get tax relief (even up to 100%) if you are out of the country for at least 365 days and do not stay in the UK for more than 62 consecutive days in the tax period. There is a difference between non-resident status and exemption from tax in the UK. If you are classed as a non-resident, you are not liable for tax on unearned income from abroad but you are still considered to be domiciled in the UK. To qualify as a non-resident you must work overseas for more than a full tax year without being in the UK for more than six months in that tax year or three months in each year if your stay abroad spans the tax year.

Many countries have agreements with the UK to prevent the double payment of tax. Credit against UK tax for payment already made abroad will be given if there is no such agreement. (See Inland Revenue form IR6 – Double Taxation Relief).

When you are teaching abroad, it is important that you think about maintaining your UK social security payments so that you qualify for the state pension. You should get form N139 from the Department of Social Security (form SA29 if you are working in the EU).

US Nationals

The U.S. income tax system (and forms) are well-known to be complex area – and for U.S. citizens working and travelling the world as teachers, this can be particularly difficult. As a U.S. citizen or resident alien, your worldwide income generally is subject to U.S. income tax regardless of where you are living. Also, you are subject to the same income tax filing requirements that apply to U.S. citizens or residents living in the United States. However, several income tax benefits might apply if you meet certain requirements while living abroad. You may be able to exclude from your income a limited amount of your foreign earned income. You also may be able either to exclude or to deduct from gross income your housing amount (more later). To claim these benefits, you must file a tax return and attach Form 2555, Foreign Earned Income. If you are claiming the foreign earned income exclusion only, you may be able to use the shorter Form 2555-EZ, Foreign Earned Income Exclusion, rather than Form 2555.

You may be able to claim a tax credit or an itemized deduction for foreign income taxes that you pay. Also, under tax treaties or conventions that the United States has with many foreign countries, you may be able to reduce your foreign tax liability.

Income Earned Abroad

You may qualify for an exclusion from tax of up to $70,000 in income earned while working abroad. However, you must file a tax return to claim the exclusion. In general, foreign earned income is income received from services you perform outside of the United States. When we use the term United States, that includes Puerto Rico, Northern Marina Islands, Republic of the Marshall Islands, Federated States of Micronesia, Guam and American Soma. While not all of these countries are part of the United States, they have special tax status. Excluded from gross earned income is your housing costs allowance which are over a certain base amount. Generally, you will qualify for these benefits if your tax home (defined below) is in a foreign country, or countries, throughout your period of bona-fide foreign residence or physical presence and you are one of the following:

1) A U.S. citizen who is a bona-fide resident of a foreign country or countries for an uninterrupted period that includes a complete tax year, or

2) A U.S. resident alien who is a citizen or national of a country with which the United States has an income tax treaty in effect and who is a bona-fide resident of a foreign country or countries for an uninterrupted period that includes an entire tax year, or

3) A U.S. citizen or a U.S. resident alien who is physically present in a foreign country or countries for at least 330 full days during any period of 12 consecutive months.

Tax Home

Generally, your tax home is the area of your main place of business, employment, or post of duty where you are permanently or indefinitely engaged to work. You are not considered to have a tax home in a foreign country for any period during which your abode (the place where you regularly live) is in the United States. However, being temporarily present in the United States or maintaining a dwelling there does not necessarily mean that your abode is in the United States. For details, see Publication 54.

A foreign country, for this purpose, means any territory under the sovereignty of a government other than that of the United States, including territorial waters (determined under U.S. laws) and air space.

Waiver of time requirements: You may not have to meet the minimum time requirements for bona-fide residence or physical presence if you have to leave the foreign country because war, civil unrest, or similar adverse conditions in the country prevented you from conducting normal business. You must, however, be able to show that you reasonably could have expected to meet the minimum time requirements if the adverse conditions had not occurred. See Publication 54 for a list of foreign countries that individuals have had to leave due to these conditions.

Travel Restrictions

If you violate U.S. travel restrictions, you will not be treated as being a bona-fide resident of, or physically present in, a foreign country for any day during which you are present in a country in violation of the restrictions. (These restrictions generally prohibit U.S. citizens and residents from engaging in transactions related to travel to, from, or within certain countries.) Also, income that you earn from sources within such a country for services performed during a period of travel restrictions does not qualify as foreign earned income, and housing expenses that you incur within that country (or outside that country for housing your spouse or dependents) while you are present in that country in violation of travel restrictions cannot be included in figuring your foreign housing amount.

Currently, these travel restrictions apply to: Cuba, Libya, and Iraq.

Exclusion Of Foreign Earned Income

If your tax home is in a foreign country and you meet either the bona fide residence test or the physical presence test, you can choose to exclude from gross income a limited amount of your foreign earned income. Your income must be for services performed in a foreign country during your period of foreign residence or presence, whichever applies. You cannot, however, exclude the pay you receive as an employee of the U.S. Government or its agencies. You cannot exclude pay you receive for services abroad for Armed Forces exchanges, officers' mess, exchange services, etc., operated by the U.S. Army, Navy, or Air Force.

Foreign Credits And Deductions

If you claim the exclusion, you cannot claim any credits or deductions that are related to the excluded income. You cannot claim a foreign tax credit or deduction for any foreign income tax paid on the excluded income. Nor can you claim the earned income credit if you claim the exclusion. Also, for IRA purposes, the excluded income is not considered compensation and, for figuring deductible contributions when you are covered by an employer retirement plan, is included in your modified adjusted gross income.

Foreign Amount Excludable

If your tax home is in a foreign country and you qualify under either the bona fide residence test or physical presence test for a tax year, you can exclude your foreign income earned during the year up to $70,000. However, if you qualify under either test for only part of the year, you must reduce ratably the $70,000 maximum based on the number of days within the tax year you qualified under one of the two tests.

Foreign Income Taxes

A limited amount of the foreign income tax you pay can be credited against your U.S. tax liability or deducted in figuring taxable income on your U.S. income tax return. It is usually to your advantage to claim a credit for foreign taxes rather than to deduct them. A credit reduces your U.S. tax liability, and any excess may be carried back and carried forward to other years. A deduction only reduces your taxable income and may be taken only in the current year. You must treat all foreign income taxes in the same way. You generally cannot deduct some foreign income taxes and take a credit for others.

Foreign Tax Credit

If you choose to credit foreign taxes against your tax liability, complete Form 1116, Foreign Tax Credit, (Individual, Estate, Trust, or Nonresident Alien Individual), and attach it to your U.S. income tax return.

Do not include the foreign taxes paid or accrued as withheld income taxes on line 55 of Form 1040.

Foreign Tax Limit

Your credit cannot be more than the part of your U.S. income tax liability allocable to your taxable income from sources outside the United States. So, if you have no U.S. income tax liability, or if all your foreign income is exempt from U.S. tax, you will not be able to claim a foreign tax credit.

If the foreign taxes you paid or incurred during the year exceed the limit on your credit for the current year, you can carry back the unused foreign taxes as credits to 2 prior tax years and then carry forward any remaining unused foreign taxes to 5 later tax years.

Foreign taxes paid on excluded income. You cannot claim a credit for foreign taxes paid on amounts excluded from gross income under the foreign earned income exclusion or the housing amount exclusion, discussed earlier.

Foreign Tax Deduction

If you choose to deduct all foreign income taxes on your U.S. income tax return, itemize the deduction on Schedule A (Form 1040). You cannot deduct foreign taxes paid on income you exclude from your U.S. income tax return.

More information. The foreign tax credit and deduction, their limits, and the carryback and carryover provisions are discussed in detail in Publication 514, Foreign Tax Credit for Individuals. For more information and to download publications and forms mentioned, visit www.irs.ustreas.gov/forms_pubs/index.html

Country Profiles

To find a particular country, see the index on the next page

Country Profiles

Country Profiles

Work the World

The following section of this guide will show you all you need to know to get a teaching job in almost any country in the world.

As a qualified English language teacher you can travel the world and find work in almost any country you choose to visit. English learners gain a valuable opportunity to interact with a native speaker and pick up all the little nuances that they would otherwise miss. This section of the Guide will prepare you for travel around the world as an EFL/ESL teacher.

Teaching English as a Foreign Language has developed into a vast global business. In many countries around the world, the development of English language skills throughout society is considered crucial to economic success. People want to learn English because they see it as a key, opening doors to new opportunities that they would not otherwise enjoy. Governments around the world are insisting that their children learn English from a very early age: Korea, Japan and Italy are just three countries that have recently announced initiatives designed to get people to learn English.

Teaching English has always had one great advantage over any other form of teaching: the travel element. The requirement for teachers in just about every country around the world leads to great opportunities for well-qualified teachers who can travel and, often, a very enjoyable lifestyle exploring countries and new cultures.

This section of the book provides thorough, in-depth profiles of all the main markets for teachers of English. We include a short description of the country and its prospects for teachers, together with details of the visa requirements, salary expectations, medical conditions, cost of accommodation and local English-language newspapers. We also include details of the main schools in the country that could offer you a job. Use this information to contact the schools with your curriculum vitae and a covering letter to find a job as a teacher.

Once you have selected the country or region where you would like to work, flip back to the 'Finding a Job' chapter earlier in this book for more details on how best to apply, how to search for jobs online and even how to travel!

Country Profiles

The countries over the following pages are organized geographically. If you already have your heart set on going to a particular country, look it up in the index at the end of this book and go straight to that section. If you have an idea of where you would like to go, but need more information before you decide, read about all of the countries in that area, because you may well find that another country is more suited to you. If you are unsure about where you would like to go, browse through all of the sections and you should get a good idea of countries that you would like to teach in.

Country-specific information

In these country sections we have tried to give you the information that you need, rather than personal opinions, so there will usually be a brief summary of the country's economic position and how that relates to the demand for English language teachers.

Next, there is information on visas and work permits, which reflects the position in the country at the time of writing. Please note that we do not recommend working illegally in any country, even though it is mentioned where teachers do tend to work without the correct paperwork. Not all countries have such a clear policy on the working rights of foreign nationals, so consult the relevant embassy for full details.

The cost of living information is intended as a guideline to enable you to establish if your local salary will be just enough to support you or will boost your savings account. Costs are described in comparison to western European and North American norms, so this information must be read in conjunction with salary information. And some teachers manage to live on tight budgets in poorer countries by adapting quickly to the local way of life.

Salaries vary enormously, even within the same town, so figures given here are averages. Where figures are quoted in UK pounds, remember that exchange rates can change very quickly, so the Sterling equivalent will also change, but the salary should still remain the same in relation to local costs.

Some employers pay teachers in US-dollars, but the vast majority of employers pay in local currency. Taxation is not set at a standard rate in some countries, so it too will vary depending on the type of institution you are working for and your contract terms.

Accommodation causes as many problems overseas as it does at home. Prices given in this section are what teachers typically pay for standard accommodation in safe areas. Obviously, luxury penthouses in chic, central areas will cost you a lot more. If you do have problems finding somewhere to live, think of alternative options, such as living with a family, which can save you money and enable you to integrate with the local community very quickly. Ask around at universities and schools – something will soon come up.

Health insurance is advisable whenever you are travelling. Travel agents, embassies and airlines should be able to provide you with up to date details of any particular health issues in the countries you are visiting.

English language newspapers are good places to look for teaching jobs and accommodation. They are also a great way to keep up to date with the local (and international) news if your grasp of the local language is not good enough to read the local papers.

At the end of each country sections, there is a list of schools to contact for jobs. The national telephone dialing code appears in brackets at the top of the list. You should also look in the 'Finding a Job' chapter earlier in this book for more details on how to apply for a job and websites listing job vacancies.

Western Europe

Demand for English as a second or foreign language is strong among the member states of the European Union. English is the working language of the EU despite efforts to promote other languages such as French, Spanish and German.

It has become more difficult for non-EU citizens to work in Western Europe. Work-related laws that are designed to make it easier for Europeans to work in member states have tightened up with regard to Americans, Canadians, and to a lesser extent, Australians.

Job prospects for native speakers from the UK are particularly good in Western Europe, but rather more difficult for Americans and Canadians with American-English qualifications (although many countries, Spain for example, have 'international' schools teaching American English). However, there are jobs available for any teacher – although schools much prefer someone with qualifications (i.e. someone who has passed at least a certificate-level course) – especially if you are flexible and don't expect to land a high-paying permanent teaching position right away. If a school wants to keep you on as a full-time and legal employee, there is usually a way to cut through the bureaucracy, even if it is time-consuming.

The traditional strong areas of the countries around the Mediterranean still attract a lot of prospective teachers, but some countries have had some economic problems recently, particularly Spain which has seen a number of its major schools shut down in the past year.

Any EU citizen can travel freely within western Europe and work in any other country – enjoying its local health system and social security benefits. However, your school will be able to advise of the local tax laws which normally require any visitor planning to work to register with the local tax authority (and often with the police, too). Non-EU citizens looking for work should contact the country's embassy and ask for current visa information. Your prospective employer is often in the best position to request (and push through) a visa application on your behalf.

AUSTRIA

It is nearly impossible for non-EU citizens to live and work in Austria on a part time, freelance basis unless married to an EU citizen or in possession of a special visa (e.g., for the arts) and nearly all teaching positions are freelance. If you have a student visa for a university in Austria, this might give you eligibility to work. There is stiff competition for teaching jobs from the expatriate community. The most commonly available work is part-time and company-based.

Embassy in UK:
18 Belgrave Mews West, London SW1X 8HU,
tel: +44 (0)20 7235 3731
www.austria.org.uk

Embassy in USA:
3524 International Court NW, Washington DC 20008
tel: +1 202 895 6700
www.austria.org

British Council office:
Schenkenstrabe 4, A -1010 Vienna
tel +43 (1) 533 2616
Www.britishcouncil.at

➲ **Visa/Work Permits:** For non-EU nationals intending to work or study in Austria, a residency permit is required. Contact the Embassy or Austrian Consulate General. US citizens may enter Austria as tourists and apply for a residency permit with the relevant authorities in Austria.

➲ **Requirements:** College Degree, experience and a TESOL/TESL Certificate

➲ **Currency:** £1 = € 1.49, $1 = € 0.76

➲ **Cost of Living:** Austria is a relatively cheap European country.

➲ **Salaries/Taxes:** Expect to earn around £10 per hour but this depends on your qualifications. If you are contracted expect to earn £600 a month. Income tax and social security deductions are around 40%.

➲ **Accommodation:** A one-bedroom apartment in Vienna or Salzburg will cost around £250 per month.

➲ **Health advice:** Insurance coverage is advisable.

➲ **English Language Media:** The Vienna Reporter

Schools in Austria (+43)

Alpha Sprachinstitut Austria
Schwarzenbergplatz 16/ Canovagasse 5, A-1010 Vienna. **Tel:** 1 503 69 69, **Fax:** 1 503 69 69-14.

American International School
Salmannsdorferstrasse 47, A-1190 Vienna. **Tel:** 1 603 02 46.

Amerika-Institut
Operngrasse 4, 1010 Vienna. **Tel:** 1 512 77 20

Austro-British Society
Wickenburgasse 19, 1080 Vienna. **Tel:** 1 431 141.

Berlitz Sprachschule
Rotenturmstrasse 1-38, 1060 Vienna.

Berufsforderungsinstitut
Kinderspitalgasse 5, 1090 Vienna.

Business Language Centre
Trattnerhof 2, 1010 Vienna.

Danube International
Guderunstrasse 184, A-1100 Vienna.

Didactica Akademie Fur Wirtschaft Und Sprachen
Schottenfledgasse 13-15, 1070 Vienna. **Tel:** 1 526 2287.

English Courses And Seminars
Salzachstr. 16, Salzburg, 5020.
Tel: 66 2629995.

English For Kids
23 Anton-Baumgartner-Str. 44, Vienna, 1230.
Tel: 1 222 66 74579.

English Institute Linzerg 27
Salzburg, 5020. **Tel:** 66 2872195.

English Institute Sprachreisen Schoenberg
31 Michaelbeuern, 5152. **Tel:** 6 274 8236.

English Language Centre
In der Hagenau 7, 1130 Vienna. **Tel/Fax:** 1 804 1869.

Graz International Bilingual School
Klusemannstrasse 25, A-8053, Graz. **Tel:** 316 1050.

inlingua School of Languages
Neuer Markt 1, A-1010 Vienna, **Tel:** 1 512 22 25 / 1 512 94 99, **Fax**: 1 513 94 56.

Innsbruck International Highschool
Schonbeg 26, A-6141, Innsbruck. **Tel:** 05225 42013992.

Institut CEF
Garnisongasse 10, 1090 Vienna. **Tel:** 1 42 04 03

International Language Services
Getreidemarkt 17, A-1060 Vienna. **Tel:** 1 585 53 47, **Fax:** 1 585 53 47.

International Montessori Preschool Vienna
Mahlerstrasse 9/13, A-1010 Vienna. **Tel**: 1 512 8733.

International Summer School Gmunden
Postfach 182, A-4810 Gmunden, **Tel:** 1 494 9994

Jelinek & Jelinek
Privatlehrinstitut, Rudolfsplatz 3, 1010 Vienna.

Kindergarten Alt Wien
Am Heumarkt 23, A-1030, Vienna.

Lizner International School
Aubrunnerweg 4, A-4040 Linz. **Tel:** 732 245 8670.

Mini Schools & English Language Day Camp
Postfach 160, 1220 Vienna. **Tel:** 227717.

Multi Lingua Language School and Translation Bureau
Hardtgasse 5/2, A-1190 Vienna.**Tel:** 1 369 15 41**, Fax:** 1 369 15 42**.**

Privat Institut Venetia Erwachsenenbildung
Große Neugasse 8/19, A-1040 Vienna. **Tel:** 1 586 82 57, **Fax:** 1 586 82 579

Salzburg International Preparatory School
Moosstrasse 106, A-5020, Salzburg. **Tel:** 662 844 485.

Sight & Sound Studio
Schubertring 12, 1031 Vienna. **Tel:** 1 512 67520.

SLL Institut Super Language Learning
Florianigasse 55, A-1080 Vienna. **Tel:** 1 408 41 84

Spidi (Spracheninstitut Der Industrie)
Lotringer Strasse 12, 1031 Vienna. **Tel:** 1 715 2506-0, **Fax:** 1 89 17431.

Sprachstudio J-J Rousseau
Untere Viaduktgasse 43, 1030 Vienna. **Tel:** 1 712 2443

Sprachinstitut Vienna
Universitatstr. 6, 1090, Vienna. **Tel:** 1422227

Sprachschule BriBou
Barichgasse 26/3, A-1030 Vienna. **Tel:** 1 710 8829.

Super Language Learning Sprachinstitut
Florianigasse 55, 1080 Vienna. **Tel:** 1 408 4184

Wiener Volkshochschulen Verband Wiener Volksbildung
Hollergasse 22, A-1150 Vienna. **Tel:** 1 891 740.

Young Austria
Alpenstrasse 108a, A-5020 Salzburg. **Tel:** 662 625 7580

BELGIUM

Currently there is little demand for teachers in Belgium. English is taught at secondary school level and school leavers usually have enough English to get by with the demands that will be made on them later in life. Some of the private language schools are struggling to survive and are often a one or two-person business. Often it is only freelance work, which is available at private language schools. This is for a number of reasons. The first is that English is an asset but not essential to work in Belgium. Also, the UK is very close for those who want to travel to the UK to learn the language and newsagents are well supplied in English magazines and the British press. Last but not least, the country has cable television and this includes a wide choice of programmes in English (British and American). Many teenagers therefore learn the language by watching British and American soaps and viewing films in the original version with subtitles in their own language.

Embassy in UK:
103 Eaton Square, London SW1W 9AW
Tel: +44 (0)20 7470 3700
www.diplobel.org/uk

Embassy in USA:
3330 Garfield Street NW, Washington DC 20008
Tel: +1 202 333 6900
www.diplobel.org/usa

British Council office:
Leopold Plaza, Rue du Trône 108 / Troonstraat 108
1050 Brussels
Tel: +32 (2) 227 08 41
www.britishcouncil.org/belgium

➲ **Visa/Work Permits:** If you're a non-EU citizen you'll need to apply for a work permit. This means you'll have to prove no EU citizen is able to do your job–it is best to leave this paperwork to your prospective employer. There are two main types of work permit (permis de travail/werkvergunning), normally you will get a B permit which is valid for only one employer and must be renewed each year. After several years you should be able to transfer this to an A permit which is valid indefinitely and allows you to work for any employer you like. Contact the Ministère de l'Emploi et du Travail for information and application forms for work permits.

➲ **Requirements:** College Degree and a TESOL/TESL Certificate

➲ **Currency:** £1 = € 1.49 $1 = € 0.76

➲ **Cost of Living:** Expect to pay a third more than in the U.S. for basic living costs. Belgium is not as cheap as it used to be.

➲ **Salaries/Taxes:** Salaries are negotiable according to experience, but on contract expect to earn around £600 a month. Employers will pay your tax and health insurance only if you are on a contract.

➲ **Accommodation:** Schools rarely assist in finding accommodation. Outside Brussels rents are quite reasonable from a minimum of £150 per month for an apartment to £400 for a small house. An apartment in Brussels costs around £200.

➲ **English Language Media:** The Bulletin (weekly)

Schools in Belgium (+32)

Access Bvba Taalbureau
Atealaan 5, 2200 Herentals. **Tel:** 14 22 61 92

Applied Language Centre
21 rue du Trone 238, 1050 Brussels.

Belgo-British Courses
Rue d'Ecosse 21, 1060 Brussels. **Tel:** 2 537 8775.

Berlitz Language Centre
Avenue Louise 306, 1050 Brussels.

Berlitz Language Centre
Place Stéphanie 10, 1050 Brussels. **Tel:** 2 512 44 04.

Berlitz Language Centre
Britselei 15, 2018 Antwerpen.

Berlitz Language Centre
Leuvenselaan 17, 3300 Tienen.

Bilingua
6 rue Renier Chalon, 1060 Brussels. **Tel:** 2 347 4534.

British Council
rue de la Charité 15, 1210 Brussels. **Tel:** 2 227 0840.

British School of Brussels
Leuvensesteenweg 19, 3080 Tervuren. **Tel:** 2 767 4700.

Brussels Language Centre
55 Rue Des Drapiers, 1050 Brussels.

Crown Language Centre
9 Rue Du Béguinage, 1000 Brussels.

The English Institute
77 Rue Lesbroussart, 1050 Brussels.

inlingua School of Languages
62 Limburgstraat, 9000 Gent.

Institute of Modern Languages and Communications
20 Av. de la Toison D'or, Bte. 21, 1060 Brussels.**Tel:** 2 512 6607.

Institut Pro Linguis SC
Place De L'eglise, 6717 Thiamont.

Linguarama
Avenue des Arts, Kunstlaan 19, 1040 Brussels. **Tel:** 2 17 9055, **Fax:** 2 17 9523.

May International
40 rue Lesbroussart, 1050 Brussels. **Tel:** 2 640 8703.

Mitchell School of English
rue Louis Hap 156, 1040 Brussels. **Tel:** 2 734 8073.

Peters School
87 rue des Deux Eglises, 1040 Brussels. **Tel:** 2 280 0021.

Phone Languages
65 rue des Echevins, 1050 Brussels. **Tel:** 2 647 4020.

Practicum
Reep 24, 9000 Gent. **Tel:** 9 223 5442.

Pro Linguis
6717 Thiaumont. **Tel:** 63 22 0462, **Fax:** 63 22 0688.

School Voor Europese Talen
28 Charlottalei, 2018 Antwerpen. **Tel:** 3 218 7370.

DENMARK

A recent study found that Denmark ranked as the number one place in the world to live in terms of its ability to provide basic social and material needs for its citizens. However, there are few jobs and prospects for non-EU teachers are particularly poor. Children of school age are decreasing in numbers and schools and universities are cutting back on staff. You may be successful if you look for work in institutes running part-time courses and evening classes, particularly in business English.

Embassy in UK:
55 Sloane Street, London SW1X 9SR
Tel: +44 (0)20 7333 0200
www.demarkemb.org

Embassy in USA:
3200 Whitehaven Street NW, Washington DC 20008
Tel: +1 202 234 4300
www.demarkemb.org

British Council office:
Gammel Mønt 12.3, 1117 Copenhagen K
Tel: +45 (33) 369 400
www.britishcouncil.dk

➲ **Visa/Work Permits:** The same criteria apply as in other EU member states.

➲ **Requirements:** College Degree and a TESOL/TESL Certificate

➲ **Currency:** £1 = 11.13 / $1 = 5.69 Danish Krone

➲ **Cost of Living:** The cost of living in Denmark is very high with a beer costing around £4 in major cities and towns and a cheap meal costing about £4-5.

➲ **Salaries/Taxes:** Salaries vary according to the institution you are working for. Taxation rates run about 55%.

➲ **Accommodation:** A room in the capital city of Copenhagen can cost around £600 per month.

➲ **Health advice:** Excellent medical facilities are widely available in Denmark. You should have medical insurance to cover any expenses you may incur.

➲ **English Language Media:** Copenhagen Post.

Schools in Denmark (+45)

Access
Hamerensgade 8, 1267 Copenhagen K. **Tel:** 14 22 6192.

Activsprog
Rosenvægets Alle 32, 2100 Copenhagen - also Odense, Ärhus & Aalborg.

Babel Sprogtræning
Vordingborggrade 18, 2100 Copenhagen.

Berlitz International
Vimmelskaftet 42a, 1161 Copenhagen.

Bis Sprogskole
Rolfsvej 14-16, 2000 Frederiksberg.

Cambridge Institute Foundation
Vimmelskaftet 48, 1161 Copenhagen.
Tel: 33 13 33 02

Centre for Undervisning
Ärhus amt, Ulla Sørensen, Vesterskovvej 4, 8660, Skanderborg. **Tel:** 86 511511, **Fax:** 86 511047

Elite Sprogcentret
Hoffmeyersvej 19, 2000 Frederiksberg.

Erhvervs Orienterede Sprogkurser
Betulavej 25, 3200 Helsinge.

European Education Centre Aps (Inlingua)
Lyngbyvej 72, 2100, Copenhagen.

FOF
Sønder Allé 9, 8000 Ärhus C. **Tel:** 88 86 12 29 55.

Frit Oplysningsforbund
Vestergrade, 5000 Odense C. **Tel:** 66 13 9813.

Linguarama
Hvilevej 7, 2900 Hellerup.

Master-Ling
Sortedam Dossering 83, 2100 Copenhagen.

FINLAND

The Finns are eager to speak English well and many spend their spare time brushing up on the language. There are plenty of opportunities to teach here as long as you are prepared to be flexible.
Finding a teaching job in Finland without qualifications will be very difficult but if you look in the bigger cities like Helsinki and Turku you may be able to get hired by a company to teach their employees. The best time to look for work is in the spring, when schools start recruiting for the following September. Before setting off and looking for work, it is advisable to contact language schools in advance as you may find it hard once you're in the country. Be prepared for taking on a lot of freelance work and the comparative instability associated with this.

Embassy in UK:
38 Chesham Place, London SW1X 8HW
Tel: +44 (0)20 7838 6200
www.finemb.org.uk

Embassy in USA:
3301 Massachusetts Ave NW, Washington DC 20008
Tel: +1 202 298 5800
www.finland.org

British Council office:
Hakaniemenkatu 2, 00530 Helsinki
Tel +358 (9) 774 3330
www.britishcouncil.fi

➲ **Visas/Work Permits:** A visa is not required for tourist or business stays up to 90 days. Any non-Scandinavian wishing to work in Finland must apply for a residence permit and a work permit before entering the country. The process usually takes 6 to 8 weeks and costs around £100.

➲ **Requirements:** College Degree, experience and a TESOL/TESL Certificate

➲ **Cost of Living:** Go to Finland with plenty of cash reserves. The cost of living is not cheap (as with all Scandinavian countries) and going out, eating and drinking will seriously dent your wallet.

➲ **Currency:** £1 = € 1.49 $1 = € 0.76

➲ **Salaries/Taxes:** Teaching rates moderate ranging from £8 - £12 per teaching hour. If you are planning to spend less than 6 months in Finland you must pay income tax at 25%.

➲ **Accommodation:** Ask if your employer will provide you with a place to live as accommodation is very expensive throughout the country.

➲ **Health advice:** There is a standard fee (relatively low) for medical services. Your insurance policy will cover it.

➲ **Tip:** As a rule, Finns are quite shy and you may find it takes some time before you start to make friends.

Schools in Finland (+358)

AAC-Opisto
Kauppaneuvoksentie 8, 00200 Helsinki. **Tel/Fax:** 9 4766 7801.

The Federation of Finnish-British Societies
Puistokatu 1bA, 00140 Helsinki, **Tel:** 639625.

IWG Kieli-Instituutti
Hämeenkatu 25 B, 33200 Tampere. **Tel:** 31 214 7573, **Fax:** 31 212 3519.

Kielipiste Oy
Kaisaniemenkatu 4A, 00100 Helsinki. **Tel:** 90 177266, **Fax:** 90 174885.

Linguarama Suomi
David Eade, Annankatu 26, 00100 Helsinki. **Tel:** 9/6803230, **Fax:** 9/603118.

Lingua-Forum Ky
Fredrikinkatu 61 A 36, 00100 Helsinki. **Tel:** 694 8426, **Fax:** 694 8417.

Lansi-Suomen opisto
Loimijoentie 280, 32700 Huittinen. **Tel:** 2 567866, **Fax:** 2 566409.

Richard Lewis Communications
Itätuulenkuja 10 B, 02100 Espo, Helsinki. **Tel:** 455 4811, **Fax:** 466592.

FRANCE

The French have become more demanding of their teachers and expect them to be experienced and professional. "Student centred" teaching methods are not in vogue and you really need to be au fait with the rules of English grammar before you take on an English language teaching job in France.

Many Americans and other non-EU nationals despair of finding a teaching job in France. But if you are determined and are not above a bit of minor rule-bending, you can work as an English language teacher in France. Reports suggest that one route for job finders is to contact all the language schools they can find in the Yellow Pages and arrange to send a CV, photo and hand-written cover letter to the relevant person. Some schools are prepared to pay foreign teachers in cash and if you can find part-time jobs at three or more schools you may find yourself rolling in Euros.

Consulate in UK:
21 Cromwell Road, London SW7 2EN
Tel: +44 (0)20 7073 1200
www.ambafrance.org.uk

Embassy in USA:
4101 Reservoir Road NW, Washington DC 20007
Tel: +1 202 944 6000
www.info-france-usa.org

British Council office:
9 rue de Constantine, 75340 Paris cedex 07
Tel +33 (1) 4955 7300
www.britishcouncil.fr

➲ **Visas/Work Permits:** A visa is not required for a tourist/business stay up to 90 days in France, Andorra, Monaco, and Corsica. Non-EU citizens must find an institute that is willing to put them on a contract before they can obtain working papers.

➲ **Requirements:** College Degree and a TESOL/TESL Certificate

➲ **Cost of Living:** Living costs in the major cities are often surprisingly high, but rent and food (particularly wine!) are cheaper than in the UK and USA.

➲ **Currency:** £1 = € 1.49 $1 = € 0.76

➲ **Salaries/Taxes:** Salaries for teachers vary immensely in France and will usually be considerably lower in the provinces than in Paris, due to the lower cost of living. Contracted teachers can expect between £500 and £750 per month, while freelancers can expect £12 per hour.

➲ **Accommodation:** Expect to pay from around £350 per month for a one-bedroom apartment in Paris.

➲ **Health advice:** You will need insurance to avoid big bills if you fall sick or have an accident.

➲ **English Language Media:** International Herald Tribune, Paris Passion, The News (monthly)

Schools in France (+33)

AABC
20 rue Gonot de Mauroy, 75009 Paris. **Tel:** 1 42 66 13 11.

Academie des Langues Appliquées
60 rue de Laxou, BP 373 65- 40 98, Nancy.

Alexandra School
32 rue Amiral de Grasse, 66130 Grasse. **Tel:** 93 36 88 01.

Alpha Formation
51 rue Saint-Ferreol, 130001 Marseille. **Tel:** 91 33 00 72.

American Centre
Belomeau, Avenue Jean-Paul Coste, Paris. **Tel:** 1 42 38 42 38.

Anglesey Language Services (ALS)
1 bis Avenue Foch, 78400 Chatou. **Tel:** 1 34 80 65 15, **Fax:** 34 80 69 91.

Arc Langue
Chemin de la Haie, 64100 Bayonne. **Tel:** 59 55 05 66.

Audio-English
44 allée de Tourny, 33000 Bordeaux **Tel:** 56 44 54 05

BEST
24 Bd Béranger, 37000 Tours. **Tel:** 47 05 55 33, **Fax:** 47 64 40 27.

British Connection International
279 rue Crequi, 69007 Lyon. **Tel:** 72 73 02 55.

British Council
9/11 rue De Constantine, 75007 Paris. **Tel:** 49 55 73 00.

British Institute (University of London)
11 rue de Constantine, 74340 Paris, Cedex 07.

Business English Service & Translation
24 Bd Béranger, 37000 Tours. **Tel:** 47 05 55 33.

BTS Language Centre
226 Route de Philippeville, 6001 Marcinelle. **Tel:** 71313076.

Centre D'Etudes Des Langues
ISF Montfoulon, 61250 Damigny. **Tel:** 33 80 84 00.

Cybée Langues
7 rue d'Artois, 75008, Paris. **Tel:** 89 18 26

Collegium Palatinium
Dept EFL/CP, Chateau de Pourtales, 61 rue Mélanie, 67000. Strasbourg. **Tel:** 88 31 01 07.

EF Corporate
3 rue de Bassano, 75016 Paris. **Tel:** 47 20 18 01, **Fax:** 1 47 20 18 80.

Executive Language Services
20 rue Sainte Croix de la Bretonnerie, 75004 Paris. **Fax:** 1 48 044 55 53.

Fontainbleau Langues et Communication
15 rue Saint-Honoré, BP 27, 77300. Fontainbleau. **Tel:** 64 22 48 96.

Formalangues
106 Blvd Haussmann, 75008 Paris. **Tel:** 45229912.

Forum
66 Rue De La Bretonnerie, 45000 Orleans. **Tel:** 38625245.

International House Nice
62 rue Geoffredo, Nice, 6000

Linguarama
6 rue Roland Garros, 38320 Eybens. **Tel:** 76 62 00 18, **Fax:** 76 25 89 60.

Metropolitan Languages
151 rue de Billancourt, 92100 Boulogne. **Tel:** 46 04 57 32. **Fax:** 46 04 57 12.

Rapid English
2 rue du Maréchal Gallieni, 76600 Le Havre.

Regency Langues
1 rue Ferdinand Duval, 75004 Paris. **Tel:** 48 04 99 97, **Fax:** 48 04 34 96.

Rothman Institute
21 Avenue du Major Général Vanier, 1000 Troyes. **Tel:** 258 03041.

Sarl Executive Language Services
25 Blvd Sébastapol, 75001 Paris.

Unilangues
1 le Parvis-Paroi Nord, 92044 Paris la Défense Cedex 41. **Tel:** 47 78 45 80, **Fax:** 49 00 03 16.

GERMANY

Many institutions in Germany are looking for qualified and experienced teachers with a specialization in an area such as business English. But even newly qualified English language teachers with Master's degrees, certificates and diplomas are being hired. Many teachers are employed on a freelance basis, often teaching in-company outside office hours. There are lots of private language schools looking for teachers.

Recent reports suggest that it is relatively easy (when compared to other EU countries) for non-EU teachers to find teaching jobs and work legally in Germany. For Americans, USIA Fulbright programme organizes for US graduates to work as teaching assistants in high schools

Embassy in UK:
23 Belgrave Square, London SW1X 8PZ
Tel: 020 7824 1300
www.german-embassy.org.uk

Embassy in USA:
4645 Reservoir Road NW, Washington DC 20007
Tel: 202 298 4000
www.germany-info.org

British Council office
Hackescher Markt 1, 10178 Berlin
Tel +49 (30) 311 0990
www.britishcouncil.de

➲ **Tip:**Materials in American English are often hard to find. Bring some books and tapes over with you from the States. Your students, who may be used to British English, will appreciate working with different materials and they may even like your accent!

➲ **Visa/Work Permits:** U.S. citizens must arrange employment and apply for a work permit before they leave. This may take a couple of months.

➲ **Requirements:** College Degree, experience and a TESOL/TESL Certificate

➲ **Cost of Living:** Contrary to popular belief, living costs in Germany are lower than you might expect, and lower than in the UK or USA.

➲ **Currency:** £1 = € 1.49 $1 = € 0.76

➲ **Salaries/Taxes:** All employees with their residence in Germany have to pay tax on their income in accordance with the German tax laws. The rates of wage or income tax are graduated progressively according to the particular wages or salary. Expect to earn £6-15 per teaching hour at a private language school and £15-20 as a freelancer. Freelancers often tend to work outside the social security and tax structures. This tends to imply high earnings but not much in the way of statutory health cover and benefits. The situation has become more complicated since 1999 when a law dating back to 1913 was resurrected by the German Social Security Administration, which requires that freelance teachers pay nearly 20% of their income into the state social security system. This law is being applied retroactively—some teachers report they have been hit with high tax bills. Many freelancers have left Germany claiming it is no longer worth their time to teach there. The situation remains uncertain at the time of writing so consult a lawyer if you have any specific questions or problems.

➲ **Accommodation:** In any of the big cities a one-bedroom apartment will cost around £500. You will usually have to pay three months rent as a deposit

➲ **Health advice:** Foreigners in Germany are placed on an equal footing with German employees when it comes to social protection. They generally have to be insured against sickness and unemployment and included in the statutory pension scheme by their employer on taking up the employment. The contributions to the social insurance scheme are borne half by the employer and half by the employee. The employee share is deducted from the gross wage and paid to the social insurance provider together with the employer's share. The employer also has to insure you against accident with his trade association: premiums for accident insurance are paid by the employer alone. In practice, many private language schools do not comply with the law and do not make contributions on your behalf. Of course, freelancers are afforded no protection and should carry their own medical insurance.

➲ **English Language Media:** Handelsblatt (English edition), Munich Found (monthly), Spotlight, World Press

Schools in Germany (+49)

American Language Academy
Charlottenstr. 65, 10117 Berlin. **Tel:** 30 20 39 78 10, **Fax:** 30 20 39 78 13.

Anglo-German Institute
Christopherstr 4, 70178 Stuttgart. **Tel:** 711 603 858 **Fax:** 711 640 9941.

ASK Sprachenschule
1 Kortumstr 71, 44787 Bochum. **Tel:** 234-12910.

Bansley College
Str Des Friedens 35, 03222 Lubbenau, Brandenburg. **Tel:** 354-244 407, **Fax:** 354 244 408.

Benedict School
Gurzenichstr 17, 50667 Koln. **Tel:** 221-212 203.

Berlitz
Friedrich-Willhelm-Strasse 30, 47051 Duisburg.

British Council
Hardenbergstr 20, 10623 Berlin. **Tel:** 3110990.

Christopher Hills School of English
Sandeldamm 12, 63450 Hanau. **Tel:** 49-6181 15015, **Fax:** 49-6181 12121.

Collegium Palatinum
Adenauerplatz 8, 69115 Heidelburg. **Tel:** 62 2146 289, **Fax:** 62-2118 2023.

Didacta
Hohenzollernring 27, 95440 Bayreuth. **Tel:** 921-27 555.

English Language Centre
Bieberer Strasse 205, 63071 Offenbach am Main. **Tel:** 69 85 87 87, **Fax:** 69 85 72 02.

English Language Institute
Sprachenchule 4, Ubersetzer Am Zwinger 14, 33602 Bielefeld. **Tel:** 521-69353.

Europa-Universitat Viadrina
Sprachenzentrum, Grosse Scharrnstr 59, 15230 Frankfurt (Oder).

FBD Schulen
Katharinenstr 18, 70182 Stuttgart. **Tel:** 711-21580.

GLS Sprachenzentrum
B Jaeshke, Pestalozzistr 886, 10625 Berlin. **Tel:** 30-313 5025.

Hallworth English Centre
Frauenstrasse 118, 89703 Ulmponau. **Tel:** 731-22668.

inlingua Spachschule
Konigstrasse 61, 47051 Duisburg. **Tel:** 203 30 53 40.

Intercom Lang Services
Muggenkampstr 38, 20257 Hamburg.

International House
Prolog Sprachenschule, Hauptstrasse 23/24, Berlin, D-10827.

International House
LGS Sprachkurse, Werderring 18, Freiburg 79098

International House
Poststrasse 51, 20354 Hamburg. **Tel:** 352041.

Knowledge Point
Hohenzollernstrasse 26, 80801 Munchen **Tel:** 89 33 3405.

Linguarama Sprachinstitut
Kant Strasse 150, 10623 Berlin. **Tel:** 30 621 81 48, **Fax:** 30 312 94 91.

Lingotek Institut
Schlueterstrasse 18, 20146 Hamburg. **Tel:** 40 459 520.

Neue Sprachschule
Rosastrass 1, 79098 Freiburg. **Tel:** 761-24810.

NSK Language and Training Services
Comeniusstr 2, 90459 Nurnburg. **Tel:** 911-441552.

Sprachstudio Lingua Nova
Thierschstrasse 36, 80538 Munich. **Tel:** 89-221 171.

Stevens English Training
Ruttenscheider strasse 68, 45130 Essen.

Vorbeck-Schule
77723 Gengenback. **Tel:** 7803 3361.

Wirtschaftwissen-schaftliche Fakultat Ingolstadt
Auf der Schanz 49, 85049 Ingolstadt.

Yes Your English Services
Altonaer Chaussee 87, 22559 Schenefeld, Hamburg. **Tel:** 408 393 2116, **Fax:** 408 393 2117

GREECE

For many years, Greece has been one of the most popular destinations for English language teachers. Of all the candidates for the Cambridge First Certificate and Cambridge Proficiency exams worldwide, one quarter are in Greece. In most parts of the country, English language teaching in state schools is inadequate for these exams and children often attend private language schools (frontisteria) as a supplement.

There are estimated to be more than 5,000 of these schools in the country. You must be a graduate to teach and a certificate, diploma or MA in TESL is desirable although not absolutely necessary. Recruitment is normally done locally in May/June or September. Those less qualified may well have more success finding work in January, as many native speakers do not return to their jobs after the holiday break.

Embassy in UK:
1a Holland Park, London W11 3TP
Tel: 020 7221 6467
www.greekembassy.org.uk

Embassy in USA:
2221 Massachusetts Avenue NW, Washington DC 20008
Tel: 202 939 5800
www.greekembassy.org

British Council office:
17 Plateia Philikis Etairias, PO BOX 3488, 10673 Athens
Tel +30 (21) 0 369 2333
Www.britishcouncil.gr

➲ **Visa/Permits:** Non-EU citizens are supposed to apply for a work permit before they come to the country, for which they will need to present an employment offer from a Greek school, at their local Greek consulate. In practice, many teachers work while they are waiting for their permits to be processed. Americans of Greek descent can obtain residency permits but should check to make sure they won't have to serve in the army.

➲ **Requirements:** College Degree and a TESOL/TESL Certificate

➲ **Cost of Living:** Greece is no longer a cheap place to live in although it is less expensive than other parts of Europe. The cost of living in Athens depends to a large extent on the price of the accommodation. You'll need around £400 per month to cover your living expenses and rent.

➲ **Currency:** £1 = € 1.49 $1 = € 0.76

➲ **Salaries/Taxes:** Expect to earn approximately £400-500 a month at a language school. You can augment your income with private lessons.

➲ **Accommodation:** Apartments in Athens can be relatively expensive starting at around £200 per month. In smaller towns, some schools offer free accommodation but check to make sure the building has a roof and is adequately heated in winter.

➲ **Health advice:** Greek hospitals are notoriously overcrowded. Make sure you have a good medical insurance plan because if you fall sick or are injured, you will want to get back home as soon as you can.

➲ **English Language Media:** Athens News, Kathimerini (English edition).

Schools in Greece (+30)

A Trechas Language Centre
20 Koundouriotou St., Keratsini. **Tel:** 1 432 0546.

Alpha Abatzolglou Economou
10 Kosma Etolou St, 54643 Thessaloniki. **Tel:** 31 830535.

A Andriopoulou
3, 28 Octobrio, Tripolis.

Anglo-Hellenic Teacher Recruitment
PO Box 263, 201 00 Corinth **Tel:** 741 53511, **Fax:** 741 86989. **Email:** info@anglo-hellenic.com

Athens College
PO Box 65005, 15410 Psychico, Athens.**Tel:** 1 6714621, **Fax:** 647 8156.

British Council
Ethnikis Amynis 9, PO Box 50007, 54013 Thessalonika. **Tel:** 31 235 236.

English Tuition Centre
3 Pythias St., Kypseli, 1136 Athens.

Enossi Foreign Languages (The Language Centre)
Stadiou 7, Syntagma, 10562 Athens. **Tel:** 3230 356.

Eurocentre
7 Solomou Street, 41222 Larissa.

Featham School of English
PO Box 12, 50 Ep. Marouli St., Rethymnon, Crete. 74100. **Tel:** 831 23428, **Fax:** 831 23922.

Hambakis Schools of English
1 Filellinon Street, Athens. **Tel:** 1 3017531/5.

Hellenic American Union
22 Massalias St., GR-106 80 Athens.

International Language Centre
35 Votsi St., 26221 Patras.

Institute of English, French, German and Greek for Foreigners
Zavitsanou Sophia, 13 Joannou Gazi St, 31100 Lefkada.

Institute of Foreign Languages
41 Epidavrou St., 10441 Athens.

Kakkos School of English
Papastratou 5, Arginio 30100. **Tel/Fax:** 641 23449.

Makris School of English
2 Pardos G Olympion St., 60100 Katerini. **Tel:** 35122859.

G. Michalopolous School of English
24E Antistasis, Alexandria, 59300 Imathias, Thessaloniki. **Tel:** 33 3 322890.

New Centre
Arkarnanias 16, 11526 Athens

Omiros Association
52 Academias St., 10677 Athens. **Tel:** 362 2887, **Fax:** 362 1833.

Peter Sfyrakis School of Foreign Languages
21 Nikiforou Foka St, 72200 Ierapetra, Crete. **Tel:** 842 28700.

Polyglosso Language School
73 Agylis St., Neos Kosmos, Athens. **Tel/Fax:** 1 9214982.

Polyglosso Language School
96 Ag.Varvaras St., Dafne, Athens. **Tel/Fax:** 1 9766542.

Protypo English Language School
22 Deliyioryi Street, Volos 38221.

School of English
8 Kosti Palama, Kavala 65302.

School of Foreign Languages
12 P Isaldari St, Xylokastro, 20400 Corinth. **Tel:** 743 24678.

SILITZIS School of Languages
42 Koumoundourou, 412 22 Larissa.

Skouras Language School
7 G. Miltiadi St., 67100 Xanthi. **Tel:** 541 26112, **Fax:** 541 23721.

Strategakis Schools
24 Proxenou Koromila St, 54622 Thessaloniki. **Tel:** 31 264276, **Fax:** 228848.

Universal School of Language
66 M. Alexandrou St, Panorama, Thessalonika 55200. **Tel:** 31 341 014

Zoula Language Schools
Sanroco Square, 49100 Corfu. **Tel:** 661 39330

IRELAND

Many language students study English in Ireland in the summer months and the country is experiencing a boom. It is difficult for non-EU citizens to find jobs in Ireland unless they are in the I.T. industry, construction or nursing.

Embassy in UK:
17 Grosvenor Place, London SW1X 7HR
Tel: 020 7235 2171

Embassy in USA:
2234 Massachusetts Ave NW, Washington DC 20008
Tel: 202 462 3939
www.irelandemb.org

➲ **Visas/Work Permits:** Ireland is a member state of the EU so you will need an EU passport to work here. Americans and Canadians of Irish descent can qualify for an Irish passport if they were born in Ireland or have parents or grandparents who were. If you do not qualify for an Irish passport your employer will have to prove that you are doing a job that another EU citizen could not do.

➲ **Requirements:** College Degree and a TESOL/TESL Certificate

➲ **Cost of Living:** Ireland can be an expensive country to live in.

➲ **Currency:** £1 = € 1.49 $1 = € 0.76

➲ **Salaries/Taxes:** Low wages for teachers and high taxation rates make it difficult for teachers to survive here.

➲ **Accommodation:** It can be very difficult to find an apartment or room in Dublin over the summer months. Rents run around £50-100 per week per person in a shared house.

➲ **Health advice:** Good medical facilities are available.

Schools in Ireland (+353)

American College Dublin
2 Merrion Square, Dublin 2. **Tel:** 1 6768939, **Fax:** 1 6768941 **Email:** pmullins@amcd.ie

Centre Of English Studies
31 Dame Street, Dublin 2. **Tel:** 353 1 6714233, **Fax:** 1 6714425 www.cesireland.ie

Cork Language Centre International
Wellington House, Wellington Road, Cork **Tel:** 21 4551661, **Fax:** 21 4551662
www.corklanguagecentre.ie

Dublin School Of English
10-12 Westmoreland Street, Dublin 2. **Tel:** 1 6773322 **Fax:** 1 6795454 www.dse.ie/dse/

English Language Institute
99, St. Stephen's Green, Dublin 2. **Tel:** 1 4752965, **Fax:** 1 4752967 www.englishlanguage.com

Galway Language Centre
The Bridge Mills, Galway. **Tel:** 91 566468, **Fax:** 91 564122.

International House Dublin
Griffith College Campus, South Circular Road, Dublin 8. **Tel:** 1 454 9260 / 454 9261, **Fax:** 1 454 9262
www.ihdublin.com

The Language Centre Of Ireland
45 Kildare Street, Dublin 2. **Tel:** 1 6716266. **Fax:** 1 6716430 www.lci.ie

Waterford English Language Centres
31 John's Hill, Waterford. **Tel:** 51 877288, **Fax:** 51 854603 www.waterfortd-net.ie

ITALY

There is a great demand for English language teachers in Italy but it can be hard to find work, especially in the major cities. Generally, teachers need to be native speakers with at least a TESL certificate. Schools are trying to save money so recruitment from overseas is much less common than it was in the past, as schools tend to recruit locally. September is the best time to arrive in Italy to start your job search. Italian schools and companies prefer to recruit locally and there is a strong demand for English teachers in the corporate sector.

Embassy in UK:
14 Three Kings' Yard, Davies Street, London W1K 2EH
Tel: 020 7312 2200
www.embitaly.org.uk

Embassy in USA:
3000 Whitehaven Street NW, Washington DC 20008
Tel: 202 612 4400
www.italyemb.org

British Council office:
Via Quattro Fontane 20, 00184 Rome
Tel +39 (06) 478 141
www.britishcouncil.it

➲ **Visas/Work Permits:** It is now easier for Americans to arrive in Italy, find a teaching job, and then apply for a work permit without having to return to the States as was once the case.

➲ **Requirements:** College Degree and a TESOL/TESL Certificate

➲ **Cost of Living:** Italian cities are generally more expensive than their American counterparts but life in rural Italy is quite cheap.

➲ **Currency:** £1 = € 1.49 $1 = € 0.76

➲ **Salaries/Taxes:** Schools usually provide 25 contact hours a week at £10-15 per hour. A typical monthly salary is around £6-800 (with overtime paid extra). Schools often provide accommodation for the first month and some even give a ticket home at the end of the contract. Tax is payable at 20% with a 10% additional social security deduction.

➲ **Accommodation:** Accommodation is easy to find but tends to be expensive. Milan is the most expensive city and a one-bedroom apartment can cost about £700 a month but accommodation can be found for £300-400. Elsewhere, rents are cheaper.

➲ **Health advice:** Insurance is advisable.

Schools in Italy (+39)

Academia Britannica
Via Bruxelles 61, 04100 Latina. **Tel:** 773 491917.

Anglo American School
Piazza S. Giovanni in Monte 9, 40124 Bologna.

Anglocentre
Via A de Gasperi 23, 70052 Bisceglie.

Arlington Language Services
cp99, 29100 Piacenza.

Bari Poggiofranco English Centre
Viale Pio XII 18, 70124 Bari.

Benedict School
Via Sauro 36, 48100 Ravenna. **Tel:** 544 38199, **Fax:** 544 38399.

The British Council
Via Manzoni 38, 20121 Milan. **Tel:** 39 2 772221.

The British Council
Palazzo del Drago, Via Quattro Fontane 20, 00184, Rome. **Tel:** 6 478141.

British Institute
Fontane 109, Rome. **Tel:** 6491979**. Fax:** 64815549**.**

British Institute of Milan
Via Marghera 45, 20149 Milan. **Tel:** 2 48011149.

British Institute of Florence
Palazzo Feroni, Via Tornalbuoni 2, Florence. **Tel:** 55 298866.

British Institute of Rome
Via 4 Fontane 109, 00184 Rome.

The British Language Centre
Via Piazzi Angolo Largo Pedrini, 23100 Sondrio. **Tel:** 342 216130.

The British Language Centre
Via Piazza Roma 3, 20038 Serengo.

The British School of Bari
Via Celentano 27, 70121 Bari. **Tel:** 080 5247335, **Fax:** 080 5247396.

Cambridge Centre of English
Via Campanella 16, 41100 Modena. **Tel:** 59 24 1004

The Cambridge School
Via S Rochetto 3, Verona. **Tel:** 458003154

The Cambridge School
Via Mercanti 36, 84100 Salerno. **Tel:** 89 228942, **Fax:** 89 252523.

Canning School
Via San Remo 9, 20133 Milan.

Centro di Lingue
Via Pozzo 30, Trento. **Tel:** 461981733**. Fax:** 461981687**.**

Centro Internazionale di Linguistica Streamline
Via Piave 34/b, 71100 Foggia. **Tel:** 88124204**.**

Cento Lingue di Vinci Antonella
Via San Martino 77, Pisa.

Centro Lingue Tradint
Via Jannozzi 8, S Donato Milanese Chandler, Viale Aventino 102, 00153 Rome.

CLM-Bell
Eugen Joa, Via Pozzo 30, 38100 Trento. **Tel:** 461 981733, **Fax:** 461 981687.

English Centre
Via Promis 8, 11100 Aosta. **Tel:** 165 235416.

The English Centre
Via P Paoli 34, 07100 Sassari, Sardinia. **Tel:** 79 238154, **Fax:** 79 238180.

The English Connection
Via Ferro 1, 30027 San Dona di Piave.

English For Business Snc
Via Magnani Ricotti, n.2, 28100 Novara. **Tel:** 321 627186 **Fax:** 321 398061.

The English Institute
Corso Gelone 82, 96100 Siracusa. **Tel:** 961 6085.

English House
Via Roma 177, 85028 Rionero, Potenza.

English Language Centre
Viale Milano 20, 21100 Varese. **Tel:** 332 282732, **Fax:** 332 236385.

The English Language Studio
Via Antonio Bondi 27, 40138 Bologna. **Tel:** 51347394**, Fax:** 51505952**.**

English School
Via dei Correttori 6, 89127 Reggio Calabria. **Tel/Fax:** 965 899535.

Eurolingue
Via Chiana 116, 00198 Roma.

European Language Institute
Via IV Novembre 65, 55049 Viareggio (LU).

Filadelfia School
Via L. Colla 22, 10098 Rivoli.

Home School
Via F. Malvotti 8, Conegliano (TV).

Inlingua
Piazza XX Settembre 36, Civitanova Marche (MC).

Inlingua
Corso Vittorio Emanuel II 68, Torino.

Inlingua
Via Leoncino 35, 37121 Verona.

Inlingua
Via Monte Piana 42, Venezia.

International British School
Via Santacaterina 146, 89100 Regio Calabria.

International House Livorno
Piazza Folgore 1, 57128 Livorno. **Tel:** 336 709682, **Fax:** 586 508060.

International House Pisa
Via Risorgimento 9, 56126 Pisa. **Tel:** 50440 40.

International House Rome
Via San Godenzo 100, 00189 Rome.

International House Turin
Via Saluzzi 60, 10125 Turin. **Tel:** 39 11 669 95.

International Language School
Via Tibullo 10, 00193 Rome. **Tel**: 687 6801, **Fax:** 683 07796.

Language Centre
Via Milano 20, 21100 Varese. **Tel:** 332 282732.

Language Centre
Via G Daita 29, 90139 Palermo.

Lb Linguistico
Centro Insegnamento, Lingue Staniere, Via Caserta 16, 95128 Catania, Sicily.

Lingua Due Villa
Pendola 15, 57100 Livorno.

Living Languages School
Via Magna Grecia, 89100 Regio Calabria. **Tel/Fax:** 39 965 330926.

London School
Fabiola Cordaro, International House, Viale Emilia 34, 90144 Palermo. **Tel:** 39 91 52 524

Lord Byron College
Via Sparano 102, 70121 Bari. **Tel:** 80 238696.

Managerial English Consultants
Via Sforza Pallavicini 11, 009193 Rome.**Tel:** 6 654 2391, **Fax:** 6 6871159.

Modern English School
Via Borgonuovo 14, 40125 Bologna. **Tel:** 51 227523, **Fax:** 51225314**.**

Multimethod
I Go Richini 8, 20122 Milan. **Tel:** 2583042**.**

Oxford School
San Marco 1513, Venice. Fax: 41 5210785.

Oxford School of English
Via S. Pertini 14, Mirano, 30035 Venice. **Tel:** 41 570 2355, **Fax:** 41 570 2390.

The Professionals
Via F Carcona 4, 20149 Milan. **Tel:** 2 48000035, **Fax:** 2 4814001.

Regency School
Via dell' Arcivescovado 7, 16121 Turin. **Tel:** 11 562 7456, **Fax**: 11 541845.

Regent International
V.U. Da Pisa 6, Milan.

Regent International
Corsa Italia 54, 21047 Saronno.

RTS Language Training
Via Tuscolana 4, 00182 Rome. **Tel:** 6 702 2730, **Fax:** 6 702 1740.

Scuola The Westminster
Via Tevere 84, Sesto Fiorentino.

Spep School
Via della Secca 1, 40121 Bologna.

Studio Linguistico Fonema
via Marconi 19, 50053 Sovigliana-Vinci. **Tel:** 571 500551.

Studio professionale Apprendimento Linguistico Programmato
Via Ferrarese 3, Bologna. **Tel:** 051 360617, **Fax:** 513 68413.

Summer Camps
Via Roma 54, 18038 San Remo. **Tel/Fax:** 184 506070.

Unimoney
Corso Sempione 72, 20154 Milan.

Victoria Language Centre
Viale Fassi 28, 41012 Capri.

Wall Street Institute
Piazza Combattento 6, 4100 Ferrara. **Tel:** 532 200231.

THE NETHERLANDS

The current employment situation for English language teachers is good. There is quite a demand for teachers of English who are native speakers and there may also be possible opportunities for qualified teachers in a new type of school. There are forty of these at present and they teach children in English and Dutch. The Dutch term for the schools is "Twee Talig Scholen." Most work opportunities available to foreign teachers tends to be teaching Business English as the levels of English in the school system are very high. Schools depending mainly on long-term freelance teachers. Self-employed teachers typically work for a number of schools or agencies and their incomes can fluctuate considerably. To find work, go through the list of language schools below as well as those in the Dutch Yellow Pages. It is likely you will have to give an observed lesson before you are given the job.

Embassy in UK:
38 Hyde Park Gate, London SW7 5DP
Tel: 020 7590 3200
www.netherlands-embassy.org.uk

Embassy in USA:
4200 Linnean Ave NW, Washington DC 20008
Tel: 202 244 5300
www.netherlands-embassy.org

British Council office:
Weteringschans 85A, 1017 RZ Amsterdam
Tel +31 (20) 550 6060
www.britishcouncil.org/netherlands

➲ **Visa/Work Permits:** The Employment of Foreign Workers Act says that an employer cannot employ a foreigner without an employment permit from the Central Employment Board. The employer must first fill a vacancy from the pool of job applicants immediately available in the Netherlands. Subsequently it will be determined if applicants from other member states of the European Community are available. Only after these two applicant pools are exhausted can subjects from other countries be eligible for an employment permit. An employment permit can be sought only after a residence permit application (VTV) has been submitted. Applications for employment permits are available from the Regional Employment Office. Applications for employment permits are processed by the Employment Office only if the form has been completed, signed by both the employer and the employee, and all required material has been attached.

➲ **Requirements:** College Degree, experience and a TESOL/TESL Certificate

➲ **Cost of Living:** The Netherlands is an expensive place to live.

➲ **Currency:** £1 = € 1.49 $1 = € 0.76

➲ **Salaries/Taxes:** Qualified teachers can earn £8-12 per lesson. Tax is payable at around 30%.

➲ **Accommodation:** Accommodation in the Amsterdam area is extremely expensive and a studio apartment could cost you over £200 a month. Also bear in mind that a place will also be very hard to find.

➲ **Health advice:** If you are a taxpayer then you will be entitled to medical coverage under the 'Ziekenfonds' system. But you will still have to pay a premium to the health insurance company of between £6 - 12 a month.

Schools in The Netherlands (+31)

All-English Institute
Iepenlaan 96, 2061 GN Bloemendaal (Haarlem area)
Tel: 023-5277394

Arcus Contracting B.V.
Diepenbrockstraat 15, 6411 TJ Heerlen **Tel:** 045-5609070

Berlitz Language Centre
Rokin 87-89, 1012 Kl Amsterdam **Tel:** 020-6221375

British Language Training Centre
NZ Voorburgwal 328, 1012 RW Amsterdam **Tel:** 020-6223634

The British School in the Netherlands
ELC, Vlaskamp 19, 2592 AA Den Haag **Tel:** 070-3338130

CBE - Centre for British English
Bergsingel 89-A, 3037 GA Rotterdam **Tel:** 010-4678705

C J Korrel
Schinkelhavenstraat 13 hs, 1075 VN Amsterdam **Tel:** 020-6627441

Communications in English
Piet Heinstraat 101, 2518 CE Den Haag **Tel:** 070-3466314

Delftse Volksuniversiteit
Ruys de Beerenbrouckstraat 31, 2613 AS Delft **Tel:** 015-2570130

De Taaltrainers
Burg. Raaymaekerslaan153, 5361 KL Grave **Tel:** 0486-421230

Double Dutch
Beringstraat 18, 7532 CP Enschede **Tel:** 053-4615822

Dutchess English Language Centre
Schoutenstraat 86, 1623 RZ Hoorn **Tel:** 0229-210581

Engelse Taalcursussen
IJsselsteinseweg 63, 3435 PB Nieuwegein **Tel:** 030-6035506

English Language Institute
Berkenweg 46, 3741 BZ Baarn **Tel:** 035-5424725

Fontys Hogeschool
Oude Markt 5, 6131 EN Sittard **Tel:** 046-4599579

Fontys Hogeschool Talencentrum
Rachelsmolen 1, 5600 AH Eindhoven **Tel:** 040-2454972

Franglais
Molenstraat 15, 2513 BH Den Haag **Tel:** 070-3611703

HES/O & O
K.P. van der Mandelelaan 22, 3062 MB Rotterdam **Tel:** 010-4532451

Int. Studiecentrum v/d Vrouw
Concertgebouwplein 17, 1071 LM Amsterdam **Tel:** 020-6761437

Instituut De Wester
Ruusbroechof 37, 1813 BC Alkmaar **Tel:** 072-5400920

Instituut Taaltip
Floralaan 19, 1211 JT Hilversum **Tel:** 035-7720211

Instituut Uithoorn
Postbus 238, 1420 AE Uithoorn **Tel:** 020-4823482

Instituut Verschoor
Piet Heinlaan 8, 3941 VG Doorn **Tel:** 0343-413319

Instituut voor Taalhantering
Postbus 368, 8000 AJ Zwolle **Tel:** 038-4449263

The Language Academy
Spuistraat 134, 1012 VB Amsterdam **Tel:** 020-5254637

Leidse Volksuniversiteit K & O
Oude Vest 45, 2312 XS Leiden **Tel:** 071-5141141

LOI Tel: correspondentie cursussen
Postbus 4200, 2350 CA Leiderdorp **Tel:** 071-5451888

Onderwijsinstituut Netty Post
Haverstraat 2, 1446 CE Purmerend **Tel:** 0299-641416

P.C.I. Talen
Bachlaan 43, 1817 GH Alkmaar **Tel:** 072-5156518

RIVO
Dorpstraat 3, 2902 BC Capelle a/d IJssel **Tel:** 010-4519464

ROC Albeda Contract Groep
Heemraadsingel 178, 3021 DL Rotterdam **Tel:** 010-4255333

ROC Da Vinci College
Groenedijk 49, Dordrecht **Tel:** 078-6572657

ROC Instituut De Factoor
Postbus 1, 8000 AA Zwolle **Tel:** 038-8508026

ROC Joke Smit School
R Vinkeleskade 62, 1071 SX Amsterdam **Tel:** 020-6647401

ROC Koning Willem I College
Vlijmenseweg 2, 5223 GW 'sHertogenbosch **Tel:** 073-6249624

Schoevers Opleidingen
Bezuidenhoutseweg 273, 2594 AN Den Haag **Tel:** 070-3491499

Schoevers Opleidingen
Eendrachtskade/Zuidzijde 12 Groningen **Tel:** 050-3182202

Schoevers Opleidingen
Javastraat 16-18, 6524 MB Nijmegen **Tel:** 024-3229742/024-3787602

Schoevers Opleidingen
Karel Doormanstraat 331, 3012 GH Rotterdam **Tel:** 010-4117855

Schoevers Opleidingen
Kromme Nieuwe Gracht 3, 3512 HC Utrecht **Tel:** 030-2368152

Schoevers Opleidingen
Markt 17, 5611 EB Eindhoven **Tel:** 040-2442832

Schoevers Opleidingen
Stadhouderskade 60, 1072 AC Amsterdam **Tel:** 020-6627156

Schoevers Opleidingen
Tuinstraat 1, 5000 JL Tilburg **Tel:** 013-5424128

Schoevers Opleidingen
Utrechtsestraat 35, 6811 LT Arnhem **Tel:** 026-4453641

The School of English
1e Wormerweg 238, 7331 MT Apeldoorn **Tel:** 055-5336769

Secradesk
Postbus 545, 7500 AM Enschede **Tel:** 053-4836420

Stichting Holtland Contract
Lammenschanspark 2, 2381 JK Leiden **Tel:** 071-5321952

Stichting IVIO
De Schans 1802, 8231 KA Lelystad
Tel: 0320-229948

Stichting Rotterdamse Volksuniv.
Heemraadsingel 275, 3023 BE Rotterdam **Tel:** 010-4761200

Suitcase
Divertimentostraat 55, 1312 EB Almere **Tel:** 036-5367482

Taalcursussen Doorn
W. Alexanderweg 117, 3945 CH Cothen **Tel:** 0343-561617

Taaltrainingscentrum Barendrecht
Maasstraat 9, 2991 AD Barendrecht **Tel:** 0180-615410

Talenlaboratorium
Rijksuniversiteit Leiden, Postbus 9515, 2300 RA Leiden **Tel:** 071-5272332

Taleninstituut Bogaers
Groenstraat 139-155, 5021 LL Tilburg **Tel:** 013-5362101

Talenpracticum Twente B.V.
Ariënsplein 2, 7511 JX Enschede **Tel:** 053-4317842

Universitair Talencentrum Nijmegen
Erasmusplein 1, 6525 HT Nijmegen
Tel: 024-612159

Utrechtse Volksuniversiteit
Nieuwe Gracht 41, 3512 LE Utrecht **Tel:** 030-2313395

Van Beers & Kooijmans vof
Rigelhof 14, 3318 CZ Dordrecht **Tel:** 078-6183998

Volksuniversiteit Amersfoort
Korte Bergstraat 5, 3811 ML Amersfoort **Tel:** 033-4656640

Volksuniversiteit Amstelland
Burg. Haspelslaan 2, 1181 NB Amstelveen **Tel:** 020-6457506

Volksuniversiteit Amsterdam
Rapenburgerstraat 73, 1011 VK Amsterdam **Tel:** 020-6261626

Volksuniversiteit Bodengraven
Groene Zoom 149, 2411 SP Bodegraven **Tel:** 0172-616924

Volksuniversiteit Enschede
Molenstraat 27, 7514 DJ Enschede **Tel:**053-4323304

Volksuniversiteit Groningen
Oude Boteringestraat 21, 9712 GC Groningen **Tel:** 050-3126555

Volksuniversiteit Haarlem
Leidsevaart 220, 2014 HE Haarlem
Tel: 023-5329453

Volksuniversiteit
Havensingel 24, 5211 TX 's Hertogenbosch **Tel:** 073-6141120

Volksuniversiteit Hilversum
Arena 301, 1213 NZ Hilversum
Tel:035-6892051/035-6892000

Volksuniversiteit Soest/Soesterberg
Di Lassostraat 65, 3766 EB Soest **Tel:**035-6022195

The Workshop
Rostocklaan 38, 7315 HM Apeldoorn **Tel:** 055-5216736

Zeeuwse Volksuniversiteit
Postbus 724, 4330 AS Middelburg **Tel:** 0118-634800

NORWAY

There is often a shortage of English language teachers in Norway but it is hard to get a job unless you are qualified and have some experience. For non-EU teachers, it is almost impossible to find a job teaching English unless you are a Norwegian citizen or you are married to one!

Embassy in UK:
25 Belgrave Square, London SW1X 8QD
Tel: 020 7591 5500
www.norway.org.uk

Embassy in USA:
2720 34th Street NW, Washington DC 20008
Tel: 202 333 6000
www.norway.org/embassy

British Council office:
Fridtjof Nansens Plass 5, 0160 Oslo
Tel +47 (22) 396 190
www.britishcouncil.no

➲ **Visas/Work Permits:** A valid passport is required. U.S. citizens may enter Norway for tourist or general business purposes without a visa for up to 90 days. Tourists who have stayed in Norway for more than 90 days without a visa are usually not permitted to re-enter the country until six months have passed. Tourists who enter without a visa cannot usually change status in Norway in order to reside or work. Travellers planning a long-term stay, marriage or employment in Norway should seek the appropriate visa before departing the United States.

➲ **Requirements:** College Degree and a TESOL/TESL Certificate

➲ **Currency:** £1 = 12.18 / $1 = 6.23 Kroner

➲ **Cost of Living:** very expensive country to live in but the social welfare system is excellent.

➲ **Health advice:** Medical facilities are widely available and of high quality, but may be limited outside the larger urban areas. The remote and sparse populations in northern Norway, and the dependency on ferries to cross fjords of western Norway, may affect transportation and access to medical facilities.

Schools in Norway (+47)

Det Internasjonale Spraksenter
Dronningensgenst 32, 0154 Oslo. **Tel:** 22 33 15 20. **Fax:** 22 33 69 30.

Norsk Sprachinsitut
Korgengst. 9, 0153 Oslo: **Tel:** 23 10 01 10. **Fax:** 23 10 01 27.

Sprakskolen As
Karl Johansgat 8, 0154 Oslo. **Tel:** 22 42 00 87. **Fax:** 22 42 32 94.

PORTUGAL

Over the last few years it has become more difficult for non-EU passport holders to find teaching jobs in Portugal. But Americans and Canadians do find work, especially in Lisbon where some schools are willing to employ unqualified native-speaking tourists.

Embassy in UK:
11 Belgrave Square, London SW1X 8PP
Tel: 020 7235 5331

Embassy in USA:
2125 Kalorama Road NW, Washington DC 20008
Tel: 202 328 8610

British Council office:
Rua Lui Fernandes 1-3, 1249-062 Lisbon
Tel: +351 21 321 4500
www.pt.britishcouncil.org

➲ **Visa/Permits:** If you are lucky enough to get work before you leave (which is advisable), your paperwork should be sorted out in advance. A visa is not required for tourist or business stays of up to 60 days.

➲ **Requirements:** College Degree and a TESOL/TESL Certificate

➲ **Cost of Living:** Reasonable compared to the UK and USA

➲ **Currency:** £1 = € 1.49 $1 = € 0.76

➲ **Salaries/Taxes:** Annual income for a teacher ranges from £8,000 to £10,000. Tax is about 20% of your gross salary.

➲ **Accommodation:**Accommodation may be provided by the institute. You will pay at least £400 per month for accommodation in Lisbon.

➲ **English Language Media:** The News (weekly)

Schools in Portugal (+351)

American Language Institute
Rua José Falcao 15-5° Esq, 4050 Oporto. **Tel:** 2 318 127, **Fax:** 2 208 5287.

Berlitz
Av Conde Valbom 6-4, 1000 Lisbon.

Big Ben School
Rua Moinho Fanares 4-1, 2725 Mem Martins.

Bristol School Group
Trav. Dr Carlos Felgueiras, 12-3°, 4470 Maia. **Tel:** 2 948 8803, **Fax:** 2 948 6460.

British Council
Rua De Sao Marcal 174, 1294 Lisbon Codex. **Tel:** 347 6141.

British Council
Rua Do Breyner 155, 4050 Oporto. **Tel:** 200 5577.

Cambridge School
Avenida da Liberdade 173-4, 1200 Lisbon. **Tel:** 352 7474, **Fax**: 353 4729.

Casa de Inglaterra
Rua Alexandre Herculano 134, 3000 Coimbra.

CENA-Cent.
Est. Norte Americanos, Rua Remedios 62 c/v, 1200 Lisbon.

Centro de Estudos IPFEL
Rua Edith Cavell 8, 1900 Lisbon.

Centro de Instrucao Tecnica
Rua Da Estefania 32-10 Dto, 1000 Lisbon.

Centro Internacional Linguas
Av Fontes P de Melo 25-1Dto, 1000 Lisbon.

Centro de Linguas de Alvide
Rua Fonte Nino, Viv Pe Americo 1, Alvide, 2750 Cascais.

Centro de Linguas Estrangeiras de Cascais
Av Marginal BI A-30, 2750 Cascais.

Centro de Linguas Intergarb
Tv da Liberdade 13-1, 8200 Albufeira.

Centro de Linguas de Quarteira
Rua Proj 25 de Abril 12, 8125 Quartiera.

Centro de Linguas de Queluz
Av Dr Miguel Bombarda 62-1E, 2745 Queluz.

Centro de Linguas de Santarem
Lg Pe Francisco N Silva, 2000 Santarem.

CIAL-Centro De Linguas
Av Republica 14-20, 1000 Lisbon. **Tel:** 3533733

Class
Rua Gen Humberto Delgado 40-1, 7540 Santiago Do Cacem.

Clube Conversacao Inglesa 3M
Rua Rodrigues Sampaio 18-3, 1100 Lisbon.

Communicate Language Institute
Praceta Joao Villaret 12B, 2675 Povoa de Sto. Adriao.

Curso de Linguas Estrangeiras
Rua Dr Miguel Bombarda 271-1, 2600 Vila Franca De Xira.

Ecubal
Lombos, Barros Brancos, Porches, 8400 Lagoa.

ELTA
Av Jose E Garcia 55-3, 2745 Queluz.

Encounter English
Av Fernao De Magalha, 4300 Porto. **Tel:** 567916.

English at PLC
Praca Luis de Camoes 26, Apartado 73, 5001 Vila Real.

English Institute Setubal
Av 22 Dezembro 88, 2900 Setubal.

The English Language Centre
Rua Calouste Gulbenkian 22-r/c C, 3080 Figueira Da Foz.

The English School of Coruche
Rua Guerreiros 11, 2100 Coruche.

Escola de Linguas de Agueda
Rua Jose G Pimenta, 3750 Agueda

Escola de Linguas de Ovar
Rua Ferreira de Castro 124-1 A/B, 3880 Ovar.

Eurocentre Instituto de Linguas
Av de Bons Amigos 4-1, 2735 Cacem.

Gab Tecnico de Linguas
Rua Hermenegildo Capelo 2-2, 2400 Leiria.

GEDI
Pq Miraflores Lt 18-lA/B, 1495 Alges.

Greenwich Instituto de Linguas
Rua 25 de Abril 560, S Cosme, 4420 Gondomar. **Tel/Fax:** 2 483 6429.

IF - Ingles Funcional
Rua Afonso Albuquerque 73-A, 2460 Alcobaca.

IF Ingles Funcional
Rua Com Almeida Henriques 32, 2400 Leiria.

IF Ingles Funcional
Ap 303, 2430 Marinha Grande. **Tel:** 44 568351

INESP
Rua Dr Alberto Souto 20-2, 3800 Aveiro.

INLINGUA
Campo Grande 30-1A, 1700 Lisbon.

INPR
Bernardo Lima 5, 1100 Lisbon.

Instituto Britanico
Rua Conselheiro Januario 119/21, 4700 Braga. **Tel:** 53 23898.

Instituto Britanico
Rua Municipio Lt B - 1 C, 2400 Leiria.

Instituto Britanico
Rua Dr Ferreira Carmo, 4990 Ponte De Lima.

Instituto Franco-Britanico
Rua 5 de Outubro 10-1Dto, 2700 Amadora.

Instituto Inlas do Porto
Rua S da Bandeira 522-1, 4000 Porto.

Instituto de Linguas
Rua Valverde 1, 2350 Torres Novas.

Instituto de Linguas do Castelo Branco
Av 1 Maio, 39 S - 1 E, 6000 Castelo Branco.

Instituto de Linguas de Faro
Av 5 de Outubro, 8000 Faro.

Instituto de Linguas do Fundao
Urb Rebordao Lt 17-r/c, 6230 Fundao.

Instituto de Linguas de Oeiras
Rua Infante D Pedro 1 e 3-r/c, 2780 Oeiras.

Instituto de Linguas de Paredes
Av Republica, Casteloes Cepeda, 4580 Paredes.

Instituto Nacional de Administracao
Centro de Linguas, Palacio Marquus de Oeiras, 2780 Oeiras.

Instituto Sintrense de Linguas
Rua Dr Almeida Guerra 26, 2710 Sintra.

Interlingua
Lg 1 de Dezembro 28, 8500 Portimao.

Interlingu
Rua Dr Joaquim Telo 32-1E, 8600 Lagos.

International House Braga
Rua dos Chaos 168, 4710 Braga. **Tel:** 53 215250, **Fax**: 53 61228.

International House Lisbon
Rua Marques Sa Da Bandeira 16, 1000 Lisbon

International Language School
Av Rep Guine Bissau, 26-A, 2900 Setubal.

ISLA
Bo S Jo de Brito, 5300 Bragan A, Manitoba,C Com Premar, l 72, 4490 Povoa De Varzim.

Know-How
Av Alvares Cabral 5-300, 1200 Lisbon.

Lancaster College
Rua C Civico, Ed A Seguradora 2, 6200 Covilha.

Lancaster College
Pta 25 Abril 35-1E, 4400 Vila Nova De Gaia. **Tel:** 2 306 495.

Language School
Rua Alm Candido Reis 98, 2870 Montijo.

Linguacoop
Av. Manuel da Maia 46-10 D, 1000 Lisbon.

Linguacultura
Rua Dr Joaquim Jacinto 110, 2300 Tomar.

Linguacultura
Lg Sto Antonio 6-1 Esq, 2200 Abrantes.

Lisbon Language Learners
Rua Conde Redondo 33-r/cE, 1100 Lisbon.

Mundilingua
Rua Dr Tefilo Braga, Ed Rubi-1, 8500 Portimao.

Mundilinguas
Rua Miguel Bombarda 34-1, 2000 Santarem.

The New Institute of Languages
Urb Portela Lt 197-5B-C, 2685 Sacavem. **Tel:** 1 943 5238.

Novo Instituto de Linguas
Rua Cordeiro Ferreira 19 C-1D, 1700 Lisbon.

PEC
Rua SA Bandeira 5385, 4000 Oporto. **Tel:** 200 5077.

PROLINGUAS
Rua Saraiva Carvalho, 84 - Pt2, 1200 Lisbon.

Royal School of Languages
Av Dr Lourenco Peixinho 92-2, 3800 Aveiro. **Tel:** 34 2956, **Fax**: 34 382870.

Tell School
Rua Soc Farmaceutica 30-1, 1100 Lisbon.

Tjaereborg Studio
Av Liberdade 166-4F, 1200 Lisbon.

Wall Street Institutes Avenidas
Av Praia de Vitoria, 71 3rd, 1000 Lisbon. **Tel:** 316 0810, **Fax**: 316 0553.

Wall Street Institutes Oporto
Rua do Campo Alegre 231, 3°, 4100 Oporto. **Tel/Fax:** 2 600 9875.

Weltsprachen-Institut
Qta Carreira 37 r/c, 2765 Sao Joao do Estoril. **Tel:** 4684032.

Whyte Institute
Lg das Almas 10-2E/F, 4900 Viana Do Castelo.

World Trade Centre - Lisbon
Av Brasil 1-5e8, 1700 Lisbon.

SPAIN

Spanish English language schools have been hit recently with economic problems and some of the major chains have closed down. However, there are still jobs available, especially to qualified native-speakers from the UK.

The best time to arrive is in September when schools are recruiting teachers for the academic year ahead.

Native English language speakers can find teaching jobs in Spain without any qualifications beyond their degrees but it will make your professional life much easier if you have some training under your belt! The maximum number of teaching hours per week allowed by law is 33, but contracted teachers can expect a challenging schedule with close to this number of hours and often at inconvenient times. You might consider working privately as an option because it is better paid than working in schools. If you're in Madrid, try putting an ad for your services in the Segundamano newspaper.

Although it is difficult for non-EU citizens to work legally in Spain, the reality is that thousands of Americans, Canadians and Australians are employed as English language teachers throughout the country.

Embassy in UK:
39 Chesham Place, London SW1X 8SB
Tel: 020 7235 5555

Embassy in USA:
2375 Pennsylvania Ave NW, Washington DC 20037
Tel: 202 452 0100

www.spainemb.org

British Council office:
Paseo del General Martinez, Campos 31, 28010 Madrid
Tel +34 (91) 337 3500
www.britishcouncil.es

➲ **Tip:** Dress formally for your interview. If you get the job, you can ask later if casual attire is acceptable while working.

➲ **Visa/Permits:** Officially, non-EU citizens require a work visa if they are staying over three months. In reality most work without contracts at least initially as work visas are difficult to get. Those wishing to stay in the long term usually apply for a work visa, involving documentation from their employer and from the US, and a return trip, a lengthy process

➲ **Requirements:** College Degree and a TESOL/TESL Certificate

➲ **Cost of Living:** Urban life in Spain is expensive but if you are selective about where you stay, then savings can be made.

➲ **Currency:**£1 = € 1.49 $1 = € 0.76

➲ **Salaries/Taxes:** A minimum monthly salary in the big cities is around £300 / $550 a month. Private sector schools pay approximately £4.50 / $8 per hour in Catalonia. Freelancers can earn around £6.50 / $12 per hour in the Valencia region and from £7 / $13 to £15 / $30 in Catalonia. Residents pay 15% tax in arrears and an additional 6% in social security payments. If you work cash in hand (which is often the case for non-EU teachers) then you should not pay any taxes.

➲ **Accommodation:** Rents often take up more than a quarter of a teacher's salary, especially in the major cities, although lower salaries in smaller towns might include accommodation. Expect to pay a minimum of £200 / $380 a month if you want to live on your own. For shared accommodation, standard rents are around £100 / $190 per month. Up to two months' deposit is payable in advance.

➲ **Health advice:** As you will probably be working as a freelancer you must have your own insurance.

➲ **English Language Media:** Barcelona Metropolitan, Costa Blanca News, Sur

Schools in Spain (+34)

Academia Andaluza de Idiomas
Crta El Punto 9, Conil, Cadiz, **Tel:** 956 44 0552.

Academia Britanica
Rodriguez Sanchez 15, 14003 Cordoba.

Academia Saint Patricks
Calle Caracuel 1, 17402 Girona.

Academia Wellington House
Guiposcoa 79, 08020 Barcelona.

Acento - The Language Company
Ruiz De Alarcon 7, 21, 41007 Sevilla.

Afoban
Alfonso Xii 30, 41002 Sevilla.**Tel:** 95 421 8974

AHIZKE/CIM
Loramendi 7, Apartado 191, Mondragon Guipuzcoa.

Alce Idiomas
Nogales 2, 33006 Oviedo. **Tel:** 85 254543.

Aljarafe Language Academy
Crta Castilleja-Tomares 83, Tomares, Sevilla.

American Institute
El Bachiller 13, 46010 Valencia.

Apple Idiomas
Aben al Abbar 6, 46021 Valencia. **Tel:** 362 25 45.

Audio Jeam
Pza Ayuntamiento 2, 46002 Valencia.

Augusta Idiomas
Via. Augusta 128, 08006 Barcelona.

Aupi
Jesus 43, 46007 Valencia.

Avila Centre of English
Bajada de Don Alonso 1, 05003 Avila. **Tel:** 18 213719, **Fax**: 213631.

Berlitz
Gran Via 80, 4º, 28013 Madrid. **Tel:** 1 542 3586, **Fax**: 1 541 2765.

Berlitz
Edif Forum 1 Mod, 3 Av Luis Morales, S/N 41018 Sevilla.

Big Ben College
Plaza Quintiliano 13, Calahorra 26500 La Rioja.

Brighton
Rambla Catalunya 66, 08007 Barcelona.

Bristol English School
Fundacion Jado 10 Bis 6, 48950 Erlandio, Vizacaya. **Tel:** 467 6332.

Britannia School of English
s 9, Bl 2 41010 Sevilla.

Britannia School
Leopoldo Lugones 3-1B, 33420 Lugones, Asturias. **Tel:** 85 26 2800.

Britannia School
Raset 22, Barcelona 08021. **Tel:** 3 200 0100

British Council
Bravo Murillo 25, 35003 Las Palmas De Gran Canaria. **Tel:** 28 36 83 00.

British Council
General San Martin 7, 46003 Valencia. **Tel:** 6 351 8818.

British Language Centre
C/Bravo Murillo 377, 28020 Madrid. **Tel:** 1 733 07 39.

Business & United Schools
Genova 4, 1D 41010 Sevilla. **Tel:** 427 3183

Callan Method School of English
Calle Alfredo Vicenti 6 bajo, 15004 La Coruna.

Cambridge School
Placa Manel Montanya, 4, 08400 Granollers. **Tel:** 3 870 2001.

Centro Atlantico
Villanueva 2 apdo, 28001 Madrid. **Tel:** 1 435 3661, **Fax:** 1 578 1435.

Centro Britanico
Republica De El Salvador 26-10m, (Edificio Simago), 15701 Santiago De Compostela, La Coruna. **Tel:** 81 59 7490.

Centro Britannico
Alfereces Provisionoles 1 1ºB, 23007 Jaen.**Tel/Fax:** 53 22 21 08.

Centro Cooperativo de Idiomas
Clavel 2, 11300 La Linea, Cadiz.

Centro Estudios Norteamericanos
Aparisi y Guijarro 5, 46005 Valencia.

Centro de Estudios de Ingles
Garrigues 2, 46001 Valencia. **Tel:** 352 21 02.

Centro De Idiomas Liverpool
Libreros 11-1º, 28801 Alcala de Henares, Madrid. **Tel:** 1 881 3184, **Fax:** 1 881 3584.

Centro De Ingles
Tejon Y Marin, S/N 14003 Cordoba.

Centro de Ingles Luz
Passage Luz 8bajo, 46010 Valencia. **Tel:** 361 40 74.

Centro Linguistico del Maresme
Virgen De Montserrat. **Tel:** 35 55 5403.

Centro Superior de Idiomas
Tuset 26, 08006 Barcelona.

Chatterbox Language Centre
Verge de l'Assumpcio 21, Barbera del Valles.

CLIC (Centro de Lenguas e Intercambio Cultural)
Santa Ana, 1141002, Sevilla. **Tel:** 34-5-437 4500/438 6007, **Fax**: 437 18 06.

Collegium Palatinum
Calle de Rodriguez San Pedro 10, 28015 Madrid. **Tel:** 1 446 2349, **Fax**: 1 593 4446.

The English Academy
Cruz 15, 11370 Los Barrios, Cadiz.

English Activity Centre
Pedro Frances 22a, 07800 Ibiza. **Tel:** 7131 5828.

The English Centre
Apdo de Correos 85, 11500 El Puerto De Santa Maria, Cadiz. **Tel:** 34 56 850560, **Fax:** 873804.

English Institute
Carrer La Mar 38, 03700 Denia. **Tel:** 6 781026.

English Language Centre
Jesus Maria 9-1d, 14003 Cordoba.

English Studies
SA Avenida de Arteijo 8-1, 15004 La Coruna.

English Way
Platero 30, San Juan De Aznalfarache, Sevilla.

Epicentre
Niebla 13, 41011 Sevilla.

Eurocentre
Puerta De Jerez 3-1, 41001 Sevilla.

Eurolingua
San Felipe 3, 14003 Cordoba.

European Language Schools
Regueiro 2, 36211 Vigo. **Tel/Fax:** 86 291748.

European Language Studies
Edificio Edimburgo, Plaza Nina, 21003 Huelva. **Tel:** 34 59 263821, **Fax:** 34 59 280778.

Fiac School
Mayor 19, 08221 Terrassa, Barcelona.

FLAC
Escola De Idiomes Moderns Les Valls, 10 2/0, 08201 Sabadell, Barcelona.

Glossa English Language Centre
Rambla De Cataluna 9, 78 20 2A, 08008 Barcelona.

Idiomas Blazek
Llinas 2, 07014 Palma de Mallorca. **Tel:** 71 457260.

Idiomas Oxford
Calvo Sotelo 8-1, 26003 Logrono.**Tel:** 4124 41332.

Idiomas Progreso
Plaza Progreso 12, 07013 Palma De Mallorca. **Tel:** 7123 8036.

Idiomaster
Los Maristas, 2 Lucena, Cordoba.

inlingua Idiomas
C/o Greforio Fernandez 6, 47006 Valladolid.

inlingua Idiomas
Maestro Falla 5, 2,12, Puerto del Rosario, 35600 Fuerteventura, Canary Islands.

inlingua Idiomas
Tomas Morales 28, 35003 Las Palmas de Gran Canaria. **Tel:** 2836 0671.

Inlingua Madrid SL
Calle Arenal 24, 28013 Madrid. **Tel:** 1 541 3246, **Fax:** 1 542 8296.

The Institute of English
Santiago Garcia 8, 46100 Burjasot.

Interlang
Pl Padre Jean de Mariana 3-2, 45002 Toledo.

International House
Trafalgar 14 Entlo, 08010 Barcelona. **Tel:** 3 268 4511.

International House
Calle Zurbano 8, 28010 Madrid. **Tel:** 1 310 13 14, **Fax:** 1 308 53 21 **Email:** ih_madrid@mad.servicom.es

International House
C/Hernan Cortes 1, 21001 Huelva. **Tel:** 959 246529.

International House
Pascual y Genis 16, 46002 Valencia. **Tel:** 352 06 42.

John Atkinson School of English
Isaac Peral 11 y 13, 11100 San Fernando.

Kensington Centros de Idiomas
Avda Pedro Mururuza 8, 20870 Elgoibar. **Tel/Fax:** 43 740236.

Language Study Centre
Corredera Baja 15 Bajo, Chiclana De La Frontera, Cadiz.

Lawton School
Cura Sama 7, 33202 Gijon, Asturias. **Tel:** 8534 9609.

Lexis
Avenida de la Constitucion 34, 18012 Granada.

Linguasec
Malaga 1, 14003 Cordoba.

London House
Baron de S Petrillo 23 bajo, 46020 Benimaclet.

Manchester School
San Bernado 81, 33201 Gijon, Alicante. **Tel:** 8535 8619. **Fax**: 8535 6932.

Modern School
Gerona 11, 41003 Sevilla.

Nelson English School
Jorge Manrique 1, Santa Cruz, Tenerife.

Ten Centro de Ingles
Caracuel 24, Jerez de la Frontera.

The New School
Calle Sant Joan 2, 2a Reus Tarragona.**Tel:** 77 330775

Number Nine English Language Centre
Sant Onofre 1, 07760 Ciudadella De Menorca, Baleares. **Tel:** 7138 4058.

Onoba Idiomas
Rasco 19-2, 21001 Huelva.

Oxford Centre
Alvaro de Bazan 16, 46010 Valencia.

Oxford House
San Jeronimo 9-11, Granada.

The Oxford School
Maron Feria 4, 41800 Sanlúcar La Mayor, Sevilla.

Piccadilly English Institute
Los Chopos 8, 14006 Cordoba.

Preston English Centre
Edif El Carmen Chapineria 3, Jerez De La Frontera, Cadiz.

Principal English Centre
Aptdo 85, Puerto De Santa Maria, Cadiz.

SALT Idiomes
Prat De La Riba, 86, 08222 Terrassa (Barcelona). **Tel:** 3 735 80 35.

San Roque School
Plaza San Roque 1, Guadalajara.

Skills
Trinidad 94, 12002 Castellon. **Tel:** 6424 2668.

Stanton School of English
Colon 26, 03001 Alicante. **Tel:** 65207581.

St Patrick's Caracuel
1 Jerez De La Frontera, Cadiz.

TELC
Av Andalucia 8, 6º, 11006 Cadiz. **Tel:** 56 271097.

The Tolkien Academy
Juan Bautista Erro 9, 20140 Andoain.

Trafalgar Idiomas
Avda Castilla 12, 33203 Gijon, Asturias. **Tel:** 85 332361.

Trinity School SL
C/ Golondrina (Plaza Jardines) 17 Bajo, 11500 Puerto De Santa Maria, Cadiz. **Tel:** 9 56 871926.

Wall Street Institute
Gutenberg 3-13, 1º 8a, 08224 Terrassa (Barcelona). **Tel:** 3 788 8566, **Fax**: 3 788 8538.

Wall Street Institute
Centro Comercial Cuesta Blanca, Local 12, 2º Planta, la Moralega, Alcobendas, 28100 Madrid. **Tel:** 1 650 7602, **Fax**: 1 650 8246.

Wall Street Institute
Av República Argentina 24 P12 D, 41011 Sevilla.

Warwick House Centro Linguistico Cultural
Lopez Gomez 18-2, Valladolid 47002.

Westfalia
Chapineria 3, Edificio El Carmen, Modulo 310, 11403 Jerez de la Frontera.

William Halstead School of English
Camilo Jose Cela 12, 11160 Barbate de Franco.

Windsor School of English
Virgen De Loreto 19-1, 41011 Sevilla.

York House
English Language Centre, Muntaner 479, 08021 Barcelona. **Tel:** 32 113200.

SWEDEN

There are opportunities for teaching EFL in Sweden but you need to be already living or studying there. Many teachers start out at adult education centres (studieförbund) such as Folkuniveristet, Studiefrämjandet, Medborgarskolan, TBV, ABF and Vuxenskolan. Unfortunately, most of these institutions are reluctant to hire teachers on a full-time basis since their student populations fluctuate from term to term There is rarely employment available during the summer months. Opportunities may exist in private language schools such as Berlitz. Many of these schools contract to teach groups or individuals at companies such as Ericsson, Volvo and Sony.

Embassy in UK:
11 Montagu Place, London W1H 2AL
Tel: 020 7917 6400
www.swedish-embassy.org.uk

Embassy in USA:
1501 M Street NW, Washington DC 20005
Tel: 202 467 2600
www.swedish-embassy.org

British Council office:
PO Box 27819, S-115 93 Stockholm
Tel +46 (8) 671 3110
Www.britishcouncil.se

➲ **Visa/Work Permits:** If you want to teach in Sweden and are not an EU citizen you must be studying there or be married to somebody working there to qualify for a work permit.

➲ **Requirements:** College Degree and a TESOL/TESL Certificate

➲ **Cost of Living:** It is very expensive to live in Sweden but the standards of living are very high. A meal for two at an inexpensive café or restaurant can range from £10-20. A bottle of strong beer in a bar costs around £3. Transportation is expensive.

➲ **Currency:** £1 = 13.49 / $1 = 6.90 Swedish Krona.(SK)

➲ **Salaries/Taxes:** Annual income can be anywhere between £10,000-20,000. This often includes four weeks paid winter and summer vacation. Salaries really depend on the teaching situation and your qualifications and experience. Studieförbund generally pay less. State & private schools pay fairly well and offer some stability. Taxation runs at 30-35% of your salary.

➲ **Accommodation:** You can expect to pay between £150-£300 or more for a one-bedroom apartment in a large city. ESOL teachers usually live in single rental apartments or shared apartments.

➲ **Health advice:** If your contract is for less than a year you will not be covered by Sweden's socialized medical programmes even though you will be paying for them through your taxes.

Schools in Sweden (+46)

ABC Engelsk o. Amerikanska Sprakundervisning
Safflegatan 7, 123 44 Farsta.

British Institute
Hagatan 3, 511348 Stockholm.. **Tel:** 8 341200, **Fax:** 8 344192.

Folkuniversitetet
Box 26152, 100 41 Stockholm. **Tel:** 8 679 2950, **Fax:** 8 678 1544.

SWITZERLAND

Tight immigration laws mean that Switzerland is a difficult place to find a teaching job on spec. School have to prove that no Swiss person could do the job which means that only highly qualified teachers stand a chance of getting a job here. If you do decide that Switzerland is the place for you, you could approach the various international schools in the country to see if they need teachers. You'll need at least two years' experience in teaching before they will consider you for employment. January and February are good months to look for jobs. Another possibility is to find work at a summer camp.

Embassy in UK:
16-18 Montagu Place, London W1H 2BQ
Tel: 020 7616 6000

Embassy in USA:
2900 Cathedral Ave NW, Washington DC 20008
Tel: 202 745 7900
www.swissemb.org

British Council office:
Sennweg 2, PO Box 532, CH 3000 Berne 9
Tel +41 (31) 301 1473
www.britishcouncil.org/switzerland

➲ **Visa/Work Permits:** You need a work permit to teach in Switzerland. Permis A allows you to work seasonally, when the local workforce is considered too small to cope with the demand. Permis B can be granted for a specific one-year contract but the employer must guarantee a minimum number of hours and prove that no Swiss national could do the job.

➲ **Requirements:** College Degree and a TESOL/TESL Certificate

➲ **Cost of Living:** Most things are expensive in Switzerland.

➲ **Currency:** £1 = 2.38 / $1 = 1.22 Swiss Francs

➲ **Salaries/Taxes:** SFR 40-70 per hour in a private language school. Tax is around 15%.

➲ **Accommodation:**Schools often help in finding accommodation, but as with everything in Switzerland – it's expensive!

➲ **Health advice:** Many international schools will take care of your health insurance.

Schools in Switzerland (+41)

The American School in Switzerland
Summer Language Programmes, 6926 Montagnola-Lugano. **Tel:** 91 993 1647.

Basilingua Sprachschule
Birsigstrasse 2, 4054 Basel. **Tel:** 61 281 3954.

Bell Language School
12 Chemin des Colombettes, 1202 Geneva. **Tel:** 00 41 22 740 20.

Berlitz
14 rue De L'Ancien-Port, 1202 Geneva. **Tel:** 22 738 3200.

Berlitz
Munzgasse 3, 4001 Basel. **Tel:** 61 261 6360, **Fax:** 61 261 6405.

Institute Le Rosey
Camp d'Eté, Route des Quatre Communes, CH-1180 Rolle.

Leysin American School
CH-1854 Leysin. **Tel:** 24 493 37 7, **Fax:** 24 494 1585.

Markus Frei Sprachenschulen
Neugasse 6, 6300 Zug. **Tel:** 710 4240, **Fax:** 710 4525.

Village Camps
1296 Coppet. **Tel:** 22 776 2059, **Fax**: 22 776 2060.

UNITED KINGDOM

British English native-speakers will have few problems in the UK since teachers are expected to teach British English; Americans will find it hard to get a job, however, for Canadians, Australian and New Zealand citizens, the prospects are much better. You will need an EU passport if you want to teach English in the UK

Embassy in USA:
3100 Massachusetts Ave, Washington DC 20008
Tel: +1 202 588 6500
www.britainusa.com

British Council main office:
10 Spring Gardens, London SW1A 2BN
Tel +44 (0) 20 7930 8466
www.britishcouncil.org

➲ **Visas/Work permits:** A work permit is extremely difficult for a non-EU citizen to obtain. You must find an employer who will apply on your behalf before you enter the country.

➲ **Requirements:** College Degree and a TESOL/TESL Certificate

➲ **Cost of Living:** The UK is an expensive country.

➲ **Currency:** US$1 = approx £0.51

➲ **Salaries/Taxes:** London salaries are not particularly good although schools in Oxford and Cambridge pay well.

➲ **Accommodation:** The cost of accommodation in London is steadily rising and public transportation is expensive and unreliable. In the centre of London expect to pay a minimum of £100 / $200 a week. In the suburbs anything from £60 / $110 - £70 / $130 is standard. Outside of London accommodation is usually a lot cheaper, especially in student towns, but a lot of schools will arrange accommodation for you often with host families.

➲ **Health advice:** While good medical services are widely available, free care under the National Health System is allowed only to EU citizens.

Schools in the UK (+44)

There are over 1000 ELT schools in the UK – we could fill this entire book (in fact, there are books just for the UK ELT market). The best way to find schools is to use one of the professional associations of schools – or the British Council. These offer accreditation to schools that meet various criteria, with over 500 accredited schools in the UK and some 5-600 schools concentrating just on teaching in the summer holidays (see Finding a Job), it is a vast market. Try some of these major schools and organisations of schools who can provide help and lists of contacts. See also the website directory, pages 60 and 272 for online search engines that will find schools in the UK

Organisations of Schools

ARELS
56 Buckingham Place, London SW1E 6AG. **Tel:** 020 7802-9200**, Fax:** 020 7630-9824.
www.arels.org.uk
Note: ARELS and BASELT have merged into EnglishUK, however both offices and websites still operate as before.

BALEAP
English Language Centre, University of Bath, Claverton Down, Bath BA2 7AY. **Tel:** (01225) 826 191, **Fax**: (01225) 323135.
www.baleap.org.uk

BASELT
Cheltenham and Gloucester College of Higher Education, Francis Close Hall, Swindon Road, Cheltenham, Glos GL50 4AZ. **Tel:** 01242 227099, **Fax**: 01242 227055.
www.baselt.org.uk
Note: ARELS and BASELT have merged into EnglishUK, however both offices and websites still operate as before.

BATQI
35 Barclay Square, Bristol BS8 1JA. **Fax:** 0117 925 1537.

British Council
10 Spring Gardens, London SW1A 2BN, **Tel**: +44 (0)20 7389 4931, **Fax**: +44 (0)20 7389 4140
www.britishcouncil.org

Central Bureau for Educational Visits and Exchanges (CBEVE)
The British Council, 10 Spring Gardens, London SW1A 2BN. **Tel:** 020 7389 4004, **Fax**: 020 7389 4426.

Department for Education and Employment (DfEE)
Mowden Hall, Darlington DL3 9BG. **Tel:** 01325 460155, **Fax**: 01325 392695

IATEFL
3 Kingsdown Chambers, Whitstable, Kent CT5 2FL
Tel: 01227 276528, **Fax**: 01227 274415
www.iatefl.org

School Groups

Bell Language Schools
The Personnel Department, Hillscross, Red Cross Lane, Cambridge CB2 2QX. **Tel:** 01223 246644, **Fax**: 01223 414080.

Berlitz
9-13 Grosvenor Street, London W1A 3BZ. **Tel:** 020 7915-0909, **Fax**: 020 7915-0222.

EF
36-38 St Aubyns, Hove, East Sussex BN3 2TD, **Tel**: 01273 201431, **Fax**: 01273 746742
www.englishfirst.com

Inlingua
Rodney Lodge, Rodney Road, Cheltenham, Gloucs. GL50 1JF. **Tel:** 01242 253171, **Fax**: 01242 253181.

International House
106 Piccadilly, London W1V 9FL. **Tel:** 020 7491-2598, **Fax**: 020 7409-0959
www.ihlondon.com

Linguarama
Oceanic House, 89 High Street, Alton, Hants GU34 1LG. **Tel:** 01420 80899, **Fax**: 01420 80856.

St Giles
154 Southampton Row, London WC1B 5JX, **Tel**: 020 7837 0404, **Fax**: 020 7837 4099
www.stgiles.co.uk

Saxoncourt (UK) Ltd.
59 South Molton Street, London W1Y 1HH. **Tel:** 020 7491 1911, www.sgvenglish.com

Eastern and Central Europe

European countries outside of the European Union (EU) are very good places to look for work, and in these countries British and North American teachers are often on a level playing field (unlike Western Europe, where it's easier for British teachers to find work). American culture is a major influence in the former Soviet Bloc, learning English is big business in the region.

Some countries even offer similar salaries to those being offered in Western Europe now–amazing, considering that the cost of living is so much lower. However, times have changed since the post-communist honeymoon, when any young American, Canadian, Australian or Brit would be offered a teaching job on the street and then would spend their time partying rather than teaching. Now schools are looking for more dedicated and qualified teachers.

Some of these countries (in particular, Bulgaria and Romania) are about to join the European Union, which will eventually make it more difficult for non-EU citizens to work. However, for the time being, governments are focusing on bringing stability to their economies. This coupled with recent growth in tourism means that these are increasingly popular teaching destinations in spite of the occasionally difficult living and teaching conditions.

Competition means that qualifications may be necessary, especially in the major cities. A number of Central and East European countries are served by voluntary and cultural organizations, including the Peace Corps and the Soros Foundation. These organizations usually offer teachers salaries at local rates along with supplements, accommodation and other benefits.

There is considerable voluntary, as well as some commercial, English language teaching activity in the Ukraine, Uzbekistan and Kazakhstan, but standards of living can be modest and conditions harsh. Opportunities can be found also in the Baltic states, especially Lithuania and Estonia, which have been progressive in implementing policies promoting free enterprise.

AZERBAIJAN

A former Soviet republic, Azerbaijan became independent in 1991. The capital, Baku, is relatively safe, although attacks on foreigners can occur, but the southern parts of the country along the border with Iran, and the western region (Nagorno-Karabakh) are not safe for travellers. The climate is generally quite warm, except in the mountains

Embassy in UK:
4 Kensington Court, London W8 5DL
Tel: 020 7938 5482
www.president.az

Embassy in USA:
927, 15th St NW, Suite 700, Washington DC 20035
Tel: (202) 842 0001
www.azembassy.com

British Council office:
1 Vali Mammadov Street, Ichari Shahar, Baku 370004
Tel +994 (12) 971 593
www.britishcouncil.org/azerbaijan

➲ **Visas/Work Permits:** You need to obtain a visa from an embassy before you enter the country, if you are intending to work, and a letter of introductrion from your employer is required.

➲ **Requirements:** College Degree and a TESOL/TESL Certificate

➲ **Currency:** £1 = 8041 / $1 = 4884 manats

➲ **Cost of Living:** Quite cheap by Western standards, but everything has to be paid in local currency. Credit cards are not generally accepted

➲ **Health advice:** there are good medical facilities in Baku, but none elsewhere. It is better to get medical insurance before you travel. Immunization against typhoid is recommmended. All water should be boiled

➲ **English language media:** Baku Sun, Baku Today, Caspian Business News

Schools in Azerbaijan (+994)

The British Council Training Centre
1 Vali Mammadov St, Baku
Tel: 994 12 972 013
Fax: 994 12 989 236
Email: enquiries@britishcouncil.az

International House Baku
Khagani 20/20, Baku 370001
Tel: 994 12 97 15 85
Email: intellect@azerin.com

Webb Academy
148 Vidadi St, Baku 37000
Tel & Fax: 994 12 973 047

BELARUS

Belarus is a destination only for the brave and adventurous. Economic and political reform in Belarus has stalled under the current government. Tourist facilities are not highly developed, and many of the goods and services taken for granted in other countries are not yet available. However, there is considerable demand for English, so it's quite easy to find a job. You need to contact a school or a University directly, that's the easiest and the quickest way. If you want to work in the private sector, there are a couple of schools in Minsk.

➲ **Visas/Work Permits:** A visa must be obtained before entering Belarus.

➲ **Requirements:** College Degree and a TESOL/TESL Certificate

➲ **Currency:** £1 = 4014 / $1 = 2170 rubles.

➲ **Salaries/Taxes:** A school teacher's average salary is about $200 per year after tax. University teachers are paid slightly more.

➲ **Accommodation:** If you work for a University, they will offer you free accommodation at a student hostel. It is easy to rent an apartment, but it will cost about £30 per month.

➲ **Cost of Living:** The cost of living is quite low. Local food is very cheap, but imported goods are expensive.

➲ **Health advice:** Belarus requires all foreign nationals entering the country to purchase medical insurance at the port-of-entry regardless of any other insurance one might have. Costs for this insurance will vary according to the length of stay. You will then be entitled to receive health care, but medical care in Belarus is limited. There is a severe shortage of basic medical supplies, including anesthetics, vaccines and antibiotics. Insurance with evacuation cover is recommended.

➲ **English Language Newspapers:** Minsk News

Schools in Belarus (+375)

The British Council
ul. Zakharova, 21, Minsk 220662. **Tel:** 17 236 79 53, **Fax**: 17 236 40 47, **Email:** Natasha.Naida@britishcouncil. org.by

International House Minsk
Room 306, Gykalo 9, Minsk 220071. **Tel:** 17 238 47 12, **Fax**: 17 231 60 07, **Email:** ih@user.unibel.by

BULGARIA

Bulgaria has been slower than most of its neighbours to embrace the West and the elected government is Communist, so most teaching opportunities are in the public sector. Bulgarians take education very seriously and, in spite of economic and political troubles, there is a surprisingly developed English language teaching structure. Secondary schools recruit foreign language teachers on yearly, renewable contracts. Apply to: Bulgarian Ministry of Education & Science, 2A Knjaz Dondoukov, 1000 Sofia, Tel: 2 988 0494. In the private sector, teaching loads are notoriously great with some having to travel between two or three schools to deliver 30 contact hours per week, often in the evenings and on Saturday mornings. In the state sector, a week's teaching is likely to consist of about twenty 40-minute lessons. Because many students make great sacrifices to pay for their courses, a Bulgarian teaching experience can be very inspiring and rewarding. On the negative side expect food shortages, power failures, transport strikes and food rationing, especially in winter.

Embassy in UK:
186-188 Queen's Gate, London SW7 5HL
Tel: 020 7584 9400
www.bulgarianembassy.org.uk

Embassy in USA:
1621 22nd St. NW, Washington DC 20008
Tel: 202 387 0174
www.bulgaria-embassy.org

British Council office:
7 Krakra Street, 1504 Sofia
Tel +359 (2) 942 4344
www.britishcouncil.org/bulgaria

➲ **Visas/Work Permits:** All non-Bulgarian citizens need an entry visa, which can be obtained from the Bulgarian Consular Service in any country. Your school should help you to organize your work permit once you are in the country.

➲ **Requirements:** College Degree, experience and a TESOL/TESL Certificate

➲ **Currency:** £1 = 2.91 / $1 = 1.49 Bulgarian Leva.

➲ **Salaries/Taxes:** Most contracted teachers are paid at local rates the equivalent of £100-200 per month. Paid airfares, furnished accommodation and holiday are generally included and there might also be a supplement paid in pounds or dollars. A highly paid Bulgarian teacher might be paid the equivalent of £90 per month. Tax deductions amount to 30-40%.

➲ **Cost of Living:** Very cheap in the provinces, reaching western levels in Sofia, but conditions are still extremely tough.

➲ **Accommodation:** Very basic accommodation is usually provided by employers.

➲ **Health advice:** Insurance is provided by state employers but it should be supplemented with a private policy.

➲ **English language media:** The Sofia Echo.

Schools in Bulgaria (+359)

Alliance for Teaching of Foreign Languages
3 Slaveikov Square, 1000 Sofia. **Tel:** 2 880238, **Fax**: 2 882 349.

American College
Tsar Osvobodital Blvd, Sofia.

Avo-3 School of English
House of Culture Sredets, 2a Krakra Street, Sofia.

ANDRA
1 Ch Smirnenski Blvd, Sofia 1421.

BELS & BELIN Centre
PO Box 37, 1421 Sofia.

British Council
7 Tulovo Street, 1504 Sofia. **Tel:** 2 946 0098.

Bulgarian Dutch College for Commerce
Trudovets Botebgrad, Sofia.

Business Private High School
Kosta Lulchev Street, Sofia.

Centre for Language Qualification
National Palace of Culture, Administrative Bldg, 2nd Floor, Room 131, Sofia.

ELIT
14 GM Dimitrov Boulevard, Sofia.

ELS
PO Box 16, Sofia 1404.

ESPA
Darwin Street, Sofia.

First Private English Language Medium School
Stara Planina Street, Sofia.

ICO Intellect
Kostur Street, 1618 Sofia.

Institute of Tourism
Park Ezero, 8000 Bourgas.

Language High School
64 Tsar Shishman St., 6000 Stara Zagora. **Tel:** 42 52975, **Fax**: 42 40188.

Language Schools of Europe
82 Tsar Simeon St., entr 3, fl 3, Sofia 1000.

MEL
First Primary School, 24 Vessela Street, Plovdiv 4000.

Meridian
22 Alabin Street, Sofia 1000.

Meridian
Macedonia Boulevard, Sofia 1606.

New School of English
13 Serdika Street, Sofia.

Pharos Ltd
2 S.Vrachanski Street, Vasrazhdane Square, Sofia.

Private High School for Law & Management
D Dimov Street, Sofia.

Private High School for Management & Business Admin
Raiko Alexiev Street, Complex Iztok, Sofia.

Private School for Banking and Business
83 Bogomil Street, Plovdiv

Yanev School
bl 147 ap 38, Complex Liulin, Sofia.

CROATIA

Croatia is being rapidly integrated into mainstream Europe and English is very much in demand in this former Yugoslav republic. With private schools operating in Zagreb, Karlovac and Varazdin, there are plenty of opportunities for teachers, especially native speakers, who are in short supply. Because there is an established English language industry, schools are reluctant to employ unqualified teachers. Jobs are often advertised abroad, but it's also possible to job-hunt on arrival. Full-time contracts are the norm and there are opportunities for private lessons.

British Council office:
Illica 12, PP 55, 10001 Zagreb
Tel +385 (1) 489 9500
www.britishcouncil.hr

➲ **Visas/Work Permits:** Anyone who intends to teach will need a work visa. This can be secured in advance when you present your contract to your local embassy, then a labour permit can be sent from an employer in Croatia.

➲ **Requirements:** College Degree, experience and a TESOL/TESL Certificate

➲ **Currency:** £1 = 10.97 / $1 = 5.61 Kuna.

➲ **Salaries and taxes:** Salaries are usually paid in German marks with tax deducted by the employer. The average rate is between £500-1100 per month, while experienced teachers can earn more.

➲ **Accommodation:** Accommodation is quite reasonable. An apartment with two or three rooms will cost around £300 per month.

➲ **Cost of Living:** Food and basic living costs are cheap, but price are higher in tourist resorts.

➲ **Health advice:** Insurance is advisable, as state facilities are under considerable strain. There are plenty of private doctors and dentists.

Schools in Croatia (+385)

Centar za strane jezike
Vodnikova 12, 4100 Zagreb *and* Trg republike 2, 58000 Split.

Class
Jankomirska 1, 41000 Zagreb.

The English Workshop
Medulinska 61, 52000 Pula.

Interlang
Krisaniceva 7, 51000 Rijeka

Lancon
Kumiciceva 10, 10000 Zagreb. **Tel:** 1 485 4985, **Fax**: 1 485 4984. **Email: lmo@lancon.hr**

Linguar Centar
Miroslava Krleze 4c, 47000 Karlovac. **Tel/Fax:** 47 621900.

Linguae
Radiceva 4, 51000 Rijeka.

Linguapax Language School
B.Jelacica 1 32100 Vinkovci. **Tel:** 32 334 827, **Fax**: 32 334 827.

LS Lukavec
Skolska 27, 41409 Donja Lomnica.

Narodno sveuciliste
Skola za strane jezike, Trg Matice Hrvatske 3, 41430 Samobor.

Octopus
Savska 13, 41000 Zagreb.

Radnicko sveuciliste
Bozidar Maslaric, 5400 Osijek.

Skola Stranih Jezika
Cesarceva 10, 42000 Varazdin. **Tel:** 42 215055, **Fax**: 42 215036.

Svjetski jezici
Varsavska 13, 41000 Zagreb.

Verba
M. Gorkog 5, 51000 Rijeka.

Vern Lingua
Senoina 28, 10000 Zagreb. **Tel/Fax:** 1 428 548.

CYPRUS

The Mediterranean island of Cyprus is divided into two sectors: Greek and Turkish. In the Greek sector of Cyprus, English is the second language. Here, it is very hard for foreigners to find teaching positions because local teachers fill most of the demand and the authorities are very strict about allowing non-Cypriots to work. In the Turkish sector, there are a number of private universities that hire native English speakers to teach in language departments.

British Council office:
3 Museum Street, CY- 1097 Nicosia
Tel +357 (22) 585 000
www.britishcouncil.org.cy

Schools in Cyprus - Greek Sector (+357)

American Academy
PO Box 112, Larnaca. **Tel:** 4 652046, **Fax**: 4 651046.

American Academy
Despinas Pattichi Street, PO Box 1867, Limassol. **Tel:** 5 337054, **Fax**: 5 387488.

American Academy
3A Michalaki Paridi Street, Nicosia. **Tel:** 2 462886, **Fax**: 2-466290.

Centre of Higher Studies
PO Box 545, 2 Evagorou Street, floor 6, Nicosia. **Tel:** 2 445970, **Fax**: 2 369171.

The English School
PO Box 3575, Nicosia. **Tel:** 2 422274.

Forum Language centre
PO Box 5567, Nicosia 1310. **Tel:** 2497766, **Email: forum1@cytanet.com.cy**

The Grammar School
Katinas Paxious Street, PO Box 1340, Limassol. **Tel:** 5 727933, **Fax**: 5 727818.

The Grammar School
PO Box 2262, Nicosia. **Tel:** 2 621744, **Fax**: 2623044

International Language Institute (ILI)
12 Richard & Verengaria Street, Limmassol. **Tel:** 371017, **Fax**: 371018.

Pascal English School
3 Costaki Panteli Street, PO Box 4746, Nicosia. **Tel:** 2 450900, **Fax**: 2 367214.

Terra Santa College
12 Lycourgos Street, PO Box 1546, Nicosia. **Tel:** 2 421100, **Fax**: 2 466767.

Turkish Sector (+90)

Cyprus International University
Ortakoy, Lefkosa, KKTC. **Tel:** 392 223 49 00. **Fax**: 392 223 49 01.

Eastern Mediterranean University
Famagusta, North Cyprus. **Tel:** 392 366 6588. **Fax**: 392 366 1317. **Email: webmaster@salamis.emu.edu.tr**

Girne American University
Yeni Liman Yolu, Girne, North Cyprus. **Tel:** 392 815 53 12. **Fax**: 392 815 11 33.

International American University
P.O.Box. 292, Kyrenia, North Cyprus. **Tel:** 392 815 6791 **Fax**: 392 815 6790.

Near East University
P.O. Box: 670, Nicosia, North Cyprus. **Tel:** 392 223 64 64. **Fax**: 392 223 64 61.

CZECH REPUBLIC

This country in the heart of Europe has long been a magnet for English language teachers (particularly from the USA and Canada). It is estimated that there are as many as 60,000 Americans in Prague alone, many drawn here by its reputation as a "party" town. But the Czech authorities have tightened up regulations recently and it is difficult to find a teaching job unless you have some kind of qualification. Although most teaching jobs are in the private sector, the Ministry of Education recruits qualified teachers (who should have at least a certificate qualification and a BA in English) for elementary, middle and high schools. The main advantage of a governmental teaching contract is that it is worth the paper it's printed on and you will probably be eligible for health insurance coverage. In the private sector, alas! This is not the case. There are disreputable schools out there that will employ teachers illegally and then take advantage of them. If you are unhappy with your school, just pick up your last salary cheque and move on – there are plenty of good schools.

Embassy in UK:
26-30 Kensington Palace Gardens, London W8 4QY
Tel: 020 7243 1115
www.czechembassy.org.uk

Embassy in USA:
3900 Spring of Freedom St NW, Washington DC 20008
Tel: 202 274 9100
www.czechembassy.org

British Council office:
Bredovsky dvur, Politickych Veznu 13, 110 00 Prague 1
Tel +420 (0)2 2199 1111
www.britishcouncil.cz

➲ **Visas/Work Permits:** In January 2000, a major revision of the Czech law regarding residency permits for foreign nationals came into force. In the old days, teachers could arrive on tourist visas (good for up to 90 days) and then apply for a residency permit. Now, American citizens must apply for a residency permit before they travel. You can still apply for a work permit while you are in the country but your employer must prove you are not taking a job that a Czech national could do.

➲ **Requirements:** College Degree, experience and a TESOL/TESL Certificate

➲ **Currency:** £1 = 41.29 / $1 = 21.12 Koruny.

➲ **Salaries/Taxes:** Teachers working in state schools earn about £180-250 per month after tax and accommodation is usually provided. You can earn more at a private language school (£300-400) as long as you are given the hours. Private tuition can be very lucrative; some business people are prepared to pay upwards of £20 an hour to learn English with a qualified and experienced tutor. The tax rate is around 30% for those working legally.

➲ **Accommodation:** It's difficult to find a room in Prague. Ask the school to provide you with accommodation. If that doesn't work, try and find people who will share with you. Check out the notices at The Globe bookstore (where you might run into another teacher in the same predicament!) or at Radnost, a popular coffee shop. If you're very lucky you might pick up a cheap apartment for around £90 a month but you'll probably have to pay at least £150 so it's a good idea to share.

➲ **Cost of Living:**If you eat fruit and vegetables that are locally produced and develop a taste for heavy Czech fare, then life will not be too expensive. Czech beer is very strong and very cheap!

➲ **Health advice:**In Prague medical facilities are good but try not to be sick in a remote region. Better still, get a job with a school that provides medical coverage for its teachers.

➲ **English language media:** The Prague Post.

Schools in the Czech Republic (+420)

Academic Information Agency (Public Sector Recruitment)
Dum zahranicnich sluzeb MSMT, Senovazne nam. 26, 111 21 Prague 1. **Tel:** 2 2422 9698, **Fax**: 2 2422 9697, **Email: aia@dzs.cz**

Agency Unitour
Libor Kivana, Senovazne 21, 110 00 Prague 1. **Tel:** 2 35 9917, **Fax**: 786 2153.

Agentura Educo
Veletrzni 24, 170 00 Prague 7. **Tel:** 2 333 70 163, **Fax**: 2 203 97 310.

Anglictina Expres
Korunni 2, 12000 Prague 2. **Tel:** 2 2423 8186, **Email: anexpres@vol.cz**

Bell School
Nedvezska 29, 100 00 Prague 10. **Tel:** 2 781 5342, **Fax**: 2 782 2961.

British Council
Narodni 10, 12501 Prague 1. **Tel:** 2 203 751.

Caledonian School
Vitavska 24, 150 00 Prague 5. **Tel:** 2 573 13650, **Email:** jobs@caledonianschool.com

California Sun School
Na Bojisti 2, Prague 2. **Tel:** 2 294 817, **Fax**: 2 822 271.

Cheb Free School
Kubelikova C.4, Cheb 350 02.

Encounter English
Azalkova 12, 100 00 Prague 10-Hostivar. **Tel:** 2 758 773.

Eminenc Language School
Vodikova 26, 110 00 Prague 1. **Tel:** 2 96226141-3, **Fax**: 2 96226144.

English House
Vysedhradska 2, 128 00 Prague 2. **Tel/Fax:** 2 29 31 41.

Et Cetera Language School
Dusni 17, 110 00 Prague 1. **Tel:** 2 231 3062, **Fax**: 2 894 484.

Euro contact - Hulesovice
Jankovskeho 52, 7 Hulesovice. **Tel:** 801 527.

European Management Institute (TEMI)
Na Truhlarce 3, 180 00 Prague 8. **Tel/Fax:** 2 68 88 052.

Foundation for a Civil Society
Jeleni 200/3, 118 00 Prague 1. **Tel:** 2 24 51 08 73, **Fax**: 2 24 51 08 75.

GEJA
Seifertova 37, Prague 3. **Tel:** 2 225 3750.

Hampson CS Ltd
Zelezna 107, Mlada Boleslav 293 01. **Tel:** 326 732732, **Fax**: 326 732732.

International House (ILC) Brno
Sokolska 1, 602 00 Brno. **Tel:** 5 41 24 0493, **Fax**: 5 41 24 59 54, **Email:** ihbrno@sky.cz

International House Prague - Akcent
Bitovska 3, 140 00 Prague 4. **Tel:** 2 420 595, **Fax**: 2 422 845, **Email:** info.accent@akcent.cz.

International Training Solutions Ltd
Thamova 24, 186 00 Prague 8. **Tel/Fax:** 2 238 0545.

Intertext Servis Karel Narvatil
Anglicka 24, 360 09 Karlovy Vary. **Tel/Fax:** 17 323 0436.

Languages at Work
Na Florenci 35, 110 000 Prague 1. **Tel/fax**: 2 248 11379, **Email:** employment@mbox.vol.cz

London School of Modern Languages
Francouzska Ulice 30, 120 00 Prague 2. **Tel:** 2 242 53 437, **Fax**: 2 242 54 259.

Prague 8 Language School
Buresova 1130, 18 000 Prague 8. **Tel:** 2 858 8028.

Pro English 90
Mala Strana, 5 Helichova (4th floor), Prague 1. **Tel:** 534 551.

St. James Language centre
Namesti Miru 15, 12000 Prague 2. **Tel.**: 2 2251 7869, **Fax:** 2 2251 7870.

Spusa Education centre
Navratilova 2, 110 00 Prague 1. **Tel:** 2 22 23 17 02, **Fax**: 2 22 23 25 30.

Statni Jazykova Skola
Skolsa 15, 116 72 Prague 1. **Tel:** 2 222 32 237, **Fax**: 2 222 32 236, **Email:** sjs@sjs.cz

ESTONIA

Estonia is a rapidly developing Baltic state that has experienced significant success in reforming its political and economic institutions since regaining independence in 1991. With about 30 universities and private schools in Estonia, there are many opportunities for teachers. Students tend to be highly motivated although they can be a little shy at first.

Embassy in UK:
16 Hyde Park Gate, London SW7 5DG
Tel: 020 7589 3428
www.estonia.gov.uk

Embassy in USA:
Suite 503, 1730 M Street NW, Washington DC20036
Tel: 202 588 0101
www.estemb.org

British Council office:
Vana Posti 7, Tallinn 10146
Tel +372 (6) 257 788
www.britishcouncil.ee

➲ **Visas/Work Permits:** No visas are required for stays up to 90 days. It's not too difficult to obtain a work permit once you have found a job, although the paperwork can be confusing.

➲ **Requirements:** College Degree, experience and a TESOL/TESL Certificate

➲ **Currency:** £1 = 23.34 / $1 = 11.94 Kroon

➲ **Salaries/Taxes:** After tax (around 26%) you should be able to take home £300-500 a month.

➲ **Accommodation:** The average rent for a two-bedroom apartment in Tallinn's suburbs is around £200. Hopefully, your school will include accommodation in the package they offer you.

➲ **Cost of Living:** Estonia is much cheaper than the UK and U.S.A. so even if you are only earning a few hundred dollars a month it should go a long way.

➲ **Health advice:** The quality of medical care in Estonia is good but can fall short of UK and USA standards. Estonia has many highly trained medical professionals, but hospitals and clinics still suffer from a lack of equipment and resources. Your school may give you a medical card entitling you to free treatment but health insurance is a good idea.

Schools in Estonia (+372)

ALF Training centre
Ravala pst. 4, Tallinn EE0001

Concordia International University
Kaluri tee 3, Viimsivald, Harjumaa EE3006. **Tel:** 6090077, **Fax**: 6090216

International House Tallinn
Pikk 69, Tallinn 10133. **Tel:** 64 10 607, **Fax**: 64 10 607, **Email:** ihte@online.ee

International Language Services
Pikk 9, Tallinn EE 0001. **Tel:** 646 4258, **Fax**: 641 2476, **Email:** ilsinfo@online.ee

Tallinn Pedagogical University
Language centre, Narva mnt. 29, Tallinn EE0101

GEORGIA

This Caucasian republic (birthplace of Josef Stalin) has a long and fascinating history. There is a shortage of native English-speaking teachers in Georgia so intrepid travellers should not have a problem finding work here, either in a private school or even with one of the big energy companies that have offices in Tbilisi.

British Council office:
34 Rustaveli Avenue, Tblisi 380008
Tel +995 (32) 250 407 988014
www.britishcouncil.org.ge

➲ **Visas/Work Permits** Tourist visas are available on entry at Tbilisi Airport. The Georgian authorities are generous with work visas.

➲ **Requirements:** College Degree and a TESOL/TESL Certificate

➲ **Currency:** £1 = 3.36 / $1 = 1.82 Lari

➲ **Salaries/Taxes:** You can expect to earn around £200 a month in a private language school. If you land a teaching job with an oil company you can make a lot more money.

➲ **Accommodation:** A bedroom in a shared apartment costs around £90-100 per month.

➲ **Cost of Living:** Locally grown products are cheap and delicious. Georgian cuisine enjoys a high reputation.

➲ **Health advice:** Medical care in Georgia is limited. There is a severe shortage of basic medical supplies, including disposable needles, anesthetics, and antibiotics. Uninsured travellers who require medical care overseas may face extreme difficulties. The U.S. Embassy advises American citizens to avoid travel to the separatist-controlled region of Abkhazia.

Schools in Georgia (+995)

Byron School of Tbilisi
2 Griboedov Street, Tbilisi 380008. **Tel:** 32 93 1578.

The English Language Centre
10 Shevchenko Street, Tbilisi 380018. **Tel/Fax:** 32 933999

International House Tbilisi
centre for Language Studies, 2, 26 May Square, Tbilisi 380015. **Tel:** 32 940 515, **Fax**: 32 001 127, **Email:** schoolih@access.sanet.ge

International Language Academy
17 Chavchavadze Ave., Tbilisi 380079. **Tel:** 32 22 0504, **Email:** ila@caucasus.net

Professional Training Group
82 Paliashvili Street, Second floor, III entrance, Tbilisi **Tel:** 32 294482/ 253547.

Public Service Language centre
8 Rustaveli Street, Tbilisi.

Tbilisi Central School of English
8 Jambuli Street, Tbilisi 380008. **Tel:** 32 92 3454.

HUNGARY

The demand for native-speaking English language teachers has slowed in recent years. Reports suggest that competition for teaching jobs in Budapest has become pretty fierce. There are numerous tales of unscrupulous school owners operating in the Hungarian capital. Better opportunities for TEFL teachers lie in the provincial centres such as Gyor, Szeged and Pecs. September is a good time to come and look for a job.

Embassy in UK:
35 Eaton Place, London SW1X 8BY
Tel: 020 7235 5218
www.huemblon.org.uk

Embassy in USA:
3910 Shoemaker St NW, Washington DC 20008
Tel: 202 362 6730
www.huembwas.org

British Council office:
Benczúr u. 26, 1068 Budapest
Tel +36 (1) 478 4700
www.britishcouncil.hu

➲ **Visas/Work Permits:** You do not need a visa for a period of up to 90 days. Hungary is very strict about work permits and everybody needs to apply for one if they are going to teach in the country. Permits must be arranged before leaving your country of residence and the applicant has to produce a labour permit sent from the employer in Hungary, stating that no Hungarian is available to do the job.

➲ **Requirements:** College Degree, experience and a TESOL/TESL Certificate

➲ **Currency:** £1 = 377 / $1 = 193 Forint

➲ **Salaries/Taxes:** Expect to earn between £3-5 an hour at a private language school which is pretty good by local standards. Taxation is very high – up to 50 % of your income. Many school owners prefer to pay foreign teachers under the table to avoid red tape. In the public sector, teachers make around the same as in the private schools (c. £180-200 per month) but accommodation is usually included.

➲ **Accommodation:** You can find an apartment for around £90-100 a month in Budapest

➲ **Cost of Living:** A decent dinner costs around £2. Annual inflation runs around 10% and salaries do not always rise to match this. Many teachers find it hard to get by on their salaries.

➲ **Health advice:** Local health insurance is available but it doesn't cover repatriation, however, healthcare facilities are good, particularly compared to other countries of central Europe.

➲ **English language media:** Budapest Sun, Budapest Week, Budapest Business Week Journal.

Schools in Hungary (+36)

American International School of Budapest
Kakukk u.1-3, 1121 Budapest. **Tel:** 396 2176.

Atalanta Business and Language School
Visegradi u.9, 1132 Budapest. **Tel**: 1 131 4954, **Fax**: 1 339 8549, **Email: atalanta.marketing@atalanta.datanet.hu**

Avalon '92 Agency
Erzsebet Krt 15, 1/19, 1073 Budapest. **Tel/Fax:** 342 6355.

Babilon Nyelvstudio
Karoly krt. 3/a IV.em, 1075 Budapest. **Tel:** 1 269 5531, **Fax**: 1 322 6023, **Email: bab@mail.datanet.hu**

Bell Iskolak
Tulipan u. 8, 1022 Budapest. **Tel:** 1 326 8457, **Fax**: 1 326 5033, **Email: bellisk@mail.matav.hu**

Belvarosi Tanoda Foundation
Molnar u.9, 1056 Budapest. **Tel:** 118 6769.

Britannica International School
Villanya ut 11-13, 1114 Budapest. **Tel:** 166 6306.

Buda Drawing School
Marvany u. 23, 1126 Budapest. **Tel/Fax:** 155 0341.

Budenz Jozsef High School Foundation
Budenz ut 20-22, 1021 Budapest. **Tel:** 176 3177.

Budapest Pedagogical Institute
Horveth Mihaly Ter 8, Budapest 8.

Business Polytechnic
Vendel u. 3/b, 1096 Budapest. **Tel:** 1 215 4900, **Fax**: 1 215 4906, **Email: titkar@mail.poli.hu**

English Teachers' Association of the National Pedagogical Institute
Bolyai u.14, 1023 Budapest.

Europai Nyelvek Studioja
Muzeum Krt. 39, 1053 Budapest. **Tel:** 1 317 1302 **Fax**: 1 266 3889, **Email: els@mail.datanet.edu**

Helle Studio
Deri Miksa u. 18, 1084 Budapest. **Tel:** 1 114 0128.

IHH A Nyelviskola
Teleki u. 18, 9022 Gyor. **Tel:** 96 315444/30 363195 **Fax**: 96 315665.

Interclub
Hungarian Language School, Szent Janos u. 16, 1039 Budapest. **Tel** 250 4640.

International Business School Budapest
Etele u. 68, 1115 Budapest. **Tel:** 203 0351.

International House Budapest
Bimbó út 7, 1022 Budapest. **Tel:** 1 212 4010, **Fax**: 1 316 2491, **Email: bp@ih.hu**

International House Eger
Mecset utca. 3, 3300 Eger. **Tel:** 36 413 770; 36 411 313, **Fax**: 36 413 770, **Email: eger@ih.hu**

IS Prospero Language School
Sas u. 25, 1051 Budapest. **Tel:** 1 331 6779, **Fax**: 1 331 2150, **Email: isprospero@mail.matav.hu**

Karinthy Frigyes Gimnazium
Thokoly utca 7, Pestlorinc, 1183 Budapest. **Tel:** 1 290 4316, **Fax**: 1 291 2367, **Email: ba@karinthy.hu**

Kulvarosi Tankor
Dregelyvar u. 6, 1158 Budapest. **Tel:** 272 2604.

Living Language Seminar
Fejer Gyorgy u 8-10, 1053 Budapest. **Tel:** 1 326 3251, **Fax**: 1 317 9655.

London Studio
Villanyi ut. 27, 1114 Budapest. **Tel:** 1 385 0177.

Mentor Nyelviskola
Ferenc krt. 37, 1094 Budapest. **Tel:** 216 0765.

Novoschool Nyelviskola
Ulloi ut 63, 1091 Budapest. **Tel:** 1 215 5480, **Fax**: 1 215 5488, **Email:** novoschool@mail.matav.hu

Oxford Nyelviskola
Ikva u. 52, Gyor. **Tel:** 96 429 882.

Vaci Street Development centre
Terez Krt. 47, 1067 Budapest. **Tel:** 1 302 2214, **Fax**: 1 353 3274, **Email:** bruce@mail.inext.hu

Western Maryland College Budapest
Villanyi ut 11-13, 1114 Budapest. **Tel:** 1 66 7366.

LATVIA

Many teachers report that Latvia is a great place to live and work in although you must like cold weather to survive here! Students are keen to learn English and really appreciate being taught by a native speaker. You won't make a lot of money in Latvia but you will feel you are making a worthwhile contribution to society.

Embassy in UK:
45 Nottingham Place, London W1M 3FE
Tel: 020 7312 0040

Embassy in USA:
4325 17th Street NW, Washington DC 20011
Tel: 202 726 8213
www.latvia-usa.org

British Council office:
5a Blaumana iela, Riga LV-1011
Tel +371 728 1730
www.britishcouncil.org/latvia

➲ **Visas/Work Permits:** No visa is required for travellers remaining up to 90 days in a half-calendar year (from January to June and from July to December). If you stay in Latvia for more than 90 days, including 180-day periods that cross over two half-calendar years, you must apply for temporary residence. Teachers who plan to remain in Latvia for more than 90 days must apply for an "educational work" visa or in-country for temporary residence.

➲ **Requirements:** College Degree and a TESOL/TESL Certificate

➲ **Currency:** £1 = 0.96 / $1 = 0.52 Lat.

➲ **Cost of Living:** Basic foods & clothing are very cheap, but specialist goods can be expensive.

➲ **Salaries/Taxes:** Pay is low – around £2-3 an hour in the private language schools.

➲ **Accommodation:** Expect very basic living conditions. Many buildings were constructed during the Soviet era and are of poor quality.

➲ **Health advice:** Foreign aid agencies and private employers sometimes provide medical insurance. Medical care in Latvia is steadily improving but remains limited.

➲ **English language media:** The Baltic Times

Schools in Latvia (+371)

Alna IT
K.Valdemara 149 - 600, Riga LV-1013. **Tel:** 7377142, **Email:** info@alnait.lv

English Language Centre Satva
Office 8, Dzirnavu Str. 79/85, Riga LV-1011. **Tel:** 7226641

Lingva Tev
Brivibas 87, Riga LV-1001. **Tel:** 7377734, **Email:** lintev@binet.lv

Mirte
Culture and Education centre, Raina bulvaris 23, k.2 Riga LV-1050. **Tel:** 7223837, **Email:** info@mirte.lv

Socialo tehnologiju augstskola
Balta 10 , Riga LV-1055. **Tel:** 2461001 **Email:** sti@mail.junik.lv

Spade un Mikelsons
Barristers' Bureau, Aspazijas bulvaris 24 - 305, Riga LV-1050. **Tel:** 7228519, **Email:** spade@mail.bkc.lv

Staltbriedis
Dzelzavas 74 - 507, Riga LV-1082. **Tel:** 2572676, **Email:** staltbriedis@mail.lv

Strombus
Kalku 24 - 4.st., Riga LV-1050. **Tel:** 7224376, **Email: strombus@com.latnet.lv**

T.U.I.
K.Barona 28, Riga LV-1011. **Tel:** 7242429, **Email:** tui@parks.lv

Valodu macibu centres
Smilsu 1/3 - 3.st., Riga LV-1050. **Tel:** 7212251, **Email:** vmc@latnet.lv

LITHUANIA

Lithuania is a country undergoing political and economic change. Tourist facilities are improving. Many goods and services are now available in the major cities, but may not be fully comparable to Western standards.

Most teachers who come here have very good experiences, as long as they are aware of the economics of teaching here before they arrive. Lithuanian schools are very keen to recruit native speakers, qualified or not, so graduates should have no problem finding a position either by contacting the Lithuanian embassy or by contacting the *Ministry of Education*, Dept. of Foreign Relations, Volvano Gatve 2/7, 2691 Vilnius, Tel: 2 622 483, Fax: 2 612 077. For US teachers, the main placement organization is the *American Partnership for Lithuanian Education*, PO Box 617, Durham, CT 06422, USA, Tel: (203) 347 7095, Fax: (203) 327 5837.

Embassy in UK:
84 Gloucester Place, London W1U 6AU
Tel: 020 7486 6401

Embassy in USA:
2622 16th Street NW, Washington DC 20009
Tel: 202 234 5860
www.ltembassyus.org

British Council office:
Business Centre 2000, Jogailos 4, 2001 Vilnius
Tel +370 (5) 264 4890
www.britishcouncil.lt

➲ **Visas/Work Permits:** You do not normally need Lithuanian visas for stays of 90 days or less. For most nationalities the procedure to obtain work permits is relatively straightforward once a position has been secured.

➲ **Requirements:** College Degree and a TESOL/TESL Certificate

➲ **Currency:** £1 = 4.84 / $1 = 2.64 Lati

➲ **Salaries/Taxes:** Salaries can be as low as £60 per month, so most teachers supplement their income with private lessons. The main private schools can pay up to £300 a month or about £5 an hour.

➲ **Accommodation:** Usually provided free with jobs arranged through the Ministry of Education. About £60 a month for an apartment in the suburbs, but cheap accommodation is difficult to find. Living with a family is the cheapest way and very easy to arrange.

➲ **Cost of Living:** Relatively low, but salaries will cover basic living costs only.

➲ **Health advice:** Insurance with evacuation cover is advised. Medical care in Lithuania is improving. Most medical supplies are now widely available.

➲ **English language media:** Baltic Times, also *Vilnius/Kaunas/Klaipeda in your pocket* are useful guides to living in these places.

Schools in Lithuania (+370)

American English School
Pylimo 20, 2001 Vilnius. **Tel/Fax**: 2 791011.

EF
Kosciuskos g.11, 2000 Vilnius. **Tel:** 2 791616, **Fax**: 2 791646.

Klaipeda International School of Languages
Zveju g. 2, 5800 Klaipeda. **Tel:** 6 311190, **Fax**: 6 311190, **Email: mark.uribe@klaipeda.omnitel.net**

Klaipeda Tourist School
Taikos pr. 69, Klaipeda. **Tel:** 34 10 83, **Fax**: 34 02 46.

Public Service Language Centre
Vilniaus St 39/6, 2000 Vilnius. **Tel:** 2 220384, **Fax**: 2 220370, **Email: lora@vlkc.vno.osf.lt**

Soros International House
Ukmerges 41, korp A, 2662 Vilnius. **Tel:** 2 72 48 79, Fax: 2 72 48 39, **Email: daiva@ihouse.osf.lt**

Siauliai Pedagogical Institute
P. Visinskio 25, 5400 Siauliai. **Tel/Fax:** 1 432592, **Email: spi@siauliai.omnitel.net**

MOLDOVA

Demand for English is very high in this tiny republic sandwiched between Romania and the Ukraine that only gained independence in 1991.

➲ **Visas/Work Permits:** U.S. citizens will need a tourist visa. The school will advise you regarding work permits. Teachers report that the rules do not seem to be enforced very strongly.

➲ **Requirements:** College Degree and a TESOL/TESL Certificate

➲ **Currency:** £1 = 22.85 / $1 = 12.35 Leu

➲ **Salaries/Taxes:** An average wage is £3 an hour, which, locally, is quite a good rate.

➲ **Cost of Living:** Very reasonable. Be aware that Moldova is basically a "cash-only" society. Credit cards are not used here very much.

➲ **Health advice:** Medical care in Moldova is limited, with severe shortages of basic medical supplies.

Schools in Moldova (+373)

ACTR-ACCELS
Stefan cel Mare Blvd. 162, oficiul 1506, Chisinau, MD-2004. **Tel:** 2 248012, **Fax**: 2 238389, **Email: accels@dnt.md**

Linguata
Stefan cel Mare 64, office 175, Chisinau. **Tel:** 2 277779 **Email: linguata@ch.moldpac.md**

Pro Didactica
13 Armeneasca St., Chisinau, MD-2012. **Tel:** 2 541994, **Email: lschool@cepd.soros.md**

Travel Teach Group
Str. P. Rares 39, Chisinau MD-2074. **Tel:** 2 294067, **Fax**: 2 321817.

POLAND

There is probably more demand for English in Poland than any other European country, due to two factors. Poland will soon become a member of the EU and is keen to attract foreign investment from the West. Secondly, the economy has been growing rapidly due to a steady process of liberalization, which has helped the export sector, which in turn increases the demand for English. In particular, the government administers English Proficiency exams, which are very important for young Poles who want to go into business. Many teachers head towards the big university cities such as Warsaw and Lodz. Most work in the private schools and supplement their income through individual private lessons. There are plenty of opportunities for English language teachers in both the public and private sectors. Public sector English teachers are placed in schools, teacher training colleges (NKJOs) and universities. Private schools are usually Polish-owned and very welcoming to native speaker teachers. They pay well and will assist with housing, health insurance and other matters.

Embassy in UK:
47 Portland Place, London W1B 1JH
Tel: 0870 7742700

Embassy in USA:
2640 16th Street NW, Washington DC 20009
Tel: 202 234 3800
www.polandembassy.org

British Council office:
Al Jerozolimskie 59, 00 697 Warsaw
Tel +48 (22) 695 5900
www.britishcouncil.pl

➲ **Visas/Work Permits:** You should not require visas for stays up to 90 days for tourist, business, or transit purposes. Americans should ensure that their passports are date-stamped upon entry. Persons planning to stay in Poland for longer than 90 days or who will be employed in Poland must obtain a visa before they arrive in the country. Polish law requires every traveller to be able to show means of support, if asked. For persons above 16 years of age, this has been defined as 100 Polish zloty per day. Regulations have been tightened up recently. All teachers need to apply for a work permit before arriving in the country. Your notarized degree, TEFL certificate (if appropriate) and promissory permit from your prospective employer should be presented to your local Polish consulate. You cannot obtain a work permit while you are in Poland.

➲ **Requirements:** College Degree, experience and a TESOL/TESL Certificate

➲ **Currency:** £1 = 5.65 / $1 = 2.89 Zlotych

➲ **Salaries/Taxes:** The average salary is around £200 per month in the state sector and at least £300 in the private sector. Some schools may offer overtime at £6-8 per hour. If you teach full-time and throw in a few private lessons you should be able to make over £600 a month which is very good money by Polish standards. The basic tax rate is 18%.

➲ **Accommodation:** A one-bedroom apartment in the centre of Warsaw will cost £200-300 per month. Studios cost £200-250. You will have to pay at least a month's rent as deposit in advance to the landlord and you might have to pay the same amount to the agent who located your apartment. It's a good idea to arrive in July or August with some spare cash and beat the September rush for accommodation. January is also a good time. Some schools offer free accommodation.

➲ **Cost of Living:** If you use public transport and adopt the local diet you can live on £5 a day including three good meals.

➲ **Health advice:** The hospitals are of variable quality. Private doctors and clinics are very cheap, as are dentists. Your school may offer you health insurance as part of your employment package. Whatever deal you have, make sure you have comprehensive coverage.

➲ **English language media:** The Warsaw Voice.

Schools in Poland (+48)

ABC
Uslugi Jezykowe, ul ZWM 6 m 29, Warsaw.

AJM
ul Klaudyny 30 m 100, 01-684 Warsaw.

American English School
PO Box 3, 00-991 Warsaw 44. **Tel/Fax:** 22 272654.

Angloschool
ul Elblaska 65/84, Warsaw.

AS
Studio Jezykow Obcych, ul M L Kinga 13, 75-016 Koszalin.

Bakalarz
Prywatne Studium Jezykowe, ul Rakowiecka 45/25, 02-528 Warsaw. **Tel:** 489 889.

Best
ul. Pestalozziego 11/13, 80-445 Gdansk. **Tel/Fax:** 58 41 2902.

Beyond 2000
Szkola Jezykow Obcych, ul Szuberta 39/5, 02-408 Warsaw.

British Council Warsaw Studium
Warsaw Technical University, ul. Filtowa 2, 00-611 Warsaw. **Tel/Fax:** 22 25 82 87.

Business and Educational English
ul Conrada 10 m 57, 01-922 Warsaw.

Centre of English Language Training
Konarskiego 2, 30-049 Krakow. **Tel:** 12 15 17 32.

Compact School of English
Spolka Cywilna, Nowogrodzka 78/24, 02-018 Warsaw.

Cosmopolitan Private Language School
Ul. Katowicka 39, 45-061 Opole. **Tel/Fax:** 77 54 87 73.

Discovery
Osrodek Nauczania Jezykow Obcych, Klub WAT-u, ul Kaliskiego 25a, 01-489 Warsaw.

The Eagle English Centre
Al Stanow Zjednoczonych 26 m 25, 03-965 Warsaw.

Elan
Mokotowska 9 m 6, Warsaw. **Tel:** 25 19 91.

Elite Language School
ul Dunikowskiego 10 m 56, 02-784 Warsaw.

English House
AWF Przedszkole, ul. Marymoncka 34, Bielany.

English House
"Domino" ul. Boguslawskiego 6a Lingwista, ul. Reymonta 25, Bielany. **Tel:** 663 5569.

English is Fun
Warchalowskiego 2/13, Warsaw.

English Language Academy
ul Narbutta 9 m 3, 02-564 Warsaw.

English Language Centre
University of Silesia, Plac Sejmu Slaskiego 1, 40-032 Katowice. **Tel:** 32 156 1269, **Fax**: 32 155 2245.

English Language Studio
ul Jadzwingow 1 m 34, 02-692 Warsaw.

English Unlimited
Podmlynska 10, 80-885 Gdansk. **Tel/Fax:** 58 31 33 73, **Email:** info@eu.com.pl.

Falaland
Osrodek Nauczania Jez Obcych, Margerytki 52, 04-908 Warsaw.

Fast
Firma Prywatna, Jezyk Angielski-Lektoraty, Ewa Maszonska-Pazdro, ul. Mickiewicza 74/58, 01-650 Warsaw.

Gama-Bell School of English
ul. sw. Krzza 16, 31-023 Krakow. **Tel:** 12 21 97 55, **Fax**: 12 21 73 79.

Greenwich School of English Poland
ul Zakroczymska 6, 00-225 Warsaw.

Human
Agencja Jezykow Obcych, ul Swietokrzyshka 20 pok 317, 00-002 Warsaw.

International House Bielsko Biala
ul. Krasinskiego 24, Bielsko Biala 43-300. **Tel:** 33 811 79 27/811 02 29, **Email:** ihbb@silesia.top.pl

International House Bydgoszcz
ul. Dworcowa 81, Bydgoszcz 85-009. **Tel:** 52 322 3515/322 3444, **Fax**: 52 322 3515, **Email:** bydgoszcz@inthouse.pl

International House Katowice
ul. Sokolska 78/80, Katowice 40-128. **Tel:** 32 58 68 96, 599 997, **Fax**: 32 253 88 33, **Email:** ihih@silesia.top.pl

International House Kielce
ul. Wesola 33, Kielce 25-353. **Tel:** 41 34 30 258/34 30 430, **Fax**: 41 34 30 261, **Email:** kielce@ih.pl

International House Koszalin
ul. Zwyciestwa 7/9, Koszalin 75-728. **Tel:** 94 3410 196, **Fax**: 94 3410 180, **Email:** hubbard.koszalin@inthouse.pl

International House Krakow
ul. Pilsudskiego 6, lp, 31-109 Krakow. **Tel:** 12 21 94 40, **Fax**: 12 21 86 52. **Email:** admin@ih.pl

International House Lodz
ul. Zielona 15, Lodz 90-601. **Tel:** 42 630 00 26, **Fax:** 42 630 00 28, **Email:** lodz@ih.pl

International House Opole
ul. Kosciuszki 17, Opole 45-062. **Tel/Fax:** 77 454 66655, **Email:** sekret@pol.pl

International House Poznan
ul. Sw. Marcin 66/72, Poznan 61 807. **Tel:** 61 851 6171, **Fax**: 61 851 61 71, **Email:** Lektor_international_house@poczta.fm

International House Torun
Ul. Legionow 15, Torun 87-100. **Tel:** 56 655 40 83, **Fax**: 56 622 50 81, **Email:** torun@inthouse.pl

International House Wroclaw
ul. Leszczynskiego 3, Wroclaw 50-078. **Tel:** 71 781 72 90, **Fax**: 71 781 72 93, **Email:** wrocdos@id.pl

International House Warsaw
ul. S. Noakowskiego 3, Warsaw 00-664. **Tel:** 22 660 71 01, **Fax**: 22 660 71 05, **Email:** monika_s@ihwarsaw.com.pl

International Language School
ul. Krakowska 51, 45-075 Opole. **Tel/Fax:** 77 53 15 97.

Junior Art-Language Studio
ul Gwardzistow 8 m 5, 00-422 Warsaw.

Kajman
ul Felinskiego 15, 1 LO pokoj 57 (parter), 01-513 Warsaw.

Konwersatorium
Jezykow Obcych, Grojecka 40a m 13, 02-320 Warsaw.

Kozlowski I Rejman
Kusocinskiego 2, 31-300 Mielec.

Langhelp
Al Jerozolimskie 23/34, Warsaw. **Tel:** 21 44 34.

Lektor Prywatna Firma Jezykowa
ul. Olawska 2, 50-123 Wroclaw. **Tel/Fax:** 71 343 25 99.

Lexis College of Foreign Languages
ul Danilowiczowska 11m 18, Warsaw.

Lingua
Studium Jezykow Obcych, ul Moniuszki 4a, Lodz.

Lingwista
ul Saska 59, 03-958 Warsaw, *and* Janowskiego 50, 02-784 Warsaw.

Mosak
ul Bonifacego 83/85 m 87, Warsaw.

MTA Language Centre
Pl. Teatrainy 6, Wroclaw. **Tel:** 71 72 13 34.

NKJO - Krakow
ul. Kanonicza 14, 31-002 Krakow. **Tel:** 12 22 79 55, **Fax**: 12 22 63 06.

Omnibus - Poznan
Pl Wolnosci 5, 61-738 Poznan. **Tel:** 52 79 08.

The Orlik Language Centre
ul Rogalskiego 2 m 49, 03-982 Warsaw.

Perfect
Naucanie Jezykow Obcych, ul Wolnosci 2 bl 13 m 14, Maciej Musiala, Zielonka.

Perfekts
Firma Oswiatowa, ul J Kaden-Bandrowskiego 2/16, 01-494 Warsaw.

Polanglo
Szkola Jezykow Obcych, ul Nowowiejska 1/3, 00-643 Warsaw. **Tel/Fax:** 22 25 37 47.

Poliglota
Biuro Jezykowych, Dzialdowska 6, 01-184 Warsaw.

Prima
Osrodek Nauczania Jezykow Obcych, ul Rozana 7 m 3, 15-669 Bialystock.

Program-Bell
ul Fredry 7, Poznan. **Tel:** 61 53 69 72.

Promotor
Szkola Jezykow Obcych, Agencja Oswiatowa, Kraszewskiego 32a, 05-800 Pruszkow.

Prymus
ul Jasna 2/4 pok 209, 00-950 Warsaw.

Pygmalion
Szkola Jezyka Angielskiego, ul Saska 78, Warsaw.

Studio Troll
Wrzeciono 1/22, Warsaw.

Studium Jezydow Obcych Project
ul Szwolezerow 5B/10, 66-400 Gorzow Wlkp. **Tel:** 95 32 05 59.

Success
ul Agawy 5/9, 01-158 Warsaw. **Tel:** 22 623 39 40, **Fax**: 22 20 55 03.

Surrey Business and Language Centres
ul Gwardzistow 20, 00-422 Warsaw.

Szansa
Prywatny Zaklad Oswiatowy, ul Agawy 5/9, 01-158 Warsaw. **Tel:** 22 623 39 40, **Fax**: 22 20 55 03.

Target
ul. Dziatwy 6/ róg Modolinska, Bialoleka. **Tel:** 660 7028, 660 7029

University of Szczecin
Al. Piastow 40B, 71-065 Szczecin. **Tel:** 91 33 31 61, **Fax**: 91 34 29 92.

Urszula
Biuro Organizacji Kursow, Dworcowa 1 pok 13, 10-431 Olsztyn

Warsaw School of Commerce
Kursy Handlowe, al Chlodna 9, Warsaw.

Warsaw Study Centre
Osrodek Jezykowo-Szkoleniowy, ul Raszynska 22, 02-026 Warsaw.

World
ul Basztowa 17, 31-143 Krakow. **Tel:** 22 91 61.

Worldwide English School
ul Stoleczna 21 paw 24, 01-530 Warsaw.

YES
ul Reformacka 8, 35-026 Rzeszow. **Tel/Fax:** 17 52 07 20.

Schools in Warsaw and its suburbs:

Mokotów ABC
ul. Wita Stwosza 57, **Tel:** 682-86-19

Columbus
ul. Kazimierzowska 16, **Tel:** 841-38-04

Greenwich
ul. Zakrzewska 25, **Tel:** 825-29-53

K i D
Osrodek Nauczania Jezyków Obcychul. Barcelonska 8, **Tel:** 39-52-04, 0-90-20-48-20

Lingwista agencja
ul. Rózana 22/24, **Tel:** 845-05-71

Lingua Nova
ul. Basniowa 3, **Tel:** 825-80-86, 668-76-74

Lingwista agencja
ul. Filtrowa 75, **Tel:** 659-50-01

Ochota English Language Centre
ul. Filtrowa 2, **Tel:** 660-55-71

Osrodek Kultury Ochoty Kursy Jezyków Obcych
ul. Grójecka 75, **Tel**: 822-48-70, 822-74-37

TARGET
ul. Bagatela 13, **Tel:** 660 70 28

Warsaw Study Centre
ul. Raszynska 22, **Tel/fax:** 822-14-12, **Tel:** 822-46-18

Praga Poludnie English House
Grochów-Kamionek, S.P. 60, ul. Zbarska 3

English House
Saska Kepa, ul. Kryniczna 12/14

Lingwista
Saska Kepa, ul. Saska, **Tel:** 617-85-83

Lingwista
ul. Miedzyborska 70, **Tel:** 810-22-01

TARGET
Saska Kepa, ul. Obronców 31, **Tel:** 660 70 28

Sródmiescie ABILITY
ul. Mazowiecka 12, **Tel:** 827-20-29

Academia de la Lengua (Language Mega Centre)
ul. Nowy Swiat 21A, **Tel:** 625-71-02, 828-10-21, 828-41-39, 646-09-20

ALBION
ul. Noakowskiego 26 m.26, **Tel/fax:** 628-89-92

BOKANO SCHOOL
ul. Chmielna 24/2, **Tel:** 826-31-17

Cambridge School of English
ul. Zakroczymska 6, **Tel:** 831-09-55

Context Language centre
ul. Hoza 61/1, **Tel:** 622 58 60

Didier Szkola Jezyka Obcych
ul. Kopernika 28 lok. 16, **Tel/fax:** 827-24-78

EMPiK Szkola Jezyków Obcych
ul. Marszalkowska 104/122, **Tel:** 826-79-17/18, 551-44-55/56/57, 826-71-04

EMPiK Szkola Jezyków Obcych
ul. Foksal 11, **Tel:** 826-79-17/18, 551-44-55/56/57, 826-71-04

EMPiK Szkola Jezyków Obcych
ul. Trebacka 3, **Tel:** 826-79-17/18, 551-44-55/56/57, 826-71-04

English Language College "Metodysci"
ul. Mokotowska 12, **Tel**: 628-53-48

Infor
ul. Sniadeckich 17, **Tel:** 629-77-90

International House
ul. Noakowskiego 3, **Tel:** 660-71-01

Greenwich
ul. Nowowiejska 5, **Tel**: 825-29-53

Konwersatorium
ul. Sempolowskiej 4, **Tel:** 622-61-62

Lang LTC
Al. Niepodleglosci, **Tel:** 825-39-40

Language Mega Centre (Academia de la Lengua)
ul. Nowy Swiat 21A, **Tel:** 625-71-02, 828-10-21, 828-41-39, 646-09-20

Lexis
Al. Marszalkowska 60 lok. 34, **Tel:** 625-53-86, 625-72-98, 622-25-23

Lingua Nova
ul. Nowowiejska 37A, **Tel**: 825-80-86, 668-76-74

Lingwista
ul. Kopernika 6, **Tel:** 827-86-85

Lingwista
ul. Ujazdowskie 37, **Tel:** 629-25-88

Lingwista
ul. Smolna 30 (Liceum/Szkola Podst.), **Tel:** 827-89-55

Loyd
pl. Powstanców Warszawy, Hotel Warsaw, **Tel:** 827 51 37.

Maart
ul. Nowy Swiat 1, **Tel:** 621-11-71

Peritia
ul. Bednarska 2/4, **Tel:** 826-95-30

PolAnglo
ul. Nowowiejska 1/3, **Tel:** 825-21-56

TARGET
ul. Polna 50, **Tel:** 660 70 28

TFLS
ul. Krucza 47 A, **Tel:** 622-20-58

Tower Language Institute
ul. Szpitalna 5 lok.16, **Tel:** 846-22-26, 828-36-99

Towarzystwo Wiedzy Powszechnej
pl. Defilad 1, Palac Kultury i Nauki, **Tel:** 656 61 75

Twins/English Conversation
ul. Krucza 17 lok. 3, **Tel:** 0604817817

Warszawskie Centrum Studenckiego Ruchu Naukowego
ul. Mokotowska 48, **Tel:** 628-82-81

Wola Cambridge School of English
ul. Bema 70, **Tel:** 632-78-86

Focus Language School
ul. Smocza 21, **Tel:** 636-44-31, 654-13-11

Lingwista
ul. Ksiecia Janusza 45/47, **Tel:** 36-17-71

Poliglota
ul. Dzialdowska 6, **Tel:** 632-02-08

Zoliborz English House
SDK, ul. Próchnika 8a, Teatr Komedia

Greenwich
ul. Gdanska 2/3, **Tel:** 833-24-31

Lingwista
ul. Gen. Zajaczka 7, **Tel:** 39-25-85

TARGET
ul. Kochanowskiego 8, **Tel:** 660 70 28

Twins/English Conversation
ul. Dyminska 6A/146 **Tel**: 39 25 84

Cambridge School of English
ul. Wasilkowskiego 6, **Tel:** 644-63-06

English House
Lasek Brzozowy 2 (Klub przez Lasku)

Lingwista
ul. Janowskiego 50, **Tel:** 643-72-79

ROMANIA

Romania is a fascinating country but teachers who come here must be prepared to live in a society that is still coming to terms with its post-Communist phase. Life can be very harsh here, especially in big industrial cities. The demand for English is not as strong in Romania as it is in other countries but a determined teacher should be able to find work without too many problems.

Embassy in UK:
Arundel House, 4 Palace Green, London W8 4QD
Tel: 020 7937 9666

Embassy in USA:
1607 23rd Street NW, Washington DC 20008
Tel: 202 332 4848
www.roembus.org

British Council office:
Calea Dorobantilor 14, 71132 Bucharest
Tel +40 (21) 307 9600
www.britishcouncil.ro

➲ **Visas/Work Permits:** You need a visa for a stay of longer than 30 days. You can hunt for jobs while visiting the country as a tourist. Your employer will apply for permits. Do not remain in Romania without a valid visa because the fines for overstaying are very high.

➲ **Requirements:** College Degree, experience and a TESOL/TESL Certificate

➲ **Currency:** £1 = 5.08 / $1 = 2.60 New Lei.

➲ **Salaries/Taxes:** A typical package for a native-speaking teacher is £200 per month with accommodation provided by the school.

➲ **Accommodation:** One of the major headaches of living in Romania is finding affordable and decent housing. Most buildings were poorly constructed and rents are very expensive.

➲ **Cost of Living:** Food and locally made products are very cheap.

➲ **Health advice:** Medical care in Romania is limited. Basic medical supplies are scarce. Doctors and hospitals often expect immediate cash payment for health services.

➲ **English language media:** Bucuresti, Nine O'Clock, In Review, Romanian Business News.

Schools in Romania (+40)

ACCESS Language Centre
Str. Tebei nr. 21, 3400 Cluj. **Tel:** 064 420480 ext 128, **Fax**: 064-420476 **Email: ovidiu@access.soroscj.ro**

Best
Colegiul I.L.Caragiale, Calea Dorobantilor 163, Sector 1 Bucharest. **Tel:** 01-322 6744.

Calepinus Foundation - Foreign Language Institute
Piata Bernady 3, 4300 Târgu Mures. **Tel:** 065-216514

Cambridge Study Centre
Liceul Vasile Alecsandri Str., N.Balcescu 41, 6200 Galati. Centrul de limbi straine FIDES, Str.Vasile Conta 7-9, sc.E, et.7, ap.168, 70139 Bucharest. **Tel:** 01-313 8080 ext.160

CLASS (Constanta Language Association)
Bd. Mircea cel Batran nr. 103, 8700 Constanta. **Tel/Fax:** 041-612877.

Computer Academy
Sala Mare A Palatului, Bucharest. **Tel:** 01-615 9710 Ext. 745/742.

Daniel - Centrul De Limba Engleza
Str. Cosminului 10, 4300 Târgu Mures. **Tel:** 065-216950.

International House Language Centre
Bd. Republicii nr. 9, 1900 Timisoara. **Tel/Fax:** 056-190593 **Email:** rodica@ihlctim.sorostm.ro

International Language Centre
Str. Moara de Foc nr. 35 et. 8, 6600 Iasi. **Tel:** 032-252850 / 252870. **Fax**: 032-252902. **Email: acolib@ilc.iasi.osf.ro**

Lexis Bucuresti
Piata Pache Protopopescu 96, Sector 2, Bucharest. **Tel:** 01-310 2629.

Linguarama
Str. Delea Veche 51, Bl.46, Et.9, Sector 2, Bucharest. **Tel:** 01-320 6156.

Open Doors School of English
Michelle Penta, str. L. Blaga nr. 4, 1900 Timisoara. **Tel/Fax:** 056- 194252.

PROSPER-ASE Language Centre
Calea Grivitei 2-2A et.2 s.4211, Bucharest. **Tel/Fax:** 01-2117800, **Tel**: 01-6505456, **Email:** prosper1@rolink.iiruc.ro

PROSPER-Transilvania Language Centre
c/o Bd.Valea Cetatii 6, bl.B2, sc.B, ap.6, 2200 Brasov **Tel:** 068-319908, **Email:** g.cusen@unitbv.ro

RALEX Linguistic Centre
Str. Buna Vestire nr. 35, 1000 Ramnicu Valcea, **Tel/Fax:** 050-740032

Soros Educational Centre Miercurea-Ciuc
Str. Florilor 9, et.3, 4100 Miercurea-Ciuc. **Tel/Fax:** 066-171799, **Email:** office@cemc.topnet.ro

'YES' Language Centre
Colegiul Pedagogic Andrei Saguna, Str. Gen. Magheru 34, 2400 Sibiu. **Tel:** 069-430809 / 434002

RUSSIA

The Western media tends to paint a distorted view of life in Russia. Although there are hardships, most teachers report that their experiences in Russia were positive and that people actually enjoy themselves quite a lot here! There is a great demand for English lessons throughout Russia at present. This demand is not confined to the major cities like Moscow and St. Petersburg. Private language schools are springing up across the country – some more reputable than others. Russians are fascinated by all things American and you will be the centre of much attention, especially in the smaller towns. But a word of warning: if your grammar and spelling are weak, you better brush up on these areas before you go. Russians are precise and thorough in their language learning and they are not impressed by teachers who do not know how to use the English language correctly. Many teachers start off by working for a school, then pick up some private lessons as the word gets around, and finally going completely freelance.

Embassy in UK:
13 Kensington Palace Gardens, London W8 4QX
Tel: 020 7229 8027
www.russianembassy.net

Embassy in USA:
2650 Winsconsin Ave. NW, Washington DC 20007
Tel: 202 298 5700
www.russianembassy.org

British Council office:
Ulitsa Nikoloyamskaya 1, Moscow 109189
Tel +7 095 782 0200
www.britishcouncil.ru

➲ **Visas/Work Permits:** A Russian visa is considered as a permission to enter and exit the country. To obtain a visa, you must be sponsored by an organization such as a school or a tour company. Only the sponsor can extend or change your visa. It is a good idea to apply for a three-month visa. After that, the school can arrange your paperwork but you will have to have an AIDS test. Be careful that the school does not decide to keep you longer than you want to stay! Be prepared for a lot of red tape as far as visas are concerned. There is no such thing as a work permit in Russia.

➲ **Requirements:** College Degree, experience and a TESOL/TESL Certificate

➲ **Currency:** £1 = 51.57 / $1 = 26.39 Rubles

➲ **Salaries/Taxes:** Russia operates on a dual currency system with both rubles and dollars in circulation. You can earn around £3500-4000 per month and as much as £30 an hour for private lessons in Moscow and St. Petersburg. These rates are lower in the provincial centres.

➲ **Accommodation:** You can find an apartment in Moscow from around £100 a month.

➲ **Cost of Living:** Moscow has a reputation as an expensive city but you will be earning far more than most people do there. If you indulge in imported western goods, then life will be expensive. Carry cash (US$) – credit cards are useless unless you plan to spend your time at expensive hotels.

➲ **Health advice:** Teachers thinking of going to Russia should be in good physical condition. It's not a good idea to entrust your health to a badly paid doctor operating in poor facilities. Get shots for rabies, diphtheria, cholera, hepatitis A and B and encephalitis before you go.

➲ **English language media:** Moscow Times, Moscow Tribune, Moscow News, St. Petersburg Press, Neva News, The Vladivostok Times.

Schools in the Russian Federation (+7)

Anglo-American School
Penkovaya ui. 5, St. Petersburg.

Benedict School
Chapligina 92, Novosibirsk 630099. **Tel:** 3832 23 24 33, **Fax**: 3832 23 59 97.

Centre for Intensive Foreign Language Instruction
Sparrow Hills, Building 2, Moscow 119899.

Go West Department of Foreign Languages
2/1 Spartakovskaya Street, Moscow. **Tel:** 95 267 4700, **Fax**: 95 267 4700.

Language Link Schools
Novoslobod-skaya ul. 5 bld.2, 101030 Moscow. **Tel/Fax:** 95 973 2154.

Link School of Languages
15/81 pl Lenina, 394000 Voronezh. **Tel:** 732 55 23 31.

LK Varioline
12-12, Klyutchevskaya St., Ekaterinburg 620109. **Tel/Fax:** 3432 46 91 39.

Marina Anglo-American School
Leninsky Prospekt 39a, Moscow 117313.

Moscow-Cambridge Project Management
Academy of National Economy, 82 Vernadsky Prospect, Moscow 117571. **Tel:** 502 234 3873, **Fax**: 95 434 6030.

Moscow International School
2nd Ulitsa Maryinoy Roshchi 2a, Moscow.

Moscow MV Lomonosov University
Sparrow Hills, Moscow 117234.

NevaLingua International Study Centre
Malaya Morskaya Str. 11, office 315, St. Petersburg 191186. **Tel:** 812 3128129, **Fax**: 812 3126211.

Polyglot International School
22 Volkov Pereulok, Flat 56, Moscow.

ProBa Language Centre
Professora Popova 5, St.Petersburg 197376. (Mail: P.O. Box 109, Lappeenranta FIN-53101 Finland.) **Tel:** 812 3252618. **Fax**: 812 3462758

Pushkin Language Centre
Pushkin. (**Contact:** Leipziger Strasse 27, 09648 Mittweida, Germany.) **Tel:** +49-3727-976228

RISC Language School
Ulitsa Dekabristov 23a, Moscow

Russian-American Centre
Tomsk Polytechnic University, Room 319, Lenin Ave 30, 634004 Tomsk. **Tel:** 3822 22 45 48, **Fax**: 3822 22 46 07.

St Petersburg University
Universitetskaya, Naberezhnaya 7/9 B-164, St Petersburg, 199164.

St Petersburg University of Humanities & Social Sciences
Dept. of Foreign Languages, 15 Fuchika St., St Petersburg 192238. **Tel:** 812 269 1925.

Sunny School
PO Box 23, Moscow 125057. **Tel:** 95 151 2500.

System-3 Language & Communication
Arnold Rubinstein, Kantemirovskaya Street 16 #531, Moscow 115522. **Tel:** 9210141, **Fax**: 3204035.

World Language Club
1106e Panfilovsky Prosp., Zelenograd 103460. **Tel:** 95 532 9121, **Fax**: 95 530 9646.

Spelling It Out in Siberia

Professor Yuri Tambovtsev explains what intrepid EFL teachers might expect to find if they come to work in the Siberian city of Novosibirsk.

Novosibirsk is situated on the great Siberian River, Ob. It is quite a young city, founded in 1893 when a crossing for the Trans-Siberian railway was built over the river. Until 1925, Novosibirsk was known as Novonikolaevsk, named after the last Russian tsar, Nicholas II. In the 1920s, the city grew rapidly as it became an important industrial and transport centre between the coalfields to the southeast and the mineral deposits in the Ural Mountains to the west. In 1930s the building of the Turko-Siberian railway south from Novosibirsk to Almaty in Kazakhstan made the city a crucial transport link between Russia and Central Asia.

Now a city of 1.5 million people, Novosibirsk is a major industrial, scientific, cultural and educational centre. One in three of the city's residents is a student. Many Russians have realized how important it is to be able to talk English because it is the language of international communication. Every Russian pupil begins learning English in the 1st grade. The pupils usually have two lessons a week. However, there are some schools which give six lessons a week. At the universities, students normally have English once a week for two years. The communicative method of teaching English is now in fashion. It is also called the "rapid" method of teaching. The pupils at schools and the students at universities are asked to talk in English during the entire lesson. A native-speaking teacher is required to talk English all the time without resorting to Russian. Although having a degree, certificate or diploma in teaching EFL/ESL will help you as a teacher, the universities and colleges will hire unqualified native speakers of English if they can prove they are able to conduct an English lesson well.

I think one problem with English teaching here in Russia is that we do not teach the students correct English phonetics. The special lessons devoted to teaching only English sounds have been abolished. Now, the students just substitute English sounds with similar Russian ones. Younger teachers of English and their students speak with a stronger Russian accent than the older teachers of English, who had a special course of practical and theoretical phonetics every day for one year of their degree course. The English language textbooks are usually British or American, so the teacher does what is in the textbook.

The salaries of young native-speaking teachers are usually 5-10 times higher than those of Russian teachers of English. Native speakers can earn £30-60 a month. One can live very well on this salary: A kilogram of meat is £2 and one can buy 4 kilograms of potatoes for under £1. There are no problems buying food, as the shops are full of goods. It is cheaper to buy food at the street markets, but you can't be sure if the food is fresh or safe to eat. Although the city is rather gray and functional, its residents are very proud of it since it has the biggest railway station, the biggest library and the biggest airport in Siberia. The local ballet company is said to be the best after Bolshoi and Mariinsky companies of Moscow and St. Petersburg respectively. The Art Gallery in Novosibirsk is certainly worth seeing as it contains original works of Shishkin, Rerickh, Repin, and Surikov. Opera, ballet and theatre tickets cost from £2-4. Universities can provide a native speaker of English with a room in a dormitory for £6-12 a month. The dormitory has showers with hot water, and places to wash clothes and a communal kitchen for cooking. Elsewhere in the city, a one-room apartment costs about £30 a month; with two rooms – £50. One can easily hire a car, but it is cheaper to use public buses or trams: 12 rides cost under £1.

During your summer vacation, you can travel by boat along the Ob river to the North to visit the Finno-Ugric aboriginal peoples of Siberia: Mansi (Voguls), Hanty (Ostjaks), and Nenets (Samoyeds). You can also visit the Altai Mountains and meet the Turkic aboriginal peoples: Altai-Kizh, Kumandin, Telengits, and Teleuts, the latter being remnants of Genghis Khan's Tatar-Mongolian hordes.

Professor Yuri Tambovtsev teaches at the Department of English and Linguistics, SIM, Novosibirsk, Russia.

SLOVAKIA

There is a shortage of English language teachers in Slovakia and native speakers with qualifications can land jobs here without problems. Many schools start recruiting in the spring for a September start but if you arrive in July or early August, you'll probably find work. If you want to teach in the state sector, you'll have to approach the Ministry of Education or one of the cultural organizations. There are fewer westerners here than in neighbouring countries.

Embassy in UK:
25 Kensington Palace Gardens, London W8 4QY
Tel: 020 7243 0803
www.slovakembassy.co.uk

Embassy in USA:
3523 International Court NW, Washington DC20008
Tel: 202 237 1054
www.slovakembassy-us.org

British Council office:
Panska 17 - P.O. Box 68, 814 99 Bratislava
Tel: +421 (2) 5443 1074
Www.britishcouncil.sk

➲ **Visas/Work Permits:** You do not need a visa for a stay of less than 30 days so you can look for a job during this time. Your employer should arrange all the necessary paperwork but you will have to take a rather unpleasant medical exam.

➲ **Requirements:** College Degree and a TESOL/TESL Certificate

➲ **Currency:** £1 = 52 / $1 = 26 Koruny.

➲ **Salaries/Taxes:** You will be paid at local rates–around £180-200 per month for a 25-hour week plus some overtime. You can earn better money with private lessons but these are hard to come by.

➲ **Accommodation:** This is usually supplied by the school who should also pay for the utilities.

➲ **Cost of Living:** Although Slovakia is cheap in dollar terms, you will be paid at local rates so basic items like shampoo and other toiletries are actually expensive. Credit cards are not used much here.

➲ **Health advice:** There are decent medical facilities but few doctors speak English. Some schools will provide you with medical coverage.

➲ **English language media:** Slovak Spectator.

Schools in Slovak Republic (+421)

Academia Vzdelavania
Gorkeho 10, 815 17 Bratislava. **Tel:** 7 367 580, **Fax**: 7 321 661.

American Language Institute
Drienova 34, PO Box 78, 820 09 Bratislava.

Foundation for a Civil Society
V Zahradach 29A, 811 03 Bratislava. **Tel:** 7 580 2491, **Fax**: 7 531 1622.

VSMU
Katedra Jazykov, Venturska 3, 813 01 Bratislava.

SLOVENIA

There are lots of opportunities for English language teachers in Slovenia. The country is more "western" than others in the region. Native-speaking teachers are usually assigned to the more advanced learners. It's easier to come here and find a job rather than applying from outside the country.

British Council office:
Cankarjevo nabrezje 27, Ljubljana 1000
Tel +386 (0) 1 200 0130
Email: info@britishcouncil.si

➲ **Visas/Work Permits:**You can stay for up to 90 days without a visa. Most teachers just stay on although some do overcome the red tape and acquire work permits. The authorities don't worry about foreign teachers unless they are perceived (rightly or wrongly) as troublemakers.

➲ **Requirements:** College Degree and a TESOL/TESL Certificate

➲ **Currency:** £1 = 357 / $1 = 182 Tolar

➲ **Salaries/Taxes:** Expect to earn $9-12 per 45-minute teaching "hour". Most teachers are paid in tolars

➲ **Accommodation:** A nice one-bedroom apartment costs around £150-200 per month or you can share a place for about £60.

➲ **Cost of Living:** You should be able to live well off your salary here.

➲ **Health advice:** Schools will not offer medical coverage so you should have your own policy. Medical care in Slovenia is of a high quality and many doctors speak English.

Schools in Slovenia (+386)

Accent On Language d.o.o.
Ljubljanska c. 36, 1230 Domzale. **Tel:** 61 712 658, **Fax**: 61 712 658.

Ambra d.o.o.
Pobeska 18, 6000 Koper. **Tel:** 66 33 128.

AS Asistent d.o.o.
Glavni trg 17 B, 2000 Maribor. **Tel:** 62 226 442, **Fax**: 62 224 432.

Athena d.o.o.
Kolodvorska c. 17, 6230 Postojna. **Tel:** 67 23 907, **Fax**: 67 23 907.

Bled
Sola za Tuje Jezike, Breda Vukelj s.p., Kajuhova c. 11, 4260 Bled. **Tel:** 64 741 780, **Fax**: 64 741 780.

Candor Dominko k.d.
Turnovse 19, 1360 Vrhnika. **Tel:** 61 756 032, **Fax**: 61 756 032.

Cenca p.o.
Masarykova c. 18, Ribnisko selo, 2000 Maribor. **Tel:** 62 306 479**.**

D and A Ita d.o.o.
Japljeva ul. 4, 1241 Kamnik.

Dialog d.o.o
Jezikovna Sola, Terceva ul. 39, 2000 Maribor. **Tel:** 62 24 779, **Fax**: 62 227 866.

Dude d.o.o
English Language centre, Slamnikarska 1, 1230 Domzale. **Tel:** 61 716 913

Eurocentre
Lilijana Durdevic s.p., Kidriceva ul. 46, 6000 Koper. **Tel:** 66 272 336, **Fax**: 66 281 653.

Europa Bled, d.o.o.
Alpska 7, 4260 Bled. **Tel:** 64 741 563, **Fax**: 64 741 563 **Email:** evropabl@perfech.si

Ekocept d.o.o
Sola za Tuje Jezike, Prezihova 8, 2230 Lenart v Slovenskih Goricah. **Tel:** 62 724 064, **Fax**: 62 724 064.

Flamingo d.o.o.
Pod Gonjami 44, 2391 Prevalje. **Tel:** 602 31 455.

Forum centre d.o.o.
Markova pot 8, 5290 Sempeter pri Gorici, **Tel:** 65 31 205, **Fax**: 65 31 205.

G.A.P.
Belokriska 22, 6320 Portoroz. **Tel:** 66 75 894 .

Hello Sempeter d.o.o
Sempeter v Savinjski dolini 34, 3311 Sempeter v Savinjski dolini. **Tel**: 63 701 795, **Fax**: 63 701 795.

ISCG Domzale
Kolodvorska c. 6, 1230 Domzale. **Tel:** 61 711 082, **Fax**: 61 712 278.

Jezikovna Sola Tabula Rasa
Trg svobode 12, 3325 Sostanj. **Tel:** 63 892 291.

Jezikovni biro Lindic
Cankarjeva ul. 10, 2000 Maribor. **Tel:** 62 221 629.

Jezikovni Studio Kotar Sonja, s.p.
Cesta na Svetino 10, 3270 Lasko, **Tel:** 63 731 105.

Kamnik
Altera d.n.o., Kettejeva 23, 1241 Kamnik. **Tel:** 61 831 117.

Lingua
Carman Cilka, s.p., Cesta Zasavskega bataljona 27, 1270 Litija. **Tel:** 61 882 137

Lingva d.o.o.
Ograde 7, 1386 Stari trg pri Lozu, Cerknica. **Tel:** 61 708 017.

Lingva p.o
Zavod za Opravljanje Izobrazevalnih in Organizacijskih Storitev, Vetrinjska ul. 16, 2000 Maribor. **Tel:** 62 24 795, **Fax**: 62 301 071.

Little England Club d.o.o.
Medvedova ul. 6, 1241 Kamnik . **Tel:** 61 832 865, **Fax**: 61 832 865.

Ljudska Univerza Celje
Cankarjeva ul. 1, 3000 Celje. **Tel:** 63 484 320, **Fax**: 63 443 181.

Ljudska Univerza Dodatno Izobrazevanje
Maistrova ul. 5, 2000 Maribor. **Tel:** 62 226 441, **Fax**: 225 017.

Ljudska Univerza Koper
Cankarjeva ul. 33, 6000 Koper. **Tel:** 64 271 258, **Fax**: 64 271 291.

Ljudska Univerza Kranj
Cesta Staneta Zagarja 1, 4000 Kranj. **Tel:** 64 222 226, **Fax**: 64 222 226.

Ljudska Univerza Nova Gorica
Cankarjeva ul 8, 5000 Nova Gorica. **Tel:** 65 22 302, **Fax**: 65 21 291.

Ljudska Univerza Ptuj
Mestni trg 2, 2250 Ptuj. **Tel:** 62 771 539, **Fax**: 62 779 849.

Ljudska Univerza Ravne Na Koroskem
Prezihova ul. 24, 2390 Ravne na Koroskem. **Tel:** 602 22 296.

Ljudska Univerza Sezana
Bazoviska c. 9, 6210 Sezana. **Tel:** 67 73 428, **Fax**: 67 32 046.

Ljudska Univerza Skofja Loka
Podlubnik 1b, 4220 Skofja Loka. **Tel:** 64 621 865.

Most d.o.o.
Kovinarska ul. 9, 8270 Krsko. **Tel**: 608 22 397, **Fax**: 608 22 397.

Multilingua d.o.o.
Ulica bratov Hvalic 16, 5000 Nova Gorica. **Tel:** 65 24 292, **Fax**: 65 24 292

New College d.o.o.
Med ogradami 11A, 5000 Nova Gorica. **Tel:** 65 23 733, **Fax**: 65 23 733.

Niansa Jezikovna Sola
Cesta talcev 4, 1230 Domzale. **Tel:** 61 712 012, **Fax**: 61 712 230.

Nista Language School
Smarska cesta 5d, 6000 Koper. **Tel:** 66 250 400, **Fax**: 66 258 440, **Email:** nista@siol.net

Ontario d.o.o.
Miklosiceva ul. 5, 2250 Ptuj. **Tel:** 62 779 108, **Fax**: 62 779 108.

Poliglot d.o.o.
Ljubljanska c. 110, 1230 Domzale. **Tel:** 61 723 089, **Fax**: 61 723 089.

Pharmagan d.o.o.
Straziska ul. 7, 4000 Kranj. **Tel:** 64 311 463, **Fax**: 64 311 463.

Prevajanje s.p.
Kalinskova ul. 31, 4000 Kranj. **Tel:** 64 340 010, **Fax**: 64 340 011.

Progress Jezikovni Tecaji d.o.o.
Seljakovo n. 33, 4000 Kranj, **Tel:** 64 310 644.

Rossana d.o.o.
Trzaska c. 30, 1370 Logatec. **Tel:** 61 743 104, **Fax**: 61 742 910.

Samba d.n.o.
Bresterniska ul. 225, 2351 Kamnica. **Tel:** 62 621 905.

Selih Kozelj d.n.o.
Studio za Ucenje, Miklosiceva 9, 3000 Celje. **Tel:** 63 24 236, **Fax**: 63 24 326.

Sibon
centre za Tuje Jezike, Ljubljanska c. 76 II, 1230 Domzale. **Tel:** 61 722 901, **Fax**: 61 722 928.

Speak It
Heintzman and Heintzman, Trubarjeva ul. 4, 1230 Domzale. **Tel:** 61 712 312, **Fax**: 61 712 312.

Spektra Jameks d.o.o.
Malgajeva 10, 3000 Celje. **Tel:** 63 443 192, **Fax**: 63 443 191.

Sports Life Pevc d.o.o.
Cesta talcev 6, 4220 Skofja Loka. **Tel:** 64 622 335, **Fax**: 64 623 121.

Studiom d.o.o.
Radvanjska c. 74, 2106 Maribor. **Tel:** 62 35 481.

Tecaj d.o.o.
Kraigherjeva ul. 19A, 2230 Lenart v Slovenskih Goricah. **Tel:** 62 723 224, **Fax**: 62 723 224.

Tonson Posrednistvo in Storitve d.o.o.
Velike Brusnice 13, 8321 Brusnice. **Tel:** 68 85 882, **Fax**: 68 85 882.

Unique d.o.o.
Jezikovno izobrazevanje, Gabrsko 69, 1420 Trbovlje. **Tel**: 601 27 442, **Fax**: 601 27 442.

Vox Vidrih Trebnje k.d.
Sola Tujih Jezikov, Simonciceva ul. 10, 8210 Trebnje. **Tel:** 68 45 158, **Fax**: 68 45 158.

Yurena d.o.o.
Ulica Marjana Kozine 49A, 8000 Novo Mesto. **Tel:** 68 341 434, **Fax**: 68 341 434.

Zavod za Izobrazevanje BAS
Grajski trg 1, 2000 Maribor. **Tel:** 62 223 344, **Fax**: 62 223 344.

Zavod za Izobrazevanje
Kopitarjeva 5, 2000 Maribor. **Tel:** 62 214 273, **Fax**: 62 225 003.

UKRAINE

There are two sides to the Ukraine: the western part is mostly Catholic and looks to Poland while the east is Orthodox. There is also the Crimea, a fascinating region on the Black Sea heavily populated by Tatars. There is a lot of interest in English language learning in the Ukraine and many schools have opened in last five years. The economic situation is still precarious and outside of Kiev, you may find daily life difficult to cope with, as food shortages still happen. It is probably better to arrange a job before arriving here.

British Council:
4/12 Vul. Hryhoriya Skovorody, Kyiv 04070
Tel +380 (44) 490 5600
Www.britishcouncil.org/ukraine

➲ **Visas/Work Permits:** To obtain a visa, you need an invitation to visit Ukraine either from a company or an individual. Once you find a job, your employer should take care of all the paperwork. Some teachers work illegally but this is not advisable and it is very hard to change your visa status from tourist to worker while in the country.

➲ **Requirements:** College Degree and a TESOL/TESL Certificate

➲ **Currency:** £1 = 9.8 / $1 = 5.33 Hyvrnia.

➲ **Salaries/Taxes:** Salaries range anywhere from £300-600 a month, which is good by local standards.

➲ **Accommodation:** Housing is usually provided by the school but you can find apartments from £100-200 pm in the suburbs and a lot more in the city centre.

➲ **Cost of Living:** Food and transport costs are very reasonable.

➲ **Health advice:** Medical facilities are limited. Insurance is essential.

➲ **English language media:** The Kiev Post, The Odessa Post.

Schools in the Ukraine (+380)

American Academy of Foreign Languages
89 Chervonoarmeyska, Office 9, 2520006 Kiev. **Tel/Fax:** 44 268 5095.

London School of English
PO Box B 158, 2420001 Kiev. **Tel/Fax:** 44 2418654. **Email: admin@lse.kiev.ua**

Odessa English Language Study centre
Pereulok Onilovoy 26/28, 270027 Odessa. **Tel/Fax:** 44 82 344 000.

Ukrainian-American Humanitarian Institute
9 Pyrogov St., 252030 Kiev. **Tel/Fax:** 44 216 0666.

Union Forum
10 Ternopiliska St., P.O. Box 10722, 290034 L'viv. **Tel/Fax:** 322 759 488.

Middle East & North Africa

Over the past decade, many of the oil-rich Gulf States offered some the best employment packages for English language teachers anywhere in the world. However, with continuing tensions and war-zones, the Middle East it is often viewed as a dangerous region - some countries are certainly too dangerous to visit, but others are still as friendly and welcoming to travellers and teachers: Arab hospitality is legendary. Although the region was dominated by Great Britain in the earlier part of the 20th century, there is a growing preference for teachers with American and Canadian accents, particularly in the Gulf. Possible drawbacks include cultural differences, and, in certain countries, restrictions on the lifestyles of female teachers. But the region, birthplace to the three great monotheistic religions (Judaism, Christianity, and Islam), is as compelling as it is complex in terms of its history and its people. Teachers willing to overcome preconceptions will be rewarded by experiences that will open their eyes to the realities of a region rich in tradition and humanity. However, the current war zones and continuing tensions involving Iraq, Iran and some other neighboring countries means this is potentially a *very* dangerous region and you should check before travelling to any of the countries in this region. Visit www.fco.gov.uk for the UK-government's advice for travellers.

The overwhelming majority of teaching positions overseas are offered by private schools–including the American-sponsored international schools in Asia and Middle East. A graduate degree is a standard requirement for employment, including the UAE, Bahrain, Lebanon, Jordan, Syria, and others.

North Africa: Morocco and Tunisia are attractive prospects for English language teachers as both countries are shaking off their Francophone past by embracing English as a working language. Algeria and Libya are basically off limits to American teachers although oil companies in Libya do employ teachers. Recent reports suggest that Canadian teachers are welcome in Tripoli although only the truly intrepid should attempt to teach in Libya.

Two good sources of information on this region are: *TESOL Arabia* website - http://tesolarabia.uaeu.ac.ae and *AMIDEAST* - www.amideast.org

BAHRAIN

A small island situated in the Arabian Gulf, Bahrain is quite liberal and women are welcome in the classroom. Employment should be arranged in advance and the only opportunities are for qualified teachers.

Embassy in UK:
30 Belgrave Square, London SW1X 8QB
Tel: 020 7201 9170

Embassy in USA:
3502 International Drive NW, Washington DC 20008
Tel: 202 342 0741
www.bahrainembassy.org

British Council office:
AMA Centre (PO Box 452), 146 Shaikh Salman Highway, Manama 356
Tel +973 261 555
www.britishcouncil.org/bahrain

➲ **Visas/Work Permits:** Passports and visas are required. Three-day and seven-day visas may be obtained upon arrival at the airport in Manama, but obtaining visas before travel is recommended. Work permits are also required. Your employer should make arrangements to provide you with one soon after your arrival.

➲ **Requirements:** College Degree, experience and a TESOL/TESL Certificate

➲ **Currency:** £1 = 0.695 / $1 = 0.377 Dinar

➲ **Salaries/Taxes:** Monthly salaries range from BHD 600-750 for teachers with certificates and BHD 600-700 for those with diplomas. Part-time teachers earn around BHD 10 per hour. There is no personal income tax.

➲ **Accommodation:** A small apartment will cost BHD 200-500 per month. There is no deposit but three months' rent is required in advance.

➲ **Cost of Living:** More expensive than other Gulf States and salaries are often lower.

➲ **Health advice:** Essential: health insurance can be obtained locally at competitive rates. Vaccinations: Hepatitis A, Polio, Typhoid.

➲ **English Language Media:** Bahrain Tribune, Gulf Daily News.

➲ **Tip:** There has been some unrest from time to time in Bahrain. Although Americans have not to date been subject to attacks, it is wise to exercise caution at all times.

Schools in Bahrain (+973)

Al Maalem centre
PO Box 20649, Manama. **Tel:** 553808. **Fax**: 554240.

Al Rawasi Academy
Mina Salman, Bahrain. **Tel:** 727442. **Fax**: 725442.

Awal Training Institute
PO Box 82110, Riffa. **Tel:** 773445. **Fax**: 771909.

Bahrain Computer & Management Institute
PO Box 26176, Manama. **Tel:** 293493.

Bahrain Training Institute
PO Box 323333, Essa Town. **Tel/Fax:** 683416.

BCMI
PO Box 26176, Manama. **Tel:** 293493. **Fax**: 290358.

The British Council
AMA Centre, (PO Box 452), Manama 356. **Tel:** 261555.

The Cambridge School of English
PO Box 20646, Manama. **Tel:** 532838.

Capital Institute
PO Box 22521, Manama. **Tel:** 740744/740088, **Fax**: 720060.

Child Development
PO Box 20284, Manama. **Tel:** 728000, **Fax**: 722636.

Daar al Merifa
PO Box 3174, Manama **Tel**: 722183, **Fax**: 722283.

Delmon Academy
PO Box 10362, Manama. **Tel:** 294400, **Fax**: 292010.

Global Institute
PO Box 11148, Manama. **Tel:** 740940, **Fax**: 720030.

Gulf Academy
PO Box 10333, Manama. **Tel:** 721700, **Fax**: 722636.

Gulf School of Languages
PO Box 20236, Manama. **Tel:** 290209, **Fax**: 290069.

Pitman IPE
PO Box 26222, Manama. **Tel:** 290028, **Fax**: 292006.

Polyglot School
PO Box 596, Manama. **Tel:** 271722, **Fax**: 722636.

University of Bahrain
PO Box 32038, Sakhir-Bahrain. **Fax:** 973 4496603.

EGYPT

English language teaching in Egypt suffers from the usual catalogue of third-world woes such as overcrowded classrooms and inadequate resources. English is nevertheless well established as the primary foreign language, and most people want to learn it and work hard at doing so. There is great enthusiasm for improving programmes in the teaching of English by means of better teacher training and the development of better methods and materials for teachers to work with. More and more schools in Egypt are starting to go online although connections are not always reliable. Foreigners will have no problem in finding a teaching job in a private language school here, especially if they are prepared to work with young learners. Employers usually recruit teachers in the spring to begin work in September. Some teachers survive on a freelance basis through advertising and contacts.

Embassy in UK:
26 South Street, London W1Y 6DD
Tel: 020 7499 2401

Embassy in USA:
3521 International Drive NW, Washington DC20008
Tel: 202 895 5400

British Council office:
192 El Nil Street, Agouza, Cairo
Tel +20 (2) 303 1514
www.britishcouncil.eg

➲ **Tip:** When teaching in Egypt, it is a good idea to keep in mind that most people come from an oral culture and do not have a tradition of reading, even in their own language, let alone English.

➲ **Visas/Work Permits:** American and Canadian citizens can obtain visitor's visas from the airport upon their arrival in Cairo. Go to one of the bank counters before the immigration windows and buy the visa stamps for your passport. They cost £10 ($19) and it is best to pay in U.S. dollars. These visas are good for one month. When they are close to expiration, you can easily get a six-month or one-year's extension. The schools obtain residency and work permits, which take around two months to process, while the teacher is working on a visitor's visa. A blood test is necessary.

➲ **Requirements:** College Degree and a TESOL/TESL Certificate

➲ **Currency:** £1 = 11.18 / $1 = 5.72 Pound

➲ **Salaries/Taxes:** Most teachers earn around £5-10 per hour in a school, Rates for private lessons range from £10-20. Monthly salaries start around EGP 2,000 and go up to EGP 7,500 depending on your qualifications and experience, and the prestige of the school. Tax is 5-7%.

➲ **Accommodation:** Do not expect Western standards of comfort and plumbing. Expect to pay around £300 a month for basic accommodations in Cairo and less elsewhere.

➲ **Cost of Living:** Imported (American) items are expensive although locally made versions of most products (varying in quality) are available at a reasonable price. Fruit and vegetables are plentiful, inexpensive, and are fresh and delicious.

➲ **Health advice:** Most foreign teachers are ill within a few weeks of arrival in Egypt although the majority of them recover. Hepatitis is a concern. Schistosomiasis, a parasitic infection, is found in fresh water in the region, including the Nile River. Do not swim in fresh water (except in well-chlorinated swimming pools.) Malaria is a risk limited to the El Faiyum area. Vaccinations: Hepatitis A, Polio, Typhoid, Malaria, Yellow Fever (if coming from an infected area).

➲ **English Language Media:** Al Ahram Weekly, Cairo Today, Egyptian Gazette, Middle East Times.

➲ **Information for Women:** You may find yourself the centre of attention as you walk down the street and the object of rude comments or occasional unwanted touching. Dress modestly and avoid public transportation (or, if taking the Cairo subway, ride the first car which is reserved for women.) Violent crime is rare.

Schools in Egypt (+20)

American Cultural centre
3 El Pharana Street, Alexandria. **Tel:** 3 483 1922

The American University in Cairo
CACE, ESD P.O. Box 2511, Cairo. **Tel:** 2 357 6870.

AMIDEAST
3 El Pharana Street, Alexandria. **Tel:** 3 483 1922.

El Manar School
Amin Fikry Street, Ramleh Station, Cairo. **Tel:** 2 4823813.

El Nasr Boys' College (EBC)
Chatby. **Tel:** 2 5976812.

El Nasr Girls' College (ECC)
Chatby. **Tel:** 2 597 1752.

El Pharaana School
El Pharaana Street, Bab Sharki. **Tel:** 2 4822229.

Heliopolis
Cairo. **Tel:** 2 4155003, **Fax**: 2 2900552.

IELP
6 Mostafa El Sherei Street, Dokki 12311, Cairo. **Tel:** 2 3365898, **Fax**: 2 3384780.

International Language Institute
2, Mohamed Bayoumi St. Off Merghany St., Heliopolis, Cairo. **Tel**: 2 291 2218.

International Language Learning Institute
34 Talaat Harb Street, 5th Floor, Cairo. **Tel:** 2 748355.

Kawmeya International School
Horreya Avenue, Bab Sharki. **Tel:** 2 4938670.

Lycee El Horreya
English Dept., Chatby. **Tel:** 2 5976968.

Port Said School
7, Taha Hussein Street, Zamalek. **Tel:** 2 3403435.

Ragab Language School
Chatby. **Tel:** 2 5976968.

Sacred Heart School
Syria School, Roushdy. **Tel:** 2 840228.

Sakkara Language School
Block #3 Zahraa El-Maadi, Maadi, Cairo. **Tel:** 2 516-5205 **Fax**: 2 516-5204.

ISRAEL & PALESTINIAN AUTHORITY

There is little demand for foreign teachers and it is difficult to find employment unless you are Jewish and are claiming your right of abode (which makes you liable for military service.) Be warned that an Israeli stamp in your passport may make travel to certain other countries in the Middle East impossible. You can ask for the stamps to be placed on removable pages. Except during periods of heightened security restrictions, most U.S. citizens may enter and exit Gaza and the West Bank on a U.S. passport with an Israeli entry stamp. It is not necessary to obtain a visitor's permit from the Palestinian Authority.

British Council office:
Crystal House, 12 Hahilazon St, Ramat Gan, 52136
Tel +972 (03) 611 3600
www.britishcouncil.org.il

➲ **Currency:** £1 = 8.18 / $1 = 4.18 New Shekels

➲ **Cost of Living:** Israel is an expensive country that has suffered frequent bouts of hyperinflation.

➲ **Salaries/Taxes:** £15-30 an hour in Israel. Taxation can go as high as 40%.

➲ **Accommodation:** Very expensive in Tel Aviv and Jerusalem, but less so in Gaza City where a comfortable apartment can be had for around £150 per month.

➲ **Health advice:** In Israel, take the same health precautions as you would at home. Note that national insurance is not available for non-Jews.

➲ **English Language Media:** Jerusalem Post, Palestine Report.

Schools in Israel & Palestinian Authority (+972)

AMIDEAST
Ahmad Abd al-Aziz Street, behind Al-Karmel High School, PO Box 1247, Gaza City, Gaza. **Tel:** 7 2824635 **Fax**: 7-286-9338. **Email: westbank-gaza@amideast.org**

AMIDEAST
1 Anata Street, Shu'fat, PO Box 19665, Jerusalem. **Tel:** 2 5811962. **Fax**: 2 5328564. **Email: westbank-gaza@amideast.org**

Anglican School
P.O. Box 191, Jerusalem 91001. **Tel**: 2 38 5220.

Arab American University
P.O. Box 239/240, Jenin, West Bank. **Tel:** 6 43 6056.

Ben Gurion University of the Negev
342 Madison Avenue, Ste 1224, NY 10173, USA.

British Council
140 Hayarkon Street, (PO Box 3302), Tel Aviv 61032. **Tel:** 3 5222194.

Walworth Barbour American International School
P.O. Box 9005, Kfar Shmaryahu, Tel Aviv. **Tel:** 9 584 255

JORDAN

Jordan is a fascinating country but teaching opportunities are limited. Many schools employ Jordanians who have qualified to teach EFL/ESL. The American Language Center employs qualified teachers on a part-time basis.

Embassy in UK:
6 Upper Phillimore Gardens, London W8 7HB
Tel: 020 7937 3685
www.jordanembassyuk.org

Embassy in USA:
3504 International Drive NW, Washington DC20008
Tel: 202 966 2664
www.jordanembassyus.org

British Council office:
Rainbow Street – off first circle, PO Box 634, Amman 11118
Tel +962 (6) 4636147
www.britishcouncil.org/jordan

➲ **Visas/Work Permits:** Employers of teachers on full-time contracts will arrange for the necessary papers. Part-time teachers must pay around £500 a year for a permit.

➲ **Requirements:** College Degree, experience and a TESOL/TESL Certificate

➲ **Currency:** £1 = 1.29 / $1 = 0.70 Dinar

➲ **Salaries/Taxes:** Salaries are generally lower than those in the Gulf States.

➲ **Health advice:** Although Jordan is malaria-free, take sensible precautions with drinking water and food. Vaccinations: Hepatitis A, Typhoid, Polio, Yellow Fever (if coming from an infected area).

➲ **English Language Media:** BYTE Middle East, Issues, Jordan Times The Star.

➲ **Cost of Living:** Prices of imported goods are high but food and travel costs are reasonable.

➲ **Salaries and taxes:** A full-time teacher at a private language school can expect to earn around £6,000 a year.

➲ **Accommodation:** Expensive. Some landlords ask for a year's rent in advance which can amount to as much as £4,500.

➲ **Tip:** Bars are few and far between in Amman. Expect to pay at least £4 for a beer in a high-class hotel.

Schools in Jordan (+962)

Al-Manhal International School
Iskan Al-Jami'a, P.O. Box 278, Jubaiha, Amman 11941. **Tel:** 6 60 3751.

American Community School
Box 310, Dahiat Al-Amir Rashid, Amman, 11831.**Tel:** 6 5813944, **Fax:** 6 5823357. **Email:** school@acsamman.edu.jo

American Language Center
Amman. **Tel:** 6 5859102/3, **Fax**: 6 5859101. **Email:** info@alc.edu.jo

Amman Baccalaureate School
PO Box 441, Sweileh, Amman, 11910. **Tel:** 6 5411572, **Fax**: 541 2603. **Email:** abs@go.com.jo

Modern American School
P.O. Box: 851819, Amman 11185. **Tel:** 6 5816861. **Fax**: 6 5816860. **Email:** **admin@ModernAmericanSchool.com**

Yarmouk Cultural centre
PO Box 960312, Amman. **Tel:** 6 671447.

Yarmouk University Language Center
P.O. Box 566, Irbid. **Tel:** 2 84 2700.

KUWAIT

Kuwait was rapidly rebuilt after the Gulf War although there are not as many English language schools these days. Some teachers say that Kuwait is a good place to earn and save money because there is not that much on offer as far as nightlife is concerned.

Others claim that Kuwait is a "party town" because it's so easy to make friends with the locals and other nationalities working in the country! There are both public and private English language schools but you will need to be qualified if you want to teach English here. Opportunities do exist to find teaching jobs after your arrival in Kuwait. Check out local newspapers when you arrive.

Embassy in UK:
2 Albert Gate, London SW1X 7JU
Tel: 020 7 590 3400

Embassy in USA:
2940 Tilden Street NW, Washington DC 20008
Tel: 202 966 0702

British Council office:
2 Al Arabi Street, Block 2, PO Box 345, 13004 Safat Mansouriya, Kuwait City
Tel +965 252 0067
www.britishcouncil.org/kuwait

➲ **Visas/Work Permits:** Entry visas are required for all visitors. Work permits are difficult to obtain and involve testing for AIDS and tuberculosis. You can find private lessons although you are supposed to have a work permit before you start teaching.

➲ **Requirements:** College Degree and a TESOL/TESL Certificate

➲ **Currency:** £1 = 0.53 / $1 = 0.29 Dinar

➲ **Salaries/Taxes:** Annual salaries start at around £20,000, which usually involves a full-time post with about 24 contact hours per week. There is no income tax.

➲ **Accommodation:** Most schools find and pay for accommodation because it is expensive. Average rent is £400-600. Usually, one month's rent is required as a deposit.

➲ **Health advice:** Unexploded bombs, mines, booby traps, and other items remain in open areas and beaches throughout Kuwait. Vaccinations: Hepatitis A, Polio, Typhoid.

➲ **English Language Media:** Kuwait Times.

➲ **Other Information:** Only limited public transport exists and taxis are very expensive. Women are permitted to drive.

Schools in Kuwait (+965)

Al-Bayan Bilingual School
PO Box 24472, Safat. **Tel:** 5315125, **Fax**: 5332836 **Email:** bbs@ncc.moc.kw

American International School
Block 11, Uhud Street, Salmiya. **Tel:** 5645083/7, **Fax**: 5645089.

American School of Kuwait
Block 7, Al-Muthanna Street and 4th Ring Road, Hawalli. **Tel:** 2664341/2655172, **Fax**: 2650438. **Email: askkwt@kuwait.net**

College of Business Studies
P.O. Box 43197, Hawalli, 32046. **Tel:** 261 4962

ELS School of Languages
PO Box 5104, Salmiya 20062. **Tel:** 5722522/2558, 5710610. **Fax**: 5733005.

ELU
The Kuwait Institute of Banking Studies, PO Box 1080, Safat 13011.

Fahaheel English School
PO Box 7209, Fahaheel 64003. **Tel:** 3711070, **Fax**: 965 371 5458.

Institute For Private Education
P.O. Box 6320, 32038 Hawalli. **Tel:** 573 7811, **Fax**: 574 2924.

Language centre
Kuwait University, PO Box 2575, Safat. **Tel:** 481 0325, **Fax**: 484 3824.

LEBANON

The UK and US governments continue to list Lebanon as a country you should not visit due to the rising tensions (which include kidnappings, violence and attacks). You should read the latest reports on www.fco.gov.uk for details of the current situation and whether it is safe to visit or travel to this otherwise beautiful country. In previous, safer times, English language teachers were in great demand and schools were advertising in the U.S. and Canada for qualified staff. Canadian citizens interested in volunteer work teaching English to Palestinian refugees in Lebanon should contact CEPAL by email at info@cepal.ca or call (613) 238-0410.

Embassy in UK:
21 Kensington Palace Gardens, London W8 4QM
Tel: 020 7727 6696

Embassy in USA:
2560 28th Street NW, Washington DC 20008
Tel: 202 939 6300

British Council office:
Sidani Street, Azar Building, Beirut
Tel +961 (1) 739 459
www.britishcouncil.org/lebanon

➲ **Currency:** £1 = 2776 / $1 = 1501 Pounds

➲ **Health advice:** Most hospitals in Beirut have modern facilities.

➲ **English Language Media:** Beirut Times, Lebanon Daily Star, Monday Morning.

Schools in Lebanon (+961)

American University of Beirut
Division of Education Programmes, 850 Third Avenue, 18th Floor New York, NY 10022-6297.

AMIDEAST
P.O. Box 70-744, Antelias. **Tel:** 4 410438/19341/2/3. **Fax**: 4 419341. **Email:** lebanon@amideast.org

AMIDEAST
Ras Beirut centre Building, 4th Floor, Surati Street, Hamra Area, Beirut. **Tel:** 1 345341, 340137. **Email:** lebanon@amideast.org

Beirut International School
Bchamoun, Beirut. **Tel:** 3 215489. **Email:** bis@bis.edu.lb

Beirut Modern School
Bir Hassan area between Golf Club & Airport Highway, Beirut. **Tel:** 1 823616. **Email:** bms95@cyberia.net.lb

British Council
Sadat/Sidani Street, Azar Building, Beirut. **Tel:** 1 740123/4/5. **Fax**: 1 739461

International College
P.O. Box 11-0236, Beirut. **Tel:** 1 867252. **Fax**: 1 362500/1

Lebanese American University
Beirut Campus, P.O. Box 13-5053, Beirut. **Tel:** 1 786456/786464. **Fax**: 1 867098.

Lebanese American University
Byblos Campus, P.O. Box 36, Byblos. **Tel:** 9 547254/547262 **Fax**: 9 944851.

Lebanese American University
Sidon Campus, P.O. Box 267, Sidon. **Tel:** 7 728724, 728725 **Fax**: 7 728726.

MALTA

A culturally diverse island in the Mediterranean, with a beautiful climate. English is one of the official languages

High Commission in UK:
Malta House, 36-38 Piccadilly, London W1V 0PQ
Tel: 020 7292 4800

Embassy in USA:
2017, Connecticut Avenue NW , Washington DC 20008
Tel: (202) 462 3611

British Council office:
Whitehall Mansions, Ta'Xbiex Seafront, Malta
Tel +356 23 23 2403
www.britishcouncil.org/malta

- **Visas/Work Permits:** no visas required for tourists or for stays of less than 3 months
- **Requirements:** College Degree and a TESOL/TESL Certificate
- Visas
- **Currency:** £1 = 0.61 / $1 = 0.33 liri
- **Health advice:** Medical care is good, and available in either public or private hospitals. It is advisable to get insurance before travelling
- **English language media:** Malta Independent, Malta Star, Malta Today, Sunday Circle, Malta Financial and Business Times

Schools in Malta (+356)

Academy Magister
L-Arkati Menaija Street, St Julian's

Alpha School of English
Arznell Street, St Paul's Bay, SPB 03

AM Language Studio
299 Manuel Dimech Street, Sliema, SLM

Bell Language Centre
Elija Zammit Street, St Julian's STJ 02

Bell Valletta
10 Sur Santa Barabara, Valletta, Vlt 06
Tel: 21 25 25 30 **Fax:** 21 25 25 29

BELS
Triq Ta'Doti, Kercem, Gozo, VCT 113

Berlitz Language Centre
Draonara Street, St Julian's

Britannia College
124, Melita Street, Valletta

Burlington English Language School
Burlington Court, Dragonara Street, St Julian's, STJ 06

Club Class Language School
1 Chealsea Court, Triq L-Imghazel, Swieqi, St Andrews

Easy School of Languages
331, St. Paul's Street, Valletta VLT 01

EC Malta
Language House, New Street off Sant Andrija Street, St Julian's, STJ 02

Education English Culture
Villa Monaco, Sliema Road, Kappara, SGN 06

EF International Language School
Lower St. Augustine Street, St Julian's

Ef Language Travel
San Michel, St George'S Road, St Julians Stj 10

Elanguest
Keating House, Ross Street, Sliema, STJ 10

English Communication School
10 St Pius Street, Sliema SLM 06

English Language Academy
9 Tower Lane, Sliema, SLM 15

European Centre of English
Paolo Court, G.Cali Street, Ta'Xbiex, MSD 14

European School of English
Edgar Bernard Street, Gzira, GZR 06

Gateway International School of English
1, Triq il-Jonju, The Village, San Gwann

GEOS Malta
Level 5, The Plaza, Bisazza Str Commercial Centre, Sliema, SLM 15

In House Malta
"Ogygia Foresteria", San Dimitri Street, Birbuba, l/o Gharb, Gozo

inLingua School
9 Guze Fava Street, Tower Road, Sliema, SLM 15

Institute of English Language
Parisio Street, Gzira, GZR 04

International House Malta
11 Birbuba St, Gharb 4020
Tel: 21 5504 90 **Fax**: 21 55004 90
Email: academyblc@mail.com

Kalypso Academy of English
University of Malta, Mgarr Road, Xewkija

La Salle Institute
105 St Thomas Street, Floriana, VLT 14

Lingua Time
243E Tower Road, Sliema, SLM 05

Link School of English
Villa Petite Fleur, 88/90 Triq it-Tiben, Swieqi

Mediterranean English Language Academy
Mountbatten Street, Blata l-Bajda, Hamrun

Melita Language School
CAK s-Sommier Street, B'Kara

MSD School of English
Josmar Triq il - Paguni, Balzan

MUT - Capital Language School
213 Republic Street, Valletta VLT 03

Nsts English Language Institute
220 St Paul Street, Valletta Vlt 07

Skylark Mediterranean School of English
29 Victor Denaro Street, Msida MSD 04

Sprachcaffe Languages
Srachcaffe Club Village, Pembroke, STJ 14

Students Travel School
c/o Mimosa Valetta Road, Attard

The Academy of English and Culture
13, Mattew Pulis Street, Sliema

The British Council
Whitehall Mansions, Ta'Xbiex Seafront

The Chamber College
Edgar Bernard Street, Gzira

The Leonardo Da Vinci Global Village
Villa Rosa, St George's Bay

Think English Language School
100, Paola Square, Paola

Unilang International School of Languages
Workers Memorial Building, South Street, Valletta VLT 11

Universal Language Services
Upper Ross Street, St Julian's

Voice School of English
PO Box 60, Msida, MSD 01

World Wide Languages Academy
128, Triq is-Sirk, Swieqi

MOROCCO

Morocco has been a magnet for travellers for centuries. Everyone from Winston Churchill to William Burroughs has ended up spending some time in this intriguing country. Moroccans are tolerant people: teachers with qualifications are welcome although conditions do vary from school to school. American Language Centers (www.aca.org.ma) operates English language schools in most major Moroccan cities. Knowledge of French is helpful but not essential. Note that institutions must ensure that at least 50% of the teachers are Moroccan.

Embassy in UK:
49 Queen's Gate Gardens, London SW7 5NE
Tel: 020 7581 5001

Embassy in USA:
1601 21st Street NW, Washington DC 20009
Tel: 202 462 7979

British Council office:
36 Rue de Tanger, BP 427, Rabat
Tel +212 (37) 760 836
www.britishcouncil.org.morocco

➲ **Visas/Work Permits:** Paper-work will be taken care of by the school after you have agreed on a contract. Be patient: It can take up to three months for your papers to come through. You will need original copies of your birth certificate, degree and teaching qualification.

➲ **Requirements:** College Degree and a TESOL/TESL Certificate

➲ **Currency:** £1 = 15.70 / $1 = 8.49 Dirham

➲ **Cost of Living:** Prices are reasonable in Morocco although a meal at a sophisticated restaurant will be expensive.

➲ **Salaries/Taxes:** 8,0000-9,000 dirhams a month is an average salary. Getting paid in dollars will help compensate for the effects of inflation.

➲ **Accommodation:** 3,000 dirhams for pleasant 2-bedroom apartment in Casablanca. Rabat is cheaper.

➲ **Health advice:** There is a small risk of malaria in Morocco. Take adequate precautions. Vaccinations: Hepatitis A, Polio, Typhoid.

➲ **English Language Media:** Morocco Times.

Schools in Morocco (+212)

American Language Center
Rue des Nations-Unies, Cité Suisse, Agadir. **Tel:** 8 821589. **Fax**: 8 848272. **Email: alcagad@marocnet**

American Language Center
2, Boulevard Mohamed V, Mohammedia. **Tel/Fax:** 3 326870.

American Language Center
4 Zankat Tanja, Rabat 10000. **Tel:** 7 761269

American Language Center
2, Rue Ahmed Hiba, BP 2136, Fes. **Tel:** 5 624850. **Fax**: 5 931608.

American Language Center
1, Place de la Fraternité, 20000 Casablanca. **Tel:** 2 277765. **Fax**: 2 207457.

American Language Center
3 Impasse du Moulin de Gueliz, Marrakesh. **Tel:** 4 447259.

American Language Center
1 rue M'sallah, Tangier. **Tel:** 9 933616.

American Language Center
2 Blvd. El Kadissa, Kenitra. **Tel:** 7 366884.

American Language Center
4bis, rue de Menton, BP 382, Meknes. **Tel:** 5 523636. **Fax**: 7 516464.

American Language centre
1, rue Maarakate Zalaka, Tetuan. **Tel:** 9 963308.

Benedict School of English
124 Ave. Hassan II, Ben Slimane. **Tel**: 3 290957. **Fax**: 3 328472.

British Centre
3, rue Brahim el Amraoui, Casablanca. **Tel:** 2 267019. **Fax**: 2 267043.

British Council
36 rue de Tanger, B.P. 427, Rabat. **Tel:** 7 76 0836, **Fax**: 7 760850.

Business & Professional English Centre (BPEC)
74 rue Jean Jaurès, Casablanca. **Tel:** 2 22.47.02.79. **Fax**: 2 22.29.68.61.

EF
20 rue du Marché, Casablanca. 2 255174. **Fax:** 2 296861.

Institut Cegis
23 Boulevard Ibnou Majid Al Bahhar, Casablanca. **Tel:** 2 308410.

International Language Centre
2 rue Tihama, Rabat. **Tel:** 7 709718.

London School of English
10 Ave. des FAR, Casablanca. **Tel:** 26 89 32.

OMAN

Oman is one of the most relaxed of the Gulf States. Alcohol is freely available and single women have an easier time here than in other Islamic countries. There are many archaeological sites to visit and the country is beautiful.

Demand is high throughout the Sultanate amongst private language institutes and higher educational establishments. Teachers should go through the Yellow Pages, approach schools directly or look at the local newspapers where jobs are occasionally advertised. Teachers with Master's degrees in TESL/TEFL are in demand. CELTA-qualified teachers can usually find hourly paid work. Opportunities for work at the more reputable end of the market for inexperienced teachers are limited. The best time to search for jobs is the early summer through to September when summer schools are always on the lookout for more teachers. The government does restrict the number of permits granted to tourists so you may have to wait until the right opportunity to teach in Oman presents itself.

Embassy in UK:
167 Queen's Gate, London SW7 5HE
Tel: 020 7225 0001

Embassy in USA:
2535 Belmont Road NW, Washington DC 20008
Tel: 202 387 1980

British Council office:
Road One, Madinat al Sultan, Qaboos West, Muscat
P.O. Box 73 Postal Code 115
Tel +968 600 548
www.britishcouncil.org/oman

➲ **Visas/Work Permits:** Your employer must sponsor you (a no-objection certificate) and the paperwork is extensive. On arrival in Oman, your employer may ask you for your passport. Although this is customary, you are not legally obliged to hand it over. Surrendering your passport could seriously affect your ability to leave the country whenever you want.

➲ **Requirements:** College Degree and a TESOL/TESL Certificate

➲ **Currency:** £1 = 0.75 / $1 = 0.348 Rial

➲ **Salaries/Taxes:** Rates of pay vary according to nationality and qualifications. You can earn about RO

3-10 per hour in institutes. Monthly: Anything from RO 500-850 at the top end. There is no tax.

➲ **Cost of Living:** The cost is living is higher than that elsewhere in the Gulf, including the UAE. Food and clothing is often imported and taxed at 17%. Supermarket costs are fairly comparable with the west overall, although imported Western luxury foods are very pricey. Petrol is cheap and you can also purchase cars at very competitive prices.

➲ **Accommodation:** In most cases, the school will arrange this for you. If you are going to work at a teacher training college you will be expected to find your own lodging. It's not difficult to find an apartment. Many teachers use an agent. Apartments cost RO 150-275 a month and utilities are RO 25-45.

➲ **Health advice:** Local medical treatment varies in quality, and it can be inadequate. While hospital emergency treatment is available, there is no ambulance service in Oman. Malaria is a concern in the interior and on the Batinah Coast. Doctors and hospitals often expect immediate cash payment for health services. Vaccinations: Yellow Fever (if coming from an infected area), Malaria.

➲ **English Language Media:** Oman Daily Observer, Times of Oman.

➲ **Tip:** Try to buy a car. You will miss much of the country if you have to rely on public transport. Visitors in possession of an international or American driver's licence may drive rental vehicles, but residents must have an Omani driver's licence. Visitors hiring rental cars are urged to ensure that the vehicle they are renting is adequately insured against loss or damage.

➲ **Other Information:** As is the case in many countries, labour laws are strictly enforced in Oman. You may be contractually obliged to work for six months or more before you qualify for a holiday. Also, don't be surprised to find that you cannot work for another employer once your original contract has expired.

Schools in Oman (+968)

American-British Academy
PO Box 372, Medinat Al Sultan Qaboos, Muscat. **Tel:** 605287/603 646. **Fax**: 603544. **Email:** abaad@gto.net.om

American International School
PO Box 202, Muscat 115. **Tel**: 600374. **Fax**: 697916.

British Council
PO Box 73, Medinat Al Sultan Qaboos, Muscat 115. **Tel:** 600548. **Fax**: 698018.

Polyglot Institute
PO Box 221, Ruwi. **Tel:** 701261, **Fax**: 794602.

Sultan Qaboos University
PO Box 50, Muscat 123. **Tel:** 513333. **Fax**: 514 455.

QATAR

Qatar is a little-known Gulf emirate. It enjoys a high standard of living and is very modern. Most schools are very well-equipped boasting the latest in teaching gadgetry. Recruitment usually starts in February, interviews take place in the spring, and teachers arrive in August to start work in September. The weather can be surprisingly cold in the winter so remember to pack a couple of sweaters.

British Council office:
93 Al Sadd Street, PO Box 2992, Doha
Tel +974 442 6193 / 4
www.britishcouncil.org/qatar

➲ **Requirements:** College Degree and a TESOL/TESL Certificate

➲ **Currency:** £1 = 7.12 / $1 = 3.64 Riyal

➲ **Health advice:** Basic modern medical care and medicines are available in the government-run Hamad General Hospital in Qatar. Serious medical problems could cost you thousands of dollars. Check with your insurer to see if you are covered in Qatar.

➲ **English Language Media:** Gulf Times.

➲ **Tip:** The British Council has a very good lending library.

➲ **Information for Women:** Qatar is very safe and you will not be hassled by guys as is the case in certain other Middle Eastern countries. Bare arms and legs in public do not cause a negative reaction but draw the line at walking down the street in a bikini. Women driving alone are sometimes tailed by male drivers.

Schools in Qatar (974)

Arizona English Language School
PO Box 7949, Doha.

ELS Language centres
P.O. Box 22678, 56 Jaber bin Ghaith, Doha.**Tel:** 699223. **Fax**: 685519. **Email:** elsdoha@qatar.net.qa

English Modern School
Al Naijah, 44 Najma Street, 44 Hamad bin Zaidoun St.

Qatar International School
P.O. Box 5697, Doha. **Tel:** 469 0552 , **Fax**: 469 0557. **Email:** info@qis.org

SAUDI ARABIA

Saudi Arabia is one of the few countries where ESL/EFL teachers can expect to be rewarded with a good salary. Rates of pay may not be rising as fast as they did ten years ago but if you want to make good money in an English language teaching job, KSA (Kingdom of Saudi Arabia) is the place to be – teachers can earn upwards of £30,000 a year. And that's tax-free.

Try to get a job in one of the big cities (Jeddah or Riyadh, for example.) Life can get a little lonely in a one-camel town although the lack of any social life might do wonders for your bank balance. Ex-pat life in the big cities can be a lot of fun if you're prepared to get involved. So, the bottom line is simple: If you're not sure you can handle new experiences (good and bad) then don't go. Most teaching contracts are for one year so take some time to check out prospective employers and find out as much as you can about them and the job you will be doing. The majority of horror stories coming from teachers in KSA come from those unfortunate enough to have worked for "cowboy" operations.

Embassy in UK:

30 Charles Street, London W1X 7PM
Tel: 020 7917 3000
www.saudiembassy.org.uk

Embassy in USA:

601 New Hampshire Ave NW, Washington DC 20037
Tel: 202 337 4076
www.saudiembassy.net

British Council office:

Tower B, 2nd floor, Al Mousa Centre, Olaya Street
P.O. Box 58012, Riyadh 11594
Tel +966 (1) 462 1818
www.britishcouncil.org/saudiarabia

➲ **Visas/Work Permits:** You must be sponsored by an employer to work in Saudi Arabia. You cannot travel to Saudi Arabia as a tourist unless you travel as part of a government-approved group. Visitors to Saudi Arabia generally obtain a meningitis vaccination prior to arrival. A medical report or physical examination is required to obtain work and residence permits. The sponsor (normally the employer) obtains work and residence permits for the employee. Residents working in Saudi Arabia generally must surrender their passports while in the Kingdom and carry a Saudi residence permit ("iqama") for identification in place of their passports.

➲ **Requirements:** College Degree and a TESOL/TESL Certificate

➲ **Currency:** £1 = 6.93 / $1 = 3.75 Riyal (SAR)

➲ **Salaries/Taxes:** Starting c. £25,000 (it can be less in the smaller towns) going up to £30,000 and more. No tax. If you are offered a salary less than £15k you should think twice.

➲ **Accommodation:**Your employer should provide accommodation and it's kismet whether you land up in a resort with pools and tennis court or sharing a trailer with four fellow teachers. If you strike out on your own, unfurnished apartments in Jeddah start around 12,000 SAR per year. Sizes range from one big room with a separate bath, up to a three-bedroom apartment, depending on where you go. The cheapest compound starts around 30,000 SAR per year and offers security.

➲ **Cost of Living:** Salaries are so high and things are so cheap you can come back with a new car and lots of other goodies. Residents in Saudi Arabia may not leave the country without obtaining an exit permit prior to leaving and an exit/reentry permit if they intend to return to Saudi Arabia. The Saudi sponsor's approval is required for exit permits. Note that it is illegal for a Saudi employer to sell the approval (known as a Certificate of No Objection) to an employee.

➲ **Health advice:** Malaria is endemic to the low-lying coastal plains of southwest Saudi Arabia, primarily in the Jizan region extending up the coast to the rural area surrounding Jeddah. Take precautions to avoid being bitten by mosquitoes if you're going to be in this area.. There are reports of cases of Rift Valley fever in southwestern Saudi Arabia (near the town of Jizan.) The virus is transmitted primarily by infected mosquitoes and by direct contact with the blood or body fluids of infected animals. Prevention of Rift Valley fever should focus on personal protective measures to limit contact with infected mosquitoes. The best prevention is appropriate use of an insect repellent containing DEET (N,N -diethylmetatoluamide) and wearing long sleeves and trousers. Vaccinations: Hepatitis A, Polio, Typhoid, Malaria, Meningitis, Yellow Fever (if coming from an infected area).

➲ **Other Information:** Alcohol, women driving, men and women intermingling – are prohibited. Experienced Saudi hands have found ways to get around these restrictions but you risk punishment (even a public beating) if you are caught doing something illegal. Married couples usually stay longer in Saudi than straight singles. There is a large and active gay teaching community in KSA.

➲ **English Language Media:** Arab News, Saudi Gazette, Riyadh Daily.

Schools in Saudi Arabia

The British Council
Al Moajil Building, 5th Floor, Dhahran Street, Mohamed Street, PO Box 8387, Daman. **Tel:** 834 3484.

Girls' College of Arts
General Presidency for Female Institute for Languages and Translation, c/o King Saud University, PO Box 2465, Riyadh 11451.

King Abdulaziz University
Jeddah. **Tel:** 966 2 695 2468, **Fax**: 966 2 695 2471.

King Fahd University of Petroleum and Minerals
English Language centre, Dhahran 31261. **Tel:** 3 860 2393, **Fax**: 3 860 2340.

Riyadh Military Hospital-Training Division
PO Box 7897, Riyadh 11159. Saudi Airlines, PO Box 167, Jeddah 21231.

Saudi Language Institute
PO Box 6760, Riyadh 11575.

SCECO - East Central Training Institute
PO Box 5190, Damman 31422.

Yanbu Industrial College
PO Box 30436, Yanbu Al Sinaiyah 21477.

SYRIAN ARAB REPUBLIC

Syrian civilization is ancient: Damascus is the oldest continually occupied city in the world and there is much to explore in this friendly country. As Syria shakes off years of Russian influence, English language teachers are in great demand but the opportunities are restricted to those who are qualified. Teachers cannot expect to be hired from abroad. Teachers must come to Syria then look into teaching options. AMIDEAST offers teaching jobs but cannot provide long-term residency. The American Language Center (ALC) does too, and offers residency. Even if regular jobs are not available it is easy to find work tutoring.

Embassy in UK:
8 Belgrave Square, London SW1X 8PH
Tel: 020 7245 9012

Embassy in USA:
2215 Wyoming Ave NW, Washington DC 20008
Tel: 202 238 6313

British Council office:
Maysaloun Street, Shalaan, P O Box 33105, Damascus
Tel +963 (11) 331 0631
www.britishcouncil.org/syria

➲ **Visas/Work Permits:** You will need an entrance visa, which can be obtained from the Syrian embassy in your country. If you decide to stay 15 days or more you must register with Syrian Immigration by your15th day in Syria. You can then extend the permit for two-week periods every two weeks. This is not a difficult process but you will need to take an AIDS test at a government-approved facility. If a teacher obtains regular work, he or she may obtain residency for up to one year. Be aware that you cannot enter Syria with an Israeli stamp in your passport.

➲ **Requirements:** College Degree and a TESOL/TESL Certificate

➲ **Currency:** £1 = 96 / $1 = 52 Pound

➲ **Cost of Living:** Food is very cheap and the quality is excellent. Small restaurants offer great "shawarma" (meat, vegetables and yogurt wrapped in pita bread) and the seafood is top-notch too. Most teachers find it is cheaper to eat out than to cook.

➲ **Salaries/Taxes:**ALC and AMIDEAST pay around £6 an hour; Syrian-owned schools pay less. Native speakers may tutor and receive upwards of £6 an hour. Foreign teachers do not have to pay taxes.

➲ **Accommodation:** A basic apartment in Damascus costs around £120 a month.

➲ **Health advice:**Basic medical care and medicines are available in the principal cities of Syria, but not necessarily in outlying areas.

➲ **English Language Media:** Syria Times.

Schools in Syria (+963)

Al Kindi English Language Centre
29 May Street, Damascus.

Al Kudssi Institute
PO Box 5296, Aleppo.

Al Razi English Language Centre
PO Box 2533, Damascus. **Tel:** 11 457301.

Alson
Damascus. **Tel:** 11 2227236.

American Language Center
PO Box 29, Rawda Circle, Damascus. **Tel:** 11 3327236. **Fax**: 11 3319327.

AMIDEAST
Wahab Bin Saad Street, Abou Roummaneh, Damascus. **Tel:** 11 3332804/3314420. **Fax**: 11-333-2804. **Email:** Syria@amideast.org

Damascus Language Center
PO Box 249, Damascus.

TUNISIA

English is replacing French as the language of business in Tunisia so there is a growing demand for EFL teachers. Teachers often find they are the first native speaker their students have ever encountered. It is best to apply to schools in Tunisia before leaving home; a little knowledge of French will help you settle in.

Embassy in UK:
29 Prince's Gate, London SW7 1QG
Tel: 020 7584 8117

Embassy in USA:
1515 Massachusetts Ave NW, Washington DC20005
Tel: 202 862 1850

British Council office:
5 Place de la Victoire, BP 229, Tunis 1015 RP
Tel +216 71 259 053
www.britishcouncil.org/tunisia

➲ **Visas/Work Permits:** You should receive a temporary work permit or a letter from your employer allowing you to work until your permit is processed. Photocopies of your degree and birth certificates will not be accepted.

➲ **Requirements:** College Degree and a TESOL/TESL Certificate

➲ **Currency:** £1 = 2.54 / $1 = 1.30 Dinar

➲ **Cost of Living:** Tunisia is a cheap country to live in unless you insist on buying imported products.

➲ **Salaries/Taxes:** Private lessons earn teachers around 15 dinars an hour. At the top end, you can expect to earn a minimum of 13,000 dinars per year. The tax rate is around 20 percent.

➲ **Accommodation:** It's easy to find somewhere to live. Expect to pay 3-400 Tunisian dinars per month, with at least one month's rent as a deposit. Accommodation is largely unfurnished.

➲ **Health advice:** Essential vaccinations: Hepatitis A, Polio, Typhoid, Yellow Fever (if coming from an infected area).

➲ **English Language Media:** Tunisia News.

➲ **Tip:** Life is pretty free and easy for an Islamic country but there's not much in the way of nightlife unless you head out to tourist areas like the island of Jerba.

Schools in Tunisia (+216)

AMIDEAST
22 rue Al Amine Al Abassi, Cité Jardins, 1002 Tunis. **Tel:** 1 790559, 790563. **Fax**: 1 791913.

English Language Training Centre
British Council, 47 Avenue Habib Bourguiba, Tunis.

Institut Bourguiba des Languages Vivantes (IBLV)
47 Avenue de la Liberté, Tunis. **Tel:** 1 282418. **Fax**: 1 780389.

TURKEY

Turkey is a unique country straddling both Europe and Asia. Although its people are Muslims, there is a secular constitution. The Turkish language is written in the Latin alphabet (like English) not in the flowing Arabic script.

Turkey represents a big market for English language teachers. Government officials estimate that there is a need for 30,000 more English language teachers in the nation's public schools. English is taught from fourth grade in the public schools. In private schools, children start learning English in kindergarten. Needs for EFL teachers extend through middle and high schools to colleges and universities.

Native English speakers are in huge demand and there are plenty of opportunities for unqualified teachers who want to work in Turkey, although their salaries will not be as high as those of qualified teachers. University programmes prefer to recruit teachers with MAs in TESOL/TEFL.

Opinions about teaching English in Turkey are mixed: some teachers love it while others hate it. A lot depends on your attitude: If you expect things to be done the same way as at home then you probably won't want to come here. But that's a shame because you'll be missing out on the opportunity to live in an amazing country, which has been at the crossroads of civilizations for centuries.

Caution: be aware before you travel to Turkey that many English language teachers leave their jobs here disillusioned by working conditions and the attitudes of their employers. Unfortunately, in some schools, the directors are not concerned about the welfare of teachers, so choose carefully before you take up an offer of a job. A lot of foreigners regard teaching English in Turkey as a paid vacation. Alcohol is freely available and a few too many teachers drink to excess and are unable to carry out their duties. So you can't always blame school directors for their negative attitudes towards potential staff members.

Although your employer may not be willing to put anything down on paper, try to get something in writing from the person that you will be working for at the school. It's no guarantee that your "contract" will be honored but waving the paper around during a dispute may help your cause. Make sure that the agreement is hammered out between you and the owner of the school. A third party probably has no authority to make any agreements involving your work. Don't rely on assurances that accommodation, medical insurance or even salary have been settled. Discuss them before you sign a contract. Promises that you will be paid in American dollars may not always be kept. Important: you can always walk out of your job and find something that suits you better. Just make sure you have some backup cash.

[Note: Although a small part of the country and its major city, Istanbul, is in Europe, Turkish culture and society is Middle Eastern – with some interesting twists!]

Embassy in UK:
43 Belgrave Square, London SW1X 8PA
Tel: 020 7393 0202

Embassy in USA:
2525 Massachusetts Ave NW, Washington DC20008
Tel: 202 612 6700

British Council office:
Esat Caddesi No: 41, Kucukesat, 06660 Ankara
Tel +90 (312) 424 1644, www.britishcouncil.org.tr

➲ **Tip:** Before you agree to work at a school, ask if you can sit in on a few lessons and watch other teachers at work. A reputable organization will allow you to do this and will appreciate your commitment. A school that refuses this reasonable request probably has something to hide.

➲ **Visas/Work Permits:** If you want to work in Turkey, you must apply for a work visa. You are not supposed to arrive as a tourist and then find a job. But some schools do employ teachers on tourist visas and send them to Greece every three months to get their visas renewed. Okay, this works for most of the time but if the authorities get suspicious, they may fine you. Keep your passport with you at all times: Teachers have spent a night in prison for failing to present their documents to the police on demand.

➲ **Requirements:** College Degree and a TESOL/TESL Certificate

➲ **Currency:** £1 = 2.78 / $1 = 1.42 New Lir

➲ **Salaries/Taxes:** Wages and conditions vary depending on the location of the school and the level you are teaching at. Rates can range anywhere from £300-500 per month at private language institutes up to £600-1200 paid monthly by the elite colleges, plus air fares and other benefits.

➲ **Accommodation:** Usually, the school arranges for your housing. Unmarried teachers will often find themselves sharing an apartment with other foreign colleagues. Some schools will pay your rent, other will not. Furniture will be basic and beds may be army-style cots. Be warned that furnished apartments are often hard to find and are expensive. Many landlords are suspicious of single men. Single females will have an easier time but they may be asked to promise not to entertain male visitors. Expect a deduction of 25% from your salary for taxes.

➲ **Cost of Living:** There is a high rate of inflation in Turkey but salaries pegged to the dollar keep pace. The cost of living is low and it should be easy to save some money.

➲ **Health advice:** In the southeastern city of Diyarbakir, there are recurring outbreaks of dysentery, typhoid fever, meningitis and other contagious diseases. Malaria is still a risk in the eastern part of the country. Turkey is one of the most dangerous countries in the world as far as traffic accidents are concerned. Vaccinations: Hepatitis A, Polio, Typhoid.

➲ **English Language Media:** Turkish Daily News.

Schools in Turkey (+90)

Active English
Ataturk Bulvari 127/701, Secan Han. Bakanlikar, 06640, Ankara. **Tel:** 312 418 7973. **Fax**: 312 4258235.

Akademi School of English
PK 234, 21001 Bahar Sokar No 2, Diyarbakir. **Tel**: 412 2242297. **Fax**: 412 2288020.

Ankara University
Rektorlugu, Beslevler, Ankara. **Tel:** 312 41234361.

Best English
Bayinder Sokak No. 53, Kizilay, Ankara. **Tel:** 312 4172536. **Fax**: 312 4176808. **Email:** bestenf@alnet.net

Beykent University
Beykent Belikdirisu, 34900, Istanbul. **Tel:** 212 8726432. **Fax**: 212 8722489.

Bilkent University
06533 Bilkent, Ankara. **Tel:** 312 2902746/2664390. **Fax**: 312 2664934.

Cambridge English
Kazim Ozalp Sok. No. 15, Kat. 4, Saskinbakkal, Istanbul. **Tel:** 216 3858431. **Fax:** 216 369 7891.

Cinar School of English
860 Sokak No. 1, Kat. 4, Konak, Izmir. **Tel:** 238 4837273. **Fax**: 238 441 1113.

CTS Language Center
Tayyareci Cemal Sk. 24/4 Sisli, Istanbul. **Tel:** 212 2660043. **Fax**: 212 2331931. **Email:** cts_language_centre@yahoo.com

Cukurova University Center for Foreign Languages
Balcali, Adana. **Tel:** 322 3386084 x2921.

Dilko English Centres
Hatboyo Caddesi No. 16, 34720 Bakirkoy, Istanbul. **Tel:** 212 5701270. **Fax**: 212 543 6123.

Dilmer LTC
Unlu Cad. 7, Keykel, Bursa. **Tel:** 224 2224673. **Fax**: 224 2231163.

Elissa English
Ihsaniye Mah 41, Sokak 48, Bandirma, Balikesir.

The English Academy
1374 Sokak No. 18/4, Selvili Is Hz. Izmir. **Tel:** 238 4462520. **Fax**: 238 4253042.

The English Centre
Rumeli Caddesi. 92, Zeki Bey Apt. 4, Osmanbey, Istanbul. **Tel:** 212 247 0983.

English Fast
Burhaniye Mah. Resmi Efendi Sok. No. 4, Beylerbeyi, Istanbul. **Tel:** 216 3187018. **Fax**: 216 318 7021.

Evrim
Cengiz Topel Caddesi 8, Camlibel, 33010 Mersin. **Tel:** 324 2339541, **Fax**: 324 2370862.

Eyuboglu Kisesi
Namik Kemal Mah., Dr. Rustem Eyuboglu Sok. 1, Umraniye 81240, Istanbul. **Tel:** 216 3291614. **Fax**: 216 3357198.

Gediz College
Seyrek Beldesi, Menemen, Izmir. **Tel:** 238 8447444. **Fax:** 238 8447441. **Email:** baysoy@gediz.kiz.tr

Genctur
Istiklal Cd. Zamabk Sok. 15/5, Taksom 80080, Istanbul. **Tel:** 212 249215. **Fax**: 212 2492554. **Email:** workcamps@genctur.com.tr

Inkur ELI
Ankara Caddesi, Yalihamam Sokak No. 8, Izmi, Kocaeli. **Tel:** 262 321 5325. **Fax**: 238 3225391.

Interlang
Russell Baulk, Zuhtu Pasa Mah, Recep Peker Cd, Sefikbey Sok No 17, Kiziltoprak, Istanbul. **Tel:** 216 4183910. **Fax**: 216 4182796.

International School
Eser Apt.A Blok Kasap, Sokak 1617, Esentepe, Istanbul.

Isik Universitesi
Buyukdere Cad., Maslak, Istanbul.

Istanbul University
Rektorlugu Beyarzit, Istanbul. **Tel:** 212 5221489.

Kent English Ankara
Mithatpasa Caddesi No. 46 Kat. 3/4/5, 06420 Kizilay, Ankara. **Tel:** 312 4343833. **Fax**: 312 4357334.

Kumlu Dersanerleri
Bursa Merkez, Basak Caddesi, Bursa. **Tel:** 224 24120465.

London Language International
Abide-i Hurriyet Caddesi, Kat. 1, Mecidiyekoy, Istanbul. **Tel:** 212 211 7445. **Fax**: 212 211 7441.

Ozel Ortadogu Lisesi
Spor Caddesi 26, Yakacik, Istanbul. **Tel:** 216 3772501. **Fax**: 216 3772502.

Ozel Tan Lisesi
Yeni Yalova Yolu 12km., 16335 Bursa. **Tel:** 224 2670072. **Fax:** 224 2670071.

Sistem English
Afiye Mah. Kibris Sehitleri Cad. 18/1, Eskisehir. **Tel:** 222 2312266. **Fax**: 222 2308681.

Swissotel Training centre
Suite Tower, Bayildim Cad. No. 2, Macka 80680, Istanbul. **Tel:** 212 259 0790. **Fax**: 212 259 0240.

UNITED ARAB EMIRATES

In many ways, the United Arab Emirates (UAE) represents what most people imagine when they think of the Gulf States. In Dubai and Abu Dhabi, vast four-wheel drive jeeps fly along well-paved freeways while the passengers stay cool in the air-conditioning. In other parts of the UAE, you can still see fishermen setting out on their boats, negotiating their way past the offshore oil derricks, and coming back with their catch. Dubai is now one of the major destination cities in the world with spectacular hotels, developments and, of course, the famous horse-racing! These contrasts add up to an unforgettable experience and many foreigners decide to stay in the UAE on a long-term basis.

Teachers of English as a second or foreign language can expect only the very best here in as far as pay and accommodation are concerned.

Embassy in UK:
30 Prince's Gate, London SW7 1PT
Tel: 020 7581 1281

Embassy in USA:
Suite 700, 1255 22nd Street NW, Washington DC20037
Tel: 202 243 2400

British Council office:
Villa no. 7, Al Nasr Street, Khalidiya, Abu Dhabi
P.O. Box 46523
Tel +971 (2) 665 9300
www.britishcouncil.org/uae

➲ **Visas/Work Permits:** Tourist visas issued readily. To live and work in the UAE, you need a sponsor employer unless you are the spouse of someone already working there. Rules about visas change at short notice, so contact your local UAE embassy before departure.

➲ **Requirements:** College Degree and a TESOL/TESL Certificate

➲ **Currency:** £1 = 6.7 / $1 = 3.6 Dirham

➲ **Salaries/Taxes:** Around £2,000-3,000 per month for teachers with a Master's degree and experience. There is no tax for foreign teachers.

➲ **Accommodation:** Usually provided, otherwise expensive. A single apartment costs 30-40,000 dirhams per year, a double 50,000-75,000.

➲ **Cost of Living:** Dubai is being touted as the "new Hong Kong." Electrical goods are very reasonable and you should try to cash in on the some of the bargains available. Tailor-made clothing is also a great buy. Food is cheap and wonderful especially if you like curries. Western food items are widely available.

➲ **Health advice:** A Health Card costs about 300 dirhams. Vaccinations: Hepatitis A, Polio, Typhoid, Malaria (in some areas).

➲ **English Language Newspapers:** Emirates News, Gulf News, Gulf Today, Khaleej Times.

➲ **Other useful information:**With the growth of tourism, there has been an increasing demand for both male and female teachers. Although the percentage of women in the UAE labour force is still low, both Abu Dhabi and Dubai are cosmopolitan, relaxed cities, without the restrictions normally associated with the Gulf.

➲ **Tip:** Don't miss out on a trip to the Indian Ocean coast: Dibbah has the best beaches and Fujairah is a charming town. There are also the mangroves to visit at Kalba.

Schools in UAE (+971)

Al Farabi Language Center
PO Box 3794, Dubai. **Tel:** 2 212556.

Al-Worood School
P.O. Box 46673, Abu Dhabi. **Tel:** 2 448855, **Fax**: 2 449732.

American University of Sharjah
College of Arts and Sciences, PO Box 26666, Sharjah. **Tel:** 6585067.

Arabic Language Center
(Dubai World Trade Centre), PO Box 9292, Dubai. **Tel:** 3086036. **Fax**: 3064089.

Dar Al Ilm School of Languages
PO Box 2550, Dubai. **Tel:** 447572, **Fax**: 4 91915.

Dubai Aviation College
PO Box 53044, Dubai. **Tel:** 824000, **Fax**: 824222.

Dubai Cultural and Scientific Institute
PO Box 8751, Dubai. **Tel:** 347747/377004, **Fax**: 347369.

ELS Language centres
P.O. Box 1496, Al Jimy Area, Al Ain. **Tel:** 3 623468. **Fax**: 3 623865 **Email:** elsaa@emirates.net.ae

ELS Language centres
Al Baten, Al Khaleej Al Arabi Street, Villa #4, P.O. Box 3079, Abu Dhabi. **Tel:** 2 669225. **Fax**: 2 653165. **Email: elsiauh@ns2.emirates.net.ae**

ELS Language centres
P.O. Box 2380, Al Garhoud Area, Villa #14, Street #20 , Dubai. **Tel:** 4 827616. **Fax**: 4 827861.**Email:** dbxels@emirates.net.ae

Institute of Australian Studies
PO Box 20183, Dubai. **Tel:** 4 666400, **Fax**: 4 623500.

International Education Institute
PO Box 524714. **Tel:** 360635, **Fax**: 348209.

International Language Institute
(Language Specialists Institute), PO Box 3253, Sharjah. **Tel:** 6 377257. **Fax**: 6 363427.

International Training Solutions
Lynne Haboubi, PO Box 4234, Dubai. **Tel/Fax:** 16 363249,

Polyglot School
PO Box 1093, Dubai. **Tel:** 223429, **Fax:** 217284.

YEMEN

Yemen is the land of fables—the Queen of Sheba came from here and for centuries Yemen was the world's source of frankincense. By all accounts, people in Yemen are keen to learn English and are very receptive to qualified and experienced language teachers.

Petroleum companies operating in Yemen employ staff to teach English to local workers. Most teachers are paid by the hour. Check out some industry websites like Canadian Occidental (www.cdnoxy.com.ye) for positions on offer. Working for a foreign company may help you overcome some of the red tape that can arise when working for a Yemeni agency. If you get a job with an oil company be sure to clue up on industry-related terminology before you start teaching.

There are a number of institutes in the country—AMIDEAST operates offices in Sana'a and Aden. However, salary rates and visa processing requirements (see below) serve as disincentives. As a result of the relative difficulty in processing expatriate teachers and the seemingly low salary rates, most teachers are not recruited from abroad, but are rather language students or spouses of expatriates or Yemenis residing in Yemen.

Teachers thinking of working in Yemen should be aware that there have been a number of kidnappings of foreigners by tribesmen hostile to the government. Most of these incidents have occurred in the lawless northern region of the country although two American tourists were abducted in southern Yemen in 1998. Teachers should travel with an armed escort between cities if they choose not to fly.

Embassy in UK:
57 Cromwell Road, London SW7 2ED
Tel: 020 7584 6607

Embassy in USA:
Suite 705, 2600 Virginia Ave NW, Washington DC20037
Tel: 202 965 4760

British Council office:
3rd Floor, Administrative Tower, Sana'a Trade Centre Algiers Street, PO Box 2157, Sana'a
Tel +967 1448 356/7
www.britishcouncil.org/yemen

➲ **Visas/Work Permits:** EFL teachers from America can only obtain a one-month tourist visa (£20). The organization hiring the teacher must then convert the tourist visa into a work permit—an expensive and time-consuming procedure. All individuals wishing to obtain a work permit must have an HIV test performed in Yemen. External test results will not be accepted.

➲ **Requirements:** College Degree and a TESOL/TESL Certificate

➲ **Currency:** £1 = 344 / $1 = 186 Rial (YER)

➲ **Salaries/Taxes:** Salaries are generally paid per hour. An average salary for native speakers is around £6/hour, which is quite good because Yemen is an inexpensive country and the average monthly income is less than £60. Foreigners do not have to pay any local tax.

➲ **Accommodation:** Rents range from £30/month to £1000/month. An average rate for a furnished, one bedroom apartment in good condition is around £100 per month. Additionally, some language institutes offer housing to expatriate teachers.

➲ **Cost of Living:** Prices in Yemen are very reasonable and you can pick up exquisitely made handicrafts at bargain rates.

➲ **Health advice:** Cases of malaria are on the rise in Yemen especially along the coasts. You should take a course of chloroquine-resistant malaria suppressants before you travel to these areas. The altitude of Sana'a (7,200 feet) and lack of adequate medical facilities can cause problems for some visitors. Doctors and hospitals often expect immediate cash payment for health services. Vaccinations: Hepatitis A, Typhoid, Polio, Yellow Fever (if coming from an infected area).

➲ **English Language Media:** Yemen Times, Yemen Observer.

➲ **Tip:** Yemeni social life revolves around chewing the leaf known as "qat" that produces a mild euphoria among users.

Schools in Yemen (+967)

Al Farouq Institute
PO Box 16927, Sana'a. **Tel:** 1 209721. **Fax**: 1 209721.

The American School
PO Box 16003, Sana'a. **Tel:** 1 417119. **Fax:** 1 415355.

AMIDEAST
162 Miswat Street, Khormaksar, Aden. **Tel:** 2 238345. **Fax**: 2 238345. **Email: aden@amideast.org**

AMIDEAST
Algiers Street 66, P.O. Box 15508, Sana'a. **Tel:** 1 206222. **Fax**: 1 206942. **Email:** yemen@amideast.org

The British Council
PO Box 2157, Sana'a. **Tel:** 1 244 121. **Fax**: 1 244120.

English Language centre
PO Box 8984, Sana'a. **Tel:** 1 248090.

Faculty of Art, English Department
PO Box 6014, Aden. **Tel:** 2 32350.

Faculty of Art
English Department, Sana'a. **Tel:** 1 200515/6.

Faculty of Education
English Department, Sana'a. **Tel:** 1 250513.

Faculty of Education
English Department, PO Box 6480, Taiz. **Tel:** 4 212191/2, **Fax**: 4 219378.

MALI
PO Box 16003, Sana'a. **Tel:** 1 241561. **Fax**: 1 414348.

Madina Institute of Technology (MIT)
Aden. **Tel**: 2 251453.

The Modern Yemen School
PO Box 13335, Sana'a. **Tel:** 1 206548.

Mohammed Ali Othman School
PO Box 5713, Taiz. **Tel:** 4 211248/9.

National Institute for Administrative Science
PO Box 102, Sana'a. **Fax:** 1 251161.

Pakistani School
Sana'a. **Tel:** 1 247830/247838.

The Pioneer School
Sana'a. **Tel:** 1 210311.

Sabaa University
PO Box 14400, Sana'a. **Tel/Fax:** 1 205749.

Sam Yemen International School
PO Box 19390, Sana'a. **Tel:** 2 69648/63303. **Fax**: 2 69649.

Sana'a International School
PO Box 2002, Sana'a. **Tel:** 1 234437. **Fax**: 1 234438.

Spectra Institute
PO Box 16101, Sana'a. **Tel**: 1 414623. **Fax**: 1 414739.

University of Science Technology
PO Box 15201, Sana'a. **Tel:** 1 207026. **Fax**: 1 207027.

Yemen-American Language Institute
c/o USIA, American Embassy, Sana'a. **Tel:** 1 203251. **Fax:** 1 203364. **Email:** info@yali.org.ye

Yemen Language centre
PO Box 1691, Sana'a. **Tel:** 1 285125. **Fax**: 1 289249.

Yemen Sudanese Language centre
PO Box 13187, Sana'a. **Fax:** 1 275407.

Africa

For many countries in Africa, the English language is part of the legacy left to them by the colonialists. In these countries, English is often taught from a young age in the state schools and people use it to communicate if there are several tribal languages spoken in one nation. Other African countries are former French or Belgian colonies, so a working knowledge of French is useful.

Most English language teachers work in Africa as volunteers but there are some jobs available at university and college institutes although you may end up earning more as a volunteer than as a "salaried" teacher. Female teachers thinking of coming to Africa should be aware that societies are much more traditional here when it comes to the role of men and women. You may find that, in a staff meeting, men sit on one side of the room and women on the other.

A number of countries within Africa are currently too dangerous to visit - due to war, violence or unrest. This includes the whole of Chad, Ivory Coast, and Somalia; and parts of CAR, Burundi, Cameroon, Congo, Ethiopia, Nigeria, Sudan, Uganda. You can find out the latest status of which country is safe to visit at www.fco.gov.uk.

Teachers who want to work as volunteers can contact *Teachers for Africa,* an organization that places experienced teachers in Malawi, Namibia, Nigeria, Ethiopia, Ghana, Guinea and Benin. Web: www.ifesh.org, Email: teachers@ifesh.org.

Note: African countries on the Mediterranean Coast are dealt with in the Middle East section.

BENIN

Benin is one of smaller West African states. Its capital city is Porto Novo; however, the city of Cotonou is the main port and largest city. There are interesting sites to visit in Benin including Ouidah, an ancient port where slaves were sent around the world, Abomey, the "royal" city, and Allada, known as the "cradle" of the voodoo religion. There is a very high demand for English language teachers in Benin. The best way to find a job here is to enter the country on a tourist visa and then start hunting around. You could also contact the *Syndicat National de l'Enseignement Primaire Public du Bénin*, SNEP, B.P. 69, Cotonou, Bénin. **Tel:** +229-303.613 **Fax:** +229-303.613.

➲ **Visas/Work Permits:** Tourist visas are valid for a 15- ($45) or 30-day ($60) period within three months of date of issue.

➲ **Requirements:** College Degree and a TESOL/TESL Certificate

➲ **Currency:** £1 = 922 / $1 = 501 CFA Franc.

➲ **Salaries/Taxes:** 4,500 CFA per hour. Foreigners do not have to pay tax unless they are registered with the OBSS – the Office of Social Security.

➲ **Accommodation:** Flats cost from 80,000 to 800,000 CFA per month.

➲ **Health advice:** Medical facilities in Benin are limited and not all medicines are available. A yellow fever certificate must be presented on entry to Benin. Cholera is a serious risk in this country and precautions are essential. Malaria is also a risk as is rabies. Health insurance is recommended.

Schools in Benin (+229)

American Cultural centre

B.P. 2012, Cotonou. **Tel:** 304553. **Fax**: 301439. **Email:** vreeland@micronet.net

Brilliant Stars International School

04 B.P. 919 Cadjehoun, Cotonou. **Tel:** 302079, **Fax**: 313701/313809.

BURKINA FASO

Although this landlocked country is one of Africa's poorest nations, there is a vibrant culture evident among its people. The Pan African Film and Television Festival of Ouagadougou (the capital city) is the largest event of its kind in Africa and draws over 100,000 visitors. Volunteers in urban and rural settings teach secondary school and university English. You can also contact the national teacher's association: Syndicat National des Enseignants du Secondaire et du Supérieur (SNESS), 01 B.P. 113, Ouagadougou 01, Burkina Faso. **Tel:** +226-362.362, **Fax**: +226-362.362.

➲ **Visas/Work Permits:** You will need both. Your employer will help you obtain them.

➲ **Currency:** £1 = 922 / $1 = 501 CFA Franc.

➲ **Health advice:** Medical facilities in Burkina Faso are limited. A Yellow fever certificate is required. Precautions against cholera, malaria and rabies are recommended.

Schools in Burkina Faso (+226)

American Language centre

01 B.P. 539, Ouagadougou-01, Burkina Faso. **Tel:** 306360/308903. **Fax**: 31 5273. **Email: alcouaga@fasonet.bf**

International School of Ouagadougou

01 B.P. 35, c/o US Embassy, Ouagadougou. **Tel:** 362143**. Fax:** 362228**. Email:** issoua@fasonet.bf

CAMEROON

There are occasional opportunities for experienced EFL teachers in Cameroon, an enchanting and spectacular country with mountains and a coastline to explore. Volunteers with VSO (see p.62) work to improve the quality of English language teaching in rural schools that have little or no resources. Unfortunately, teachers report that the security situation in Cameroon is getting worse and that foreigners should be very careful when travelling.

British Council office:

Immeuble Christo, Avenue Charles de Gaulle, BP 818, Yaoundé
Tel +237 2 21 16 96/20 31 72
www.britishcouncil.org/cameroon

➲ **Visas/Work Permits:** Visas are granted for up to 12 months. You will also need a residence permit, which costs £500.

➲ **Requirements:** College Degree and a TESOL/TESL Certificate

➲ **Currency:** £1 = 926 / $1 = 501 CFA Franc.

➲ **Salaries/Taxes:** Teachers can earn £550-800 a month with a housing and transport allowance.

➲ **Health advice:** Medical facilities in Cameroon are limited. Sanitation levels are low, even in the best hospitals. Not all medicines are available. Travellers are advised to bring their own supplies. Malaria prophylaxis and vaccination against hepatitis A and B, tetanus, diphtheria, polio, typhoid, and meningococcal meningitis are recommended.

Schools in Cameroon (+237)

American School of Douala
BP 1909, Douala. **Tel:** 421437. **Fax**: 421437. **Email:** adminasd@camnet.cm

American School of Yaoundé
Dept of State, 2520 Yaounde Place, Washington DC, 20521-2520, USA. **Tel:** 230421. **Fax:** 236011. **Email:** asoy@camnet.cm

The British Council
Immeuble Christo, Rue Drouot, B.P 12,801 Douala. **Tel:** 425145. **Fax:** 425170. **Email:** ELC.Douala@bc-douala.douala1.com

The British Council
Immeuble Soms, Avenue Charles de Gaulle, B.P 818 Yaoundé. **Tel:** 211696. **Fax:** 215691. **Email:** ELC.Yaounde@bc-yaounde.iccnet.cm

Ministry of Higher Education
2, Avenue du 20 Mai, Yaoundé. **Tel:** 22 13 70

CENTRAL AFRICAN REPUBLIC

The CAR, as it known, is a developing country that it is full of potential. Interesting places to visit include the Dzanga-Sangha National Park, a primeval rain forest in the southwestern region of the country. There is a very strong demand for English language teaching in the CAR. Experienced teachers will have no problems in finding a job but they should apply from outside the country to avoid disappointment. Unfortunately, the political and social situation deters most teachers from working here and it can be a very dangerous country that features high on the list of countries to avoid at www.fco.gov.uk

➲ **Visas/Work Permits:** As rules change so quickly, it is best to check with the embassy.

➲ **Currency:** £1 = 926 / $1 = 501 CFA Franc.

➲ **Salaries/Taxes:** Expect to be paid $150-300 a month but there are problems with getting paid in government contracts in a country where civil servants haven't had any money for almost two years.

➲ **Accommodation:** An apartment in the capital, Bangui, will cost anywhere between £150-300 a month.

➲ **Cost of Living:** Very cheap

➲ **Health advice:** Medical facilities are limited, and the quality of acute care is unreliable. Sanitation levels are low. Many medicines are not available. Travellers are advised to bring their own properly-labeled supplies.

Schools in Central African Republic (+236)

Martin Luther King Cultural centre
c/o American Embassy, Avenue David Dacko, PO Box 924, Bangui. **Tel:** (236) 610200. **Fax**: 614494. **Email:** gazembetig@hotmail.com

DEM. REPUBLIC OF CONGO

The situation in this country (formerly known as Zaire) has been improving as it recovers from civil war. However, it is still a dangerous country and features on the list of countries to avoid at www.fco.gov.uk

Schools in Dem. Rep. Congo (+243)

American School of Kinshasa
c/o American Embassy, Kinshasa, APO AE 09828. **Tel:** 88 46619.

Congo-American Language Institute (CALI)
B.P. 8622, Kinshasa 1. **Tel:** 8843979. **Fax**: 88 46592.

ERITREA

Eritrea is recovering from a recent war with neighbouring Ethiopia and conditions are very harsh. Many programmes have had to be suspended at various times because of instability in the region; it is still a dangerous country and features on the list of countries to avoid at www.fco.gov.uk

British Council office:
Lorenzo Tazaz Street No 23, PO Box 997, Asmara
Tel +291 (1) 123 415
www.britishcouncil.org/eritrea

➲ **Salaries/Taxes:** Rates of pay are very low and may be supplemented by your agency.

➲ **Accommodation:** Housing will be provided by your employer.

Schools in Eritrea (+291)

Asmara International Community School
PO Box 4941, Asmara, Eritrea **Tel:** 1 161705. **Fax**: 1 161705 **Email:** aics@eol.com.er

The British Council
Lorenzo Ta'azaz Street, No. 23, PO Box 997, Asmara, **Tel**. 1 23415. **Email:** britcoun@eol.com.er

ETHIOPIA

It is not easy to find a job teaching English in Ethiopia. Most private language schools employ Ethiopians who can speak English. A number of western aid organizations work in the country and it may be possible to land a job through one of them. Volunteers need to have had at least two years' teaching experience as well as a certificate or an MA. VSO Canada sends teachers to mainstream educational programmes and technical training institutions aiming to improve the standard of English (the medium of instruction) of the students. However, be aware that this is still a dangerous country and features on the list of countries to avoid at www.fco.gov.uk

Embassy in UK:
17 Princes Gate, London SW7 1PZ
Tel: 020 7589 7212

Embassy in USA:
3506 International Drive NW, Washington DC20008
Tel: 202 364 1200

British Council office:
PO Box 1043, Artistic Building, Adwa Avenue, Addis Ababa
Tel +251 (1) 550 022
www.britishcouncil.org/ethiopia

➲ **Visas/Work Permits:** You have to get a single-entry visa and you must get an exit visa each time you leave the country, which makes travel difficult.

Schools in Ethiopia (+251)

The British Council
Adwa Avenue, Artistic Building, PO Box 1043, Addis Ababa. **Tel:** 1 55 00 22. **Fax**: 251-1-55 25 44. **Email: Solomon.Hailu @bc-addis.bcouncil.org**

International Community School
P.O. Box 70282, Addis Ababa.

Sandford English Community School
P.O. Box 30056 MA, Addis Ababa. **Tel:** 1 55 2275. **Fax**: 1 55 1945.

University of Addis Ababa
P.O. Box 1176, Addis Ababa.

GHANA

There is much to explore in Ghana from the coastal slave forts recalling a tragic period in the country's history to the rainforests of the interior. Depending on which level you intend to teach you could get in touch with the educational authorities. Try: The General Secretary, Ghana National Association Of Teachers, GNAT Headquarters, P.O. Box 209, Accra, Ghana. Tel: +233-21-221.515 / 221.576, Fax:+233-21-224.877 for more information. For volunteer teaching opportunities in Ghana, see page 62. More limited openings can be found at a handful of language institutes affiliated with colleges and universities. To teach in Ghana you have to be a certified teacher, or serving with a volunteer organization, or a tenured lecturer.

High Commission in UK:
13 Belgrave Square, London SW1X 8PN
Tel: 020 7235 4142

Embassy in USA:
3512 International Drive NW, Washington DC20008
Tel: 202 686 4520

British Council office:
11 Liberia Road, PO Box GP 771, Accra
Tel +233 (21) 683 068
www.britishcouncil.org/ghana

➲ **Visas/Work Permits:** The school will take care of this for you.

➲ **Currency:** £1 = 16551 / $1 = 8947 Cedi

➲ **Salaries/Taxes:** Volunteers can expect to earn around £30-200 a month. Teachers at universities or language institutes can earn anywhere from £1000-2000.

➲ **Health advice:** Facilities are limited, particularly outside Accra. Be aware that evidence of and/or assurances from insurance companies will not be accepted as settlement of medical expenses.

➲ **English Language Media:** Ghanaian Chronicle, Ghanaian Times, Graphic.

Schools in Ghana (+233)

Centre of Language and Professional Studies
P.O. Box 4501, Accra.

Institute of Languages
P.O. Box M67, Accra. **Tel:** 21 221052.

The Language Centre
University of Ghana, Legon. **Tel:** 21 500381.

KENYA

English is the working language of Kenyan schools and there is a shortage of qualified and experienced teachers. Although you won't make a lot of money here, it is a beautiful country with much to see and do.

High Commission in UK:
45 Portland Place, London W1N 3AS
Tel: 020 7636 2371

Embassy in USA:
2249 R Street NW, Washington DC 20008
Tel: 202 387 6101

British Council office:
ICEA Building, Kenyatta Avenue, PO Box 40751 Nairobi
Tel +254 (2) 334 855/6
www.britishcouncil.org/kenya

➲ **Visas/Work Permits:**You are supposed to have a work permit before you arrive but teachers have picked up casual work while on tourist visas.

➲ **Requirements:** College Degree and a TESOL/TESL Certificate

➲ **Currency:** £1 = 136 / $1 = 69 Shillings

➲ **Salaries/Taxes:** Teachers are usually paid at local rates, which allows for a simple lifestyle.

➲ **Accommodation:** Outside Nairobi, housing can be very basic.

Schools in Kenya (+254)

American Universities Preparation & Learning centre
P.O. Box 14842, Nairobi. **Tel:** 2 741764. **Fax**: 2 741690.

The Language centre
P.O. Box 14245, Ndemi Close, off Ngong Road, Nairobi. **Tel:** 2 569531. **Fax**: 2 568207.

MADAGASCAR

Madagascar is a large island situated in the Indian Ocean. Many unique species of animals are threatened with extinction here as sapphire mining and other destructive activities go unchecked.
Teachers interested in working at the English Teaching programme in Antananarivo should contact the programme upon arrival in the city. Please do not send applications in advance.

➲ **Visas/Work Permits:** Visas should be obtained in advance, although airport visas are available in Antananarivo, the capital.

➲ **Currency:** £1 = 3233 / $1 = 1748 Ariary

➲ **Health advice:** The hospital infrastructure is minimal and does not meet basic sanitary norms.

Schools in Madagascar (+261)

American School of Antananarivo
US Dept of State/Antananarivo, Washington DC 20521-2040, USA. **Tel:** 202242039. **Fax**: 202234539.

English Teaching Program
4 Lalana Dr., Razafindratandra Ambohidahy, Antananarivo. **Tel:** 22220238. **Fax**: 221397.

MALI

Mali is a vast land of savannas, steppes, and deserts. For the most of the last four millennia, it was one of the great centres of African civilization. During the middle ages, the Malian empire was famous throughout much of Africa and Europe; Timbuktu was a centre of Islamic study and its library was one of the largest in the world. The Peace Corps has been active in Mali for over 30 years and it enjoys a high reputation among the people of Mali.

Schools in Mali (+223)

American International School of Bamako
Dept of State - Bamako, Washington DC, 20521-2050, USA. **Tel:** 224738. **Fax**: 220853.

MOZAMBIQUE

For many years, Mozambique was in the grip of a bloody civil war. Although the situation has improved a lot since then, teachers should be warned that parts of country and the capital, Maputo, are unsafe. The civil war has severely disrupted education: The overall literacy rate is estimated at 30-40% and only 40% of school-age children actually attend school. Volunteers with VSO Canada are involved in English language teaching in rural elementary schools where there are sometimes more than 50 children in a classroom. Teachers report that life outside the capital is much more relaxed and that the weekends are worth waiting for because there are miles of empty beaches and fascinating countryside to explore. You will need to speak some Portuguese.

British Council office:
Rua John Issa, 226, P.O. Box 4178, Maputo
Tel +258 (1) 310 921
www.britishcouncil.org/mozambique

➲ **Visas/Work Permits:** You will need a visa, which must be obtained in advance. Travellers arriving without a visa are subject to a fine of £300. The authorities will impose a fine of £60 per day for each day travellers overstay their visas. Once your contract has been agreed, the school will apply for a work permit, which will take about four months to process. You are not supposed to work for the school during this period although this rule is commonly ignored.

➲ **Currency:** £1 = 33932 / $1 = 18342 Metical.

➲ **Salaries/Taxes:** Approximately £10 / $19 an hour. Tax is about 17%.

➲ **Accommodation:** Rents are expensive in Maputo. Typical rates are £600 / $1100 per month shared among three teachers. The upside is that the accommodation is clean and modern.

➲ **Cost of Living:** Mozambique is an expensive country where you might be able to save only a little of your salary – if you're not keen on going out and partying every night.

➲ **Health advice:**Facilities are minimal and many medicines are unavailable. Maputo's Sommerschield Clinic, which requires payment in hard currency, can provide general, non-emergency services. Many rural health centres were forced to close during the conflict with the MNR rebels. Yellow fever vaccination certificate is required. Cholera is a serious risk and precautions are essential. Malaria and rabies are risks.

Schools in Mozambique (+258)

Brumag
Av. Mártires Mueda 563, Maputo. **Tel:** 496199.

Contacto
R. Tchamba 64, Maputo. **Tel:** 492479.

Oikos
R. General Pereira D'Eca, Maputo. **Tel**; 497850.

Instituto De Linguas
Av. Ahmed Sekou Touré, **Tel:** 425684.

Lynden Language School
PO Box 456, 1746 Rua de Resistencia 3rd floor, Maputo. **Tel:** 417569.

Maputo International School
Rua de Nachingwea, CP 4152, 389, Maputo. **Tel:** 492131/492195. **Fax:** 492382.

SENEGAL

EFL teachers can find work on a casual basis in Senegal but there are no private language schools as such. If you want to work in the public sector, you could try the national teachers' union: *Syndicat des Professeurs du Sénégal* (SYPROS), Bourse du Travail CNTS, 7 Avenue du Président Lamine Guèye, B.P. 937, Dakar, Sénégal. **Tel:** +221-821.8381 / 821.0491, **Fax**: +221-821.7771).

Embassy in UK:
39 Marloes Road, London W8 6LA
Tel: 020 7937 723

Embassy in USA:
2112 Wyoming Ave NW, Washington DC 20008
Tel: 202 234 0540

British Council office:
34-36 Boulevard de la République, Immeuble Sonatel
BP 6238, Dakar
Tel: +221 822 2015/822 2048
www.britishcouncil.org/senegal

Schools in Senegal (+221)

American English Language programme **(AELP)**
c/o USIS, B.P. 49, Dakar. **Tel:** 211634. **Fax:** 222345.

The British Council
34-36 Bd de la République, Dakar.**Tel:** 822 20 15 / 822 20 48. **Fax**: 821 81 36.

SOUTH AFRICA

The demand for English language instruction is very strong in post-apartheid South Africa. There are many private language institutes, especially in and around Cape Town that cater to an international student market. In rural South Africa, the Peace Corps operates a programme where volunteers advise local elementary school teachers on English language lesson planning. Qualified teachers are welcomed here but you will be competing for jobs with South African citizens who do not require work permits.

High Commission in UK:
South Africa House, Trafalgar Square, London WC2N 5DP
Tel: 020 7451 7299
www.southafricahouse.com

Embassy in USA:
3051 Massachusetts Ave NW, Washington DC 20008

Tel: 202 238 4400

British Council office:
Ground Floor, Forum 1, Braampark, 33 Hoofd Street, Braamfontein, Johannesburg 2001
Tel +27 (11) 718 4300
www.britishcouncil.org/southafrica

➲ **Visas/Work Permits:** Teachers need a work permit, which costs around £90. Your employer must be able to prove that the position was made available to South Africans before a foreign national took up the offer.

➲ **Requirements:** College Degree and a TESOL/TESL Certificate

➲ **Currency:** £1 = 13.65 / $1 = 6.98 Rand

➲ **Salaries/Taxes:** Teachers earn from £5-10 an hour and taxation ranges from 20-30%.

➲ **Accommodation:** There is a high standard of accommodation in South Africa and rents are expensive if you are earning a local salary.

➲ **English Language Media:** The Star, Weekly Mail.

Schools in South Africa (+27)

Cape Communication Centre
37 Riebeck Street, Cape Town 8001. **Tel:** 21 419 1967

Cape Studies Language School
100 Main Road Seapoint, P.O.Box 4425, Cape Town 8000. **Tel:** 21 439 0999. **Fax**: 21 439 3130

Cape Town School of International Languages
66 Main Rd, Claremont, Cape Town 7708. **Tel:** 21 674 4117.

Good Hope Studies
4 Mount Road, Rondebosch, Cape Town 7700. **Tel:** 21 686 0060. **Fax**: 21 686 0061.

inlingua Language Training Centre
Kinellan House, 14 Portswood Rd., Cape Town 8000. **Tel:** 21 4190494. **Fax**: 21 4190725.

INTERLINK School of Languages
P.O. Box 694, Seapoint, Cape Town 8060. **Tel:** 21 4399 834 **Fax**: 21 4343 267.

Language & Travel
Mail: P.O. Box 868, Green Point, Cape Town 8051, 2 Braemar Road, Green Point 8005. **Tel:** 21 439 3101. **Fax**: 21 439-0616.

Language Lab International House
4th Floor Aspern House, 54 De Korte Street, Braamfontein, Johannesburg 2001. **Tel:** 11 339 1051. **Fax**: 11 403 1759.

One World Language School
P. O. Box 7888, Roggebaai, Cape Town 8012. **Tel/Fax:** 21 4231833.

Table Mountain Language Centre
7 Mountain Road, Claremont, Cape Town 7700. **Tel**: 21 674 49 14. **Fax**: 21 671 09 97.

Tygerberg Language and Leisure Centre
P.O. Box 2483, Bellville 7530. **Tel:** 21 948 3460.

TANZANIA

You can find work as a volunteer teacher and there are sometimes opportunities in colleges and universities.

British Council office:
Samora Avenue / Ohio Street, PO Box 9100, Dar Es Salaam
Tel +255 (22) 211 6574
www.britishcouncil.org/tanzania

➲ **Visas/Work Permits:**Employers make the necessary arrangements.

Schools in Tanzania (+51)

International Languages Orientation Services
Oysterbay, Karume Road, P.O. Box 6995, Dar es Salaam. **Tel:** 667159. **Fax**: 112752.

TOGO

There are limited opportunities for EFL teachers in this small West African nation. Contact the teachers' union: Fédération des Syndicats de l'Education Nationale, Tel: +228-222.547

Schools in Togo (+228)

English Language Program
Centre Culturel Américain, Rues Pelletier et Vauban, B.P. 852, Lomé. **Tel:** 212166.

Australia & New Zealand

Over the past few years, the prospects in both Australia and New Zealand have changed. Many English language schools have closed down and it is no longer quite as easy to get a job as it used to be. These ups and downs in the local ELT industry began when, in Australia, the recession has eased and the government introduced new 12-month visitor visas (allowing temporary work) giving young teachers the perfect excuse to travel around this fabulous country ... and earn money teaching. New Zealand, too, has had a change in government policy that impacted on English language teachers: the rules for immigration have been relaxed, resulting in an influx of immigrants – all keen to learn English. As a result, New Zealand is suffering something of a shortage of teachers. Only US citizens will find a problem with either country: they cannot take advantage of the 12-month young traveller visa and report that finding work (and a school willing to sponsor them through a work visa) is very tough.

AUSTRALIA

There are good job prospects for teachers in Australia, with a good increase in the numbers of (mostly Asian, due to its location) students visiting Australia to learn English. The majority of the language schools are in Sydney (which has over 70 schools); around two-thirds of the schools are accredited and prefer teachers with experience and a recognised certificate (visit the accreditation website, www.neasaccred.com for more details).

Australia has changed its visa policy to around a dozen countries, allowing young (18-30) teachers to stay for 12 months and teach in temporary jobs. US citizens report that it is hard to get a job without a work permit which is difficult for US citizens to obtain (however, Canadians are included in the 12-month working-holiday visa scheme).

High Commission in UK:
Australia House, Strand, London WC2B 4LA
Tel: 020 7379 4334
www.australia.org.uk

Embassy in USA:
1601 Massachusetts Ave NW, Washington DC20036
Tel: 202 797 3000
www.austemb.org

British Council office:
PO Box 88, Edgecliff, Sydney NSW 2027
Tel: (2) 9326 2022
www.britishcouncil.org/au.htm

➲ **Visas:** British, American and Canadian citizens require visas before they can enter Australia as tourists, students or workers. Tourists need a tourist visa (now generally replaced by the Electronic Travel Authority, issued by major travel agents). All other work visas are assessed by the high commission or embassy – the basic reasoning is that any job opportunity must be made available to Australian citizens, so you will have to show that your work cannot be carried out by an existing resident.

➲ **Requirements:** College Degree and a TESOL/TESL Certificate

➲ **Currency:** £1 = 2.49 / $1 = 1.27 Dollars

➲ **Health:** Health-care standards are very high and if you are granted a resident permit for more than six months, you can join the national medical and hospital service, Medicare.

➲ **Tax:** all employees must register for tax – find out how at www.ato.gov.au

Schools In Australia (+61)

acl (Australian Centre for Languages)
157-161 Gloucester Street, The Rocks, New South Wales 2000, **Tel**: +61 2 9252 3788, **Fax**: +61 2 9252 3799, **Email:** enquiry@acl.edu.au , **Web:** www.acl.edu.au

Adelaide Institute of TAFE English Language Centre
120 Currie Street, Adelaide, South Australia 5000, **Tel**: +61 8 8207 8279, **Fax**: +61 8 8207 8283,**Web:** www.tafe.sa.edu.au

Anutech Education Centre at the Australian National University
Cnr Barry Drive and North Road, Canberra, Australian Capital Territory 2601, **Tel**: +61 2 6125 5000, **Fax**: +61 2 6125 5664, **Email:** edu.enquiries@anutech.com.au, **Web:** www.anutech.com.au/ELC

ASPECT ILA Sydney
98-104 Goulburn Street, Sydney, New South Wales 2000, **Tel**: +61 2 9283 8055, **Fax**: +61 2 9283 9055, **Email:** aspectsydney@educate.com , **Web:** www.aspectworld.com

Australian College of English
Level 1, 237 Oxford Street, Bondi Junction, New South Wales 2022, **Tel**: +61 2 9389 0133, **Fax**: +61 2 9389 6880, **Email:** info@ace.edu.au, **Web:** www.ace.edu.au

Australian Pacific College
189 Kent Street, Sydney, New South Wales 2000, **Tel**: +61 2 9251 7000, **Fax**: +61 2 9251 7575, **Email:** info@apc.edu.au, **Web:** www.apc.edu.au

Billy Blue English School
Level 9, Northpoint, 100 Miller Street, North Sydney, New South Wales 2060, **Tel**: +61 2 9955 1122, **Fax**: +61 2 9957 1811, **Email:** info@billyblue.com.au **Web:** www.billyblue.com.au

Cambridge English Language Centre
Level 3, 4 Cross Street, Hurstville, New South Wales 2220, **Tel**: +61 2 9579 6244, **Fax**: +61 2 9570 9565, **Email:** cambridgecollege@tpg.com.au **Web:** www.camb.aust.com

Central Queensland University Language Centre
Bruce Highway, North Rockhampton, Queensland 4702, **Tel**: +61 7 4930 6577, **Fax**: +61 7 4930 6321 **Web:** www.language.cqu.edu.au

Centre for English Teaching
University of Sydney, 88 Mallett Street, Camperdown, New South Wales 2006, **Tel**: +61 2 9351 0706, **Fax**:

+61 2 9351 0710, **Email:** info@cet.usyd.edu.au, **Web:** www.usyd.edu.au/cet

Centre for English Language Teaching
35 Stirling Highway, Crawley, West Australia 6009, **Tel**: +61 8 9380 3539, **Fax**: +61 8 9380 1077, **Email:** celt@ecel.uwa.edu.au, **Web:** www.celt.uwa.edu.au

EF International Language Schools
Sydney, 5-7 Young Street, Sydney, New South Wales 2000, **Tel**: +61 2 9247 7668, **Fax**: +61 2 9247 7691, **Email:** ils.sydney@ef.com, **Web:** www.ef.com

ELICOS
University of Canberra, University Drive, Bruce, Australian Capital Territory 2601, **Tel**: +61 2 6201 2102, **Fax**: +61 2 6201 5089, **Web:** www.slie.canberra.edu.au

ELS Language Centres
Charles Sturt University, Boorooma Street, Wagga Wagga, New South Wales 2678, **Tel**: +61 2 6933 2602, **Fax**: +61 2 6933 4018, www.csu.edu.au/division/internat/els/index.htm

Embassy CES Brisbane
119 Charlotte Street, Brisbane, Queensland 4000, **Tel**: +61 7 3238 1500, **Fax**: +61 7 3238 1670, **Email:** w.day@sga.edu.au, **Web:** www.embassyces.com

Embassy CES Melbourne
399 Lonsdale Street, Melbourne, Victoria 3000, **Tel**: +61 3 9670 3788, **Fax**: +61 3 9670 9356, **Email:** info@taylorscollege.com, **Web:** www.embassyces.com

International House Sydney
(Waratah Education Centre), 22 Darley Road, Manly, New South Wales 2095, **Tel**: +61 2 9976 2422, **Fax**: +61 2 9976 2433, **Web:** www.ihsydney.com

International House Queensland English Language College
130 Mcleod Street, Cairns, Queensland 4870, **Tel**: +61 7 4031 3466, **Fax**: +61 7 4031 3464, **Email:** admin@ihqld.com, **Web:** www.ihqld.com

Northern Territory University Centre for Access and ESL
Ellengowan Drive, Casuariana, Northern Territory 0909, **Tel**: +61 8 8946 6160, **Fax**: +61 8 8946 7144, **Email:** uni-info@ntu.edu.au, **Web:** http://mindil.ntu.edu.au

Shane Global Village Sydney
Level 12, 222 Pitt Street, Sydney, NSW 2000 **Tel:** +61 (2) 9283 1088 **Fax**: +61 (2) 9283 1760 **Web:** www.sgvenglish.com

TAFE English Language Centre, Northern Sydney
213 Pacific Highway, St Leonards, New South Wales 2065, **Tel**: +61 2 9942 0086, **Fax**: +61 2 9906 7423, **Email:** Telc@tafensw.edu.au , **Web:** www.tafensw.edu.au

TAFE International Education Centre, Liverpool
Cnr Bigge and Moore Streets, Liverpool, New South Wales 2170, **Tel**: +61 2 9827 5197, **Fax**: +61 2 9827 5198, **Web:** www.tafensw.edu.au/swsit/

Universal English College
Level 12, 222 Pitt Street, Sydney, New South Wales 2000, **Tel**: +61 2 9283 1088, **Fax**: +61 2 9283 1760, **Email:** enquiries@uec.edu.au, **Web:** www.uec.edu.au

University of NSW Institute of Languages
22 King Street, Randwick, New South Wales 2031, **Tel**: +61 2 9385 0339, **Fax**: +61 2 9399 5420, **Email:** Institute.Languages@unsw.edu.au, **Web:** www.lang.unsw.edu.au/elicos/

University of Western Sydney English Language Centre
Building 16, Room 7, Campbelltown Campus, Campbelltown, New South Wales 2560, **Tel**: +61 2 4620 3028, **Fax**: +61 2 4620 3782, **Email:** l.simon@uws.edu.au, **Web:** http://uwselc.uws.edu.au

Victoria University English Language Institute
Level 5, 301 Flinders Lane, Melbourne, Victoria 3000, **Tel**: +61 3 9248 1175, **Fax**: +61 3 9248 1298, **Email:** eli@vu.edu.au, **Web:** www.vu.edu.au/eli/

NEW ZEALAND

New Zealand is experiencing a good demand for English teachers – thanks to a rush of immigrants who want to learn English (tempted by the favourable exchange rate and easing of immigration requirements). There are not enough teachers and schools currently will offer to help visitors get a work visa; schools prefer teachers with a CELTA or Trinity certificate and at least one year's experience.

Auckland and Christchurch are the two main centres for schools and the bulk of students are from China, Korea and Japan. The only drawback is for US citizens hoping to teach in New Zealand – like Australia, it's hard for US citizens to get a work permit. A good source for jobs are the websites www.english-schools.co.nz and www.teachnz.gov.nz

High Commission in UK:
New Zealand House, 80 Haymarket, London SW1Y 4TQ
Tel: 020 7930 8422
www.nzembassy.com

Embassy in USA:
37 Observatory Circle NW, Washington DC20008
Tel: 202 328 4800
www.nzemb.org

British Council office:
44 Hill Street, PO Box 1812, Wellington 1
Tel (4) 495 0897
www.britishcouncil.org.nz

➲ **Visas:** British citizens can enter NZ without a visa and can stay in the country for up to six months, if they can prove that they have enough money and a return ticket. UK, Irish, Canadian, and Japanese citizens aged 18-30 can use the Working Holiday visa scheme to take temporary work whilst they stay in NZ for up to 12months. US citizens can stay for up to 90 days without a visa.

➲ **Requirements:** College Degree and a TESOL/TESL Certificate

➲ **Currency:** £1 = 2.82 / $1 = 1.44 Dollars

Schools in New Zealand (+ 64)

Auckland English Academy
PO Box 11 241, Ellerslie, Auckland, New Zealand, **Tel**: +64 9-3793777, **Fax**: +64 9-358-3383, **Email:** learn@english.co.nz, **Web:** www.english.co.nz

Auckland Institute of Studies
PO Box 2995, Auckland, New Zealand, **Tel**: +64 9-815 3772, **Fax**: +64 9-815 1802, **Web:**www.ais.ac.nz

Canterbury Language Institute
PO Box 4425, Christchurch, New Zealand, **Tel**: +64 3-365 1920, **Fax**: +64 3-365 2296, **Email:** can.lancol@xtra.co.nz

Dynaspeak English
12 O'Connell Street, PO Box 106-052, Auckland 1001, New Zealand, **Tel**: +64 9 377 2434, **Fax**: +64 9 377 2811, **Web:** www.dynaspeak.co.nz

Edenz Colleges
Ideal House, 17 Eden St, Newmarket. Auckland, **Tel**: +64 9 815 3772, **Fax**: +64 9 522 1211, **Email:** edenz@edenz.com, **Web:** www.edenz.com

ILA South Pacific - Auckland
PO Box 9107, Auckland, New Zealand, **Tel**: +64 9 307 6507, **Fax**: +64 9 379 5749, **Email:** principal@ila.co.nz, **Web:**www.ila.co.nz

ILA South Pacific - Christchurch
PO Box 25170, Christchurch, New Zealand, **Tel**: +64 3 379 5452, **Fax**: +64 3 379 5957, **Email:** ilachch@ila.co.nz, **Web:**www.ila.co.nz

LSNZ Christchurch & Queeenstown
PO Box 911, Queenstown, New Zealand, **Tel**: +64 3 442 6625, **Fax**: +64 3 442 6676, **Web:**www.languageschool.co.nz

Modern Age English Language School
PO Box 2137, Tauranga, New Zealand, **Tel**: +64 7-576 8831, **Fax**: +64 7-578 5883, **Web:**www.modernage.co.nz

New Horizon College of English
PO Box 66, Napier, New Zealand, **Tel**: +64 6-835 6423, **Fax**: +64 6-835 6523, **Email:** office@nhce.ac.nz, **Web:** www.nhce.ac.nz

Shane Global Village English Centre
Level 2, 138 Queen Street. PO Box 105-347, Auckland., **Tel**:+64 (9) 309 2201, **Email:** auckland@sgvenglish.com, **Web:** www.sgvenglish.com

Shore English School of Language New Zealand
PO Box 33-614, Takapuna, Auckland, **Tel**: +64 9-486 1896, **Fax**: +64 9-486 0648, **Email:** shoreng@nznet.gen.nz, **Web:**www.shore-english.co.nz

Southern Lakes English School
PO Box 405, Queenstown, New Zealand, **Tel**: +64 3 442-7810, **Fax**: +64 3 442-7815, **Email:** slec@xtra.co.nz, **Web:** www.slec.co.nz

Wellington College of Languages
PO Box 9546, Wellington, New Zealand, **Tel**: +64 4 382 8938, **Fax**: +64 4 385 1518, **Email:** wcl@xtra.co.nz, **Web:** www.wcl.co.nz

Worldwide School of English
PO Box 1802, Auckland, New Zealand, **Tel**: +64 9 302 5288, **Fax**: +64 9 302 5299, **Email:** info@wwse.co.nz, **Web:** www.wwse.co.nz

Asia

The English language is in great demand in Asia, particularly in China (including Hong Kong), Korea, Japan, Taiwan, and Thailand. In some countries in South East Asia (notably Singapore) the local government has introduced legislation to teach core subjects in English – students must learn maths and science in English.

China has seen an explosion of English language schools over the past few years - with native Chinese speakers learning English and learning to teach English with equal enthusiasm. Korea also offers a fast-changing EFL landscape, but is mostly dominanted by American-English language schools; by comparison, Japan and Thailand offer established school chains, networks for teachers and associations to help teachers.

The region has, mostly, recovered from the Asian economic crisis at the start of the 2000s; money has returned for education, and proficiency in English is widely regarded as essential for a successful career. For EFL teachers, levels of pay, the cost of living and working conditions vary considerably between the developed nations and the developing countries. Some teachers work long hours to live well and save a lot of money, whereas others supplement their income with their savings and use their spare time to travel. Whatever your goals, there is no doubt that you will have the experience of a lifetime should you choose to teach in Asia.

BANGLADESH

Bangladesh is one of the most impoverished countries in the world. There are opportunities for EFL teachers here, mostly volunteer work with aid organizations and the Peace Corps.

Schools in Bangladesh (+880)

The British Council Teaching Centre
754b Satmasjid Road, Dhanmondi, Dhaka 1205. **Tel:** 9116171/9116545. **Fax**: 8116554. **Email:** dto@TheBritishCouncil.net

Regional Resources Centre
SESDC Building, Teachers Training College, Bakulia, Chittagong. **Fax:** 031 612307.

CAMBODIA

Cambodia is for the adventurous EFL teacher who enjoys getting off the beaten track. The country is still rebuilding after a long civil war and some areas are not safe, including the capital, Phnom Penh, where a number of teachers from Oklahoma were kidnapped and later released. Crime waves are sporadic so take the same precautions as you would in any major western city. Salaries are low when compared to the richer Asian nations. The better language schools (offering better rates of pay!) will want to see evidence of TESL certification but there are street corner schools who will take on an unqualified native speaker. Most schools employ teachers on a term-to-term basis so you are only guaranteed work for 10 weeks at a time. There is a lot of demand for English in Cambodia as the country slowly integrates with its neighbours. Few Cambodians can afford private lessons.

➲ **Visas/Work Permits:** You can arrive in Cambodia and purchase a one-month working visa for £14 / $30. The working visa can be renewed without leaving the country. You don't have to prove you are actually working for someone when you arrive. Your school will not reimburse you for the cost of your visa.

➲ **Requirements:** High School Diploma and a TESOL/TESL Certificate

➲ **Requirements:** High School Diploma and a TESOL/TESL Certificate

➲ **Currency:** £1 = 7455 / $1 = 4030 Riel

➲ **Salaries/Taxes:** Schools pay around £6-10 / $10-20 an hour if you are qualified. Tax is 6%.

➲ **Accommodation:** Rents have been forced up artificially high as Phnom Penh plays host to large groups of overpaid foreign aid workers. Flats are difficult to find and cost anywhere from £100-170 / $190-320 a month.

➲ **Cost of Living:** The presence of well-paid foreign aid workers means that prices are sky-high in Phnom Penh. You can live on £6-10 / $11-20 a day if you rule out drinking at a smart hotel with the UN gang. Personal cheques and credit cards are not widely accepted within Cambodia, although a number of banks in Phnom Penh accept Visa cards for cash advances. Use dollars and/or Thai baht.

➲ **Health advice:** The Cambodian medical system was virtually wiped out during the Khmer Rouge period, 1975-1979, and is being rebuilt slowly. Medical facilities and services in Cambodia are not up to western standards. Get shots for Hepatitis A, Malaria, Typhoid, and Polio. Make sure you have medical insurance that will guarantee and pay for evacuation in the case of an emergency.

➲ **English Language Media:** The Cambodia Daily, Phnom Penh Post.

Schools in Cambodia (+855)

ACE
46 Street 214, Sangkat Boeung Raing, Khan Daun Penh, Phnom Penh. **Email:** ace.idpcam@bigpond.com.kh

ACE Battembang
Rumcheak Village, Khum Ratanak, Svay Po, Battembang. **Tel:** 012-833602.

ACE Siem Reap
House 0675, Group 20, Wat Bo Village, Siem Reap, **Tel:** 012-833601.

ACE Kompong Cham
House 034, Group 1, Village 7, Sangkat Kompong Cham. **Tel:** 012-833600.

CHINA

There is a huge demand for native-speaking English language teachers (both American and British English speakers), especially in the enterprise zones near Hong Kong. The demand for teachers is led both by private schools in China and by companies eager to take advantage of new international business opportunities. There is a need for ESP teachers, such as English for business, for tourism, for medical purposes and for law, the latter because many foreign firms regard cultural and legal differences between Chinese and external cultures as a major cause of failed western investments. Another area that is booming is teaching English for young learners but these positions are usually advertised by word of mouth and require an extrovert personality and infinite patience. Unqualified EFL teachers can find work here but given the demands of teaching in China a TESL certificate is highly recommended. Also, holders of more advanced qualifications like an MA in TESOL will be paid at higher rates because they are classified as Foreign Experts rather than Foreign Teachers who are usually under 25 and have a BA. Job seekers can contact the nearest Chinese embassy or State Bureau of Foreign Experts, Friendship Hotel, 3 Bai Shi Qiao Rd., 100873 Beijing. Tel: +86 10-849-888, x 83500, fax: +86 10-831-5832 for more information and should also check out the TEFLChina Teahouse site: www.teflchina.com– an excellent source of information about teaching English in China. There is also a listing of teaching jobs in China at www.teach-in-china.com

Embassy in UK:
49 Portland Place, London W1N 4JL
Tel: 020 7631 1430
www.chinese-embassy.org.uk

Embassy in USA:
2300 Connecticut Ave NW, Washington DC 20008
Tel: 202 328 2500
www.china-embassy.org

British Council office:
4/f Landmark Building Tower 1, 8 North Dongsanhuan Road, Chaoyang District, 100004 Beijing
Tel +86 (10) 6590 6903
www.britishcouncil.cn

➲ **Visas/Work Permits:** If you negotiate your job from home, then your employer or the agency will take care of the necessary paperwork. It is possible to go to China on a tourist visa and then start job hunting. Once you have a job, the school will send you to the Public Security Bureau (PSB) so that your visa can be changed. To obtain a visa, negative HIV test results are required for teachers who plan to stay nine months or more. Some teachers go to Hong Kong and secure six-month multiple entry visas from there.

➲ **Requirements:** College Degree and a TESOL/TESL Certificate

➲ **Currency:** £1 = 15.29 / $1 = 8.27 Yuan Renminbi (RMB)

➲ **Salaries/Taxes:** Minimum salaries are RMB 2,500 per month for bachelors' degree holders and RMB 3,000 per month for masters' degree holders. Private tutors can charge around RMB 150-200 an hour in the big cities. There is no tax.

➲ **Accommodation:** Your housing should be provided by the employer. Conditions can be harsh by Western standards although foreign teachers usually receive the best accommodation that the institution can offer.

➲ **Cost of Living:** It is cheaper in the North than the South, excluding Beijing.

➲ **Health advice:** Although most schools provide health insurance, which covers 80% of the cost, you should have your own insurance. The quality of medical care in China varies. Competent, trained doctors and nurses are available in major metropolitan centres, but many do not speak English. Hospital accommodation is spartan, and medical technology is not up-to-date. Doctors and hospitals often expect immediate cash payment for health services and may not accept cheques or credit cards.

➲ **English Language Media:** Beijing Scene, China Daily, Shanghai Daily Star.

Schools in China (+86)

AES
102 Chang Jiang Liu, Dalian, 116001. **Tel:** (411) 281-7932. **Fax**: (411) 281-3318. **Email:** future@mail.dlptt.ln.cn

American International School of Guangzhou
Box 212, Ti Yu Dong Post Office, 510620 Guangzhou. **Tel:** (20) 3881.0001/2711/2712,2713; **Fax**: (20) 3881.1102; **Email**: asg@gitic.com.cn; **Website:** www.concept.se/user/aisg/ aisg.html

ASM Overseas Corp.
Asian Games Garden, Building No.2-6A. No.12 Xiaoying Lu, Chao Yang Dist., Beijing 100101. **Tel:** (10) 6497 4451. **Fax**: (10) 6497 4872.

Capital Language School
101 Dong Yuan Lu, Zhongshan 528400.**Tel:** (760) 8335067. **Fax**: (760) 8335067.

CDI Career Development Institute
110 Bloor Street West, Suite 202, Toronto, Ontario M5S 2W7, Canada. **Tel:** 1 416 964 6707, **Fax**: 1 416 964 0520.

China International Education League
Suite C 4th floor, Zao Fong World-trade Building, Jiangsu Road 369, Shanghai 200050. **Tel:** (21) 52400111 **Fax**: (21) 52400101.

China Teaching Program
Western Washington University, Old Main 530A, Bellingham, WA 98225-9047. **Tel:** (360) 650-3753, **Fax**: (360) 650-3753.

Chinese Education Association for International Exchange
37 Damucang Hutong, Beijing 100816. **Tel:** (10) 660-20-731, **Fax**: (10) 660-16-56.

Colorado China Council
4556 Apple Way, Boulder, CO 80301. **Tel:** (303) 443 1108, **Fax**: (303) 443-1107.

Concordia International School Shanghai
Plot S2, Huang Yang Rd., P.O. Box 206-005, Pudong, 201206 Shanghai; **Tel:** (21) 5899.0380; **Fax**: (21) 5899.1685; **Email:** sandras@ciss.com.cn; **Website:** www.ciss.com.cn

Fountain School
Ji Da Guo Mao, Zhuhai, Guangdong 519015. **Email:** wuvoka@mailcity.com

The International School of Beijing
c/o American Embassy Beijing, PSC461, Box 50, FPO AP 96521-0002; **Tel:** (10) 6437.7119; **Fax**: (10) 6437.6989; **Email:** isb-info@isb.bj.edu.cn; **Website:** www.isb.bj.edu.cn

New China Education Foundation
1587 Montalban Drive, San Jose, CA 95120. **Tel**: (408) 268 0418.

Qingdao International School
Middle Sector Song Ling Rd., Stone Old Man National Tourism Spot, Qingdao, 266101. **Tel:** (532) 889-1000; **Fax**: (532) 889-8950l; **Email:** gbapk@hotmail.com **Website:** www.qischina.com

Shanda International House
29-3 South Road Shanda, Jinan 250100, Shandong Province.**Tel:** +61 7 40 313466, **Fax**: +61 7 40 313 464.

Shanghai American School
50 Ji Di Lu, Zhudi, Minhang, 201107 Shanghai. **Tel**: (21) 6221.1445 **Fax**: (21) 6221.1269. **Email:** info@saschina.org; **Website:** www.saschina.org

Shanghai Changning International School
Jiangsu Lu 155, Shanghai 200050. **Tel:** (21) 6252-3688; **Fax**: 21-6212-2330; **Email:** scis@uninet.com.cn, **Website:** www.uninet.com.cn/userhome/scis/index.htm

Shekou International School
c/o Arco China, Inc., P.O. Box 260888, Plano, TX 75026-0888. **Tel:** (755) 669.3669; **Fax**: (755) 667.4099; **Email:** szejones@public.szptt.net.cn

TEDA International School
Tianjin; 6 Mu Nan Dao, 300050 Tianjin; **Tel:** (22) 2311.4772. **Fax**: (22) 2331.2263; **Email:** tist@public1.tpt.tj.cn

Tianjin International School
Xing Guo School, Xi Yuan Road, Ti Yuan Bei, He Xi District, 300061 Tianjin. **Tel:** (22) 283.53200. **Fax**: (22) 283.52455. **Email:** tmis@public1.tpt.tj.cn

Western Academy of Beijing
P.O. Box 8547, 7A Bei Si Huan Dong Lu, Chao Yang Qu, Beijing 100102. **Tel:** (10) 6437.5935. **Fax**: (10) 6437.5936. **Email:** wabinfo@ns.wab.senet.gov.cn

Wuhan Guanghua School
Jiangxia District, Wuhan, Hubei 430205. **Tel:** (27) 8794-4154, **Fax**: (27) 8794-5564. **Email:** doralei@netease.com

Wuhan International School
c/o Holiday Inn Tian An Wuhan 868 Jie Fang Da Dao Wuhan 430022. **Tel:** (27) 8576-8863; **Fax**: (27) 8584-5353. **Email**: wis@teacher.com

Xiamen International School
Jiu Tian Hu, Xinglin District, 361022 Xiamen; **Tel**: (592) 625.6581 **Fax**: (592) 625.6584. **Email:** JDFisch47@yahoo.com

Yang-En University
Quanzhou, Fujian 362014. **Tel:** (595) 208 7600, Ext. 202. **Fax**: (595) 208 2017. **Email:** yeu@yeu.qz.fj.cn

Yew Chung Shanghai International School (SIS)
11 Shui Cheng Road, 200336 Shanghai; **Tel:** (21) 6242.3243; **Fax**: (21) 6242.7331; **Email:** ycsis@public.sta.net.cn

Yunnan Institute of the Nationalities
Foreign Affairs Office, Kunming 650031, Yunnan. **Tel:** (871) 515 4308.

HONG KONG

The former British colony is now a Special Administrative Region within China. There are plenty of well-paid jobs here for EFL teachers although it is more difficult to obtain a working visa than it was in the past. The NET (Native-speaker English Teacher) program, operated by the Hong Kong Department of Education, allows qualified native speaker teachers to teach English in the public schools for 12 months. The scheme pays teachers anywhere from £1300-3500 / $2400-6500 per month plus a 15% gratuity payable on the completion of the contract. For more information about NET on the web, visit www.info.gov.hk/ed/english/teacher/native_eng_teacher/home.htm but be sure to confirm pay and other benefits before you decide to go.

Embassy information:

see China

British Council office:

3 Supreme Court Road, Admiralty, Hong Kong
Tel +852 291 35100
www.britishcouncil.org.hk

➲ **Visas/Work Permits:** U.S. citizens can obtain tourist visas for stays of up to 90 days. To get a work permit, you need to be sponsored by an employer. One way to do this is to arrive in Hong Kong, land a job, and then leave the country while your application is being processed. If you work illegally, be aware that teachers have been deported in the past for overstaying their visas.

➲ **Requirements:** College Degree and a TESOL/TESL Certificate

➲ **Currency:** £1 = 15.18 / $1 = 7.77 HK Dollars

➲ **Salaries/Taxes:** Teachers earn anywhere from HK$12,000 and up per month in the public schools to HK$50,000 at the top end in the private sector. Freelancers make from HK$300-750 or more per hour for private tuition.

➲ **Accommodation:** Expensive if you are staying in Hong Kong itself, less so if you live on one of the other islands like Lamma, where a shared apartment costs around HK$2,500-3,000 a month.

➲ **Cost of Living:** Expensive, until you learn where the best deals are.

➲ **Health advice:** Medical facilities are excellent. Your employer should provide medical coverage for you.

➲ **English Language Media:** Hong Kong Magazine, Hong Kong Standard, South China Morning Post.

Schools in Hong Kong (+852)

Berlitz

1 Pacific Place, Central. **Tel:** 2826 9223, **Fax**: 2 525 9757.

English Language Club

Ka Nin Wah Commercial Building, Hennessy Road. **Tel:** 2 838 0036.

First Class Language Centre

22a Bank Tower, 351-353 King's Road, North Point. **Tel**: 5 887 7555.

Hong Kong English Club

Ground floor, 41 Carnarvon Road, Tsimshatsui, Kowloon. **Tel**: 2 366 6961.

HK Institute of Languages

Rm 501 30-32 D'Aguliar St, Central. **Tel:** 2 877 6160.

Island International High School

20 Borrett Road, Hong Kong.

Pasona Education

2/F One Hysan Ave, Causeway Bay. **Tel:** 2 577 8002.

The British Council

English Language Centre, 3 Supreme Court Road, Admiralty. **Tel:** 2 913 5100, **Fax**: 2 913 5102.

The Centre for Professional and Business English

7th Floor, Core B, Hong Kong Polytechnic University, T.S.T. **Tel:** 2 766 5111.

Venture Living Languages Ltd

1A 163 Hennessy Road, Hong Kong. **Tel:** 2 507 4985, **Fax**: 2 858 1764.

INDONESIA

Indonesia is undergoing economic and political turmoil at present and the country is not as stable as it used to be. EFL teachers report that although life goes on as usual, especially in the big cities like Jakarta, Surabaya and Medan, demand for English lessons is slowing due to financial instability. If you decide to go to Indonesia, try getting a job with one of the larger and more reputable schools. Do not go unless you are offered a work visa, otherwise you will be working illegally and the authorities deal very harshly with offenders. For US teachers, anti-American sentiment is running high in some areas and caution should be exercised if you are travelling in Indonesia.

British Council office:

S. Widjojo Centre, Jalan Jenderal Sudirman Kav 71, Jakarta 12190
Tel +62 (21) 252 4115
www.britishcouncil.or.id

➲ **Visas/Work Permits:** Tourist visas are granted for up to two months. Some teachers do "visa runs" to Singapore or Darwin, Australia, when their visas run out but these trips are expensive, especially now that the rupiah is worth so little. Your school should arrange a work permit for you. Vacation time is usually given at Christmas and one week at the end of Ramadan.

➲ **Requirements:** High School Diploma and a TESOL/TESL Certificate

➲ **Currency:** £1 = 17706 / $1 = 9063 Rupiah.

➲ **Salaries/Taxes:** Teachers are paid in rupiah. Your contract should guarantee that you are paid between 6-8 million rupiah a month. If the amount is less than that, you will find it difficult to live in Indonesia. There is a mandatory tax of £90 / $170 each month imposed on all foreigners.

➲ **Accommodation:** You should receive free or subsidized housing as part of your employment package. Rents are generally cheaper outside of Jakarta.

➲ **Cost of Living:** Food, clothing and other basics are reasonable.

➲ **Health advice:** Make sure your school offers you insurance coverage that offers protection against sickness and that will evacuate you in the event of full scale rioting.

➲ **English Language Media:** Indonesian Times, Indonesian Observer, Jakarta Post.

Schools in Indonesia (+62)

American English Language Training (AELT)
Jalan R.S. Fatmawati 42a, Keb Baru, Jakarta Selatan 12430. **Tel:** 21 769 1001, **Fax**: 21 751 3304.

American Language centre
Jl. Panglima Polim Raya No. 101, Kebayoran Baru, Jakarta 12160. **Tel:** (21) 722-8957, **Fax**: (21) 722-8960.

Centre for Language Training (CLT)
Soegijapranata Catholic University, Jl. Menteri Supeno 35, Semarang 50241. **Tel:** 24 316028, **Fax**: 24 415429.

English Education Centre (EEC)
Jalan Let. Jend S.Parman 68, Slipi, Jakarta 11410. **Tel:** 21 532 3176, **Fax**: 21 532 3178.

Executive English Programmes (EEP)
Jalan Wijaya VIII 4, Kebayoran Baru, Jakarta Selatan 12160. **Tel:** 21 720 1896.

ELS International
Jalan Tanjung Karang 7 c-d, Jakarta Pusat. **Tel:** 21 323 211.

English First Bogor
Jl. Raya Pajajaran No.8, Bogor 16151.**Tel:** 251 324374, **Fax**: 251 379296.

EF English First Kelapa Gading
Jl.Raya Bulevar Barat Blok LA 1 no 8 - 9 Kelapa Gading Permai Jakarta 14240. **Tel:** 21 451 5858. **Fax**: : 21 451 5959.

EF English First Lampung
Jl. Diponegoro No. 146, Bandar Lampung 35127. **Tel**: 721 263154, **Fax**: 721 263154.

EF English First (Ujung Pandang)
Jl. Slt. Hasanuddin 27, Ujung Pandang 90113. **Tel:** 411 315288, **Fax**: 411 315251.

English Language Training International (ELTI)
Complex Wijaya Grand Centre, Blok F 84A & B, Jalan Wijaya II, Jakarta Selatan 12160. **Tel:** 21 720 6653, **Fax**: 21 720 6654.

Indonesia-Australia Language Foundation (IALF)
Wisma Budi, Suite 503, Jalan HR Rasuna Said Kav c-6, Kuningan, Jakarta Selatan 12940. **Tel:** 521 3350.

International Language Programmes (ILP)
Jalan Raya Pasar Minggu No. 39A, Jakarta 12780. **Tel**: 21 798 5210, **Fax**: 21 798 5212.

International Language Studies (ILS)
Jalan Ngemplak 30, Arbengan Plaza B-34, Surabaya 60272. **Tel:** 31 531 9644, **Fax**: 31 531 9645.

I-Pro International Programmes
Jl Ngagel 179 -183, K1, Surabaya 60245. **Tel:** 31 566 2914/5, **Fax**: 31 566 7979.

Logo Education Centre (LEC)
Jelan H.Z. Arifin 208A, Medan, Sumatra Utara. **Tel**: 61 534991, **Fax**: 61 552385.

Maple Leaf English Language Centre
Jl. Raya Darmo Permai II / 32, Surubaya. **Tel:** 31 710722. **Fax**: (031) 725476

Oxford Course Indonesia (OCI)
Jalan Cempaka Putih Tengah 33C-2, Jakarta Pusat 10510. **Tel:** 21 424 3224, **Fax**: 21 425 4041.

Professional Training Services (PTS)
Wisma BII, Jalan Permuda No. 60-70, Surabaya 60271. **Tel/Fax:** 31 531 0006.

School for International Communication
Jalan Taman Pahlawan 194, Purwakarta. **Tel:** 264 201204.

School for International Training (SIT)
Jalan Sunda 3, Menteng, Jakarta Pusat 10350. **Tel:** 21 390 6920, **Fax**: 21 335671.

Strive International
Setiabudi 1 Building, Jalan HR Rasuna Said, Jakarta 12920. **Tel:** 21 521 0690, **Fax**: 21 521 0692.

The British Institute (TBI)

Plaza Setiabudi 2, Jalan HR Rasuna Said, Jakarta 12920. **Tel:** 21 525 6750, **Fax**: 21 520 7574.

Triad English Centre

Jalan Purnawarman 76 Bandung- 40116. **Tel:** 62 22 431309, **Fax**: 22 431149.

JAPAN

Despite the recent economic crisis, Japan is still a popular destination for English teachers, qualified or not, because of the number of jobs available and the favourable salaries. Japan's commitment to learning English was recently demonstrated in very clear terms by Prime Minister Keizo Obuchi's recent statement that, "Achieving world-class excellence demands that all Japanese acquire a working knowledge of English." But it must be noted that the standard pay of 250,000 yen per month for an EFL teacher has hardly changed in the last few years despite rising inflation so teaching contracts in Japan are not as lucrative as they once were.

After-hours schools where students cram for exams, including English language, are known as juku. An eikaiwa is an English conversation school. Unqualified teachers can find work in private schools giving conversation classes or through Japan's Exchange and Teaching (JET) programme (see below). There are also chains of schools, such as AEON, GEOS and NOVA employing thousands of trained and untrained teachers. The best time to come to Japan and look for a job is in March just before the start of the school year on April 1. The big chains of schools recruit teachers year-round. If you negotiate your employment with a recruiter outside of Japan remember that the agreement you make is not binding. It is the contract with the school in Japan that governs the terms of your employment. Also, be aware that you may run into problems if you decide to go off and work for another school once you are in Japan. Your first employer may not be happy that you are leaving after so much of the school's money and time has been invested in you. If you work outside of Tokyo, you may need an International Drivers' Licence if the school provides you with a car. It can take up to six months to get acclimatized to Japan. Prepare yourself before you go: look at the ELT News website for regularly updated information about teaching in Japan (www.eltnews.com).

The JET Program

The Japan Exchange and Teaching programme is a Japanese government scheme aimed at improving English language teaching and international relations in schools. Although 34 countries are involved in the program, the majority of the almost 6,000 participants are from the US. Anyone under 35, with a degree and an interest in Japan, can apply. The salary and conditions are excellent – you get about £20,000 / $38,000 for the year, free return airfare, 3-day orientation and you work in conjunction with a Japanese teacher – one participant described it as the best job in the world! Applications must be in by early December for the following summer.
Contact: www.jetprogramme.org

Embassy in UK:

101-104 Piccadilly, London W1V 9FN
Tel: 020 7465 6500
www.embjapan.org.uk

Embassy in USA:

2520 Massachusetts Ave NW, Washington DC20008
Tel: 202 238 6700

British Council office:

1-2 Kagurazaka, Shinjuku-ku, Tokyo 162-0825
Tel +81 (3) 3235 8031
www.britishcouncil.org/japan

➲ **Visas/Work Permits:** The opportunity to obtain a working visa to teach in Japan is generally available to native English speakers with a valid college diploma in any field. The sponsoring employer in Japan must first obtain a "Certificate of Eligibility" from the Japanese Ministry of Justice, Immigration Bureau, for the prospective teacher. With this certificate, you can apply for a work visa at the Japanese Embassy (or at a Japanese Consulate) prior to entering Japan. If you try to enter into Japan as a tourist under the 90-day visa waiver programme to work, you may be denied entry, but you can enter to look for work. If you enter Japan as a tourist and subsequently obtain an offer to teach, you must leave the country, taking the "Certificate of Eligibility" to a Japanese Embassy/Consulate in the U.S. or a neighbouring country (usually Korea), to obtain a work visa, and then re-enter Japan on the correct visa. If you are from Canada, Australia or New Zealand, you can teach in Japan on a working holiday visa, which can be obtained in your own country. You must be between the ages of 18-30 to go on the working holiday programme. The visa is for six months, renewable only once. After that you have to find a sponsor or a legal visa to continue working in Japan.

➲ **Requirements:** College Degree and a TESOL/TESL Certificate

➲ **Currency:** £1 = 230 / $1 = 118 Yen.

➲ **Salaries/Taxes:** Salaries begin at around 250,000 yen per month with airfares and end-of-contract bonuses offered to more qualified candidates. If you are teaching privately, expect to earn 3,000-6,000 yen per lesson. Salaries are usually based on an hourly wage with a guarantee of a certain minimum number of hours per month. Many teachers hold side jobs as private tutors outside of their classroom workplaces. Income tax in Japan is approximately 10% of your gross income. Watch out for deductions by employers for accommodation and other expenses, which, if you are not careful, can leave you with about half your salary left to live on. It should be possible for you to save a fair amount of money after living in Japan for a few months.

➲ **Accommodation:** Many schools provide free or subsidized housing as part of the employment package. If a school is taking a large percentage of your pay packet (more than 25%) then you should question the contract. If you try to rent on your own you should be aware that housing is much smaller, more expensive and complicated to rent than in the west. Rents range from 60,000-100,000 Yen per month in Tokyo, but in smaller towns you can pay a quarter of that. Landlords can be reluctant to rent to foreigners, so it's worth getting somebody to help. There are agents who can do this, otherwise you might have to pay as much as £3000 / $5,700 in advance as a security deposit ('key money'). Schools that recruit overseas often include accommodation assistance as a benefit.

➲ **Cost of Living:** The cost of living, especially in cities, is very high. Don't be surprised if a simple night out with some friends costs you $100 or more! Supporting a family in Japan, according to middle class American standards, is well beyond the reach of most English teachers. However, salaries are high. It is possible to save money, especially outside Tokyo.

➲ **Health advice:** Most teachers are covered by the Japanese national health insurance plan, which does not pay for treatment outside of Japan. While recently on the rise, health care costs remain relatively low in Japan. The self-employed, employees of small businesses, and others not covered by employer plans can apply to the government for low-cost National Health Insurance, which provides coverage similar to workplace-based insurance. Teachers should obtain private, supplemental health insurance that includes worldwide medical evacuation and care. Vaccinations recommended are Polio and Typhoid.

➲ **English Language Media:** Japan Times, Kansai Time Out, The Asahi Evening News, The Daily Mainichi.

Schools in Japan (+82)

A Ring English Conversation
3-9-12-304 Nishi Shinagawa, Shinagawa-ku, Tokyo, **Tel:** 0120-750510.

Academy of Language Arts
Tobundo Bldg. No.2 5F, 2-16 Agebacho, Shinjuku-ku, Tokyo. **Tel:** 03-3235-0071. **Fax**: 03-3235-0004. **Website:** www.kbic.ardour.co.jp /~newgenji/ala/

ACE English School
6-6-22 Hachimancho, Takahmashi Aichiken, 444-1302. **Tel:** 566 52 3235. **Fax**: 566 52 3235.

AEON Inter-Cultural
Shinjuku I-Land Tower Bldg. 12F, 6-5-1 Nishishinjuku, Shinjuku-ku, Tokyo 163-1377. **Tel:** 03-5381-1500. **Fax**: 03-5381-1501. **Email:** aejedu@corp.aeonet.co.jp)

Ai Language Studio
Hiroo Stone Bldg. 5F, 1-3-17 Hiroo, Shibuya-ku, Tokyo. **Tel:** 03-5421-3572. **Fax**: 03-5421-3573. **Website:** http://plaza24.mbn.or.jp/~als/ **Email:**tokim@bd.mbn.or.jp

AIT Foreign Language Centre
3-20-1-401, Tamagawa, Setagaya-ku, Tokyo. **Tel**: 03-3700-8110. **Fax**: 03-3700-8889. **Website:** www2.tky.3web.ne.jp/~ai **Email:** aitcorp@tky2.3web.ne.jp

Alpha Language School
3-4-11 Mita, Minato-ku, Tokyo. **Tel:** 03-3769-4945. **Fax**: 03-3769-4946. **Email:**ala@a2.mbn.or.jp

American Academy
4-1-3 Kudan Kita Chiyoda-Ku, Tokyo 102.

America Eiko Gakuin
Misono-cho 5-2-21, Wakayama City 640.

American School of Business
1-17-4 Higashi Ikebukuro, Toshima-Ku, Tokyo 170.

Azabu Academy
401 Shuwa-Roppongi Building, 3-14-12 Roppongi, Minato-ku, Tokyo 106.

Berlitz Schools of Languages (Japan) Inc.
Kowa Bldg. 1,5f, 11-41, Alasaka 1-chome, Minato-ku, Tokyo 107.

Bernard Group
2-8-11 Takezono, Tsukuba City, lbaraki-Ken, 305.

CA English Academy
2nd Floor Kotohira Building, 9-14 Kakuozan-dori, Chikusa-ku, Nagoya.

Cambridge English School
Dogenzaka 225 Building, 2-23-14 Dogenzaka, Shibuya-ku,Tokyo 150.

Cambridge School of English
Kikumura 91 Building1-41-20 Higashi, Ikebukuro Toshima-ku, Tokyo 170.

Canadian School of Languages
3-9-7-202 Kameari, Katsushika-ku, Tokyo. **Tel:** 03-3601-7648.

Canning
The, 2-9 Kanda Surugadai, Chiyoda-ku, Tokyo. **Tel:** 03-3295-5500

CEC Foreign Language centre
Takahashi Bldg. 3F, 1-44-3 Fuda, Chofu City, Tokyo. **Tel:** 0424-87-3623. **Fax**: 0424-87-3664.

Celc English School
1-3-5 Mejirodai, Hachioji City, Tokyo. **Tel:** 0426-66-5041 **Fax**: 0426-66-5042. **Website:** http://member.nifty.ne.jp/celc/index.html

CIC English Schools
Kawamoto Building, Imadegawaagaru Nishigawa Karasuma-dori, Kamigyo-ku, Kyoto.

Cosmopolitan Language Institute
1-8-9 Yaesu, Chou-ku, Tokyo. **Tel:** 03-3273-7878. **Fax**: 03-3273-7870.

Cosmopolitan Language Institute
Yashima B Building 4f, 1-8-9 Yesu Chuo-ku,Tokyo 104.

Daigakushorin International Language Academy
AM Bldg., 9 Rokubancho, Chiyoda-ku, Tokyo. **Tel:** 03-3264-2131. **Fax**: 03-3264-2133.

David English House
2-3f Nakano Building, 1-5-17 Kamiyacho, Naka-ku, Hiroshima 730.

Den House International Language School
Meguro Ekimae School, Tamagawa Bldg. 3F, 2-16-5 Kami Osaki, Shinagawa-ku, Tokyo. **Tel:** 03-3280-4255.

Den House International Language School
Tsukahara Bldg. 2F, 7-19-9 Okusawa, Setagaya-ku, Tokyo. **Tel:** 03-3704-7636. **Fax**: 03-3704-7696.

Dual International School
ILYA Bldg. 3F, 1-1 Yotsuya-dori, Chikusa-ku, Nagoya, 464-0819. **Tel:** 052-783-3031. **Fax**: 052-783-3035. **Email:** dual6@spice.or.jp

Educational Information Institute
Level 7, 2-15-22 Shinjuku, Shinjuku-ku, Tokyo. **Tel:** 03-5366-7201. **Fax**: 03-5366-7203. **Email:** eiijabli@mx2.ttcn.ne.jp

EEC Foreign Languages Institute
Shikata Building 2f, 4-43 Nakazald-Nishi 2-chrome, Kita-ku, Osaka 530.

Ell Educational Information Institute
S2 Bldg. 7F, 2-15-22 Shinjuku, Shinjuku-ku, Tokyo. **Tel:** 03-3352-5280. **Fax**: 03-3352-5871

ELEC Eigo Kenkyujo (The English Language Education Council)
3-8 Kanda Jimbo-cho, Chiyoda-ku,Tokyo 101.

English Plus
Bell Maison Minami Azabu 1F, 1-3-17 Minami Azabu, Minato-ku, Tokyo. **Tel:** 03-5238-0111. **Fax**: 03-5238-0300.

English Studio
Meguro School, 5-12-2 Shimo Meguro, Meguro-ku, Tokyo. **Tel:** 03-3712-3313. **Fax**: 03-3710-3386. **Email:**e-studio@tkc.att.ne.jp

English Studio
Mita School, 4-2-14 Mita, Minato-ku, Tokyo. **Tel**: 03-3798-5829. **Fax**: 03-3769-9397. **Email:** e-studio@tkc.att.ne.jp

English Studio
Setagaya School, 2-31-11 Oyamadai, Setagaya-ku, Tokyo. **Tel:** 03-5758-2491. **Fax**: 03-5758-2492. **Email:** e-studio@tkc.att.ne.jp

English Telephone Club
1-19-28 Minami Oizumi, Nerima-ku, Tokyo. **Tel:** 03-3924-3853. **Fax**: 03-3978-0990. **Website:** www.aoki.com/job/etc

Eurocentres Japan
2-12 Kanda Surugadai, Chiyoda-ku, Tokyo. **Tel:** 03-3295-5441. **Fax**: 03-3295-5443. **Email:** euro-tyo@orange.ifnet.or.jp

Evergreen Language School
1-21-18 Yutenji, Meguro-ku, Tokyo. **Tel:** 03-3713-4958. **Fax**: 03-3719-4383. **Website:** www.evergreen.gr.jp

Executive Gogaku Centre (Executive Language Centre)
1 Kasumigaseki Building, 12F, 3-2-5 Kasumigaseki, Chiyoda-ku,Tokyo 100.

English Circles/EC Inc.
President Building 3rd Floor, South-1, Chuo-Ku, Sapporo 060. **Tel:** 011 221 0279.

FCC (Fukuoka Communication Centre)
Dai Roku Okabe Building, 5f, Hakata Eki Higashi, 2-4-17 Hakata-ku, Fukuoka 812.

F L Centre (Foreign Language Centre)
1 Iwasaki Building, 3f, 2-19-20 Shibuya-ku,Tokyo 150.

Four Seasons Language School
Aoyamashorin 2F, 4-32-11 Sanarudai, Hamamatsu-shi 432. **Tel:** 534 48 1501.

Fukuoka International School
3-18-50 Momochi, Sawara-ku, 814-0006 Fukuoka. **Tel:** 92- 841.7601, **Fax**: 92-841-7602, **Email:** adminfis@fka.att.ne.jp

Galaxy Language Academy
Noda Bldg., 2-12-1 Kami Osaki, Shinagawa-ku, Tokyo. **Tel:** 03-3473-3251. **Fax**: 03-3473-3334. **Website:** www.jade.dti.ne.jp/~galaxy **Email:** galaxy@jade.dti.ne.jp

Gateway Gakuin Rokko
Atelier House, 3-1-15 Yamada-cho, Nada-ku, Kobe.

GEOS Language System
Shin Osaki Bldg. 4F, 1-6-4 Osaki, Shinagawa-ku, Tokyo. **Tel:** 03-5434-0200.

Global English School
1887-86 Kubo-cho, Matsusaka, Mie 515-0044. **Tel**: 598-60-0873. **Fax**: 598-29-8730. **Email:** recruit@ges-hiro.co.jp

Gregg Int'l School of Language
1-14-16 Jiyugaoka, Meguro-ku, Tokyo. **Tel:** 03-3718-0552.

Gunma Language centre **& Language Academy**
840, Munadaka, Gunma 370-3521. **Tel:** 273-732310. **Fax**: 273-730846. **Email:** glc@mail.wind.co.jp

Hearts English School
2146-3 Shido, Shido-cho, Okawa-gun, Kagawa 769-21. **Tel:** 878 943557, **Fax**: 947227.

Hiroshima International School
3-49-1, Kurakake, Asakita-ku, 739-1743 Hiroshima. **Tel:** 82-843-4111, **Fax**: 82-843-6399, **Email:** hisadmin@orange.ocn.ne.jp

ICA Kokusai Kaiwa Gakuin (International Conversation Academy)
l Mikasa 2 Building, 1-16-10 Nishi.

IF Foreign Language Institute
7f Shin Nakashima Building, 1-9-20 Nishi Nakashima, Yodogawa-ku, Osaka.

inlingua School of Languages
Kyodo Bldg. 9F, 2-4-2 Nihombashi Muromachi, Chuo-ku, Tokyo. **Tel:** 03-3231-0167. **Fax**: 03-3245-0086. **Email:** inlingua@fa.mbn.or.jp

INTECO Business English School
4-2-25 Akasaka, Minato-ku, Tokyo. **Tel:** 03-5570-2821 **Fax**: 03-5570-4323 **Website:** www2.gol.com/users/inteco/ **Email:** inteco@gol.com

Jelly Beans English School
16-908 Shinyashiki cho, Kagoshima City 892-0838. **Tel:** 90-2511-9733. **Fax**: 81-99-224-7595.

Kains English School in Gakko
1-5-2 Ohtemon Chuo-ku, Fukuoka 810.

KICC English and Computer School
1-8-13 Horai, Shingu, Wakayama 647. **Tel:** 735-28-2230. **Fax**: 735-28-2234. **Email:** kiccjob@kiccworld.com

King's Road School of English
Akasaka 4-3-9, Takaike Bldg 3F, Minato-ku, Tokyo 107. **Tel:** 03 3589 8666, **Fax**: 03 3589 7868.

Kyoto English Centre
Sumitomo Seimei Building, Shijo-Karasuma Nishi-iru Shimogyo-ku, Kyoto.

Kobe Language Centre
3-18 Wakinoharnacho, 1-chome, Chuo-ku, Kobe 651. **Tel:** 78 2614316.

Language Education Centre
7-32 chome Ohtemachi Nakaku, Hiroshima-shi 730.

London Study Centre
1-1 Tamanuki, Ohyama, Ohtama, Fukushima 969-1301. **Tel:** 0243 48 4286. **Fax**: 0243 48 4286. **Email:** londoneigakuin@hotmail.com

Luna International
1-2-9 Asahi, Matsumoto-shi 390-0802. **Tel:** 263-34-4481. **Fax:**263-34-4481.

Matty's School of English
3-15-9 Shonan-takatori, Yokosuka 234. **Tel:** 468 658717.

Michigan English School
Enya Bld. 2F, 7-34 Sanae-cho, Gifu-city, Gifu, 500-8304. **Tel:** 058 251 9421. **Fax**: 058 251 9850. **Email: michigan@ktroad.ne.jp**

MIL Language Centre
3.4F Eguchi Bldg, 1-6-2 Katsutadai, Yachiyo-shi, Chiba-ken 276. **Tel:** 474 857555.

Miyazaki College
2520-88, Hongo-kitakata, Miyazaki 880-0925. **Tel:** 985-52-7568. **Fax**: 985-52-7568.

Mobara English Institute
618-1 Takashi, Mobara-shi, Chiba-ken 297. **Tel:** 475 224785.

Mukogawa Women's University
Edagawa-cho 4-16, Nishinomiya-city, Hyogo-ken 545-0002. **Tel:** 0798-47-6436. **Fax**: 06-6719-2093.

NOVA Group
Harajuku Carillon Bldg. 4F, 1-8-9 Jingumae, Shibuya-ku, Tokyo 150-0001. **Tel:** 03-3478-3475. **Fax**: 03-3478-3665.)

Op-Net English Institute
Torinose 222-1, Iwata-shi, Shizuoka-ken 438-0072. **Tel:** 0538 34 8686. **Fax**: 0538 36 1217. **Email:** opnet1@peace.ne.jp

Ph.D. Language centre
743 Aramaki-machi, Maebashi-shi, Gunma-ken 371-0044. **Tel:** 027 233 8442. **Fax**: 027 233 8442.

Plus Alpha
2-25-20 Denenchofu, Ota-Ku, Tokyo 145.

Queens School of English
3f Yuzuki Bldg, 4-7-14 Minamiyawata, Ichikawa 272.

Pegasus Language Services
Sankei Building, 1-7-2 Otemachi Chiyoda-ku, Tokyo 100.

REC School of Foreign Language
Nijojo-mae Ebisugawasagaru Higashihorikawa-dori, Nakagyo-ku, Kyoto.

Royal English Language Centre
4-31-3-2 Chyo Hakata-ku, Fukuoka 82.

Seido Language Institute
12-6 Funado-cho, Ashiya-shi, Kyoto.

Stanton School of English
12 Gobancho Chiyoda-ku, Tokyo 102.

Sumikin-Intercom, Inc
Andrew Vaughan, 7-28 Kitahama 4-Chome, Chou-ku, Osaka 541 **Tel:** 06 220 5500, **Fax**: 06 220 5713.

Sun Eikaiwa School
6f Cherisu Hachoubori Building, 6-7 Hachoubori Naka-ku, Hiroshima-shi 730.

Toefl Academy
1-12-4 Kundankita, Chiyoda-ku, Tokyo 102. **Tel:** (3) 2303500.

Tokyo English Centre
(TEC) 7-9 Uguisudai-cho, Shibuyaku, Tokyo 150.

Tokyo Language Centre
Tatsunama Building, 1-2-19 Yaesu, Chuo-Ku, Tokyo 103.

Tokyo YMCA College of English
7 Kanda Mitoshiro-Cho, Chiyoda-ku, Tokyo T-101.

Trident School of Languages
1-5-31 Imaike, Chikusa-ku, Nagoya 464.

Westgate HR Department
1173-1 Hagiwara, Gotemba Shizuoka 412-0042. **Fax:** 550-84-0770

World Language School Inc
Tokiwa Soga Ginko Building 4f 1-22-8 Jinnan, Shibuya-ku, Tokyo 171.

KOREA (SOUTH)

A favourite destination for teachers of English over the last 15 years, South Korea has witnessed rapid economic growth and foreign investment, and even if the economic situation is not as favourable as in the past, there are still many opportunities. Korean is the major language, but some Mandarin Chinese is spoken. English is not widely spoken but is much sought after. Crime is not a major problem but be aware that nationalist sentiments are on the increase, particularly in view of recent economic difficulties and sales to foreign companies. As the number of teachers applying to go to Korea has decreased you should be able to find a job easily and, with the right qualifications, you can secure some very favourable terms and conditions. Many teachers are able to build up significant savings while working in Korea.

In the past, many travellers took casual teaching jobs on tourist visas. However, the government has cracked down on illegal workers. Anyone found working without a work permit will be deported and any companies found employing illegal workers risk being closed down.

One popular type of school is known as the hagwon. These are privately-run schools where Koreans learn English, usually on a daily basis before they go to work in the morning or in the evenings. If you teach in one of these schools, you'll be expected to teach 20-30 hours a week and you'll have free time during the day. Finding a job in a hagwon should not be a problem. Try and get a reference from a teacher already working at the hagwon you are thinking of working for. Also, if you can have a Korean-speaker help you go through your contract, it may help you in the event of a dispute—the Korean language contract takes precedence over the English language version and the two do not always match up. Another place to find a teaching job in Korea is in a chaebol. These are the large corporations that dominate business in Korea (like Hyundai). Many run their own in-house English language programmes. Expect to work at least 30 hours a week from early morning until late at night. If you have an MA in TESOL you could look for a job in a university language institute. The institutes offer better pay and conditions than hagwons or the chaebol – so the staff turnover is low. Do not confuse working in a university foreign language institute with teaching in a university department. Departmental teaching involves conversational English classes with large numbers of university students for about 10-15 hours a week. Private lessons offer teachers the chance to supplement their incomes but you should know that working for another school other than your official employer or teaching privately is illegal. Some teachers have been imprisoned and/or deported for giving private lessons. If you do teach private students insist on payment in advance for the lessons.

In 1995, the Korean government started the EPIK (English programme in Korea) program, which sends English language instructors to work in Korean middle and high schools. EPIK participants are paid according to their qualifications anywhere from 1.6-2 million won a month. Airfares and accommodation are included in the package. In 2000, 392 teachers took part in the programme. Reactions from participants have been mixed — some teachers were unhappy at being placed in rural schools far from the big cities and contact with other foreigners. But most old Korea hands agree that it is a good way to be introduced to the country's culture and people. Contact your closest Korean consulate or embassy for more information. The website is http://cc.knue.ac.kr/~epik/off.htm

Embassy in UK:

60 Buckingham Gate, London SW1E 6AJ
Tel: 020 7227 5505

Embassy in USA:

2450 Massachusetts Avenue NW, Washington DC20008
Tel: 202 939 5600

British Council office:

Joongwhoo Building, 61-21 Taepyungro 1 Ka, Choong-ku, Seoul, Korea100-101
Tel +82 (2) 3702 0600
www.britishcouncil.or.kr

➲ **Visas/Work Permits:** In order to work legally in Korea, one must first obtain the appropriate employment visa. The Korean government tightly controls employment visas and sometimes teachers have been unable to obtain visas. A person who wishes to work in Korea must obtain the visa outside Korea. One can, however, come to Korea on a tourist visa, obtain sponsorship documents, and apply for the visa in a nearby country. Depending on the job and other factors, it can take between one week and two months to obtain the appropriate visa. A teacher arriving in Korea with a teaching visa must register with Korean Immigration and obtain a residence certificate and re-entry permit within 90 days of entry. Most English instructors are granted an E-2 visa (conversation instructor), an E-1 visa (professor at educational institution higher than a junior college), or an E-5 visa (professional employment with a public relations firm or corporation). Once you have your permit, do not allow your employer to keep your passport even if he or she says that you are obliged to do so – you are not!

➲ **Requirements:** College Degree and a TESOL/TESL Certificate

➲ **Currency:** £1 = 1806 / $1 = 925 Won.

➲ **Salaries/Taxes:** The average salary at a hagwon is equivalent to about £1,000 / $1,900 per month. You can earn £1,200 / $2,200 or more at a chaebol. University departments pay c. £1,000 / $1,900 per month while the foreign language institutes pay the highest salaries. It is possible to earn £30 / $57 per hour with private tuition. Taxation rates depend on how much you are actually making – they range from 10-40% of your income. There is also a residency tax of 7.5%. If your employer tries to deduct a "security deposit" from your wages for any reason, refuse (politely but firmly).

➲ **Accommodation:** Your employer should meet your accommodation needs otherwise renting an apartment is an expensive and complicated process. You will probably find yourself sharing a small apartment with some fellow teachers. Apartments vary in quality.

➲ **Cost of Living:** Korea is not a cheap place to live in — especially if you are fond of sampling the nightlife. This the reason why so many teachers risk deportation lured by the extra cash of private lessons.

➲ **Health advice:** Your employer should cover your insurance costs.

➲ **English Language Media:** Korea Herald, Korea Times.

Schools in South Korea (+81)

All Nations Community Church

Yuksamdong Youngdong APT 28-105, Yuksamdong 765, Seoul 135-082. **Tel:** (02) 555-0691.

ANC Language Institute
2nd Fl., Taeyoung Bldg., Daechi-dong, Kangnam-gu, Seoul. **Tel:** (02) 508-0081.

ANS Recruiters
Seocho-gu, Seocho-dong 1575-2, Soryong Bldg. 5th Fl. Seoul, Korea 137-070. **Tel:** (02) 3472-0020. **Fax:** (02) 3472-2173.

ARC Language Education
636-1 Shinsa-dong, Kangnam-gu, Seoul. **Tel:** (02) 511-9311.

BCM Institute (Konkuk University)
3nd Fl. Daeyang Bldg., 8-3 Noyu 1-dong, Kwangjin-gu, Seoul. **Tel:** (02) 499-0515. **Fax**: (02) 464-0516.

BCM Institute (Kuro)
1123-3 Kuro 3-dong, Kuro-gu, Seoul. **Tel:** (02) 830-2121. **Fax**: (02) 830-2124.

BCM Institute (Pundang)
4th Fl. Kwangrim Praza, Sohyon-dong, Pundang-gu, Kyonggi-do. **Tel:** (031) 708-9090. **Fax**: (031) 708-3161.

BCM Institute (Sanggye)
6th Fl. Kyongwon Bldg., Sanggye-dong, Nowon-gu, Seoul. **Tel:** (02) 933-8656. **Fax:** (02) 933-4525.

BCM Institute (Songpa)
11th Fl. Taemun Bldg., 174-2 Sokchon-dong, Songpa-gu, Seoul. **Tel:** (02) 412-2335. **Fax**: (02) 412-2350.

Berlitz Korea company
2nd Fl., Sungwoo Academy Bldg., 1316-17 Socho-dong, Socho-gu, Seoul. **Tel:** (02) 3481-5324.

Berlitz Language Centre (Ulchiro)
16th Fl. Ileun Jungkwon Bldg., 198 Ulchiro 2-ga, Chung-gu, Seoul. **Tel:** (02) 775-1441. **Fax**: (02) 775-8088.

Chungmoon Language Institute
18-4 Kwanchol-dong, Chongno-gu, Seoul. **Tel:** (02) 738-5151.**Fax**: (02) 738-6150.

CPI Foreign Language Academy
#2-24, Mullae-dong, Yongdungpo-gu, Seoul. **Tel:** (02) 723-8882. **Fax**: (02) 2631-0928. **Email:** sscyber@sscyber.com

Daewon Language School
3nd Fl. Hyundae Chonghap Sangga, 615 Shinsa-dong, Kangnam-gu, Seoul. **Tel:** (02) 3445-7221. **Fax**: (02) 3445-7791.

Ding Ding Dang English School
1275-3 Bummul-dong Soosung-gu, Taegu 706-100. **Fax**: (53) 782-6434. **Email:** dings@thrunet.com

E2 Foreign Language Institute
1F Lotte Hagwon Sangga, 63 Yong Ho Dong, Changwon, Kyungnam, 641-041. **Tel:** (55) 261-0888 **Fax:** (55) 261-0797 **Email:** mcleod@chollian.dacom.co.kr

ESS Language Institute
38-1, 1-ka Kwangbok-dong, Jung-ku, Pusan 600-131. **Tel:** (51) 246-3251. **Fax**: (51) 241-1988.

FSFA Institute of Junior English
Gu-wol Plaza 2F, 342-256, Gu-wol dong, Nam-dong Gu, Inchon. **Tel:** (32) 472-8636. **Fax**: (32) 472-8637. **Email:** DarrenFSFA2000@yahoo.com

Global Christian School
243-5 Pogwang-dong, Yongsan-gu, Seoul. **Tel:** (02) 797-0234. **Fax**: (02) 797-0401.

Good Teacher's Centre
271-19, Dong In-3Ga, Jung-Gu, Taegu 700-423. **Tel:** (53) 427-8291. **Fax**: (53) 427-8295. **Email:** gta@chollian.dacom.co.kr

JS Language School
1-190, Munwha-dong, Jung-ku, Daejon. **Tel:** (42) 252-3500. **Fax**: (42) 254-6452. **Email:** js@englishcampus.com

Jungjin Language Institute 1
5th Fl., Kyungdong Bldg., 130-1 Hyehyun 1-ga, Chung-gu, Seoul. **Tel:** (02) 753-5243.

Kangrung High School
886 Kyo-1 Dong, Kangrung City 210-101. **Tel:** (016) 297-2095. **Fax**: (031) 641-5118.

Kid's College
261-2 Yangjae-dong Seocho-gu, Seoul 137-130. **Tel:** (2) 529-3831. **Fax**: (2) 529-7809. **Email:** kidscollege@hotmail.com

Korea Foreign Language Institute
16-1 Kwanchol-dong, Chongno-gu, Seoul. **Tel:** (02) 739-8000. **Fax**: (02) 739-0602.

Korea Herald Institute (Hoehyondong)
9th Fl. Korea Herald Bldg., 1-12 Hoehyon-dong 3-ga, Chung-gu, Seoul. **Tel:** (02) 727-0271. **Fax**: (02) 727-0668.

Korea Herald Institute (Suhadong)
65-1 Suha-dong, Chung-gu, Seoul. **Tel:** (02) 757-0419. **Fax**: (02) 319-3936.

Latt
3nd Fl. Shinwon Bldg., Chungjongno 3-ga, Sodaemun-gu, Seoul. **Tel:** (02) 363-3291. **Fax**: (02) 313-5620.

L.A. Wilshire Language School Korea
4th Fl., Kyewon Bldg., 32-7 Kwanchol-dong, Chongno-gu, Seoul. **Tel:** (02)723-6600.

LovEnglish
Duckhwa Building, Room 401, 444-17, Seogyo-dong, Mapo-gu, Seoul 121-210. **Tel:** (02) 3141-3766. **Fax**: (2) 332-2245. **Email:** mhkim@lovenglish.com

Mido Foreign Language School
Mido Sang-Ga Bldg., 3rd Floor, 311, Dae-Chi Chong, Kang-nam Ku, Seoul 135-280. **Tel:** (2) 562-6714. **Fax**: (2) 563-1205. **Email:** mido1@uriel.net

Mokdong SLP
7F Whasung Plaza BL 909-5, Mok-dong Yangchun-ku, Seoul. **Tel:** (2) 654-7331. **Fax**: (2) 654-7335. **Email:** perryr@nuri.net

Naeja English Academy
19-2 Hyoja-dong, Chunchon City, Kangwon Province. **Tel:** (361) 242-0505. **Fax**: (361) 251-0594. **Email:**naeja@hanmail.net

Oedae Junior English Institute
4th Fl. 286 Wolpyong-dong, So-gu, Taejon. **Tel:** (042) 487-0545. **Fax**: (042) 487-0572.

Pagoda Foreign Language Institute (Abgujung)
2nd Fl. Dongwon Bldg., 598 Shinsa-dong, Kangnam-gu, Seoul. **Tel:** (02) 3442-4000. **Fax**: (02) 515-4030.

Pagoda Foreign Language Institute (Chongno)
56-6 Chongno 2-ga, Chongno-gu, Seoul. **Tel:** (02) 2274-4000. **Fax**: (02) 2278-7538.

Pagoda Foreign Language Institute (Kangnam)
816 Yoksam-dong, Kangnam-gu, Seoul. **Tel:** (02) 501-4002. **Fax**: (02) 3452-0203.

Pusan Foreign School
1492-8, Chung-Dong, Haeundae-Gu, Pusan 612-010. **Tel:** (51) 747-7199. **Fax**: : (51) 747-9196. **Email:** pfs@chollian.net

Seoul Academy
Young dong P.O. Box 85, Seoul 135-600. **Tel :** (02) 555-2475.

Sisa English Institute
19-23 Kwanchol-dong, Chongno-gu, Seoul. **Tel:** (02) 734-2442. **Fax**: (02) 720-4931.

Sisa Language Institute
1099-1 Sanpon, Kunpo-dong, Kyonggi-do. **Tel:** (031) 398-6445.

Sogang University
1-1 Sinsu-dong, Mapo-gu, Seoul 121-742. **Tel:** (2) 716-1230. **Fax**: (2) 705-8733. **Email:** johng@ccs.sogang.ac.kr

Space Corporation
2nd floor, 656-502, 1Dong, 1Ga, Sung Su-Dong Sung Dong-Gu, Seoul 133-821. **Tel:** (2) 2205-0213. **Fax**: (2) 2205-0382.

World Language School
19-16 Kumnaro-1 ka, Dung-ku, Kwangju 501-021. **Tel:** (62) 228-1723. **Fax**: (62) 226-3562. **Email:** worldedu@hotmail.com

YBM
Sisa Yongasa 5-3 Kwanchul-dong, Chongno-gu, Seoul. **Tel:** (02) 2276-0509. **Fax**: (02) 2271-0172.

YES English School
Daewon Bldg., 5th Fl., Daechi-dong 599, Kangnam-gu, Seoul 135-281. **Tel:** (2) 553-8880. **Fax**: (2) 553-5764. **Email:** yescho@nuri.net

LAOS

Unlike other southeastern Asian countries, Laos is only just starting to open up to the West. As a recently joined member of the Association of Southeast Asian Nations (ASEAN), which uses English as the language of communication, the Lao P.D.R. government has made the study of English a national priority. The national curriculum was recently rewritten to include an emphasis on English learning, and members of the government are eager to study English. Most English language teaching is done by volunteers but there are limited opportunities for teachers at private language schools in Vientiane and possibility of work at universities.

➲ **Visas/Work Permits:** You can obtain a tourist visa (good for two weeks) at the border or at the airport in Vientiane. A one-month visa offered by a sponsor can be extended. Work permits are hard to acquire and the government is strict about illegal workers.

➲ **Requirements:** College Degree and a TESOL/TESL Certificate

➲ **Currency:** £1 = 19147 / $1 = 10350 Kip

➲ **Salaries/Taxes:** You can make anywhere from $1,200-2,000 a month less income tax of 10%.

➲ **Accommodation:** The school should assist you with accommodation.

➲ **Cost of Living:** Laos is much cheaper than neighbouring Thailand to live in but some basic items are hard to find.

➲ **Health advice:** Medical facilities and services are severely limited and do not meet Western standards. The blood supply is not screened for HIV or AIDS.

Visitors should be aware that there have been a number of bombings in public places and to be vigilant at all times.

Schools In Laos (+856)

Lao American Language Centre
152 Sisangvone Road, Saysettha, Ban Naxay, P.O. Box 327, Vientiane. **Tel:**21 41 4321, **Fax:** 21 41 3760.

National University of Laos
Dept. of English, Dong Dok Campus, Vientiane.

Vientiane International School
P.O. Box 3180, Vientiane. **Tel:** 21-313606, **Fax:** 21-315008, **Email: VIS@panlaos.net.la**

Vientiane University College
P.O. Box 4144, Vientiane. **Tel:** 21 414873, **Fax:** 21 414 346.

MALAYSIA

Malaysia offers a relatively high standard of living, a stunning natural environment, and easy access to many of the other countries in the region. It is difficult for foreign teachers to find a job here because of strict labour laws and current economic problems. English is widely spoken throughout the country.

British Council office:
PO Box 10539, 50916 Kuala Lumpur
Tel +60 (3) 2723 7900
www.britishcouncil.org/malaysia

➲ **Visas/Work Permits:** Most teachers go to Malaysia on a tourist visa (Valid for 90 days) and then look for jobs. The number of work permits allocated to foreigners are limited: teachers have to have an MA, teaching experience and be between 26-55 years old to qualify for one. Be prepared to spend up to a year doing "visa runs" to Singapore, Thailand or Indonesia if you are working without a permit.

➲ **Requirements:** High School Diploma and a TESOL/TESL Certificate

➲ **Currency:** £1 = 6.9 / $1 = 3.5 Malaysian ringgit.

➲ **Salaries/Taxes:** Local salaries are reasonable if you are not planning to send money home. Teachers earn around 2,500-4,000 ringgits per month. Income tax is 5% (once you have a permit) but the rate leaps to 30% if you spend less than 165 consecutive days in the country.

➲ **Accommodation:** Rents are very high for Asia. Expect to share accommodation with others.

➲ **Cost of Living:** Reasonable with lots of products available in Kuala Lumpur.

➲ **Health advice:** The standard of medical care is very high in the major cities where credit cards are usually accepted.

➲ **English Language Media:** The Borneo Post (Sabah and Sarawak), The Daily Express (Sabah), The New Straits Times, The Malay Mail, The Sarawak Tribune, The Star, The Sun.

Schools in Malaysia (+60)

Bangsar English Language Centre
60-1 Jalan Ma'arof Bangsar Baru, 59100 Kuala Lumpur. **Tel:** 3 282 3166/68, **Fax**: 3 282 5578.

Centre for Language and Communications Studies
University Malaysia Sarawak, 94300 Kota Samarahan, Samarahan, Sarawak.

Centre for Promoting Language and Knowledge
University Malaysia Sabah, Temporary Campus Likas Bay, Locked bag 2073, 88999 Kota Kinabalu, Sabah.

The English Language Centre
1st Floor, Lot 2067, Block 10, K.C.L.D., Jalan Keretapi, PO Box 253, 93150 Kuching, Sarawak.

International Tuition School
PO Box 3062, 93760 Kuching, Sarawak. **Tel:** 82 480780, **Fax**: 82 416250.

IPOH Learning and Training Institute
6 Jalan Tambun, 30350 Ipoh, Perak. **Tel:** 5 255 3067, **Fax**: 5 255 7136.

Kinabulu Commercial College
3rd & 4th Floors, Wisma Sabah, Kota Kinabulu, Sabah.

Kinabulu International School
PO Box 12080, 88822 Kota Kinabalu, Sabah.

Lodge Preparatory School
Jalan Tabuan Jaya, 93550 Kuching, Sarawak.

University Malaysia - Sarawak
94300 Kota Samarahan, Sarawak. **Tel:** 82 671000, **Fax**: 82 672315.

MONGOLIA

Teachers who have experienced Mongolia report that the country has a harsh but fascinating beauty. The capital, Ulaanbaatar, suffers from pollution but visiting the surrounding steppes is a must do. There are many opportunities to teach English here, including volunteer workers (notably the Peace Corps). Most of the teaching involves young learners in the public schools system. Flights to and from Mongolia should be arranged by your organization.

➲ **Visas/Work Permits:** It is common practice for a school to reimburse you for the cost of your visa when you have completed your contract. Visas may be obtained at the international airport in Ulaanbaatar and at train stations on the Russian and Chinese borders. Two photographs and a $50 processing fee are required.

➲ **Requirements:** College Degree and a TESOL/TESL Certificate

➲ **Currency:** £1 = 2360 / $1 = 1280 Tugrik.

➲ **Salaries/Taxes:** Some schools in Ulaanbaatar pay around $3.50 for a 45-minute lesson based on 20-24 teaching hours per week. At other schools, like those in smaller cities like Erdenet, expect around $100 a month, which is more than adequate to live on.

➲ **Accommodation:** Simple accommodation (including utilities) should be provided by your employer.

➲ **Cost of Living:** Very reasonable if you develop a taste for yak's milk!

➲ **Health advice:** Medical facilities in Mongolia are very limited, and some medicines are unavailable. Infectious diseases, such as plague and meningococcal meningitis, are present at various times of the year. Your school should provide medical cover.

Schools in Mongolia (+976)

Battsetseg Namnansuren
Box-346, Central Post Office, Ulaanbaatar-13. **Tel**: 9611-5597. **Fax**: 1-459732.

Eagle English Language School
Erdenet 976-35. **Tel:** 9935-1450. **Fax**: 35-73978. **Email:** rinchin@erdnet.mn

International School of Ulaanbaatar
P.O. Box 49/564, Ulaanbaatar. **Tel:** 1-452-959. **Fax**: 1-358-375. **Email:** inschool@magicnet.mn

Mongol Caravan, Inc.
P.O. Box 191, Clifton, NJ 07011. **Tel:** (973) 594-0655. **Fax**: (209) 729-4674. **Email:** mc@mongolcaravan.com

Santis Educational Services
P.O. Box 1174, Ulaanbaatar 210657. **Fax:** 32-3276. **Email:** santis@magicnet.mnw

MYANMAR

The current government restrictions on foreigners make it very difficult to find a teaching job in Myanmar (formerly Burma). Access to the Internet is strictly monitored by the present regime. The U.S. government has forbidden American companies to make new investments in the country while human rights abuses continue in Myanmar. General dissatisfaction with public universities has led to the opening of large numbers of private schools, many of which teach English. If the political situation improves, Myanmar could develop into a leading destination for EFL teachers.

➲ **Visas/Work Permits:** A single entry visa is good for a stay of 28 days.

➲ **Requirements:** College Degree and a TESOL/TESL Certificate

➲ **Health advice:** Medical facilities in Myanmar are inadequate for even routine medical care. There are few trained medical personnel because the universities were closed for several years and have recently reopened.

Schools in Myanmar (+951)

The British Council
78 Kanna Road, P.O. Box 638, Yangon. **Tel:** (951) 254658. **Fax**: 245345 **Email:** malcolm.jardine@bc-burma. bcouncil.org

NEPAL

Nepal is an impoverished country with great educational needs: the adult literacy rate in Nepal in 1995 was just 41% for men and a meagre 14% for women. It is possible to land a job teaching English at a private language school in Kathmandu, the backpacking capital of the world, but most teachers will not make enough money to survive unless they have funds of their own to fall back on.

There are aid agencies and Christian groups working with schools throughout and you might find a job in Nepal by taking this route.

British Council office:
PO Box 640, Lainchaur, Kathmandu

Tel +977 (1) 410 798
www.britishcouncil.org/nepal

➲ **Visas/Work Permits:** A tourist visa is extendible to a maximum period of 150 days in one visa year (January to December). Working for a school will allow you to remain in the country for up to a year.

➲ **Currency:** £1 = 129 / $1 = 70 Rupees

➲ **Salaries/Taxes:** You can earn around £60 / $110 a month in a private language school but remember that Nepal is a very poor country — you are not here for the money!

➲ **Accommodation:** You can find a small apartment for $150-200 per month. Volunteers will have housing provided for them.

➲ **Cost of Living:** Estimate £300-350 / $570-650 per month as a comfortable amount.

➲ **Health advice:** Medical care is extremely limited and is generally not up to Western standards. Serious illnesses often require evacuation to the nearest adequate medical facility (in Singapore, Bangkok or New Delhi). Illnesses and injuries suffered while on trek in remote areas often require rescue by helicopter. The cost is typically £1,800 - 6,000 / $3,300 - 11,000. Medical insurance is essential.

➲ **English Language Media:** Kathmandu Post, The Rising Nepal.

Schools in Nepal (+977)

Debbie's English Language Institute
Jamal, Kathmandu. **Tel:** 1-248140.

Friendship Club Nepal
P.O. Box 11276, Maharajgunj, Kathmandu. **Tel:** 1-427406. **Email: fcn@ccsl.com.np**

Grahung Kalika
(Walling Village Development Committee), Bartung, Ward 2, Syngja. **Fax:** 1-262878.

Insight Nepal
P.O. Box 489, Zero K.M. Pokhara, Kaski. **Email:** insight@clcexp.mos.com.np

Universal Language & Computer Institute
Putalisadak, Kathmandu. **Tel:** 1-419443.

PAKISTAN

It is best to arrange a job before coming to Pakistan, as there are few private language schools in the country. There is a small demand for TEFL teachers although most children learn English in the public schools. Do not take on private lessons unless your contract stipulates that you may do so. The authorities will not be pleased if you are caught working illegally.

High Commission in UK:
935 Lowndes Square, London SW1X 9JN
Tel: 020 76649220

Embassy in USA:
2315 Massachusetts Avenue, Washington DC20008
Tel: 202 939 6200

British Council office:
House 1, Street 61, F-6/3, P.O. Box 1135, Islamabad
Tel +92 (51) 111 424 424
www.britishcouncil.org/pk

➲ **Visas/Work Permits:** paperwork will be taken care of by your employer.

➲ **Requirements:** College Degree and a TESOL/TESL Certificate

➲ **Salaries/Taxes:** A TEFL teacher can earn £10,000-12,000 / $19,000-22,000 a year.

➲ **Accommodation:** This should be provided by your employer.

➲ **Health advice:**Adequate medical care is available in major cities in Pakistan, but is limited in rural areas. Your employer should provide you with medical insurance.

Schools in Pakistan (+92)

Pakistan American Cultural Center (PACC)
Head Office, 11 Fatima Jinnah Road, Karachi. **Tel:** 5670516/5670513/515305.

Pakistan American Cultural Center (PACC)
Hyderabad centre, Hospital Road, Hyderabad. **Tel:** (0221) 619854/618854.

Pakistan American Cultural Center (PACC)
Karimabad centre, 54-C, Block 8, Federal B Area, Karimabad. **Tel: 6363277.**

Pakistan American Cultural Center (PACC)
Lahore centre, 25 E-3, Gulberg III, Lahore. **Tel:** (042) 5756438 / 5763644.

Pakistan American Cultural Center (PACC)
Peshawar centre, c/o St. John's School, 1, Sir Syed Rd., Peshawar. **Tel:** (0521) 279734.

Pakistan American Cultural Center (PACC)
Quetta centre, Sadiq Shaheed Park, Arbab Mohammad Khair Road, Mali Bagh, Quetta. **Tel:** (081) 826970 / 830889.

SINGAPORE

This wealthy city-state is popular with British and Australian EFL teachers. The demand here is for well-qualified teachers with a lot of experience and it is harder to find a job here than in other Asian countries.

High Commission in UK:
9 Wilton Crescent, London SW1X 8SP
Tel: 020 7235 8315
www.mfa.gov.sg/london

Embassy in USA:
3501 International Place, Washington DC20008
Tel: 202 537 3100

British Council office:
30 Napier Road, 258509
Tel +65 6473 1111
www.britishcouncil.org/singapore

- **Visas/Work Permits:** U.S. citizens can stay for 90 days without a visa.
- **Requirements:** College Degree and a TESOL/TESL Certificate
- **Currency:** £1 = 3.01 / $1 = 1.54 Dollar (S$).
- **Salaries/Taxes:** An average salary in Singapore is S$2,500 a month.
- **Accommodation:** Apartments start at around S$1,200 a month.
- **Cost of Living:** You can get by on S$2,000 a month.
- **English Language Media:** Straits Times.

Schools in Singapore (+65)

Advanced Training Techniques (ATT)
Tanglin Shopping Centre, 19 Tanglin Road, Singapore 247909. **Tel:** 235 5222, **Fax**: 738 1257.

American College
25 Paterson Road, Singapore 0923. **Tel:** 235 9537.

Canadian International School
5 Toh Tuck Road, Singapore 2159. **Tel**: 467 1732.

Coleman Commercial and Language Centre
Peninsula Plaza, Singapore. **Tel:** 336 3462.

Corrine Private School
Selegie Complex 04-277, Selegie Road, Singapore 188350. **Tel:** 339 5564.

Dimensions Language Centre
50 East Coast Road, Singapore 0238.

Dover Court Preparatory School
Dover Road, Singapore 0513. **Tel:** 775 7664.

ILC Language and Business Training Centre
545 Orchard Road, 11-07 Far East Shopping Centre, Singapore 238882. **Tel:** 338 5415.

inlingua School of Languages
1 Grange Road, 04-01 Orchard Building, Singapore 239693. **Tel:** 737 6666.

International School
21 Preston Road, Singapore 0410. **Tel:** 475 4188.

Language Teaching Institute of Singapore
30 Orange Grove Road, Singapore 258352. **Tel:** 737 9044.

Linguamedia
230 Orchard Road, 04-230 Faber House, Singapore, 238854. **Tel:** 235 2620. **Fax**: 235 2080.

Morris Allen Study Centre
1 Newton Road 02-47/49, Goldhill Plaza, Singapore 307943. **Tel:** 253 5737.

St. Francis Methodist School
11 Mount Sophia, Singapore 228461. **Tel** 334 8944. **Fax:** 337 1825. **Email:** stfrancm@mbox5.signet.com.sg

TAIWAN

Prospects for English language teachers are very good in Taiwan, especially for native speakers of American English. Private schools (bushibans) pay well and jobs are easy to find. Unqualified travellers can find work here as long as they have a degree. The school year begins on August 1 and the winter semester starts on February 1. Local papers start carrying ads for English language teaching jobs in the spring. You should bring between £750-1500 / $1,300-$2,200 to cover you for about five weeks until you get your first full salary cheque. If you pre-arrange a teaching job before you travel to Taiwan, you may find that you are making less money than fellow teachers who arrived here and then found a job. But remember that you will not have to go through all the hassles of getting the right permits which can be very complicated since the authorities in Taiwan cracked down on freelance teachers working without the correct papers.

British Council office:
Education and Cultural Section, 7F-1, Fu Key Building, No 99 Jen Ai Road, Section 2, Taipei 100
Tel +886 (2) 2192 7050
www.britishcouncil.org.tw

➲ **Visas/Work Permits:** Most teachers go to Taiwan on a tourist visa. Try to get a 60-day visa, as your paperwork will take 3-6 weeks to process: If your visa runs out before your paperwork is processed you will have to leave the country to obtain another resident visa. Many teachers go to Hong Kong, Singapore or Thailand to do this. If you enroll for Chinese language lessons at a government-recognized school, you can renew your visa without having to leave the country. To teach legally in Taiwan, the government requires that all foreign teachers obtain a Work Permit and an Alien Resident Certificate. To be eligible for these, you must: possess a bachelor's degree or higher, a contract of employment and a health certificate from a government-approved hospital. You will be expected to pay for the health exam and the processing fees.

➲ **Requirements:** College Degree and a TESOL/TESL Certificate

➲ **Currency:** £1 = 63 / $1 = 33 New Dollar (NT$)

➲ **Salaries/Taxes:** Monthly wages for teachers in language schools usually average between NT$30,000 and NT$70,000 net depending on the number of classes taught, and the number of students in each class. An entry-level salary for an assistant professor is about NT$73,000 per month (including a $10,000/month bonus for those with PhDs); for a lecturer, the figure is about NT$52,000 per month. Taxes are 20% until you have worked in Taiwan for 183 days in the same calendar year. The rate then drops to 10%. If you have been in the country for more than 183 days you are eligible for a tax refund in the next year.

➲ **Cost of Living:** A quick lunch costs NT$100.

➲ **Health advice:** Teachers working for the larger and more reputable schools are usually covered under group plans where they might make a contribution (around 30%) from their pay packets. Medical insurance is essential for freelancers unless you have a residency permit in which case you will be covered under the state system.

➲ **English Language Media:** China News, China Post.

Schools in Taiwan (+886)

American Children School
Chung Shan North Road, Sec. 6, #34, Taipei. **Tel:** 2-2831-6664.

American Culture University
Hsu Chang Street, #40, 7F, Taipei. **Tel:** 2-2311-8889.

Bao Bei Yuan English School
Jian Guo South Road, Sec. 1, Lane 270, #28, Taipei.. **Tel:** 2-2754-8066.

Bei Song Language School
Chung Hsiao East Road, Sec. 6, #436, Taipei. **Tel:** 2-2788-7778.

Big Bird English
Hsin Yi Road, Sec. 4, #181, 2F, Taipei. **Tel:** 2-2754-8405.

Big Byte English School
Hsin Hai Road, Sec. 2, #35, Taipei. **Tel:** 2-2369-2868.

Canadian-American Language Schools
2 Chien Hua Street, 2nd Floor, Hsinchu, 300. **Tel:** 3-562-0535. **Fax**: 3-561-6905. **Email:** employment@can-am.org

Chang Chun English
Ji Long Road, Sec. 2, #260, 2F, Taipei. **Tel:** 2-2735-2617.

Cheng Dan Language Centre
Hang Chou Road, Sec. 1, Lane 71, #9, Taipei. **Tel:** 2-2381-5147.

Ching Shan English
Hsin Sheng South Road, Sec. 3, #98, 3F, Taipei. **Tel:** 2-2366-1855.

David's English Centre
Nan Jing East Road, Sec. 2, #100, 3F, Taipei. **Tel:** 2-2522-4004.

Disney English
Hoping East Road, Sec. 2, Lane 96, Taipei. **Tel**: 2-2732-8969.

Dong Han English
Hoping East Road, Sec. 2, #100, 4F, Taipei. **Tel:** 2-2732-2771.

English Bridge School
Ching Hsing Road, Lane 202, No. 8, 2F, Taipei. **Tel**: 2-2933-5118.

Everyday English
Chung Hsiao East Road, Sec. 4, #177, 4F, Taipei. **Tel**: 2-2741-8989.

Gauden Language School
48 Jing Shiou East Road, Yuan Lin, Changhwa, 510. **Tel:** No: 4- 8368176. **Fax:** 4-8350431. **Email:** gauden@ms11.hinet.net

Giraffe English Centre
Shi Pai Road, Sec. 1, #160, 1F, Taipei. **Tel:** 2-2822-5815.

Gram English Centre
4th fl., 116 Yung Ho Rd., Sec. 2, Yung Ho City 234, Taipei County. **Tel:** 2-2927-2478. **Fax**: 2-2926-2183. **Email:** gram@ms7.hinet.net

Great Time English
Min Chuan East Road, Sec. 3, Lane 60, #10, Taipei. **Tel:** 2-2506-3033.

Han Bang Language School
Chung Yuan Road, #17, 2F, Taipei. **Tel:** 2-2531-6835.

He Jing Language School
Dun Hua South Road, Sec. 1, Lane 100, Alley 5, #37, 2F, Taipei. **Tel:** 2-2731-2822.

Hers English School
Wo Long Street, #1, 11F-2, Taipei. **Tel:** 2-2378-0186.

Hess Educational Organization
English Human Resources Department, 419 Chung Shan Rd., Section 2, Chung-Ho City 235, Taipei County. **Tel:** 2-3234-6188 ext. 1053/ 1052. **Fax**: 2-2222-9499. **Email:** hesswork@hess.com.tw

Jya Ying English
Jwang Jing Road, Lane 325, #37, Taipei. **Tel**: 2-2723-5259.

Kid Castle
Min Chuan Rd., No.98, 8F, Taipei. **Tel:** 2-2218-5996 ext.319/341.

Lang Wen English Centre
Jin Jou Street, #16, 3F, Taipei. **Tel:** 2-2551-8956.

Lian Hsin English
Fu Hsing North Road, #92, 7F, Taipei. **Tel:** 2-2773-6161.

Little Forest
4th Floor 121-5 Da Chang 2nd Rd., Kaohsiung. **Tel:** 7-395-6736. **Fax**: 7-733-4288. **Email:** jyy@usa.net

Melody English centre
San Min Area, Bei Ping 2nd Street, #6, 6F, Kaohsiung. **Tel:** 7-322-7492.

Min Sheng English School
Min Sheng East Road, Sec. 5, #226-1, Taipei. **Tel**: 2-2769-9243.

Ming Tai Language School
Chung Hua Road, Sec. 1, #178, Taipei. **Tel:** 2-2311-0712.

Pai Han Language School
Bo Ai Road 1st Road, #43, Kaohsiung. **Tel:** 7-316-1518.

Sen Yuan Language Centre
Jian Guo Road, #258, Kaohsiung. **Tel:** 7-395-6736.

Timothy Language Institute
Cheng Kung Road, Lane 213, No. 97. Tainan County, Jiali. **Tel:** 6- 723-2819.

Tom's English
Shwun Hsing Road, #95, Hualien. **Tel:** 38-822-2059.

Wen Tzao English
Yong Fu Road, Sec. 2, #83, 7F, Tainan. **Tel:** 6-228-2818.

Woodpecker English School
Bo Ai Road, #282, Pingtung. **Tel:** 8-733-8691.

Ya Ya
Chwei Yang Road, #199, 2F, Chiayi. **Tel:** 5-223-0358.

Twinkle Language Centre
66 Hsyn Hai Road, Tso Ying District 813, Kaohsiung. **Tel:** 7-557-1883. **Email:** twinklelanguagecentre@yahoo.com

THAILAND

There are many opportunities to teach English in Thailand, particularly in the universities. Most teachers find work in Bangkok, which is the economic, cultural and educational hub of the country. Although there was a slump in English language teaching because of the Asian economic crisis, the situation is improving. English language schools are looking for qualified teachers — the market for native-speaking beach bums has dried up.

Embassy in UK:
29 Queens Gate, London SW7 5JB
Tel: 020 7589 2944

Embassy in USA:
1024 Wisconsin Avenue, Suite 401, Washington DC20007
Tel: 202 944 3600

British Council office:
254 Chulalongkorn Soi 64, Siam Square, Phayathai Road, Pathumwan, Bangkok 10330
Tel +66 (2) 252 5480
www.britishcouncil.org/th

➲ **Visas/Work Permits:** In order to get a work permit you need at least a bachelors' degree and some form of TEFL certification. Before you can apply for a work permit you need a Non-Immigrant 'B' visa. These are three-month visas that initially allow you to be employed in Thailand. To get the visa, you need a copy of a letter from your prospective employer clearly stating that you have been offered a job. For private companies you need a copy of the said company's registration documents, if you're working for a government establishment this isn't required, and two passport sized photos. Embassies usually take two days to process the visa. Most Thai embassies work on the system that you apply for visas in the morning and collect them the next afternoon. Once you have found a job and can prove to the Immigration officials that you are applying for a work permit you can extend the three-month visa for another nine months.

➲ **Requirements:** College Degree and a TESOL/TESL Certificate

➲ **Currency:** £1 = 68 / $1 = 35 Baht

➲ **Salaries/Taxes:** There is a wide range of salaries for teaching English in Thailand. Working a typical 20-25 hour week can earn you anywhere from as little as 10,000 baht a month for an unqualified teacher in a small provincial school to 60,000 baht or more for a TEFL teaching professional in a well-established school or college in Bangkok or another large city. Teachers at language schools pay around 25-35,000 baht per month in tax. Income earned by a U.S. resident from teaching in Thailand is exempt from income tax if the following conditions are met: (1) The teacher is a resident of the U.S.A. before he/she comes to teach in Thailand; (2) The teacher teaches in a university, college or other recognized educational institution in Thailand; (3) It must be his/her first visit for teaching in Thailand; and (4) He/she does not teach in Thailand for more than two years.

➲ **Accommodation:** In most cases, teachers are expected to find their own accommodation. Studios in Bangkok cost around 5,000 baht a month while a one-bedroom apartment is anywhere from 8-18,000 baht.

➲ **Cost of Living:** You can get a three-course meal for two people without alcohol for under 100 baht. If you can't do without imported items you will spend a lot of money because they are heavily taxed.

➲ **Health advice:** Basic coverage for a single teacher will cost you at least 15,000 baht.

➲ **English Language Media:** Bangkok Post, The Nation.

Schools in Thailand (+66)

Adanac Community Language Centre
Suite 605 Bangkok Tower, 2170 New Petchburi Rd., Bangkok 10320. **Tel/Fax:** 3080787-90

American University Alumni (AUA) Language Center
24 Rajadamnern Road, Chiang Mai 50200. **Tel:** 53-211-973. **Fax**: 53-211-973. **Email:** aualanna@loxinfo.co.th

Anna's Language School & Travel Services
116 Phisitgaranee Road, Patong Beach, Phuket 83150. **Tel:** 296372. Fax: 296423 **Email:** annatran@samart.co.th

Arundel Language Institute (A.L.I.)
61/67 Soi 10/3 Pechkasem Rd, Hat Yai, Songkla, 90110. **Tel:** 74 345 341-2. **Fax**: 74 345 342. **Email:** ali@loxinfo.co.th

Asia Institute of English
59 Sukhumvit 56, Bangchak , Prakanong, Bangkok 10250. **Tel:** 741- 5970-2. **Email:** AIE@a-net.net.th

Bangkok Patana School
2/38 Soi Lasalle, Sukhumvit 105, 10260 Bangkok. **Tel**: 2 398 0200. **Fax**: 2 399 3179. **Email:** patanarc@loxinfo.co.th/patreg@loxinfo.co.th

The Bell Educational Trust Thailand
204/1 Ranong 1 Road, Samsen, Dusit, Bangkok 10300. **Tel:** 2 241 0356. **Fax**: 2 668 2124. **Email:** gbradd@loxinfo.co.th

Berlitz Silom
191 Silom Complex Building, 22F Silom Road, Bangkok 10500. **Tel:** 2 231 3652. **Fax:** 2 231 3656.

Berlitz Sukhumvit
Times Square Bldg, 14F, 2246 Sukhumvit Road, Bangkok 10110. **Tel:** 2 250 0950. **Fax**: 2 250 0907.

Bright Future Language Centre
7th floor, Room 107, Sibunruang 2 Building, 1/7 Convent Road, Silom, Bangkok 10500. **Tel:** 2 631 0497. **Fax**: 2 631 0498. **Email:** bright@idn.co.th

Cambridge World International English Language Schools
296-300 Sukhumvit Road, Paknam.

Charansanitwong School of Business (CSB)
Charansanitwong Soi 41 Charansanitwong Rd.,

Bangkoknoi
Bangkok 10700. **Tel:** 8824676 **Fax**: 4333647. **Email:** NewCambridge@excite.com

Company English Language Centre of Excellence
Prince of Songkla University Wichitsongraram Rd, A. Kathu Phuket 83120. **Tel**: 202556 / 7. **Fax**: 202558 **Email:** h&t@ratree.psu.ac.th

English and Computer College (ECC)
430/17-24 Chula Soi 64, Siam Square, Bangkok 10330. **Tel:** 2-253-3312; **Email:** eccthai@comnet3.th.

EF - English First
Suite 701, 7th Fl. Sibunruang 2 Bldg. 1/7 Convent Rd., Silom, Bangrak, Bangkok 10500. **Tel:** 2-631-0497/8. **Fax**: 2-631-0499. **Email:** teach@efenglish.com

English Language Centre
Assumption University, Ramkhamhaeng 24 Road, Huamark, Bangkapi, Bangkok 10240. **Fax:** 2 318 7159. **Email:** elc@au.edu

English Language Instructors
PO Box 604 Amphur Muang, Phuket 83000. **Tel:** 282230.

Fun English Language Programme
Pichaya Suksa School, Pakkret, Nonthaburi. **Tel:** 1-313-8319. **Fax**: 2-373-8938. **Email:** ayatan@mozart.inet.co.th

Fun English Language Programme
Yamsaard School, Lard-Praow Road, Bangkok. **Tel:** 1-313-8319. **Fax**: 2-373-8938. **Email:** ayatan@mozart.inet.co.th

In Company Ltd.
206/1 Nawanakorn, Pathumthani 12120. **Tel:** 2 909 1569. **Email:** patricks@samart.co.th

inlingua International School of Languages
Central Chidlom Tower, 22 Ploenchit Road, Patumwan, 10330 Bangkok. **Tel:** 2 254 7028-30. **Fax**: 2 254 7098. **Email:** inlingua.bkk@inlingua.com

International Community School
72 Soi Prong Jai, Sathorn, 10120 Bangkok. **Tel:** 2 679 7175. **Fax**: 2 287 4530. **Email:** icsbkk@loxinfo.co.th

International School Bangkok
39/7 Soi Nichada Thani, Samakee Road, 11120 Pakkret. **Tel:** 2 583 5401-19. **Fax**: 2 583 5431-2. **Email:** reyadik@isb.ac.th

Kaplan
Bangkok Bank Siam 5th floor, 394 Rama 1 Road, Siam Square Soi 5, Bangkok 10330. **Tel:** 2 658-3991 / 2. **Fax**: 2 658-4354.

NAVA Language School
183/105-106 Phang-Nga Rd, A. Muang, Phuket 83000. **Tel:** 238398/9 **Fax**: 238397. **Email:** nava@phuket.a-net.net.th

NAVA Language Schools
34 Yosawadi Bldg, Ground floor, Paholyothin 7 Road (Soi Aree) Samsen-nai, Payathai, Bangkok. **Tel:** 617 1390-3. **Email:** nava3@navaed.co.th

PEP Kindergarten
973 Ladproaw 48, Bangkok 10310. **Tel:** 2 277 8170. **Fax**: 2 276 2916. **Email:** maes2@hotmail.com

P.E.T.S.
160/1-2 Praram 6 Muang Trang 92000. **Tel:** 075 215556. **Email:** ffracs@hotmail.com

Rajabhat Institute Surin
186 Surin-Prasart Road, Muang District, Surin 32000. **Tel:** 44 511 604. **Fax**: 44 511 631. **Email:** intanin@risurin.nepnet.ac.th

St. Stephen's International School
107 Viphavadi Rangsit Road, Ladyao Chatuchak, Bangkok 10900. **Tel:** 2 513 0270. **Fax**: 2 513 0265. **Email:** richard@ocean.co.th

Sarasas Witaed Suksa School
300 Moo 10 Pracha Utid Rd., Naiklongbangplakot, Samut Prakarn,Bangkok 10270. **Tel:** 2 815 7540. **Fax**: 2 815 7541. **Email:** saraswitaed@hotmail.com

Siam Computer & Language School
2 Soi 7, Phang-Nga Road Amphur Muang, Phuket 83000. **Tel:** 219914/5.

Taiyo Co. Ltd.
3F, Prasarnmit Plaza, Sukhumvit Road, Soi 23Klong Toey, Bangkok 10110. **Tel:** 2 6641630-2. **Fax**: 2 6641630. **Email:** taiyobkk@yahoo.com

Training Creativity Development (TCD)
399/7 Soi Thongloh 21, Sukhumvit Soi 55, Bangkok 10110. **Tel/Fax:** 2 391 5670.

Universal International School
49 Moo 4, Thanarat Road, Tambol Nongnamdaeng, Amphur Pak Chong, Nakorn Ratchasima 30130. **Tel**: 44 328 334. **Fax**: 44 313 519.

Teaching in Thailand

Well, how about Thailand? You may well have seen the TV and print ads extolling the virtues of Thailand, 'The Land of Smiles', and Bangkok, 'The City of Angels'. New Year's Resolution No. 1 "I Really Want To Teach Somewhere Hot & Sunny". But is life as an English language teacher as idyllic as these slogans suggest? Well, I've been here for three and a half years so it can't be too bad. Hopefully the following article will equip you with enough basic information to ascertain whether you are ready for Thailand and whether Thailand is ready for you.

Thailand has never been colonized, unlike most South East Asian countries. While Thais are justifiably proud of this fact, there is actually a long history of English teaching in Thailand and an English teaching methodology has developed over a long period of time. Indeed, in the majority of Thai schools and colleges, English is taught in the same manner as every other subject–learning by rote.

Average students in Malaysia, Singapore and the Philippines usually have a much higher standard of English than their peers in Thailand. In an effort to correct this imbalance and make up lost ground on fellow ASEAN countries, the government has recently introduced new legislation to try to replace the old "chalk and talk" teaching methods with student centred activities. Although this idea is commendable, schools and individual teachers have been rather too quick to point out obstacles rather than channelling their thoughts constructively. As with many good ideas, implementation is often delayed indefinitely and/or the original proposal/law altered beyond all recognition or scrapped entirely. According to the local press, the Education Reform Bill 2000 appears, unfortunately, to be heading for oblivion.

With respect to the availability of jobs in Thailand, teaching in Thailand equates to teaching in Bangkok. I'd estimate that over 90% of the foreign teachers working in Thailand are working in Bangkok. It's best to think of Bangkok and the rest of Thailand as two separate entities when it comes to teaching opportunities, salaries, job prospects, standard of living and so on.

If you have a notion of coming to Thailand, heading down to the beaches of Phuket or Samui or up to the hills in the North and getting a job teaching to finance your stay in paradise – forget it. Job vacancies and opportunities are very limited in these areas.

It is possible to find work if you have the right connections but, if you're reading this and not already there doing the job, then you don't have the right connections. If you prefer the quieter life, then both Chiang Mai, in the north and Hat Yai in the south offer some possibilities – however the competition, particularly in Chiang Mai can be fierce, and the pay is often much lower than in Bangkok.

My first impressions of Bangkok? It's a sprawling, polluted, overcrowded city with one foot in the past and another, attempting, to stride into the future. But once you've been here a while you discover that the city does have quite a few redeeming qualities: It's much safer than any western city – it's very rare to hear of any muggings or see any fights. Friendly people – Thais are more willing that most to make you feel at home in their country. Their natural, relaxed lifestyle makes life here less hectic and stressful than say Japan, Taiwan or Hong Kong.

Cheap food: 20 to 30 baht will buy you a simple meal at a roadside food stall – of which there are thousands in Bangkok. Dining out is a way of life here – very few people cook their evening meals, preferring to eat with their friends after work or to buy something on the way home.

The Skytrain, an elevated electric railway opened in December 1999 and, although underused by Thais, is very popular amongst the teaching community. If you're lucky enough to live and teach on or near either of the two lines your commuting time will be measured in minutes rather than hours.

Nightlife: Let's be honest, one of the big attractions of Bangkok is its well-known and varied nightlife. For many teachers this is the deciding factor in the choice whether to come to Bangkok or go elsewhere. There's something for everyone from arty, stylish bars and bistros to the delights of Patpong and Nana Plaza.

Phillip Williams, Academic Director at the Asia Institute of English, sums up life in the City of Angels, "Bangkok is just about as teacher-friendly as a city can be. It offers a relatively low cost of living, affordable entertainment and, most importantly teachers are always in great demand."

The vast majority of Thai students have been taught by using the rote learning method. Anything the teacher writes on the board is automatically copied into your book. The information passes straight from the eye to the hand, bypassing the brain entirely. Students rarely question the teacher, that's tantamount to having doubts about the teacher's knowledge and, as we all know, what your teacher says is always correct. This results in the average Thai student being able to cope with a multiple choice exam, which requires the memorization of facts, but having a great deal of difficulty with any exam that requires a written answer, especially one requiring a personal opinion or giving a reason why something happens.

Historically the Thai people are farmers and the economy largely based on agriculture. During labour intensive periods, planting and harvest time, families will pool their resources to help their neighbours. Nowadays, the attitude on lending a helping hand is carried over into schools, colleges and the workplace. Weaker students are given a lot of encouragement (usually in the form of the correct answers) by their classmates. Teachers give students every opportunity to pass an exam; this manifests itself in the form of multiple retakes until the student passes. Work colleagues help each other produce reports for their bosses to ensure that even those staff who would be considered a liability stay in a job. Therefore, if you are teaching a class, never expect Thai students to work on their own, everything must be done in pairs or groups. If you specify that a certain report or piece of homework must be done individually, you could count on at least 70% still being done as a group effort.

By now you may feel that I am being very negative about Thai students, but that's not so. I'm just trying to make you aware of the differences between western students and Thai students. Knowing about these differences before you start teaching will save you a great deal of head shaking and hand wringing. On the plus side the students respect their teachers, Thais are talkative but you won't have any verbal or physical abuse from any students. As I said earlier, your students will enjoy a laugh and a joke; they won't expect you to be serious.

From an academic point of view, teaching ability isn't really valued or regarded as high as it should be. There are many well-qualified TEFL/ESL teachers who have left Thailand in despair. The reason being is that their abilities aren't recognized by fellow Thai teachers and/or their students. Succeeding in Thailand isn't a matter of having a list of academic qualifications as long as your arm. What is important is that you can understand your students and know what they expect from you. For many career teachers, especially those used to teaching more dedicated students in, say, Japan, this aspect of Thai life can be particularly frustrating.

Anyone who is governed by absolute obedience to rules and regulations, timetables, strict adherence to contracts, the need to be involved in decisions that affect you and the belief that displays of anger will get conflicts resolved won't enjoy their time here.

Get used to "going with the flow," "rolling with it" and any number of similar phrases that could be used to describe the notion that if something goes to plan then that's an unexpected bonus and not something to be taken for granted.

Thai students want to be entertained as well as educated. Therefore the successful teacher has to bear in mind what I like to call the 4 "Ss" of Thai society. These are:

1. Sanook (Fun): If it's not sanook then why do it? Makes sense really, doesn't it. Listening to a teacher drone on, copying from a board, mindless choral repetition and so on aren't considered fun activities. First thing to remember then is to make your lessons fun, include a couple of games and keep things moving along – don't get bogged down – it's not sanook.

2. Suay (Beautiful): All the best things in life are suay. Outward appearance is of the utmost importance to Thais, especially personal appearance. While teachers in the west may be used to teaching classes in jeans and a t-shirt, that kind of attire is frowned upon here. Get a haircut; you'll need to wear a shirt, tie, trousers and polished shoes. Female teachers should stick to knee length skirts and non-revealing blouses. Suay carries over into any homework or reports you'll give your students. Style is always placed above substance, trying to convince your students otherwise will result in a look of disbelief upon their faces.

3. Suparb (Politeness): This covers everything from the way you dress and speak to having respect for your elders. Everyone knows their place in the social structure and people address each other accordingly. Elders are addressed as pee: those younger than you as nong. Elders, bosses, teachers and monks are all wai'd (use the traditional Thai greeting with them – hold your hands as though praying and bow your head slightly.) This is just the tip of the politeness iceberg. A couple of things for teachers new to Thailand should remember are: (1) never perch on the edge of a desk (as I remember my old high school math teacher doing every lesson) and (2) don't touch your students, especially if you're a male teacher and have female students.

4. Serious – Okay, it's an English word but all Thais know it. This is something you shouldn't be. It's almost as bad as being boring. God help you if you are boring and serious because no one else will. In a practical sense, try to treat everything with the same lack of importance, for example if someone doesn't do their homework, give them more time, if they copy – no problem so long as they only copy a little bit. You'll find your students are only serious around exam time – and then for only a very brief period.

In a nutshell, dress the part, clean, smart and professional – if you look like a teacher Thais will treat you like a teacher (i.e., with respect.) Be friendly and polite, stay calm, don't raise your voice, have fun, laugh at yourself and laugh with your students. This isn't a licence to spend all class playing games but it does allow you to be creative and not be ruled by the constraints of a rather dull course book.

You've read some background information (I hope!), now ask yourself this question: "Why do I want to go to Thailand to teach English as a foreign language? Here are three possible scenarios:

(1) You plan on backpacking around the world and you need to earn some cash during your trip. The large language schools are always looking for teachers, qualified or not. This fact, coupled with the reality that Thai students are fun loving and pretty forgiving towards their teachers means that it's not hard to get work, even if you only plan to stay a few months. You will be on a relatively low hourly rate but a couple of months work will enable you to save enough to pay for a month chilling out on Koh Pa Ngan.

(2) You've finished university and it's time to take a year out before starting life as a corporate slave. Ideally you'd like to spend a year experiencing life in a country with a different culture and climate from what you're used to. However you also have the reality of a student loan to face, so you need to make sure that your overseas experience pays it's way and doesn't drain your bank account further. If you have the presence of mind to, at least, read a couple of books on TEFL, then armed with a copy of your degree, you should have no trouble in finding work once you arrive. A better scenario would be if you took some form of TEFL course before you arrived, it doesn't have to be a Cambridge or Trinity College course – which are quite expensive. Any evidence that you had the presence of mind to take a TEFL qualification will show employers that you are serious about teaching and will ensure a better rate of pay and more job opportunities than if you'd arrived totally unprepared.

(3) You are an experienced teacher. If you're financially minded then you will probably want to avoid Thailand. Sure, there are well paying jobs available at international schools and private universities, but on the whole the availability of these jobs is pretty limited. Even if you worked as the Director of Studies at a language school you might find that your 10+ years of experience only results in a salary a few thousand Baht higher than a new teacher with little or no experience. In general, teaching experience isn't well rewarded financially. However, if you're looking for a less stressful atmosphere than say, Japan or Korea (where experience is more prized), then Thailand offers the chance to earn a decent salary in a more easy going atmosphere.

A recent report concluded that the market for English language teaching was far from saturated and would continue to grow. The demand for both qualified and unqualified teachers would therefore remain high for the foreseeable future. But what about the conditions for teachers in language schools? Back to Phillip Williams, who, having been here for over 10 years, has noticed the impact the Internet is having on the standard of teaching and the attitude of schools towards their teachers. "What is very worrying for the poor quality schools is the number of times their name gets dragged through the mud on the Internet. Their dirty laundry is hung out for all to see."

Schools now realize that they are being held accountable for their actions. Phillip continues, "Many schools are waking up to the fact that proper teachers want proper jobs. They want a competitive salary, access to modern teaching materials, a worthwhile training programme and an employer who values their contribution to the education of Thailand's new generation."

So, what are you waiting for . . . . ?

Ian McNamara teaches EFL at Assumption College, Bangkok and, when time allows, is also the webmaster of Ajarn – for those interested in life in Bangkok (www.ajarn.com).

VIETNAM

After the first flush of foreign investors rushing to do business in Vietnam, the atmosphere is more cautious as people realize that the country cannot embrace capitalism overnight. There is a lot of bureaucracy in Vietnam and this extends into the English language teaching market. Demand for English fluctuates from year to year: early in 2000, EFL teachers reported a big downturn in student numbers enrolling in established schools in Ho Chi Minh City (Saigon). Students were moving to cheaper schools that had just opened and competition among EFL schools was intensifying. Although unqualified travellers can find jobs here, they will be paid very badly. It's much better to apply for jobs in Vietnam as a qualified EFL teacher.

British Council office:
40 Cat Linh Street, Dong Da, Hanoi
Tel +84 (4) 843 6780
www.britishcouncil.org/vietnam

➲ **Visas/Work Permits:** A tourist visa is valid for 30 days. This can be extended three more times for one-month periods. After this time, you have to leave Vietnam. Some teachers apply for six-month, multiple-entry extendable business visas through travel agencies, which should charge no more than $60-80 for the service. You then have to leave the country and pick up your business visa at the Vietnamese embassy in the country you choose to go to — usually Cambodia.

➲ **Requirements:** High School Diploma and a TESOL/TESL Certificate

➲ **Currency:** £1 = 31400 / $1 = 16100 Dong

➲ **Salary:** £520-750 / $950-1400 per calendar month, depending on the school and your experience. Be prepared to work longer hours in Vietnam (30 or more hours a week) than in other Asian countries. Overtime rates are around £7.50-10 / $12-18 per hour.

➲ **Accommodation:** You can get nice "shophouse" accommodation in Ho Chi Minh City for £120-240 / $220-450 a month. Hanoi is more expensive. Make sure your landlord has a permit to rent to foreigners or you could face eviction.

➲ **Cost of Living:** There are two sets of prices in Vietnam: one for locals and another (higher) for foreigners. Teachers who have lived for some time in the country still expect to pay 20-30% more than the Vietnamese. But remember that foreign English language teachers are paid far more than the average Vietnamese worker. You can live well on $500 a month. Sharpen your bargaining skills – it's expected of you to haggle over prices.

➲ **Health advice:** Avoid local hospitals if you can help it. Membership of a foreign-run medical service costs around £100 / $190 a year. You can get evacuation insurance for about £300 / $570 a year.

➲ **English Language Media:** Vietnam News.

Schools in Vietnam (+84)

American English School
357 Dien Bien Phu, District 3, Ho Chi Minh City. **Tel**: 8 8324226. **Fax**: 8342484.

Apollo Education Centre
67 Le Van Huu, Hanoi. **Tel:** 4-9432051. **Fax**: 4-9432052.

Duong Minh Language School
132c Phan Dang Luu, Phu Nhuan District, Ho Chi Minh City.

English 2000
70 Ding Tien Hoang, District 1, Ho Chi Minh City.

ELT Lotus
8 Nguyen Van Trang, District 1, Ho Chi Minh City.

Hanoi ELTE
10 Hang Chao St, Hanoi. **Fax:** 4 7332300.

Hollywood English centre
87B Tran Din Xu, District 1, Ho Chi Minh City.

Interaction Language School
59 Dinh Tien Hoang, District 1, Ho Chi Minh City.

New Star ELT centre
217 Doi Can Street, Cong Vi, Ba Dinh, Hanoi. **Tel:** 4 762 6696. **Fax**: 4 762 6699.

T.E.S.T. International
552 Dien Bien Phu, District 10, Ho Chi Minh City. **Tel:** 8 8334676.

Top Globis Training centre
3-7 P. Tan Phu, District 7, Ho Chi Minh City.

VATC (Vietnamese American Training College)
44 Truong Dinh St, District 3, Ho Chi Minh City. **Tel**: 8 8293971.

Latin America

Central and South American countries are attractive destinations for many American and Canadian language teachers, simply because a majority of these countries are not too far away and there are plenty of job opportunities.

Although teachers in some countries like Brazil, Mexico and Argentina can earn a good salary dollars, Latin America is not really a good destination for graduates trying to pay off their loans. However, teaching English in Latin America gives foreigners the opportunity to experience life in a vibrant and fascinating culture.

Mexico is probably the easiest place to find a job, and there are also well-paid positions in Brazil, Argentina, Colombia and Brazil. There are fewer opportunities in Peru, Chile, Venezuela, Uruguay, Ecuador and Bolivia, but even in these countries, there are prestigious schools, willing to pay the premium for native speakers. Even unqualified teachers will find jobs in most of these countries however; to get the better jobs you will need to have either a TESOL-related MA or a recognized certificate. Opportunities in Belize, Dominican Republic, the eastern Caribbean, Guatemala, Guyana, Haiti, Honduras, Nicaragua and Panama are mainly for qualified teachers through aid programmes. Salaries are usually paid in local currency and might not cover living costs, but some include supplements paid in foreign currency.

In South America, the best time of year to find a job is just before the beginning of the academic year in January or February, although many institutes operate year-round and are happy to recruit teachers at any time. In addition to jobs at universities, English-medium schools, bi-national centres and language schools, companies encourage their employees to learn and often employ in-company teachers.

Only the larger schools with an international presence recruit overseas, mainly because work permits require so much paperwork, so most teachers find work by approaching schools in places they want to stay in. It also helps if you send your CV beforehand and dress professionally when you go looking for work: you can usually revert to jeans and a t-shirt after you've got the job.

ARGENTINA

Following the economic and political uncertainties caused by the massive devaluation of the local currency, Argentine schools cannot pay much for teachers – but, what you are paid covers local costs (food and accommodation) since these too have been re-valued. Unqualified teachers are having a harder time finding work. Nowadays, there is plenty of competition for jobs teaching English from local English language graduates and English-speaking immigrants.

Nevertheless, a qualified teacher should have no problem finding a job here. October is a good time to start writing to potential employers who will want teachers who can start work in the following spring. Teachers who want to pick up a job while they're in the country should arrive in January. If you don't like big cities, consider Mendoza or Cordoba as alternatives to Buenos Aires. You could even travel out to Patagonia and chance your luck there.

Embassy in UK:
65 Brook Street, London W1Y 1YE
Tel: 020 7318 1300
www.argentine-embassy-uk.org

Embassy in USA:
1600 New Hampshire Ave NW,Washington DC20009
Tel: 202 238 6400
www.embajadaargentina-usa.org

British Council office:
Marcelo T de Alvear 590 4th Floor, C1058AAF, Buenos Aires
Tel +54 (11) 4311 9814/7519
www.britishcouncil.org.ar

➲ **Visas/Work Permits:** Most teachers work on tourist visas, which they renew every three months. For a longer-term contract, you need a work permit, which is arranged by your employer. Securing the correct permit can take up to eight months, so plan ahead.

➲ **Requirements:** TESOL/TESL Certificate

➲ **Currency:** £1 = 5.9 / $1 = 3.1 Pesos.

➲ **Salaries/Taxes:** Institutes pay on an hourly basis

➲ **Accommodation:** If you are paying for accommodation yourself, a one-bedroom apartment can cost £300 / $600 per month. Teachers often live in simple, family-owned hotels until they have saved some deposit money.

➲ **English Language Media:** Buenos Aires Herald.

➲ **Tip:** Many institutes expect teachers to provide their own supplies including copies of materials, blank tapes, videos, a tape player and books.

➲ **Health advice:** Local medical care is expensive, so private insurance is recommended. You should be protected against Hepatitis A; in some northern areas Malaria and Yellow Fever occur.

Schools in Argentina (+54)

ALICANA
San Martin 2293, 3000 Sanata Fe. **Tel:** 42 53 7567, **Fax**: 42 55 2026

American English Institute
Córdoba 392 3ºD, Buenos Aires. **Tel:** 4312-8646, **Fax**: 4312-8462

American English Institute
Gral.M. Belgrano 179, Paraná, Entre Rios. **Tel:** 0343 431-5018

American Training Co.
Viamonte577, Piso 7, 1053 Buenos Aires. **Tel:** 1 311 3699, **Fax**: 1 315 3573

AMICANA
Chile 985, 5500 Mendoza. **Tel:** 61 23 6271, **Fax**: 61 29 8702, **Email:** amicana@impsat.com.ar

ARICANA
Buenos Aires 934, 2000 Rosario. **Tel:** 41 21 7664, **Fax**: 41 21 9179, **Email:** aricana@interactive.com.ar

Asociacion Comodoro Rivadavia de Intercambio Cultural Argentino Norteamericano (ACRICANA)
Escalada 1567, 9000 Comodoro Rivadavia. **Tel:** 0297 446-6311. **Fax**: 0297 447-3956. **Email:** acricana@infovia.com.ar

Asociacion Mendocina de Intercambio Cultural Argentino Norteamericano (AMICANA)
Chile 987, 5500 Mendoza. **Tel**: 0261 423-6271. **Fax**: 0261 429-8702. **Email:** amicana2@impsat1.com.ar

Asociacion Paranense de Intercambio Cultural Argentino Norteamericano (APICANA)
Córdoba 256, 3100 Parana. **Tel/Fax:** 0343 431-9678. **Email:** apicana@arnet.com.ar

Asociacion Rosarina de Intercambio Cultural Argentino Norteamericano (ARICANA)
Buenos Aires 934, 2000 Rosario . **Tel:** 0341 421-7664. **Fax**: 0341 421-9179. **Email:** aricana@arnet.com.ar

Instituto Cultural Argentino Norteamericano (ICANA)
Maipú 672, 1006 Buenos Aires. **Tel:** 011 4322-3855/4557. **Fax**: 011 4322-2106. **Email:** secretaria@bcl.edu.ar

Instituto Dean Funes de Intercambio Cultural Argentino Norteamericano (IDFICANA)
9 de Julio 177, 5200 Dean Funes, Córdoba. **Tel:** 03521 420-738. **Fax**: 03521 421-001/20899

Instituto de Intercambio Cultural Argentino Norteamericano (IICANA)
Dean Funes 726, 5000 Cordoba. **Tel: 0**351 421-4026/425-4216/426-3941. **Fax**: 0351 423-7858. **Email:** IICANA@onenet.com.ar

ATICANA
Salta 581, 4000 San Miguel de Tucuman. **Tel:** 81 31 0616, **Fax**: 81 30 3070, **Email:** aticana@starnet.net.ar

Berlitz
Av. de Mayo 847, 1er Piso, 1084 Buenos Aires. **Tel:** 11 4342 0202 **Email:** info@berlitz.com.ar

Berlitz
Av. Del Libertador 15231, 1640 Acassuso. **Tel**: 11 4747 1871 **Fax**: 11 4732 3096, **Email:** info@berlitz.com.ar

British Council
Marcalo T De Alvera 590 (4th Floor), 1058 Buenos Aires. **Tel:** 54 1 311 9814.

Brooklyn Bridge
Reconquista 715, 6 C y D, 1003 CF, Buenos Aires. **Tel**: 1 313 1652

CAIT
Maipú 863 3ºC, Buenos Aires. **Tel/Fax:**4311-44/4314-2583

Centum Servicios de Idiomas
Bartlome Mitre, Piso 4, 1036 Buenos Aires. **Fax:** 1 328 5150/2385

English Studies
Catalina Hansen, Rodriguez Pena 238 40 A - 1020, CF, Buenos Aires. **Tel:** 1 371 5352, **Fax**: (21) 720393.

ICANA
Maipu 672, 1006 Buenos Aires. **Tel:** 1 322 3855, **Fax**: 1 322 2106, **Email:** icana@bcl.edu.ar

IELI
Alberti 6444, San Jose de la Esquina, Santa Fe 2185.

Instituto Cultural Argentino-Britanico
Calle 12, No 1900, La Plata.

Instituto ELT
Isabel Gonzalez Bueno, Soler 458 (1714) Ituzaingo, Buenos Aires. **Tel/Fax:** (541) 624 0148.

International House
IH Belgrano, Arcos 1830, 1428 CF, Buenos Aires. **Tel:** 541 785 4425.

Landmark
Marcelo T. de Alvear 2331 1ºD, Buenos Aires. **Tel/Fax:** 4825-3806, **Email:** landmark@arnet.com.ar

Language Network
Billinghurst 2476 piso 2, Buenos Aires. **Tel:** 4805-6586, **Fax**: 4805-8452.

Liceo Superior de Cultura Inglesa
Italia 830, Tandil, 7000 Pica de Buenos Aires.

Northside School of English
G W Seminario, Velez Sarsfield 56, 1640, Martinez, Buenos Aires. **Tel/Fax:** 798 5150/793 5469.

St John's School
Recta Martinoli 3452, V Belgrano 5417 Cordoba, Pica de Cordoba.

TOKEN
Av.de Mayo 1260-5°Q, Capital Federal. **Tel:** 4383-0751/15-4979-3929 **Email:** houses@infovia.com.ar

BOLIVIA

Bolivia is currently enjoying a rare period of political and economic stability although there has been a recent increase in street protests because of popular dissatisfaction with government policies. Bolivians are keen to learn English and travelling teachers are attracted here because it is cheap to stay and culturally fascinating.

Embassy in UK:
106 Eaton Square, London SW1W 9AD
Tel: 020 7235 4248

Embassy in USA:
3014 Massachusetts Ave NW, Washington,DC20008
Tel: 202 483 4410

British Council office:
Avenida Arce 2708 (esq.Campos), Casilla 15047, La Paz
Tel: +591 (2) 2431 240
www.britishcouncil.org/bolivia

➲ **Visas/Work Permits:** All nationalities need a tourist visa to enter. Work permits are obtained by employers for contracts of at least one year in duration. They cost around £240 / $450 and you may be asked to make a small payment (around £12-30 / $22-57) towards the cost.

➲ **Requirements:** TESOL/TESL Certificate

➲ **Currency:** £1 = 14.91 / $1 = 8.06 Bolivianos

➲ **Salaries and taxes:** As low as £2 / $4 per hour in a language school or £6/ $11 per hour for private classes.

➲ **Accommodation:** Cheap and easy to find. Some schools will supply accommodation.

➲ **Cost of Living:** Bolivia is an inexpensive country to live in.

➲ **Health advice:** La Paz is situated 13,000 feet above sea level and many teachers report that it takes a while to become acclimatized to the altitude. Medical facilities, even in La Paz, are not adequate to handle serious medical conditions, such as cardiac problems. You should be vaccinated against Hepatitis A, Malaria, and Yellow Fever.

School in Bolivia (+591)

Centro Boliviano Americano
Calle 25 de Mayo N-0365, Casilla 1399, Cochabamba. **Tel:** 42 21288, 22518. **Fax**: 42 51225. **Email:** suarezmj@llajta. nrc.bolnet.bo

Centro Boliviano Americano
Parque Iturralde Zenon 121, Casilla 12024, La Paz. **Tel:** 2 430107 **Fax**: 2 431342. **Email:** cbalp@datacom-bo.net

Centro Boliviano Americano
c/ Cochabamba 66, Casilla 510, Santa Cruz de la Sierra. **Tel:** 3 342299 **Fax**: 3 350188 **Email:** cbascz@datacom.bo.net

Centro Boliviano Americano
Calle Calvo No. 331, Casilla No. 380, Sucre, Chuquisaca. **Tel:** 64 41608. **Fax**: 64 41608. **Email:** cba@nch.bolnet.bo

Colegio Ingles Saint Andrews
Av. Las Relamas, La Florida, La Paz. **Tel:** 2 792484.

Colegio San Calixto
C/Jenaro Sanjines 701, La Paz. **Tel:** 2 35 5278.

Colegio San Ignacio
Av. Hugo Ernest 7050, Seguencoma, La Paz. **Tel:** 2 783720.

Pan American English Centre
Edificio Avenida, Avenida 16 de Julio 1490, 7º piso, Casillo 5244, La Paz. **Tel/Fax:** 2 340796.

Tito's Place
P.O. Box 3112, Cochabamba. **Tel:** 4 289308, **Fax**: 4 288836. **Email:** mahergar@comteco.entelnet.bo

BRAZIL

English language teaching is booming in Brazil and it is still quite easy to find a job here. Many parents are keen to send their children to private schools to learn English. There are around 60 bi-national centres specifically catering for Brazilians who want to learn English.

Embassy in UK:
32 Green Street, London W1Y 4AT
Tel: 020 7499 0877
www.brazil.org.uk

Embassy in USA:
3006 Massachusetts Ave NW, Washington DC20008
Tel: 202 238 2700
www.brasilemb.org

British Council office (see also appendix):
Ed.Centro Empresarial Varig, SCN Quadra 04 Bloco B Torre Oeste Conjunto 202, Brasilia – DF, 70710-926
Tel +55 (0) 61 327 7230
www.britishcouncil.org/brazil

➲ **Visas/Work Permits:** It is very difficult to obtain a work visa so some schools are willing to hire teachers who are in the country on a tourist visa although this is illegal. It is your responsibility to arrange for a work visa, which can take up to two months to get. You need to be qualified and have at least two years' experience to qualify. It is possible to arrive, find a job, apply for a visa and pick it up in a neighbouring country, although this is not popular with the bureaucrats. Otherwise, the process should be initiated through the Brazilian consulate at home.

➲ **Requirements:** TESOL/TESL Certificate

➲ **Currency:** £1 = 4.19 / $1 = 2.14 Reals.

➲ **Salaries/Taxes:** Teachers are paid for 13 months and salaries are higher in cities. Full-time salaries start at around £220 / $400 a month, plus basic accommodation. Incomes can be supplemented with private classes starting at £6 / $11 per hour. Experienced teachers with a Masters degree can earn up to £1500 / $2700 a month. Tax varies from 13-25% and there is a statutory social security deduction.

➲ **Accommodation:** Better schools will either provide lodging for free or offer a subsidy. Otherwise, expect to pay between a third and half of your salary. Apartments are unlikely to be furnished.

➲ **Cost of Living:** Food is cheap but rent is moderately expensive.

➲ **Health advice:** Insurance can be taken out locally but it does not cover dental treatment.

➲ **English Language Media:** Brazil Herald.

Schools in Brazil (+55)

Britannia Schools
Rua Garcia D'Avila 58, Ipanema, Rio De Janeiro RJ 22421-010. **Tel:** 21 511 0143, **Fax**: 21 511 0893.

Britannia Special English Studies
Rua Dr Timoteo, 752 Moinhos De Vent, Porto Alegre RS.

Britannia Juniors
Rua. Barao da Torre 599 - Rio de Janeiro, CEP 22411-003, RJ. **Tel:** 21 239-8044, **Fax**: 21 259 6197.

Britannia Executive School
Rua Barao De Lucena 61, Botofogo, Rio de Janeiro 22260.

Britannic International House
Rua Hermogenes de Morais 178, Recife PE 50610-160. **Tel:** 81 445 5564, **Fax**: 81 445 5481.

British House
Rua Tiradentes 2258, Centro, Pelotas RS 96060-160. **Tel:** 532 27 3139, **Fax:** 532 27 5604.

Cambridge Sociedade Brasileira de Cultura Inglesa
Rua Piaui 1234, Londrina 86020-320 PR. **Tel:** 43 324 1092, **Fax**: 324 0314.

Casa Branca
Rua Ma-chado De Assis 37, Boqueirao, Santos SP. **Tel/Fax:** 13 233 5258.

CEBEU
Av Marechal Rondon, 745, Centro, Ji-Paraná RO. **Tel**: 69 422 3100, **Fax**: 69 422 3100.

CEL-LEP
Av. dos Tajuras 212, Sao Paulo SP 05670-000. **Tel:** 11 212 6183, **Fax**: 11 210 0698.

CEMID
Rua Santa Luzia, 799 conj. 401, Rio de Janeiro - RJ 20030-040. **Tel:** (21) 262-1793 **Fax**: (21) 240-9271

Centro Britanico
Rua Joao Ramalho 344, Sao Paulo SP 05008-011. **Tel:** 11 622984, **Fax**: 11 872 9483.

Centro Cultural Brasil Estados Unidos
Avenida T 5, No. 441, 74230-040 Goiania GO. **Tel:** 62 833 1313, **Fax**: 62 833 1308.

Centro Cultural Brasil-Estados Unidos
Avenida Julio de Mesquita, 606 13025-061 Campinas – SP. **Tel:** 192 51-3664/52-6668. **Fax**: 192 52-6888

Centro Cultural Brasil-Estados Unidos
Rua Amintas de Barros, 99, Edifico Itatiaia Centro Caixa Postal 3328, 80060-200 Curitiba – PR. **Tel:** 41 320-4777. **Fax**: 41 238-2822

Centro de Enseñanza PLI
Rua de Octubro, 1234 Conj 4, Porto Alegre, RS 90000.

Centro De Cultura Inglesa
Av Guapore 2.236, Cacoal RO, CEP 78 975-000. **Tel**: (69) 441-2833, **Fax**: 441-5346.

Cultura Inglesa
Rua do Progresso 239, Recife PE 50070-002. **Tel:** 81 423 6266, **Fax**: 81 231 2318.

Cultura Inglesa
Av. 17 de Agosto 233, Recife PE 52060-090. **Tel:** 81 268 6938, **Fax**: 81 268 6787.

Cultura Inglesa
Rua Visconde Albuquerque 205, Recife PE 50610-090. **Tel:** 81 228 6649, **Fax**: 81 228 6649.

Cultura Inglesa
Rua Visconde de Inhauma 980, Ribeirao Preto SP 14010-100. **Tel:** 16 610 6616, **Fax**: 16 610 9868.

Cultura Inglesa
Rua Paul Pompeia 231, Rio de Janeiro RJ 22080-000. **Tel:** 21 287 0990, **Fax**: 21 267 6474.

Cultura Inglesa
R. Plinio Moscoso 945, Salvador BA 40155-020. **Tel:** 71 247 9788, **Fax**: 71 245 3287.

Cultura Inglesa
Av. Sao Sebastiao 848, Santarem PA 68005-090. **Tel:** 91 522 7247.

Cultura Inglesa
Rua Sao Sebastiao 1530, Sao Carlos SP 13560-230. **Tel:** 16 272 2276, **Fax**: 16 272 9875.

Cultura Inglesa
Av. Tiradentes 670, Sao Joao del Rei MG 36300-000. **Tel:** 32 371 7035, **Fax**: 32 371 2764.

Cultura Inglesa
Av. Brig. Faria Lima 2000, Sao Paulo SP 01452-002. **Tel:** 11 870 4955, **Fax**: 11 813 1945.

Cultura Inglesa
R. Joao Pinheiro 808, Uberlandia MG 38400-58. **Tel:** 34 235 2786, **Fax**: 34 236 9250.

Cultura Inglesa
Praca Rosalvo Ribeiro 110, Maceio AL 57021-57. **Tel:** 82 221 8055, **Fax**: 82 223 3785.

Cultura Inglesa
R. Eng Mario de Gusmao 603, Maceio AL 57035-000, **Tel:** 82 231 8687. **Fax**: 82 231 8132.

Cultura Inglesa
Rua Natal 553, Manaus AM 69005-000. **Tel:** 92 611 1635, **Fax**: 92 611 1635.

Cultura Inglesa
Av. Rio Branco 741, Maringa PR 87015-380. **Tel/Fax**: 44 225 1518.

Cultura Inglesa
Rua Acu 495, Natal RN 59020-110. **Tel:** 84 211 6070, **Fax**: 84 211 4559.

Cultura Inglesa
R. Silvio Henrique Braune 15, Nova Friburgo RJ 28625-050. **Tel:** 245 22 5392, **Fax**: 245 22 5392.

Cultura Inglesa
R. Eduardo de Moraes 147, Olinda PE 53030-250. **Tel/Fax:** 81 429 2281.

Cultura Inglesa
Av. Bernardo Vieira de Melo 2101, Jaboatao dos Guararapes PE 54 410-010. **Tel:** 81 361 3458, **Fax**: 813610467.

Cultura Inglesa
Rua Paula Xavier 501, Ponta Grossa PR 84010-430. **Tel/Fax:** 42 223 2735.

Cultura Inglesa
Rua Quintino Bocaiuva 1447, Porto Alegre RS 90570-010. **Tel/Fax:** 51 333 4033.

Cultura Inglesa
Rua Mamanguape 411, Recife PE 51020-50. **Tel:** 81 326 1908, **Fax**: 81 326 7618.

Cultura Inglesa
Alameda Julia da Costa 1500, Curitiba PR 80730-070. **Tel:** 41 222 7339, **Fax**: 41 224 1024.

Cultura Inglesa
Rua Ponto Grossa 1565, Dourados MS 79824-160. **Tel:** 67 421 7147, **Fax**: 67 421 5753.

Cultura Inglesa
Rua Conde de Porto Alegre 59, Buque de Caxias RJ 25070-350. **Tel:** 21 671 2543, **Fax**: 21 671 4346.

Cultura Inglesa
Rua Rafael Bandeira 335, Florianapolis SC 88015-450. **Tel/Fax:** 48 224 2696.

Cultura Inglesa
Rua Ana Bilhar 171, Fortaleza CE 60160-110. **Tel:** 85 244 3784, **Fax**: 85 224 4665.

Cultura Inglesa
Rua Marechal Deodoro 1326, Franca SP 14400-440. **Tel/Fax:** 16 722 0011

Cultura Inglesa
Rua 86 No. 7, Golania GO 74083-330. **Tel:** 62 241 4516, **Fax**: 62 241 2582.

Cultura Inglesa
Rua 20 778, Ituiutaba MG 38300-000. **Tel:** 34 268 1681, **Fax**: 34 261 5975.

Cultura Inglesa
Av. Rio Grande do Sul 1411, Joao Pessoa PB 58030-021. **Tel:** 83 224 7005, **Fax**: 83 224 9479.

Cultura Inglesa
Rua Dr. Joao Colin 559, Joinville SC 89204-004. **Tel:** 47 433 7603, **Fax**: 47 433 4512.

Cultura Inglesa
Av. dos Andras 536, Juiz de Fora MG 36036-000. **Tel:** 32 215 5169, **Fax**: 32 215 9659.

Cultura Inglesa
Av. Barao de Maruim 761, Aracaju SE 49015-020. **Tel/Fax:** 79 224 7360.

Cultura Inglesa
Rua Almeida Campos 215, Araxa MG 38180-00. **Tel/Fax:** 34 661 3275.

Cultura Inglesa
Rua Virgilio Malta 1427, Bauru SP 17040-440. **Tel/Fax:** 142 23 3016.

Cultura Inglesa
R. Fernandes Tourinho 538, Belo Horizonte MG 30112-011. **Tel:** 31 221 6770, **Fax**: 31 225 1791.

Cultura Inglesa
Rua Mal. Floriano Peixoto 433, Blumenau SC 89010-000. **Tel/Fax:** 47 326 7272.

Cultura Inglesa
SEPS 709/908 Conjunto B DF 70390-89. **Tel:** 61 244 5650, **Fax**: 61 244 8571.

Cultura Inglesa
Av. Guapore 2236, Cacoal RO 78975-000. **Tel:** 69 441 2833. **Fax**: 69 441 4547.

Cultura Inglesa
Rua Lino Gomes da Silva 53, Campina Grande PB 58107-613. **Tel/Fax:** 83 322 4658.

Cultura Inglesa
R. Humberto de Campos 419, Campo Grande MS 79020-060. **Tel:** 67 751 0272, **Fax**: 67 751 0570.

Cultura Inglesa
Av. Agamenon Magalhaes 634, Caruaru PE 55000-000. **Tel/Fax:** 81 721 4749.

Cultura Inglesa
Rua Antonio Ataide 515, Vila Velha ES 29100-290. **Tel:** 27 229 5194, **Fax**: 27 229 3206.

Cultura Inglesa
Av Tiradentes 670, 36300 Sao Joao Del Rei MG. **Tel**: 32 371-4377, **Fax**: 32 371-4377.

Cultura Inglesa de Londrina
Rua Goias 1507, Centro, Londrina PR 86020-340. **Tel/Fax:** 43 323 7700.

ELC
Rua Sa e Souza 655, Boa Viagem, Recife PE, 51030-350. **Tel:** 81 342 0351.

English Forever
Rua Rio Grande Do Sul 356, Pituba, Salvador-BA 41830-140. **Tel:** 71 240 2255, **Fax**: 71 248 8706, **Email:** forever@svn.com.br

George Otto
Av. Vereador de Campo 525, Sala 4, Ibiuna - SP 18150-000. **Tel:** (15) 249-1159 **Fax**: (15) 249-1159

Global Team
Caixa Postal 6504, Campinas - SP 13084-970. **Tel**: (19) 287-4427 **Fax**: (19) 287-4427

Independent British Institute
SHCGN 703 Area Especial, s/no Brasilia DF 70730-700. **Tel:** 61 322 8373, **Fax**: 323 5524, **Email:** ibi@nutecnet.com.br

Instituto Britanico
Rua Deputado Carvalho Deda 640, 49025-070 Salgado Filho, Aracaju SE, Brazil. **Tel:** 79 23 2791, **Fax**: 79 27 2645.

Instituto da Lingua Inglesa
Av. do CPA 157, Cuiaba MT 78008-000. **Tel/Fax:** 65 624 1197.

International House-Matriz
The School House, Rua 4, 80, Goiania GO 74110-140, Goais, SE.

International House Goiania
Rua 04 nº 80, Setor Oeste, Goiania GO 74110-140. **Tel:** 62 224 0478, Fax: 62 223 1846.

Liberty English Centre
Rua Amintas De Barros 1059, Curitiba PR. **Tel:** 41 263 3586, **Fax**: 41 262 1738.

New Vision English
Av. Ibirapuera, 2249 São Paulo - SP 01248-001. **Tel**: (11) 241-4199 **Fax**: (11) 241-4199 **Email:** newvision@newvision.com.br

PBF - Pink and Blue – Freedom
Rua Santos, 54 Joinville - SC 89202-46. **Tel:** (47) 433-1471 **Fax**: (47) 433-1471 **Email:** pbfjlle@netville.com.br

Sharing English
Rua Souza de Andrade 56, Recife PE 52050-300. **Tel/Fax:** 81 421 2286.

Schütz & Kanomata
Rua Galvão Costa, 85 Santa Cruz do Sul – RS. **Tel:** (51) 711-2248 **Fax**: (51) 715-3366 **Email:** english@viavale.com.br

Skill
Av. Indianópolis, 3356 São Paulo - SP 04062-003. **Tel:** (11) 5589-9535 **Fax**: (11) 5589-9447 **Email:** fruggie@ibm.net

Sky Cursos e Traduções SC Ltda
Av. Senador Virgílio Távora, 867 - Meireles Fortaleza - CE 60170.250. **Tel:** (85) 224-7879 **Fax**: (85) 224-7879 **Email:** sky@fortalnet.com.br

Soc Bras de Cultura Inglesa
Rua Fernandes Tourinho, 538 - Savassi BH-MG 30112.000, **Tel:** 031 221 6770.

System 2000
Rua Deputado Jose Lajes 491, Ponta Verde, Maceio AL 57035-330. **Tel:** 82 231 7808, **Fax**: 82 327 4946.

St. Peter's English School
Rua Berilo Guimaraes, 182 Centro Itabuna, Bahia.

The English Office
Rua D. Eponina Afonseca, 116 São Paulo - SP 04720-010. **Tel:** (11) 546-0796 **Fax**: (11) 546-0796

Universitas
Rua Goncalves Dias 858, Belo Horizonte MG 30140-091. **Tel:** 031 261 1477, **Fax**: 031 261 5138.

UP Language Consultants
Av. Brig. Faria Lima, 1912, 18m, Jd. Paulistano, Calcerter Sao Paulo - SP 01451-907. **Tel:** (11) 211-9990 **Fax**: (11) 212-5366 **Email:** up@ensino.net

Upper English
Rua 09 de Julho 2143, Sao Carlos SP 13560-590. **Tel**: 16 271 8146, **Fax**: 16 271 5431.

Vision English Escola de Inglês S/C Ltda.
Av. Ibirapuera, 2249 São Paulo - SP 04029-100.**Tel:** (11) 241-4199 **Fax**: (11) 241-4199 **Email:** newvision@newvision.com.br

CHILE

The government has placed improving Chile's school system and developing foreign export markets at the top of its economic agenda, so learning English is an educational priority.

Although it is impossible to work in state schools without a qualification from a Chilean university, there are many private schools and in-company opportunities, especially in Santiago.

Foreigners coming to Chile with the intention of teaching English should preferably have both qualifications and experience. Teachers are trained to a high standard in Chile and competition is tough.

Embassy in UK:
12 Devonshire Street, London W1N 2DS
Tel: 020 7580 1023

Embassy in USA:
1732 Massachusetts Ave NW, Washington DC20008
Tel: 202 785 1746
www.chile-usa.org

British Council office:
Eliodoro Yáñez 832, Providencia 6640356, Casilla 115 Correo 55, Santiago
Tel +56 (2) 410 6900
www.britcoun.cl

➲ **Visas/Work Permits:** Applications for visas and work permits should be made before arrival. It is possible to obtain a work permit after arrival but you have to show that you intend to stay for at least one year. Working on a tourist visa is possible although illegal.

➲ **Requirements:** (A degree is recommended) TESOL/TESL Certificate

➲ **Currency:** £1 = 1024 / $1 = 524 Pesos

➲ **Salaries/Taxes:** For contracted teaching positions, a minimum wage for overseas teachers is set by the government at £250 / $460 per month. Rates vary from £5-7 / $9-13 per hour for qualified teachers. Private lessons pay £10 / $19 an hour and up. The tax rate is around 10%

➲ **Accommodation:** Schools offer little help, but it's easy to stay in hostels, which offer long-term accommodation with good facilities. Rooms can cost as little as £80 / $150 per month but average around £150 / US$300. Apartments can cost up to £250 / US$470 per month and landlords expect a month's rent as deposit and a month in advance.

➲ **Cost of Living:** Low but higher than much of Latin America.

➲ **English Language Media:** News Review.

➲ **Health advice:** Medical care is generally good, but it may not meet western standards. Supplemental medical insurance, which includes specific overseas coverage, including provisions for medical evacuation, is highly recommended, as in-country medical evacuations from outlying areas to Santiago cost £1,500 / $2,700 or more.

Schools in Chile (+56)

Abbey Road Cultural De Idiomas
Gorostiaga 608 P. 2, Iquique. **Tel:** 57 428737 **Fax**: 57 427147 **Email:** Abbeyroa.Dcult001@chilnet.cl

Acpen Academy
Av. Lib. B. O'higgins 949 P. 21 Of. 2103, Santiago. **Tel:** 2 6724460 **Fax**: 2 6724460 **Email:** acpenaca.demy001@chilnet.cl

American Language Institute
Lord Cochrane 286, Santiago, **Tel:** 2 6955487 **Fax**: 2 6955487 **Email:** american.langu001@chilnet.cl

Anglo-American International School
San Sebastian 2975, Santiago. **Tel:** 231 1771.

Antofagasta British School
Pedro Leon Gallo 723, Casilla 1, Antofagasta. **Tel:** 241 368.

Berlitz
Padre Mariano 305, Santiago. **Tel:** 2 2361557 **Fax**: 2 2361563 **Email:** Berlitz_Chile@entelchile.net

British Council
Eliodora Yanez 832, Casilla 115-Correo 55, Santiago. **Tel:** 22361199.

British School
Waldo Seguel 454, Casilla 379, Punta Arenas. **Tel:** 223 381, **Fax**: 248 447.

British High School
Los Gladiolos 10281, Santiago. **Tel:** 217 2204.

Burford Institute
Av. Pedro De Valdivia 511, Santiago. **Tel:** 2 2744603 **Fax**: 2 2235944 **Email:** info@burford.cl

Cie
Orella 1089, Antofagasta. **Tel:** 55 268339 **Fax**: 55 268339 **Email:** Capacita.Cioni001@chilnet.cl

Centro Chileno Canadiense Idiomas Ltda.
Av. Luis Thayer Ojeda 0191 P. 6 Of. 601, Santiago. **Tel:** 2 3341090 **Fax**: 2 3341089 **Email:** centroch.ileno001@chilnet.cl

Colegio Charles Darwin
Manantiales 0314, Punta Arenas. **Tel:** 212 671.

Colegio Dunalastair
Av. Las Condes 11931, Santiago. **Tel:** 215 2666.

Colegio Ingles George Chaytor
Callejon Ingles 4B, Temuco. **Tel:** 211 301.

Colegio Ingles De Talca
12 Norte 5/6 Oriente, Talca. **Tel:** 226692.

Colegio St George
Av. Americo Vespucio Norte 5400, Santiago.

Colegio Del Verbo Divinio
Av. Presidente Errazuriz 4055, Santiago. **Tel:** 228 6076.

Colegio Villa Maria Academy
Av. Presidente Errazuriz 3753, Santiago. **Tel:** 228 3398.

Computacion E Ingles Intercom Ltda.
Av. Pedro Montt 2053, Valpo. **Tel:** 32 254436 **Fax**: 32 254436 **Email:** Intercom.Pescu001@chilnet.cl

Craighouse
El Rodeo 12525-La Dehesa, Casilla 20007-Correo 20, Santiago. **Tel:** 242 4011, **Fax**: 215 7400.

Fischer English Institute
Cirujano Guzman 49, Santiago, **Tel:** 2 2356667 **Fax**: 2 2359810 **Email:** Fischeri.Nstit001@chilnet.cl

Grange School
Av. Principe De Gales 6154, Casilla 51-Correo 12, Santiago. **Tel:** 277 1181, **Fax**: 277 0946.

Greenhouse School
Ines De Suarez 1500, Temuco. **Tel:** 240 840.

Henry Renna Instituto
Mark Twain 7241, Santiago. **Tel:** 2 2199044 **Fax**: 2 2193013 **Email:** Henryren.Nains002@chilnet.cl

Instituto Chileno-Britanico De Cultura
Baquedano 351, Casilla 653, Arica. **Tel:** 58 238399, **Fax**: 58 231960.

Instituto Chileno-Britanico De Cultura
3 Norte 824, Casilla, Vina Del Mar. 929. **Tel:** 32 971061, **Fax**: 32 686656.

Instituto Chileno-Norteamericano De Cultura
Moneda 1467, Santiago. **Tel:** 2 238 6107, **Fax**: 2 698 1175, **Email:** infocent@hood.ichn.cl

Instituto Chileno Norteamericano
Caupolican 315, Casilla 612, Concepcion. **Tel:** 41 248 589, **Fax**: 41 233 851 (Also In Antofagasta And Valparaiso.)

Instituto De Ingles Impact S.A.
Rosa O'higgins 259, Santiago. **Tel:** 2 2125609, **Fax**: 2 2116165

Instituto De Ingles Lorbeth
Cabo Arestey 2468 Of., Santiago. **Tel:** 2 6956018, **Fax**: 2 6956018,

International Preparatory School
Pastor Fernandez 16001, El Arrayan, Santiago. **Tel**: 215 1094.

Iquique English College
Jose Joaquin Perez 419, Iquique. **Tel:** 411 2409.

Let's Do English
Villa Vicencio 361, Office 109, Santiago.

Lincoln International Academy
Camino San Antonio 55- Las Condes, Santiago. **Tel:** 217 1907.

Linguatec
Av. Los Leones 439, Santiago. **Tel:** 2 2334356, **Fax**: 2 2341380, **Email:** Michelle@Linguatec.cl

Lm Language centre
Agustinas 853 Of. 301, Santiago. **Tel:** 2 6396419, **Fax**: 2 6396419

Mackay School
Vicuna Mackenna 700, Casilla 558, Vina Del Mar. **Tel:** 832 574 **Fax**: 832 419.

Mayflower School
Las Condes 12167, Santiago. **Tel:** 217 1085.

Nido De Aquilas School
Nido De Aquilas 14515, Casilla 16211-Providencia, Santiago. **Tel:** 216 6842.

Polyglot
Villavicencio 361 Of. 102, Santiago. **Tel:** 2 6398078, **Fax**: 2 6322485, **Email:** info@polyglot.cl

Redland School
Camino El Alba 11357, Las Condes, Santiago. **Tel:** 2 214 1265, **Fax**: 2 214 1020.

St Gabriel's School
Av. Bilbao 3070-Providencia, Santiago. **Tel:** 204 1915, **Fax**: 225 2136.

St John's School
Pedro De Valdivia 1783, Casilla 284, Concepcion. **Tel:** 331 670, **Fax**: 340 809.

St Margaret's School
5 Norte 1351, Casilla 392, Vina Del Mar. **Tel:** 977 000, **Fax**: 975 220.

St Peter's School
Calle Libertad 575, Vina Del Mar. **Tel**: 971 993.

Santiago College
Los Leones 584, Casilla 130-D, Santiago. **Tel:** 2 238 1813, **Fax**: 2 238 0755.

Thewhla's English School
Las Camelias 2854, Santiago.

Tronwell
Av. Apoquindo 4499 P. 1-2-3, Santiago. **Tel:** 2 2461040, **Fax**: 2 2289739, **Email:** tronwell.sa001@chilnet.cl

Wenlock School
Carlos Pena Otaegui 10880, La Foresta, Los Dominicos, Santiago. **Tel**: 212 8982, **Fax**: 212 9226.

Wessex School
Colo-Colo 222-Clasificador 43, Concepcion. **Tel:** 315436.

Windsor School
Av. Francia Esq. Simpson, Casilla 530, Valdivia.

COLOMBIA

Although Colombia has a reputation for violence, the reality is that, except for a few "no-go" zones, the country is no more dangerous than others in the region. In fact, many teachers report that Bogota and even Medellin are fun cities to live and work in as long as you don't stray into the "bad" areas. Schools welcome native English speakers because Colombia's image has deterred many of them from travelling here to teach. There is a network of Colombian-American Cultural centres and many of the private schools are American-owned. Bilingual high schools offer good rates of pay and working conditions.

This means that most EFL teachers, including unqualified ones looking for short-term positions while travelling, can get a job pretty easily. However, you may have to settle for a low wage initially, and it might be only just enough to live on. If you want to stay more than a few months, it is advisable to get qualified first so that you can get a job in one of the more reputable schools and build up contacts to secure more lucrative private classes.

Consulate in UK:
15-19 Great Titchfield Street, London W1P 7FB
Tel: 020 7495 4233

Embassy in USA:
2118 Leroy Pl NW, Washington DC20008
Tel: 202 387 8338
www.colombiaemb.org

British Council office:
Calle 87 No 12 - 79, Bogota
Tel +57 (1) 618 7680
www.britishcouncil.org.co

➲ **Visas/Work Permits:** Teachers must have a work permit before they travel and, unusually, a private language school cannot intervene directly with the Ministry of Foreign Affairs to secure this visa. Only those employed by approved institutions can secure the necessary work visa and the application process is associated with red tape, delays and expense, so make sure that your school is one of these institutions. If you want to insure that your visa is legal, your degrees and transcripts must be validated with a rubber stamp from the Colombian consulate.

➲ **Requirements:** TESOL/TESL Certificate

➲ **Currency:** £1 = 4436 / $1 = 2290 Pesos

➲ **Salaries/Taxes:** Many schools offer no more than £2 / $4 per hour and unqualified teachers can expect to earn as little as £1.50 / $3. Average monthly salaries are £300 / $570 and it is possible to earn £525 / $950. Many institutions pay teachers 14 times per year and, with a normal work visa, teachers will pay tax of 10% or more. Private lessons can earn you up to £12 / $20 an hour or more.

➲ **Accommodation:** Expensive. In Bogota and most other cities, a one-bedroom apartment in a safe area costs about £200 / $380 per month and administration fees are often charged on top.

➲ **Cost of Living:** The cost of living is expensive as inflation is still high. However, once established, teachers can have a good standard of living.

➲ **Health advice:** Travellers to Bogota may require some time to adjust to the altitude (8,600 feet), which can adversely affect blood pressure, digestion and energy levels

Schools in Colombia (+57)

Academia Ingles Para Niños
Calle 106 No 16-26, Bogota.

Advanced Learning Service
Transversal 20 No 120-15, Bogota.

American School Way
Dg. 110 No. 42-96 Casa 4, Bogota. **Tel:** 1 6198356

Aprender Ltda
Calle 17 No 4-68 Ql. 501, Bogota.

Aspect Language Schools Ltda.
Cl. 90 No. 11-44 Of. 301, Bogota. **Tel:** 1 2569827

BBC De Londres
Calle 59 No 6-21, Bogota.

Babel
Avenida 15 No 124-49 Cf. 205, Bogota.

Berlitz
Tr. 26 No. 146-60 L. 201, Bogota. **Tel:** 1 6275294

Berlitz de Colombia S.A.
Tr. 26 No. 146-60 L. 201, Medellin. **Tel:** 4 6250554

Best English Services Today
Dg. 110 No. 42-85 Int. 2 Ap. 103, Bogota. **Tel:** 1 2156380

Bi-Cultural Ltda.
Cl. 98 No. 10-32, Bogota. **Tel**: 1 6231478

Bi Cultural Institute
Avenida 7 No 123-97 Of. 202, Bogota.

Boston School of English Ltda
Carrera 43 No 44-02, Barranquilla.

Británico Americana
Cr. 40 No. 105A-46, Bogota. **Tel:** 1 2180045

The British Council
Calle 87 No 12-79, Bogota.

Business Language Centre Ltda.
Carrera 49 No 15-85, Medellin.

California English Institute
Cl. 46 No. 8-10 P. 2, Bogota. **Tel:** 1 2456429

California Institute Of English
Carrera 51 No 80-130, Barranquilla.

Carol Keeney
Carrera 4 No 69-06, Bogota.

Ceico
Calle Siete Infantes, San Diego, Cartagena.

Centro Anglo Frances
Carrera 11 No 6-12, Neiva.

Centro Audiovisual De Ingles Chelga
Calle 137 No 25-26, Bogota.

Centro Colombo Andino
Calle 19 No 3-16 Of. 203, Bogota.

Centro Colombo Americano
Cr. 45 No. 53-24, Medellin. **Tel:** 4 513 4444

Centro Cultural Colombo Americano
Carrera 43, No. 51-95, Apt. Aereo 2097, Barranquilla.

Centro de Idiomas Teachers Inc.
Cr. 15 No. 75-35, Bogota. **Tel:** 1 5456712

Centro de Inglés Lincoln
Cl. 49 No. 9-37, Bogota **Tel:** 1 2852224

Centro De Idiomas Winston Salem
Calle 45 No 13-75, Bogota.

Centro De Idiomas Winston Salem
Transversal 74 No C2-33 Laureies, Medellin.

Centro De Idiomas Winston Salem
Ave-nida La Ceste No 10-27, Santa Teresita, Cali.

Centro De Idiomas y Turismo De Cartagena
Popa Calle 30 No 20- 177, Cartagena.

Centro De Lengua Inglesa
Calle 61 No 13-44 Of. 402, Bogota.

Centro De Lenguas Modernas
Carrera 38 No 69 C 65, Barranquilla.

Centro Educativo Bilingüe English centre Envigado
Cr. 40 No. 38 S-55 Int. 201-202-203, Medellin. **Tel:** 4 3312211

Clever Training Corp
Av. Cr. 15 No. 122-51 Of. 509-510, Bogota. **Tel:** 1 2143239

Coningles Ltda
Cl. 71 No. 13-56, Bogota. **Tel:** 1 3128646

Darlitz English Institute Ltda.
Cr. 11A No. 69-89, Bogota. **Tel**: 1 3101780

Dynamic English
Cl. 119 No. 13A-84, Bogota. **Tel:** 1 3102277

Easy English
Carrera 45 A No 34 Sur 29 Torre No 4, Portal Del Cerro , A. A. 80511, Envigado, Medellin.

Eccles
Cr. 13 No. 55-21 P. 3, Bogota. **Tel:** 1 2550552

EF Educación Internacional
Cr. 38 No. 10A-40, Bogota. **Tel:** 1 2664275

EF Learn a Language
Cl. 76 No. 9-66, Medellin. **Tel**: 4 3172180

El Centro Inglés
Cl. 11A No. 43D-14 P. 3, Medellin. **Tel:** 4 3114888

El Centro Inglés
El Poblado Carrera 10 A.No 36-39, Medellin.

Elci
Cr. 18 No. 89-16, Bogota. **Tel:** 1 2183535

ELS Language centres
Cl. 118 No. 25-52, Bogota. **Tel:** 1 6299774

English For Infants (John Dewey)
Diagonal 110 No 40-85, Bogota.

English Hard Practice for School and Company Improvement
Cl. 34 S No. 50-10, Bogota. **Tel:** 1 2701142

English Language & Culture Institute (ELCI)
Calle 90 No 10-51, Bogota.

English Systems
Cr. 13 No. 32-51 Of. 807, Bogota. **Tel:** 1 3381361

English Zone Institute
Cl. 70A No. 31-29, Bogota. **Tel:** 1 3108559

Esquela De Inglés
Calle 53 No 38-25, Barranquilla.

Escuela De Idiomas Berlitz
Calle 83 No 19-24, Bogota.

Escuela Mundial del Idioma Inglés
Cl. 34 No. 64A-11, Medellin. **Tel:** 4 2353266

First Class English
Cr. 12 No. 93-78 P. 4, Bogota. **Tel:** 1 6238380

FLS Learning centres
Cl. 106 No. 16-50, Bogota. **Tel:** 1 2140031

Genelor International
Avenida 78 No 20-49 Piso 20, Bogota.

I.C.L.
Calle 119 No 9a-25, Bogota.

Idiomas-Munera-Cros Ltda
Carrera 58 No 72-105, A.A. 52032, Barranquilla.

Ingles Cantando Y Jugando
Calle 106 No 16-26, Bogota.

Inlingua
Cr. 77B No. 47-70, Medellin. **Tel:** 4 2509746

Instituto Anglo Americano De Idiomas
Carrera 16a No 85-34 Of. 204, Bogota.

Instituto Bridge Centro De Idiomas
Carrera 65 No 49 A 09, Cali.

Instituto De Inglés Thelma Tyzon
Carrera 59 No 74-73, Barranquilla.

Instituto De Lenguas Modernas
Carrera 41 No 52-05, Baranquilla.

Instituto Electronico De Idiomas
Carrera 6 No 12-64 Piso, Bogota.

Instituto Experimental De Atlantico
"jos Celestino Mutis", Calle 70 No 38-08, Barranquilla.

Instituto Meyer
Calle 17 No 10-16 Piso 80, Bogota.

Interlingua
Cr. 18 No. 90-38, Bogota. **Tel:** 1 2180942

International Language Institute
Carrera 5a No 21-35, Neiva.

International Language Institute Ltda
Carrera 11 No. 65-28 Piso 3, Bogota. **Tel:** 571 235-8152/72, **Fax**: 310-2892.

International Language Institute Ltda
Carrera 13 No 5-79 Castillogrande, Cartagena.

International Study centre
Cl. 116 No. 22-45 Of. 503, Bogota. **Tel**: 1 2165830

International System
Transversal 6 No 51 A 33, Bogota.

K.O.E De Columbia
Calle 101 A No 31-02, Bogota.

Life Ltda.
Transversal 19 No 100-52, Bogota.

Lubigon English Academy
Cl. 48D No. 67A-14 , Medellin. **Tel:** 4 2602840

Oxford Centre
A.A. 102420, Santate de Bogota.

Royal English Club
Tr. 42A No. 100-59, Bogota. **Tel:** 1 2718887

Stanton School of English
Cl. 57 No. 18-22 Of. 503, Bogota. **Tel:** 1 3479017

The Better English centre
Cr. 73B No. 5C-19, Bogota. **Tel:** 1 2640535

The British Council
Cl. 91 No. 21-55, Bogota. **Tel:** 1 6103077

The British Council
Cr. 42 No. 16A S-41, Medellin. **Tel:** 4 3131867

Total English Solutions
Cr. 36A No. 57A-15, Bogota. **Tel:** 1 2215233

Wall Street Institute
Cr. 9 No. 70-09, Bogota. **Tel:** 1 3460566

Way's English School
Calle 101 No 13 A 17, Bogota.

Winston - Salem
Cl. 45 No. 13-75, Bogota. **Tel:** 1 3128646

World English centre
Cr. 16 No. 53-46 P. 2, Bogota. **Tel:** 1 5471950

Worldwide Languages
Tr. 19A No. 123-49, Bogota. **Tel:** 1 6190762

COSTA RICA

Costa Rica is one of the more stable of the Central American republics and a favourite haunt of "gringo" tourists. There are many private language schools, especially in San José, an unattractive city of around five million people. The rest of the country is beautiful with volcanoes and beaches to explore.

Consulate in UK:
14 Lancaster Gate, London W2 3LH
Tel: 020 7706 8844

Embassy in USA:
2114 S Street NW, Washington DC 20008
Tel: 202 234 2945

➲ **Visas/Work Permits:** Work visas are very difficult to obtain and teachers usually work on tourist visas.

➲ **Currency:** £1 = 1011 / $1 = 517 Colon

➲ **Salaries/Taxes:** Salaries are not high (c. £2.50-3 / $4-5 an hour for part-timers; £400-600 / $700-$1100 per month full-time is about average) but you can earn enough to get by. Private tutors can earn $15 an hour.

➲ **Accommodation:** £180-250 / $350-460 a month for a shared apartment. Assistance is often provided by employers.

➲ **Cost of Living:** One of the most expensive Central American countries, but prices are still low compared to the UK or USA.

➲ **Health advice:** Medical facilities are available, but may be limited outside urban areas. Be aware that on both the Caribbean and Pacific coasts, currents are swift and dangerous. Several drownings occur each year, and there are no lifeguards.

Schools in Costa Rica (+506)

American International School
Bosques de Doña Rosa, Ciudad Cariari, Heredia. **Fax**: 239-0625.**Email:** brown@cra.ed.cr

Berlitz
Santa Ana 2000, Edificio A, Planta Baja, Santa Ana, San José. **Tel**: 204 7555, **Fax**: 204 7444.

Blue Valley School
P.O. Box 2050-1000, San José. **Fax:** 253-7708

Centro Cultural Costarricense Norteamericano
Apt.1489-1000, San José, Calle Los Negritos, Barrio Dent. **Tel:** 225 6433, **Fax**: 224 1480, **Email:** centrcr@sol.racsa.co.cr

Country Day School
P.O. Box 8-6170-1000, San José. **Fax:** 289-6798

ELS International
Apt 6495-1000, San Jose. **Tel:** 261 4242, **Fax**: 261 3212, **Email:** acallen@uicr.ac.cr

Escuela Britanica
P.O. Box 8184-1000, San José. **Fax:** 234-7833.

Instituto Britanico
PO Box 8184-1000, San José. **Tel:** 256 0256, **Fax**: 253 1894

Instituto de Inglés USA
Apt 418-1000 C1 A6, San José

International Christian School
P.O. Box 3512-1000, San José. **Tel:** 236-7879. **Fax**: 235-1518.

Lincoln School
P. O. Box 1919-1000, San José. **Tel:** 235 7733, **Fax**: 236-1706.

Marian Baker School
P.O. Box 4269-1000, San José. **Fax:** 273-4609

Monteverde Friends School
P.O. Box 10165-1000, San José. **Fax:** 645-5219.

World Education Forum
PO Box 383-4005, San Antonio de Belén, Heredia. **Tel:** 239 2245, **Fax**: 239 2254.

ECUADOR

Even though the current economic situation is very difficult and many Ecuadorians are leaving the country, there are hundreds of language schools in Ecuador, so it is easy for English teachers to find work, whether they are qualified or not. Until recently, Ecuador was generally considered the best starting point for teachers planning to work their way round South America. Although this may no longer be the case, it is more relaxed than many of its neighbours and still offers plenty of opportunities, especially in the main cities of Quito, Guayaquil and the beautiful Cuenca. Terms and conditions vary and most recruitment is done locally. Many expatriates have set up their own schools, but to find work in them it helps to have personal contacts and qualifications. Some private language schools will often take on native speakers with no qualifications.

➲ **Visas/Work Permits:** It is illegal to work on a tourist visa but many teachers do so anyway. Visitors from the US and Canada are entitled to stay for up to 90 days on a tourist visa. To extend the visa, you have to prove that you have independent income not resulting from work in the country. If you intend to work legally, you have to produce a return air ticket. A tourist visa cannot be exchanged for any other kind of visa once in the country, so if you get a job while you are there, you have to go to neighbouring country, usually Colombia, to get the necessary paperwork. Most employers will help teachers to get a one-year cultural exchange visa, as long as they commit themselves to staying for a reasonable length of time.

➲ **Requirements:**TESOL/TESL Certificate

➲ **Currency:** £1 = 1.85 / $1 = 1 US Dollar

➲ **Salaries/Taxes:** The monthly rate for unqualified teachers starts at about £215 / $370, rising to £300 / $570 in more reputable schools. The top schools pay about £500-600 / $950-1100 a month and only employ qualified teachers. Everyone is taxed at 5-8%, regardless of whether they are working legally or not.

➲ **Accommodation:** Wide range available. Expect to pay £150 / $220 a month for an apartment in Quito. The usual deposit is two months' rent.

➲ **Cost of Living:** Figure on basic living expenses of around £80 / $150 per month.

➲ **Health advice:** Malaria and dengue fever are prevalent in Guayaquil and the coastal region. Cases of cholera and yellow fever have also been reported. Medical care is available but it varies in quality and generally is below UK or U.S. standards.

➲ **English Language Media:** Q Magazine, Inside Ecuador.

Schools in Ecuador (+593)

American Language School
Carchi 904 y Velez, Guayaquil. **Tel:** 452926.

Benedict
9 De Octubre 1515, Y Orellana, Quito.

Benedict
Datiles y La Primera, CC Urdesa, Guayaquil.

Centro De Estudios Interamericanos
Casilla 597, Cuenca.

Centro Ecuatoriano NorteAmericano
Luis Urdeneta y Cordoba, Guayaquil, **Tel:** 564536.

Experimento De Convivencia International Del Ecuador
Les Embleton, Hernando de la Cruz 218 y Mariana de Jesús, Quito. **Tel:** 2 551937/550179, **Fax**: 2 55 0228.

Fulbright Commission
Almagro 961 y Colon, Quito. **Tel:** 2 562999.

International Benedict Schools of Languages
PO Box 09-01-8916, Guayaquil. **Tel:** 4 444418, **Fax**: 4 441642.

Lingua Franca
Edificio Jerico, 12 De Octubre 2449 y Orellana, Casilla 17-2-68, Quito. **Tel:** 2 546075, **Fax**: 2 500734.

Quito Language and Culture Centre
Republica De El Salvador, 639 Y Portugal, Quito.

EL SALVADOR

The country is slowly recovering from the long and bloody civil war. English lessons are very popular and experienced native speakers are genuinely respected here. Many English teachers are Salvadorans who have been deported from the States who are not teachers by profession. An MA in TESOL or a certificate will enhance your wage-earning capacity. You can live quite well on a tight budget in El Salvador and, unlike Costa Rica the country is not full of "gringos."

➲ **Visas/Work Permits:** You will be given a free tourist card on your flight in. This is good for 30-90 days after which you can pop across the border and then return. Each school has its own policy regarding work permits but most are not interested in red tape and will employ teachers on tourist visas.

➲ **Currency:** £1 = 16.37 / $1 = 8.75 Colon

➲ **Salaries/Taxes:** You can earn about £300 / $570 per month or £6 / $10 an hour if you freelance. Foreigners are liable for tax unless they are being paid "under the table."

➲ **Accommodation:** In San Salvador, you can pay anywhere from £30 / $60 a month for a two-bedroom house in a rough area to £600 / $1100 for an apartment in a tony neighbourhood. Utility bills are around £12-15 / $20-30 per month.

➲ **Health advice:** Dengue fever is a big problem. Medical care is limited. Emergency services, even in San Salvador, are very basic. Ambulance services are not staffed by trained personnel and lack life-saving necessities such as oxygen. Doctors in the major hospitals are generally well-trained, often in U.S. hospitals, but nursing and support staff are not up to UK or U.S. standards.

Schools in El Salvador (+503)

Academia Europea
911 Ave Nte., San Salvador. **Tel:** 263 4355. **Fax**: 263 4430. **Email:** jobs@euroacad.edu.sv

Escuela Superior De Idiomas
Ave. La Campilla 226, Colonia San Benito, San Salvador. **Tel:** 264 1256.

GUATEMALA

This is a popular destination for EFL teachers drawn here by the colourful indigenous culture. Some teachers recommend getting out of the capital city and working in Antigua or Quetzaltenango.

➲ **Visas/Work Permits:** Tourist visas are granted for 90 days. It is possible to obtain a work permit through the Guatemalan consulate in your country if you have a job offer, your last two bank statements showing you have funds to live on when you arrive and two photos. But most teachers just turn up and work on their tourist visas.

➲ **Currency:** £1 = 14.39 / $1 = 7.78 Quetzal

➲ **Salaries/Taxes:** Salaries are very low (around £3 / $5 an hour) and if you have a work permit, you will pay tax at around 30%.

➲ **Health advice:** Health care facilities are good in Guatemala City but are more limited elsewhere.

Schools in Guatemala (+502)

American School of Guatemala
Aptdo Postal 83, 01901 Guatemala City. **Tel:** 369 0791.

Colegio Americano de Guatemala
11 Calle 15-70, Zona 15, Vista Hermosa.

Colegio Ingles Americano
0 Calle 19-70, Zona 15, V.H. II , Guatemala City.

Colegio Internacional Montessori
Carr. el Salv. Km. 13.5 finca la Luz, Guatemala City. **Tel:** 2 3641921.

Instituto Guatamelteco-Americano
Ruta 1, 4-05, Aptdo. 691, Guatemala City. **Tel:** 331 0022, **Fax**: 332 3135

Modern American English School
c/ de los Nazarenos 16, Antigua. **Tel:** 932 3306, **Fax**: 932 0217.

HONDURAS

Demand for native-speaking English teachers is high in Honduras both in private language schools and in the state colleges and universities. Students at bilingual school learn Spanish and English. The country is still recovering from the effects of Hurricane Mitch and many roads are washed out.

➲ **Visas/Work Permits:** After you have a job offer, you must apply for a residency permit from outside Honduras although you can look for work as a tourist. The process can take anywhere from four to six months.

➲ **Currency:** £1 = 34 / $1 = 18 Lempira

➲ **Salaries/Taxes:** Rates of pay vary from £2-5 / $3-$8 an hour.

➲ **Accommodation:** Your employer will probably offer you accommodation but be warned that many apartment blocks in the country are not finished and that you may be without water and electricity for periods of time.

➲ **Cost of Living:** Cheap.

➲ **Health advice:** Doctors are generally well trained; but support staff and facilities generally are not up to UK or U.S. standards, and facilities for certain surgical procedures are not available. Drink bottled water.

Schools in Honduras (+504)

American School
c/o American Embassy-Tegucigalpa Honduras, Dept of State, Washington DC 20521-3480: **Tel:** 32 4696

Centro Cultural Sampedrano
3 Calle, entra 3A y 4A Avenida 20 , Apartado Postal 511, San Pedro Sula, Cortes.

Escuela Anglo-Hondurena
Colonia La Reforma, Edificio Club Arabe, Tegucigalpa. **Tel:** 236 8814.

Harris Communications
Ave. Circunvalación, Centro Comercial La Careta, San Pedro Sula **Tel/Fax:** 504 557-3416.

Instituto Hondureno de Cultura Americana
2 Avenida entre 5 y 6 Calles No. 520, Apartado 201, MDC Comayaguela , Tegucigalpa.

JAMAICA

American teachers are sometimes employed in the public school system. There are limited opportunities in private EFL schools.

➲ **Visas/Work Permits:** You can apply for a work permit when you are in Jamaica.

➲ **Currency:** £1 = 130 / $1 = 67 Dollars

➲ **Salaries/Taxes:** You can make J$150 an hour in a school and at least J$200 for private lessons. Income tax is 25%.

➲ **Accommodation:** It is hard to find cheap accommodation in Kingston.

➲ **Cost of Living:** Prices are quite high in Kingston but cheaper elsewhere.

➲ **Health advice:** Ambulance service is limited both in the quality of emergency care, and in the availability of vehicles in remote parts of the country.

Schools in Jamaica (+809)

Language Training Centre Ltd
24 Parkington Plaza, Kingston 10. **Tel:** 926 03756.

Target English Associates
9a Duquesnam Ave., Kingston 10. **Tel:** 929 2473.

MEXICO

Mexico is a popular destination for all ELT teachers, particularly American ESL teachers – after all, they can always walk home if something goes wrong! Increasing trade with the U.S. and Canada means that more Mexican companies require their employees to speak (American) English. Teachers are usually recruited locally because schools are finding it increasingly difficult to pay recruitment costs, but it is possible to organize a job from outside the country.

Embassy in UK:
8 Halkin Street, London SW1X 7DW
Tel: 020 7235 6393

Embassy in USA:
1911 Pennsylvania Ave NW, Washington DC 20006
Tel: 202 728 1600
www.embassyofmexico.org

British Council office:
Lope de Vega 316, Col. Chapultepec Morales, Mexico City, 11570 DF
Tel +52 (55) 5263 1900
www.britishcouncil.org.mx

➲ **Visas/Work Permits:** Tourist visas are available for up to six months. Work permits are usually granted to American and Canadian teachers. Make sure your school provides you with all the necessary paperwork. There are reports of immigration inspectors posing as English students trying to catch out foreigners teaching on tourist visas. Keep your tourist card or other papers on you at all times in case an official asks to see them.

➲ **Requirements:** TESOL/TESL Certificate

➲ **Currency:** £1 = 21.1 / $1 = 10.7 New Peso

➲ **Salaries/Taxes:** Teachers with an MA in TESOL can make £600 / $1100 a month or more, which is very good considering that Mexican teachers in the state system receive about £300 / $570. Tax is about 25%.

➲ **Cost of Living:** It's quite expensive to live in Mexico City but much cheaper elsewhere. Meals can be anywhere from £1-£6 depending on where you go, Estimate spending about £20-£30 / $35-$60 a month on food if you eat out once or twice a week.

➲ **Accommodation:** Rents vary across the country. Cheap rooms (£30 / $60 per month) can be found in most towns but they tend to be in noisy and sometimes dangerous areas. If you're lucky you may find a nice apartment for around £100 / $190 a month. Of course, you will pay a lot more in the big cities.

➲ **Health advice:** Health facilities in Mexico City are excellent. Care in more remote areas is limited. Serious medical problems requiring hospitalization and/or medical evacuation home can be very costly so insurance is essential.

➲ **English Language Media:** The Mexico City Times, The News.

Schools in Mexico (Mexico City)

Academia De Idiomas Tepeyac
Misterios 764 Col. Tepeyac Insurgentes C.P.07020, Ciudad De Mexico, Df. **Tel:** 5 7810046

American Tem
Eduardo Molina 440 Col. Nueva Atzacoalco C.P.07420, Ciudad De Mexico, Df. **Tel:** 5 7531720

Angloamericano
Campos Eliseos 111 Col. Polanco, C.P.11560, Ciudad De Mexico, Df. **Tel**: 5 6586700

Aprendizaje Inst. A.C.
Montevideo 284 P-2 Col. Lindavista, C.P.07300, Ciudad De Mexico, Df. **Tel:** 5 5860196, 5 5861776

Berlitz
Reforma 132 Col. Juarez C.P.06600, Ciudad De Mexico, Df. **Tel:** 5 7050832, 5 7056695

Berlitz
Bosque Ciruelos 194, 2do. Piso Col. Bosques De Las Lomas C.P.11700 Ciudad De Mexico, Df. **Tel:** 5 5969190, 5 5969160

Berlitz
Pafnucio Padilla #17 P-1, Plaza Satelite Col. Satelite C.P.53100 Ciudad De Mexico, Df. **Tel:** 5 5620190, 5 3934981 **Email:**Centro.Satelite@Berlitz.com.mx

Berlitz
Sn Jeronimo 240 Col. Pedregal De San Angel C.P.01900 Ciudad De Mexico, Df. **Tel:** 5 6161953, 5 5504359

Berlitz
Montevideo 365 Esq. Payta Col. Lindavista C.P.07300 Ciudad De Mexico, Df. **Tel:** 5 7521210, 5 7544025.

Berlitz
Lomas Verdes 825 43 Col. Lomas Verdes Nauc C.P.53120 Ciudad De Mexico, Df. **Tel:** 5 3436497.

Berlitz
Del Hueso 480 Loc 11 Col. Los Girasoles C.P.04920 Ciudad De Mexico, Df. **Tel:** 5 6797146, 5 6790438

Berlitz
Insurgentes Sur 1261 Col. Extrem-adura Insurgentes C.P.03740 Ciudad De Mexico, Df. **Tel:** 5 6117468, 5 6117551

Britt Servicios Linguisticos
Berlin 27 B Col. Juarez C.P.66000 Ciudad De Mexico, Df. **Tel:** 5 5466160

Britt Servicios Linguisticos
Luz Avi&On 2011 4 Col. Narvarte C.P.03020 Ciudad De Mexico, Df. **Tel:** 5 6873571

Centro Cultural Benjamin Franklin Sc
Parroquia 828 Col. Del Valle C.P.03100 Ciudad De Mexico, Df. **Tel:** 5 5347747

Ef
Alfredo Musset 228 Col. Polanco C.P.11550 Ciudad De Mexico, Df. **Tel:** 5 5143333, 55313156

Endicott College
Ibsen 43 P-8 Col. Polanco C.P.11560 Ciudad De Mexico, Df. **Tel**: 5 2801225

English For Executives
Viveros De Asis 19 Col. Viveros De La Loma C.P.54080 Ciudad De Mexico, Df. **Tel:** 5 3615234, 5 3979980

English On Site
Giotto 29, Col. Mixcoac C.P.04300 Ciudad De Mexico, Df. **Tel:** 5 6112761

Harmon Hall
San Luis Potosi 196 Col. Roma C.P.06700 Ciudad De Mexico, Df. **Tel:** 5 5847599

Harmon Hall
Puebla 319 0670 Col. Roma C.P.06700 Ciudad De Mexico, Df. **Tel:** 5 2118019, 5 2112020

Harmon Hall
Rojo Gomez 1457 Col. San Pablo C.P.09360 Ciudad De Mexico, Df. **Tel:** 5 6853258, 5 6852181

Ingles Sin Barreras Sa De Cv
Medellín 184, Col. Roma C.P.06700 Ciudad De Mexico, Df. **Tel:** 5 5649739

Instituto Anglo Mexicano De Cultura Ac
Av Insurgentes Sur 710 Col. Del Valle C.P.03100 Ciudad De Mexico, Df. **Tel:** 5 5360742, 55236641

Instituto Anglo Mexicano De Cultura Ac
Rio Nazas 116 Col. Cuauhtemoc C.P.54124 Ciudad De Mexico, Df. **Tel:** 5 2085640, 5 2085799, 5 7051866, 5 5116926

Instituto Hamer Sharp
M Avila Camacho 291 Loc-A-29-30-31 Col. Lomas De Sotelo C.P.11220 Ciudad De Mexico, Df. **Tel:** 5 5806834, 5 5800341, 5 5802676

Instituto Mexico Norteamericano De Idiomas Sc
Av Chapultepec 474 Col. Roma Nte C.P.06700 Ciudad De Mexico, Df. **Tel:** 5 2116758, 5 2867988

Instituto Superior De Idiomas Sa
Viena 71-301 Col. Del Carmen Coyoacan C.P.04100 Ciudad De Mexico, Df. **Tel:** 5 6593533

Interlingua
Montevideo 405 073 Col. Lindavista C.P.07300 Ciudad De Mexico, Df. **Tel:** 5 5864398, 5 5863275, 5 7541209

Interlingua
Medicos 3 Piso 2 Col. Zp Em Frac Cd Satelite C.P.53100 Ciudad De Mexico, Df. **Tel:** 55720697, 55623713

Interlingua, Genova
33 Piso 6 Col. Juarez C.P.06600 Ciudad De Mexico, Df. **Tel**: 5 2080737, 5 5336706, 5 2081919

Interlingua
J Balmes 11 D Piso 2 Mezzanine Col. Los Morales Polanco C.P.11510 Ciudad De Mexico, Df. **Tel:** 5 5572875

Interlingua
Av Insurgentes Sur 1971 L-280 Col. Guadalupe Inn C.P.01020 Ciudad De Mexico, Df. **Tel:** 5 6627208, 5 6627207, 5 6613787, 5 6614166

La Villa English Academies
Calz De Guadalupe 156-A Col. Ex-Hipodromo De Peralvill C.P.06220 Ciudad De Mexico, Df. **Tel:** 5 5176876, 5 5374401

Prompt English
Viveros De Asis 13-200 Col. Viveros De La Loma C.P.54080 Ciudad De Mexico, Df. **Tel:** 5 3628523

Quick Learning
Arellano 14 Piso 2 Col. Cd Satelite Edo De Me C.P.53100 Ciudad De Mexico, Df. **Tel:** 5 5626036, 5 5626852

Quick Learning
Av Division Del Norte 2646 Col. Churubusco Contry Club C.P.04210 Ciudad De Mexico, Df. **Tel:** 5 6055934

Quick Learning
Liverpool 143-102 Col. Juarez C.P.06600 Ciudad De Mexico, Df. **Tel:** 5 5330180

Quick Learning
Av Molina Enriquez 4231 P-1 Col. Viaducto Piedad C.P.08200 Ciudad De Mexico, Df. **Tel:** 5 7406466, 5 7406569, 5 7406576

Quick Learning
Insurg Sur 586 601 Col. Del Valle C.P.03100 Ciudad De Mexico, Df. **Tel:** 5 6829970, 5 6829962

Quick Learning
Rancho Toyocan 61 Loc 5 Y 6 - R Col. Los Girasoles C.P.04929 Ciudad De Mexico, Df. **Tel:** 5 6842021

Quick Learning
Calle Sur 71 No 4231 Col. Viaducto Piedad C.P.08200 Ciudad De Mexico, Df. **Tel:** 5 7406677

Quick Learning
Calendula 111 Piso 3 Col. Xotepingo C.P.04610 Ciudad De Mexico, Df. **Tel:** 5 6104246

Quick Learning
Jinetes 18 Col. Las Arboledas C.P.53139 Ciudad De Mexico, Df. **Tel**: 5 3792054

Schola
Lamartine 138 Col. Polanco C.P.11570 Ciudad De Mexico, Df. **Tel:** 5 5453804

Servicio Linguistico Empresarial
Sn Fco 657 A Piso 12 - B Col. Del Valle C.P.03100 Ciudad De Mexico, Df. **Tel:** 5 6874449, 5 5236380, 5 6691066

USA
Egipto 206 Loc-A Col. Claveria C.P.02080 Ciudad De Mexico, Df. **Tel:** 5 3410092, 5 3410211

Wall Street Institute
Ipsen 4343 8 Piso Col. Polanco C.P.11580 Ciudad De Mexico, Df. **Tel:** 5 2800809

Wave Comunicacion De Mexico
Insurg Cto 51 204 Col. San Rafael C.P.06470 Ciudad De Mexico, Df. **Tel:** 5 5350124

Schools (on outskirts of Mexico City)

Academia Comercial Linguistica
Montiel 205 Col. Lindavista C.P.07300 Ciudad De Mexico, Df. **Tel:** 5 5776380

Active English Institute
Playa Langosta 311 Col. Reforma Ixtaccihuatl C.P.08840 Ciudad De Mexico, Df. **Tel:** 5 5322881

American British English Sc
Londres 40-10 P-1 Col. Juarez C.P.06600 Ciudad De Mexico, Df. **Tel**: 5 5145856, 5 5143348

Angloamericano De Satelite
Circunvala-cion Ote 8 Deps 308 Col. Satelite C.P.0 1030 Ciudad De Mexico, Df. **Tel:** 5 5721737

Angloamericano Satelite
Vito Alessio Robres 233 Col. Florida C.P.01030 Ciudad De Mexico, Df. **Tel:** 5 5727838

Centro Angloamericano Sur
Alessio Robles 233 Col. Florida C.P.01030 Ciudad De Mexico, Df. **Tel:** 5 6580889

Centro De Aprendizaje Linguistico
M Escobedo 446 Col. Polanco Reforma C.P.11550 Ciudad De Mexico, Df. **Tel:** 5 2503851

Centro De Capacitacion En Lenguas Extranjeras
Coru&A 201 P-1 Col. Viaducto Piedad C.P.08200 Ciudad De Mexico, Df. **Tel:** 5 5305050

Centro De Convivencia Linguistica
O Del Ajusco 2 Col. Zp Em Frac Los Pirules C.P.54040 Ciudad De Mexico, Df. **Tel:** 5 3791310

Centro De Idiomas Dinamicos
Luz Savi&On 13 - 902 Col. Del Valle C.P.03100 Ciudad De Mexico, Df. **Tel:** 5 5438280

Centro De Idiomas Dinamicos
L Savi&On 13 301 Col. Del Valle C.P.03100 Ciudad De Mexico, Df. **Tel:** 5 5434534, 5 6692259, 5 6829805

Centro De Idiomas Sn Angel
Insurgentes Sur 2047 P-A Col. Sn Angel C.P.01000 Ciudad De Mexico, Df. **Tel:** 5 6617422, 5 6617423

Centro De Idiomas Tamesis
Plaza De La Constitucion Mz 10 Lt 44 Col. Plaza Aragon C.P.57139 Ciudad De Mexico, Df. **Tel:** 5 7104056

Centro De Idiomas Y Diversificados
Nuevo Leon 253 305 Col. Escandon C.P.11800 Ciudad De Mexico, Df. **Tel:** 5 5163049

Centro De Ingles Abraham Lincoln
Ermita Izt 326 Col. Cacama C.P.09080 Ciudad De Mexico, Df. **Tel:** 5 5814949

Centro Especializado De Servicios Linguisticos
Cuitlahuac 2725 Col. Popular C.P.02840 Ciudad De Mexico, Df. **Tel:** 5 3412253

Centro Linguistico De Ingles
Yucatan 189 Ph Col. Tizapan Sn Angel C.P.01090 Ciudad De Mexico, Df. **Tel:** 5 6164315

Centro Linguistico Lucerna
Tulipan 40 Col. Ciudad Jardin C.P.04370 Ciudad De Mexico, Df. **Tel:** 5 5495846

Centro Linguistico Valle Dorado
Atenas 137 Col. Valle Dorado C.P.54020 Ciudad De Mexico, Df. **Tel:** 5 3706808

Centro Universitario De Idiomas
Dr Galvez 38 Col. San Angel C.P.01000 Ciudad De Mexico, Df. **Tel:** 5 5503808

Chanor Company
Bosques De Duraznos 65-205 Col. Bosques De Las Lomas C.P.11 700 Ciudad De Mexico, Df. **Tel:** 5 5964052

Cima
Saltillo 91-B Col. Hipodromo C.P.0 6100 Ciudad De Mexico, Df. **Tel:** 5 516 4605, 5 5165516, 5 5167138

Comercio En Mexico Del Centro Internacional
Ave Insurgentes Sur No 421 B-301 Col. H Condesa C.P.06140 Ciudad De Mexico, Df. **Tel**: 5 2643347

Complete Language Services
Barranca Del Muerto 525 P-6 Col. Merced Gomez C.P.01600 Ciudad De Mexico, Df. **Tel:** 5 6643622

Converse International School Of Languages
Melchor Ocampoloc-H4-6 193 Col. Veronica Anzures C.P.11590 Ciudad De Mexico, Df. **Tel:** 5 2609002

Cosmo Educacion
Insurg Sur 1694 Col. Florida C.P.01030 Ciudad De Mexico, Df. **Tel:** 5 6618367

Cosmo
Anatole France 13 Altos Col. Chapultepec Polanco C.P.11560 Ciudad De Mexico, Df. **Tel:** 5 2820380, 5 2821021, 5 2821279

Desarrollo De Idiomas Empresariales
Av Nuevo Leon 270 P-2 Col. Hipodromo C.P.06100 Ciudad De Mexico, Df. **Tel:** 5 2719024

Easy English Institute
Ave San Isidro 288-302 Col. San Isidro Cacahulton C.P.0 2720 Ciudad De Mexico, Df. **Tel:** 5 3523951

EF Servicios En Educacion Internacional
Londres No 188 Mezzanine Col. Juarez C.P.06600 Ciudad De Mexico, Df. **Tel:** 5 5251216

Escuela Internacional De Idiomas
Ailes 144 Col. Sn Mateo. C.P.53240 Ciudad De Mexico, Df. **Tel:** 5 3733951

Escuela Mexicana Canadiense De Ingles
Miguel A De Quevedo 928 - M Col. Parque Sn Andres C.P.04330 Ciudad De Mexico, Df. **Tel:** 5 6593252

Escuela Mexicana Canadiense De Ingles
M Angel De Quevedo 928 Col. Rosedal Coyoacan C.P.04330 Ciudad De Mexico, Df. **Tel:** 5 6594112

Especialistas En Idiomas
Arellano 14 P-1 Col. Cd Satelite C.P.53100 Ciudad De Mexico, Df. **Tel:** 5 3939271

Focus On Language
Colina De Aconitos 15 8 Col. Boulevares C.P.53140 Ciudad De Mexico, Df. **Tel:** 5 5723406

Gpo Educativo Angloamericano
Vito Alessio Robles 233 Col. Florida C.P.01030 Ciudad De Mexico, Df. **Tel:** 5 6589656, 5 6589540, 5 6589632

Harmon Hall
Eugenia No 106 Col. Narvarte C.P.03020 Ciudad De Mexico, Df. **Tel:** 5 5238359

Harmon Hall Del Valle
Eugenia 1606 Col. Narvarte C.P.03020 Ciudad De Mexico, Df. **Tel:** 5 6874987

Harmon Hall Toreo
Av Rodolfo Gadna No 6 5do Piso Col. Lomas De Sotelo C.P.53390 Ciudad De Mexico, Df. **Tel:** 5 3958300, 5 3958400

Ibi Instituto Britanico
Campos Eliseos 204 P-4 Col. Polanco C.P.11560 Ciudad De Mexico, Df. **Tel:** 5 2812465

Idiomas Empresariales Marva
Gante 15 P-2 Loc-213 Col. Centro C.P.06010 Ciudad De Mexico, Df. **Tel:** 5 5102997, 5 5129242

Guadalajara

American English centre
Av Federalismo Sur 415 Col. Centro C.P.44100 Guadalajara, Jalisco. **Tel:** 3 6131667

Ayusa Internacional Mexico
Lopez Co-tilla 1713 202 Y 204 Col. Guadalajara C.P. 44140 Guadalajara, Jalisco. **Tel:** 3 6153688

Berlitz
Plabo Neruda #2914 Col. Providencia C.P.44630 Guadalajara, Jal-isco. **Tel:** 3 412074, 3 6414048, 3 641 2768.

Berlitz
Vallarta 1550 Loc. 1 Esq. Marsella Col. Sector Juarez C.P.44140 Guadalajara, Jalisco. **Tel:** 3 6303987

Centro Cultural Linguistico
Universo 746 Jardines Del Col. Jardines Del Bosque Guad C.P.44520 Guadalajara, Jalisco. **Tel:** 3 1216863

Centro De Desarrollo Linguistico
Mariano Otero No 3677 Col. La Calma C.P. 45070 Guadalajara, Jalisco. **Tel:** 3 6342713

Centro Linguistico De Guad-Alajara
S H C 68 Nº 46 Desp 201 Col. Santa Teresita C.P.44600 Guadalajara, Jalisco. **Tel:** 3 6166626

Connections
Francisco Frejes 222 Col. Rojas Ladron De Guevara C.P.44650 Guadalajara, Jalisco. **Tel:** 3 6160708, 3 6157451, 3 6161077

EF
Circun Agustin Ya&Ez 2559-2 Col. Arcos Del Sur C.P.44100 Guadalajara, Jalisco. **Tel:** 3 6301835

Harmon Hall
Lopez Mateos Sur Esq Chimalhuacan Col. Cd Del Sol C.P.45050 Guadalajara, Jalisco. **Tel:** 3 1212435

Imac Ingles Total
Donato Guerra 180 Col. Alcalde Barranquitas C.P.44270 Guadalajara, Jalisco. **Tel:** 3 6132805

Ingles Individual
Agustin Ya&Ez 2583 Loc-106 Col. Barrera C.P.44150 Guadala-jara, Jalisco. **Tel:** 3 6165798, 3 6303393

Instituto Anglo Mexicano De Cultura
Av Americas No 1301 Col. Col Providencia Norte C.P.05308 Guadalajara, Jalisco. **Tel:** 3 8170506

Instituto Luis Silva
Eje Pte 644 Col. El Santuario C.P.44200 Guadalajara, Jalisco. **Tel:** 3 6134392

Interlingua
Mexico 3065 Piso 2 Col. Vallarta San Lucas Zp 6 C.P.44690 Guadalajara, Jalisco. **Tel:** 3 6150401

International Language Design
Madero 710 Col. Centro C.P.44100 Guadalajara, Jalisco. **Tel:** 3 8251226

Kaao Centro De Idiomas
Av Lapizlazuli 2515 A Col. Residencial Victoria C.P. 44580 Guadalajara, Jalisco. **Tel:** 3 6327949

Lenguaje World Institud
Santuario 830 Col. Juan Diego C.P.44510 Guadalajara, Jalisco. **Tel**: 3 1219833

Linguas
Patria 1398 Desp 203 Col. Villa Universitaria C.P.45110 Guadalajara, Jalisco. **Tel:** 3 6402272

USA English Institute
Victoriano Aguer-os 1474 Col. Sect Juarez Cp 44150 C.P. 44150 Guadalajara, Jalisco. **Tel:** 3 6302255

Vancouver Language Centre
Avenida Vallarta 1151 Col. Americana C.P.44100 Guadalajara, Jalisco. **Tel:** 3 8254271

Wall Street Institute
Ave Vallarta 1530 Col. Centro C.P.44100 Guadalajara, Jalisco. **Tel:** 3 6162803, 3 6166615, 3 6169124

Aprendizaje Instantaneo Chap
Av Del Sur 1974 Sj Col. Obrera Centro C.P.44140 Guadalajara, Jalisco. **Tel:** 3 8252059

Centro Cultural Norteamericano
Sjc 18 N 172 A S Col. Zp 1 Gua C.P.44100 Guadalajara, Jalisco. **Tel:** 3 6144416

Idiomas SA
Ave Mexico 3065 Piso 2 Col. Residencial San Jorge C.P.44690 Guadalajara, Jalisco. **Tel:** 3 6152451

Idiomas SA
Shc 7 3065 Piso 2 Col. Residencial San Jorge C.P.44690 Guadalajara, Jalisco. **Tel:** 3 6300118

Instituto Cultural Mexicano Norteamericano De Guadalajara
Enrique Diaz De Leon 300 Col. Americana C.P.44170 Guadalajara, Jalisco. **Tel:** 3 8258834

Instituto Anglo Mexicano De Cultura
Manuel Acu&A 2674-102 Col. Lomas De Guevara C.P.44680 Guadalajara, Jalisco. **Tel:** 3 6414017

Instituto Anglo Mexicano De Cultura
Manuel Acu&A 2674-102 Col. Lomas De Guevaraia C.P.44680 Guadalajara, Jalisco. **Tel:** 3 6414009

Universidad De Guadalajara
Esc Superior De Lenguas Modernas, Apdo. Postal 2-416, 44280 Guadalajara, Jalisco.

Monterrey

Academia De Ingles Prof Gil Moreno
Galeana Sur 418 Col. Centro C.P.64000 Monterrey, Nl. **Tel:** 83 423844

Berlitz
Padre Mier Pte 241 Pte. Col. Centro C.P.64000 Monterrey, Nl. **Tel:** 83 441761, **Fax**: 83 455781

Berlitz
Rio Mississippi 101-B Col. Del Valle C.P.66220 Monterrey, Nl. **Tel:** 83 561605, 83 561612

English Comunication System
De Los Leones Po 2389 A Col. Las Cumbres C.P.64610 Monterrey, Nl. **Tel:** 83 001878

English Now
Av Constitucion 1251 Pte Col. Centro C.P.67460 Monterrey, Nl **Tel:** 83 449175

Harmon Hall
Pablo Moncayo 117 Piso 2 Col. Col De San Jeronimo C.P.64630 Monterrey, Nl. **Tel:** 83 477775, 83 477774

Instituto Mexicano Norteamericano De Relaciones Culturales
Hidalgo Pte 768 Col. Centro C.P.64000 Monterrey, Nl. **Tel**: 83 401583

Ingles En Siete Semanas
Washington 2910 Pte Col. Obispado C.P.64610 Monterrey, Nl. **Tel:** 83 477735

Interlingua
Av Jose Vasconcelos Or 109 Col. Del Valle C.P.66220 Monterrey, Nl. **Tel:** 83 786222, 83 786500

Quick Learning
Gonzalitos 916 Col. Mitras Centro C.P.64460 Monterrey, Nl. **Tel:** 83 468718

Teaching Of International Languages
San Lorenzo 165 Col. Vista Hermosa C.P.64620 Monterrey, Nl. **Tel**: 83 465655

Boston English Institute
Matamoros Pte 1174 Col. Centro C.P.64610 Monterrey, Nl. **Tel:** 83 401981

British American School Sc
P Mier Pte 421 Col. Centro C.P.64000 Monterrey, Nl. **Tel:** 83 423887

British American School Sc
Priv Rhin Y Priv Sevilla 651 Col. Centro C.P.64000 Monterrey, Nl. **Tel:** 83 401993

Centro Educativo Bilingue La Paz Sc
Hidalgo Pte 380 Col., C.P. 64000 Monterrey, Nl. **Tel:** 83433406

Cosmo Educacion
Rio Colorado Ote 240 Loc-10 Col. Del Valle C.P.66220 Monterrey, Nl. **Tel:** 83 562935, 83 562925, 83 359221

Desarrollo De Ingles Rapido
Av Pino Suarez 853 Nte Col. Centro C.P.64000 Monterrey, Nl. **Tel:** 83 721222

English Through The Arts
Av Jose Vascon-celos 1501 Pte Loc-5 Y 6 Col. San Pedro C.P.66220 Monterrey, Nl. **Tel:** 83 386713

Harmon Hall Hermosillo A C
Av Eug-enio Garza Sada 3755 Col. Alfonso Reyes C.P.64730 Monterrey, Nl. **Tel:** 83 691924

Interlingua
Av Jose Vasconcelos Or 109 Col. Del Valle C.P.66220 Monterrey, Nl. **Tel:** 83 786244

Liceo Anglofrances De Monterrey
Rio De La Plata Ote 114 Col. Jardines Del Valle C.P.66220 Monterrey, Nl. **Tel:** 83 353576

New English System
Dr Coss 843 Sur Loc-E Col. Centro C.P.64000 Monterrey, Nl. **Tel:** 83 421222

Puebla

Berlitz
39 Ote 3302 Loc 36, 37,38 Col. Centro C.P.72400 Puebla, Pue. **Tel:** 22 498838, 22 498850, **Fax**: 22 498 475, **Email:** Centro.Puebla@Berlitz.Com.Mx

English Training centre
Calle 27 Sur 111 Col. Centro C.P.72000 Puebla, Pue. **Tel**: 22 303243, 22 313848

Harmon Hall
Juarez 2108 Col. La Paz C.P. 72160 Puebla, Pue. **Tel**: 22 424160, 22 424158

Harmon Hall
Calle 51 Pte 505 Loc 133 Col. Plza America C.P.72000 Puebla, Pue. **Tel:** 22 435701, 22 435487

Personal English
9 Pte No 1308 1er Piso Col. Centro C.P.72000 Puebla, Pue. **Tel:** 22 465880

Instituto Anglo Frances
Calle 25 Oriente 1020 Col. Bellavista C.P.72500 Puebla, Pue. **Tel:** 22 432748

Instituto Anglo Mexicano De Cultura Ac
C Comercial Plaza Crystal Zona A Loc 9-1 Col. C.P.72440 Puebla, Pue. **Tel:** 22 405 122

Instituto Angloamericano De Puebla
Clle 5 Pte No 1710 Col. Centro C.P.72000 Puebla, Pue. **Tel:** 22 328885

Interlingua
Reforma Sur 231 R Col. La Paz Pue C.P.72160 Puebla, Pue. **Tel:** 22 493168

Langtrac
Juarez 1905 Desp 10 Col. La Paz C.P.72160 Puebla, Pue. **Tel:** 22 329245, 22 420339

Lewis & Lopez Language Services For Business
7 Poniente 309 Desp 201 Col. Col Centro C.P.72080 Puebla, Pue. **Tel:** 22 422116

Quick Learning
Blvd 5 De Mayo Nº 2509 Col. Del Carmen C.P.72000 Puebla, Pue. **Tel:** 22 430075

Total English
Av Juarez 1310 P-4 Col. Cen-tro C.P.72000 Puebla, Pue. **Tel:** 22 466020

English Training centre
Calle 7 Pte 1905 7 Col. La Pue C.P.72000 Puebla, Pue. **Tel:** 22 323874

Idiomas SA De Cv
Reforma Sur 231 R Col. La Paz Pue C.P.72160 Puebla, Pue. **Tel:** 22 49317

PERU

Spanish is the first language in Peru, and there is a lot of interest in learning English. Many ELT teachers also take advantage of the Spanish schools to learn or improve their Spanish. The country has some of the world's most important archaeological remains, and people leaving the country may be searched to make sure they are not exporting illegal treasures.

Embassy in UK:
52 Sloane Street, London SW1X 9SP
Tel: 020 7838 9223
www.peruembassy-uk.com

Embassy in USA:
1700 Massachusetts Avenue NW, Washington DC 200036
Tel: (202) 833 9860
www.peruemb.org

British Council Office:
Torre Parque Mar (Floor 22)
Av Jose Larco 1301
Miraflores
Tel: 51 1 617 3060
www.britishcouncil.org.pe

➲ **Visas/Work Permits:** Visitors from the UK and US are entitled to stay for 90 days as tourists without a visa. If you intend to work it is necessary to obtain a visa before entering the country.

➲ **Requirements:** TESOL/TESL Certificate

➲ **Currency:** £1 = 6.27 / $1 = 3.21 nuevos soles

➲ **Health advice:** You should get vaccinations against yellow fever, smallpox and malaria if you are intending to visit the tropical north of the country

Schools in Peru (+51)

Cambridge School
La Alameda de Los Molinos 730-28, La Encantada de Villa, Chorrillos, Lima 9
Tel: (1) 254 0163
Fax: (1) 254 0149
Email: camoff@amauta.rcp.net.pe

Colegio Franklin Delano Roosevelt
Av. La Palmeras 325, Urb. Camacho, La Molina, Lima 12
Tel: (1) 435 0890
Fax: (1) 436 0927
Email: fdr@amersol.edu.pe

Colegio Peruano Britanico
Av. Via Lactea 445, Monterrico
Tel: (1) 436 0151
Fax: (1) 436 1006
Email: postmast@copebrit.edu.pe

Davy College
Av. Hoyos Rubio 2684, Cajamarca
Tel: (44) 827 501
Fax: (44) 827 502
Email: postmaster@davycollege.edu.pe

Hiram Bingham Scool
Paseo La Castellana 919, Higuereta, Lima 33
Tel: (1) 448 1222
Fax: (1) 479 0430
Email: dpringle@hirambingham.edu.pe

Markham College
Augusto Angulo 291, Miraflores, Lima 18
Tel: (1) 241 7677
Fax: (1) 447 5295
Email: postmaster@markham.edu.pe

Newton College
Av. Elias Aparicio 240, Las Lagunas de La Molina, Lima 12
Tel: (1) 368 0163
Fax: (1) 479 0430
Email: college@newton.edu.pe

San Silvestre School
Av. Santa Cruz 1251, Miraflores, Lima 18
Tel: (1) 241 3334
Fax: (1) 445 5075
Email: postmaster@staff.sansil.edu.pe

Sir Alexander Fleming School
Av. America Sur 3701, Trujillo
Tel: (44) 280 395
Fax: (44) 284 440
Email: admissions@fleming.edu.pe

URUGUAY

Rapidly developing economy, with few political problems. There is some street crime in Montevideo, and especially along the beach resorts

Embassy in UK:
140 Brompton Road, London SW3 1HY
Tel: 020 7589 8835

Embassy in USA:
2715 M St NW Washington DC 20007
Tel: (202) 331 1313
www.embassy.org/uruguay

➲ **Visas/Work Permits:** no visas required for short stays, but you will need to show a return ticket; work permits needed if you are intending to work

➲ **Currency:** £1 = 45.7 / $1 = 24.7 pesos

➲ **Health advice:** Medical care is reasonable, but you should get insurance before travelling

Schools in Uruguay (+598)

London Institute (International House)
Av. Brasil 2831, Montevideo 11300
Tel: (2) 709 6774
Fax: (2) 709 0297
Email: araz@netgate.com.uy

VENEZUELA

Venezuela is a well-developed reasonably rich country, and life in the major cities is subject to urban crime like anywhere else

Embassy in UK:
1 Cromwell Rd, London SW7 2HR
Tel: 020 7584 4206
www.venezlon.demon.co.uk

Embassy in USA:
1099, 30th St NW, Washington DC 20007
Tel: (202) 342 2214
www.embavenez-us.org

British Council office:
Piso 3, Torre Credicard, Av Principal El Bosque, Caracas
Tel: +58 (212) 952 9965
www.britishcouncil.org.ve

➲ **Visas/Work Permits:** you need to get a tourist entry card (on the plane) before you enter the country. You also need to show your return ticket. For a longer stay (up to 180 days) you will need a visa and a letter of introduction from your employer

➲ **Currency:** £1 = 4194 / $1 = 2147 bolivares

➲ **Health advice:** Medical care is good in the capital Caracas, but less good in country areas. You will probably have to pay cash to get treatment. Vaccinations against cholera and yellow fever are recommended

Schools in Venezuela (+58)

Wall Street Institute
Avda. Francisco de Miranda, Torre Lido, Piso 11, Torre C, Ofic. 111C, 113C El Rosal, Caracas
Tel: (212) 953 7473
Fax: (212) 374 65 03
Email: wsiven@cantv.net

North America

Finding a job in the USA is not as straight-forward as in other countries. If you are not a US-resident, you'll find that the Homeland Security laws introduced following the September-11 tragedy make it far harder to get a visa and work permit. In addition, in the USA, each state has its own requirements for teacher qualifications. The schools within a state must also be registered with their state education body – and each state, again, has its own requirements for schools operating in their area. To send out your CV, you need a a list of ESL schools (and there are hundreds in the US). The simplest route to getting a list is to use the Internet. See p28 for job-websites. Alternatively, visit the state's main website for a listing of schools within that state. In Canada, each province has its own requirements for teacher qualification and certification; provinces also have their own requirements for any school operating in the area.

Job seekers in the field of teaching ESL have a number of resources to turn to help them in their search for work in the U.S. and Canada. *ESL Magazine* and *Language Magazine* often feature display advertisements for ESL teaching positions from schools, colleges and school districts in Canada and the U.S. Local newspapers usually carry advertisements for teaching jobs and *The Chronicle of Higher Education* usually contains positions in colleges and universities for ESL teachers. Try *Education Week* for jobs teaching young learners (the K-12 sector).

You can also find some U.S. and Canadian teaching jobs through the Placement E-Bulletin offered by the professional teachers' association, TESOL, Inc. (http://careers.tesol.org) – the service is free of charge to TESOL members. Jobseekers should also check out the annual TESOL convention where employers are often recruiting for new staff. Another cybersource for ESL jobs is the useful TESLJB-L listserv. Go to http://lc.ust.hk/~teach/temployment.html for information on how to sign on.

On the Internet, visit Tom Riedmiller's site (www.uni.edu/riedmill) for a good overview of ESL job sites. The job listings at Dave's ESL Café site (www.eslcafe.com) sometimes contain positions in Canada and the U.S. although it's mostly adverts for other countries.

If you are interested in teaching in the K-12 sector (Kindergarten to 12th grade – young learners), you might try a search of a local school district's site: many of them feature a jobs section. Also, most universities and colleges offer a recruitment section although many of these are not regularly updated and the deadlines for job applications can be months or even years past due.

Career Development

Teachers can also be learners! Career development is a natural progression if you want to expand your horizons within the profession or in related fields

Diploma/Certificate Qualifications *See p.178*

Diploma courses usually take 8 weeks full-time or a year part-time. With this qualification, you can normally get a job anywhere round the world if you have one.

Master's Degree Courses *See p.193*

Masters' programmes offer a way to study the more theoretical approach to English language teaching. In the US, they are not always post-experience qualifications, whereas outside the US applicants will be expected to have an undergraduate degree and two or more years experience.

Specialist Qualifications *See p.208*

English as a second language is a field that continues to specialize as it matures as a discipline. This section covers Computer Assisted Language Learning and English for Specific Purposes.

Doctorates *See p.217*

This is the ultimate academic qualification in English language teaching. A Ph.D. is generally research-based whereas the Ed.D. is often geared towards those who want to be a school administrator.

Publishing *See p.220*

Teachers often find that they are creating their own materials for classroom use. Over the years, these materials are honed until they are fully developed. This is when you should consider getting these materials published.

School Management / Ownership *See p.226*

As teachers work their way up the career ladder, some are drawn to school management. Others go on to be school owners.

Diploma and Master's Courses

To further your career you will certainly need a higher qualification. One way to do this is to take a Diploma course such as those run by the University of Cambridge or Trinity College London, or a Master's degree.

There is an acute shortage of highly qualified teachers around the world, so having a Diploma under your belt means that you can be guaranteed a job anywhere. It proves that you have both a high level qualification and several years' experience in the classroom. Because the University of Cambridge and Trinity College London Diplomas have both theoretical exams and observed teaching practice, most schools prefer these.

The two best-known qualifications are: Diploma in English language teaching to adults (DELTA) from University of Cambridge ESOL and the Licentiate Diploma in Teaching English to speakers of other languages (LTCL.DipTESOL) from Trinity College London.

In order to enrol on a Cambridge DELTA course, you must have at least two years' post-qualification teaching practice The requirements for Trinity are similar. Each course takes eight weeks full-time, although there are part-time courses and many schools and colleges offer distance learning options for part of the course.

There are also university diploma courses which are usually one year in length. They are open to both native speakers and non-native speakers with at least one year's teaching experience. University diploma courses tend to be more theoretically oriented than the University of Cambridge ESOL or Trinity College London Diplomas.

Diploma/Certificate Courses

This section lists Diploma courses from the main schools around the world. The main courses are: **DELTA** (Diploma in English Language Teaching to Adults, from University of Cambridge ESOL) and **LTCL DipTESOL** (Licentiate Diploma in TESOL, from Trinity College London). Many of the schools and universities listed provide the Diploma course by distance learning.

ARGENTINA (+54)

International House
Pacheco De Melo 2555, 1425 Buenos Aires
Tel: 11 4805 6393
Fax: 11 4805 6393
Email: ihrecoleta@international-house.com.ar
Website: www.international-house.com.ar
Courses offered DELTA

AUSTRALIA (+61)

Australian TESOL Training Centre (ATTC)
PO Box 82, Bondi Junction, New South Wales 2022
Tel: 29389 0249
Fax: 29389 7788
Email: lynnev@ace.edu.au
Website: www.attc.nsw.edu.au
Courses offered DELTA

International House, Sydney
Level 3, 89 York Street, Sydney, NSW 2000
Tel: 29279 0733
Fax: 0 29279 4544
Email: jtennant@ihsydney.com
Website: www.training.ihsydney.com
Courses offered DELTA

Phoenix English Language Academy
223 Vincent Street, West Perth 6005
Tel: 89227 5538
Fax: 89227 5540
Email: pjones@phoenixela.com.au
Website: www.phoenixela.com.au
Courses offered DELTA

Australian TESOL Training Centre,
PO Box 82, Bondi Junction NSW 2022.
Tel: 02 9389 0249.
Fax: 02 9389 7788.
Courses Offered: DELTA

AUSTRIA (+43)

Berufsförderungsinstitut Wien,
Kinderspitalgasse 5, A-1090 Vienna.
Tel: 1 4043 5114.
Fax: 1 4043 5124.
Email: bfi.dion@bfi-wien.or.at
Website: www.bfi-wien.or.at/bfiwien/
Courses Offered: DELTA

BAHRAIN (+973)

The British Council
P.O. Box 452, Manama 356.
Tel: 261555.
Fax: 258689.
Courses Offered: DELTA

BULGARIA (+359)

AV0-3 School Of English
2a Krakra Street, 1504 Sofia
Tel: 2 944 3 032
Fax: 2 943 3 943
Email: binnie@avo-3.com
Website: www.avo-3.com
Courses offered: DELTA

BRAZIL (+55)

SBCI, Sao Paulo
Rua Ferreira de Araujo 741, 3° andar - Pinheiros, São Paulo - SP, CEP 05428-002
Tel: 11 3039 0551
Fax: 11 3039 0562
Website: www.culturainglesasp.com.br
Courses offered DELTA

CANADA (+1)

Oxford Seminars
131 Bloor Street West, Suite 200-390, Toronto, Ontario M5S 1R8
Tel: 416-924-3240
Toll Free: 1-800-779-1779
Email: info@oxfordseminars.com
Website: www.oxfordseminars.com
Courses Offered: TESOL/TESL Certificate

Oxford Seminars
10405 Jasper Avenue, Suite 16-21, Edmonton, Alberta T5J 3S2
Tel: 780 428 8700
Toll Free: 1-800-779-1779
Email: info@oxfordseminars.com
Website: www.oxfordseminars.com
Courses Offered: TESOL/TESL Certificate

Shane Global Village English Centre
Westcoast English Language Centre Ltd
888 Cambie Street, Vancouver BC, V6B 2P6
Tel: 604 684 2112
Fax: 604 648 2124
Email: teachertraining_van@sgvenglish.com
Website: www.sgvenglish.com
Courses offered: DELTA

CYPRUS (+357)

Forum Language Centre
PO Box 25567, Nicosia 1310
Tel: 2 2317 029
Fax: 2 249 7766
Email: peterl@forum.ac.cy
Website: www.forum.ac.cy
Courses offered DELTA

CZECH REPUBLIC (+42)

Akcent International House
Bítovská 3, 140 00 Praha 4
Tel: 02 6126 1675
Fax: 02 6126 1880
Email: dana@akcent.cz
Website: www.akcent.cz
Courses offered DELTA

EGYPT (+20)

The British Council
192 Sharia El Nil, Agouza, Cairo
Tel: 2 301 8341
Fax: 2 344 3076
Email: Louise.Greenwood@britcoun.org.eg
Website: www.britishcouncil.org.eg
Courses offered DELTA

GREECE (+30)

Anglo-Hellenic Teacher Recruitment,
PO Box 263, 201 00 Corinth.
Tel: 741 53511.
Fax: 741 86989.
Courses Offered: Lic. Dip.TESOL

CELT Athens
77 Academias Street, 106 78 Athens
Tel: 1 330 2406
Fax: 1 330 1455
Email: celtath@otenet.gr/celt@celt.gr
Website: www.celt.gr
Courses offered DELTA

Profile
Frantzi 4 and Kallirois Street, 117 45 Athens
Tel: 1 922 2065
Email: profile-educinst@ath.forthnet.gr
Website: www.profiletraining.net
Courses offered DELTA

Study Space
Tsimiski 86, Thessaloniki 54622
Tel: 31 269 697
Fax: 31 269 697
Email: studyspa@compulink.gr
Website: www.compulink.gr/studyspa
Courses offered DELTA

HUNGARY (+36)

International House, Budapest
PO Box 92, 1276 Budapest
Tel: 1345 7046
Fax: 1 316 2491
Email: ttraining@ih.hu
Website: www.ih.hu
Courses offered DELTA

IRELAND (+353)

English in Dublin
Clifton House, Lower Fitzwilliam Street, Dublin, 2
Email: cliftonh@indigo.ie
Tel: 1 4732101
Fax: 1 661 5200
Courses Offered: LTCL.Dip.TESOL

Language Centre
O'Rahilly Building, UCC, The National University of Ireland, Cork.
Tel: 3 21 4902043.
Fax: 3 21 4903223.
Course Length: 9 months' intensive
Course Fees: IEP1750.
Start dates: January / October
Courses Offered: DELTA.

University College Cork
The Language Centre, College Road, Cork
Tel: 2 1490 2962 / 2043
Fax: 2 1490 3223
Email: rmasterson@langcent.ucc.ie
Website: www.ucc.ie
Courses offered DELTA

ITALY (+39)

British Council, Milan
Via Manzoni 38, 20121 Milano
Tel: 02 772221
Fax: 02 781119
Email: simon.creasy@britishcouncil.it
Website: www.britishcouncil.it/milan
Courses offered DELTA

British Council, Naples
Via Morghen, 31, NAPLES 80129
Email: BCNaples@britishcouncil.it
Courses Offered: DELTA
Course Length: 6 months
Start Dates: June, September

British Institute of Florence Language Centre
Palazzo Strozzino, Piazza Strozzi 2, 50123 Firenze
Tel: 055 267 782 05
Fax: 055 267 782 23
Email: sellis@britishinstitute.it
Courses offered DELTA

The Cambridge School of English
Via Rosmini 6, 37123 Verona
Tel: 45 800 3154
Fax: 45 801 4900
Email: info@cambridgeschool.it
Website: www.cambridgeschool.it
Courses offered: DELTA

International House
Via Jannozzi 6, 20097 S. Donato, Milan.
Tel: 02 527 9124.
Fax: 02 5560 0324.
Courses Offered: DELTA.

International House, Rome
Via Marghera 22, ROMA 185
Email: tiziana.di.dedda@dilit.it
Courses Offered: DELTA
Course Length: FT:12 weeks; PT 4 months

Lord Byron College
Via Sparano 102, 20121 Bari
Tel: 08 052 32 686
Fax: 08 052 41 349
Email: johncredico@lordbyroncollege.com
Website: www.lordbyroncollege.com
Courses offered DELTA

JAPAN (+81)

British Council, Tokyo
2-1 Kagurazaka, Shinjuku-ku, Tokyo 162-0825
Tel: 3 3235 8011
Fax: 3 3235 0049
Email: sheena.palmer@britishcouncil.or.jp

Website: www.uknow.or.jp
Courses offered DELTA

KOREA (+82)

Yonsei Institute Of Language Research & Education
Yonsei University, 134 Shinch'on-Dong
Sodaemun-Gu, Seoul 120-749.
Tel: 2 361 3462.
Fax: 2 393 4599.
Courses Offered: DELTA.

MEXICO (+52)

British Council, Mexico
Lope de Vega 316, Col. Chapultepec Morales (Polanco), 11570 Mexico D.F.
Tel: 52 63 1900 (General) 1982 / 1965
Fax: 52 63 1960 / 1660
Email: bcmexico@britishcouncil.org.mx
Website: www.britishcouncil.org/mexico
Courses offered DELTA

International House
Calle Montecito 38, Piso 20 Despacho 4, Colonia Nápoles, 03810 Mexico DF
Tel: 5 448 0223
Fax: 5 488 0223
Courses Offered: DELTA

NEW ZEALAND (+64)

Capital Language Academy
PO Box 1100, Wellington.
Tel: 4 385 0600.
Fax: 4 385 6655.
Courses Offered: DELTA.

Christchurch College of English
PO Box 31-212, Christchurch, New Zealand
Tel: 3 343 3790
Fax: 3 343 3791
Email: pauline.taylor@ccel.co.nz
Website: www.ccel.co.nz
Courses: DELTA

EDENZ EDENZ College
PO Box 10-222 Dominion Road, Mt Eden, Auckland, 1030
tel: 0064 9 522 1211
fax: 0064 9 522 1511
Courses Offered: LTCL.Dip.TESOL.

ILA South Pacific Limited
21 Kilmore Street, PO Box 25170, Christchurch.
Tel: 3 3795452.
Fax: 3 3795373.
Courses Offered: DELTA

Languages International
PO Box 5293, Auckland 1036.
Tel: 9 309 0615.
Fax: 9 377 2806.
Courses Offered: DELTA
Course Length: 7 months
Start dates: April

Unitec School Of Languages
Private Bag 92025, Auckland.
Tel: 9 849 4180.
Fax: 9 815 2906.
Courses Offered: DELTA

POLAND (+48)

ELS - Bell Gdansk
ul. Matejki 6, 80-238Gdansk
Tel: 58 344 7312
Fax: 58 344 0784
Email: ziolkowska@elsbell.pl
Website: www.bellschools.pl
Courses offered DELTA

ELS - Bell (Warsaw)
ul. Nowy Swiat 2, 00-497 Warszawa
Tel: 58 344 7312
Fax: 58 344 0784
Email: ziolkowska@elsbell.com
Website: www.bellschools.pl
Courses offered DELTA

International House, Katowice
ul. Leszczynskiego 3, 50-078 Wroclaw
Tel: 71 781 7290
Fax: 71 781 7290 ext 11
Email: ttcentre@ih.com.pl
Website: www.ih.com.pl
Courses offered DELTA

International House, Kraków
ul. Pitsudskiego 6, 31-109 Kraków
Tel: 12 422 6482
Fax: 12 430 1000
Email: admin@ih.pl
Website: www.ih.pl
Courses offered DELTA

International House, Wroclaw
ul. Leszczynskiego 3, 50-078 Wroclaw
Tel: 71 781 7290
Fax: 71 781 7290 ex. 111
Email: ttcentre@ih.com.pl
Website: www.ih.com.pl
Courses offered DELTA

UEC Bell School Of English
Plac Trzech Krzyzy 4/6, 00-499 Warsaw.
Tel: 22 625 4792.
Fax: 22 629 5865.
Courses Offered: DELTA

PORTUGAL (+35)

International House
Rua Marquês Sá da Bandeira, 16, 1050-148 Lisbon.
Tel: 1 21 315 1496
Fax: 1 21 353 0081
Email: ttraining@ihlisbon.com
Website: www.international-house.com
Courses offered DELTA

RUSSIA (+7)

BKC - International House, Moscow
ul. Tverskaya 9a, Building 4, 103009 Moscow
Tel: 095 234 0314
Fax: 095 234 0316
Email: t-training@bkc.ru
Website: www.bkc.ru
Courses offered DELTA

Language Link
Novoslobodskaya ul. 5/2, 103030 Moscow.
Tel: 095 238 0225.
Fax: 095 234 0703.
Courses Offered: DELTA.

SEBIA (+381)

Embassy CES / Rainbow School of English
Kolo Srpskih Sestara 8, 21000 Novi Sad
Tel: 21 363
Email: infouk@embassyces.com
Courses offered: DELTA

SOUTH AFRICA (+27)

International House, Johannesburg
4th Floor, Aspern House, 54 De Korte Street, Braamfontein 2001, Johannesburg
Tel: 11 339 1051
Fax: 11 403 1759
Email: laurencek@ihjohannesburg.co.za
Website: www.ihjohannesburg.co.za
Courses offered DELTA

SPAIN (+34)

The British Council Examination Services
P. del General Martinez Campos 31
Madrid, 28010
Tel: 91 337 3529
Fax: 91 337 3586
Website: www.britishcouncil.org
Courses Offered: LTCL.Dip.TESOL.

British Language Centre
Calle Bravo Murillo 377, 2°, 28020 Madrid
Tel: 91 733 0739
Fax: 91 314 5009
Email: ted@british-blc.com
Website: www.british-blc.com
Courses offered DELTA

Campbell College
Teacher Training Centre, Calle Pascual y Genís, 14 pta 4, 46002 Valencia
Tel: 96 352 4217
Fax: 96 351 5402
Email: campbell@cpsl.com
Website: www.cpsl.com/campbell
Courses offered DELTA

International House, Barcelona
Trafalgar 14
08010 Barcelona
Tel: 93 268 4511 / 93 268 3304

Fax: 93 268 0239
Email: training@bcn.ihes.com
Website: www.ihes.com/bcn
Courses offered DELTA

International House, Madrid
Calle Zurbano 8, 28010 Madrid
Tel: 91 319 7224
Fax: 91 308 5321
Email: ttraining@ihmadrid.es
Website: www.ihmadrid.es
Courses offered DELTA

International House, Seville
Calle Menez Nunez 7, 41001 Seville
Tel: 954 50 03 16
Fax: 954 50 08 36
Email: mike@clic.es
Website: www.clic.es
Courses offered DELTA

Language Institute
Calle Oliva 11-20, Pontevedra, Galicia.
Tel/Fax: 986 871978.
Courses Offered: LTCL.Dip.TESOL.

Oxford TEFL
Avda Diagonal 402 Principal , Barcelona, Spain 8037
Tel: 93 458 0111
Fax: 93 458 6638
Website: www.oxfordtefl.com
Courses Offered: LTCL.Dip.TESOL.

SWITZERLAND (+41)

Klubschule Migros
Bahnhofplatz 2, 9001 St. Gallen
Tel: 71 228 1609
Fax: 71 228 1601
Email: lee.walker@ksmos.ch
Website: www.klubschule.ch
Courses offered DELTA

TURKEY (+90)

Bilkent University
Teacher Training Unit, School of English Language, 06533 Bilkent, Ankara
Tel: 312 290 5190
Fax: 312 266 4320
Email: elif@bilkent.edu.tr
Website: www.bilkent.edu.tr/~busel
Courses offered DELTA

Eastern Mediterranean University
School of Foreign Languages, Magosa, PO Box 95 Mersin 10
Tel: 392 630 2163
Fax: 392 365 0785
Email: stephen.pike@emu.edu.tr
Website: www.emu.edu.tr
Courses offered DELTA

International Training Institute (I.T.I.)
Kallavi Sok 7-9 Kat: 4, Istiklal Cad, Galatasaray Istanbul
Tel: 212 243 28 88
Fax: 212 245 31 63
Email: iti_ist@yahoo.com
Website: www.itistanbul.org
Courses offered DELTA

UNITED ARAB EMIRATES (+971)

Abu Dhabi Men's College
Higher College of Technology, P.O. Box 25035, Abu Dhabi
Tel: 2 4048 353
Fax: 2 4048 367
Email: phil.quirke@hct.ac.ae
Website: www.admc.hct.ac.ae/admcinternet/
Courses offered DELTA

Dubai Men's College
PO Box 15825, Dubai
Tel: 4 608 5502
Fax: 4 269 1369
Email: suzanne.McMahon@hct.ac.ae
Website: www.hct.ac.ae
Courses offered DELTA

Higher Colleges of Technology - Al Ain
Al Ain Men's College, PO Box 17155, Al Ain
Tel: 382 0888
Fax: 782 0099
Email: norman.williams@hct.ac.ae
Website: www.hct.ac.ae
Courses offered DELTA

Higher Colleges Of Technology - Abu Dhabi
PO Box 25026, Abu Dhabi.
Tel: 2 681 4600 / 5654.
Fax: 2 681 5833 / 6579.
Courses Offered: DELTA.

Higher Colleges of Technology - Sharjah
Sharjah Men's College, PO Box 7946, S.A.I.F. Zone, Sharjah
Tel: 6558 5222
Fax: 6558 5252
Email: lesley.dick@hct.ac.ae
Website: www.hct.ac.ae
Courses offered DELTA

UNITED KINGDOM (+44)

Aberdeen College
Gallowgate Centre, Gallowgate, Aberdeen, AB25 1BN
Website: www.abcol.ac.uk
Email: b.ilett@abcol.ac.uk
Tel: 01224 612000
Fax: 01224 612001
Courses Offered: LTCL.Dip.TESOL

Anglo-Continental School of English
29-35 Wimborne Rd, Bournemouth, England BH1 4PT
Tel: 01202 557414
Fax: 01202 556156
Website: www.anglo-continental.com
Courses Offered: LTCL.Dip.TESOL

Barnet College
CELTA Courses, LBCE Department
Wood Street, Barnet, EN5 4AZ
Tel: 0208 275 2828
Fax: 0208 441 5236
Email: fran.linley@barnet.ac.uk
Website: www.barnet.ac.uk
Courses offered DELTA

Bell Language School
Bowthorpe Hall, Norwich, NR5 9AA
Tel: 01603 745615
Fax: 01603 747669
Email: info.norwich@bell-centres.com
Website: www.bell-centres.com
Courses offered DELTA

Bell Language School
South Road, Saffron Walden, CB11 3DP
Tel: 01799 582100
Fax: 01799 582199
Email: Bruce.Milne@bell-centres.com
Website: www.bell-schools.ac.uk
Courses offered DELTA

Bournemouth & Poole College of FE
Room 345, Lansdowne, Bournemouth, England BH1 3JJ
Tel: 01202 205851
Website: www.bpc.ac.uk
Courses Offered: LTCL.Dip.TESOL

Brasshouse Language Centre
50 Sheepcote Street, Birmingham, B16 8AJ
Tel: 0121 303 0114
Fax: 0121 303 4782
Website: www.birmingham.gov.uk/brasshouse
Courses offered DELTA

City of Bristol College
Department of Languages, Cabot House, Brunel Centre, Ashley Down, Bristol, BS7 9BU
Tel: 0117 312 5163
Fax: 0117 312 5180
Email: emmanuel.raud@cityofbristol.ac.uk
Website: www.cityofbristol.ac.uk
Courses offered: DELTA

Brooklands College
Heath Road, Weybridge, KT13 8TT
Website: www.brooklands.ac.uk
Email: uover@brooklands.ac.uk
Tel: 01932 847029
Fax: 01932 797800
Courses Offered: LTCL.Dip.TESOL

Canterbury Christ Church University College
North Holmes Road, Canterbury, CT1 1QU
Tel: 01227 458459
Email: english@cant.ac.uk
Website: www.cant.ac.uk/io
Courses Offered: LTCL.Dip.TESOL

Chester English College/English in Chester
9-11 Stanley Place, Chester, CH1 2LU
Tel: 0 1244 318913
Fax: 0 1244 320091
Email: study@english-in-chester.co.uk

Website: www.english-in-chester.co.uk
Courses offered DELTA

City of Bath College
Avon Street, Bath, BA1 1UP
Tel: 0 1225 312191
Fax: 0 1225 444213
Email: bulld@citybathcoll.ac.uk
Website: www.citybathcoll.ac.uk
Courses offered DELTA

City College Manchester
Fielden Centre, 141 Barlow Moor Road, West Didsbury, Manchester, M20 2PQ
Website: www.ccm.ac.uk
Email: aspencer@ccm.ac.uk
Tel: 0161 957 1660
Fax: 0161 434 0443
Courses Offered: LTCL.Dip.TESOL

Ealing, Hammersmith, West London College
Gliddon Road, London W14 9BL
Freephone: 44 800 980 2175
Fax: 0 20 8563 8247
Email: cic@hwlc.ac.uk
Website: www.hwlc.ac.uk
Courses offered DELTA

Eastbourne School of English
8 Trinity Trees, Eastbourne, BN21 3LD
Tel: 0 1323 721759
Fax: 0 1323 639271
Email: delta@esoe.co.uk
Website: www.esoe.co.uk
Courses offered DELTA

Embassy CES
Palace Court, White Rock, Hastings, TN34 1JY
Tel: 0 1424 720100
Fax: 0 1424 720323
Email: training@embassyces.com
Website: www.embassyces.com
Courses offered DELTA

English First - EF
26 Wilbraham Road, Manchester, England M14 6JX
Tel: 0161 256 1400
Website: www.ef.com
Courses: LTCL.Dip.TESOL

English Language Centre
5 New Street, York, YO1 8RA
Tel: 0 1904 423594
Fax: 0 1904 672200
Email: english@btinternet.com
Website: www.elcyork.com
Courses offered DELTA

English Language House
300 Saxon Gate West, Milton Keynes, MK9 2ES
Tel: 0 1908 694 357
Fax: 0 1908 694 355
Email: info@englishlanguagehouse.co.uk
Website: www.englishlanguagehouse.co.uk
Courses offered DELTA

Golders Green Teacher Training Centre
11 Golders Green Road, London NW11 8DY
Email: teachertraining@ggcol.fsnet.co.uk
Tel: 020 8731 0963
Fax: 020 8455 6528
Courses Offered: LTCL.Dip.TESOL

Guildford College Of Further & Higher Education
Stoke Park, Guildford, GU1 1EZ
Tel: 01483 448500
Fax: 01483 448600
Courses Offered: LTCL.Dip.TESOL

inlingua
iTTR, Rodney Lodge, Rodney Road, Cheltenham, GL50 1HX
Tel: 01242 253 171
Email: info@inlingua-cheltenham.co.uk
Website: www.inlingua-cheltenham.co.uk
Courses: LTCL.Dip.TESOL

Institute for Applied Language Studies
Univ. of Edinburgh, 21 Hill Place, Edinburgh EH8 9DP
Tel: 0 131 650 6200
Fax: 0 131 667 5927
Email: IALS.enquiries@ed.ac.uk
Website: www.ials.ed.ac.uk
Courses offered DELTA

International House
106 Piccadilly, London W1J 7NL
Tel: 0 20 7518 6928
Fax: 0 20 7518 6921

Email: fernando.nonohay@ihlondon.co.uk
Website: www.ihlondon.com
Courses offered DELTA, also Distance-DELTA (www.thedistancedelta.com)

International House
14-18 Stowell Street, Newcastle upon Tyne, NE1 4XQ
Tel: 0 191 238 9551
Fax: 0 191 238 1126
Email: info@ihnewcastle.com
Website: www.ihnewcastle.com
Courses offered DELTA

International House - The Distance DELTA
106 Piccadilly, London W1V 9FL
Tel: +44 (0) 20 7518 6999
Fax: +44 (0) 20 7518 6998
Email: admin@thedistancedelta.com
Website: www.thedistancedelta.com
Courses offered DELTA

ITTC United Kingdom Ltd
Nortoft Road, Charminster, Bournemouth, BH8 8PY
Tel: 0 1202 397 721
Fax: 0 1202 309 662
Email: tefl@ittc.co.uk
Website: www.ittc.co.uk
Courses offered DELTA

Kingston College
Kingston Hall Road, Kingston-upon-Thames, KT1 2AQ
Website: www.kingston-college.ac.uk
Email: kise@kingston-college.ac.uk
Tel: 020 8546 2151
Fax: 020 8268 2900
Courses Offered: LTCL.Dip.TESOL

King's College English Language Teaching Centre
English Language Centre, King's College London, The Strand, London WC2R 2LS
Tel: 0 207 848 1604
Fax: 0 207 848 1601
Email: christopher.sciberas@kcl.ac.uk
Website: www.kcl.ac.uk
Courses offered DELTA

Langside College
50 Prospecthill Road, Glasgow, G42 9LB
Website: www.langside.ac.uk
Email: tfoster@langside.ac.uk
Tel: 0141 649 4991
Fax: 0141 632 5252
Courses Offered: LTCL.Dip.TESOL

Language Link
181 Earls Court Road, Earls Court, London SW5 9RB
Tel: 0 20 7370 4755
Fax: 0 20 7370 1123
Email: languagelink@compuserve.com
Website: www.languagelink.co.uk
Courses offered DELTA

Language Project
27 Oakfield Road, Clifton, Bristol, BS8 2AT
Website: www.languageproject.co.uk
Email: info@languageproject.co.uk
Tel: 0117 9090911
Fax: 0117 907 7181
Courses Offered: LTCL.Dip.TESOL

Languages Training and Development
Suite 2, Waterloo House , 58-60 High Street , Witney, OX28 6RJ
Tel: 01993 708 637
Fax: 01993 862791
Website: www.ltdoxford.com
Courses Offered: LTCL.Dip.TESOL

Manchester Academy of English
St Margaret's Chambers, 5 Newton Street, Manchester, England M1 1HL
Tel: 0161 237 5619
Fax: 0161 237 9016
Website: www.manacad.co.uk
Courses Offered: LTCL.Dip.TESOL

Northbrook College
Modern Languages Centre, Littlehampton Road, Goring-by-Sea, Worthing, BN12 6NU
Website: www.nbcol.ac.uk
Email: s.scowen@nbcol.ac.uk
Tel: 01903 606243
Fax: 01903 606207
Courses Offered: LTCL.Dip.TESOL

Nottingham Trent University
Nottingham Language Centre, Newton Building, Room 519, Burton Street, Nottingham, NG1 4BU
Tel: 0 115 941 8418
Fax: 0 115 848 6513

Email: linda.taylor@ntu.ac.uk
Website: www.nlc.ntu.ac.uk
Courses offered DELTA

Oxford Cherwell College
Oxpens Road, Oxford, OX1 1SA
Tel: 01865 269268
Fax: 01865 269412
Email: Steven_Haysham@oxfordcollege.ac.uk
Website: www.oxfordcollege.ac.uk
Courses offered: DELTA

Oxford College of Further Education
1 Oxpens Road, Oxford, OX1 1SA
Tel: 0 1865 269268
Fax: 0 1865 269412
Email: Steven_Haysham@oxfordcollege.ac.uk
Website: www.oxfordcollege.ac.uk
Courses offered DELTA

Oxford House College
28 Market Place, Oxford Circus, London W1W 8AW
Tel: 0 20 7580 9785
Fax: 0 20 7323 4582
Email: teachertraining@oxfordhouse.co.uk
Website: www.oxfordhousecollege.co.uk
Courses offered DELTA

Reading College
Crescent Road, Reading, RG1 5RQ
Tel: 0118 967 5442
Fax: 0118 967 5441
Courses Offered: LTCL.Dip.TESOL

Reading College and School of Arts & Design
Crescent Road, Reading, RG1 5RQ
Tel: 0 118 967 5411
Fax: 0 118 967 5441
Email: baberm@reading-college.ac.uk
Website: www.reading-college.ac.uk
Courses offered DELTA

Regency School of English
Royal Crescent, Ramsgate, CT11 9PE
Email: regency.school@btinternet.com
Tel: 01843 591212
Fax: 01843 850035
Courses Offered: LTCL.Dip.TESOL

St George International
76 Mortimer Street, London W1N 7DE
Website: www.stgeorges.co.uk
Email: teflenq@stgeorges.co.uk
Tel: 020 7299 1700
Fax: 020 7299 1711
Courses Offered: LTCL.Dip.TESOL

St Giles College
51 Shepherds Hill, London, N6 5QP
Tel: 0208 340 0828
Fax: 0208 348 9389
Email: edtrust@stgiles.co.uk
Website: www.tefl-stgiles.com
Courses offered: CELTA, DELTA

Sandwell College
Smethwick Campus, Crocketts Lane, Smethwick, B66 3BU
Website: www.sandwell.ac.uk
Email: gill.clarke@sandwell.ac.uk
Tel: 0121 253 6238
Fax: 0121 253 6322
Courses Offered: LTCL.Dip.TESOL

Saxoncourt Teacher Training
Nightingale House, 56-70 Queensway, Petts Wood
Bromley, BR5 1DH
Tel: 01689 873 355
Fax: 01689 872 100
Email: jshepheard@sgvenglish.com
Website: www.sgvenglish.com
Courses offered DELTA

Shane English School
59 South Molton Street, London W1K 5SN
Website: www.sgvenglish.com
Email: awhitehead@sgvenglish.com
Tel: 07491 1911
Fax: 07493 3657
Courses Offered: LTCL.Dip.TESOL

Sheffield Hallam University, TESOL Centre
36 Collegiate Campus, Sheffield, S10 2BP
Email: tesol@shu.ac.uk
Tel: 0114 225 2240
Fax: 0114 225 2280
Website: www.shu.ac.uk
Courses Offered: LTCL.Dip.TESOL

Sidmouth International School
May Cottage, Sidmouth, EX10 8EN
Email: efl@sidmouth-int.co.uk

Tel: 01395 516754
Fax: 01395 579270
Website: www.sidmouth-int.co.uk
Courses Offered: LTCL.Dip.TESOL

Solihull College
Blossomfield Road, Solihull, Birmingham, B91 1SB
Tel: 0 121 678 7173
Fax: 0 121 678 7276
Email: pat.morris@Solihull.ac.uk
Website: www.solihull.ac.uk
Courses offered DELTA

Stanton Teacher Training Courses Ltd
Stanton House, 167 Queensway, London W2 4SB
Tel: 0 207 221 7259
Fax: 0 207 792 9047
Email: study@stanton-school.co.uk
Website: www.stanton-school.co.uk
Courses offered DELTA

TLI Ltd. Teacher Training
Jewel and Esk Valley College, 24 Milton Road East, Edinburgh, England EH15 2PP UK
Tel: +441316201163
Website: www.tlieurope.com
Courses Offered: LTCL.Dip.TESOL

Universal Language Services
43 - 45 Cambridge Gardens, Hastings, TN34 1EN
Website: www.unversallanguageservices.com
Email: enquiries@universallanguageservices.com
Tel: 01424 438025
Fax: 01424 438050
Courses Offered: LTCL.Dip.TESOL

University of Bath
English Language Centre, Claverton Down, Bath BA2 7AY,
Tel: 0 1225 323139
Fax: 0 1225 323135
Email: mlssfw@bath.ac.uk
Website: www.bath.ac.uk/elc/delta.htm
Courses offered DELTA

University of Durham
The Language Centre, Elvet Riverside, New Elvet, Durham, DH1 3JT
Tel: 0 191 374 3716
Fax: 0 191 374 7790
Email: CELTA.training@durham.ac.uk
Website: www.durham.ac.uk/language.centre
Courses offered DELTA

University College London (UCL)
Language Centre, 136 Gower Street, London, WC1 6BT
Tel: 0207 679 7722
Fax: 0207 383 3577
Email: a.salisbury@ucl.ac.uk
Website: www.ucl.ac.uk
Courses offered: DELTA

Waltham Forest College
Forest Road, London E17 4JB
Tel: 0 208 501 8178
Fax: 0 208 501 8001
Email: lawlor@waltham.ac.uk
Website: www.waltham.ac.uk
Courses offered DELTA

Warwickshire College
Thornbank Centre, Warwick New Road, Leamington Spa, CV32 5JE
Tel: 0 1926 318 165
Fax: 0 1926 427 317
Email: bbuxton@warkscol.ac.uk
Website: www.warkscol.ac.uk
Courses offered DELTA

Westminster Kingsway College
Castle Lane, London SW1E 6DR
Tel: 0 207 963 8515
Fax: 0 207 233 8509
Email: annecandan@westking.ac.uk
Courses offered DELTA

Windsor Schools TEFL
21 Osborne Road, Windsor, SL4 1RU
Website: www.windsorschools.co.uk
Email: info@windsorschools.co.uk
Tel: 01753 858995
Fax: 01753 831726
Courses Offered: LTCL.Dip.TESOL

USA (+1)

Oxford Seminars
12335 Santa Monica Blvd, Suite 337, Los Angeles, California 90025
Tel: 310-820-2359

Toll Free: 1-800-779-1779
Email: info@oxfordseminars.com
Website: www.oxfordseminars.com
Courses Offered: TESOL/TESL Certificate

Oxford Seminars

244 5th Avenue, Suite J262, New York, New York 10001-7406
Tel: 212-213-8978
Toll Free: 1-800-779-1779
Email: info@oxfordseminars.com
Website: www.oxfordseminars.com
Courses Offered: TESOL/TESL Certificate

St. Giles Language Teaching Center

One Hallidie Plaza (Suite 350), San Francisco, CA 94102
Tel:1 415 788 3552
Fax:1 415 788 1923
Email: info@stgiles-usa.com
Website: www.stgiles-usa.com
Courses offered: DELTA

VIETNAM (+84)

International Language Academy (ILA) Vietnam

402 Nguyen Thi Minh Khai Street, District 3, Ho Chi Minh City
Tel: 8 929 0100 (Ext. 124)
Fax: 8 929 0070
Email: andrew@ilavietnam.com
Website: www.ilavietnam.com
Courses offered: DELTA

Master's Degree Courses

A Master's Degree in Teaching English as a Second or Foreign Language is, for many teachers, a key qualification for those who want to work to cover the theory of teaching – or work in schools and universities.

For many prospective teachers, the decision to embark on a Master's degree programme represents a crucial investment of time and money in the career of their choice. Studying for a postgraduate degree need not leave you in debt for the rest of your natural born days. Many students can earn money while they study by working as a Graduate Teaching Assistant. A number of universities and colleges offer the opportunity for students to earn their degrees on a part-time basis, so you can work while you study. Most institutions offer scholarships and grants: make it a point to inquire about when you contact a school or university.

Master's degree programmes vary in terms of admission requirements and length of the course. Prospective students whose first language is not English will need to submit their TOEFL score for consideration by the admission office of the institution where they hope to study. M.A. programmes are usually flexible and you can work at your own pace within the criteria for time set by the university or college: a programme can take anywhere from one to five years or more to complete.

There are many different titles for Master's degree programmes. These often reflect the historical development of the department with which they are associated. There are three main types of Master's degree in TESL/TEFL: (1) Master of Arts (M.A.), (2) Master of Science (M.Sc./M.S.) and (3) Master of Education (M.Ed.). In general there is little difference between all three degrees although earning a M.Ed., which is awarded by a School of Education, may not involve writing a thesis. A MAT (Master of Arts in Teaching) focuses more on the practical aspects of teaching than the other types of Master's degrees.

An Applied Linguistics degree is at the other end of the scale in that these programmes usually concentrate more directly on the theory of language acquisition.

The benefits associated with a Master's degree are numerous. Although there are institutions that will hire teachers who have not earned a postgraduate degree, the reality is that many employers, especially in higher education, insist that candidates have a Master's degree in TESL, TEFL, TESOL or Applied Linguistics. A teacher with a Master's degree usually enjoys a greater earning power than a colleague who lacks the qualification. Above all, embarking on a Master's programme reinforces a teacher's decision that he or she has made teaching English as a second or foreign language a serious career choice.

Choosing a Master's Degree Course

If you plan to make teaching English as a second or foreign language your major career decision, then you should be looking at studying for a Master's Degree in TEFL / TESL or TESOL. The first thing to do when choosing a Master's programme that's right for you is to ask questions: What kinds of courses are offered in the programme? Are they more practical or theoretical in scope?

If you are interested in going directly into teaching after getting your master's degree, you should stick to a practical curriculum. If, on the other hand, if you are interested in pursuing a Ph.D., then look for a more theoretically based curriculum.

What do the faculty members specialize in? Ask the programme director for details on what the faculty is researching, what courses they teach, and even about possible internships or assistantships. This will give you an indication of what the focus of your studies will be. If your interests are similar to theirs, you have a good match.

If you are interested in teaching in the public school system, find out if the courses lead to or satisfy provincial certification (in Canada) or state certification (in the US), and if they satisfy the requirements in any other province or state where you are considering certification.

General guidelines:

Ask for a reference or two from the programme director. Ask recent graduates if they are satisfied with the education they received and whether their courses have actually met the needs of their current jobs.

Examine the job placement record. What percentage of graduates get jobs after they receive their degrees? Does the school offer placement assistance or career counselling?

English for Special Purposes Programmes

Other professions look to university TESOL master's programmes to develop instructors for teaching English for special purposes (ESP) in specialized fields such as medicine, law, computer science, engineering, and business. Master's programmes must provide their students with the expertise needed to teach a wide range of skills from basic English for those working at entry level jobs to accent reduction courses for professionals seeking to improve their oral communication skills. Additionally, ESP instructors need a familiarity with the content subject matter and the ability to collaborate with professors and other professionals in various disciplines.

Teaching English as a Foreign Language Master's programmes continue to fulfill their traditional role of providing teachers for positions overseas teaching English to college and university students, professional adults, and young learners. With the worldwide demand for English instruction, American students wishing to teach abroad and international students planning to teach English in their own countries seek out programmes in Teaching English as a Foreign language (TEFL).

These programmes feature methodology courses that address frequent concerns of English as a Foreign Language (EFL) teachers such as curriculum design for large class sizes, limited budget and materials, and resources for authentic language input and natural language acquisition.

TESOL Doctoral Study

TESOL master's programmes are the training ground for prospective doctoral candidates who conduct research, design and implement university degree programmes, and teach at the master's and doctoral levels in institutions of higher education. To prepare students for these academic roles, TESOL master's programmes typically offer a combination of courses which cover the theory and practice of second language teaching, curriculum development, and the fundamentals of linguistics, second language acquisition, and research.

Master's Degree Courses

There are opportunities to study for a Master's degree in a wide range of countries around the world. These selected schools and universities offer full-time intensive courses and, more often, distance or part-time study.

AUSTRALIA (+61)

Macquarie University
Sydney, NSW 2109.
Tel: 2 9850 9647.
Fax: 2 985 9352.
Email: lingpgo@ling.mq.edu.au
Website: www.ling.mq.edu.au/researchdegrees
Start Dates: March, July.
Course length: 2-3 semesters FT.
Admission Requirements: First degree + teaching experience.
Programmes: M.A. and PhD in Applied Linguistics.

University of Melbourne
Department of Language, Literacy and Arts Education, Room 513, Doug McDonell Building, Victoria, 3010.
Tel:: 3 9344 8377.
Fax: 3 8344 5104.
Programmes: Master of TESOL.

Monash University
Normanby House, 100 Normanby Road, Clayton, Victoria 3168.
Programmes: M.A. TESOL.
Course length: 2 years FT; 4 years PT.

Queensland University of Technology
School of Culture & Language Education, Kelvin Grove Campus, Victoria Park Road, Kelvin Grove, Brisbane, Queensland 4059.
Tel: 7 3864 3242.
Fax: 7 3864 3988.
Programmes: M.Ed. TESOL.

University of South Australia
GPO Box 2471, Adelaide, South Australia 5001.
Start Dates: February.
Tel: 8 8302 6611.
Fax: 8 8302 2466.
Programmes: M.Ed. TESOL.

Tasmania University
English Language Centre, P.O. Box 1214, Launceston, Tasmania 7250.
Programmes: M.Ed. TESOL.
Course length: 18 months FT.
Admission Requirements: Degree.

Victoria University
PO Box 14428, Melbourne, Victoria, 8001.
Tel: 3 9688 4000.
Fax: 3 9689 4069.
Programmes: M.A. in TESOL and Literacy.

CANADA (+1)

Brock University
Graduate Studies Office, St. Catharines, Ontario L2S 3A1.
Tel: (905) 688-5550, x4467.
Fax: (905) 988-5488.
Email: ekoop@spartan.ac.brocku.ca
Programmes: M.Ed. TESL.
Course length: 1 year.
Grant Availability: Possible Teaching/Research Assistantships.
Admission Requirements: Undergraduate degree with an average of 75% in the last 10 credits.

Carleton University
School of Linguistics and Applied Language Studies,1125 Colonel By Drive, Ottawa K1S 5B6.
Tel: (613) 520-2802.
Fax: (613) 520-6641.
Email: linguistics@carleton.ca
Programmes: M.A. Applied Language Studies.
Course length: Usually 4 terms FT depending on which option student decides to complete (thesis, research essay or all courses).
Start Dates: January, May, September.
Admission Requirements: Honors degree with a B - B+ average.

Concordia University
School of Graduate Studies
Office of the Registrar, 1455 de Maisonneuve Boulevard West, Montreal, Quebec H3G 1M8
Tel: 1-514-848-2837
Website: www.concordia.ca
Programmes: M.A. - Applied Linguistics

McGill University
Faculty of Graduate Studies, Second Language Education, Education Bldg., Rm. 431A, 3700 McTavish St. Montreal, Quebec H3A 1Y2
Tel: 1-514-398-6982
Website: www.education.mcgill.ca
Programmes: M.Ed.-Second Language Education

Memorial University of Newfoundland
Faculty of Education, Teaching and Learning Studies, St. John's, Newfoundland, A1B 3X8
Tel: 1-709 -737 -8587
Website: www.mun.ca
Programmes: M.Ed.-Teaching and Learning Studies

Trinity Western University
School of Graduate Studies, Teaching English as a Second or Other Language, 7600 Glover Road, Langley, BC V2Y 1Y1
Tel: 1-604-888-7511
Website: www.twu.ca
Programmes: M.A. - TESOL

University of Calgary
Faculty of Graduate Studies, Graduate Division of Educational Research, Teaching English as a Second Language, Education Tower 940, 2500 University Drive, NW, Calgary, Alberta T2N 1N4
Tel: 1-403-220-5675
Website: www.ucalgary.edu
Programmes: M.Ed

York University
Room 544 South, Ross Building, 4700 Keele Street, Toronto, Ontario M3J 1P3.
Tel: (416) 736-2100 x88270.
Fax: (416) 736-5483.
Email: gradling@yorku.ca
Programmes: M.A. in Theoretical & Applied Linguistics.
Course length: 1 year FT.
Grant Availability: Teaching Assistantships, Graduate Assistantships, Scholarships.
Start Dates: September.

CHINA (HONG KONG +852)

Chinese University of Management
Tat Chee Avenue, Kowloon, Hong Kong.
Tel: 2788 9400.
Fax: 2788 9020.
Programmes: M.A. Degree in TESL.

City University of Hong Kong School of Graduate Studies
5/F Cheng Yick Chi Building, Tat Chee Avenue, Kowloon.
Tel: 2788 9076.
Fax: 2788 9940.
Programmes: M.A. Degree in TESL.

EGYPT (+20)

The American University in Cairo
113 Sharia Kasr el-Aini - PO Box 2511, Cairo, 11511.
Tel: 2-357-5530.
Fax: 2-355-7565. (In the U.S.: **Tel:** (212) 730-8800. **Fax:** (212) 730-1600)
Programmes: M.A. Degree in TEFL.
Course length: 2 years FT.
Start Dates: January, September.

IRELAND (+353)

The National University of Ireland Language Centre
O'Rahilly Building, UCC, Cork.
Tel: 21 4902043.
Fax: 21 4903223.
Programmes: M.A. Degree in Applied Linguistics.
Course length: FT 1 year.
Start Dates: October.

Admission Requirements: Relevant primary degree at Second Class Honors standard or equivalent, English to IELTS level 7.0 if not a native speaker.

MALAYSIA (+60)

Universiti Malaya Faculty of Languages and Linguistics Education

Pantai Valley, 59100 Kuala Lumpur.
Tel: 3 756-0022.
Fax: 3 755-2975.
Programmes: M.A. Degree in TESOL.

MEXICO (+52)

Universidad de las Américas

Departamento de Lenguas, Sta. Catarina Mártir, Puebla, 72820.
Tel: 22 29 31 05.
Fax: 22 29 31 05
Email: eckep@mail.udlap.mx
Programmes: M.A. in Applied Linguistics.
Course length: 2 years FT.
Aid Availability: A limited number of teaching assistantships is available in the Department of Languages. Mexican students can obtain financial aid from the Ministry of Public Education of Mexico, while participants from the United States are able to procure college loans from institutions in the U.S.

PHILIPPINES (+63)

University of the East

2219 C.M. Recto Avenue, 1008 Manila.
Tel: 2-741-95-01.
Programmes: M.A. Degree in Education (TESOL).

SOUTH AFRICA (+27)

Rhodes University

Faculty of Arts, P.O. Box 94, Grahamstown, 6140.
Tel: 461 318111.
Fax: 461 25049.
Programmes: M.A. TESOL.
Course length: 2 years.

THAILAND (+66)

Assumption University

A Building 3rd floor, Ramkhamhaeng Soi 24 Road Huamark, Bangkok 1.
Tel: 2-3004543-62.
Fax: 2-7191521.
Programmes: M.A. TESL.

Chiang Mai University Graduate School

Chang Mai, 50200.
Tel: 53-221699.
Fax: 53-892231.
Programmes: M.A. TEFL.

Payap University Graduate School

Chiang Mai, 50000.
Programmes: M.A. TEFL.

UNITED KINGDOM (+44)

Aston University

Aston Triangle, Birmingham B4 7ET.
Tel: 121 359 3611.
Fax: 121 359 6350.
Email: lsu@aston.ac.uk
Website: www.les.aston.ac.uk/tesol
Programmes: M.Sc./Diploma TESOL, M.Sc./Diploma Teaching ESP. Both distance learning

Canterbury Christ Church University

North Holmes Road, Canterbury, Kent, CT1 1QU
Tel: 01227 458459
Fax: 01227 781558
Email: english@cant.ac.uk
Website: www.cant.ac.uk/io
Courses offered: MA TESOL

Middlesex University

White Hart Lane, London N17 8HR
Tel: 020 8411 5898
Email: admissions@mdx.ac.uk
Website: www.ilrs.mdx.ac.uk/lang/eng_non.htm
Courses: MA TEFL with Applied Linguistics

Norwich Institute of Language Education (NILE)

PO Box 2000, Norwich, NR2 1LE
Tel: 01603 664473
Fax: 01603 664493

Email: registrar@nile-elt.com
Website: www.nile-elt.com
Courses offered: MA language education, range of courses for professional development

Oxford Brookes University

Tel: 01865 483874
Fax: 01865 484377
Email: icels@brookes.ac.uk
Website: www.brookes.ac.uk/icels
Courses: MA Applied Linguistics, MA ELT

Sheffield Hallam University

TESOL Centre, School of Education, Collegiate Campus, Sheffield S10 2BP.
Tel: 114 225 2240.
Fax: 114 225 2280.
Programmes: M.A. TESOL.
Course length: 12 weeks DL phase, + 6 weeks intensive direct contact phase, + 32 weeks research phase.
Admission Requirements: PG diploma in TESOL or equivalent.
Start Dates: March or October each year for direct contact phase in June/July or January/February.

University of Bath

English Language Centre, University of Bath, BA2 7AY
Tel: 01225 383024
Fax: 01225 383135
Email: english@bath.ac.uk
Website: www.bath.ac.uk/elc
Courses offered: MA in ELT

University of Birmingham

CELS, University of Birmingham, Edgbaston, Birmingham, B15 2TT
Tel: 0121 414 3239
Fax: 0121 414 3298
Email: cels@bham.ac.uk
Courses offered: MA Applied Linguistics, MA ESP, MA TEFL/TESL,

University of Bristol

Graduate School of Education, 35 Berkeley Square, Bristol BS8 1JA.
Tel: 117 9287040.
Fax: 0117 927250.
Email: janet.woolways@bristol.ac.uk
Course length: 12 months full-time; up to five years part-time; can also be done in term blocks: October to December or January to March.
Start Dates: October FT; January, October PT. Summer Session: Presessional available for those who have either been out of full-time education for many years, or who feel that one month of academic preparation would benefit them. Usually held throughout September.
Admission Requirements: First degree, at least one year fulltime teaching experience; preferably a professional qualification in TESL/TEFL.
Specializations: Specialist centre for Language in Education, including English as an additional language; off shore teaching can be provided for closed groups.
Courses: M.Ed. TEFL, EdD in TESOL, MPhil/PhD in TESOL

University Of Durham

Department Of Linguistics And English Language, Elvet Riverside II, New Elvet, Durham DH1 3JT.
Tel: 191 374 2641.
Fax: 191 374 2685.
Programmes: M.A. Linguistics, M.A. Language Acquisition, M.A. Applied Linguistics, M.A. Applied Linguistics (ELT), M.A. Applied Linguistics, (ELT, CALL And Educational Technology), M.A. Applied Linguistics (ESP), M.A. Applied Linguistics (ESOL), M.A. Applied Linguistics (ELT And Materials Development), M.A. Applied Linguistics (FLT),M.A. Applied Linguistics (Translation).

University of Edinburgh

Faculty of Education, Moray House, Holyrood Road, Edinburgh, EH8 8AQ, Scotland.
Tel: 131 651 6329.
Fax: 131 651 6409.
Programmes: M Ed (TESOL) (DL available).
Course length: 1 year FT; 6 years maximum DL.
Start Dates: October Full-time; Continuous DL.
Admission Requirements: Degree + 3 years' relevant experience.

University of Exeter

School of Education and Lifelong Learning
Tel: 01392 264815
Fax: 01392 264902
Email: ed-cpd@exeter.ac.uk
Website: www.ex.ac.uk/sell
Courses: MPhil in TESOL, MEd in TESOL, EdD in TESOL, Pgcert in TESOL

University of Leicester
21 University Road, Leicester, LE1 7RF.
Tel: 116 2523675.
Fax: 116 2523653.
Programmes: M.A. Applied Linguistics and TESOL (DL).

University of London
Institute of Education, 20 Bedford Way, London WC1 0AL.
Tel: 020 7612 6536
Fax: 020 7612 6534.
Course length: 1 year FT; 2 years PT.
Start Dates: October.
Admission Requirements: A good Honours class degree, and at least two years teaching experience
Programmes: M.A. TESOL.

University of Luton
Vicarage Street, Luton, LU1 3JU.
Tel: 1582 489019.
Fax: 1582 743466.
Programmes: M.A. Applied Linguistics (TEFL).
Course length: 1 year. **Start Dates:** October.

University of Manchester
Faculty of Education, University of Manchester, Oxford Rd, Manchester, M13 9PL
Tel: 0161 275 3463
Email: education.enquiries@man.ac.uk
Website: www.man.ac.uk/langlit
Courses offered: MEd in ELT, MEd in Educational Technology and ELT

University of Newcastle upon Tyne
Language Centre, Level 4, Old Library Bldg., Newcastle upon Tyne, NE1 7RU.
Tel: 191 222 5621
Fax: 191 222 5239
Programmes: M.A. in Media Technology for TEFL, M.A. in Linguistics for TESOL.
Course length: 12 months FT for both programmes, 24 months PT (Linguistics).
Requirements: 2 years FT teaching experience.
Start Dates: September.

University of Sunderland
School of Humanities & Social Sciences, Forster Building, Chester Road, Sunderland SR1 3SD.
Tel: 191 515 2000.
Programmes: M.A. TESOL.

University of Surrey
English Language Institute, Guildford, Surrey, GU2 5XH.
Tel: 1483 259910
Fax: 1483 259507.
Programmes: M.A. Linguistics (TESOL) (DL). M.Sc. English Language Teaching Management (DL).
Course length: 27 months.
Start Dates: March, October.
Admission Requirements: 1st degree and at least two years full time teaching experience.

University of Warwick
CELTE, Coventry, CV4 7AL
Tel: 024 7652 3200
Fax: 024 7652 4318
Email: celte@warwick.ac.uk
Website: www.warwick.ac.uk/celte
Courses offered: MA in ELT, MA in ESP, Diploma in ELT and Administration, research degrees of PhD and EdD

University of York
EFL Unit, York, Y010 5DD
Tel: 01904 432483
Fax: 01904 432481
Email: efl2@york.ac.uk
Website: www.york.ac.uk/ltc/efl
Courses offered: MA in teaching young learners (distance course)

UNITED STATES OF AMERICA (+1)

Adelphi University
1 South Avenue, Garden City, NY 11530
Tel: (516) 877-3000
Website: www.adelphi.edu/

University of Alabama
UAB Station, Birmingham, AL 35294
Website: http://main.uab.edu/
Courses: MA TESOL

American University
4400 Massachusetts Avenue NW, Washington DC 20016-8001
Tel: 202-885-6000
Fax: 202-885-1025
Website: www.american.edu/
Email: afa@american.edu

Courses: MA TESOL, MIP TESOL, w/Peace Corps., MAT in ESOL

University of Arizona
Tucson, AZ 85721
Tel: (520) 621-2211
Website: www.arizona.edu
Courses: MA ESL

Arizona State University
University Drive and Mill Avenue, Tempe, AZ 85287
Tel: (480) 965-9011
Website: www.asu.edu/
Email: information@asu.edu
Courses: MA TESL

Azusa Pacific University
901 E. Alosta Ave.Azusa, CA 91702
Tel: (626) 969-3434
Website: www.apu.edu/
Courses: MA TESOL

Ball State University
Muncie, IN 47306.
Tel: 1-800-482-4278.
Website: www.bsu.edu/
Email: ASKUS@bsu.edu.
Courses: MA in TESOL

Biola University
13800 Biola Avenue, La Mirada, California 90639-0001
Tel: 1-800-OK-BIOLA 562-903-4752
Website: www.biola.edu/
Courses: MA TESOL

Boston University
121 Bay State Road Boston, MA 02215
Website: www.bu.edu/
Courses: Med TESOL

Bowling Green State University
Bowling Green, OH 43402
Website: www.bgsu.edu/
Courses: MATESL programme

Brigham Young University
Provo, Utah 84602
Tel: (801) 378-INFO or (801) 378-4636
Website: www.byu.edu
Courses: TESOL MA

University of California
One Shields Avenue, Davis, CA 95616
Tel: (530) 752-1011
Website: www.ucdavis.edu/
Courses: MA with concentration in applied linguistics.

University of California, LA
405 Hilgard Avenue, Box 951361 Los Angeles, CA, 90095-1361
Tel: (310) 825-4321
Website: www.ucla.edu/
Courses: MA Applied Linguistics & TESL
Course length: 2 years full-time

Cal. State Univ, Dominguez Hills
1000 E. Victoria Street, Carson, California 90747
Tel: (310) 243-3696
Website: www.csudh.edu/
Courses: MA English with TESL

Cal. State Univ, Fresno
5241 N. Maple Avenue, Fresno, CA 93740-8027
Tel: (559) 278-4240
Website: www.csufresno.edu/
Email: webmaster@csufresno.edu
Courses: MA Linguistics/ESL Emphasis

Cal. State Univ, Fullerton
P.O. Box 34080 Fullerton, CA 92834
Tel: (714) 278-2011
Website: www.fullerton.edu/
Courses: MS Education with TESOL
Course length: 30 semester units, full or part-time

Cal. State Univ, Hayward
25800 Carlos Bee Blvd, Hayward, CA 94542
Tel: (510) 885-3000
Website: www.csuhayward.edu/
Courses: MA English – TESOL option.
Course length: 6 quarters, full or PT
Aid Availability: Grants and fellowships

Cal. State Univ, Los Angeles
5151 State University Drive, Los Angeles, California 90032-4226
Tel: (323) 343-3000
Website: www.calstatela.edu/
Courses: MA TESOL
Course length: 45 quarter units, full or part-time

Aid Availability: Scholarships, assistantships, fee waivers
Other: Practical teacher-training programme. Also offered in Buenos Aires, Argentina.

Cal. State Univ, Northridge
18111 Nordhoff Street, Northridge, California 91330
Tel: (818) 677-1200
Website: www.csun.edu/
Courses: MA Linguistics with TESOL

Cal. State Univ, Sacramento
6000 J Street, Sacramento, CA 95819.
Tel: (916) 278-6011
Website: www.csus.edu/
Courses: MA TESOL
Other: Pedagogical focus. Joint programme with the Peace Corps is offered.

Cardinal Stritch University
6801 N. Yates Rd., Milwaukee, WI 53217-3985.
Tel: (414) 410-4000
Website: www.stritch.edu/
Courses: MEd Professional Development with ESL

Carson-Newman University
Jefferson City, TN, 37760.
Tel: (865) 475-9061
Fax: (865) 471-3502
Website: www.cn.edu/
Courses: MAT-ESL
Course length: 12-18 mths

Central Connecticut St. University
1615 Stanley St., New Britain, CT 06050
Tel: 1-860-832-CCSU
Website: www.ccsu.edu/
Courses: M.S. in TESOL; Certification in TESOL.

Central Michigan University
Mount Pleasant, Michigan 48859
Tel: (517) 774-4000
Website: www.cmich.edu/
Courses: MA TESOL
Other: Teaching practice at university language centre.

Central Missouri State Universtiy
Department of Linguistics
P.O. Box 800, Warrensburg MO 64093
Tel: 1-800-SAY-CMSU (1-800-729-2678)
Website: www.cmsu.edu
Courses: MA-TESL, MA TESL

Central Washington University
400 E. 8th Avenue, Ellensburg, WA 98926
Tel: (509) 963-1211
Website: www.cwu.edu/
Courses: MA English with TESL/TEFL

University of Cincinnati
2624 Clifton Avenue, Cincinnati, OH 45221
Tel: (513) 556-6000
Website: www.uc.edu
Courses: MEd Literacy Education with TESL

University of Colorado, Boulder
Boulder, CO 80309
Tel: (303) 492-1411
Website: www.colorado.edu
Courses: MA Linguistics
Other: Primarily an MA in linguistics with some TESL options.

University of Colorado, Denver
P.O. Box 173364, Denver, CO 80217-3364
Tel: 303-556-2400
Website: www.cudenver.edu/
Courses: MA English/Applied Linguistics (ESL)
Other: Teaching practice available at the program's campuses in Beijing and Moscow.

Colorado State University
Fort Collins, CO 80523
Tel: (970) 491-1101
Website: www.colostate.edu/
Courses: MA TESL

Colombia Intl. University
Website: www.columbia.edu/
Courses: MA TEFL/Intercultural Studies

University of Delaware
Newark, DE 19716
Tel: (302) 831-2000
Website: www.udel.edu/
Courses: MA ESL
Other: Teaching practice locally in ESL and bilingual classes, and option of teaching in Panama.

East Carolina University
East Fifth Street, Greenville, NC 27858-4353
Tel: (252) 328-6131
Website: www.ecu.edu/
Courses: MA English with TESL

Eastern College
521 Lancaster Avenue, Richmond, KY, 40475
Tel: (859) 622-1000
Website: www.eku.edu/
Courses: MA Multicultural Education.
Other: Strong spiritual emphasis.

Eastern Mennonite University
1200 Park Road, Harrisonburg, VA 22802-2462
Tel: (540) 432-4000
Fax: (540) 432-4444
Website: www.emu.edu/
Courses: MA Education, minor in TESL

Eastern Michigan University
Ypsilanti, MI, USA 48197
Tel: (734) 487-1849
Website: www.emich.edu/
Courses: MA in TESOL, Graduate Certificate in TESL/Summer 2 Summer, Undergrad minor in TESOL, K-12 Endorsement in TESL.

Evergreen State College, WA
2700 Evergreen Parkway NW, Olympia, WA 98505
Tel: (360) 866-6000
Website: www.evergreen.edu/
Courses: MA Teaching w/ Bilingual Education and TESL.

Fairfield University
1073 North Benson Road, Fairfield, CT 06430
Tel: (203) 254-4000
Website: www.fairfield.edu/
Courses: MA TESOL

Fairleigh Dickinson University
Florham-Madison Campus, 285 Madison Avenue, Madison, NJ 07940
Tel: (973) 443-8500
Courses: MAT ESL

University of Findlay
1000 North Main Street, Findlay, OH 45840-3695
Tel: (800) 472-9502 or (419) 422-8313
Website: www.findlay.edu/
Courses: MA TESOL and Bilingual Education

Florida Atlantic University
Tel: 1-800-299-4FAU
Website: www.fau.edu/about/campuses/
Email: admisweb@fau.edu
Courses: Master's In Curriculum and Instruction with ESOL.

Florida International University
University Park Campus
11200 S.W. Eighth Street, Miami, FL 33199
Tel: 305-348-2000
Website: www.fiu.edu/
Courses: M.Sc TESOL.

Florida State University
Tallahassee, FL 32306
Tel: 850-644-2525.
Website: www.fsu.edu/
Courses: Multilingual/Multicultural Education

Fordham University
New York City
Website: www.fordham.edu/
Courses: MS Education.

Fresno Pacific University
Tel: 559-453-2039 Toll-Free: 1-800-660-6089
Fax: 559-453-2007
Website: www.fresno.edu/
Email: ugadmis@fresno.edu
Courses: TESOL M.A. and TESOL Certificate

Georgetown University
37th and O Street NW, Washington DC 20057
Tel: (202) 687-0100
Website: www.georgetown.edu/
Courses: MAT TESL

Grand Canyon University
3300 West Camelback Road, Phoenix, AZ 85017
Tel: 1-800-800-9776
Courses: MA with TESL major

University of Hawaii, Manoa
2444 Dole Street, Honolulu, HI 96822
Tel: (808) 956-8111
Website: www.uhm.hawaii.edu/

Courses: MA ESL
Other: The first and largest Department of ESL at a US university.

Hofstra University
Hempstead, NY 11549-1000
Tel: 1-800-HOFSTRA
Website: www.hofstra.edu/
Courses: MS Education: TESL, MEd TESL

University of Houston
4800 Calhoun Rd, Houston, Texas 77204
Tel: (713) 743-2255
Website: www.uh.edu/
Courses: MA Applied English Linguistics

University of Houston, Clear Lake
2700 Bay Area Blvd, Houston, TX 77058
Tel: (281) 283-7600
Website: www.cl.uh.edu/
Courses: MS Multicultural Studies with ESL endorsement

Hunter College, City Univ, NY
Website: www.hunter.cuny.edu/
Courses: MA Education (TESOL)

Univ. Illinois, Urbana-Champaign
1401 West Green Street, Urbana, IL 61801
Tel: (217) 333-1000
Website: www.uiuc.edu/
Email:iuforum@uiuc.edu
Courses: MA TESL

Illinois State University
Website: www.uillinois.edu/
Courses: MA Writing with TESOL
Other: Can lead to an Illinois credential in TESOL. Teaching internships and jobs on offer in Korea and Thailand.

Indiana State University
Terre Haute, IN 47809
Tel: (812) 237-6311
Website: www-isu.indstate.edu/
Courses: MA English with Applied Linguistics
Other: Exchange programmes with institutions in China and Japan.

Indiana University
107 S. Indiana Ave, Bloomington, IN 47405-7000
Tel: (812) 855 4848
Website: www.indiana.edu/
Courses: MA Applied Linguistics with TESOL

Indiana Univ. Of Pennsylvania
Indiana, PA 15705-0001.
Tel: (724) 357-2100.
Website: www.iup.edu/
Courses: MA English (TESOL)

University of Iowa
107 Calvin Hall, Iowa City, Iowa 52242
Tel: 1-800-553-IOWA
Courses: MA Linguistics with TESL

University of Kansas
Lawrence, KS 66045
Tel: 785/864-2700
Website: www.ukans.edu/
Courses: MA TESL

Long Island University
700 Northern Boulevard, Brookville, New York 11548-1326
Website: www.liu.edu/
Courses: MS Education with TESL
Course length: 3 semesters, full or p/t

Univ. of Maryland, Balt. County
1000 Hilltop Circle, Baltimore, MD 21250
Tel: (410) 455-1000
Website: www.umbc.edu/
Courses: MA ESOL/Bilingual Instructional Systems Development (w/ K-12 cert)

Univ. of Maryland, College Park
College Park, MD 20742
Tel: (301) 405-1000
Website: www.umcp.umd.edu/
Courses: MEd TESOL

Marymount University
2807 N. Glebe Road, Arlington, VA, 22207-4299
Tel: 703-522-5600 or 800-548-7638
Website: www.marymount.edu/
Courses: MEd ESL

Univ. Massachusetts, Amherst
Massachusetts, 01003.
Tel: (413) 545-0111
Website: www.umass.edu/
Courses: MEd with ESL

Univ. Massachusetts, Boston
100 Morrissey Boulevard, Boston, Ma 02125-3393
Tel: 617-287-5000
Website: www.umb.edu/
Courses: MA Bilingual or ESL Studies

University of Memphis
Office of Admissions, 229 Administration building, Memphis, TN 38152-3370
Tel: 901 678-2169 or 800-669-2678
Website: www.memphis.edu/
Courses: MA English with ESL
Other: Summer programme in Czech Republic and internships in China.

Meredith College
3800 Hillsborough Street, Raleigh, NC, 27607-5298
Tel: 919-760-8600 Fax: 919-760-2828
Website: www.meredith.edu/
Courses: MEd with ESL

Michigan State University
East Lansing, Michigan
Tel: 517-355-1855
Website: www.msu.edu/
Courses: MA TESOL

University of Montana
Website: www.umt.edu/
Courses: MA in TESOL

Montclair University
1 Normal Avenue, Upper Montclair, N.J. 07043
Tel: (973) 655-4000
Website: www.montclair.edu/
Courses: MA Applied Linguistics with TESL

Monterey Int. Of Intl. Studies
425 Van Buren Street, Monterey, CA 93940
Tel: (831) 647-4100
Fax: (831) 647-4199
Website: www.miis.edu/
Courses: MA in TESOL, MA in TFL (Teaching Foreign Language), Peace Corps Masters' International

Mount Vernon College
800 Martinsburg Road, Mount Vernon, Ohio 43050
Tel: (740) 397-9000 or (800) 782-2435
Website: www.mvnc.edu/
Courses: MA TESOL

National-Louis University
Chicago Campus, 122 S. Michigan Avenue, Chicago, IL 60603
Tel: 800-443-5522 (x 5151)
Website: www.nl.edu/
Courses: MEd with ESL

University of Nevada, Reno
Reno, Nevada 89557
Tel: (775) 784-1110
Website: www.unr.edu/
Courses: M.A. in TESOL

University of New Hampshire
Durham, NH 03824
Website: www.unh.edu/
Courses: MA in English Language and Linguistics (with ESL specialization), MAT w/TESL

University of New Mexico
Albuquerque, NM 87131
Tel: (505) 277-0111
Website: www.unm.edu/
Courses: MA Education with TESOL

New Mexico State University
Las Cruces, NM 88003-8001
Tel: 505-646-0111
Website: www.nmsu.edu/
Courses: MA Curriculum & Instruction with TESOL
Other: Teaching practice in Mexico

College of New Rochelle, NY
Website: http://cnr.edu/
Courses: MS TESOL

New York State Univ., Albany
1400 Washington Avenue, Albany, NY 12222
Tel: (518) 442-3300
Website: www.albany.edu/
Courses: MA TESOL

New York State Univ., Buffalo
15 Capen Hall, Buffalo, NY 14260-1660

Tel: 1-888-UB-ADMIT or (716) 645-6900
Website: www.buffalo.edu/
Courses: MEd TESOL

New York State Univ., Stony Brook
Stony Brook, NY 11794
Website: www.sunysb.edu/
Courses: MA TESOL

New York University
70 Washington Square South, New York, NY 10012
Tel: (212) 998-1212
Website: www.nyu.edu/
Courses: MA TESOL

Univ. Of North Carolina, Charlotte
9201 University City Blvd, Charlotte, NC 28223-0001
Tel: (704) 687-2000
Website: www.uncc.edu/
Courses: MEd

University of North Texas
PO Box 311277, Denton, TX 76203
Tel: (940) 565-2000 or (800) 735-2989
Website: www.unt.edu/
Courses: MA English with ESL

Northeastern Illinois University
5500 North St. Louis Avenue, Chicago, Illinois
Tel: 60625-4699, 773/583-4050
Website: www.neiu.edu/
Courses: MA Linguistics with TESL

Northern Arizona University
Tel: 1-888-MORE-NAU
Website: www.nau.edu/
Courses: MA TESL
Programme length: 4 semesters, full or p/t

Northern Illinois University
DeKalb, IL 60115
Tel: (815) 753-1000
Website: www.niu.edu/
Courses: MA English with TESOL
Other: Teaching opportunities in China.

Northern Iowa University
1227 West 27th Street, Cedar Falls, IA 50614
Tel: (319) 273-2311
Website: www.uni.edu/
Courses: MA TESOL
Other: International programme with internships in Moscow and St. Petersburg, Russia.

Notre Dame College
1500 Ralston Avenue, Belmont, California 94002-1997
Tel: (650) 593-1601
Fax: (650) 508-3660
Website: www.cnd.edu/
Courses: MEd TESL

Nova Southeastern University
3301 College Avenue, Fort Lauderdale, Florida 33314
Tel: 800-541-6682
Courses: MS Education with TESOL

Ohio State University
Columbus, Ohio 43210
Tel: 614-292-OHIO
Website: www.ohio-state.edu/
Courses: MA TESOL

Ohio University
Department of Linguistics, Gordy Hall 383, Athens, OH 45701-2979
Tel: (740) 593-4564
Fax: (740) 593-2967
Website: www.ohiou.edu/
Email: lingdept@ohio.edu
Courses: MA Applied Linguistics and TESOL

Oklahoma State University
Office of Academic Affairs, 101 Whitehurst, Stillwater, OK 74078
Tel:(405) 744-5627
Fax: (405) 744-5495
Website: http://osu.okstate.edu/
Courses: MA English with TESL
Programme length: 4 semesters, full or p/t

University of Oregon
Eugene, OR 97403
Tel: (541) 346-1000
Website: www.uoregon.edu/
Courses: MA in Linguistics with concentration in Applied Linguistics/SLA

Our Lady of the Lake Univ., S.A.
411 S.W. 24th Street, San Antonio, TX 78207-4689
Tel: 800-436-OLLU

Website: www.ollusa.edu/
Courses: MEd with ESL

University of Pennsylvania
3451 Walnut, Philadelphia PA 19104
Tel: 215-898-5000
Website: www.upenn.edu/
Courses: MSEd TESOL

Pennsylvania State University
201 Shields Building, Box 3000 , University Park, PA 16804-3000
Tel: (814) 865-5471
Website: www.psu.edu/
Courses: MA TESL

University of Pittsburgh
Pittsburgh, PA 15260
Tel: (412) 624-4141
Website: www.pitt.edu/
Courses: MA Linguistics with TESOL certificate

Portland State University
PO Box 751, Portland, Oregon 97207
Tel: (503) 725-3000
Website: www.pdx.edu/
Courses: MA TESOL

University of Puerto Rico
PO Box 364984, San Juan, PR 00936-4984
Tel: (787) 250-0000
Website: www.upr.clu.edu/
Courses: BA in TESS, MEd. in TESL,and ED.D. in Curriculum and Teaching in TESL

Queen's College of City Univ, N.Y.
65-30 Kissena Blvd,.Flushing, New York 11367
Tel: (718) 997-5000
Website: www.qc.edu/
Courses: MA applied Linguistics, MS Education with TESOL

Radford University
East Norwood St., Radford, Va. 24142
Tel: 540- 831-5000
Website: www.runet.edu/
Courses: MS Education with curriculum/instruction and TESL

Rhode Island College
600 Mount Pleasant, Providence, RI 02908-1991
Tel: (401) 456-8000
Website: www.ric.edu/
Courses: MEd TESL

Rhode Island University
Kingston, RI 02881
Tel: 401-874-1000
Website: www.uri.edu/
Courses: MA TESL

University of Rochester
Rochester, New York 14627
Tel: (716) 275-2121
Website: www.rochester.edu/
Courses: MS TESOL

State Michael's College
1 Winooski Park, Colchester, VT 05439
Tel: 802-654-2000
Website: www.smcvt.edu/
Courses: MA TESL/TEFL

Sam Houston State University
1803 Ave I, Huntsville, TX 77341
Tel: (936) 294-1111
Website: www.shsu.edu/
Courses: MEd with Bilingual Education

University of San Francisco
2130 Fulton Street, San Francisco, CA 94117
Tel: 415-422-5555
Website: www.usfca.edu/
Courses: MA TESL

San Francisco State University
1600 Holloway Avenue, San Francisco, CA 94132
Website: www.sfsu.edu/
Courses: MA English with TESOL

San Jose State University
One Washington Square, San José, CA 95192
Tel: 408-924-1000
Website: www.sjsu.edu/
Courses: M.A. TESOL

College of Santa Fe
1600 St. Michael's Drive, Santa Fe NM 87505
Tel: 800-456-2673 or 505-473-6011

Website: www.csf.edu/
Courses: MA with TESL (focus on at-risk youth)

School for International Training
Kipling Road, P.O. Box 676, Brattleboro, Vermont 05302-0676
Tel: (802) 257-7751
Fax: (802) 258-3248
Website: www.sit.edu/
Email: info@sit.edu
Courses: Master of Arts in Teaching (TESOL, French, Spanish), U.S. Public School Certification option, Bilingual-Multicultural Education Endorsement option, also Master International programme with US Peace Corps option., Master of Arts in Teaching Concentrations : English to Speakers of Other Languages (ESOL), Spanish, French
Further options available in the Academic Year MAT Program: Vermont State Certification and Bilingual-Multicultural Education Endorsement

Seattle University
900 Broadway, Seattle WA 98122-4340
Tel: (206) 296-6000
Website: www.seattleu.edu/
Courses: MA/MEd TESOL
Other: Focus on teaching the adult learner

Seattle Pacific University
3307 Third Avenue West, Seattle, WA 98119-1997
Tel: (206) 281-2000
Website: www.spu.edu/
Courses: MA TESOL

Seton Hall University
400 South Orange Avenue, South Orange, New Jersey 07079
Tel: (973) 761-9000
Website: www.shu.edu/
Courses: MA ESL

Shenandoah University
1460 University Drive, Winchester, VA 22601
Tel: 800-432-2266
Website: www.su.edu/
Courses: MSEd TESOL

Simmons College
300 The Fenway, Boston, Massachusetts, 02115-5898
Tel: 617-521-2000 or 800-345-8468
Fax: 617-521-3190
Website: www.simmons.edu/
Courses: MATESL

Soka University
California campus, 26800 West Mulholland Hwy, Calabasas, California 91302
Tel: (818) 880-6400
Fax: (818) 880-9326
Courses: MA Second/Foreign Language Education with TESOL
Other: Internships and post-graduation working opportunities at affiliating university and high schools in Japan.

University of South Carolina
Columbia, South Carolina, 29208
Tel: (803) 777-7000
Website: www.sc.edu/
Courses: MA Linguistics

University of South Florida
Tel: (813) 974-3350 or 1-877-USF-BULL
Website: www.usf.edu/
Courses: MA Applied Linguistics

Southeast Missouri State Univ.
One University Plaza, Cape Girardeau, MO 63701
Website: www.semo.edu/
Courses: MA TESOL

Southern Illinois Univ., Carbondale
Carbondale, Illinois 62901
Tel: 618.453.4381
Website: www.siuc.edu/
Courses: MA TESOL

University of Southern Maine
P.O. Box 9300, Portland, ME 04104-9300,
Tel: (207) 780-4141 or 1-800-800-4USM
Website: www.usm.maine.edu/
Courses: MSEd Literacy Education with ESL

University of Southern Mississippi
Multiple Campuses
Website: www.usm.edu/
Courses: MA Teaching of Languages (TESOL)

Syracuse University
Syracuse, NY 13244
Tel: (315) 443-1870

Website: www.syr.edu/
Courses: MA Linguistics with ESL

Teachers College, Columbia Univ.
525 West 120th St. New York, 10027
Tel: 212-678-3000
Website: www.tc.columbia.edu/
Courses: MA TESOL

Temple University
1801 North Broad Street, Philadelphia, PA 19122
Tel: 215-204-7000
Website: www.temple.edu/
Courses: EdM TESOL, MEd TESOL

University of Texas, Arlington
701 South Nedderman Drive, Arlington, TX 76019
Tel: (817) 272-2222
Website: www.uta.edu/
Courses: MA Linguistics with TESOL

University of Texas, Austin
University of Texas at Austin, Austin, Texas 78712
Tel: (512) 475-7348.
Website: www.utexas.edu/
Courses: MA Foreign Language Education with TESL/TEFL

University of Texas, El Paso
500 West University Avenue, El Paso, Texas 79968
Website: www.utep.edu/
Courses: MA Linguistics with Applied or Hispanic Linguistics

University of Texas, Pan American
Website: www.panam.edu/
Courses: MA ESL

University of Texas, San Antonio
6900 N. Loop 1604 West, San Antonio, TX 78249
Tel: 210-458-4011
Website: www.utsa.edu/
Courses: MA with ESL
Programme length: 3-4 semesters, full or p/t

Texas Tech University
Website: www.ttu.edu/
Courses: MA Applied Linguistics
Programme length: 4 semesters, full or p/t

Texas Woman's University
304 Administration Dr., Denton, TX 76201
Tel: (940)TWU-2000
Website: www.twu.edu/
Courses: MEd with ESL
Programme length: 2 years, full or p/t

University of Toledo
Toledo, Ohio 43606-3390
Tel: (419) 530-4242
Website: www.utoledo.edu/
Courses: MA-Ed. in ESL; MA in English with Concentration in ESL (these are two different degrees; both are offered)

U.S. International University
Admissions Office, 10455 Pomerado Road, San Diego, CA 92131
Website: www.usiu.edu/
Courses: MEd with TESOL

University of Utah
Salt Lake City, UT 84112
Tel: (801) 581-7200
Website: www.utah.edu/
Courses: MA Bilingual Education

University of Washington
Website: www.washington.edu/
Courses: MA TESL

Washington State Univ., Vancouver
Vancouver, Canada
Website: www.vancouver.wsu.edu/
Courses: Master of Education programme with ESL Endorsement (K-12 emphasis)

West Chester University
West Chester, PA 19383
Tel: (610) 436-1000
Website: www.wcupa.edu/
Courses: Master's & Certificate in TESL (Teaching English as a Second Language)

West Virginia University
Morgantown, WV 26506
Tel: (304) 293-0111
Website: www.wvu.edu/
Courses: Master of Arts in Foreign Languages (TESOL area of emphasis)

Western Kentucky University
1 Big Red Way, Bowling Green, KY 42101-3576
Tel: (270) 745-0111.
Website: www.wku.edu/
Courses: MA English with TESL

Wheaton College
501 College Avenue, Wheaton, IL 60187
Tel: (630) 752-5000
Website: www.wheaton.edu/
Courses: MA Intercultural Studies with TESL
Other: Leads to an Illinois State credential in TESOL. Teaching practice may be undertaken locally or at universities in Russia, China, Korea or Romania.

Wichita State University
1845 Fairmount, Wichita, Kansas, 67260.
Website: www.wichita.edu/
Courses: MA Curriculum and Instruction with TESOL
Other: Leads to Kansas State endorsement in TESOL. Opportunities for fieldwork in Mexico and Cyprus.

William Paterson University
300 Pompton Road, Wayne, NJ 07470
Tel: (973) 720-2000
Website: www.wpunj.edu/
Courses: MEd. in Education with a concentration in Bilingual/ESL

Wright State University
3640 Colonel Glenn Highway, Dayton, Ohio 45435-0001
Tel: (937) 775-3333
Website: www.wright.edu/
Courses: Undergraduate or graduate TESOL certificate. Undergraduate BA in English – emphasis in TESOL, MA in TESOL, Graduate endorsement in TESOL

Youngstown State University
One University Plaza, Youngstown, OH 44555
Tel: 877-GO-TO-YSU (877-468-6978)
Website: www.ysu.edu/
Courses: M.A. English, TESOL concentration

VENEZUELA

Universidad Central de Venezuela
Facultad de Humanidades y Educación, Ciudad Universitaria, Los Chaguaramus, Caracas, 1050.
Programmes: M.A. Degree in TEFL.

Learning to teach with CALL

Computer technology is now playing an increasingly influential role in education. Language education has been no exception to this trend and in recent years growing numbers of teachers are undertaking graduate level courses in the field of CALL. Mark Peterson examines the choices and courses for Computer Assisted Language Learning teaching qualifications

In the first part of this article, I will identify the major factors that teachers must consider before embarking on such a course of study. In the second part, a number of CALL courses in various countries will be examined in order to assist teachers in the task of selecting a course suitable to their needs and future career goals.

The number of CALL focused training programmes has increased rapidly in recent years. In contrast to a few years ago, teachers considering entering a CALL course are faced with a larger number of possible options. However before embarking on a course potential applicants should consider the following questions:

If your goal is to improve your knowledge of computer technology, ESL pedagogy and make you a better teacher then a CALL course may be for you. If you think that graduating from such a course will make you more marketable you may be disappointed. While it is certainly true that the demand for computer literate language teachers is increasing, the number of posts is limited. Moreover the financial returns from undertaking what is a demanding course of study may not justify for some, the high costs of such a programme.

Most Masters degree programmes in CALL will usually require applicants to hold a first degree and some previous teaching experience in the language teaching field. However there are exceptions and negotiation over entry requirements may be possible in some cases.

To many mid-career teachers or those with circumstances that make undertaking a residential based MA impossible, distance learning may be a possibility. However candidates should first ensure that any distance based course is accredited internationally. British universities appear to be taking the lead in this area of course provision; distance based MA's have been common in the UK for many years. However it is fair to say that suspicion regarding the validity of distance based degrees remains in some conservative quarters of academe.

Graduate programmes are expensive and CALL courses are no exception to this trend. Potential applicants should compare and contrast course fees and the cost of living in North America with the UK and Australia. Potential applicants should also note that the level of course fee is often determined by an applicant's nationality and residency status. For example, a British applicant for a Master degree resident in the UK usually qualifies for reduced rate "home" fees, while applicants from overseas are usually required to pay a higher "overseas" fee.

Various sources of financial help are available including competitive bursaries, home government support and grants from private charitable institutions.

Masters degrees generally take two years to complete. But in some cases it is certainly possible to finish in a shorter period of time. Burn out may be a factor in many accelerated MA programmes. Full time doctoral degrees generally take three or more years to complete. Part time doctoral degrees may take several years.

There is a great deal of variety in course content within CALL postgraduate degree programmes. In the author's view, a CALL degree should include a through grounding in applied linguistics, instructional design and practical computer skills. These skills should include the use of software tools and authoring packages. Other relevant areas of study would be hypermedia design; concordancing and computer mediated communication (CMC).

Any potential applicant to a masters programme would do well to find out as much as possible about the faculty who will teach them. CALL experts of long experience are still relatively rare so it pays to take the time to check if the faculty in your programme have the appropriate backgrounds in educational technology and applied linguistics.

Potential students with little or no background in computers may expect to experience a degree of technostress in the initial stages of the course. This may be alleviated to some degree, if applicants already possess basic computer skills including familiarity with the more popular word processing, email and web browser software. Being prepared may make the course more enjoyable and beneficial in the long term.

Assessment methods vary between degree programmes, but in most cases assessment is based on course work and examinations.

In the US most CALL based degrees are given the degree classification of MA whereas in the UK by contrast, many CALL degrees are classed as M.Sc. or M.Ed.

Although the job market in EFL is growing, finding a suitable post where CALL skills can be fully utilized isn't easy. One motivation for those taking CALL courses is that they provide transferable skills. Many graduates of CALL programmes have launched alternative careers in areas such as web design, instructional technology and software consultancy.

Embarking on an advanced degree is a major undertaking in terms of time and money. Selecting an appropriate course depends largely on an individuals needs and circumstances.

While obtaining a Master's degree in the CALL field does not guarantee employment, it does provide opportunities in what is an expanding and increasingly important area of language education.

Mark Peterson holds an M.Sc. degree in TESOL and CALL and is currently a faculty member at Japan Advanced Institute of Science and Technology.

CALL courses

M.Sc. in CALL and TESOL

University of Stirling, Scotland

www.stir.ac.uk/celt/courses/MSC.HTM#CALL

This course was established in 1993 by John Higgins, one of the founders of CALL as a distinct discipline, and was the first CALL postgraduate programme to be launched in the UK. The programme combines a solid grounding in applied linguistics with the essential elements in CALL theory and methodology.

The M.Sc. may be completed in one year and is made up of the following components: Autumn Semester, Teaching of skills or Second Language, Acquisition, Syntax and Grammar, Classroom Observation (for those with no ELT experience), Introduction to CALL concepts and materials, CALL and communications. In addition to the above requirements, students must also complete a 12,000 to 15,000 word dissertation on an aspect of CALL research. A research based Ph.D. degree in CALL is also available at the same institution.

M.Ed. (Master of education) course in educational technology and ELT

University of Manchester

Centre for English Language Studies in Education

www. man.ac.uk/CELSE/centre/techELT.htm

This course is offered in two modes. Fully distance learning based or through a combination of distance learning and summer attendance. In order to graduate students are required to undertake a total of six course modules, three of which must be in the area of educational technology. In addition, students are required to submit a dissertation of between 12,000 and 20,000 words on a research topic of their choice.

MA in Media Technology for Teaching English as a Foreign Language

University of Newcastle upon Tyne

www.ncl.ac.uk/langcen/postgrad/media.html

This course provides students with a grounding in language teaching methodology, computer assisted language learning, video in ELT and media centre management. This course may be completed in 12 months of full-time study. Assessment is by course work.

MA in Computer Assisted Language Learning

University of Melbourne, Australia

www.hlc.unimelb.edu.au/callmast_ arts.htm

The above programme takes the form of a research based Masters degree. In order to graduate a candidate must submit a 30,000 word CALL based thesis by research. It may be undertaken on a full time or part-time basis. The University of Queensland Australia runs a Master's degree programme in applied linguistics, computers, technology and language learning (www.cltr.uq.edu.au/mactll.html). This programme is composed of the following courses: Structure of Language, Applied Linguistics and Language Teaching, Second Language Acquisition, Second Language Research, Second Language Teaching, Literacy or Issues in Language programme

Development, Portfolio and Synthesis, Principles of Computer-Enhanced Language Learning. This programme can be completed over 3 semesters full time or 6 semesters part time.

MA in applied linguistics (includes a major CALL element)

University of Waikato New Zealand

Department of General and Applied Linguistics

www.waikato.ac.nz/ling/

Some of the courses in this degree programme may be undertaken through distance learning. M.Phil. and Ph.D. courses in CALL are also available.

MA in TESOL with a CALL specialization

Iowa State University

www.engl.iastate.edu/tesling/ma.html

This programme is designed for those seeking training as a CALL specialist in an ESL programme. Core course requirements include the study of the following subjects: Sociolinguistics, Grammatical Analysis, Second Language Acquisition Methods for Teaching ESL, Second Language Testing Practicum in Teaching ESL or Seminar in Teaching English Composition, Computer Assisted Language Learning Instructional Technology.

In order to graduate, MA candidates must submit a research-based thesis. The University of South Florida runs a Ph.D. programme in interdisciplinary education (www.coedu.usf.edu/deptseced/forlanged/phdbro.html) with an emphasis on SLA and instructional technology. Core elements of this programme include courses in the areas of applied linguistics and SLA. In addition courses are available in the areas of educational technology described below: Authoring Systems and Languages, Programming Languages for Education, Applications for Computers as Educational Tools, Instructional Design for Computers, Interactive Media and Interactive Video, Telecommunications, Sound Processing and Animation. Candidates are also required to complete a dissertation. The State University of New York at Stony Brook has established a Doctor of Arts programme in Foreign Language Instruction (http://ccmail.sun- .edu/llrc//llrc3.html). Included in this programme are graduate level courses in educational technology with an emphasis on CALL and language centre management. Students on this course have access to the Language Learning and Research centre at Stony Brook.

MA in teaching English as foreign language

Ohio University in Athens Ohio

Department of Linguistics

http://cscwww. cats.ohiou.edu/linguistics/dept/

This degree programme is composed of courses in language teaching theory and methodology, curriculum development and evaluation. A practicum is also provided to give candidates supervised teaching experience.

Five CALL courses are available as part of the above degree programme. Entry requirements include a bachelor’s degree with a minimum GPA of 3.0. This course requires two years of full time study.

Teaching ESP (English for Specific Purposes)

Some EFL teachers find that their career becomes focused on teaching a particular aspect of English language skills geared to help learners in specific situations or jobs. Teachers of ESP (English for Specific Purposes) are usually in high demand, not only in schools in English-speaking countries but all over the world.

Teaching ESP (English for Specific Purposes) can be an interesting and lucrative career development. Much of the world's demand for English comes from the business sector: companies are prepared to pay top rates for training to ensure that their workforce is effective and competitive. In particular, company executives are almost universally required to know Business English, the world's working language. Other common areas of specialization in ESP are English for Law, English for Medical Purposes and English for Tourism. Less well studied in terms of student numbers but equally important areas are English for Engineering, English for Shipping, Aviation English, English for Engineering and English for the Military.

Teaching positions in the ESP require specialized knowledge of the key language used in the field and an understanding of the industries and professions that the language derives from. Some teachers come from a background (legal or medical, for example) that will enable them to tutor in a specialist field without too much trouble. Quite often, teachers with a specialist background like law are able to teach students from their homes. Other teachers find teaching ESP hard going especially when they have little confidence in their own knowledge of the industry or specialization they are teaching in. There are courses available for teachers who do not have experience in a particular field but who do want teach it.

Obtaining a job teaching ESP may require a little lateral thinking on your part. For example, if you'd like to work in the oil industry, why not contact some of the big oil companies and find out what their hiring practices are. If you want to teach English for Military Purposes, contact the relevant branch of the armed forces. Integration of commands within NATO means that the military are always on the lookout for a few good men (and women) to teach English.

Teaching ESP can be a fascinating and rewarding experience. Some ESP teachers have taught classes to oilrig workers on derricks in the Gulf of Siam. Others have worked with Russian cosmonauts in training in closed cities in Siberia. Be aware, however, that many teachers are turned off by the very thought of teaching anything with a commercial aspect. If you want to succeed as an ESP teacher, you must at least display a passing interest in your area of specialization.

Business English

One of the most important areas of specialization in teaching English as a second or foreign language is Business English. Many students are interested in improving their English skills in the areas of conducting business meetings, making presentations, understanding discussions at formal meetings, reading technical journals, and receiving and interacting effectively with American and other English-speaking business people.

Some schools specialize in teaching Business English as a means of attracting better-paying clients. Teachers working with these classes often find they are dealing with mixed-ability students whose needs may not be all the same. After all, an accountant has to learn a different vocabulary than that required by a sales person. Of course, schools with larger numbers of Business English students can divide classes based on ability and needs. Another factor to be aware of when teaching Business English is that there are very few textbooks and published materials to rely on. Most major publishers feel that it is not worth investing in textbooks for teaching Business English (and ESP in general) because it is too small a market to be profitable. Business English teachers are often forced to create their own materials for use in the classroom using a variety of sources such as professional journal articles and college textbooks.

If you are employed in a corporate environment, remember that you will be expected to dress accordingly. You may find yourself discussing students' progress with managers, directors and human resources personnel so it pays to have a professional appearance and attitude.

There are a number of ESP examinations – mostly for students, but with one or two designed for teacher-training certification (from LCCIB). The following are the main developers of business English examinations:

Cambridge ESOL (UCLES)

Business English Certificates (BEC)

Certificate in English for International Business and Trade (CEIBT)

University of Cambridge Local Examinations Syndicate

Syndicate Buildings, 1 Hills Road, Cambridge, CB1 2EU

Tel: 01223 553311

Fax: 01223 460278 / 553068

Website: www.ucles.org.uk

The Chauncey Group (part of ETS)

TOEIC (Test of English for International Communication)

The Chauncey Group

664 Rosedale Road, Princeton, NJ 08540-2218 USA

Tel: +1 609.720.6500

Fax: +1 609.720.6550

Email: info@chauncey.com

Website: www.toeic.com

LCCI Examinations Board

English for Business (EFB)
Spoken English for Industry and Commerce (SEFIC)
Written English for Tourism (WEFT)
English for the Tourism Industry (EFTI) - Group Award (WEFT + SEFIC)
Practical Business English (PBE)
English for Commerce (EFC)
LCCI Examinations Board
Athena House, 112 Station Road, SIDCUP, Kent DA15 7BJ, UK
Tel: +44 (0)20 8302 0261
Fax: +44 (0)20 8302 4169
Website: www.lccieb.com

Pitman

English for Business Communications
English for Office Skills
Pitman Qualifications, 1 Giltspur Street, London EC1A 9DD
Tel: +44 (0)20 7294 3500
Fax: +44 (0)20 7294 2403
Email: info@pitmanqualifications.co.uk
Website: www.pitmanqualifications.com

Certificates for Teachers of Business English

There are few specialist certificates for teachers who want to gain a qualification in this specialist field. The standard CELTA, Cert.TESOL or other certificate courses (see chapter 1) include sections on teaching business English, as do diplomas, MAs and Ph.Ds. The main two certificates for teachers are issued by LCCIEB (London Chamber of Commerce and Industry Examinations Board (below).

LCCIEB: CertTEB (Certificate in Teaching English for Business)

Administered in conjunction with ARELS (www.arels.org.uk), this exam is aimed at teachers with little or no previous experience of teaching business English, and develops the basic practical skills for working in this area.

FTBE (Further Certificate for Teachers of Business English)

A supplementary qualification for teachers of general English wishing to broaden their skills into teaching business English. The exam aims to develop management skills and understanding of business practice. LCCIEB has a worldwide network of centres.

LCCI Examinations Board
International House, Siskin Parkway East, Middlemarch Bus. Park, Coventry, CV3 4PE, UK
Tel: 08707 202909
Fax: +44 (0)24 76516505
Website: www.lccieb.com
Email: customerservice@ediplc.com

English For Academic Purposes

Students whose first language is not English who are planning to study at a college or university where instruction is in English often require specialist help if they are going to succeed on their courses. In particular, these students need to learn how to write about academic subjects including all the relevant conventions of presentation and style that will be expected of them as they submit written work like essays and theses. Students taking EAP (English for Academic Purposes) classes often do so while taking part in an IEP (Intensive English Program) programme at the university they plan to graduate from. Specialist tutors, usually from an academic background themselves, are in demand to teach EAP on these courses. One of the easiest ways to find a position in teaching EAP is to apply through the university that you are studying at or have graduated from although you will probably need to have a Master's degree if your job application is to be considered seriously.

Young Learners

Another area of teaching ESL/EFL that is becoming increasingly specialized is the teaching of English to young learners. Teachers have realized that approaches to the teaching of a second language to children requires specific training if they are to be successful. Teachers with an interest in this area may want to follow a course in order to qualify as a specialist. There are several certificate courses on offer and a Master's degree in Teaching English to Young Learners is offered on a two-year distance-based programme by the University of York in the UK. Teachers with a specialization in TEYL (Teaching English to Young Learners) are especially welcome in Asian countries like Korea and Japan where the teaching of young learners is considered a priority.

Many schools and colleges offer certificate-level courses that prepare teachers for the discipline of teaching young learners. The best-known international qualifications are the CELTYL (Certificate in ELT for Young Learners) from University of Cambridge ESOL and the Cert.TEYL (Certificate in Teaching English to Young Learners) from Trinity College, London. We list the schools and colleges offering these courses in the section on Certificate courses. These courses are normally taken as an add-on module to a standard CELTA or Cert.TESOL certificate course).

Cultural Studies

In the U.S., the teaching of ESL can be just one component of an overall programme in "American Culture" or "American Studies." If you have a degree or qualification in sociology, cultural studies or a related field, you may be able to find a job on one of these programmes where both your language teaching skills and your specialist knowledge can be put to good use.

Studying for a Doctorate

As the English language teaching field becomes more saturated with teachers who hold a Master's degree, many observers predict that doctorates will become a prerequisite for those who want to hold jobs on the top rungs of the professional ladder. Mary Ellen Butler-Pascoe evaluates TESOL doctoral programmes.

While there has been a continuous professional dialogue on various aspects of the TESOL master's programmes, similar attention has not been afforded to TESOL doctoral programmes. This in large part has been a result of the master's degree serving as a terminal degree for most ESL instructors and many directors and teacher trainers. Since the 1960s, when master's degrees in TESOL were rare, the TESOL profession has gained worldwide respect as the discipline continues to define and assert itself in local school districts and institutions of higher education. Master's programmes have proliferated over the past two decades with over 200 now being offered in Canada and the United States alone. Considerably fewer colleges and universities offer doctoral degrees in TESOL or related fields.

What are the characteristics of these doctoral programmes? Where are they located both geographically and institutionally? In what disciplines do they offer degrees? What courses does a typical curriculum include and most importantly in what areas can these programmes better meet the needs of TESOL practitioners? A recent study of these programmes sought to shed light on these questions.

Of the twenty-nine doctorate-granting institutions surveyed, the degree programmes were housed in seventeen different academic departments with the most frequent placement being in the Department of Linguistics, followed by the Departments of English and Education. Eighteen institutions granted Ph.D. degrees and six awarded Ed.D. degrees, indicating the research emphasis of the majority of the programmes. Three institutions provided a choice of either a Ph.D. or Ed.D. and two offered a Doctorate of Arts degree. Within the departments, degrees were granted in a wide range of disciplines with the greatest number in Linguistics followed closely by Education/ Curriculum and Instruction. Only three departments offered a degree major specifically in TESOL though several of the various other degrees offered TESOL/ESL as a concentration option. The standard admissions requirements included a master's degree with acceptable GPAs, letters of recommendation, the Graduate Record Exam (GRE) or in two cases the Miller's Analogy Test (MAT), and the TOEFL for non-native English-speaking applicants. Only two programmes specifically listed teaching experience as a prerequisite for admissions.

For degree completion, twenty-seven of the twenty-nine programmes required comprehensive exams and one required a qualifying paper; nineteen demanded knowledge of a foreign language, and twenty-three required a dissertation/thesis. Only six required practice teaching experience. The length of time for degree completion ranged from two to seven years with an average of 4.02 years of study. Interestingly, 45% of the graduates were non-native English speakers.

While there was a great variety in course nomenclature, for purposes of clarity, similar course titles were grouped together under common headings. Of the twenty-five different course types identified in the curriculum of the various universities, the one most frequently offered was first and/or second Language Acquisition. The second most prominent course was Teaching Methods with over three-fourths of the programmes listing courses specifically in TESOL Methodology/Teaching ESL. Others in the top five headings included Research Methodology focusing on research designs of both quantitative and qualitative methods, Linguistics, and Testing and Assessment.

Heading the list of least frequently provided courses was ESL Administration with only four programmes listing courses in any aspect of programme administration, design, or evaluation. Courses in Language Policy and Planning were similarly scarce, and only eight offered courses in technology media or allowed electives in technology. A typical programme profile incorporating the most common characteristics of all the doctoral programmes would be a Ph.D. degree programme in Linguistics administered by the Department of Linguistics. Students would be predominantly native-English speakers who would be required to have knowledge of a foreign language, pass comprehensive exams, conduct a research study and author a dissertation in order to complete a degree in approximately four years. The curriculum would include courses in second language acquisition and TESOL Methods with no student teaching required.

Fortunately, there are several combinations of location, degree, discipline, concentration, curriculum, and programme requirements from which students may choose a programme. But a review of the current programmes does raise questions as to how well they serve the needs of the ESL practitioners who are teaching in all types of educational settings and directing programmes around the world.

The TESOL profession has conducted a long campaign to establish ESL as a legitimate academic discipline. To this end it has established standards for ESL and teacher training programmes with the latest being the standards for the Intensive English Programmes. To a large extent the profession has overcome the notion that any native English speaker could intuitively teach ESL as evidenced by the now routine requirement of a master's degree for adult education, community college and university level ESL instruction, and teaching abroad. For the thousands of teachers who have obtained master's degrees in recent years, it would be a natural next step in their professional development and in the struggle to constantly enhance the status of the TESOL profession for more practitioners to seek a doctoral degree.

Many obstacles such as lack of time, financial resources, professional incentives and encouragement hinder aspirations for a doctorate, but there are additional negative factors influencing that decision that are within the control of TESOL faculties, university administrators, and even the prospective students themselves. One could start by looking at the doctoral curriculum.

An obvious group of candidates for doctoral study would be programme directors or directors of study in a school. "Doctorate in TESL or related field preferred" is commonly seen in job announcements for ESL director positions, yet only four universities listed doctoral courses in ESL programme design and implementation, programme evaluation, or administrative leadership. Despite the importance of language policy and planning to TESOL local and worldwide issues, just five programmes provided study in this area. Leaders in second language teaching need the knowledge and skills to infuse technology into the language curriculum and to manage the technology that is becoming an increasingly important part of the language learning process. While the role of English as the de facto language of the Internet means that the ESL/EFL field will continue to grow in prominence, it also suggests a need to study the impact the power of English and technology might have on languages in non-English-speaking societies. In what ways do our doctoral programmes assist students in gaining the expertise to use computer technology in language teaching? The answer does not appear to be in the curriculum of most TESOL doctoral programmes. Few programmes offered courses in the educational uses of technology and of those only four included courses specifically in technology for language teaching. Looking at recent journals, it is obvious that an area of great concern to the profession is the apparent dichotomy between classroom teaching and research. TESOL professionals have warned that too many classroom teachers view research as the domain of the university research community and fail to see its relevance to their own teaching. Stronger teacher-researcher ties through collaborations between university faculty and classroom teachers have been urged. The solution argued by the teacher-researcher movement is to recognize teachers as researchers with the advantages of insider perspective and applicability. Perhaps the simplest way to ensure connections between university research and classroom practice is for more classroom teachers to join the university research community as doctoral students. The crucial questions on teaching practices and the language learning process will not be answered until ESL teachers assume a more prominent role in the research process. TESOL doctoral programmes face the challenge of facilitating that outcome.

As TESOL continues to mature, more positions will demand a doctorate, therefore making it incumbent upon universities to analyze the needs not only of traditional doctoral students seeking a university teaching or research position, but also the needs of classroom teachers from a wide array of instructional settings.

Getting Published

Dr. Lin Lougheed, the author of more than 40 English language teaching textbooks, offers his advice on becoming a successful writer in this market, highlighting the best ways to getting your lesson plans in print.

So you want to contribute to the overpublished field of ESL/EFL? Don't give up your day job teaching. You'll need it if you like to eat and if you want a good place to test your materials. A textbook is simply an organized set of daily lesson plans well printed and illustrated on good quality paper. You have these lesson plans, don't you? You've been repurposing textbooks for years and creating your own supplementary materials. Everyone loves them. Your students' proficiency rates have skyrocketed because of the quality of your exercises. Your lazy colleagues are always praising them as they photocopy them to use in their own classes.

The path to becoming an author is simple: have an idea; write it down; teach it; revise it; send it to a publisher. At that point, the process becomes more complicated so let's analyze the easy steps first.

The idea

To publish your fabulous activities you'll need to convince someone in the publishing world that your materials have universal applicability and universal appeal. Universal is the key word. You don't want your materials to be unique. You want them to stand out, but you don't want them to be so different from everything else on the market that no one (especially the sales reps who will have to push your books) can understand how to use them.

Write it down

You have a brilliant idea how to teach a particular grammar point or to reinforce vocabulary. You want to develop a lesson around this idea. Look at some of the books you are using. Most lessons have three parts: Warm Up, Teach, Review.

When you begin to put your idea on paper, think in terms of a class session. How will you prepare the students, teach the lesson, and review the material in 45 minutes?

As you refine your lesson, keep in the back of your mind how you plan to expand this one lesson into enough lessons to fill a book.

Teach it

Actually, have someone else teach it. You have been experimenting with your materials in class. You know what will work or how to make it work. Let someone else interpret your perfect lesson. It would be ideal if you could observe your colleague teaching your materials. Then you can gauge the reactions of the students to the material and the way your colleague adapts the material to his or her teaching style.

Revise it

Was your audience confused? Did they not follow your structure? Was the teacher supplementing the material with too many explanations? Or was it just perfect? Remember your materials have to work outside your own classroom. They will undoubtedly be used by total strangers in a variety of teaching situations. They must have universal appeal.

At this stage, you might even want to post your material on a teacher's bulletin board on the Web or email it to colleagues around the world. Don't worry about someone stealing your idea. I send entire manuscripts to teachers in countries that don't even have copyright protection agreements. What is important to me is their feedback. If they want to use some messy, unformatted materials in their classroom, they can be my guest. Remember there is very little that has not already been done in EFL/ESL publishing.

Send it to a publisher

You have written, taught, revised, taught, and revised again your materials. You are now ready to have your work published. Your material can be published as a book to be downloaded from the Internet or can be published in book form and sold in bookstores. You can do either yourself or you can find a publisher to do it for you.

Whatever the method, you will have to do a market survey of your material. You should know–and your publisher will want to learn from you–why your book should be on the market.

You will put this explanation in a proposal that provides a brief description of your book, a list of the outstanding features, a survey of the competition, a discussion of the target audience, production considerations, and projected time frame for completing the book. You will also want to include a table of contents and a few of the completed chapters along with any reviewers' comments.

The survey of the competition is a very important part of your proposal. A publisher will need to know where your book fits and how it will be positioned. This detailed analysis will help the publisher see right away how your book can be set apart from the competition. You can't discuss the outstanding features of your book without pointing out the specific best selling titles that lack those features.

The survey of the competition will also help the publisher target the right market and determine whether it fills a hole in his or her publishing plan. A publisher might be desperate for a book on email correspondence for Japanese students in middle school with limited English proficiency or the publisher may feel that this topic would not be a good match for its marketing and sales force.

Production issues are also important to consider. Does your book require a lot of art work or photos? Would an audio CD or an on-line component be necessary? Any publisher has to look at the bottom line. What will it cost to produce this material? How much can I sell it for? How many copies can I sell? This information is also important to you. The cheaper it is, the better the sales; the better the sales, the larger the royalty cheque.

Time is money. When are you going to finish this work?

Think about your own expenditure of time. Is it worth the year writing, the year revising, the year in production? Could you not do something more productive (i.e. earn more money) doing something else with your time?

That's a question I ask myself constantly. Is it worth it? I wrote my first book when I had a Fulbright scholarship to Sri Lanka. I remember a Sri Lankan professor asking me, "Aren't you a little young to be writing a book?" I wasn't too young to write it, but I was too naïve to publish it properly. I left it in the public domain. Similarly, my second book was written while I was on a Fulbright to Tunisia. That book too ended up in the public domain.

My third book, The Great Preposition Mystery, was also in the public domain. It's been licensed around the world and is still in print after 20 years. I don't get a dime, but I do have great name recognition in Angola.

These three books gave me a track record as an author and made it easier for me to get my manuscripts considered by publishers. Even after 20 years of writing EFL/ESL textbooks, the process for me is the same as for any potential author. I still start with an idea that I test and revise and test and revise. I still submit proposals and competition surveys. Once a book is published, I am still revising for subsequent printings and new editions.

It's difficult for me to find time to start new projects because the old ones are still demanding attention. They need to be revised to meet the demands and expectations of new users and to keep current users on their toes. Keeping my readers and myself interested is a challenge in itself and one that makes each day at the computer worthwhile. Besides, I always make time for something new.

*Dr.Lin Lougheed has written over 40 EFL/ESL textbooks and has produced a daily radio show in China, Dr. Lougheed's Business English. You can find out more about Lin's books at his website***:** *www.lougheed.com.*

Finding a Publisher

So you're trying to find a publisher interested in your work. What next? Andy Martin, Mona Scheraga, and Tina Carver explain the process.

You've done it in class. You've done it in more than one class. Your techniques work. The students are learning year after year and it's your material and your methods that are creating success. Now, how are you going to share it.

A great place to find publishers is at one of the big English language shows: the TESOL Convention in the USA, the TESL Canada conference, or the IATEFL show in the UK. If you can't get to a show, get as many publishers' catalogues as you can. Study them to see who is publishing the kind of material you have to offer. Talk to the salespeople from these publishing companies when they visit your institution. They know what your colleagues are asking for. Then make the transition from teacher's lesson plans to author's presentation. Write a prospectus–a brief summary of what you want to publish, including any research and/or experience that validates your premise. Include what market/target audience you have in mind, such as elementary, community college, etc.; the skills to be taught, the pedagogy involved, ancillary materials to be included, and reasons why your work is so necessary. Be sure to include a brief CV. Work up a Table of Contents and at least one chapter, including any exercises and an answer sheet so the publisher has an idea of the format you have in mind. Check the competition so you can discuss what makes your work different, special, and unique. Write up a competitive analysis indicating what the competition is, what its strengths and weaknesses are and how your work compares.

Finally, your sample package is ready, accompanied by your cover letter, which outlines what is included and together with your resume or bio data.

Get the name of the acquisitions editor at each publisher you submit your work to so you have a person to connect with. Talk to people in the field and get an idea of the marketing practices of different publishers; what kinds of royalties/flat fees are being paid; what kinds of materials sell best (series versus individual books, CD-ROMs, etc.). Get out to conferences and do as many workshops as possible presenting your materials so you can keep refining them.

You've got a publisher who's interested in your idea or manuscript and you've been assigned an acquisitions editor. Your editor receives the sample package. Now it's ready for review. The best reviewers are usually those who are currently teaching the equivalent course for which your project is designed and who are accustomed to evaluating materials, either as textbook committee members or as experienced publishers' reviewers. The editor sends along the review package and guide questionnaires that will give you feedback on your materials. This feedback will enable you and the editor to evaluate your sample and perhaps redirect the writing, or make major changes (for example, adding/eliminating units, changing exercise formats, redesigning the flow of lessons).

So, you've passed the trial by fire, the reviewing stage, discussed the reviews with your editor and have agreed on changes. It's time to publish your masterpiece! After the editor's supervisor approves the project, a contract is offered.

Negotiation is an important part of the process. Before you sign, read your contract carefully, suggesting any changes you'd like to make. Feel good about what you're doing and whom you're working with. The bottom line for contract negotiations is this: Is it fair to both you and the publisher? No publisher is out to "get" an author on a contract.

The financials of publishing are complex and royalties are figured as an expense of publishing the material, along with expenses of development, production, and publishing. It's important to understand that, as an author, if you receive 10%, the publisher's profit is not 90%. In fact, educational publishing survives on a very small margin of profit. Don't expect to get rich from an EFL textbook

Once the contract is signed, the work is yours to finish on time in a form appropriate for final review. When this work is complete, the editor can "accept" the manuscript. Now the fun begins.

In most publishing houses, the manuscript heads for development and you are assigned an editor (sometimes called a Development Editor or DE). It is the editor's job to work with you to look over the manuscript with an eagle eye. It has to be checked for consistency, that it ensures a variety of activities, that the exercises all work, that the tape script is accurate. Together, you must finalize the art and photo specs to get the manuscript in tip-top shape for the copy editor. The copy editor scrutinizes the manuscript for consistency, for punctuation, misspellings and for minute errors that have evaded detection. You will be sent the manuscript with copy edit queries and for approval of changes made.

Next step: Production. Here, your manuscript, art specification notes and sketches are thrown into the crucible. With the magic of computer programmes, art on CD ROMS, and the talents of the production editor, the whole hodgepodge becomes beautiful pages.

Your job is to review the pages and make corrections only–no changes at this stage, heaven forbid! It's necessary for you to meet schedules and support the production editor so the process of making the work on time and in an excellent fashion proceeds smoothly.

Voilà! After many sleepless nights, the work is in your hands. Your work, now your publication, is ready to move into the next phase, marketing and sales.

Now you've got a publisher, signed a contract, been through reviews, edits, production and design, but this means nothing if the book does not sell! Here's what's got to happen: the book has to be positioned, pre-sold, packaged, promoted and possibly piloted.... Hmmm, a lot of "p's" here. Then there's advertising, journal reviews, direct mail, following up on leads, and committee presentations. Sampling is critical, it can be done at your basic sales calls, telemarketing, book fairs, and convention presentations. We mustn't forget Internet sales, and if warranted, author tours, and in-service training. Yeah, there's a lot of stuff to be done. We don't usually cover all the bases, but there's always some combination of these sales and marketing elements. It all depends on the sales potential of the book. Basically marketing/sales can be split into pre- and post-publication activities. Let's take a peek...

- Positioning - what is the best potential market for the book?
- Sales forecasting - Estimating how many copies a book will sell and how much money it will bring in to the publishing company
- Packaging consulting - Working with editorial and design to come up with the most attractive cover for the book
- Pre-selling - Getting the word out to build excitement and interest about a book before it's published. Word-of-mouth, conferences, brochures, and marketing focus group can do this.
- Piloting - Having a teacher, or several, try-out the finished books in their classes in the hope that if they like it the whole school or district will adopt the text.
- Advertising - Promoting the book in the company catalogue, brochures and space ads in professional publications such as English Teaching Professional (www.etprofessional.com).
- Journal reviews - Mailing the new book to the review editors of professional ESL journals in the hope that they will have the book reviewed (positively) in their journal.
- Direct mail - Sending out a brochure by mail to all the English language professionals on a rented mailing list, or from the company's own database.
- Following up leads - Contacting professionals upon the recommendation of the author who provides a list of contacts in the field.
- Presenting to textbook committees - Meeting with committees at colleges or schools to convince them to adopt the new text.
- Sampling/Complimentary/Examination copies (perhaps the single most important selling technique) - Giving away free copies of the book to potential users to try out with their students.
- Calling on English language teachers/decision makers - Meeting with teachers, coordinators, chair people, etc., to present the book, let them know it's available and making sure they have a sample.
- Telemarketing - Calling potential customers on the phone to try to get them to agree to accept a sample book.
- Book fairs - Displaying the book along with other new and best selling titles at a meeting of all the ESL teachers in a school or department with the goal of leaving copies of the book behind as samples.
- Presenting at conventions - Conducting commercial presentations and giving away sample copies at local state, regional and national ESL conventions.
- The Internet - Promoting a book at the company website and soliciting sample requests and/or actual on-line sales.
- Author Tours - For really big books/series, sending the Author to different parts of the United States and/or other countries to present the book.
- In-service Training - When a book/series is adopted by a very large program, the publisher will provide free training of the English language teachers on how to use the book.

Mona Scheraga (ESL author, freelance writer, reviewer, and trainer extraordinaire) wrote "Finding a Publisher." "Publishing the Book" was written by Tina Carver, ESL publisher, editorial director, and best-selling author. "Selling the Book" was written by Andy Martin.

Running a School

One of the many opportunities for teachers is the possibility of opening or running your own school. Here, Barbara Stipek discusses the possibility of doing just that – and the freedom of designing your own programmes.

Many of you who are teaching English, or will soon be teaching English, may also be wondering about possible areas of advancement within the field. The most common pathway is via promotion into administrative positions. Another possibility is to open a new school, design a programme or in some way test your entrepreneurial skills. Whichever path you choose, you are about to embark on a wonderful adventure with many rewards.

The best way to become a first-choice candidate for promotions is to show a high degree of interest in the programme where you are teaching. Make it known that you are available for extra work or special projects (paid or unpaid!) so that you can a) gain valuable administrative experience and b) show everyone how terrific you are.

I once hired a teacher who was so eager to advance into administration that the first time I asked him if he'd like to help out on a project, he said "yes!" before I even finished the question. He consistently made himself useful and showed a great willingness to learn. He was enthusiastic and performed competently, so when openings came available, he was the first person we thought of. Eventually, he was working full-time in administration.

Looking back on my early administrative positions, I realize this was a time of great personal and professional development for me. It was a time to solidify my skills in administration and management, and I found I had an insatiable appetite for learning what made a programme good. I spent a lot of time talking to other ESL administrators and studying criteria used by external agencies to judge strong programmes. I started finding many commonalties among programmes that I felt to be academically sound, such as:

- Participation of site directors and teachers in writing curricula and course descriptions that are shared throughout the school system.
- Centralization of financial, marketing, and business affairs so that site personnel can focus on academic matters.
- Establishment of clear forms of communication so that information can be shared between all of the sites and administrative offices.
- Participation in the greater ESL community through conference attendance, presentations, and board memberships.

The most important thing to remember is that you don't know everything. You can always learn something from others—even if it's what NOT to do. You can also find wonderful camaraderie with fellow administrators, which is important because, once you've been promoted out of teaching, the teachers you used to work with won't be your peer group any longer. You will be tested in your management skills as you learn to define your new role and keep the loyalty of your teaching staff.

If you find yourself as the director of the programme where you are currently teaching, there are a few important "don'ts" to keep in mind:

- Don't pretend to know something that you really don't—you'll miss wonderful opportunities to learn new things.
- Don't try to be "one of the guys" with your teaching staff—they know you are now responsible for their jobs, and they'll see through the act
- Don't be too surprised if there is some resentment among the staff—especially if you're supervising anyone who was passed over for your promotion

After I'd been in programme administration for awhile, I started recognizing my own entrepreneurial tendencies. While I thoroughly enjoyed the technical and managerial phases of my career—and would certainly feel comfortable in any teaching or administrative position—there was something particularly exhilarating about this period. I wanted to open new programmes and have the authority—and the responsibility— to do what needed to be done to be successful. I believe that there are certain qualities that indicate an "entrepreneurial bent", including:

- Higher than average risk tolerance.
- Ability to recognize talent around you and to empower it.
- Being able to listen to others and learn new things.
- Strong desire to see your vision become a reality.
- High level of comfort with change.

It is important to do some self-assessment to decide what type of person you are and what path is right for you. If security is an important issue for you, the entrepreneurial route is probably not the best way to go.

Clearly the English language field is filled with opportunities for teachers who wish to pursue administrative careers or entrepreneurial ventures. Keep your eyes and ears open, show enthusiasm and a willingness to learn, and most of all, follow your dreams.

Barbara Stipek is the former owner of IEI, a group of ten intensive English programmes located on university campuses across the United States. In 1998, she sold the company to Sylvan Learning Systems, Inc.

School Management

Some teachers make their way up to management level. These positions require organizational and people skills – talents you will have developed during your teaching career.

It is not difficult to find teachers and former teachers who have taken over managerial roles at language schools all over the world. Some accept their new responsibilities with a pang of regret for their glory days in front of the blackboard. A manager with teaching experience should be sympathetic to the needs of staff members but will also be aware of the realities of what can go on both in and outside the classroom.

Many teachers find themselves becoming managers in an incremental process. When a teacher is appointed or chooses of his or her own free will to be in charge of examination matters or of materials procurement, the initial steps towards a managerial role have already been taken.

This role can develop as a teacher becomes involved with the creation and implementation of school or departmental policies, compliance with accreditation bodies, staff recruitment and teacher training programmes.

Some teachers naturally develop an interest in the commercial aspects of language teaching. It is not too difficult to understand that a school needs to recruit students in order to survive. Some of criteria normally associated with the private sector such as "customer satisfaction" and "performance" have even seeped into the current administrative thinking that governs the public schools system.

To be an effective manager, teachers need to be aware of basic business principles. Familiarity with the theory and practice of marketing methods is a definite asset as is managerial ability.

A head for figures and an acquaintance with accounting will also help the school or department manager deal with professionals such as bookkeepers and accountants, not that the manager has to be an expert in any of these fields.

If you decide to manage and/or own a school in a country that is not your homeland be sure that you are familiar with local laws regarding business operations and employment regulations. Although most schools operated by native English speakers run without major problems, there have been reports of foreign-owned institutions falling foul of local labour laws and other regulations and having to pay sometimes harsh penalties.

Teachers who are serious about a career in school management could consider taking one of the specialist courses that exist for managers. There is a M.Sc. degree course (taught by distance) in English Language Teaching Management offered by the University of Surrey. Core Modules include: ELT Management, Syllabus Design, Business and Financial Management in ELT, Human Resource Management, Language Teaching Methodology, Marketing Research Methods and Quantitative Techniques for Business Language Testing.

If you are interested in the issues that surround ESL/EFL school management, you may be interested in joining the ELT Management SIG (Special Interest Group) operated under the auspices of IATEFL.

Further Reading:

Impey, G. and Underhill N. An ELT Manager's Handbook, Heinemann ELT, 1994, Oxford.

White R. V. et. al. Management in English Language Teaching, Cambridge University Press, 1991, Cambridge.

Appendix

Essential contacts to help you plan your travel, find a job, source ideas for your lessons, find a supplier, join an association, subscribe to a magazine or find a publisher

Teacher Associations

New teachers have much to learn during their training periods. Developing professional bonds with other teachers and networking as much as you can will help you to stay informed early in your career.

AsiaTEFL

Organisation with over 5000 members who work as ELT teachers in the countries within the asian region (mostly within south-east asia).
Website: www.asiatefl.org

IATEFL

Based in the UK, but with regional offices and special-interest groups around the world., IATEFL (The International Association of Teachers of English as a Foreign Language) offers members an annual international conference with a programme of talks, workshops and exhibits as well as hundreds of regional meetings and SIG-sessions.

IATEFL
Darwin College, University of Kent
Canterbury, Kent
CT2 7NY, UK
Website: www.iatefl.org

MLA

The MLA (Modern Language Association) serves the interests of teachers of language and literature.

Modern Language Association
26 Broadway, 3rd Floor
New York, NY 10004-1789
Tel: (646) 576-5000
Fax: (646) 458-0030
Website: www.mla.org

NABE

NABE (National Association for Bilingual Education) is the only national organization exclusively concerned with the education of language-minority students in American schools. Many of these students are English language learners who need to preserve their first languages while they acquire a second one. NABE News, the Association's official news magazine, is published eight times annually and distributed free of charge to its members.

National Association for Bilingual Education
1220 L Street NW, Suite 605
Washington DC 20005-4018
Tel: (202) 898-1829
Fax: (202) 789-2866
Website: www.nabe.org

NAFSA

NAFSA promotes the exchange of students and scholars to and from the United States. There is a professional section of NAFSA called ATESL (Administrators and Teachers in English as a Second Language). These international teachers have responsibilities in the teaching and/or management of programmes, which teach English to speakers of other languages.
NAFSA: Association of International Teachers
1307 New York Avenue NW, Eighth Floor,
Washington DC 20005-4701
Tel: (202) 737-3699
Fax: (202) 737-3657
Website: www.nafsa.org

TESOL

The largest professional association of teachers of English as a second or foreign language is called TESOL: Teachers of English to Speakers of Other Languages, Inc. Based in Alexandria, Virginia, TESOL numbers some 16,500 members (Source: The World Almanac 2000). TESOL provides a variety of useful services to members including the bimonthly newsletter, TESOL Matters, which is included with the basic level of membership. Other benefits of TESOL membership include access to an insurance programme and the opportunity to apply for a number of awards and grants including the TESOL Professional Scholarships. During the summer months, the TESOL Academies are held at venues across the U.S. These events are designed to "increase your effectiveness as a teacher, administrator, curriculum planner, computer specialist, or ESL resource specialist." The annual TESOL Convention offers the prospect of workshops and lectures designed to inform and stimulate teachers. It is also the showcase of ESL/EFL teaching materials.

TESOL: Teachers of English to Speakers of Other Languages, Inc.
700 South Washington Street, Suite 200
Alexandria, VA 22314
Tel. 703-836-0774
Fax: 703-836-7864
Website: www.tesol.org

TESOL Affiliates

Organized on a state or regional basis, the TESOL affiliates serve members on a local level. Some are larger than others (California-based CATESOL is the biggest) but all are active in the promotion of ESL teaching. Nearly all of them hold an annual convention where teachers can attend professional development seminars, join in networking groups and visit exhibits by ESL/EFL book publishers, software producers and other businesses. There are also international associations affiliated with TESOL such as MEXTESOL in Mexico and KOTESOL in Korea. If you want to know if there is a local TESOL affiliate in your area, go to www.tesol.org/isaffil/affil/index.html

Country-specific Teacher Associations

Almost every country has its own teacher association (often with many regional groups and smaller special-interest groups too!). These all help to provide information, news, support and newsletters (with the latest job opportunities). There are 100s of country-specific assocations and 1000s of special interest groups, however most are affiliates of the two biggest worldwide groups: IATEFL and TESOL. Almost every association organises its own conferences, talks and shows – great places to learn new techniques and find out about the latest materials on offer - the selection below is a good place to start but, if your country is not listed, find out more from www.iatefl.org or www.tesol.org.

American Council on the Teaching of Foreign Languages (ACTFL)
www.actfl.org

Asociación De Profesores de Inglés de América Latina
www.aplial.net

Association for the Advancement of Computing in Education (AACE)
www.aace.org

Australian Council for TESOL Associations
www.tesol.org.au

Arizona-TESOL
www.az-tesol.org

BALEAP - British Association of Lecturers in English for Academic Purposes
www.baleap.org.uk

BASELT - British Association of State English Language Teaching
www.baselt.org.uk/

BC TEAL - British Columbia Teachers of English as an Additional Language
www.vcn.bc.ca/bcteal/

British Institute of English Language Teaching (BIELT)
www.bielt.org

Business English Special Interest Group of IATEFL
www.besig.org

California TESOL
www.catesol.org

Carolina TESOL
www.intrex.net/cartesol/

CASLT - Canadian Association of Second Language Teachers
www.caslt.org

English Language Teachers' Associations of Stuttgart and Frankfurt/Main, Germany
www.eltas.de

English Language Teachers Contacts Scheme (ELTECS)
www.britishcouncil.org/english/eltecs/

EUROCALL - European Association for Computer Assisted Language Learning
www.hull.ac.uk/cti/eurocall.htm

Fukuoka JALT
www.kyushu.com/FukuokaJALT.html

Georgia TESOL
www.gatesol.org

IATEFL - International Association of Teachers of English as a Foreign Language
www.iatefl.org.uk

IATEFL Greece
www.iatefl.gr

IATEFL Hungary
www.iatefl.hu

IATEFL Poland
www.iatefl.pl

Indiana TESOL
www.intesol.org

JALT (Japan association of language teachers)
www.jalt.org

JALT Testing & Evaluation Special Interest Group
www.jalt.org/test/

Japan Association of College English Teachers (JACET)
www.jacet.org

Japan Association of Language Teaching (JALT)
www.jalt.org

JETAA - The Japan Exchange and Teaching Alumni Association
www.jet.org

Korea Teachers of English to Speakers of Other Languages (KOTESOL)
www.kotesol.org

MIDITESOL - MidAmerica Teachers of English to Speakers of Other Languages
www.midtesol.org

Michigan TESOL
www.mitesol.org

NALDIC - National Association of language development in the curriculum
www.naldic.org.uk

NATECLA - National Association for Teaching English and other Community Languages to Adults
www.natecla.org.uk

Oregon TESOL
www.ortesol.org

TESOL - Teachers of English to Speakers of Other Languages
www.tesol.edu

TESOL Greece
www.tesolgreece.com

TESOL Spain
www.tesol-spain.org

TESOL Ukraine
www.tesol-ua.org

The Computer Assisted Language Instruction Consortium
www.calico.org

TESL Canada
www.tesl.ca

VATE - Victorian Association for Teaching of English
www.vate.org.au

Websites & Contacts

Plan your career, travel the world, get a job – just some of the uses of the mass of excellent websites on the Internet. You'll find the sites and contacts on the following pages help you get a visa, plan your classes, work out your teacher training, travel around the world and study the latest methodology.

Good Places To Start

Dave's ESL Café
www.eslcafe.com

EFLweb
www.eflweb.com

English Teaching Professional magazine
www.etprofessional.com

ESL Magazine
www.eslmagazine.com

IATEFL
www.iatefl.org

Internet TESL Journal
www.aitech.ac.jp/~iteslj

Linguistic Funland
www.linguistic-funland.com

Oxford Seminars
www.oxfordseminars.com

Randall's ESL Cyber Listening Lab
www.esl-lab.com

TESL
www.tesl.com

TESOL organisation
www.tesol.org

Web Directories

These sites include directories of useful websites – a good place to start looking for materials or other information online.

Business English Links for ESL
www.geocities.com/kurtracy/

English as a Second Language Page
www.rong-chang.com/

English as a Second/Foreign Language for Kids
www.eslkid.com/

English Grammar Links for ESL Students
www.gl.umbc.edu/~kpokoy1/grammar1.htm

English Language Teaching Web (ELTWEB)
www.eltweb.com/

ESL Café's Web Guide
www.eslcafe.com/search/index.html

ESL Forum
www.eslforum.net/

ESLdirectory.com
www.esldirectory.com/

GlobalStudy
www.globalstudy.com

Internet TESL Journal web directory
www.iteslj.org

Australian TESOL Websites Page
www.tesol.org.au/links.htm

English Language Teaching & Learning Resources
www.bernieh.com.ar/

Bilingual and ESL Education Related Resources
http://jan.ucc.nau.edu/~jar/BME.html

Carnegie Library's ESL/EFL Links
www.carnegielibrary.org/subject/education/esl.html

CATESOL's ESL Resources
www.catesol.org/resource.html

Online Resources for ESL Students and Teachers
www.clpccd.cc.ca.us/cc/maj/lahum/esl/resources.html

Bilingual / ESL Resources
www-rcf.usc.edu/~cmmr/BEResources.html

General English Links
www.englishnetlinks.homestead.com

ESL Webring
www.eslwebring.com

ESL-EFLworld Directory
www.esl-eflworld.com/

ESLoop
www.esloop.org/

IATEFL's BEsig - Links Page
www.besig.org/pages/links.htm

JALT Web Links
Http://jalt.org/jalt_e/main/materials_link/linkster.php

Schools

The web provides almost too much information about schools: almost every school has a website and it can be hard to compare them. Use one of these directories to help select a school according to type of course, level, experience and location.

ApplyESL.com
www.applyesl.com

Global Study
www.globalstudy.com

ESL Directory
www.esldirectory.com

Language Courses Comparison
www.languagecourse.net

eduPASS: US schools
www.edupass.org/english

Study in the USA
www.studyusa.com

Accredited schools in the UK
www.101schools.co.uk

Abracadebra ESL: Canadian schools
www.abracadabraesl.com

GotoEd - Study English Abroad (outside the US)
www.gotoed.com

Hyper Study: Australia and New Zealand
www.hyperstudy.com

American Cultural Exchange
www.cultural.org/map.htm

Canadian Association of Private Language Schools
www.capls.com

Directory of ESL Programme Websites Around the World
www.globalstudy.com/esl/

Education and Homestay in New Zealand
www.studentstay.com

EF Education: chain of schools around the world
www.ef.com

EFL Directory - EFL Courses
www.europa-pages.co.uk/uk/tefl.html

Study in the USA
www.studyusa.com

International Language Schools
www.aspectworld.com

Global Studies: UK schools
www.globalstudies.co.uk

go2study
www.go2study.com/english/index.htm

TESL courses in Thailand
www.langserv.com

Bookshops

KEL (Argentina)
www.ediciones-kel.com

SBS (Argentina)
www.sbs.com.ar

Disal (Brazil)
www.disal.com.br

Liv. Martins Fontes (Brazil)
www.martinsfontes.com.br

Special Book Services (Brazil)
www.sbs.com.br

The English Centre (Canada)
www.theenglishcentre.ca

Books and Bits S.A. (Chile)
www.booksandbits.cl

SBS (Chile)
www.sbs.cl

Algoritam (Croatia)
www.algoritam.hr

Bohemian Ventures (Czech Rep.)
www.venturesbooks.com

Mega Books International (Czech Rep.)
www.megabooks.cz

Nakladatelstvi Fraus (Czech Rep.)
www.fraus.cz

English Center (Denmark)
www.englishcenter.dk

Middle East Observer (Egypt)
www.meobserver.com.eg

Accendo (Estonia)
www.accendo.ee

Allecto Bookshop (Estonia)
www.allecto.ee

The Academic Bookstore (Finland)
www.akateeminen.com

Attica (France)
www.attica-langues.com
Bookshop Stäheli Ltd. (Germany)
www.staehelibooks.ch
Kosmos Floras Bookshops (Greece)
www.floras.gr
International Books (Ireland)
www.interbooksirl.com
Overseas Book Service (Italy)
www.overseasbookservice.com
Nellie's Group (Japan)
www.nellies.co.jp
English Resource (Japan)
www.englishresource.com
Sanseido Bookstore (Japan)
www.books-sanseido.co.jp
Kyobo Book Centre (Korea)
www.kyobobook.co.kr
Kungman Book Center (Korea)
www.kyobobook.co.kr
Delti Bookstore (Mexico)
www.delti.com.mx
CenterCom (Russia)
www.centercom.ru
Slovak Ventures (Slovakia)
www.venturesbooks.com
DZS (Slovenia)
www.dzs.si
The English Book Centre (Sweden)
www.engbookcen.se
The Uppsala English Bookshop (Sweden)
www.ueb.se
Bergli Books (Switzerland)
www.bergli.ch
Staeheli (Switzerland)
www.staehelibooks.ch
Hans Stauffacher (Switzerland)
www.stauffacher.ch
Caves Educational Training (Taiwan)
www.cettw.com
BEBC (UK)
www.bebc.co.uk
Cambridge International Book Centre (UK)
www.cibc.co.uk
English Book Centre (UK)
www.ebcoxford.co.uk
English Language Bookshop (UK)
www.elb-brighton.com
KELTIC (UK)
www.keltic.co.uk

LCL International Bookshop (UK)
www.lclib.com
Alta Book Center Publishers (USA)
www.altaesl.com
Delta Systems Co. Inc. (USA)
www.delta-systems.com
World of Reading, Ltd. (USA)
www.wor.com

Magazines

Magazines and journals are a teacher's best friends; they can provide practical information, lesson-plans and photocopiable material as well as a forum for technical discussions and analysis of methodology.

EL Gazette
www.elgazette.com
English Teaching Professional
www.etprofessional.com
ESL Magazine
www.eslmag.com
Hands-on English
www.handsonenglish.com/
IATEFL Newsletter
www.iatefl.org
IT's magazines
www.its.com
Language Magazine
www.languagemag.com
Mary Glasgow Magazines
www.link2english.com
Modern English Digest
www.ModernEnglishDigest.net
Modern English Teacher
www.onlineMET.com
Spotlight
www.spotlight.de
TESOL Journal
www.tesol.org/pubs/magz/tj.html
TESOL Quarterly
www.tesol.org/pubs/magz/tq.html

Publishers

ABAX ELT Publishers
www.abax.co.jp

Adams & Austen Press Publishers
www.aapress.com.au

Alta Book Center Publishers
www.altaesl.com

Barron's
www.barronseduc.com/english-language-arts.html

Beaumont Publishing
www.beaumont-publishing.com

Cambridge University Press
www.cup.cam.ac.uk

Crown House Publishing
www.crownhouse.co.uk

Delta Publishing
www.deltabooks.co.uk

Dymon Publications
www.dymonbooks.com

DynED International
www.dyned.com

EFL Press
www.EFLPress.com

Encomium Publications, Inc.
www.encomium.com

Express Publishing ELT Books
www.expresspublishing.co.uk

Full Blast Productions
www.fullblastproductions.com

Garnet Education
www.garneteducation.com

Georgian Press
www.georgianpress.co.uk

Griffith Books
www.griffith-books-ltd.sagenet.co.uk

HarperCollins Publishers
www.harpercollins.com

Hodder and Stoughton
www.madaboutbooks.com

Houghton Mifflin
www.hmco.com

JAG Publications
www.jagpublications-esl.com

John Benjamins Publishing
www.benjamins.com/jbp/index.html

Keyways Publishing
www.keywayspublishing.com

Longman English language Teaching
www.longman-elt.com

Macmillan Heinemann English Language Teaching
www.mhelt.com

Marsall Cavendish ELT
www.mcelt.com

New Readers Press
www.newreaderspress.com/main.html

Oxford University Press
www.oup.co.uk

Pro Lingua Associates
www.ProLinguaAssociates.com

Publishing Choice
www.publishingchoice.com

Richmond Publishing
www.richmondelt.com

Summertown Publishing
www.summertown.co.uk

The McGraw-Hill Companies
www.mhhe.com/catalogs/hss/esl/

Heinle & Heinle Thomson Learning
www.heinle.com

Prentice Hall Regents
www.phregents.com

Addison-Wesley
www.awl.com

Conferences

There are thousands of conferences, shows and lectures around the world each year: the sister magazine of this yearbook, English Teaching Professional, holds its conference every year in Londo (www.etplive.com); every association has its own. the biggest are run by IATEFL (www.iatefl.org.uk) and TESOL (www.tesol.org). Use one of these sites that catalogues them all.

ELT Events Calendar in Japan
http://eltcalendar.com

English Teaching Professional
www.etprofessional.com

ESL Magazine
www.eslmag.com

TESOL Online's Conference Calendar
www.tesol.org/isaffil/calendar/

Discussion and Mailing Lists

Discuss ideas and chat with colleagues and peers about materials, courses, schools or work – these discussion boards and emailing lists are dedicated to ELT.

Sign-up for TESL Mailing Lists
www.linguistic-funland.com/tesllist.html

Bulletin Board for EFL teachers working in Germany
http://pub95.ezboard.com/belt.html

CALLNews mailing list
http://listserv.cddc.vt.edu/mailman/listinfo/callnews

ELTASIA-L mailing list for ESL teachers in Asia
http://eltasia.com/

English Teaching Professional - discussion forum
www.etprofessional.com

ESL Café's Discussion Center for Teachers
www.eslcafe.com/discussion/#teacher

JALTTALK - Japan
www.jalt.org/jalt_e/main/jaltcall_main.shtml

Discussion Forums for Teachers
www.eslpartyland.com/tdisc.htm

Modern English Teacher - discussion forum
www.onlineMET.com

NETEACH-L - using technology in the classroom
www.ilc.cuhk.edu.hk/english/neteach/main.html

TESL-L Discussion List
www.hunter.cuny.edu/~tesl-l/

The ESL/Language ChatBoard
www.teachers.net/mentors/esl_language/

Free-ESL Discussion Forums
www.free-esl.com/teachers/forums/default.asp

Lesson Plans

Lesson plans are an essential part of every teacher's kit – trying to make teaching academic subjects such as grammar interesting, relevant to the student and easy to understand. The web has masses of free lesson plans that you can download and use as a basis for your own work. Try these sites for ideas and free plans.

EFL4U Lesson Plans
www.efl4u.com

ESLFLOW
www.eslflow.com

ESL Classroom Handouts
www.englishclub.net/handouts

Ideas for the ESL Classroom
www.eslcafe.com/ideas

ESL Teachers Guide
http://humanities.byu.edu/elc/Teacher/TeacherGuideMain

Karin's ESL PartyLand
www.eslpartyland.com/teach3.htm

askERIC Lesson Plans
Http://ericir.syr.edu/Virtual/Lessons

Games in ESL Classroom
http://eslsv001.esl.sakuragaoka.ac.jp/teachers/BR/games/Games.html

Free Instant Lessons
www.english-to-go.com

English Lessons
www.nwrel.org/sky/

PIZZAZ!
http://darkwing.uoregon.edu/~leslieob/pizzaz.html

TEFL Farm
www.teflfarm.com

TEFL.net
www.tefl.net

Lessons and Lesson Plans from The Internet TESL Journal
http://iteslj.org/Lessons/

Boggle's World ESL Lesson Plan Archive
http://bogglesworld.com/lessons/archive.htm

Bradley's Worksheets
www.bradleys-english-school.com/worksheets/nfindex.html

Free English Lessons
www.cerbranetics.com/english.html

Churchill House School of English Lanuage
www.churchillhouse.com/english/downloads.html

CNN Newsroom Daily Classroom Guide
http://Learning.turner.com/newsroom/index.html

Free Lesson
www.english-to-go.com/english/

ESL Games and Activities
www.etanewsletter.com/games.htm

Teachers Teaching Teachers
www.etni.org.il/teacteac.htm

Activities for Summer School ESL
http://everythingesl.net/lessons/summerschool_esl.php?ty=print

Classroom ESL Games
http://genkienglish.net/games.htm

Classroom Materials

Wordsmyth Glossary Maker
www.wordsmyth.net/foundry/glossary.html

Wordsmyth Vocabulary Quiz Generator
www.wordsmyth.net/foundry/vocabquiz.shtml

Course for Adults
http://iteslj.org/Lessons/Vorland-4units/

Crosswords and Word Searches for Young Children
http://abcteach.com/EasyPuzzles/kidsTOC.htm

Boggle's World
http://bogglesworld.com/
Community ESOL
www.communityesol.org.uk/
EFL Club Resource Box
www.eflclub.com/9resourcebox/resourcebox.html
What's Wrong - Intensive Reading
http://esl.about.com/homework/esl/library/lessons/nbl wrong.htm
Grammar: Nature and Teaching
www.gabrielatos.com/Grammar.htm
Pronunciation: /r/ and /l/
www.csulb.edu/~linguag/ali/r_and_l.html
Word Search Factory
www.schoolhousetech.com/wordsearch.html
Classroom Materials Generators
www.teach-nology.com/web_tools/materials/
Free handouts for EFL teachers
www.handoutsonline.com/
Simple English Grammar Exercises
www.theenglishprofessor.com/freeworksheets.htm

Teaching Tips and Ideas

If you're faced with your first class or if you are stuck for ideas after teaching the same subject to the thousandth student, these sites can help with fresh ideas and tips.

75 ESL Teaching Ideas
http://iteslj.org/Techniques/Houston-TeachingIdeas.html
Tips for Teachers
http://2merediths.org/esl/teachertips.htm
Conversation starters for students
www.languageimpact.com/articles/rw/conv_starters.htm
Teacher's Tips
www.developingteachers.com/tips/currenttip.htm
FAQ about Teaching ESL Students (For Mainstream Teachers)
www.fis.edu/eslweb/esl/students/teanotes/
Forty Helpful Hints & Tips
www.handsonenglish.com/40tips.html
Hints and Pointers
http://genkienglish.net/general.htm
Practical Teaching Ideas on ESL
www.ncte.org/teach/esl.shtml
Survival Guide for New Teachers
www.ed.gov/pubs/survivalguide/
Tips for Teaching Grammar
www.ateg.org/grammar/tips.htm

Travel advice

www.1000traveltips.org
www.canuckabroad.com
www.gapyear.com
www.goabroad.com
www.globetrotters.co.uk
www.journeywoman.com
www.lonelyplanet.com
www.roughguides.com
www.mapsworldwide.co.uk
www.timeout.com
www.vtourist.com
www.worldtimeserver.com

Official travel advice

British Foreign and Commonwealth office
www.fco.gov.uk
British passport enquiries
www.ukpa.gov.uk
Canadian Consular Affairs
www.voyage.gc.ca
US Department of State
www.travel.state.gov
NHS Direct
www.knowhow.co.uk

Visas

Visas for Australia
www.australia.org.uk
Visas for Canada
www.cic.gc.ca/english/visit/
Visas for USA
www.usembassy.org.uk

Emailing & cybercafés

Cybercafes.com
Visit before you go to find your local cybercafé – you'll be able to keep in touch with friends and family cheaply and quickly by emailing them while you're away. This website contains a list of over 4,000 Internet cafés around the globe.

Finances

Mastercard
www.mastercard.com/atm
Visa
www.visa.com/pd/atm
American Express
www.americanexpress.com

Currency Conversion
www.xe.com
Western Union Money Transfer
www.westernunion.com

Travel health

National Centre for Disease Control
www.cdc.gov/travel
Travel health
www.tmvc.com.au/info10.html
TripPrep
www.tripprep.com
World Health Organization
www.who.int/ctd

Travel Agencies

Ebookers
www.ebookers.com
Expedia
www.expedia.com
STA Travel
www.statravel.co.uk
Trailfinders
www.trailfinders.co.uk
Intl Student Travel
www.istc.org
Studenttravel.com
www.studenttravel.com
Thomson travel
www.austravel.com

Ferry operators and ports

Brittany Ferries
www.brittanyferries.com
Hoverspeed
www.hoverspeed.co.uk
P&O Stena Line
www.posl.com
Stena Line
Www.stenaline.co.uk

Bus

Buslines (Australia)
www.buslines.com.au
Greyhound (USA)
www.greyhound.com
Ticabus (Central America)
www.ticabus.com/Eindex/htm

Train

BudgetTravel
www.budgettravel.com
EuroRailways
www.eurorailways.com
Public Transport (in North America)
www.geocities.com/capitolhill/5355/
Australian Train Routes
www.gsr.com.au
RailServe
www.railserve.com
TrainWeb
www.trainweb.com/indiarail

British Council

The British Council network has offices in just about every country; these should be your first point of contact: the offices provide information on visas, grants, jobs and local economic conditions as well as English-language resources, contacts with local schools and advice for teachers hoping to get established in the country. Main Website: www.britishcouncil.org

AFGHANISTAN

The British Council
15th Street Roundabout
Wazir Akbar Khan
P.O. Box 334
Kabul
Tel: +93 (0)70 102 302
Fax: +93 (0)70 102 250
Email: richard.weyers@britishcouncil.org

ALBANIA

The British Council
Rruga Ded Gjo Luli 3/1
Tirana
Tel: +355 (4) 240856/40857
Fax: +355 (4) 240858
Email: info@britishcouncil.org.al
Website: www.britishcouncil.org.al

ALGERIA , Algiers

The British Council
Hotel Hilton International 7th floor
Pins Maritimes, Palais des Expositions
El Mohammadia
Algiers
Tel: +213 (21) 230 068
Fax: +213 (21) 230 751
Email: rachida.benyahia@fco.gov.uk
Website: www.britishcouncil.org/algeria

ARGENTINA, Buenos Aires

The British Council
Marcelo T de Alvear 590 4th Floor
C1058AAF
Buenos Aires
Tel: +54 (11) 4311 9814/7519
Fax: +54 (11) 4311 7747
Email: info@britishcouncil.org.ar
Website: www.britishcouncil.org.ar

ARMENIA

The British Council
Baghramian Avenue 24, Yerevan, 375019
Tel: +374 (1) 55 99 23
Fax: +374 (1) 55 99 29
Email: info@britishcouncil.am
Website: www.britishcouncil.am

AUSTRALIA, Sydney

The British Council
PO Box 88
Edgecliff, Sydney NSW 2027
Tel: +61 (2) 9326 2022
Fax: +61 (2) 9327 4868
Email: enquiries@britishcouncil.org.au
Website: www.britishcouncil.org/australia

AUSTRIA, Vienna

The British Council
Schenkenstrabe 4
A -1010 Vienna
Tel: +43 (1) 533 2616

Fax: +43 (1) 533 261685
Email: bc.vienna@britishcouncil.at
Website: www.britishcouncil.at

AZERBAIJAN, Baku

The British Council
1 Vali Mammadov Street
Ichari Sheher
Baku AZ1000
Tel: +994 (12) 971 593
Fax: +994 (12) 989 236
Email: enquiries@britishcouncil.az
Website: www.britishcouncil.org.az

BAHRAIN , Manama

The British Council
AMA Centre (PO Box 452)
146 Shaikh Salman Highway
Manama 356
Tel: +973 261 555
Fax: +973 241 272
Email: bc.enquiries@britishcouncil.org.bh
Website: www.britishcouncil.org/bahrain

BANGLADESH, Dhaka

The British Council
5 Fuller Road
PO Box 161, Dhaka 1000
Tel: +880 (2) 861 8905 – 7
Fax: +880 (2) 861 3375
Email: Dhaka.Enquiries@bd.britishcouncil.org
Website: www.britishcouncil.org/bangladesh

Bangladesh, Chittagong

The British Council
77/A East Nasirabad
Chittagong
Tel: +880 (31) 657884
Fax: +880 (31) 657 881
Email: Chittagong.Enquiries@bd.britishcouncil.org

BELGIUM, Brussels

The British Council
Leopold Plaza
Rue du Trône 108 / Troonstraat 108
1050 Brussels
Tel: +32 (2) 227 08 41
Fax: +32 (2) 227 08 49
Email: enquiries@britishcouncil.be
Website: www.britishcouncil.org/belgium

BOLIVIA, La Paz

The British Council
Avenida Arce 2708 (esq.Campos)
Casilla 15047
La Paz
Tel: +591 (2) 2431 240
Fax+591 (2) 2431 377
Email: information@britishcouncil.org.bo
Website: www.britishcouncil.org/bolivia

BOSNIA & HERZEGOVINA , Sarajevo

The British Council
Ljubljanska 9, Sarajevo 71 000
Tel: +387 (0) 33 250 220
Fax: +387 (0) 33 250 240
Email: British.Council@britishcouncil.ba
Website: www.britishcouncil.ba

BOTSWANA, Gaborone

British High Commission Building
Queen's Road
The Mall
P.O Box 439, Gaborone
Tel: +267 395 3602
Fax: +267 395 6643
Email: general.enquiries@britishcouncil.org.bw
Website: www.britishcouncil.org/botswana

BRAZIL, Brasilia

The British Council
Ed.Centro Empresarial Varig
SCN Quadra 04 Bloco B
Torre Oeste Conjunto 202
Brasilia – DF
70710-926
Tel: +55 (0) 61 2106 7500
Fax: +55 (0) 61 2106 7599
Email: brasilia@britishcouncil.org.br
Website: www.britishcouncil.org.br

BRAZIL, Curitiba

The British Council
Rua Presidente Faria, 51
Conjunto 204
Curitiba PR 80020-290
Tel: / **Fax:** +55 41 238 2912
Email: curitiba@britishcouncil.org.br

BRAZIL, Recife

The British Council
Av. Domingos Ferreira, 4150
Boa Viagem
51021-040 Recife PE,
P.O. Box 4079
Tel: +55 81 2101 7500
Fax: +55 81 2101 7599
Email: recife@britishcouncil.org.br

BRAZIL , Rio de Janeiro

The British Council
Rua Jardim Botanico, 518/1 andar
Jardim Botanico,
22461 – 000 Rio de Janeiro RJ
Tel: +55 21 2105 7500
Fax: +55 21 2105 7598
Email: riodejaneiro@britishcouncil.org.br
Regional Director - Mr Sital Dhillon

BRAZIL, Rio de Janeiro

The British Council
Management Centre
Av. Rio Branco 80/4 andar
Centro 20040-070
Rio de Janeiro – RJ
Tel: +55 (21) 2242 1223
Fax: +55 (21) 2221 0515

BRAZIL , São Paulo

The British Council
Centro Brasileiro-Britanico
Rua Ferreira Araujo, 741 – 3 andar, Pinheiros
05428 002, Sao Paulo – SP
Tel: +55 (0) 112126 7500
Fax: +55 (0) 11 2126 7575
Email: saopaulo@britishcouncil.org.br

BRUNEI DARUSSALAM, Bandar Seri Begawan

The British Council
Level 2, Block D,
Yayasan Sultan Haji Hassanal Bolkiah
Jalan Pretty, B.S.B.
Brunei Darussalam BS8711
Tel: +673 223 7742
Fax: +673 223 7392
Email: all.enquiries@bn.britishcouncil.org
Website: www.britishcouncil.org/brunei

BULGARIA, Sofia

The British Council
7 Krakra Street
1504 Sofia
Tel: +359 (0) 2 942 4344
Fax: +359 (0) 2 942 4222
Email: bc.sofia@britishcouncil.bg
Website: www.britishcouncil.org/bulgaria

BURMA, Myanmar

The British Council
78 Kanna Road,
(PO Box 638)
Yangon
Myanmar
Tel: + 95 (0)1 254658/256290
Fax: + 95 (0)1 245345
Email: enquiries@britishcouncil.org.mm
Website: www.britishcouncil.org/burma

CAMEROON, Yaoundé

The British Council
Immeuble Christo
Avenue Charles de Gaulle BP 818
Yaoundé
Tel: +237 2 21 16 96/20 31 72
Fax: +237 2 21 56 91
Email: bc-yaounde@britishcouncil.cm
Website: www.britishcouncil.org/cameroon

CAMEROON, Douala

The British Council
Rue Joffre, Akwa
BP 12801
Douala
Tel: +237 343 49 66
Fax: +237 342 51 70
Email: bc-douala@bc-douala.iccnet.cm

CANADA, Ottawa

The British Council
80 Elgin Street
Ottawa
Ontario K1P 5K7
Tel: +1 613 364 6233
Fax: +1 613 569 1478
Email: ottawa.enquiries@ca.britishcouncil.org
Website: www.britishcouncil.org/canada

CANADA, Montreal

The British Council
1000 ouest rue de La Gauchetiere
Bureau 4200
Montreal
Quebec H3B 4W5
Tel: +1 514 866 5863
Fax: +1 514 866 5322
Email: montreal.enquiries@ca.britishcouncil.org

CHILE, Santiago

The British Council
Eliodoro Yáñez 832
Providencia 750-0651
Casilla 115 Correo 55
Santiago
Tel: +56 (2) 410 6900
Fax: +56 (2) 410 6929
Email: info@britishcouncil.cl
Website: www.britishcouncil.cl

CHINA, Beijing

The British Council
4/f Landmark Building Tower 1
8 North Dongsanhuan Road
Chaoyang District,
100004
Beijing
Tel: +86 (10) 6590 6903
Fax: +86 (10) 6590 0977
Email: enquiry@britishcouncil.org.cn
Website: www.britishcouncil.org/china

CHINA, Hong Kong

The British Council
3 Supreme Court Road
Admiralty
Tel: +852 291 35100
Fax: +852 291 35102
Email: info@britishcouncil.org.hk
Website: www.britishcouncil.org.hk

CHINA, Shanghai

The British Council
1 Floor Pidemco Tower
318 Fu Zhou Lu
Shanghai 200001
Tel: +86 (21) 6391 2626
Fax: +86 (21) 6391 2121
Email: bc.shanghai@britishcouncil.org.cn

CHINA, Chongqing

The British Council
Room 5-7 28 Floor Metropolitan Tower
No 68 Zou Rong Road
Yuzhong District
Chongqing 400010
Tel: +86 (23) 6373 6888
Fax: +86 (23) 6373 7898
Email: bc.chongqing@britishcouncil.org.cn

CHINA, Guangzhou

The British Council
1001 Main Tower
Gungdong International Hotel
339 Huanshi Dong Lu
Guangzhou 510098
Tel: +86 (20) 8335 1316
Fax: +86 (20) 8335 1321
Email: bc.guangzhou@britishcouncil.org.cn

COLOMBIA, Bogotá

The British Council
Calle 87 No 12 - 79
Bogota
Tel: +57 (1) 618 7680
Fax: +57 (1) 218 7754
Email: info@britishcouncil.org.co
Website: www.britishcouncil.org/colombia

CROATIA , Zagreb

The British Council
Illica 12
PP 55
10001 Zagreb
Tel: +385 (0) 1 4899 500
Fax: +385 (0) 1 4833 955
Email: zagreb.info@britishcouncil.hr
Website: www.britishcouncil.hr

CUBA, Havana

The British Council
Calle 34 no 702, esq.a 7ma Avenida,
Miramar,
La Habana
Tel: +53 (0) 7 204 1771/2
Fax: +53 (0) 7 204 9214
Email: information@cu.britishcouncil.org
Website: www.britishcouncil.org/cuba

CYPRUS, Nicosia

The British Council
3 Museum Street
CY- 1097 Nicosia
Tel: +357 (22) 585000
Fax: +357 (22) 677 257
Email: enquiries@britishcouncil.org.cy
Website: www.britishcouncil.org.cy

CZECH REPUBLIC, Prague

The British Council
Bredovsky dvur, Politickych veznu 13,
110 00 Prague 1
Tel: +420 221 991 111
Fax: +420 224 933 847
Email: info.praha@britishcouncil.cz
Website: www.britishcouncil.cz

CZECH REPUBLIC, Brno

The British Council
Trida Kpt. Jarose 13
602 00 Brno
Tel: +420 545 210 174
Fax: +420 545 211 065
Email: info.brno@britishcouncil.cz

CZECH REPUBLIC, Pilsen

The British Council
Purkynova 27
301 36 Plzen
Tel: +420 377 220 180
Fax: +420 377 220 181
Email: info.pilsen@britishcouncil.cz

DENMARK, Copenhagen

The British Council
Gammel Mønt 12.3
1117 Copenhagen K
Tel: +45 (33) 369 400
Fax: +45 (33) 369 406
Email: british.council@britishcouncil.dk
Website: www.britishcouncil.org/denmark

EGYPT, Cairo

The British Council
192 El Nil Street
Agouza, Cairo
Tel: +20 (2) 303 1514
Fax: +20 (2) 344 3076
Email: british.council@britishcouncil.org.eg
Website: www.britishcouncil.org/egypt

EGYPT, Alexandria TC

The British Council
9 Batalsa Street
Bab Sharki
Alexandria
Tel: +20 (3) 486 0199
Fax: +20 (3) 484 6630
Email: british.council@britishcouncil.org.eg

EGYPT, Heliopolis TC

The British Council
4 El Minya Street
Off Nazih Khalifa Street
Heliopolis
Tel: +20 (2) 452 3395-7
Fax: +20 (2) 258 3660
Email: british.council@britishcouncil.org.eg

ERITREA, Asmara

The British Council
Lorenzo Tazaz Street No 23
PO Box 997 Asmara
Tel: +291 (1) 123 415/120 529
Fax: +291 (1) 127 230
Email: information@ britishcouncil.org.er
Website: www.britishcouncil.org/eritrea

ESTONIA, Tallin

The British Council
Vana Posti 7
Tallinn 10146
Tel: +372 (0) 625 7788
Fax: +372 (0) 625 7799
Email: british.council@britishcouncil.ee
Website: www.britishcouncil.org/estonia

ETHIOPIA, Addis Ababa

The British Council
PO Box 1043
Artistic Building
Adwa Avenue, Addis Ababa
Tel: +251 (1) 550 022
Fax: +251 (1) 552 544
Email: bc.addisababa@et.britishcouncil.org
Website: www.britishcouncil.org/ethiopia

FINLAND, Helsinki

The British Council
Hakaniemenkatu 2
00530 Helsinki
Tel: +358 (9) 774 3330
Fax: +358 (9) 701 8725
Email: office@britishcouncil.fi
Website: www.britishcouncil.fi

FRANCE, Paris

The British Council
9 rue de Constantine
75340 Paris cedex 07
Tel: +33 (1) 4955 7300
Fax: +33 (1) 4705 7702
Email: information@britishcouncil.fr
Website: www.britishcouncil.fr

GEORGIA, Tbilisi

The British Council
34 Rustaveli Avenue, Tbilisi 380008
Tel: +995 32 250 407/988014
Fax: +995 32 989591
Email: office.bc@ge.britishcouncil.org
Website: www.britishcouncil.org.ge

GERMANY, Berlin

The British Council
Hackescher Markt 1
10178 Berlin
Tel: +49 (0) 30 311 0990
Fax: +49 (0) 30 311 09920
Email: bc.berlin@britishcouncil.de
Website: www.britishcouncil.de

GHANA, Accra

The British Council
11 Liberia Road
PO Box GP 771
Accra
Tel: +233 21 683 968
Fax: +233 21 683 062
Email: infoaccra@gh.britishcouncil.org
Website: www.britishcouncil.org/ghana

GHANA, Kumasi

The British Council
Bank Road
PO Box KS 1996
Kumasi
Tel: +233 51 23462
Fax: +233 51 26725
Email: infokumasi@gh.britishcouncil.org

GREECE, Athens

The British Council
17 Kolonaki Square
Plateia Philikis Etairias
10673 Athens
Tel: +30 210 369 2333
Fax: +30 210 363 4769
Email: customerservices@britishcouncil.gr
Website: www.britishcouncil.org/greece

GREECE, Thessaloniki

The British Council
Ethnikis Amynis 9
P. O. Box 50007
54013 Thessaloniki
Tel: +30 2310 378 300
Fax: +30 2310 282 498
Email: general.enquiries@britishcouncil.gr

HUNGARY, Budapest

The British Council
Benczúr u. 26
1068 Budapest
Tel: +36 (1) 478 4700
Fax: +36 (1) 342 5728
Email: information@britishcouncil.hu
Website: www.britishcouncil.hu

INDIA, New Delhi

The British Council
17 Kasturba Gandhi Marg
New Delhi 110 001
Tel: +91 11 2-3711401
Fax: +91 11 2-3710717
Email: delhi.enquiry@in.britishcouncil.org
Website: www.britishcouncil.org/india

INDONESIA, Jakarta

The British Council
S. Widjojo Centre
Jalan Jenderal Sudirman Kav 71
Jakarta 12190
Tel: +62 (21) 252 4115
Fax: +62 (21) 252 4129
Email: information@britishcouncil.or.id
Website: www.britishcouncil.org/indonesia

INDONESIA, Surabaya

The British Council
Jalan Cokroaminoto 12A, 3rd Floor
Surabaya 60264
Tel: +62 (31) 568 9958
Fax: +62 (31) 568 9957
Email: tony.saputra@britishcouncil.or.id

IRAN, Tehran

The British Council
Qolhak
Dr Shariati Avenue
Opposite Bonbaste Elahieh
19396 13661 Tehran
Tel: +98 (021) 200 1222
Fax: +98 (021) 200 7604
Email: enquiries@britishcouncil.org.ir
Website: www.britishcouncil.org/iran

IRELAND, Dublin

The British Council
Newmount House
22/24 Lower Mount Street
Dublin 2
Tel: +3531 676 4088
Fax: +3531 676 6945
Email: tom.farrell@ie.britishcouncil.org
Website: www.britishcouncil.org/ireland

ISRAEL, Tel Aviv

The British Council
12 Hahilazon Street, Ramat Gan, Tel Aviv 52136
Tel: +972 (0) 3 611 3600
Fax: +972 (0) 3 611 3640
Email: bcta@britishcouncil.org.il
Website: www.britishcouncil.org.il

ISRAEL, Nazareth

The British Council
P O Box 2545
Nazareth 16121
Tel: +972 (0)4 657 9570
Fax: +972 (0)4 655 0436
Email: Jane.Shurrush@britishcouncil.org.il

ISRAEL , West Jerusalem TC, i

The British Council
Hagmidal, Gan Ha'technologi
Jerusalem 96951
Tel: +972 (0)2 6403900
Fax: +972 (0)2 6403919
Email: bcwjlm@britishcouncil.org.il

ITALY, Rome

The British Council
Via Quattro Fontane 20
00184 Rome
Tel: +39 (06) 478 141
Fax: +39 (06) 481 4296
Email: studyandcultureuk@britishcouncil.it
Website: www.britishcouncil.org/italy

ITALY, Bologna TC

The British Council
Corte Isolani 8
Strada Maggiore 19
40125 Bologna
Tel: +39 (051) 225 142
Fax: +39 (051) 224 238
Email: enquiries.bologna@britishcouncil.it

ITALY , Milan TC

The British Council
Via Manzoni 38
20121 Milan
Tel: +39 (02) 772 221
Fax: +39 (02) 781 119
Email: enquiries.milan@britishcouncil.it

ITALY , Naples

The British Council
Via Morghen 36
80129 Naples
Tel: +39 (081) 578 5817
Fax: +39 (081) 556 2585
Email: enquiries.naples@britishcouncil.it

JAMAICA, Kingston

The British Council
28 Trafalgar Road
Kingston 10
Tel: +1 876 929 7090
Fax: +1 876 960 3030
Email: bcjamaica@britishcouncil.org.jm
Website: www.britishcouncil.org/caribbean

JAPAN, Tokyo

The British Council
1-2 Kagurazaka
Shinjuku-ku

Tokyo 162-0825
Tel: +81 (0) 3 3235 8031
Fax: +81 (0) 3 3235 8040
Email: enquiries@britishcouncil.or.jp
Website: www.britishcouncil.org/japan

JAPAN, Kyoto TC

The British Council
Karasuma Chuo Building 8F
659 Tearaimizu-cho, Nishikikoji-agaru
Karasuma-dori
Nakagyo-ku, Kyoto 604-8152
Tel: +81 (0) 75 229 7151
Fax: +81 (0) 75 229 7154
Email: enquiries@britishcouncil.or.jp

JAPAN, Nagoya

The British Council
NHK Nagoya Broadcasting Centre Building 6F
1-13-3, Higashi Sakura
Higashi-ku Nagoya-shi, Aichi 461-0005
Tel: +81 (0) 52 963 3671
Fax: +81 (0) 52 963 3670
Email: enquiries.nagoya@britishcouncil.or.jp

JAPAN, Osaka

The British Council
Dojima Avanza 4F
1-6-20 Dojima, Kita-ku
Osaka-shi, Osaka 530-0003
Tel: +81 (0) 6 6342 5301
Fax: +81 (0) 6 6342 5311
Email: enquiries.osaka@britishcouncil.or.jp

JORDAN, Amman

The British Council
First Circle , Jebel Amman
PO Box 634, Amman 11118
Tel: +962 6 4636147
Fax: +962 6 465 6413
Email: bcamman@britishcouncil.org.jo
Website: www.britishcouncil.org.jo

KAZAKHSTAN, Almaty

The British Council
Republic Square 13, 480013 Almaty, Kazakhstan
Tel: +7 3272 72 0111
Fax: +7 3272 72 0113
Email: general@kz.britishcouncil.org
Website: www.britishcouncil.kz

KAZAKHSTAN, Astana

The British Council
1st Floor, Renco Building
62 Cosmonaut Street, Chubary District
473000 Astana
Tel: +7 3172 971179
Fax: +7 3172 971180
Email: general@kzbritishcouncil.org

KENYA , Nairobi

The British Council
ICEA Building
Kenyatta Avenue
PO Box 40751
Nairobi
Tel: +254 (0) 20 334 855
Fax: +254 (0) 20 339 854
Email: information@britishcouncil.or.ke
Website: www.britishcouncil.org/kenya

KOREA, Seoul

The British Council
Hungkuk Life Insurance Building 4F
226 Shinmunro 1-ga
Jongro-gu, Seoul 110-786
Tel: +82 (0) 2 3702 0600
Fax: +82 (0) 2 3702 0660
Email: info@britishcouncil.or.kr
Website: www.bckorea.or.kr

KUWAIT, Kuwait City

The British Council
2 Al Arabi Street, Block 2
PO Box 345, 13004 Safat
Mansouriya, Safat 13004
Tel: +965 251 5512
Fax: +965 252 0069
Email: bc.enquiries@kw.britishcouncil.org
Website: www.britishcouncil.org/kuwait

LATVIA, Riga

The British Council
5a Blaumana iela
Riga LV-1011
Tel: +371 728 1730
Fax: +371 750 4100
Email: mail@britishcouncil.lv
Website: www.britishcouncil.lv

LEBANON, Beirut

The British Council
Sidani Street
Azar Building
Beirut
Tel: +961 (1) 740 123
Fax: +961 (1) 739 461
Email: general.enquiries@lb.britishcouncil.org
Website: www.britishcouncil.org/lebanon

LIBYA, Tripoli

The British Council
24th Floor,
Burj al Fatah,
PO Box 4206, Tripoli
Tel: + 218 21 335 1473/5
Fax: +218 21 335 1471
Email: info.libya@britishcouncil-ly.org
Website: www.britishcouncil.org/libya

LITHUANIA, Vilnius

The British Council
Jogailos 4
LT - 2001 Vilnius
Tel: +370 5 264 4890
Fax: +370 5 264 4893
Email: mail@britishcouncil.lt
Website: www.britishcouncil.lt

MACEDONIA, Skopje

The British Council
Bulevar Goce Delcev 6
P.O. Box 562
1000 Skopje
Tel: +389 2 313 5035
Fax: +389 2 313 5036
Email: info@britishcouncil.org.mk
Website: www.britishcouncil.org/macedonia

MALAWI, Lilongwe

The British Council
Plot No 13/20 City Centre
P O Box 30222
Lilongwe 3
Tel: +265 (0) 1 773 244
Fax: +265 (0) 1 772 945
Email: Info@britishcouncil.org.mw
Website: www.britishcouncil.org/malawi

MALAYSIA, Kuala Lumpur

The British Council
PO Box 10539
50916 Kuala Lumpur
Tel: +60 (0) 3 2723 7900
Fax: +60 (0) 3 2713 6599
Email: kualalumpur@britishcouncil.org.my
Website: www.britishcouncil.org.my

MALAYSIA, Kota Kinabalu

The British Council
1st Floor EONCMG Building
1 Lorong Sagunting
88000 Kota Kinabalu
Sabah
Tel: +60 (88) 222 059
Fax: +60 (88) 238 059
Email: sabah@britishcouncil.org.my

MALAYSIA, Penang TC

The British Council
3 Weld Quay
10300 Penang
Malaysia
Tel: +60 (4) 263 0330
Fax: +60 (4) 263 3262
Email: penang.info@britishcouncil.org.my

MALTA, Valletta

The British Council
Whitehall Mansions, Ta'Xbiex Seafront
Tel: +356 23 238403
Fax: +356 23 238402
Email: information@britishcouncil.org.mt
Website: www.britishcouncil.org/malta

MAURITIUS, Rose Hill

The British Council
Royal Road, PO Box 111, Rose Hill
Tel: +230 454 9550
Fax: +230 454 9553
Email: general.enquiries@mu. britishcouncil.org
Website: www.britishcouncil.org/mauritius

MEXICO, Mexico City

The British Council
Lope de Vega 316
Col. Chapultepec Morales
Mexico City
11570 DF

Tel: +52 (55) 5263 1900
Fax: +52 (55) 5263 1940
Email: bcmexico@britishcouncil.org.mx
Website: www.britishcouncil.org/mexico

MOROCCO, Rabat

The British Council
36 Rue de Tanger
BP 427, Rabat
Tel: +212 (37) 760 836
Fax: +212 (37) 760 850
Email: bc@britishcouncil.org.ma
Website: www.britishcouncil.org/morocco

MOROCCO, Casablanca

The British Council
87 Boulevard Nador, Quartier Polo
20500 Casablanca
Tel: +212 (0)22 520990
Fax: +212 (0)22 520964
Email: casa.info@britishcouncil.org.ma

MOZAMBIQUE, Maputo

The British Council
Rua John Issa, 226
P.O. Box 4178
Maputo
Tel: +258 (1) 355 000
Fax: +258 (1) 321 577
Email: General.enquiries@britishcouncil.org.mz
Website: www.britishcouncil.org/mozambique

NAMIBIA, Windhoek

The British Council
1-5 Fidel Castro Street
Windhoek
Tel: +264 (61) 226 776
Fax: +264 (61) 227 530
Email: general.enquiries@britishcouncil.org.na
Website: www.britishcouncil.org/namibia

NEPAL, Kathmandu

The British Council
PO Box 640
Lainchaur
Kathmandu
Tel: +977 1 4410 798
Fax: +977 1 4410 545
Email: general.enquiry@britishcouncil.org.np
Website: www.britishcouncil.org.np

NETHERLANDS, Amsterdam

The British Council
Weteringschans 85A
1017 RZ Amsterdam
Tel: +31 (20) 550 6060
Fax: +31 (20) 620 7389
Email: information@britcoun.nl
Website: www.britishcouncil.org/netherlands

NEW ZEALAND, Wellington

The British Council
44 Hill Street
PO Box 1812
Wellington 1
Tel: +64 (4) 924 2880
Fax: +64 (4) 473 6261
Email: enquiries@britishcouncil.org.nz
Website: www.britishcouncil.org/nz

NEW ZEALAND, Auckland

The British Council
151 Queen Street
Private Bag 92014
Auckland
Tel: +64 (0) 9 373 4478
Fax: +64 (0) 9 373 4479
Email: enquiries@britishcouncil.org.nz

NIGERIA, Abuja

The British Council
Plot 2935
IBB Way
Maitama
P.M.B. 550
Garki
Abuja, FCT
Tel: +234 (9) 413 7870-7
Fax: +234 (9) 413 0902
Email: maureen.ideozu@ng.britishcouncil.org
Website: www.britishcouncil.org/nigeria

NIGERIA, Lagos

The British Council
11 Alfred Rewane Road
Ikoyi
PO Box 3702
Lagos
Tel: +234 (1) 269 2188
Fax: +234 (1) 269 2193
Email: yinka.ijabiyi@ng.britishcouncil.org

NIGERIA, Kano

The British Council
10 Emir's Palace Road
(PMB 3003)
Kano
Tel: +234 (64) 646652
Fax: +234 (64) 632500
Email: Kano.info@ng.britishcouncil.org

NORWAY, Oslo

The British Council
Fridtjof Nansens Plass 5
0160 Oslo
Tel: +47 (22) 396 190
Fax: +47 (22) 424 039
Email: british.council@britishcouncil.no
Website: www.britishcouncil.no

OMAN, Muscat

The British Council
Road One
Madinat al Sultan, Qaboos West, Muscat
P.O. Box 73 Postal Code 115
Tel: +968 600 548
Fax: +968 699 163
Email: bc.muscat@om.britishcouncil.org
Website: www.britishcouncil.org/oman

PAKISTAN, Islamabad

The British Council
House 1, Street 61, F-6/3
P.O. Box 1135
Islamabad
Tel: +92 (51) 111 424 424
Fax: +92 (51) 111 425 425
Email: info@britishcouncil.org.pk
Website: www.britishcouncil.org/pakistan

PAKISTAN , Karachi

The British Council
20 Bleak House Road
P.O. Box 10410
Karachi 75530
Tel: +92 (21) 111 424 424
Fax: +92 (21) 111 425 425
Email: info@britishcouncil.org.pk

PALESTINIAN TERRITORIES, East Jerusalem

The British Council
31 Nablus Road
PO Box 19136
East Jerusalem 97200
Tel: +972 (2) 628 7111
Fax: +972 (2) 628 3021
Email: british.council@ps.britishcouncil.org
Website: www.ps.britishcouncil.org

PALESTINIAN TERRITORIES, Gaza

The British Council
Al-Nasra Street, Al-Rimal
Gaza City
PO Box 355
Tel: +972 (0)8 282 2274
Fax: +972 (0)8 282 0512
Email: british.council@ps.britishcouncil.org

PERU, Lima

The British Council
Torre Parque Mar (Floor 22)
Av Jose Larco 1301
Miraflores
Lima 18
Tel: +51 (1) 617 3060
Fax: +51 (1) 617 3065
Email: bc.lima@britishcouncil.org.pe
Website: www.britishcouncil.org/peru

PHILIPPINES, Manila

The British Council
10th Floor, Taipan Place
Emerald Avenue
Ortigas Centre
Pasig City 1605
Tel: +63 (2) 914 1011 - 14
Fax: +63 (2) 914 1020
Email: britishcouncil@britishcouncil.org.ph
Website: www.britishcouncil.org.ph

POLAND, Warsaw

The British Council
Al Jerozolimskie 59
00 697 Warsaw
Tel: +48 (22) 695 5900
Fax: +48 (22) 621 9955
Email: bc.warsaw@britishcouncil.pl
Website: www.britishcouncil.org/poland

POLAND, Krakow

The British Council
Rynek Glowny 26

31-007 Krakow
Tel: +48 (12) 428 5930
Fax: +48 (12) 428 5940
Email: bc.krakow@britishcouncil.pl

PORTUGAL, Lisbon

The British Council
Rua Luis Fernandes 1-3
1249-062 Lisbon
Tel: +351 21 321 45 00
Fax: +351 21 347 61 51
Email: lisbon.enquiries@pt.britcoun.org
Website: www.pt.britishcouncil.org

PORTUGAL, Coimbra

The British Council
Rua de Tomar 4
3000-401 Coimbra
Tel: +351 239 823 700
Fax: +351 239 836 705
Email: Coimbra.Enquiries@pt.britcoun.org

PORTUGAL, Porto

The British Council
Rua do Breyner 155
4050-126 Porto
Tel: +351 22 207 3060
Fax: +351 22 207 3068
Email: Porto.Enquiries@pt.britcoun.org

PORTUGAL, Parede

The British Council
Rua Dr Camilio Dionisio Alvares
Lote 6
2775-177 Parede
Tel: +351 21 458 7370
Fax: +351 21 457 9918
Email: parede.enquiries@pt.britcoun.org

PORTUGAL, Cascais

The British Council
Edificio Sao Jose
1 Piso, Sala 102
Av Combatentes da Grande Guerra
2750-326 Cascais
Tel: +351 21 482 0714
Fax: +351 21 482 0722
Email: Cascais.Enquiries@pt.britcoun.org

QATAR, Doha

The British Council
93 Al Sadd Street
PO Box 2992
Doha
Tel: +974 442 6193 / 4
Fax: +974 442 3315
Email: general.enquiries@qa.britishcouncil.org
Website: www.britishcouncil.org/qatar

ROMANIA, Bucharest

The British Council
Calea Dorobantilor 14
010572 Bucharest
Tel: +40 21 307 9600
Fax: +40 21 307 96 01
Email: bc.romania@britishcouncil.ro
Website: www.britishcouncil.ro

RUSSIA, Moscow

The British Council
Ulitsa Nikoloyamskaya 1
Moscow 109189
Tel: +7 095 782 0200
Fax: +7 095 782 0201
Email: bc.moscow@britishcouncil.ru
Website: www.britishcouncil.ru

RUSSIA, St Petersburg

The British Council
46 Fontanka River Embankment
St Petersburg 191025
Tel: +7 812 118 5060
Fax: +7 812 118 5061
Email: bc.stpetersburg@britishcouncil.ru

SAUDI ARABIA, Riyadh

The British Council
Tower B, 2nd floor
Al Mousa Centre, Olaya Street
P.O. Box 58012
Riyadh 11594
Tel: +966 (1) 462 1818
Fax: +966 (1) 462 0663
Email: Enquiry.riyadh@sa.britishcouncil.org
Website: www.britishcouncil.org/saudiarabia

SAUDI ARABIA, Jeddah

The British Council
3rd Floor

Farsi Centre,
Waly Al Ahd St.
PO Box 3424, Jeddah 21471
Tel: +966 2 657 6200
Fax: + 966 2 657 6123
Email: enquiry.jeddah@sa.britishcouncil.org

SAUDI ARABIA, Dammam

The British Council
2nd Floor, Al-Waha Mall
First Street
P.O. Box 8387, Dammam 31482
Tel: +966 3 826 9036 / 9831
Fax: +966 3 826 8753
Email: enquiry.dammam@sa.britishcouncil.org

SENEGAL, Dakar

The British Council
34-36 Boulevard de la Republique
Immeuble Sonatel
BP 6238
Dakar
Tel: +221 822 2015/822 2048
Fax: +221 821 8136
Email: postmaster@britishcouncil.sn
Website: www.britishcouncil.org/senegal

SERBIA, Belgrade

The British Council
Terazije 8/I, POB 248,
11000 Belgrade
Tel: + 381 (11) 3023 800
Fax: + 381 (11) 3023 898
Email: info@britishcouncil.org.yu
Website: www.britishcouncil.org/yugoslavia

SIERRA LEONE, Freetown

The British Council
Tower Hill
PO Box 124
Freetown
Tel: +238 (22) 222 223
Fax: +238 (22) 224 123
Email: enquiry@sl.britishcouncil.org
Website: www.britishcouncil.org/sierraleone

SINGAPORE, Singapore

The British Council
30 Napier Road
258509 Singapore
Tel: +65 6473 1111
Fax: +65 6472 1010
Email: enquiries@britishcouncil.org.sg
Website: www.britishcouncil.org.sg

SLOVAKIA, Bratislava

The British Council
Panska 17 - P.O. Box 68
814 99 Bratislava
Tel: +421 (2) 5443 1074 / 5443 1185
Fax: +421 (2) 5443 4705
Email: info.bratislava@britishcouncil.sk
Website: www.britishcouncil.org/slovakia

SLOVENIA, Ljubljana

The British Council
Tivloi Center, Tivolska 30
Ljubljana 1000
Tel: +386 (0) 1 300 2030
Fax: +386 (0) 1 300 2044
Email: info@britishcouncil.si
Website: www.britishcouncil.si

SOUTH AFRICA, Pretoria

The British Council
Suite 1, Sanlam Gables
1209 Schoeman Street
Hatfield, Pretoria 0083
Tel: +27 (12) 431 2400
Fax: +27 (12) 431 2415
Email: information@britishcouncil.org.za
Website: www.britishcouncil.org/southafrica

SOUTH AFRICA, Johannesburg

The British Council
Ground Floor
Forum 1, Braampark
33 Hoofd Street
Braamfontein
Johannesburg 2001
Tel: +27 (11) 718 4300
Fax: +27 (11) 718 4400
Email: information@britishcouncil.org.za

SPAIN, Madrid

The British Council
Paseo del General Martinez
Campos 31
28010 Madrid
Tel: +34 91 337 3500

Fax: +34 91 337 3573
Email: Madrid@britishcouncil.es
Website: www.britishcouncil.es

SPAIN, Barcelona

The British Council
Calle Amigo 83
08021 Barcelona
Tel: +34 93 241 9700
Fax: +34 93 202 3168
Email: barcelona@britishcouncil.es

SPAIN, Bilbao

The British Council
Avenida Lehendakari Aguirre 29 - 2º
Deusto
48014 Bilbao
Tel: +34 94 476 3650
Fax: +34 94 476 2016
Email: bilbao@britishcouncil.es

SPAIN, Segovia

The British Council
Avenida Padre Claret 3
40003 Segovia
Tel: +34 921 43 48 13
Fax: +34 921 43 48 13
Email: aoife.buckley@britishcouncil.es

SPAIN, Valencia

The British Council
Avinguda de Catalunya, 9
46020 Valencia
Tel: +34 96 339 29 80
Fax: +34 96 369 13 89
Email: Valencia@britishcouncil.es

SRI LANKA, Colombo

The British Council
49 Alfred House Gardens
PO Box 753
Colombo 3
Tel: +94 (0)11 2581171
Fax: +94 (0)11 2587 079
Email: enquiries@britishcouncil.lk
Website: www.britishcouncil.lk

SUDAN, Khartoum

The British Council
14 Abu Sin Street
PO Box 1253
Central Khartoum
Tel: +249 183 780 817 / 777310
Fax: +249 183 774 935
Email: info@sd.britishcouncil.org
Website: www.britishcouncil.org/sudan

SWEDEN, Stockholm

The British Council
PO Box 27819
S-115 93 Stockholm
Tel: +46 (8) 671 3110
Fax: +46 (8) 663 7271
Email: info@britishcouncil.se
Website: www.britishcouncil.org/sweden

SWITZERLAND, Berne

The British Council
Sennweg 2
PO Box 532
CH 3000 Berne 9
Tel: +41 (31) 301 1473
Fax: +41 (31) 301 1459
Email: britishcouncil@britishcouncil.ch
Website: www.britishcouncil.ch

SYRIA , Damascus

The British Council
Maysaloun Street
Shalaan
P O Box 33105
Damascus
Tel: +963 (11) 331 0631
Fax: +963 (11) 332 1467
Email: general.enquiries@sy.britishcouncil.org
Website: www.britishcouncil.org/syria

TAIWAN, Taipei

British Council
2F-1, 106 XinYi Rd
Section 5
Taipei 110
Tel: +886 (0) 2 8722 1000
Fax: +886 (0) 2 8786 0985
Email: enquiries@britishcouncil.org.tw
Website: www.britishcouncil.org/taiwan

TANZANIA, Dar Es Salaam

The British Council
Samora Avenue / Ohio Street

PO Box 9100
Dar Es Salaam
Tel: +255 (22) 211 6574
Fax: +255 (22) 211 2669
Email: info@britishcouncil.or.tz
Website: www.britishcouncil.org/tanzania

THAILAND, Bangkok

The British Council
254 Chulalongkorn Soi 64
Siam Square
Phayathai Road
Pathumwan
Bangkok 10330
Tel: +66 (2) 252 5480
Fax: +66 (2) 253 5312
Email: info@britishcouncil.or.th
Website: www.britishcouncil.or.th

THAILAND, Chiang Mai

The British Council
198 Bumrungraj Road
Chiang Mai 50000
Tel: +66 (53) 242 103
Fax: +66 (53) 244 781
Email: info@britishcouncil.or.th

TRINIDAD (Caribbean), Port of Spain

The British Council
19 St Clair Avenue,
St Clair, PO Box 778
Port of Spain, Trinidad
Tel: +1 868 628 0565
Fax: +1 868 622 2853
Email: candice.fabien@britishcouncil.org.tt
Website: www.britishcouncil.org/caribbean

TUNISIA, Tunis

The British Council
c/o British Embassy
5 Place de la Victoire
BP 229
Tunis 1015 RP
Tel: +216 71 259 053 / 351 754
Fax: +216 71 353 411
Email: info@tn.britishcouncil.org
Website: www.britishcouncil.org.tn

TURKEY, Ankara

The British Council
Karum Is Merkesi
C Blok No 437 Kat 5
Kavaklidere/ Ankara
Tel: +90 312 455 3600
Fax: +90 312 455 3636
Email: bc.ankara@britishcouncil.org.tr
Website: www.britishcouncil.org.tr

TURKEY, Istanbul

The British Council
PO Box 16
81690 Besiktas
Istanbul
Tel: +90 212 355 5657
Fax: +90 212 355 5658
Email: bc.istanbul@britishcouncil.org.tr

TURKEY, Izmir

The British Council
C/o Izmir Ekonomi Universitesi
Ismet Kaptan Mahallesi
Sakarya cad. No 156
Balçova 35330 Izmir
Tel: +90 238 295 00 52
Fax: +90 238 295 00 27
Email: bc.izmir@britishcouncil.org.tr

UGANDA , Kampala

The British Council
Rwenzori Courts
Plot 2 and 4a
Nakasero Road
PO Box 7070
Kampala
Tel: +256 (0)41 234 725/730
Fax: +256 (0)41 254 853
Email: info@britishcouncil.or.ug
Website: www.britishcouncil.org/uganda

UK, London

The British Council
10 Spring Gardens
London SW1A 2BN
Tel: +44 (0) 20 7930 8466
Fax: +44 (0) 20 7839 6347
Email: firstname.surname@britishcouncil.org
Website: www.britishcouncil.org

UKRAINE, Kyiv

The British Council
4/12 Vul. Hryhoriya Skovorody
Kyiv 04070
Tel: +380 (44) 490 5600
Fax: +380 (44) 490 5605
Email: enquiry@britishcouncil.org.ua
Website: www.britishcouncil.org.ua

UNITED ARAB EMIRATES, Abu Dhabi

The British Council
Villa no. 7, Al Nasr Street
Khalidiya
Abu Dhabi
P.O. Box 46523
Tel: +9712 665 9300
Fax: +9712 666 4340
Email: information@ae.britishcouncil.org
Website: www.britishcouncil.org/uae

UNITED ARAB EMIRATES, Dubai

The British Council
Tariq bin Zaid Street
(near Rashid Hospital)
P.O. Box 1636
Dubai
Tel: +9714 337 0109
Fax: +9714 337 0703
Email: information@ae.britishcouncil.org

USA, Washington

British Embassy
3100 Massachusetts Avenue NW
Washington DC 20008-3600
Tel: +1 202 588 6500 / 7830
Fax: +1 202 588 7918
Email: studyintheuk@us.britishcouncil.org
Website: www.britishcouncil-usa.org

UZBEKISTAN, Tashkent

British Council Information Centre
11 Kounaev Street
Tashkent 700031
Tel: +998 (71) 120 6752
Fax: +998 (71) 120 6371
Email: bc-tashkent@britishcouncil.uz
Website: www.britishcouncil.org/uzbekistan

VENEZUELA, Caracas

The British Council
Piso 3, Torre Credicard
Av. Principal El Bosque
Chacaito, Caracas
Tel: +58 (212) 952 9965
Fax: +58 (212) 952 9691
Email: bc-venezuela@britishcouncil.org.ve
Website: www.britishcouncil.org.ve

VIETNAM, Hanoi

The British Council
40 Cat Linh Street
Dong Da
Hanoi
Tel: +84 (0)4 843 6780
Fax: +84 (0)4 843 4962
Email: bchanoi@britishcouncil.org.vn
Website: www.britishcouncil.org/vietnam

VIETNAM, Ho Chi Minh City

The British Council
25 Le Duan
District 1
Ho Chi Minh City
Tel: +84 (0)8 823 2862
Fax: +84 (0)8 823 2861
Email: bchcmc@britishcouncil.org.vn

YEMEN, Sana'a

The British Council
3rd Floor, Administrative Tower
Sana'a Trade Centre
Algiers Street
PO Box 2157
Sana'a
Tel: +967 1448 356/7
Fax: +967 1448 360
Email: britishcouncil@ye.britishcouncil.org
Website: www.britishcouncil.org/yemen

ZAMBIA, Lusaka

The British Council
Heroes Place
Cairo Road, (PO Box 34571)
Lusaka
Tel: +260 (1) 223 602 / 228 332
Fax: +260 (1) 224 122
Email: info@britishcouncil.org.zm
Website: www.britishcouncil.org/zambia

ZIMBABWE, Harare

The British Council
Corner House, Samora Machel Avenue
P. O. Box 664
Harare
Tel: +263 4 775 313/4
Fax: +263 4 756 661
Email: general.enquiries@britishcouncil.org.zw
Website: www.britishcouncil.org.zw

Canadian Embassies

When preparing to teach and travel overseas it is vital to contact the your local embassy or consulate within the country you are planning to visit to find out information on visas, restrictions and a variety of other information.

Afghanistan

Embassy of Canada
Address: Street No. 15, House No. 256, Wazir Akbar Khan, Kabul, Afghanistan
Tel: 93 (0) 799 742 800
Fax: 93 (0) 799 742 805
Email: kabul@international.gc.ca
Website: www.canada-afghanistan.gc.ca

Albania

Consulate of Canada
Address: Rruga "Dervish Hima," Kulla, No. 2, Apt. 22, Tirana, Albania
Tel: 355 (4) 257 274 / 25 7275 / 355 (68) 20 29364
Fax: 355 (4) 257 273
Email: canadalb@canada.gov.al

Algeria

Embassy of Canada
Address: 18, rue Mustapha-Khalef, Ben Aknoun, Alger, Algeria
Postal Address: P.O. Box 48, Alger-Gare, 16000 Alger, Algeria
Tel: 213 (0) 70-08-30-00
Fax: 213 (0) 70-08-30-70
Email: alger@international.gc.ca
Website: www.algeria.gc.ca

Angola

Consulate of Canada
Address: Rua Rei Katyavala 113, Luanda, Angola
Tel: 244 (222) 448-371,- 377, -366
Fax: 244 (2) 449-494
Email: consul.can@angonet.org

Argentina

Embassy of Canada
Address: 2828 Tagle, C1425EEH Buenos Aires, Argentina
Postal Address: P.O. Box 1598, C1000WAP, Buenos Aires, C1000WAP Correo Central, Argentina
Tel: 54 (11) 4808-1000
Fax: 54 (11) 4808-1012
Email: bairs@international.gc.ca
Website: www.dfait-maeci.gc.ca/bairs

Armenia

Consulate of Canada
Address: 1 Amirian Street, Hotel Marriott, Suite 306, Yerevan, Armenia
Tel: 374 (10) 56-79-90
Fax: 374 (10) 56-79-90

Australia

High Commission of Canada
Address: Commonwealth Avenue, Canberra ACT 2600, Australia
Tel: 61 (2) 6270-4000
Emergency toll-free to Ottawa: 0011-800-2326-6831
Fax: 61 (2) 6270-4081
Email: cnbra@international.gc.ca
Website: www.dfait-maeci.gc.ca/australia

Consulate of Canada
Address: Level 50, 101 Collins Street, Melbourne, Victoria 3000, Australia
Tel: 61 (3) 9653-9674
Emergency toll-free to Ottawa: 0011-800-2326-6831

Consulate of Canada
Address: 3rd Floor, 267 St. George's Terrace, Perth, Western Australia 6000, Australia
Tel: 61 (8) 9322-7930
Emergency toll-free to Ottawa: 0011-800-2326-6831
Fax: 61 (8) 9261-7706

Consulate General of Canada
Address: Level 5, Quay West, 111 Harrington Street, Sydney, New South Wales 2000, Australia
Tel: 61 (2) 9364-3000
Emergency toll-free to Ottawa: 0011-800-2326-6831
Fax: 61 (2) 9364-3098
Email: sydny@international.gc.ca

Austria

Embassy of Canada
Address: Laurenzerberg 2, 1010 Vienna, Austria
Tel: 43 (1) 531-38-3000
Emergency toll-free to Ottawa: 00-800-2326-6831
Fax: 43 (1) 531-38-3905
Email: vienn@international.gc.ca
Website: www.kanada.at

Bahamas

Consulate of Canada
Address: Shirley Street Shopping Plaza, Nassau, Bahamas
Postal Address: P.O. Box SS-6371, Nassau, Bahamas
Tel: 1 (242) 393-2123, -2124
Emergency toll-free to Ottawa: 1-881-949-9993
Fax: 1 (242) 393-1305
Email: cdncon@batelnet.bs

Bahrain

Consulate of Canada
Address: Al Jasrah Tower, 12th Floor, Building No. 95, Road 1702, Block 317, Diplomatic Area, Manama, Kingdom of Bahrain
Postal Address: P.O. Box 2397, Manama, Bahrain
Tel: (973) 17 536270
Emergency toll-free to Ottawa: 800-00-732
Fax: (973) 17 532520
Email: canadabh@batelco.com.bh

Bangladesh

High Commission of Canada
Address: House CWN 16/A, Road 48, Gulshan, Dhaka, Bangladesh
Postal Address: P.O. Box 569, Dhaka, 1000, Bangladesh
Tel: 880 (2) 988-7091/2/3/4/5/6/7
Fax: 880 (2) 882-3043
Email: dhaka@international.gc.ca
Website: www.bangladesh.gc.ca

Barbados

High Commission of Canada
Address: Bishop's Court Hill, St. Michael, Barbados
Postal Address: P.O. Box 404, Bridgetown, Barbados
Tel: 1 (246) 429-3550
Emergency toll-free to Ottawa: 1-888-949-9993
Fax: 1 (246) 437-7436
Email: bdgtn@international.gc.ca
Website: www.bridgetown.gc.ca

Belgium

Consulate of Canada
Address: Sint Pietersvliet, 15, B-2000 Anvers, Belgium
Tel: 32 (03) 220-0211
Emergency toll-free to Ottawa: 00-800-2326-6831
Fax: 32 (03) 220-0204

Embassy of Canada
Address: 2, avenue de Tervueren, 1040 Brussels, Belgium
Tel: 32 (02) 741-0611
Emergency toll-free to Ottawa: 00-800-2326-6831
Fax: 32 (02) 741-0619
Email: bru@international.gc.ca
Website: www.ambassade-canada.be

Belize

Consulate of Canada
Address: 80 Princess Margaret Drive, Belize City, Belize
Postal Address: P.O. Box 610, Belize City, Belize
Tel: 501 223-1060
Fax: 501 223-0060
Email: cdncon.bze@btl.net

Benin

Canadian Cooperation Support Unit
Address: behind the postal triage at the airport, Lot 2371, Zone Djomèhountin, Cotonou, Benin
Postal Address: P.O. Box 04-1124, Cotonou, Benin
Tel: 229 21-30-24-79
Fax: 229 21-30-05-32
Email: secretariat@uapbenin.net

Bhutan

Canadian Cooperation Office
Address:
Postal Address: P.O. Box 201, Thimphu, Bhutan
Tel: 975 (2) 322-109 or 332-615
Tel. after hours: 975 17110040
Fax: 975 (2) 332-614
Email: canada@druknet.net.bt

Bolivia

Consulate of Canada
Address: Calle Victor Sanjinez 2678, Edificio Barcelona, 2nd Floor, Plaza España (Sopocachí), La Paz, Bolivia
Tel: 591 (2) 241-5141, 5021,4517
Fax: 591 (2) 241-4453
Email: lapaz@international.gc.ca

Bosnia and Herzegovina

Embassy of Canada
Address: 4, Grbavicka 71 000, Sarajevo, Bosnia and Herzegovina
Tel: 387 (33) 222-033
Fax: 387 (33) 222-044
Email: sjevo@international.gc.ca

Botswana

Consulate of Canada
Address: Vision Hire Building, Queens Road, Plot 182, Gaborone, Botswana
Postal Address: P.O. Box 882, Gaborone, Botswana
Tel: 267 30-4411
Fax: 267 30-44-11
Email: canada.consul@info.bw

Brazil

Consulate of Canada
Address: Rua Rio Grande do Norte, 1164-conj. 502, Belo Horizonte, Minas Gerais, 30130-131, Brazil
Tel: 55 (31) 3261-1017
Emergency toll-free to Ottawa: 0-800-891-6614
Fax: 55 (31) 3261-1071

Embassy of Canada
Address: Setor de Embaixadas Sul, Avenida das Naçoes lote 16, 70410-900 Brasilia D.F., Brazil
Postal Address: P.O. Box 00961, Brasilia D.F, 70359-900, Brazil
Tel: 55 (61) 3424-5400
Emergency toll-free to Ottawa: 0-800-891-6614
Fax: 55 (61) 3424-5490
Email: brsla@international.gc.ca
Website: www.dfait-maeci.gc.ca/brazil

Consulate of Canada
Address: Av. Atlântica 1130, 4th Floor, Copacabana, 22021-000 Rio de Janeiro,
Tel: 55 (21) 2543-3004
Emergency toll-free to Ottawa: 0-800-891-6614
Fax: 55 (21) 3873-4843, 2275-2195
Email: rio@international.gc.ca
Website: www.dfait-maeci.gc.ca/brazil

Consulate General of Canada
Address: Centro Empresarial Nações Unidas - Torre Norte, 16th floor, Avenida Nações Unidas, 12901 São Paulo, SP, Brazil
Postal Address: P.O. Box CEP 04578-000, São Paulo,
Tel: 55 (11) 5509-4321
Emergency toll-free to Ottawa: 0-800-891-6614
Fax: 55 (11) 5509-4260
Email: spalo@international.gc.ca
Website: www.dfait-maeci.gc.ca/brazil

Brunei Darussalam

High Commission of Canada
Address: 5th Floor, Jalan McArthur Building, No. 1, Jalan McArthur, Bandar Seri Begawan, Brunei Darussalam
Tel: 673 (2) 22-00-43
Fax: 673 (2) 22-00-40
Email: bsbgn@international.gc.ca
Website: www.dfait-maeci.gc.ca/Brunei

Bulgaria

Consulate of Canada
Address: 9 Moskovska Street, 1000 Sofia, Bulgaria
Tel: 359 (2) 969-9710
Fax: 359 (2) 981-6081
Email: consular@canada-bg.org

Burkina Faso

Embassy of Canada
Address: rue Agostino Néto, Ouagadougou, Burkina Faso
Postal Address: P.O. Box 548, Ouagadougou 01, Province du Kadiogo, Burkina Faso
Tel: 226 50-31-18-94
Fax: 226 50-31-19-00
Email: ouaga@international.gc.ca
Website: www.ouagadougou.gc.ca

Burma

Australian Embassy
Address: 88 Strand Road, Rangoon, Burma (Myanmar)
Tel: 95 (1) 251810
Fax: 95 (1) 246159, 246160

Burundi

Consulate of Canada
Address: 4708, Boulevard de l'UPRONA, Bujumbura
Postal Address: P.O. Box 7112, Bujumbura, Burundi
Tel: 257 24-58-98
Fax: 257 24-58-99
Email: consulat.canada@usan-bu.net

Cambodia

Australian Embassy
Address: Villa 11, RV Senei Vannavaut Oum (Street 254), Daun Penh District, Phnom Penh, Cambodia
Tel: 855 (23) 213 470
Fax: 855 (23) 213 413
Email: pnmpn@international.gc.ca
Website: www.phnompenh.gc.ca

Cameroon - Douala

Consulate of Canada
Address: a/s PRO-PME, 1726, avenue Charles de Gaulle, Douala, Cameroon
Postal Address: P.O. Box 2373, Douala, Cameroon
Tel: 237 343-2934
Fax: 237 342-31-09

Cameroon - Yaoundé

High Commission of Canada
Address: Immeuble Stamatiades, Place de l'Hôtel de Ville, Yaoundé, Cameroon
Postal Address: P.O. Box 572, Yaoundé, Cameroon
Tel: 237 223-2311
Fax: 237 222-1090
Email: yunde@international.gc.ca

Cayman Islands

Consulate of Canada
Address: 24 Huldah Avenue, George Town, Cayman Islands
Postal Address: P.O. Box 10102 SMB, George Town, Cayman Islands
Tel: (345) 949-9400
Emergency toll-free to Ottawa: 1-888-949-9993
Fax: (345) 949-9405
Email: cdncon.cayman@candw.ky

Central African Republic

Consulate of Canada
Address: Cabinet ARC
Postal Address: P.O. Box 514, Bangui, Central African Republic
Tel: 236 61-30-39
Email: consulatbangui@yahoo.fr

Chad

Consulate of Canada
Address: Rue 5041, Porte 964, Quartier Moursal, N'Djamena, Chad
Postal Address: P.O. Box 6013, N'Djamena, Chad
Tel: 235 53-42-80
Tel. after hours: 235 27-30-27
Email: honconca@intnet.td
Email: nigel.whiteho@intnet.td

Chile

Consulate of Canada
Address: José Toribio Medina 146, Apt. 601, Antofagasta, Chile
Tel: 56 (55) 24-7652
Emergency toll-free to Ottawa: 800-201-670
Fax: (2) 652-3916
Email: ca.consul.antofa@emol.com

Consulate of Canada
Address: Caupolicán 245, Chiguayante, Concepción, Chile
Postal Address: P.O. Box Casilla 425, Concepción, Chile
Tel: 56 (41) 236-97-05
Emergency toll-free to Ottawa: 800-201-670
Fax: 56 (41) 236-81-85
Email: ca.consul.concep@emol.com

Embassy of Canada
Address: Nueva Tajamar 481, Torre Norte, 12th Floor, Las Condes, Santiago, Chile
Postal Address: P.O. Box Casilla 139, Correo 10, Chile, Chile
Tel: 56 (2) 652-3800
Emergency toll-free to Ottawa: 800-201-670
Fax: 56 (2) 652-3916
Email: stago@international.gc.ca
Website: www.santiago.gc.ca

China

Embassy of Canada
Address: 19 Dong Zhi Men Wai Street, Chao Yang District, Beijing 100600, People's Republic of China
Tel: 86 (10) 6532-3536
Emergency toll-free to Ottawa: 10800-1400125; 00800-2326-6831
Fax: 86 (10) 6532-5544
Email: bejing-cs@international.gc.ca
Website: www.beijing.gc.ca

Consulate of Canada
Address: Room 1705, Metropolitan Tower, Wu Yi Lu, Yu Zhong District, Chongqing 400010, People's Republic of China
Tel: 86 (23) 6373-8007
Emergency toll-free to Ottawa: 10800-1400125; 00800-2326-6831
Fax: 86 (23) 6373-8026
Email: chonq@international.gc.ca
Website: www.chongqing.gc.ca

Consulate General of Canada
Address: Suite 801, China Hotel Office Tower, Liu Hua Lu, Guangzhou 510015, People's Republic of China
Tel: 86 (20) 8666-0569
Emergency toll-free to Ottawa: 10800-1400125; 00800-2326-6831
Fax: 86 (20) 8667-0267
Email: ganzu@international.gc.ca
Website: www.guangzhou.gc.ca

Consulate General of Canada
Address: 14th Floor, One Exchange Square, Central, Hong Kong SAR, People's Republic of China
Postal Address: P.O. Box 11142, Central, Hong Kong SAR, China
Tel: 85 (2) 2810-4321
Emergency toll-free to Ottawa: 001-800-2326-6831
Fax: 85 (2) 2810-6736
Email: hkong@international.gc.ca
Website: www.dfait-maeci.gc.ca/hongkong/

Consulate General of Canada
Address: American International Centre, West Tower, Suite 604, 1376 Nanjing Xi Lu, Shanghai 200040, People's Republic of China
Tel: 86 (21) 6279-8400
Emergency toll-free to Ottawa: 10800-1400125; 00800-2326-6831
Fax: 86 (21) 6279-8401
Email: shngi@international.gc.ca
Website: www.shanghai.gc.ca

Colombia

Embassy of Canada
Address: Cra. 7, No. 115-33, Bogotá, Colombia
Postal Address: P.O. Box 110067, Bogotá, Colombia
Tel: 57 (1) 657-9800
Emergency toll-free to Ottawa: 01-800-919-0114
Fax: 57 (1) 657-9912
Email: bgota@international.gc.ca
Website: www.dfait-maeci.gc.ca/bogota

Consulate of Canada
Address: Edificio Centro Ejecutivo Bocagrande, Carrera 3, No. 8-129, Oficina No. 1103, Cartagena, Colombia
Tel: 57 (5) 665-5838
Emergency toll-free to Ottawa: 01-800-919-0114
Fax: 57 (5) 665-5837
Email: honcartagena@enred.com

Costa Rica

Embassy of Canada
Address: La Sabana Executive Business Centre, Building No. 5, 3rd Floor, behind the Contraloría General de la República, San José, Costa Rica
Postal Address: P.O. Box 351-1007, San José, Costa Rica
Tel: 506 242-4400
Emergency toll-free to Ottawa: 0-800-015-1161
Fax: 506 242-4410
Email: sjcra@international.gc.ca
Website: www.dfait-maeci.gc.ca/sanjose

Côte d'Ivoire

Embassy of Canada
Address: Immeuble Trade Center, 23, avenue Noguès, Le Plateau, Abidjan, Côte d'Ivoire
Postal Address: P.O. Box 4104, Abidjan 01, Côte d'Ivoire
Tel: 225 20-30-07-00
Fax: 225 20-30-07-20
Email: abdjn@international.gc.ca
Website: www.dfait-maeci.gc.ca/abidjan

Croatia

Embassy of Canada
Address: Prilaz Gjure Dezelica 4,10000 Zagreb, Croatia
Tel: 385 (1) 488-1200,1211

Fax: 385 (1) 488-1230
Email: zagrb@international.gc.ca
Website: www.dfait-maeci.gc.ca/canadaeuropa/croatia

Cuba

Consulate of Canada
Address: Hotel Atlantico, Suite # 1, Guardalavaca, Holguín, Cuba
Tel: 53 (24) 30-320
Fax: 53 (24) 30-290
Email: honconcanada.hog@enet.cu

Embassy of Canada
Address: Calle 30, nº 518, Esquina a7a, Miramar, Havana, Cuba
Tel: 53 (7) 204 2516
Fax: 53 (7) 204 2004
Email: havan@international.gc.ca
Website: www.havana.gc.ca

Consulate of Canada
Address: Calle 13E, Avenida Primera y Camino del Mar, Varadero, Mantanzas, Cuba
Tel: 53 (45) 61-2078
Fax: 53 (45) 66-7395
Email: vra.honcon@enet.cu

Curaçao

Consulate of Canada
Address: Maduro and Curiel's Bank, N.V., Plaza Jojo Correa 2-4, Willemstad (Punda), Curaçao, Netherlands Antilles
Postal Address: P.O. Box 305, Willemstad, Curaçao
Tel: 599 (9) 466-1115, 466-1121
Fax: 599 (9) 466-1122, 466-1130

Cyprus

Consulate of Canada
Address: 1 Lambousa Street, 1095 Nicosia, Cyprus
Postal Address: P.O. Box 22115-1517, Nicosia, Cyprus
Tel: 357 (2) 2775-508
Emergency toll-free to Ottawa: 8009-6082
Fax: 357 (2) 2779-905
Email: info@consulcanada.com.cy

Czech Republic

Embassy of Canada
Address: Muchova 6, 160 00 Prague 6, Czech Republic
Tel: 420 27210-1800
Fax: 420 27210-1890
Email: prgue@international.gc.ca
Website: www.dfait-maeci.gc.ca/prague

Democratic Republic of Congo

Embassy of Canada
Address: 17, Pumbu Avenue, Commune de Gombe, Kinshasa, Democratic Republic of Congo
Postal Address: P.O. Box 8341, Kinshasa 1, Democratic Republic of Congo
Tel: 243 895-0310/0311/0312
Tel. after hours: (243) 81-700-5188
Fax: 243-997-5403 / 243-81-301-6515
Email: kinshasa@international.gc.ca
Website: www.congo.gc.ca

Denmark

Embassy of Canada
Address: Kr. Bernikowsgade 1, 1105 Copenhagen K, Denmark
Tel: 45 33 48 32 00
Fax: 45 33 48 32 20
Email: copen@international.gc.ca
Website: www.canada.dk

Consulate of Canada
Address: Air Greenland, Nuuk Airport, 3900 Nuuk, Greenland
Postal Address: P.O. Box 1012, Nuuk, 3900, Denmark
Tel: 299 31-16-47
Fax: 299 32-02-88
Email: gohtsh@greenlandair.gl

Djibouti

Consulate of Canada
Address: Place Lagarde, Djibouti, Djibouti
Postal Address: P.O. Box 1188, Djibouti, Djibouti
Tel: 25 (3) 35-38-59, 35-59-50
Fax: 25 (3) 35-00-14
Email: georgalis@intnet.dj

Dominican Republic

Consulate of Canada
Address: Calle Virginia E. Ortea, Edificio Isabel de Torres, Suite 311-C,
Puerto Plata, Dominican Republic
Tel: (809) 586-5761
Emergency toll-free to Ottawa: 1-888-156-3102
Fax: (809) 586-5762
Email: pplat.canada@verizon.net.do

Embassy of Canada
Address: Capitán Eugenio de Marchena No. 39, La Esperilla, Santo Domingo, Dominican Republic
Postal Address: P.O. Box 2054, Santo Domingo 1
Tel: (809) 685-1136; 1-200-0012 (toll-free within Dominican Republic - service sans frais en République dominicaine)
Emergency toll-free to Ottawa: 1-888-156-3102
Fax: (809) 682-2691
Email: sdmgo@international.gc.ca
Website: www.santodomingo.gc.ca

Ecuador

Consulate of Canada
Address: 810 Avenida General Córdova and Victor Manuel Rendón, Edificio Torres de la Merced, 21st Floor, Guayaquil, Ecuador
Tel: 593 (4) 256-3580, 231-4561
Fax: 593 (4) 231-4562
Email: consulc1@gye.satnet.net

Embassy of Canada
Address: Avenida 6 de diciembre, No. 2816, and Paul Rivet, Josueth Gonzalez Building, 4th Floor, Quito, Ecuador
Postal Address: P.O. Box 17-11-6512 (CC1), Quito, Ecuador
Tel: 593 (2) 250-6162, 223-2114
Fax: 593 (2) 250-3108
Email: quito@international.gc.ca
Website: www.quito.gc.ca

Egypt

Embassy of Canada
Address: 26 Kamel El Shenawy Street, Garden City, Cairo, Egypt
Postal Address: P.O. Box 1667, Cairo, Egypt
Tel: 20 (2) 791-8700
Fax: 20 (2) 791-8862
Email: cairo@international.gc.ca
Website: www.cairo.gc.ca

El Salvador

Embassy of Canada
Address: Centro Financiero Gigante, Torre A, Lobby 2, Alameda Roosevelt y 63 Ave. Sur, Colonia Escalón, San Salvador, El Salvador
Tel: 503 2279-4655, -4657, -4659
Fax: 503 2279-0765
Email: ssal@international.gc.ca
Website: www.sansalvador.gc.ca

Eritrea

Consulate of Canada
Address: Abeneh Street 745, House No. 152/154, Tiravolo, Eritrea
Postal Address: P.O. Box 3962, Asmara, Eritrea
Tel: 291 (1) 18-64-90, 18-19-40
Fax: 291 (1) 18-64-88
Email: mkcca1@yahoo.com

Estonia

Office of the Embassy of Canada
Address: Toom Kooli 13, 2nd Floor, 10130 Tallinn, Estonia
Tel: 372 627 3310; -3311
Fax: 372 627 3312
Email: allinn@canada.ee
Website: www.dfait-maeci.gc.ca/dfait/missions/baltiks/

Ethiopia

Embassy of Canada
Address: Old Airport Area, Nifas Silk Lafto K.K. Kebele 4, House #122
Addis Ababa, Ethiopia
Postal Address: P.O. Box 1130, Addis Ababa, Ethiopia
Tel: 251 (0) 11-371-3022
Fax: 251 (0) 11-371-3033
Email: addis@international.gc.ca
Website: www.dfait-maeci.gc.ca/world/embassies/ethiopia/

Fiji

Consulate of Canada
Address:
Postal Address: P.O. Box 10690, Nadi Airport, Nadi
Tel: (679) 6722-400
Tel. after hours: (679) 9924 999 (mobile)
Fax: (679) 672 1936; 679 672 4489
Email: vyases@connect.com.fj

Finland

Embassy of Canada
Address: Pohjoisesplanadi 25B, 00100 Helsinki
Postal Address: P.O. Box 779, Helsinki, 00101
Tel: 358 (9) 228-530
Emergency toll-free to Ottawa: 990-800-2326-6831; +800-2326-6831
Fax: 358 (9) 601-060
Email: hsnki@international.gc.ca
Website: www.canada.fi

France

Embassy of Canada
Address: 35, avenue Montaigne, 75008 Paris, France
Tel: 33 1-44-43-29-00
Emergency toll-free to Ottawa: 00-800-2326-6831
Fax: 33 1-44-43-29-86
Email: paris@international.gc.ca
Website: www.amb-canada.fr

Consulate of Canada
Address: 30, avenue Émile Zola, 59800 Lille, France
Tel: 33 3-20-14-05-78
Emergency toll-free to Ottawa: 00-800-2326-6831
Fax: 33 3-20-14-36-96
Email: consulatcanadalille@wanadoo.fr

Consulate of Canada
Address: 21, rue Bourgelat, 69002 Lyon, France
Tel: 33 4-72-77-64-07
Fax: 33 4-72-77-65-09
Email: consulatcanadalyon@wanadoo.fr

Consulate of Canada
Address: 10, rue Lamartine, 06000 Nice, France
Tel: 33 4-93-92-93-22
Fax: 33 4-93-92-55-51
Email: cancons.nce@club-Website.fr

Consulate of Canada
Address: 16, rue Jacques Debon, Saint-Pierre, Saint-Pierre et Miquelon, France
Postal Address: P.O. Box 4370, 97500, Saint-Pierre, Saint-Pierre et Miquelon, 97500, France
Tel: 508 41-55-10
Fax: 508 41-55-10
Email: consulat.canada@cheznoo.com

Consulate of Canada
Address: 10, Jules de Rességuier, 31000 Toulouse, France
Tel: 33 5-61-52-19-06
Fax: 33 5-61-55-40-32
Email: consulat.canada.toulouse@wanadoo.fr

French Polynesia

Consulate of Australia
Address: Service Mobil, Motu Uta
Postal Address: P.O. Box 9068, Papeete (Tahiti), 98715, French Polynesia
Tel: (689) 46 88 06
Fax: (689) 43 39 26

Gabon

Consulate of Canada
Address: Résidence Saint-Georges, Pont de Gué-Gué
Postal Address: P.O. Box 4037, Libreville, Gabon
Tel: 241-44 29 65
Fax: 241-44-29-64
Email: conhongab@gmail.com

Germany

Embassy of Canada
Address: Leipziger Platz 17, 10117 Berlin, Germany
Tel: 49 (30) 20 31 20
Emergency toll-free to Ottawa: 00-800-2326-6831
Fax: 49 (30) 20 31 24 57
Email: brlin-cs@international.gc.ca
Website: www.berlin.gc.ca

Consulate of Canada
Address: Benrather Strasse 8, 40213 Düsseldorf, Germany
Tel: 49 (211) 17 21 70
Fax: 49 (211) 35 91 65
Email: ddorf@international.gc.ca

Consulate of Canada
Address: Ballindamm 35, 5th Floor, 20095 Hamburg, Germany
Tel: 49 (40) 46 00 27 0
Fax: 49 (40) 46 00 27 20
Email: hmbrg@international.gc.ca

Consulate of Canada
Address: Tal 29, 80331 Munich, Germany
Tel: 49 (89) 21 99 57 0
Fax: 49 (89) 21 99 57 57
Email: munic@international.gc.ca

Consulate of Canada
Address: Lange Strasse 51, 70174 Stuttgart, Germany
Tel: 49 (711) 22 39 67 8
Fax: 49 (711) 22 39 67 9
Email: hcons.stuttgart@consulates-canada.de

Ghana

High Commission of Canada
Address: 42 Independence Avenue, Accra, Ghana
Postal Address: P.O. Box 1639, Accra, Ghana
Tel: 233 (21) 21-15-21; 22-85-55
Fax: 233 (21) 21-15-23; 77-37-92
Email: accra@international.gc.ca
Website: www.accra.gc.ca

Greece

Embassy of Canada
Address: 4 Ioannou Ghennadiou Street, 115 21 Athens, Greece
Tel: 30 (210) 727-3400
Fax: 30 (210) 727-3480
Email: athns@international.gc.ca
Website: www.athens.gc.ca

Consulate of Canada
Address: c/o Hotel Palace, 12 Tsimiski Street, 546 24 Thessaloniki, Greece
Tel: 30 (2310) 256-350
Fax: 30 (2310) 256-351
Email: yakoumis_canada@profinet.gr

Guatemala

Embassy of Canada
Address: Edyma Plaza Building, 8th Floor, 13 Calle 8-44, Zona 10, Guatemala City, Guatemala
Postal Address: P.O. Box 400, Guatemala City, Guatemala
Tel: 502 2363-4348
Fax: 502 2365-1216
Email: gtmla@international.gc.ca
Website: www.guatemala.gc.ca

Guyana

High Commission of Canada
Address: High and Young Streets, Georgetown, Guyana
Postal Address: P.O. Box 10880, Georgetown, Guyana
Tel: 592 227-2081/2/3/4/5
Fax: 592 225-8380
Email: grgtn@international.gc.ca
Website: www.georgetown.gc.ca

Haiti

Embassy of Canada
Address: Delmas Road, between Delmas 75 and 71, Port-au-Prince, Haiti
Postal Address: P.O. Box 826, Port-au-Prince, Haiti
Tel: 509 249-9000
Tel. after hours: 558-0479
Fax: 509 249-9920, 249-9921, 249-9922, 249-9928
Email: prnce@international.gc.ca
Website: www.port-au-prince.gc.ca

Holy See (Vatican City)

Canadian Embassy to the Holy See
Address: Via della Conciliazione 4/D, 00193 Rome
Tel: 39 (06) 6830-7316, 7386, 7398
Fax: 39 (06) 6880-6283
Email: vatcn@international.gc.ca
Website: www.vatican.gc.ca

Honduras

Office of the Embassy of Canada
Address: Centro Financiero BANCO UNO, 3rd Floor, Boulevard San Juan Bosco, Colonia Payaquí, Tegucigalpa, Honduras
Postal Address: P.O. Box 3552, Tegucigalpa
Tel: 504 232-4551
Fax: 504 239-7767
Email: tglpa@international.gc.ca

Hungary

Embassy of Canada
Address: Budapest 1027, Ganz utca 12-14,
Tel: 36 (1) 392-3360
Fax: 36 (1) 392-3390
Email: bpest@international.gc.ca
Website: www.budapest.gc.ca

Iceland

Embassy of Canada
Address: 14 Tungata, 101 Reykjavik, Iceland
Postal Address: P.O. Box 1510, 121 Reykjavik
Tel: 354 575-6500
Fax: 354 575-6501
Email: rkjvk@international.gc.ca
Website: www.reykjavik.gc.ca

India

Consulate General of Canada
Address: SCO 54-56, Sector 17A, Chandigarh 160 017, India
Tel: 91 (172) 505-0300
Fax: 91 (172) 505-0323
Email: CHADG-G@international.gc.ca

Consulate of Canada
Address: 18 (Old 24), 3rd floor YAFA Tower, Khader Nawaz Khan Road, Nungambakkam, Chennai, 600 034
Tel: 91 (44) 2833-0888
Fax: 91 (44) 5215-9393
Email: cheni@gocindia.org

Consulate of Canada
Address: c/o RPG Enterprises, Duncan House, 31, Netaji Subhas Road, Calcutta 700 001, West Bengal, India
Tel: 91 (33) 2242-6820
Fax: 91 (33) 2242-6828
Email: ccklkta@rpg.in

Consulate General of Canada
Address: 6th floor, Fort House, 221 Dr. D.N. Road, Mumbai, 400 001
Tel: 91 (22) 6749 4444
Fax: 91 (22) 6749 4454
Email: mmbai@international.gc.ca

High Commission of Canada
Address: 7/8 Shantipath, Chanakyapuri, New Delhi 110021, India
Postal Address: P.O. Box 5207, New Delhi, India
Tel: 91 (11) 4178-2000, -2100
Fax: 91 (11) 4178-2023
Email: delhi.consular@international.gc.ca
Website: www.india.gc.ca

Indonesia

Consulate General of Australia
Address: Jalan Hayam Wuruk, No. 88B, Tanjung Bungkak, Denpasar, Bali, Indonesia
Postal Address: P.O. Box 3243, Denpasar, Bali, Indonesia
Tel: 62 (361) 283-011, 283-241
Emergency toll-free to Ottawa: 008-800-105-171
Fax: 62 (361) 282-281

Embassy of Canada
Address: World Trade Centre, 6th Floor, Jl. Jend Sudirman, Kav. 29, Jakarta 12920, Indonesia
Postal Address: P.O. Box 8324/JKS.MP, Jakarta, 12084, Indonesia
Tel: 62 (21) 2550-7800
Fax: 62 (21) 2550-7811
Email: jkrta@international.gc.ca
Website: www.jakarta.gc.ca

Consulate of Canada
Address: c/o Wisma Dharmala, 2nd Floor Suite 2, Jl. Panglima Sudirman 101-103, Surabaya 60271
Tel: 62 (31) 546-3419
Fax: 62 (31) 546-3420
Email: canada@sby.prima.net.id

Iran

Embassy of Canada
Address: 57 Shahid Javad-e-Sarfaraz (Darya-E-Noor), Ostad Motahari Avenue, Tehran, Iran
Postal Address: P.O. Box 11365-4647, Tehran, Iran
Tel: 98 (21) 8873-2623, -2624, -2625, -2626
Fax: 98 (21) 8873-3202
Email: teran@international.gc.ca
Website: www.iran.gc.ca

Ireland

Embassy of Canada
Address: 65 St. Stephen's Green, Dublin 2, Ireland
Tel: 353 (1) 417-4100
Tel. after hours: 353 (1) 478-1476
Fax: 353 (1) 417-4101
Email: dubln@international.gc.ca
Website: www.dublin.gc.ca

Israel, the West Bank and Gaza - Ramallah

Representative Office of Canada
Address: 12 Mahfal Street, Ramallah, West Bank
P.O. Box 2286, Ramallah, West Bank
P.O. Box 18604, Jerusalem 91184
Tel: 972 (2) 295-8604
Fax: 972 (2) 295-8606
Email: rmlah@international.gc.ca

Embassy of Canada
Address: 3/5 Nirim Street, Tel Aviv 67060, Israel
P.O. Box 9442, Tel Aviv, 67060, Israel
Tel: 972 (3) 636-3300
Fax: 972 (3) 636-3383
Email: taviv@international.gc.ca
Website: www.telaviv.gc.ca

Italy

Embassy of Canada
Address: Via Zara 30, Rome, Italy 00198
Tel: 39 (06) 85 444 2911
Fax: 39 (06) 85 444 2912
Email: rome.citizenservices@international.gc.ca
Website: www.rome.gc.ca

Consulate General of Canada
Address: Via Vittor Pisani 19, Milan, Italy 20124
Tel: 39 (02) 6758-1
Emergency toll-free to Ottawa: 00-800-2326-6831
Fax: 39 (02) 6758-3900
Email: milan@international.gc.ca

Consulate of Canada
Address: Via Carducci 29, Naples, Italy 80121
Tel: 39 (081) 401-338
Fax: 39 (081) 410-4210
Email: cancons.nap@tiscali.it

Consulate of Canada
Address: Riviera Ruzzante 25, Padova, Italy 35123
Tel: 39 (049) 876-4833
Fax: 39 (049) 878-1147
Email: consolato.padova@canada.it

Consulate of Canada
Address: Via Panfilo Mazara 26, Sulmona (Aquila), Italy 67039
Tel: 39 (0864) 212-341
Fax: 39 (0864) 210-158
Email: consolato.canadese@tin.it

Jamaica

High Commission of Canada
Address: 3 West Kings House Road, Kingston, 10, Jamaica
Postal Address: P.O. Box 1500, Kingston, 10, Jamaica
Tel: (876) 926-1500
Fax: (876) 511-3493
Email: kngtn@international.gc.ca
Website: www.kingston.gc.ca

Consulate of Canada
Address: 29 Gloucester Street, Montego Bay, Jamaica
Tel: (876) 952-6198
Fax: (876) 952-3953
Email: cancon@cwjamaica.com

Japan

Consulate of Canada
Address: FT Building, 9F, 4-8-28 Watanabe-Dori, Chuo-ku, Fukuoka-shi, Fukuoka-ken 810-0004, Japan
Tel: 81 (92) 752-6055
Fax: 81 (92) 752-6077
Email: fkoka@international.gc.ca
Website: www.dfait-maeci.gc.ca/ni-ka/contacts/fukuoka

Consulate of Canada
Address: c/o Chugoku Electric Power Co. Inc., 4-33 Komachi, Naka-ku, Hiroshima-shi, Hiroshima-ken 730-8701, Japan
Tel: 81 (82) 246-0057
Fax: 81 (82) 246-0057

Consulate of Canada
Address: Nakato Marunouchi Building, 6F, 3-17-6 Marunouchi, Naka-ku, Nagoya-shi, Aichi-ken 460-0002, Japan
Tel: 81 (52) 972-0450
Fax: 81 (52) 972-0453
Email: ngoya@international.gc.ca
Website: www.dfait-maeci.gc.ca/ni-ka/contacts/nagoya

Consulate of Canada
Address: Daisan Shoho Building, 12th Floor, 2-2-3, Nishi-Shinsaibashi, Chuo-ku, Osaka 542-0086, Japan
Tel: 81 (6) 6212-4910
Fax: 81 (6) 6212-4914
Email: osaka@international.gc.ca
Website: www.dfait-maeci.gc.ca/ni-ka/contacts/osaka

Consulate of Canada
Address: Tokyo Tatemono Sapporo Building 2F, 20 Kita-7 Nishi-2, Kita-ku, Sapporo 060-0807, Japan
Tel: 81 (11) 726-2863
Fax: 81 (11) 726-2863

Embassy of Canada
Address: 3-38 Akasaka 7-chome, Minato-ku, Tokyo 107-8503, Japan
Tel: 81 (3) 5412-6200
Fax: 81 (3) 5412-6289
Email: tokyo@international.gc.ca
Website: www.tokyo.gc.ca

Jordan

Embassy of Canada
Address: Pearl of Shmeisani Building, Shmeisani, Amman, Jordan
Postal Address: P.O. Box 815403, Amman, 11180, Jordan
Tel: 962 (6) 520-3300
Fax: 962 (6) 520-3396
Email: amman@international.gc.ca
Website: www.amman.gc.ca

Kazakhstan

Embassy of Canada
Address: 34 Karasai Batir Street (Vinogradov Street), Almaty, 480100, Kazakhstan
Tel: 7 (3272) 501151, 501153
Fax: 7 (3272) 582493
Email: almat@international.gc.ca
Website: www.dfait-maeci.gc.ca/canadaeuropa/kazakhstan

Kenya

High Commission of Canada
Address: Limuru Road, Gigiri, Nairobi, Kenya
Postal Address: P.O. Box 1013, 00621, Nairobi, Kenya
Tel: 254 (20) 366-3000
Fax: 254 (20) 366 3900
Email: nrobi@international.gc.ca
Website: www.nairobi.gc.ca

Kiribati

Australian High Commission
Address: Bairiki, Tarawa, Kiribati
Postal Address: P.O. Box 77, Tarawa, Kiribati
Tel: 686 21-184
Fax: 686 21-904

Kuwait

Embassy of Canada
Address: 24, Al Mutawakel Street, Block 4, Da'aiyah, Kuwait City, Kuwait
Postal Address: P.O. Box 25281, Safat, Kuwait City, 13113, Kuwait
Tel: 965 256-3025
Fax: 965 256-0173
Email: kwait@international.gc.ca
Website: www.kuwait.gc.ca

Kyrgyz Republic

Consulate of Canada
Address: 189 Moskovskaya Avenue, 720010 Bishkek, Kyrgyz Republic
Tel: 996 (312) 65-05-06
Fax: 996 (312) 65-01-01
Email: canada_honcon@infotel.kg

Laos

Australian Embassy
Address: J. Nehru Street, Phone Xay, Vientiane, Laos
Postal Address: P.O. Box 292, Vientiane, Laos
Tel: 856 (21) 413 600
Fax: 856 (21) 413 601

Latvia

Embassy of Canada
Address: 20/22 Baznicas St, 6th Floor, Riga, Latvia, LV-1010
Tel: 371 781-3945
Fax: 371 781-3960
Email: canembr@bkc.lv
Email: riga@international.gc.ca
Website: www.dfait-maeci.gc.ca/canadaeuropa/baltics

Lebanon

Embassy of Canada
Address: 43 Jal El Dib Highway, 1st Floor, Coolrite Building, Jal El Dib, Beirut, Lebanon
Postal Address: P.O. Box 60163, Jal El Dib, Lebanon
Tel: 961 (4) 713-900
Fax: 961 (4) 710-595
Email: berut@international.gc.ca
Website: www.beirut.gc.ca

Lesotho

Consulate of Canada
Address: c/o Sechaba Consultants, 3 Orpen Road, Maseru, Lesotho
Postal Address: P.O. Box 1191, Maseru, 100, Lesotho
Tel: 266 (22) 316-555
Fax: 266 (22) 310-437
Email: canada@lesoff.co.za

Libya

Embassy of Canada
Address: Great Al-Fateh Tower Building, Tower 1, 7th Floor, Tripoli, Libya
Postal Address: P.O. Box 93392, Al-Fateh Tower Post Office, Tripoli, Libya
Tel: 218 (21) 335-1633
Fax: 218 (21) 335-1630
Email: trpli@international.gc.ca
Website: www.libya.gc.ca

Lithuania

Office of the Embassy of Canada
Address: Jogailos St. 4, 7th Floor, Vilnius 11116
Tel: 370 (5) 249-0950
Fax: 370 (5) 249-7865
Email: vilnius@canada.lt
Website:
www.dfait-maeci.gc.ca/canadaeuropa/baltics/

Luxembourg

Consulate of Canada
Address: 15, rue Guillaume Schneider, L-2522 Luxembourg, G.D. Luxembourg
Tel: 352 2627-0570
Emergency toll-free to Ottawa: 800-23679
Fax: 352 2627-0670
Email: canada@pt.lu
Website: www.canada.lu

Macedonia, Former Yugoslav Republic of

Consulate of Canada
Address: Partizanska odredi 17-a, 1st Floor, 1000 Skopje, Former Yugoslav Republic of Macedonia
Tel: 389 (2) 3225-630
Fax: 389 (2) 3220-596
Email: honcon@unet.com.mk

Madagascar

Consulate of Canada
Address: c/o QIT Madagascar Minerals Ltd., Villa 3H, Lot II J 169, Ivandry, Antananarivo Madagascar
Postal Address: P.O. Box 4003, Antananarivo, 101, Madagascar
Tel: 261 (20) 22-425-59, 22-423-22
Fax: 261 (20) 22-425-06
Email: consulat.canada@wanadoo.mg

Malawi

Consulate of Canada
Address: Accord Centre, M Chipembere Highway, Blantyre-Limbe, Malawi
Postal Address: P.O. Box 51146, Blantyre-Limbe
Tel: 265-(1) 645-441, 641-612, 643-277
Fax: 265-1 643-446
Email: kokhai@malawibiz.com

Malaysia

High Commission of Canada
Address: 17th Floor, Menara Tan & Tan, 207 Jalan Tun Razak, 50400 Kuala Lumpur, Malaysia
Postal Address: P.O. Box 10990, Kuala Lumpur, 50732, Malaysia
Tel: 6 (03) 2718-3333
Fax: 6 (03) 2718-3399
Email: klmpr-cs@international.gc.ca
Website: www.kualalumpur.gc.ca

Consulate of Canada
Address: 3007 Tingkat Perusahaan 5, Prai Industrial Park, 13600 Prai, Penang, Malaysia
Tel: 6 (04) 390-6000
Fax: 6 (04) 390-6000
Email: tyt@lbsb.com

Mali

Embassy of Canada
Address: Immeuble Séméga, Route de Koulikoro, Bamako, Mali
Postal Address: P.O. Box 198, Bamako, Mali
Tel: 223 221-2236
Fax: 223 221-4362
Email: bmako@international.gc.ca
Website: www.bamako.gc.ca

Malta

Consulate of Canada
Address: Demajo House, 103 Archbishop Street, Valletta, Malta
Tel: 356 2552-3233
Fax: 356 2552-3233
Email: canhcon@demajo.com

Mauritania

Consulate of Canada
Address: Centre Commercial Abass, avenue Charles de Gaulle, îlot "O", n 34, 1 étage, bureau 2, Nouakchott, Mauritania
Postal Address: P.O. Box 428, Nouakchott, Mauritania
Tel: (222) 529 26 97/8
Fax: (222) 292-698
Email: j.taya@mr.refer.org

Mauritius

Consulate of Canada
Address: 18 Jules Koenig Street, c/o Blanche Birger Co. Ltd., Port Louis, Mauritius
Postal Address: P.O. Box 209, Port Louis, Mauritius
Tel: 230 212-5500
Fax: 230 208-3391
Email: canada@intnet.mu

Mexico

Consulate of Canada
Address: Centro Comercial Marbella, local 23. Prolongación Farallón s/n, Esq. Miguel Alemán, 39690, Acapulco, Guerrero México
Tel: 52 (744) 484-1305, 481-1349
Emergency toll-free to Ottawa: 001-800-514-0129
Fax: 52 (744) 484-1306
Email: acapulco@canada.org.mx

Consulate of Canada
Address: Plaza Caracol II, 3er piso, Local 330, Boulevard Kukulcán km 8.5, Zona Hotelera, 77500 Cancún, Quintana Roo, Mexico
Tel: 52 (998) 883-3360/1
Fax: 52 (998) 883-3232
Email: cancun@canada.org.mx

Consulate of Canada
Address: Hotel Fiesta Americana, Local 31, Aurelio Aceves 225, Colonia Vallarta Poniente, 44100 Guadalajara, Jalisco, Mexico
Tel: 52 (33) 3615-6270, 6215, 6266, 3616-5642
Emergency toll-free to Ottawa: 001-800-514-0129
Fax: 52 (33) 3615-8665
Email: mxicogjara@international.gc.ca

Consulate of Canada
Address: De la Colina # 567, Fracc. El Cid, Mazatlan, Sinaloa 82110, Mexico
Postal Address: P.O. Box 614, Mazatlán, Sinaloa, 82110, Mexico
Tel: 52 (669) 913-73-20
Fax: 52 (669) 914-66-55
Email: mazatlan@canada.org.mx

Embassy of Canada
Address: Calle Schiller No. 529 (Rincón del Bosque), Colonia Bosque de Chapultepec, 11580, Mexico, D.F., Mexico
Postal Address: P.O. Box 105-05, Mexico, Distrito Federal, 11580, Mexico
Tel: 52 (55) 5724-7900 Ext.: 3322
Tel. after hours: 01-800-706-2900
Fax: 52 (55) 5724-7943
Email: mxico@international.gc.ca
Website: www.mexico.gc.ca

Consulate General of Canada
Address: Edificio Kalos, Piso C-1, Local 108-A, Zaragoza 1300 Sur y Constitución, 64000 Monterrey, Nuevo Léon, Mexico
Tel: 52 (81) 8344-32-00, 27-53,29-06, 29-61, 8345-9105, 9045
Fax: 52 (81) 8344-30-48
Email: mxicomntry@international.gc.ca

Consulate of Canada
Address: Pino Suarez 700, Local 11B, Multiplaza Brena, Colonia Centro, 68000 Oaxaca, Oaxaca
Postal Address: P.O. Box Apartado Postal 29, Succ C,, Oaxaca, 68050 Oaxaca, Mexico
Tel: 52 (951) 513-3777
Fax: 52 (951) 515-2147
Email: oaxaca@canada.org.mx

Consulate of Canada
Address: Edificio Obelisco Local 108, Avenida Francisco Medina Ascencio No. 1951, Zona Hotelera Las Glorias, Puerto Vallarta, Jalisco, Mexico
Postal Address: P.O. Box 48300, Mexico
Tel: 52 (322) 293-0098/9
Fax: 52 (322) 293-2894
Email: vallarta@canada.org.mx

Consulate of Canada
Address: Plaza José Green, Local 9, Boulevard Mijares s/n, Colonia Centro, 23400 San José del Cabo, Baja California Sur, Mexico
Tel: 52 (624) 142-4333
Fax: 52 (624) 142-4262
Email: loscabos@canada.org.mx

Consulate of Canada
Address: Germán Gedovius No. 10411-101, Condominio del Parque, Zona Río, 22320 Tijuana, Baja California Norte, Mexico
Tel: 52 (664) 684-04-61
Fax: 52 (664) 684-03-01
Email: tijuana@canada.org.mx

Micronesia

Australian Embassy
Address: H&E Enterprises Building, Kolonia, Pohnpei, Micronesia
Postal Address: P.O. Box S, Kolonia, Pohnpei, Micronesia
Tel: 691 320-5448
Fax: 691 320-5449

Monaco

Consulate of Canada
Address: Palais de la Scala, 1, avenue Henry Dunant, bureau 1178, 98000 Monaco
Tel: 377 97 70 62 42
Fax: 377 97 70 62 52
Email: consul-canada@monte-carlo.mc

Mongolia

Consulate of Canada
Address: Bodi Tower, 7th floor, Sukhbataar Square, Sukhbaatar District, Ulaanbaatar, Mongolia
Postal Address: P.O. Box 1028, Central Post Office, Ulaanbaatar-13, Mongolia
Tel: 976 (11) 328-285
Fax: 976 (11) 328-289
Email: can_honcon@mongolnet.mn

Morocco

Embassy of Canada
Address: 13 bis, rue Jaafar Assadik, Agdal-Rabat, Morocco
Postal Address: P.O. Box 709, Agdal-Rabat, Morocco
Tel: 212 (37) 68 74 00
Fax: 212 (37) 68 74 30
Email: rabat@international.gc.ca
Website: www.rabat.gc.ca

Mozambique

High Commission of Canada
Address: 1138 Kenneth Kaunda Avenue, Maputo, Mozambique
Postal Address: P.O. Box 1578, Maputo, Mozambique
Tel: 258 (21) 492-623
Fax: 258 (21) 492-667
Email: mputo@international.gc.ca
Website: www.maputo.gc.ca

Namibia

Consulate of Canada
Address: Suite 1118, Sanlam Centre, Independence Ave, Windhoek, Namibia
Tel: 264 (61) 251 254
Fax: 264 (61) 251 686
Email: canada@mweb.com.na

Nepal

Canadian Cooperation Office
Address: Lazimpat, Nepal
Postal Address: P.O. Box 4574, Kathmandu, Nepal
Tel: 977 (1) 4415-193, -389, -391, -861, 4426-885, 4425-669
Fax: 977 (1) 4410-422
Email: cco@canadanepal.org

Netherlands

Embassy of Canada
Address: Sophialaan 7, 2500 GV, The Hague, The Netherlands
Postal Address: P.O. Box 30820, The Hague, Netherlands
Tel: 31 (70) 311-1600
Fax: 31 (70) 311-1620
Email: hague@international.gc.ca
Website: www.canada.nl

New Caledonia

Australian Consulate General
Address: Immeuble Foch, 7th Floor, 19 Avenue du Maréchal Foch, Nouméa, New Caledonia
Postal Address: P.O. Box 22, Nouméa, 98845, New Caledonia
Tel: 687 272-414
Fax: 687 278-001

New Zealand

High Commission of Canada
Address: Level 11, 125 The Terrace, Wellington
Postal Address: P.O. Box 8047, Wellington
Tel: 64 (4) 473-9577
Fax: 64 (4) 471-2082
Email: wlgtn@international.gc.ca
Website: www.wellington.gc.ca

Nicaragua

Office of the Embassy of Canada
Address: De los Pipitos, 2 blocks west 25, Nogal Street,Bolonia, Managua, Nicaragua
Postal Address: P.O. Box 25, Managua, Nicaragua
Tel: 505 (2) 68-0433,-3323
Fax: 505 (2) 68-0437
Email: mngua@international.gc.ca

Niger

Office of the Embassy of Canada
Address: Sonara II Building, 8th Floor, Avenue du Premier Pont, Niamey, Niger
Postal Address: P.O. Box 362, Niamey, Niger
Tel: 227 75-36-86, -87
Fax: 227 75-31-07
Email: niamy@international.gc.ca

Nigeria

High Commission of Canada
Address: 15 Bobo Street, Maitama, Abuja, Nigeria
Tel: 234 (9) 413-9910
Fax: 234 (9) 413-9932
Email: abuja@international.gc.ca
Website: www.nigeria.gc.ca

Deputy High Commission of Canada
Address: 4 Anifowoshe Street, Victoria Island, Lagos
Tel: 234 (1) 262-2512, -2513, -2515
Fax: 234 (1) 262-2517
Email: lagos@international.gc.ca
Website: www.nigeria.gc.ca

Consulate of Canada
Address: 15 Ahoada Street, Rumuibekwe Housing Estate, Port Harcourt, Nigeria
Tel: 234 (8) 461-0434; 461-1601
Fax: 234 (8) 461-0899
Email: phconsul2004@yahoo.com

North Korea

Swedish Embassy
Address: Munsudong, Daehak Street, Taedonggang District, Pyongyang, Democratic People's Republic of Korea
Tel: 850 (2) 381-7908
Fax: 850 (2) 381-7663
Email: ambassaden.pyongyang@foreign.ministry.se
Website: www.sweden.gov.se/sb/d/4189/l/en/pd/4189/e/3647

Norway

Consulate of Canada
Postal Address: P.O. Box 2439, Solheimsviken, 5824, Bergen Norway, Bergen, 5824, Norway
Tel: 47 55-29-71-30
Emergency toll-free to Ottawa: 00-800-2326-6831
Fax: 47 55-29-71-31
Email: honconbergen@canada.no

Embassy of Canada
Address: Wergelandsveien 7, 0244 Oslo, Norway
Tel: 47 22-99-53-00
Emergency toll-free to Ottawa: 00-800-2326-6831
Fax: 47 22-99-53-01
Email: oslo@international.gc.ca
Website: www.canada.no

Pakistan

High Commission of Canada
Address: Diplomatic Enclave, Sector G-5, Islamabad
Postal Address: P.O. Box 1042, Islamabad, Pakistan
Tel: 92 (51) 227-9100, 208-6000
Tel. after hours: 92 (51) 227-9113
Fax: 92 (51) 227-9110
Email: isbad@international.gc.ca
Website: www.islamabad.gc.ca

Consulate of Canada
Address: c/o Beach Luxury Hotel, Room 120, Moulvi Tamiz Uddin Khan Road, Karachi 0227, Pakistan
Tel: 92 (21) 561-0685, 1031/2/3/4/5/6/7
Fax: 92 (21) 561-0673/4
Email: honcon@khi.comsats.net.pk

Consulate of Canada
Address: 102-A Siddiq Trade Centre, 1st floor, 72 Main Boulevard, Gulberg, Lahore, Pakistan
Tel: 92 (42) 578-1763, -1966
Email: canconlhr@yahoo.com
Email: canconlhr@brain.net.pk

Panama

Embassy of Canada
Address: World Trade Center, Calle 53E, Marbella, Galería Comercial, Piso 1, Panama City, Panama
Postal Address: P.O. Box Apartado 0832-2446, Estafeta World Trade Center, Panama City, Panama
Tel: 507 264-9731, 7115
Fax: 507 263-8083
Email: panam@international.gc.ca
Website: www.panama.gc.ca

Papua New Guinea

Australian High Commission
Address: Godwit Road, Waigani, NCD, Port Moresby, Papua New Guinea
Tel: 675 325-9333
Fax: 675 325-9239

Paraguay

Consulate of Canada
Address: Prof. Ramírez No. 3 at Juan de Salazar (between Perú and Padre Pucheu), Asunción, Paraguay
Tel: 595 (21) 227-207
Fax: 595 (21) 227-208
Email: honconpy@telesurf.com.py

Peru

Embassy of Canada
Address: Calle Libertad 130, Miraflores, Lima 18, Peru
Postal Address: P.O. Box 18-1126, Miraflores Post Office, Lima, 18, Peru
Tel: 51 (1) 444-4015
Fax: 51 (1) 242-4050
Email: lima@international.gc.ca
Website: www.lima.gc.ca

Philippines

Consulate of Canada
Address: 45-L Andres Abellana Street, Cebu City 6000, Philippines
Tel: 63 (32) 256-3320
Fax: 63 (32) 255-3068
Email: canada-consulate-cebu@mozcom.com

Embassy of Canada
Address: 6th, 7th, and 8th Floors, RCBC Plaza Tower 2, 6819 Ayala Ave., 1200 Makati City, Manila,
Tel: 63 (2) 857-9000, 857-9001
Fax: 63 (2) 843-1082, 810-4299
Email: manil@international.gc.ca
Website: www.manila.gc.ca

Poland

Embassy of Canada
Address: ul. Matejki 1/5, 00-481 Warsaw, Poland
Tel: 48 (22) 584-3100
Fax: 48 (22) 584-3192, 584-3101
Email: wsaw@international.gc.ca
Website: www.dfait-maeci.gc.ca/warsaw

Portugal

Consulate of Canada
Address: Rua Frei Lourenço de Santa Maria No. 1, 1st Floor, Faro, Portugal
Postal Address: P.O. Box 79, Faro, 8001-957, Portugal
Tel: 351 289-80-3757
Emergency toll-free to Ottawa: 800-819-826
Fax: 351 289-88-0888
Email: consul.faro.canada@mail.net4b.pt

Embassy of Canada
Address: Avenida da Liberdade 196-200, 3rd Floor, 1269-121 Lisbon, Portugal
Tel: 351 21316-4600
Fax: 351 21316-4693
Email: lsbon@international.gc.ca
Website: www.portugal.gc.ca

Consulate of Canada
Address: Rua Antonio Jose de Almeida, No. 27, 1st Floor, 9500-053 Ponta Delgada, Sao Miguel, Azores
Tel: 351 296-281488
Fax: 351 296-281489
Email: canadapdl@mail.telepac.pt

Romania

Embassy of Canada
Address: 1-3, Tuberozelor St. 011411, Bucharest, sector 1, Romania
Postal Address: P.O. Box 117, Post Office No. 22, Bucharest, Romania
Tel: 40 (21) 307-5000
Fax: 40 (21) 307-5010
Email: bucst@international.gc.ca
Website: www.bucharest.gc.ca

Russia

Embassy of Canada
Address: 23 Starokonyushenny Pereulok, Moscow, 119002 Russia
Tel: 7 (495) 105-6000
Fax: 7 (495) 105-6004
Email: mosco@international.gc.ca
Website: www.moscow.gc.ca

Consulate General of Canada
Address: 32B Malodetskoselski Prospekt, St. Petersburg, 198013 Russia
Tel: 7 (812) 325-8448
Fax: 7 (812) 325-8393
Email: spurg@international.gc.ca

Consulate of Canada
Address: 306-46 Verhneportovaya street, Vladivostok 600003, Russia
Tel: 7 (4232) 49-11-88
Fax: 7 (4232) 49-11-88
Email: cbcrfe@mail.ru
Email: cbcrfe@rambler.ru

Rwanda

Office of the Embassy of Canada
Address: 1534 Akagera Street, Kigali, Rwanda
Postal Address: P.O. Box 1177, Kigali, Rwanda
Tel: 250 5 73210
Fax: 250 5 72719
Email: kgali@international.gc.ca

Saint Martin/Sint Maarten

Consulate of Canada
Address: 11A Green Star Shell Road, Dawn Beach, Sint Maarten
Tel: (599) 543-6261
Tel. after hours: (599) 520-5202
Fax: (599) 543-6291
Email: canadacon@caribserve.net

Saint Vincent and the Grenadines

Address:
Emergency toll-free to Ottawa: 1-881-949-9993
See: Barbados

Samoa

Australian High Commission
Address: Fen Gai Ma Leata Building, Beach Road, Tamaligi, Apia, Samoa

Postal Address: P.O. Box 704, Apia, Samoa
Tel: 68 (5) 234-11, 236-13, 252-32
Fax: 68 (5) 231-59

Saudi Arabia

Consulate of Canada
Address: Madina Road, Ali Reza Tower, 11th Floor, Jeddah, Saudi Arabia
Postal Address: P.O. Box 9484, Jeddah, 21413, Saudi Arabia
Tel: 966 (2) 653-0597, 653-0434
Fax: 966 (2) 653-0538
Email: canada.consulate.jeddah@nazergroup.com

Embassy of Canada
Address: Diplomatic Quarter, Riyadh, Saudi Arabia
Postal Address: P.O. Box 94321, Riyadh, 11693, Saudi Arabia
Tel: 966 (1) 488-2288
Fax: 966 (1) 488-1997
Email: ryadh@international.gc.ca

Senegal

Embassy of Canada
Address: corner of Galliéni and Brière-de-l'Isle Streets, Dakar, Senegal
Postal Address: P.O. Box 3373, Dakar, Senegal
Tel: 221 889-47-00
Fax: 221 889-47-20
Email: dakar@international.gc.ca
Website: www.dakar.gc.ca

Serbia

Embassy of Canada
Address: Kneza Milosa 75, 11000 Belgrade, Serbia
Tel: 381 (11) 306-3000
Tel. after hours: 381 (11) 306-3050
Fax: 381 (11) 306-3042
Email: bgrad@international.gc.ca
Website: www.dfait-maeci.gc.ca/canada-europa/serbia/

Singapore

High Commission of Canada
Address: One George Street, #11-01Singapore 049145
Postal Address: P.O. Box 845, Robinson Road, Singapore, 901645, Singapore
Tel: 65 6854-5900
Fax: 65 6854-5912
Email: spore@international.gc.ca
Website: www.singapore.gc.ca

Slovakia

Office of the Embassy of Canada
Address: Carlton Court Yard & Savoy Buildings, Mostova 2, 811 02 Bratislava, Slovakia
Tel: 421 (2) 5920-4031
Fax: 421 (2) 5443-4227
Email: brtsv@international.gc.ca

Slovenia

Consulate of Canada
Address: Maple Leaf, c/o Slovenijales Business Centre, Dunajska 22, 1511 Ljubljana, Slovenia
Tel: 386 (1) 430-3570
Fax: 386 (1) 430-3575
Email: canada.consul.ljubljana@siol.net

Solomon Islands

Australian High Commission
Address: Corner Hibiscus Avenue and Mud Alley, Honiara, Solomon Islands
Tel: 677 21561
Fax: 677 23691

South Africa

High Commission of Canada
Address: 19th Floor, South African Reserve Bank Building, 60 St. George's Mall, Cape Town 8001
Postal Address: P.O. Box 683, Cape Town, 8000, South Africa
Tel: 27 (21) 423-5240
Fax: 27 (21) 423-4893
Email: cptwn@international.gc.ca
Website: www.dfait-maeci.gc.ca/southafrica

Consulate of Canada
Address: 25/27 Cypress Avenue, Stamfordhill, Durban 4001, South Africa
Postal Address: P.O. Box 712, Durban, 4000
Tel: 27 (31) 303-9695
Fax: 27 (31) 303-9694
Email: vnaidu@trematon.co.za
Website: www.dfait-maeci.gc.ca/southafrica

High Comission of Canada Trade Office
Address: Cradock Place, 1st Floor, 10 Arnold Road, off Cradock Avenue, Rosebank, Johannesburg 2196
Postal Address: P.O. Box 1394, Parklands, Johannesburg, 2121, South Africa
Tel: 27 (11) 442-3130
Fax: 27 (11) 442-3325
Email: jobrg@international.gc.ca

High Commission of Canada
Address: 1103 Arcadia Street, Hatfield, Pretoria 0083
Postal Address: P.O. Box Private Bag X13, Hatfield, Pretoria, 0028, South Africa
Tel: 27 (12) 422-3000
Fax: 27 (12) 422-3052
Email: pret@international.gc.ca
Website: www.pretoria.gc.ca

South Korea

Consulate of Canada
Address: c/o Dongsung Chemical Corporation, 472 Shin Pyung-dong, Saha-gu, Busan 604-721, South Korea
Tel: 82 (51) 204-5581
Emergency toll-free to Ottawa: 001-800-2326-6831; 008-800-2326-6831; 002-800-2326-6831
Fax: 82 (51) 204-5580

Embassy of Canada
Address: 10th and 11th Floors, Kolon Building, 45 Mugyo-Dong, Chung-Ku,
Seoul 100-170, Korea
Postal Address: P.O. Box 6299, Seoul, 100-662, South Korea
Tel: 82 (2) 3455-6000
Fax: 82 (2) 755-0686, 3455-6123
Email: seoul@international.gc.ca
Website: www.seoul.gc.ca

Spain

Consulate of Canada
Address: Elisenda de Pinos, 10, 08034 Barcelona,
Tel: 34 (93) 204-2700
Fax: 34 (93) 204-2701
Email: bcncon@sefes.es

Embassy of Canada
Address: Goya Building, 35 Nuñez de Balboa, 28001 Madrid, Spain
Postal Address: P.O. Box 587, Madrid, 28080, Spain
Tel: 34 (91) 423-3250
Emergency toll-free to Ottawa: 00-800-2326-6831
Fax: 34 (91) 423-3251
Email: mdrid@international.gc.ca
Website: www.canada-es.org

Consulate of Canada
Address: Horizonte Building, Plaza de la Malagueta 2, 1st Floor, 29016, Málaga, Spain
Postal Address: P.O. Box 99, Málaga, 29080, Spain
Tel: 34 (95) 222-3346
Emergency toll-free to Ottawa: 00-800-2326-6831
Fax: 34 (95) 222-9533
Email: cancon@microcad.es

Sri Lanka

High Commission of Canada
Address: 6 Gregory's Road, Cinnamon Gardens, Colombo 7, Sri Lanka
Postal Address: P.O. Box 1006, Colombo, 7, Sri Lanka
Tel: 94 (11) 522-6232 (5-CANADA), 532-6232
Fax: 94 (11) 522-6299
Email: clmbo-cs@international.gc.ca
Website: www.srilanka.gc.ca

Sudan

Embassy of Canada
Address: 29 Africa Road, Block 56, Khartoum 1, Sudan
Tel: 249 (183) 56 36 70/72/73/74
Fax: 249 (183) 56 36 71
Email: khrtm@international.gc.ca

Suriname

Consulate of Canada
Address: Wagenwegstraat 50, boven, Paramaribo
Postal Address: P.O. Box 1449, Paramaribo, Suriname
Tel: 59 (7) 424-527, -575
Fax: 59 (7) 425-962
Email: cantim@sr.net

Sweden

Consulate of Canada
Address: Marieholmsgatan 1, 415 02, Gothenburg
Tel: 46 (31) 707 4288
Emergency toll-free to Ottawa: 00-800-2326-6831
Fax: 46 (31) 154 099
Email: got@canadianconsulate.se

Consulate of Canada
Address: Canadian Oil Co. AB, Verkstadsgatan 4, 233 51 Svedala, Sweden
Tel: 46 (40) 40 23 51
Emergency toll-free to Ottawa: 00-800-2326-6831
Fax: 46 (40) 40 31 12

Embassy of Canada
Address: Tegelbacken 4, 7th Floor, Stockholm, Sweden
Postal Address: P.O. Box 16129, Stockholm, 10323, Sweden
Tel: 46 (8) 453-3000

Emergency toll-free to Ottawa: 00-800-2326-6831
Fax: 46 (8) 453-3016
Email: stkhm@international.gc.ca
Website: www.stockholm.gc.ca

Switzerland

Embassy of Canada
Address: Kirchenfeldstrasse 88, 3005 Bern, Switzerland
Postal Address: P.O. Box 3000, Bern, 6, Switzerland
Tel: 41 (31) 357-3200
Fax: 41 (31) 357-3210
Email: bern@international.gc.ca
Website: www.bern.gc.ca

Permanent Mission of Canada to the Office of the United Nations, Consular Section
Address: 5, avenue de l'Ariana, 1202 Geneva, Switzerland
Tel: 41 (22) 919-9200
Fax: 41 (22) 919-9233
Email: genev@international.gc.ca

Syria

Consulate of Canada
Address: Al Sabil Street No. 2, Al Rabat Avenue, Sector 12, Aleppo, Syria
Postal Address: P.O. Box 1250, Aleppo, Syria
Tel: 963 (21) 268-4160
Fax: 963 (21) 268-4100
Email: mhismail@net.sy

Embassy of Canada
Address: Lot 12, Mezzeh Autostrade, Damascus, Syria

Postal Address: P.O. Box 3394, Damascus, Syria
Tel: 963 (11) 611-6692, 611-6851
Fax: 963 (11) 611-4000
Email: dmcus@international.gc.ca
Website: www.damascus.gc.ca

Taiwan

Trade Office of Canada
Address: 13th Floor, 365 Fu Hsing North Road, Taipei, 105, Taiwan,
Tel: 886 (2) 2544-3000
Emergency toll-free to Ottawa: 00-800-2326-6831
Fax: 886 (2) 2544-3590
Email: tapei@international.gc.ca
Website: www.canada.org.tw

Tanzania

High Commission of Canada
Address: 38 Mirambo Street, Corner Garden Avenue, Dar es Salaam, Tanzania
Postal Address: P.O. Box 1022, Dar es Salaam, Tanzania
Tel: 255 (22) 216 3300
Fax: 255 (22) 211 6897
Email: dslam@international.gc.ca
Website: www.daressalaam.gc.ca

Thailand

Embassy of Canada
Address: 15th Floor, Abdulrahim Place, 990 Rama IV, Bangrak, Bangkok 10500, Thailand
Postal Address: P.O. Box 2090, Bangkok, 10501
Tel: 66 (2) 636-0540
Fax: 66 (2) 636-0555
Email: bngkk@international.gc.ca
Website: www.dfait-maeci.gc.ca/bangkok

Consulate of Canada
Address: 151 Super Highway, Tambon Tahsala, Muang, Chiang Mai 50000, Thailand
Tel: 66 (53) 850-147, 242-292
Fax: 66 (53) 850-147, 850-332
Email: cancon@loxinfo.co.th

Timor-Leste (East Timor) - Dili

Australian Embassy
Address: Avenida dos Martires da Patria, Dili, East Timor
Tel: 670 332-2111
Fax: 670 332-2247

Tonga

Australian High Commission
Address: Salote Road, Nuku'Alofa, Tonga
Tel: 676 23-244
Fax: 676 23-243

Trinidad and Tobago

High Commission of Canada
Address: Maple House, 3-3A Sweet Briar Road, St. Clair, Port of Spain, Trinidad and Tobago
Postal Address: P.O. Box 1246, Port of Spain, Trinidad and Tobago
Tel: 1 (868) 622-6232 (6-CANADA)
Emergency toll-free to Ottawa: 1-800-387-3124
Fax: 1 (868) 628-2581

Email: pspan@international.gc.ca
Website: www.portofspain.gc.ca

Tunisia

Embassy of Canada
Address: 3, rue du Sénégal, Place d'Afrique, Tunis-Belvédère, Tunisia
Postal Address: P.O. Box 31, Tunis-Belvédère, 1002, Tunisia
Tel: 216 71-104-000
Fax: 216 71-104-191
Email: tunis@international.gc.ca
Website: www.tunis.gc.ca

Turkey

Embassy of Canada
Address: Cinnah Caddesi No. 58, Çankaya 06690, Ankara, Turkey
Tel: 90 (312) 409-2700
Emergency toll-free to Ottawa: 00800-14-220-0149
Fax: 90 (312) 409-2712
Email: ankra@international.gc.ca
Website: www.dfait-maeci.gc.ca/ankara

Consulate of Canada
Address: Istiklal Caddesi No 373/5, Beyoglu, 80050 Istanbul, Turkey
Tel: 90 (212) 251-9838
Emergency toll-free to Ottawa: 00800-14-220-0149
Fax: 90 (212) 251-9888

Uganda

Consulate of Canada
Address: Jubilee Insurance Centre, 14 Parliament Avenue, Kampala, Uganda
Postal Address: P.O. Box 20115, Kampala, Uganda
Tel: 256 (41) 258-141, 256 (31) 260-511
Fax: 256 (41) 349-484
Email: canada.consulate@utlonline.co.ug

Ukraine

Embassy of Canada
Address: 31 Yaroslaviv Val Street, Kyiv 01901, Ukraine
Tel: 380 (44) 590-3100
Fax: 380 (44) 590-3134
Email: kyiv@international.gc.ca
Website: www.kyiv.gc.ca

Consulate of Canada
Address: 2 Bohomoltsa Street, Suite 4, Lviv 79005, Ukraine
Tel: 380 (322) 97-1772
Fax: 380 (322) 97-8154
Email: oksmyr@link.lviv.ua

United Arab Emirates

Embassy of Canada
Address: Abu Dhabi Trade Towers (Abu Dhabi Mall), West Tower, 9th Floor
Postal Address: P.O. Box 6970, Abu Dhabi, United Arab Emirates
Tel: 971 (2) 694-0300
Emergency toll-free to Ottawa: 800-014-0145
Fax: 971 (2) 694-0399
Email: abdbi@international.gc.ca
Website: www.abudhabi.gc.ca

Consulate of Canada
Address: Bank Street Building, Suite 701, Bur Dubai, United Arab Emirates
Postal Address: P.O. Box 52472, Dubai, United Arab Emirates
Tel: 971 (4) 314-5555
Emergency toll-free to Ottawa: 800-014-0145
Fax: 971 (4) 314-5556/5557
Email: dubai@international.gc.ca

United Kingdom

Consulate of Canada
Address: Unit 3, Ormeau Business Park, 8, Cromac Avenue, Belfast, Northern Ireland BT7 2JA, United Kingdom
Postal Address: P.O. Box 405, Belfast, BT9 5BL, United Kingdom
Tel: 44 (2891) 272060

Consulate of Canada
Address: 55 Colmore Row, Birmingham, B3 2AS, England, U.K.
Tel: 44 (121) 236-6474
Emergency toll-free to Ottawa: 00-800-2326-6831
Fax: 44 (121) 214-1099

Consulate of Canada
Address: c/o St John Cymru Wales, Beignon Close, Ocean Way, Cardiff, Wales, CF24 5PB, United Kingdom
Tel: 44 (0) 2920- 449635
Emergency toll-free to Ottawa: 00-800-2326-6831
Fax: 44 (0) 2920-449645
Email: dan.clayton-jones@talk21.com

Consulate of Canada
Address: Burness, 50 Lothian Road, Festival Square, Edinburgh, United Kingdom, EG3 9WJ
Tel: 44 (131) 473 6320
Emergency toll-free to Ottawa: 00-800-2326-6831
Fax: 44 (131) 473 6321
Email: canada.consul@burness.co.uk

High Commission of Canada
Address: Canada House, Consular Services, Trafalgar Square, London, SW1Y 5BJ, England, U.K.
Tel: 44 (20) 7258-6600
Emergency toll-free to Ottawa: 00-800-2326-6831
Fax: 44 (20) 7258-6533
Email: LDN@international.gc.ca
Website: www.london.gc.ca

United States of America

Consulate General of Canada
Address: 1175 Peachtree Street N.E., 100 Colony Square, Suite 1700, Atlanta, Georgia 30361-6205, U.S.A.
Tel: (404) 532-2000
Emergency toll-free to Ottawa: 1-888-949-9993
Fax: (404) 532-2050
Email: atnta@international.gc.ca
Website: www.atlanta.gc.ca

Consulate General of Canada
Address: 3 Copley Place, Suite 400, Boston, Massachusetts 02116, U.S.A.
Tel: (617) 262-3760
Emergency toll-free to Ottawa: 1-888-949-9993
Fax: (617) 262-3415
Email: bostn@international.gc.ca
Website: www.boston.gc.ca

Consulate General of Canada
Address: 1 HSBC Center, Suite 3000, Buffalo, New York 14203-2884, U.S.A.
Tel: (716) 858-9500
Emergency toll-free to Ottawa: 1-888-949-9993
Fax: (716) 852-4340
Email: bfalo@international.gc.ca
Website: www.buffalo.gc.ca

Consulate General of Canada
Address: Two Prudential Plaza, 180 North Stetson Avenue, Suite 2400, Chicago, Illinois 60601, U.S.A.
Tel: (312) 616-1860
Emergency toll-free to Ottawa: 1-888-949-9993
Fax: (312) 616-1877
Email: chcgo@international.gc.ca
Website: www.chicago.gc.ca

Consulate General of Canada
Address: St. Paul Place, 750 North St. Paul Street, Suite 1700, Dallas, Texas 75201-3247, U.S.A.
Tel: (214) 922-9806
Emergency toll-free to Ottawa: 1-888-949-9993
Fax: (214) 922-9815
Email: dalas@international.gc.ca
Website: www.dallas.gc.ca

Consulate General of Canada
Address: World Trade Center, 1625 Broadway, Suite 2600, Denver, CO 80202, U.S.A.
Tel: (303) 626-0640
Emergency toll-free to Ottawa: 1-888-949-9993
Fax: (303) 572-1158
Email: denvr-g@international.gc.ca
Website: www.denver.gc.ca

Consulate General of Canada
Address: 600 Renaissance Center, Suite 1100, Detroit, Michigan 48243-1798, U.S.A.
Tel: (313) 446-4747
Emergency toll-free to Ottawa: 1-888-949-9993
Fax: (313) 567-2164
Email: dtrot@international.gc.ca
Website: www.detroit.gc.ca

Consulate General of Canada
Address: 550 South Hope Street, 9th Floor, Los Angeles, California 90071-2327, U.S.A.
Tel: (213) 346-2700
Emergency toll-free to Ottawa: 1-888-949-9993
Fax: (213) 620-8827
Email: lngls@international.gc.ca
Website: www.losangeles.gc.ca

Consulate General of Canada
Address: 200 South Biscayne Boulevard, Suite 1600, Miami, Florida 33131, U.S.A.
Tel: (305) 579-1600
Emergency toll-free to Ottawa: 1-888-949-9993
Fax: (305) 374-6774
Email: miami@international.gc.ca
Website: www.miami.gc.ca

Consulate General of Canada
Address: 701 Fourth Avenue South, Suite 900, Minneapolis, Minnesota 55415-1899, U.S.A.
Tel: (612) 333-4641

Emergency toll-free to Ottawa: 1-888-949-9993
Fax: (612) 332-4061
Email: mnpls@international.gc.ca
Website: www.minneapolis.gc.ca

Consulate General of Canada
Address: 1251 Avenue of the Americas, Concourse Level, New York, New York 10020-1175, U.S.A.
Tel: (212) 596-1628
Emergency toll-free to Ottawa: 1-888-949-9993
Fax: (212) 596-1666/1790
Email: cngny@international.gc.ca
Website: www.newyork.gc.ca

Consulate of Canada
Address: Home Mortgage Plaza, 268 Ponce de Leon, Suite 802, San Juan, Puerto Rico 00918, U.S.A.
Tel: (787) 759-6629
Emergency toll-free to Ottawa: 1-866-600-0184
Fax: (787) 294-1205

Consulate of Canada
Address: 402 W.Broadway, 4th Floor, San Diego, CA 92101, U.S.A.
Tel: (619)-615-4286
Emergency toll-free to Ottawa: 1-888-949-9993
Fax: (619) 615-4287
Website: www.dfait-maeci.gc.ca/can-am/san_diego

Consulate General of Canada
Address: 580 California Street, 14th floor, San Francisco, CA 94104, U.S.A.
Tel: (415) 834-3180
Emergency toll-free to Ottawa: 1-888-949-9993
Fax: (415) 834-3189
Email: sfran@international.gc.ca
Website: www.dfait-maeci.gc.ca/san_francisco

Consulate General of Canada
Address: 1501-4th Ave, Suite 600, Seattle, Washington 98101, U.S.A.
Tel: (206) 443-1777
Emergency toll-free to Ottawa: 1-888-949-9993
Fax: (206) 443-9662
Email: seatl@international.gc.ca
Website: www.seattle.gc.ca

Embassy of Canada
Address: 501 Pennsylvania Avenue N.W., Washington, D.C. 20001, U.S.A.
Tel: (202) 682-1740
Emergency toll-free to Ottawa: 1-888-949-9993
Fax: (202) 682-7738
Email: wshdc-outpack@international.gc.ca
Website: www.washington.gc.ca

Uruguay

Embassy of Canada
Address: 749 Plaza Independencia, app.102, Montevideo, Uruguay
Tel: 598 (2) 902-2030
Fax: 598 (2) 902-2029
Email: mvdeo@international.gc.ca
Website: www.montevideo.gc.ca

Uzbekistan

Consulate of Canada
Address: 56, U. Nasir Street, Apt. 39-40, 100100, Tashkent, Uzbekistan
Tel: 998 (71) 362-9205
Fax: 998 (71) 120-7270
Email: antal@rol.uz

Vanuatu

Australian High Commission
Address: KPMG House, Port Vila, Vanuatu
Tel: 678 22-777
Fax: 678 23-948

Venezuela

Embassy of Canada
Address: Consular Section, Avenida Francisco de Miranda con Avenida Sur Altamira, Altamira, Caracas
Postal Address: P.O. Box 62-302, Caracas, 1060-A
Tel: 58 (212) 600-3000, 3042, 3043
Fax: 58 (212) 263-4981
Email: crcas@international.gc.ca
Website: www.caracas.gc.ca

Consulate of Canada
Address: Calle Ortega, Edificio Monseratto, #12-100, apartment 2-B, Porlamar, Estado Nueva Esparta
Tel: 58 (295) 264-1684
Fax: 58 (295) 249-1071

Vietnam

Embassy of Canada
Address: 31 Hung Vuong Street, Hanoi, Vietnam
Tel: 84 (4) 734 5000
Fax: 84 (4) 734 5049
Email: hanoi@international.gc.ca
Website: www.hanoi.gc.ca

Consulate General of Canada
Address: 10th Floor, The Metropolitan, 235 Dong Khoi Street, District 1, Ho Chi Minh City, Vietnam
Tel: 84 (8) 827-9899
Fax: 84 (8) 827-9936
Email: hochi@international.gc.ca

Yemen

Consulate of Canada
Address: Yemen Computer Co. Ltd., Building 4, 11th Street off Haddah Street, Sanaa, Yemen
Postal Address: P.O. Box 340, Sanaa, Yemen
Tel: 967 (1) 20 88 14
Fax: 967 (1) 20 95 23
Email: canconsulye@hotmail.com

Zambia

High Commission of Canada
Address: 5199 United Nations Avenue, Lusaka
Postal Address: P.O. Box 31313, Lusaka, Zambia
Tel: 260 (1) 25 08 33
Fax: 260 (1) 25 41 76
Email: lsaka@international.gc.ca
Website: www.zambia.gc.ca

Zimbabwe

Embassy of Canada
Address: 45 Baines Avenue, Harare, Zimbabwe
Postal Address: P.O. Box 1430, Harare, Zimbabwe
Tel: 263 (4) 252-181/2/3/4/5
Fax: 263 (4) 252-186/7
Email: hrare@international.gc.ca
Website: www.harare.gc.ca

US Embassies

When preparing to teach and travel overseas it is vital to contact the embassy of the country, or the U.S. embassy within the country, you are planning to visit to find out information on visas, restrictions and a variety of other information.

Afghanistan

Embassy
The Great Masoud Road, Kabul
Tel: 93 2 2300436
Fax: 93 2 2301364

Albania

Embassy
Rr Elbasanit 103, Tirana
Tel: +355 4247285
Fax: +355 423 2222

Angola

Embassy
Rua Houari Boumeddienne #32
Luanda, Angola
C.P. 6468
Tel: 244-2-445-481
Fax: 244-2-446-924

Consulate
Rua Major Kanhangulo #132-136
Luanda, Angola
C.P. 6468
Tel: 244-2-371645
Fax: 244-2-390-515
Email: luandaconsular@yahoo.com

Argentina

Embassy
Avenida Colombia 4300
1425 Buenos Aires
Tel: 54-11-5777-4533
Fax: 54-11-5777-4240

Armenia

Embassy
18 Baghramyan Avenue, Yerevan. 375019
Tel: (3741) 52-0791
Fax: (3741) 52-0800
Email: usinfo@uas.am

Australia

Embassy
Moonah Place, Yarralumla, Canberra, ACT 2600
Tel: (02)-6214-5600
Fax: (02) 6214-5970
Email: info@usembassy-australia.state.gov

Sydney Consulate
MLC Centre, Level 59
19-29 Martin Place
Sydney NSW 2000
Tel: (612) 9373-9200

Melbourne Consulate
U.S. Consulate General, 553 St Kilda Road, VIC 3004
Tel: (03) 9526-5900

Perth Consulate- General
16 St. George's Terrace, 13th Floor, Perth, WA 6000
Tel: (08) 9202 1224

Austria

Embassy
Boltzmanngasse 16
A-1090 Vienna
Tel: +43 (1) 31339-0
Fax: +43 (1) 310-0682
Email: embassy@usembassy.at

Consular Agency Salzburg
Alter Markt 1
A-5020 Salzburg
Tel: (0662)848 776
Fax: (0662) 849 777

Azerbaijan

Embassy
83 Azadlyg Prospecti, AZ 1007 Baku
Tel: (994-12) 980 335
Fax: (994-12) 656 671

Bahrain

Embassy
Manama, Bahrain, Bldg 979, Road 3119, Block 321, Zinj.
Tel: (973) 273 300.

Bangladesh

Embassy
Madani Avenue
Dhaka 1212, Bangladesh
Tel: [880] (2) 885 55 00
Fax: [880] (2) 882 37 44
Email: dhaka@pd.state.gov

Barbados

Embassy
P.O. Box 302
Bridgetown, Barbados, W.I.

From the U.S.:
(Department Name)
CMR 1014
APO AA 34055

Tel: (246) 436-4950
Fax: (246) 429-5246
Consular Tel: (246) 431-0225
Consular Fax: (246) 431-0179

Belarus

Embassy
46 Starovilenskaya Street, Minsk 220002
Tel: +375-17 210 1283
Fax: +375-17 234 7853

Belgium

Embassy
Regentlaan 27 / Boulevard du Régent
B-1000 Brussels
Tel: +32-2-508-2111
Fax: +32-2-511-2725

Bolivia

Embassy
Avenida Arce 2780,
Casilla 425
Tel: (591) 2 216 8000
Fax: (591) 2 216 8111
Email: Ipzirc@state.gov

Bosnia & Herzegovina

Embassy
Alipasina 43, 71000 Sarajevo
Tel: +387-33 445-700
Fax: +387-33 659-722
Email: bhopa@state.gov

Botswana

Embassy
PO Box 90 Gaborone
Tel: (267) 395 3982
Fax. (267) 395 6947
Email: consulargaboro@state.gov

Brazil

Rio de Janeiro Consulate
Avenida Presidente Wilson, 147
20030-020 Rio de Janeiro,RJ
Tel.: 55 21 2292-7117
Fax: 55 21 2262-1820

São Paulo Consulate
Rua Henry Dunant, 700
Chacara Santo Antonio
04709-110 São Paulo - SP
Tel: 55 11 5186 7000
Fax: 55 11 5186 7199

Bulgaria

Embassy
Consular Section: 1 Kapitan Andreev Street, Sofia 1421
Tel: (359) (2) 963 2022
Email: irc@usembassy.bg

Burkina Faso

Mailing from the U.S.A.:
AmEmbassy Ouagadougou
Department of State
2440 Ouagadougou Place
Washington DC 20521-2440

International Mailing:
Ambassade des Etats-Unis d'Amérique
01 B.P. 35 Ouagadougou 01
Burkina Faso

Tel: (226) 50-30-67-23
Fax: (226) 50-31-23-68
Email: amembouaga@state.gov

Burundi

Embassy
BP 1720 Ave des Etats-Unis, Bujumbura
Tel: (257) 22-34-54
Fax: (257) 22-29-26

Cambodia

Embassy
16, Street 228 (between streets 51 and 63)
Phnom Penh, Cambodia
Tel: (855-23) 216 436
Fax: (855-23) 216 437

Cameroon

Embassy
Rue Nachtigal
P.O. Box 817, Yaoundé, Cameroon
Tel: (237) 223-40-14
Fax: (237) 223-07-53

Embassy Branch Office in Douala
Tel: (237) 342-53-31; (237) 342-03-03
Fax: (237) 342-77-90

Canada

Embassy
490 Sussex Drive
Ottawa, Ontario K1N 1G8
Tel: 613.238.5335

Consular Service
490 Sussex Drive
Ottawa, Ontario K1N 1G8
Tel: 613 238 5335
Fax: 613 688 3082

Halifax Consulate
1969 Upper Water Street
Purdy's Wharf Tower II, suite 904
Halifax, Nova Scotia, B3J 3R7
Tel: 902 429-2485

Montréal Consulate
1155 St. Alexandre Street, Montréal, Québec, H3B 1Z1
Tel: 514 398-9695

Mailing address:
PO Box 65, Station Desjardins, Montréal, Québec H5B 1G1

Québec City Consulate
2 Place Terrasse Dufferin, B.P. 939
Québec City, Québec, G1R 4T9
Tel: 418 692-2095

Toronto Consulate
360 University Avenue, Toronto, Ontario
M5G 1S4
Tel: : 416 595-1700

Calgary Consulate
615 Macleod Trail SE, Calgary, Alberta
T2G 4T8
Tel: 403 266-8962

Vancouver Consulate
1095 West Pender Street, Vancouver, British Columbia, V6E 2M6
Tel: 604/685-4311

Winnipeg Consulate
860-201 Portage Avenue, Winnipeg, Manitoba
R3B 3K6
Tel: 204 940 1800

Chad

Embassy
BP 413 Ave Felix Eboué, N'Djamena

Tel: (235) 51 70 09
Fax: (235) 51 56 54

Chile

Embassy
Andres Bello 2800, Las Condes,
Santiago, Chile
Tel: (56-2) 238-2600
Fax: (56-2) 330-3710

China

Embassy
3 Xiu Shui Bei Jie, Beijing, China 100600
Tel: (86-10) 6532-3431

Shenyang (Consulate)
No.52, 14 Wei Road, Heping District, Shenyang,
Liaoning, P.R.C 110003
Tel: (86-24) 2382-1198
Fax: (86-24) 2382-2374

Shanghai (Consulate)
1469 Huai Hai Zhong Lu, Shanghai 200031
Tel: (86-20) 6433-6880
Fax: (86-20) 6433-4122

Guangzhou (Consulate)
1 Shamian South Street, Guangzhou 510133
Tel: (86-20) 8121-8000
Fax: (86-20) 8121-9001

Colombia

Embassy
Calle 22D-Bis # 47-51
Bogotá, Colombia
Tel: (571) 315-0811
Fax: (571) 315-2197

Democratic Republic of Congo

Embassy
310 Ave des Aviateurs, Kinshasa
Tel: (243) 81 225 5872
Fax: (243) 88 43467

Costa Rica

Embassy
Calle 120 Avenida O
Pavas, San José, Costa Rica
Tel: (506) 519-2000
Fax: (506) 519-2305

Côte d'Ivoire

Embassy
Rue Jesse Owens,
B.P. 1712,
Abidjan 01, République de Côte d'Ivoire
Tel: (225) 20-21-09-79
Fax: (225) 20-22-32-59

Croatia

Embassy
Thomasa Jeffersona 2, 10010, Zagreb
Tel: 385 (1) 661 2200

Cuba

US Interests Section
U.S. Interests Section of the Embassy of Switzerland,
Havana, Cuba
Calzada between L & M Streets, Vedado, Havana
Tel: (53-7) 33-3551/59
Fax: (53-7) 833 33-1084

Cyprus

Embassy
P.O.Box 24536,
Metochiou & Ploutarchou Street, Engomi, 2407
Nicosia, Cyprus
Tel: 357-22-393939
Fax: 357-22-780944

Czech Republic

Embassy
Trziste 15, 118 01 Praha 1, Czech Republic
Tel: (420-2) 5753-0663

Denmark

Embassy
Dag Hammarskjölds Allé 24,
2100 København Ø.
Tel. +45 35 55 31 44
Fax: +45 35 43 02 23

Dominican Republic

Embassy
César Nicolás Penson esquina Máximo Gomez
Tel: (809) 221-2171
Fax: (809) 685-6959

Ecuador

Embassy
Av. Patria and Av. 12 de Octubre,
Quito, Ecuador
Tel: (593) 22 562 890
Fax: (593) 22 502 052

Egypt

Embassy
5 Latin America Street, Garden City, Cairo, Egypt
Tel: [20] 2-795-7371
Fax: [20] 2-797-3200

El Salvador

Embassy
Final Boulevard Santa Elena, Antiguo Cuscatlán, La Libertad, El Salvador
Tel: (011) [503] 278-4444
Consular Fax: (011) [503] 278-5522

Estonia

Embassy
Kentmanni 20, 15099 Tallinn
Tel: (372) 668 8100
Fax: (372) 668 8134,
Email: USASaatkond@state.gov

Consular Fax: (372) 668 8267
Email: visaTallinn@state.gov

Ethiopia

Embassy
P.O.Box 1014 , Addis Ababa , Ethiopia
Tel. 251-1-174000
Fax: 251-1-174001
Email: usemaddis@state.gov

European Union

U.S. Mission to the European Union
27, Boulevard du Régent, B-1000 Brussels, Belgium
Tel: 32-2-508-2774
Fax: 32-2-512-5720

Finland

Embassy
Itäinen Puistotie 14 B,
FIN-00140, Helsinki
Tel: +358-9-616 250
Email: webmaster@usembassy.fi

France

Embassy
2, ave Gabriel, 75008 Paris
Cedex 08
Tel: (33) 1- 43 12 22 22
Fax: (33) 1- 42 66 97 83

Gabon

Embassy
Blvd. Du Bord de Mer B.P. 4000
Libreville, Gabon
Tel: (241) 76.20.03
Fax: (241) 74.55.07

Georgia

Embassy
25 Atoneli Street, Tbilisi, 0105
Tel: (995-32) 98-99-67
Fax: (995-32) 93-37-59

Germany

Embassy
Neustädtische Kirchstr. 4-5
10117 Berlin
Tel: (49) (030) 8305-0

Düsseldorf (Consulate General)
Willi-Becker-Allee 10, 40227,
Düsseldorf
Tel: (0211) 788 – 8927
Fax: (0211) 788-8938

Frankfurt (Consulate General)
Siesmayerstraße 21
60323 Frankfurt
Tel: (49) (69) 7535-0
Fax: (49) (69) 7535-2277

Hamburg (Consulate General)
Alsterufer 27/28
20354 Hamburg
Tel: (040) 411 71-100
Fax: (040) 4132 7933

Leipzig (Consulate General)
Wilhelm-Seyfferth-Straße 4
04107 Leipzig
Tel: (49) (341) 213-84-0

Munich (Consulate General)
Königinstraße 5

80539 Munich
Tel: (49)(89) 2888-0
Fax: (49)(89) 280-9998

Ghana

Embassy
PO Box 194, Accra
Tel: (233) 21-775348
Fax: (233) 21-776008
Email:accracon@ghana.com

Greece

Embassy
91 Vassilissis Sophias Avenue,
101 60 Athens
Tel: (30-210) 721-2951
Public Affairs Section
8 Makedonon Street, 115 21 Athens,
Tel: (30-210) 363 8114
Fax: (30-210) 364 2986

Thessaloniki (Consulate General)
43 Tsimiski , 7th Floor
546 23 Thessaloniki
Tel: (0030) 2310 242 905
Email: amcongen@compulink.gr

Guatemala

Embassy
Avenida Reforma 7-01, Zona 10
Guatemala Ciudad
Tel: (502) 2331-1541/55
Fax: (502) 2334-8477

Haiti

Embassy
5, Boulevard Harry S Truman
Port-au-Prince, Republic of Haiti
Tel. (509) 222-0200
Fax. (509) 223-9038

Honduras

Embassy
Avenida La Paz
Apartado Postal No. 3453
Tegucigalpa, Honduras
Tel: (504) 236-9320
Fax: (504) 236-9037

Hong Kong

Hong Kong (Consulate-General)
26 Garden Road, Hong Kong
Tel: (852) 2523-9011
Fax: (852) 2845-1598

Hungary

Embassy
Szabadság tér 12., H-1054
Budapest
Tel: (36-1) 475-4400
Fax: (36-1) 475-4764

Iceland

Embassy
Laufásvegur 21, Reykjavik
Tel: : (354) 562-9100
Consular Fax: (354) 562-9110

India

Embassy
Shantipath, Chanakyapuri
New Delhi 110021
Tel:(11) 419-8000
Fax:(11) 419-0017
8:30am-5:30pm Monday-Friday

Calcutta (Consulate)
U.S. Mission Calcutta
5/1 Ho Chi Minh Sarani,
Calcutta 700-071
Tel: 91-33-2282-3611;
Fax: 91-33-2282-2335.

Indonesia

Embassy
Jl. Merdeka Selatan 4-5, Jakarta 10110, Indonesia
Tel: (62-21) 3435 9000
Fax: (62-21) 3436-9922

Ireland

Embassy
42 Elgin Road, Ballsbridge, Dublin 4
Tel: 353-1-668-8777
Fax: 353-1-668-9946

Israel

Embassy
1 Ben Yehuda Street,
Tel Aviv, Israel

Tel: (972) 3 510 3822
Fax: (972) 3 510 3828

Italy

Embassy
Via Vittorio Veneto 119/A - 00187
Roma
Tel: (+39) 06.4674.1
Fax: (+39) 06.4882.672

Milan Consulate General
Via Principe Amedeo 2/10
20121 Milano
Tel: 39 02 290 351
Fax: 39 02 2900 1165

Florence Consulate General
Lungarno Vespucci 38
50123 Firenze
Tel: 39 055 266 951
Fax: 39 055 284 088

Naples Consulate General
Piazza della Republica
80122 Napoli
Tel: 39 081 583 8111
Fax: 39 081 7611 869

Jamaica

Embassy
2 Oxford Road
Kingston 5
Jamaica, West Indies
Tel: (876) 935 6053
Fax: (876) 929 3637
Email: opakgn@state.gov

Japan

Embassy
1-10-5 Akasaka, Minato-ku, Tokyo 107-8420
Tel: (03)3224-5000
Fax: (03)3505-1862

Consulate General Osaka
11-5 Nishitenma 2-chome,
Kita-ku, Osaka 530-8543
Tel: (06)6315-5900
Fax: (06)6315-5914

Consulate General Naha
2564 Nishihara,
Urasoe-shi, Okinawa 901-2101
Tel: (098)876-4211
Fax: (098)876-4243

Consulate General Sapporo
Kita 1-jo, Nishi 28-chome,
Chuo-ku, Sapporo 064-0821
Tel: (011)641-1115
Fax: (011)643-1283

Consulate Fukuoka
5-26 Ohori 2-chome,
Chuo-ku, Fukuoka 810-0052
Tel: (092)751-9331
Fax: (092)713-9222

Consulate Nagoya
Nishiki SIS Bldg. 6F, 10-33, Nishiki
3-chome, Naka-ku,
Nagoya 460-0003
Tel: (052)203-4011
Fax: (052)201-4612

Jordan

Embassy
P.O. Box 354, Amman 11118
Jordan
Tel: 962-6-592-0101
Fax: 962-6-592-0121

Kazakhstan

Embassy
99/97 Furmanova Street at the Corner of Aiteke Bi and Furmanova, Almaty
Tel: +7 (3272) 63-39-21
Fax: +7 (3272) 50-62-69

Kenya

Embassy
United Nations Avenue
P.O. Box 606
00621 Nairobi
Tel: 254-2-3636000
Fax: 254-2-537810

Korea

Embassy
32 Sejongno, Jongno-gu
Seoul 110-710
American Citizen Services Fax: 82-2-397-4101

Kuwait

Embassy
Bayan, Al-Masjed Al-Aqssa St.
PO Box 77, Safat, 13001Kuwait
Tel: 539-5307/8
Fax: 538 0282

Laos

Embassy
19 Rue Bartholonie, Vientiane, Lao P.D.R
Tel: (856) 21-212966
Fax: (856) 21-213045

Latvia

Embassy
7 Raina Blvd.,
Riga LV1510, Latvia
Tel: +371-7036 200
Fax: +371-7820 047

Lebanon

Embassy
Awkar, P.O. Box: 70-840
Beirut – Lebanon
Tel: (04) 542600
Fax: (04) 544136

Lithuania

Embassy
Akmenu 6
2600 Vilnius, Lithuania
Tel.: (370-5) 266 5500
Fax: (370-5) 266 5510
Monday-to-Friday: 8:30 a.m. - 5:30 p.m.

Macedonia

Embassy
Consular Section
Bul.Ilinden B.B.
1000 Skopje
Tel: 389-2 31 16 180
Fax: 389-2 31 17 103

Madagascar

Embassy
14, Rue Rainitovo, Antsahavola, Antananarivo 101
B.P. 620, Antsahavola
Tel: (261-20) 22-212-57
Fax: (261-20) 22-345-39

Malaysia

Embassy
376 Jalan Tun Razak, 50400 Kuala Lumpur.
Tel: 603-2168-5000
Fax: 603-2142-2207

Mali

Embassy
Rue Rochester NY, Bamako
Tel: (223) 222 5470
Fax: (223) 222 3712

Malta

Embassy
Development House, 3rd Floor
St. Anne Street, Floriana, Malta. VLT 01
Tel: (356) 2561 4000
Fax: (356) 2124 3229

Mauritius

Embassy
4th Floor, Rogers House, John Kennedy Ave
Port Louis, Mauritius
Tel: (230) 202 4400
Fax: (230) 208 9534
Email: usembass@intnet.mu

Mexico

Embassy
Paseo de la Reforma 305
Col. Cuauhtémoc
06500 Mexico, D.F.

Mail from the U.S.
American Embassy Mexico
P.O. Box 9000
Brownsville, TX 78520-9000

Tel. From the U.S.: (011-52) 55-5080-2000
Tel. From Mexico: (01) 55-5080-2000
Fax From the U.S.: (011-52) 55-5511-9980
Fax From Mexico: (01) 55-5511-9980

Guadalajara (Consulate)
Progreso 175, Guadalajara, Jalisco C.P. 44100
Tel: (33) 3268 2100
Fax: (33) 3826-6549

Juarez (Consulate)
Av. Lopez Mateos 924 Nte.

Ciudad Juarez, Mexico
U.S. Mailing Address:
P.O. Box 10545
El Paso, Tx. 79995
Tel: (52) (656) 611-3000
Fax: (52) (656) 616-9056

Monterrey (Consulate)
Ave. Constitucion 411 Pte.
Monterrey NL 64000 Mexico
Tel: (8) 345-2120
Office Hours: Monday-Friday 08:00am-5:00pm

Tijuana (Consulate)
Tapachula No. 96
Colonia Hipódromo
22420 Tijuana, B.C., México
Tel: (664) 622-7400
Fax: (664) 622 7625

Micronesia

Embassy
P.O. Box 1286
Kolonia, Pohnpei , FSM 96941
Tel: (691) 320-2187
Fax. (691) 320-2186
Email: usembassy@mail.fm

Moldova

Embassy
103 Mateevici Street,
Chisinau MD-2009
Tel: 373-2-233772
Fax: 373-2-233044

Mongolia

Embassy
P.O. Box 1021
Ulaanbaatar-13
Tel: 976-11-329095
Fax: 976-11-320776

Morocco

Embassy
2 Avenue de Mohamed El Fassi
Rabat, Morocco
Tel: (212)(7)-76-22-65
Fax: (212)(7)-76-56-61

Mozambique

Embassy
Av. Kenneth Kaunda, 193; P.O. Box 783, Maputo
Tel: (258 1) 492797
Fax: (258 1) 490114

Namibia

Embassy
14 Lossen Street, Private Bag 12029,
Windhoek
Tel: 264 61 221601
Fax: 264 61 229792
Email: HealyKC@state.gov

Netherlands

Embassy
Lange Voorhout 102
2514 EJ The Hague
The Netherlands
Tel: +31 70 310-9209
Fax: +31 70 361-4688
Website: www.usemb.nl

Amsterdam (Consulate)
Museumplein 19
1071 DJ Amsterdam
Tel: 020-575 5309
Fax: 020-575 5310

Netherlands Antilles

Consulate-General
J B Gorsiraweg #1, Willemstad AN, Curacao
Netherlands Antilles
Tel: (011) 59-99 461-3066
Fax: (011) 59-99 461-6489
Email: info@amcongencuracao.an

New Zealand

Embassy
29 Fitzherbert Terrace, Thorndon
(Or P.O. Box 1190)
Wellington, New Zealand
Tel: 644 462 6000
Fax: 644 499 0490

Auckland (Consulate)
3rd Floor Citibank Building
23 Customs Street
Auckland, New Zealand
Or Private Bag 92022

Auckland
Tel: (649)303-2724
Fax: (649)366-0870

Nicaragua

Embassy
Apartado Postal 327
Kilometro 4 ½ Carretera Sur
Managua, Nicaragua
Tel: 505-268-0123
Fax: 505-268-9943

Niger

Embassy
BP 11201, rue des Ambassades
Niamey, Niger
West Africa
Tel. (227) 73-31-69
Fax: (227) 73-55-60
Email: NiameyPASN@state.gov

Nigeria

Embassy
7 Mambilla Street, Abuja
Tel: (234) 9 523 0916
Fax: (234) 9 523 0353
Email: usabuja@state.gov

Consulate
2, Walter Carrington Crescent, Victoria Island, Lagos
Tel: (234) 1 261-0050
Fax: (234) 1 261-9856
Email: uslagos@state.gov

Norway

Embassy
Drammensveien 18
0244 Oslo, Norway
Tel: (+47) 22 44 85 50
Email: oslo@usa.no

Oman

Embassy
P.O. Box 202, Code 115
Madinat Al-Sultan Qaboos, Muscat
Sultanate of Oman
Tel: (968) 698-98
Fax: (968) 699-18
Email: aemctira@omantel.net.om

Pakistan

Embassy
Diplomatic Enclave, Ramna 5
Islamabad, Pakistan
Tel: (92) 51-2080-0000
Fax: (92) 51-2276427

Lahore (Consulate)
50, Empress Road
Lahore, Pakistan
Tel: (92) 42-6365530
Fax: (92) 42-6365177

Karachi (Consulate)
8, Abdullah Haroon Road
Karachi, Pakistan
Tel: (92) 21-5685170
Fax: (92) 21-5683089

Peshawar (Consulate)
11, Hospital Road
Peshawar, Pakistan
Tel: (92) 91-279801-3
Fax: (92) 91-276712

Panama

Embassy
Apartado 0816-02561
Zona 5 Panama
Tel: (507) 207 7000
Fax: (507) 227 1964
Email: panamaweb@pd.state.gov

Paraguay

Embassy
1776 Mariscal Lopez Ave.
Asunción
Tel: (595) 21 213-715
Fax: (515) 21 213-728

Peru

Embassy
Avenida La Encalada cdra. 17 s/n
Surco, Lima 33, Peru
Tel: (51-1) 434-3000
Fax: (51-1) 618 2397

Philippines

Embassy
1201 Roxas Boulevard, Ermita 1000, Manila
Tel: (63-2) 528 6300
Fax: (63-2) 522 4361

Poland

Embassy
Al. Ujazdowskie 29-31, 07-00-540 Warsaw
Tel.: +48-22 504 2000
Fax: +48-22 504 2688

Portugal

Embassy
Av. das Forças Armadas
1600-081 LISBOA
Apartado 4258
1507 LISBOA CODEX
Tel: 351-21-727-3300
Fax: 351-21-727-9109

Qatar

Embassy
PO Box 2399
Doha, Qatar
Tel: 488-4101
Fax: 488 4298
Email: usembdoh@qatar.net.qa

Romania

Embassy
Tudor Arghezi, 7-9,
Bucharest, Romania
Tel: (40-21) 210-4042
Fax: (40-21) 210-0395

Russia

Embassy
Bolshoy Deviatinsky Pereulok, 8
Moscow, 121099 Russia
Tel: 7 095 728 5000
Fax: 7 095 738 5090

St. Petersburg (Consulate)
Furshtadtskaya ul., 15
191028 St. Petersburg, Russia
Tel: (812) 331 2600
Fax: (812) 331 2852

Saudi Arabia

Embassy
P.O. Box 94309
Riyadh 11693
Saudi Arabia
Tel: (966-1) 488-3800
Fax: (966-1) 488-7360

Jeddah (Consulate)
P.O. Box 149, Jeddah 21411
Jeddah - Saudi Arabia
Tel: (966-2) 667-0080
Fax: (966-2) 660-2567

Dhahran (Consulate)
P.O. Box 38955, Dhahran Airport 31942
Saudi Arabia
Tel: (966-3) 330-3200
Fax: (966-3) 330-0464

Senegal

Embassy
Ave. Jean XXIII x Rue Kleber
BP. 49 Dakar, Senegal
Tel: (221) 823.42.96
Fax: (221) 822.29.91
Email: usadakar@state.gov

Singapore

Embassy
27 Napier Road
Singapore 258508
Tel: (65)-6476 9100
Fax: (65)-6476 9340

Slovak Republic

Embassy
PO Box 309, 814 99 Bratislava
Tel: +421-5443 3338
Fax: +421-5443 0096

Slovenia

Embassy
Presernova 31
1000 Ljubljana,
Slovenia
Tel: +386 (1) 200-5500
Fax: +386 (1) 200-5555

South Africa

Embassy
PO Box 9536
Pretoria 0001
877 Pretorius St
Pretoria
Tel: (27-12) 431-4000
Fax: (27-12) 342-2299

Spain

Embassy
Serrano, 75, 28006 Madrid, Spain
Tel: (34) 91587 2200
Fax: (34) 91587 2303

Barcelona (Consulate)
Paseo Reina Elisenda de Montcada, 23
08034 Barcelona, Spain
Tel: (34) 93280 2227
Fax: (34) 93280 6175

Sri Lanka

Embassy
210, Galle Road,
Colombo 3, Sri Lanka
Tel: (94-1) 448-007
Fax: (94-1) 437-345

Sweden

Embassy
Dag Hammarskjölds Väg 31,
SE-115 89 Stockholm
Tel: (46) 8 783 5300

Switzerland

Embassy
Jubiläumsstrasse 93, 3001 , Bern
Tel.: 031/357-7011
Fax: 031/357-7344

Tanzania

Embassy
PO Box 9123, 686 Old Bagamoyo Road
Msasani, Dar-es-Salaam
Tel: 255 22 266 8001
Fax: 255 22 266 8238
Email: embassyd@state.gov

Thailand

Embassy
120/22 Wireless Road
Bangkok 10330
Tel: (66) 2 205-4000
Fax: (66) 2 254-1171

Chiang Mai (Consulate)
387 Wichayanond Road,
Chiang Mai 50300, Thailand
Tel: (66-53) 252-629
Fax: (66-53) 252-633

Togo

Embassy
Angle Rue Kouenou et Rue 15 Béniglato
B.P. 852, Lomé
Tel: (228) 221 2994
Fax: (228) 221 79 52

Trinidad & Tobago

Embassy
15 Queen's Park West
Port of Spain, Trinidad & Tobago
Tel: (868) 622-6371/6
Fax: (868) 628-5462

Turkey

Embassy
Public Affairs Section
Paris Caddesi 32 Kavaklidere
06540 Ankara, Turkey
Tel: (90-312) 468-6102
Fax: (90-312) 468-6145

Istanbul (Consulate)
Kaplicalar Mevkii Sokak No 2
Istinye 34460, Istanbul,
Tel: 212 335 9000

Turkmenistan

Embassy
9 1984 Street, Ashgabat, Turkmenistan
744000
Tel: (99312) 35 0045
Fax: (993-12) 39 2614

Ukraine

Embassy
Yuriya Kotsubinskoho 10,
Kyiv 04053, Ukraine.
Tel: (380) 44 490 4000
Fax: (380) 44 490 4085
Consular Fax: (380) 44 216 3393

United Arab Emirates

Embassy
P.O. Box 4009, Abu Dhabi,
United Arab Emirates
Tel: 9712 443 6691

United Kingdom

Embassy
24 Grosvenor Square
London W1A 1AE
United Kingdom
Tel: 020 7499-9000

Uruguay

Embassy
Lauro Muller 1776
Montevideo 11200
Tel: (598-2) 418-7777
Fax: (598-2) 418-8611

Uzbekistan

Embassy
82, Chilanzarskaya Street
Tashkent, 700115, Uzbekistan
Tel:(+998-71)120-54-44
Fax:(+998-71)120-63-35
Email: consul_tashkent@yahoo.com

Venezuela

Embassy
Calle F con Calle Suapure
Urb. Colinas de Valle Arriba
Caracas 1080, Venezuela
Tel: (58-212) 975-6411
Email: consularcaracas@state.gov

Vietnam

Embassy
7 Lang Ha Street
Ba Dinh District
Hanoi, Vietnam
Tel: (844) 7721 500
Fax: (844) 7721 510
Email: hanoicons@fpt.vn

Ho Chi Minh City (Consulate)
4 Le Duan Blvd., District 1,
Ho Chi Minh City
Tel: (84-8) 822 9433
Fax: (84-8) 824 5571

Yemen

Embassy
Sa'awan Street
PO Box 22347, Sana'a, Yemen
Tel: 967-1-303155

Zambia

Embassy
Public Affairs Section
COMESA Building, Ben Bella Road,
PO Box 32053
Lusaka
Tel: (260-1) 227 993
Fax: (260-1) 226 523
Email: paslib@zamnet.zm

Zimbabwe

Embassy
172 Herbert Chitepo Avenue, Harare
Tel: 263- 4-703169/378/478
Fax: 263- 4-796488
Email: consularharare@state.gov

Index

PASCAL
PROGRAMS IN SCIENCE AND ENGINEERING

PREFACE

This book is meant to serve two purposes. First, it is an instant software library which contains 112 Pascal programs that can be recalled whenever needed. The programs are primarily related to mathematics and engineering, especially electrical engineering. They are the result of hundreds of hours of work, and we invite you to take full advantage of them.

The second purpose of this book is to serve as a guide to the newer Pascal programmer by demonstrating various techniques for programming in Pascal. In many cases similar tasks are programmed in several ways to show that there is more than one correct way to write a program. In addition, the programs are modular and therefore can be easily incorporated into your own programs.

Programs that design circuits will require some additional documentation, such as circuit diagrams, but most programs are meant to stand alone. Information not included in the programs is usually included in the writeup given with each program listing.

All of the programs in this book were written and tested on the Apple II Plus computer with Apple UCSD Pascal. This makes most of the programs very portable. For cases in which the program states "Uses Transcend," check your version of Pascal to see what should be used.

All programs start with the same header. This asks:

S(creen or P(rinter?

Entering P or p will send only the output report to the printer. All other output, including program prompts, will go to the screen.

All of the programs are set up so that they can be completely rerun. This feature is incorporated to accommodate the slowness of the Pascal operating system. If you enter some data and decide to save the results, the program can be run again, and the option of screen or printer can be changed.

Variables are named to make the programming as understandable as possible.

In the sample runs, if more than one set of data is used—as it is when the answer is "Y" to the question "Another set of data? (Y/N)"—the above header and the program title would normally be repeated. To spare you this repetition, however, the header is not shown in the examples and the title is displayed only once.

The listings are set up to work on the normal 40-column screen of the Apple II Plus. If you are so equipped, during key-in this can be easily changed to 80 columns.

Happy computing.

JULES H. GILDER and J. SCOTT BARRUS

EQUIPMENT NEEDED

To use the programs in this book on an Apple computer, you will need the following equipment.

Apple II Plus with 64K RAM or 48K RAM plus Language Card

Apple Pascal Language system

Apple Disk Drive (2 recommended)

Display monitor

Printer (optional)

Because the programs in this book were prepared on an Apple II computer, they are likely to run unmodified only on the same equipment. If you own a different system, you should be able to make use of the programs, but you should also expect to have to make changes to the listings here before they can be compiled and run on your computer.

At the time of this writing, UCSD Pascal has been implemented on all of the following processors. (Source: Softech Microsystems.)

6502	68000	9900
PDP-11	LSI-11	6809
Z80	8080	8085
8086	8088	

If your computer uses one of the above processors, chances are there is a preconfigured version of UCSD Pascal available.

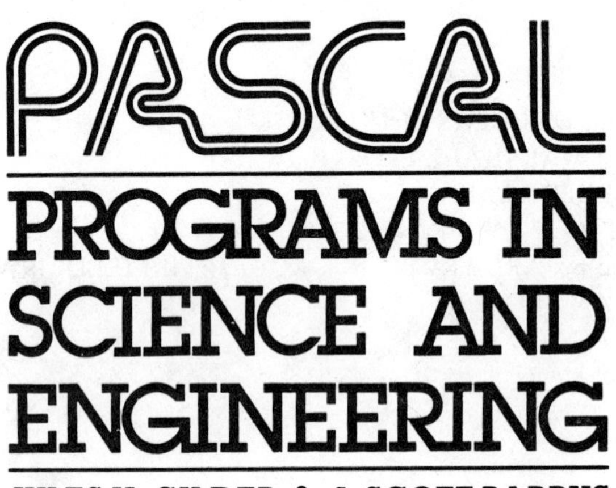
PASCAL
PROGRAMS IN SCIENCE AND ENGINEERING
JULES H. GILDER & J. SCOTT BARRUS

HAYDEN BOOK COMPANY, INC.
Rochelle Park, New Jersey

Acquisitions Editor: DOUGLAS McCORMICK
Production Editor: MARSHALL E. OSTROW
Art Director: JIM BERNARD
Compositor: ART, COPY, & PRINT COMPANY
Printed and bound by: ARCATA BOOK GROUP/FAIRFIELD GRAPHICS DIVISION
Text design: SUSAN BROREIN
Cover design: DELGADO DESIGN
Cover photo: KEN KARP

Library of Congress Cataloging in Publication Data

Gilder, Jules H., 1947-
Pascal programs in science and engineering.

Includes bibliographical references.
1. PASCAL (Computer program language) 2. Mathematics—Computer programs.
3. Electrical engineering—Computer programs. 4. Engineering mathematics—
Computer programs. I. Barrus, J. Scott. II. Title. QA76.73.P2G54 1983
510'.28'5425 83-12648 ISBN 0-8104-6265-6

Printed in the United States of America

1	2	3	4	5	6	7	8	9	PRINTING
83	84	85	86	87	88	89	90	91	YEAR

CONTENTS

1 GENERAL MATHEMATICS

Most home computers come with a wide variety of built-in mathematical functions. There are, however, many occasions when these are not enough and more capabilities are needed. The programs in this chapter are in part designed to extend your machine's capabilities.

In some of the programs (e.g., extended factorial and improved combinations) the computer is programmed to handle numbers whose size would ordinarily cause an overflow. By carefully analyzing this programming technique, it is possible to learn how it can be used in other applications.

In addition to the probability-related programs (1.1 to 1.5), this section also shows how the special mathematics of vectors can be handled (1.6 and 1.7). Pascal has both Log (base 10) and Ln (natural log) capabilities. A simple program (1.8) will provide the capability to work with logs of any base.

Two programs in this chapter deal with coordinate systems. The first (1.9) provides the capability of translating and/or rotating a given coordinate system. This is handy if you want to convert data or measurements from one frame of reference to another. The second coordinate-related program is 1.10. This one converts data from Cartesian (rectangular) coordinates to polar coordinates, and vice versa.

The next program is one that works on numbers. Program 1.11 finds the maximum and minimum of a given set of numbers, an absolute necessity if any data plotting is to be done.

If you've ever had to draw an object on a piece of paper accurately, you will quickly realize the value of the dimension scaling program (1.12). And if you want to do some plotting, the histogram program (1.13) produces attractive output. As given, it is used to plot functions. With very little effort, however, it can be used to plot individual data points. Try to do this by combining 1.13 with the min/max locator program.

Did you ever have three points through which you wanted to draw a circle? Try to do it. Draw three points on a piece of graph paper and try to draw a circle that will pass through all of them. It is not easy. Now enter the location of those three points into program 1.14. It will tell you the radius of the circle and where the center is located. The remaining programs produce normally distributed numbers, find any root of any number, and convert any number into a rational fraction.

1.1 FACTORIAL

The factorial of a number is an operation frequently used in probability calculations. It is arrived at by evaluating:

$$n! = n(n-1)(n-2)\ldots 1$$
$$5! = 5(4)(3)(2)(1) = 120$$

The notation n! is often used to designate the factorial operation and is read "n factorial." There is one special case, for 0!, which is defined as being equal to 1. The factorial calculation is carried out in *Procedure Fact.* The variable "factor" is initialized to one. If this were not done, Apple Pascal could initialize the number to anything. The program loops through the calculation *factor:= factor* ✱ *variable* through *number* cycles. If *number* were equal to zero, the program would loop through once and *factor* would be equal to 1.

```
(*=========================================*)
(*                                         *)
(* Title:   Factorial                      *)
(* Program Summary: Program computes       *)
(*          the factorial of a given       *)
(*          number.                        *)
(*                                         *)
(*=========================================*)


Program FACTORIAL;

 Var
   number,
   variable                 : Integer;
   factor                   : Integer[35];
   answer                   : Char;
   out                      : Text;
   cont                     : Char;

 Procedure Outputs;
   begin
      page(output);
      repeat
         gotoxy(Ø,6);
            Write('S(creen or P(rinter? ');
            readln(answer);
      until answer in ['s','S','p','P'];
   end; { outputs }
```

```
Procedure Title;
  begin
     page(output);

     Gotoxy(15,5);
     Writeln('Factorial');
     Writeln;Writeln;
     Writeln('This program computes the factorial of');
     Writeln('a given number. Operation is limited ');
     Writeln('by the largest number your computer can');
     Writeln('handle.  Only integers are valid.');
  end; { title }

Procedure Fact;
  begin
     factor:=1;
       for variable:= 1 to number do
          factor:= factor * variable;
  end; { factor }


Procedure Input;
  begin
     Write('Enter number: ');
     Readln(number);
  end; { input }

Procedure Printout;
  begin
     case answer of
       's','S' :  begin
                    Writeln(number,' ! = ',factor);
                  end;
       'p','P' :  begin
                    rewrite(out,'printer:');
                    Writeln(out,number,' ! = ',factor);
                    Writeln(number,' ! = ',factor);
                    close(out);
                  end;
     end; { case of }
  end; { printout }


 begin { main program }
    Repeat
      Outputs;
        Title;
           Writeln;Writeln;
        Input;
           Writeln;Writeln;
        Fact;
        Printout;
      Writeln;Writeln;
      Write('Do you have another number?(Y/N) ');
      Readln(cont);
    until cont in ['n','N'];
 end.
```

Factorial

This program computes the factorial of
a given number. Operation is limited
by the largest number your computer can
handle. Only integers are valid.

Enter number: 9

9 ! = 36288Ø

Do you have another number? (Y/N) Y

Enter number: 35

(system overflow error)

1.2 EXTENDED FACTORIAL[1]

A major problem with the previous program is that the limit of the computer's number-handling ability is quickly reached. Most home computers become overloaded with a calculation whose results produce a number larger than 10^{39}. This program contains a separate routine that factors the answer into powers of 10 so that much greater capability is offered (last part of *Procedure Fact*).

As in the previous program, the factorial is calculated in *Procedure Fact*. Thus numbers such as 121! can easily be evaluated, whereas before they would have overloaded the computer.

```
(*==========================================*)
(*                                          *)
(* Title:   Extended Factorial              *)
(* Program Summary: Program computes        *)
(*          the factorial of a given        *)
(*          number.  This program is        *)
(*          like Factorial, but it will     *)
(*          accept larger numbers.          *)
(*                                          *)
(*==========================================*)


Program EXTENDFACTORIAL;

 Var
   expo,
   variable,
   number                   : Integer;
   factor                   : Real;
   answer                   : Char;
   out                      : Text;
   cont                     : Char;

 Procedure Outputs;
   begin
     page(output);
     repeat
       gotoxy(Ø,6);
       Write('S(creen or P(rinter? ');
       Readln(answer);
     until answer in ['s','S','p','P'];
   end; { outputs }

 Procedure Title;
   begin
     page(output);
      gotoxy(11,5);
      Writeln(' Extended Factorial ');
      Writeln;Writeln;
      Writeln('This program will perform factorial');
      Writeln('calculations whose results are much');
      Writeln('larger than 1Ø^39, a value that ');
      Writeln('overloads most home computers. Just');
      Writeln('enter the number when asked.');
   end; { title }
```

[1]Adapted from W. M. Bunker, "Computer Program Extends Computation of Factorials," *Electronic Design*, (April 1971), p. 86.

```
Procedure Fact;
  begin
    expo:=0;
    factor:=1;
       for variable:=1 to number do
           begin
             factor:= factor*variable;
               if factor >=1E5 then
                  begin
                     factor:= factor*1E-5;
                     expo:=expo+5;
                  end;
           end;
  end; { fact }

Procedure Input;
  begin
    Write('Enter number : ');
    Readln(number);
  end; { input }

Procedure Printout;
  begin
    case answer of
      's','S' : begin
                  Writeln(number,' ! = ',factor,' times');
                  Writeln;
                  Writeln('10 to the ',expo,' power.');
                end;
      'p','P' : begin
                  Rewrite(out,'printer:');
                  Writeln(out,number,' ! = ',factor,'times');
                  Writeln(number,' ! = ',factor,' times');
                  Writeln(out);Writeln;
                  Writeln(out,'10 to the ',expo,' power.');
                  Writeln('10 to the ',expo,' power.');
                  Close(out);
                end;
    end; { case of }
  end; { printout }

begin { main }
   Repeat
     Outputs;
       Title;
         Writeln;Writeln;
       Input;
         Writeln;Writeln;
       Fact;
       Printout;
         Writeln;Writeln;
         Write('Do you have another number?(Y/N) ');
         Readln(cont);
   until cont in ['n','N'];
end.
```

Extended Factorial

This program will perform factorial calculations whose results are much larger than 10 39, a value that overloads most home computers. Just enter the number when asked.

Enter number: 35

35 ! = 1.03331 times
10 to the 40 power.

Do you have another number? (Y/N) Y

Enter number: 121

121 ! = 8.Ø9429 times
1Ø to the 2ØØ power.

Do you have another number? (Y/N) N

1.3 PERMUTATIONS

In some probability problems, it is often necessary to calculate the number of ways that n objects can be arranged in order of groups of r objects. In figuring this out, we see that there are n ways of choosing the first object, n–1 ways of choosing the second object, and so forth. This continues until the last object of the group is reached. For this object, there are n–r+1 ways of choosing it. The different ways that these objects can be arranged are called permutations (P). Mathematically, the number of permutations can be calculated by multiplying the number of ways each object can be chosen by the next.

$$P = n(n - 1)(n - 2)(n - 3) \ldots (n - r + 1)$$

This expression can be written more simply as:

$$P = n!/(n - r)!$$

The term n! is called "n factorial" and is equal to $n(n - 1)(n - 2) \ldots 1$ so 3! equals 3(2)(1) or 6.

The program uses the factorial approach to calculate the number of permutations. The factorials are calculated in *Procedure Factorial.* After both factorials are calculated, the larger is divided by the smaller to get the final result.

The program contains two error-detection tests, both of which are located in *Procedure Error.* These tests will determine if an invalid input entry is made.

```
(*==========================================*)
(*                                          *)
(* Title:   Permutations                    *)
(* Program Summary: Program computes        *)
(*          the permutations of "N"         *)
(*          objects taken "R" at a          *)
(*          time.                           *)
(*                                          *)
(*==========================================*)


Program PERMUTATIONS;

 Var
   Answer,
   Cont                   : Char;
   out                    : Text;
   Variable,
   NumObjects,
   NumGroups,
   Temp                   : Integer;
```

```
  Permut,
  FactObject,
  FactGroup,
  Factor                    : Integer[35];


Procedure Outputs;
  begin
    page(output);
    repeat
      gotoxy(Ø,5);
      Write('S(creen or P(rinter? ');
      Readln(answer);
    until answer in ['s','S','p','P'];
  end; { outputs }

Procedure Title;
  begin
    page(output);
    gotoxy(14,5);
    Writeln('Permutations');
    Writeln;Writeln;
    Writeln('This program computes the permutations');
    Writeln('of "N" objects taken "R" at a time');
  end; { title }

Procedure Input;
  begin
    Write('Enter number of objects: ');
    Readln(NumObjects);
    Write('Enter number per group: ');
    Readln(NumGroups);
  end; { input }

Procedure Error;
  begin
    if NumObjects< NumGroups then
       begin
         Writeln(chr(7),'Input error! The number of objects is less');
         Writeln('than the number of groups.');
         Input;
       end;
    if Numgroups < Ø then
       begin
         Writeln(chr(7),'Input error! The number of groups is less ');
         Writeln('than zero.');
         Input;
       end;
  end; { error }

Procedure Factorial;
  begin
    factor:=1;
      for variable:= 1 to temp do
        factor:= factor*variable;
  end; { Factorial }


Procedure Printout;
 begin
  case answer of
  's','S' : begin
              Writeln('There are ',permut,' permutations of');
              Writeln;
              Writeln(NumObjects,' objects taken ',Numgroups,' at a time.');
              end;
  'p','P' : begin
              Rewrite(out,'printer:');
              Writeln(out,'There are ',permut,' permutations of');
              Writeln('There are ',permut,' permutations of');
              Writeln(out);Writeln;
              Writeln(out,NumObjects,' objects taken ',NumGroups,' at a time.');
              Writeln(NumObjects,' objects taken ',NumGroups,' at a time.');
              Close(out);
              end;
  end; { case of }
end; { printout }
```

```
begin { main }
  Repeat
     Outputs;
        Title;
          Writeln;Writeln;
        Input;
        Error;
          Writeln;Writeln;
            temp := NumObjects;
          Factorial;
            FactObjects := Factor;
            temp := NumObjects - NumGroups;
          Factorial;
            FactGroup := Factor;
              Permut := FactObject DIV FactGroup;
          Writeln;Writeln;
        Printout;
          Writeln;Writeln;
          Write('Do you have another permutation? (Y/N) ');
          Readln(cont);
  until cont in ['n','N'];
end.
```

```
Permutations

This program computes the permutations
of "N" objects taken "R" at a time.

Enter number of objects: 10
Enter number per group: 3

There are 720 permutations of
10 objects taken 3 at a time.

Do you have another permutation? (Y/N) N
```

1.4 COMBINATIONS

In the permutations program, the number of ways n objects could be arranged in order into groups of r was calculated. Sometimes the order of the objects in a group is not important—for example, the number of ways 12 people can be divided into groups of three. Another application is found in the error-detection and correction codes for communication systems.

The number of ways items can be grouped into groups of r objects, where order is not important, is called the "combination."

The combination of 12 objects taken three at a time is equal to

$$C = 12!/(3!(12 - 3)!) = 220$$

The generalized formula for calculating combinations is

$$C = n!/(r!(n - r)!)$$

The difference between this calculation and the one for permutations is the extra r! term in the denominator.

Once again, the factorial calculation is carried out by *Procedure Factorial.* The

results of the three calls of this routine are used to calculate the combinations. As in the permutations program, error-detection tests are included.

```
(*========================================*)
(*                                        *)
(* Title:  Combinations                   *)
(* Program Summary: Program computes      *)
(*         the combination of "N"         *)
(*         objects taken "R" at a         *)
(*         time.                          *)
(*                                        *)
(*========================================*)


Program Combinations;


  var
    answer,
    cont              : Char;
    out               : Text;
    NumObjects,
    NumGroups,
    Temp,
    Variable          : Integer;
    FactObject,
    FactGroup,
    FactDiff,
    Combine,
    Factor            : Integer[35];

  Procedure Outputs;
    begin
       page(output);
        repeat
          gotoxy(Ø,5);
          Write('S(creen or P(rinter ? ');
          Readln(answer);
        until answer in ['s','S','p','P'];
    end; { outputs }

  Procedure Title;
    begin
       page(output);
        gotoxy(14,5);
        Writeln('Combinations');
        Writeln;Writeln;
        Writeln('This program computes the combinations');
        Writeln('of "N" objects taken "R" at a time');
    end; { title }

  Procedure Factorial;
    begin
       factor:= 1;
        for variable := 1 to temp do
          factor := factor * variable
    end; { factorial }

  Procedure Input;
    begin
       Write('Enter number of objects: ');
       Readln(NumObjects);
       Write('Enter number of groups: ');
       Readln(NumGroups);
    end; { input }

  Procedure Error;
    begin
       if NumObjects < NumGroups then
           begin
             Writeln(chr(7),'Input Error! Number of objects less');
             Writeln('than number of groups');
             Input;
           end;
```

```
       if NumGroups < 0 then
           begin
             Writeln(chr(7),'Input Error! Number of groups less');
             Writeln('then zero');
             Input;
           end;
    end; { error }

  Procedure Printout;
   begin
   case answer of
   's','S' : begin
             Writeln('There are ',combine,' combinations of');
             Writeln;
             Writeln(NumObjects,' objects taken ',NumGroups,' at a time');
             end;
   'p','P' : begin
             Rewrite(out,'printer:');
             Writeln(out,'There are ',combine,' combinations of');
             Writeln('There are ',combine,' combinations of ');
             Writeln(out);Writeln;
             Writeln(out,NumObjects,' objects taken ',NumGroups,' at a time');
             Writeln(NumObjects,' objects taken ',NumGroups,' at a time');
             Close(out);
             end;
   end; { case of }
  end; { printout }

begin  { main }
   Repeat
      Outputs;
        Title;
          Writeln;Writeln;
        Input;
        Error;
          Writeln;Writeln;
           Temp := NumObjects;
            Factorial;
              FactObject := Factor;
           Temp := NumObjects - NumGroups;
            Factorial;
              FactDiff := Factor;
           Temp := NumGroups;
            Factorial;
              FactGroup := Factor;
           Combine := FactObjects DIV FactDiff DIV FactGroup;
        Printout;
          Writeln;Writeln;
          Write('Do you have another combination?(Y/N)');
          Readln(cont);
    until cont in ['n','N'];
 end.
```

Combinations

This program computes the combinations
of "N" objects taken "R" at a time.

Enter number of objects: 12
Enter number of groups: 3

There are 220 combinations of
12 objects taken 3 at a time.

Do you have another combination? (Y/N) N

1.5 IMPROVED COMBINATIONS[1]

The problem with the previous program is that it is limited in the magnitude of the numbers used to calculate combinations. If the number of objects is greater than or equal to 34, the computer will indicate that the number is too large and give an overflow error message.

This program uses a method of calculation that prevents the overflow problem, unless the final answer itself is too large for the computer. First, the program determines if the number of objects n is less than twice the number of objects per group k. If it is, the program sets I = 1 and calculates:

$$\binom{k+I}{k} = \binom{k+1}{k} = k + 1$$

Then for I = 2,3,4,5, . . . ,n − k, the following expression is recursively computed:

$$\binom{k+I}{k} = \binom{k+I-1}{k} \left(\frac{k+I}{I}\right)$$

If n is equal to or greater than twice k, the following recursive calculation is made for values of j = 2,3,4,. . .,k :

$$\binom{n}{j} = \left(\frac{n}{j-1}\right) \left(\frac{n+1-j}{j}\right)$$

For the case where k is equal to 1, the combination is simply n.

In the following listing, the variable *Objects* is n, and *Groups* is k as described above.

```
(*==========================================*)
(*                                          *)
(* Title:    Improved Combinations          *)
(* Program Summary: Program computes        *)
(*           combinations without           *)
(*           causing an overflow,           *)
(*           unless the final answer        *)
(*           itself is too large for        *)
(*           the computer ("N" as large     *)
(*           as 120 can be used).  It       *)
(*           computes the combinations      *)
(*           of N objects taken R at a      *)
(*           time.                          *)
(*                                          *)
(*==========================================*)

Program Improved_Combinations;

  var
    answer,
    cont               : Char;
    out                : Text;
    Objects,
```

[1]Adapted from R. Lambert, "Perform Large-Value Computations on a Computer," *Electronic Design.*

```
    Groups,
    Stepper,
    Difference       : Integer;
    Factorial,
    Combine,
    Combination      : Real;

  Procedure Outputs;
    begin
       page(output);
        repeat
          gotoxy(0,5);
          Write('S(creen or P(rinter ? ');
          Readln(answer);
        until answer in ['s','S','p','P'];
    end; { outputs }

  Procedure Title;
    begin
       page(output);
        gotoxy(9,5);
        Writeln('Improved Combinations');
        Writeln;Writeln;
        Writeln('This program will calculate combinations');
        Writeln('without causing an overflow unless');
        Writeln('the final answer is too big for');
        Writeln('the computer.  "N" as large as 120');
        Writeln('can be used.');
    end; { title }

  Procedure Factorlo;
    begin
        Combination := Factorial * ( Groups + Stepper )/ Stepper;
        Factorial := Combination;
    end; { factorlo }

  Procedure Factorhi;
    begin
        Combination := Factorial * (( Objects +1 - Stepper)/ Stepper);
        Factorial   := Combination;
    end; { factorhi }


  Procedure Input;
    begin
       Write('Enter number of objects: ');
       Readln(Objects);
       Write('Enter number of groups: ');
       Readln(Groups);
    end; { input }


  Procedure Printout;
   begin
   case answer of
   's','S' : begin
             Writeln('There are ',combine,' combinations of');
             Writeln;
             Writeln(Objects,' objects taken ',Groups,' at a time');
             end;
   'p','P' : begin
             Rewrite(out,'printer:');
             Writeln(out,'There are ',combine,' combinations of');
             Writeln('There are ',combine,' combinations of ');
             Writeln(out);Writeln;
             Writeln(out,Objects,' objects taken ',Groups,' at a time');
             Writeln(Objects,' objects taken ',Groups,' at a time');
             Close(out);
             end;
    end; { case of }
  end; { printout }

begin  { main }
   Repeat
      Outputs;
        Title;
          Writeln;Writeln;
```

```
      Input;
        Writeln;Writeln;
      If Objects  < 2* Groups then
        begin
          Difference := Objects - Groups;
            factorial := Groups + 1;
         for Stepper := 2 to Difference do
              factorlo;
            combine:= combination;
        end { if }
       else
        begin
          factorial := Objects;
         for Stepper := 2 to Groups do
              factorhi;
            combine := combination;
         end; { else }
            Printout;
        Writeln;Writeln;
        Write('Do you have another combination?(Y/N)');
        Readln(cont);
   until cont in ['n','N'];
end.
```

Improved Combinations

This program will calculate combinations
without causing an overflow unless
the final answer is too big for the
computer. "N" as large as 12Ø
can be used.

Enter number of objects: 12Ø
Enter number of groups: 23

There are 2.69ØØ3E24 combinations of
12Ø objects taken 23 at a time.

Do you have another combination? (Y/N) N

1.6 SCALAR PRODUCT OF TWO VECTORS

The scalar product of two vectors, also called the "dot product," is defined as the product of the magnitude of the two vectors multiplied by the cosine of the angle between them. The result of this calculation is a scalar (not vector) quantity.

While this approach can be used to calculate the scalar product, a much simpler method is to use the following formula:

$$A * B = (XA)(XB) + (YA)(YB) + (ZA)(ZB)$$

where

$$A = XA + YA + ZA$$
$$B = XB + YB + ZB$$

This law is used to perform the calculation in *Procedure Product* where A is *Vect1* and B is *Vect2*. An interesting feature of the scalar product is that if neither of the

two vectors is zero but the scalar product is zero, the two vectors are perpendicular to one another.

```
(*=========================================*)
(*                                         *)
(* Title:  Scalar Product of Two Vectors   *)
(* Program Summary: Program computes       *)
(*         the scalar product of two       *)
(*         vectors.  This is defined as    *)
(*         the product of the magnitude    *)
(*         of two vectors multipled by     *)
(*         the cosine of the angle         *)
(*         between them.                   *)
(*                                         *)
(*=========================================*)


Program Scalar_Products;


  var
    answer,
    cont             : Char;
    out              : Text;
    Vect1X,
    Vect1Y,
    Vect1Z,
    Vect2X,
    Vect2Y,
    Vect2Z           : Integer;
    Scalar           : Integer[35];

  Procedure Outputs;
    begin
       page(output);
        repeat
          gotoxy(Ø,5);
          Write('S(creen or P(rinter ? ');
          Readln(answer);
        until answer in ['s','S','p','P'];
    end; { outputs }

  Procedure Title;
    begin
       page(output);
        gotoxy(13,5);
        Writeln('Scalar Product');
        Writeln;Writeln;
        Writeln('This program calculates the scalar');
        Writeln('product of two vectors.');
        Writeln;
        Writeln('Use a space between data entries.');
    end; { title }

  Procedure Product;
    begin
       scalar:= Vect1X * Vect2X + Vect1Y * Vect2Y + Vect1Z * Vect2Z;
    end; { product }

  Procedure Input;
    begin
       Write('Enter vector 1 (x,y,z): ');
       Readln(Vect1X,Vect1Y,Vect1Z);
       Write('Enter vector 2 (x,y,z): ');
       Readln(Vect2X,Vect2Y,Vect2Z);
    end; { input }


  Procedure Printout;
    begin
      case answer of
      's','S' : Writeln('The scalar product is ',scalar);
      'p','P' : begin
                  Rewrite(out,'printer:');
```

```
                    Writeln(out,'The scalar product is ',scalar);
                    Writeln('The scalar product is ',scalar);
                    Close(out);
                   end;
         end; { case of }
      end; { printout }

   begin  { main }
      Repeat
         Outputs;
            Title;
              Writeln;Writeln;
            Input;
              Writeln;Writeln;
            Product;
            Printout;
              Writeln;Writeln;
              Write('Do you have another scalar product?(Y/N)');
              Readln(cont);
      until cont in ['n','N'];
   end.
```

```
Scalar Product


This program calculates the scalar
product of two vectors.

Use a space between data entries.

Enter vector 1 (x,y,z): 2 3 4
Enter vector 2 (x,y,z): -3 5 Ø

The scalar product is 9.

Do you have another scalar product? (Y/N) N
```

1.7 CROSS PRODUCT OF TWO VECTORS

Unlike the scalar product of two vectors, the cross product of two vectors is itself a vector. The magnitude of this vector is defined as the product of the magnitude of the two original vectors multiplied by the sine of the angle between them.

The direction of the cross product vector is perpendicular to the plane that contains the two original vectors. The value of the cross product vector components can be calculated by using the formulas:

$$X = Y1 * Z2 - Z1 * Y2$$
$$Y = Z1 * X2 - X1 * Z2$$
$$Z = X1 * Y2 - Y1 * X2$$

where the original vectors are

$$A1 = X1 + Y1 + Z1$$
$$A2 = X2 + Y2 + Z2$$

These calculations are performed in *Procedure Product.*

```
(*=========================================*)
(*                                         *)
(* Title:  Cross Product of Two Vectors    *)
(* Program Summary: Program computes       *)
(*         the cross product of two        *)
(*         vectors.  The magnitude is      *)
(*         the magnitude of two vectors    *)
(*         multiplied by the sine of       *)
(*         the angle between them.         *)
(*                                         *)
(*=========================================*)

Program Cross_Products;


  var
    answer,
    cont              : Char;
    out               : Text;
    Vect1X,
    Vect1Y,
    Vect1Z,
    Vect2X,
    Vect2Y,
    Vect2Z            : Integer;
    CrossX,
    CrossY,
    CrossZ            : Integer[35];

  Procedure Outputs;
    begin
       page(output);
        repeat
          gotoxy(Ø,5);
          Write('S(creen or P(rinter ? ');
          Readln(answer);
        until answer in ['s','S','p','P'];
    end; { outputs }

  Procedure Title;
    begin
       page(output);
        gotoxy(14,5);
        Writeln('Cross Product');
        Writeln;Writeln;
        Writeln('This program calculates the cross');
        Writeln('product of two vectors.');
        Writeln;
        Writeln('Use spaces with data entries instead');
        Writeln('of using commas');
    end; { title }

  Procedure Product;
    begin
      CrossX := Vect1Y * Vect2Z - Vect1Z * Vect2Y;
      CrossY := Vect1Z * Vect2X - Vect1X * Vect2Z;
      CrossZ := Vect1X * Vect2Y - Vect1Y * Vect2X;
    end; { product }

  Procedure Input;
    begin
       Write('Enter vector 1 (x,y,z): ');
       Readln(Vect1X,Vect1Y,Vect1Z);
       Write('Enter vector 2 (x,y,z): ');
       Readln(Vect2X,Vect2Y,Vect2Z);
    end; { input }


  Procedure Printout;
    begin
      case answer of
      's','S' : begin
                  Writeln('The Cross Product is: ');
                  Writeln;
                  Writeln('X = ',CrossX,'    Y = ',CrossY,'    Z = ',CrossZ);
                end;
```

```
        'p','P' : begin
                  Rewrite(out,'printer:');
                  Writeln(out,'The Cross Product is: ');
                  Writeln('The Cross Product is:');
                  Writeln(out);Writeln;
                  Writeln(out,'X = ',CrossX,'     Y = ',CrossY,'     Z = ',CrossZ);
                  Writeln('X = ',CrossX,'    Y = ',CrossY,'     Z = ',CrossZ);
                  Close(out);
                  end;
        end; { case of }
      end; { printout }

   begin  { main }
     Repeat
       Outputs;
         Title;
           Writeln;Writeln;
         Input;
           Writeln;Writeln;
         Product;
         Printout;
           Writeln;Writeln;
           Write('Do you have another cross product?(Y/N)');
           Readln(cont);
     until cont in ['n','N'];
   end.
```

```
Cross Product

This program calculates the cross
product of two vectors.

Use spaces with data entries instead
of using commas.

Enter vector 1 (x,y,z): 2 3 4
Enter vector 2 (x,y,z): -3 5 Ø

The Cross Product is:
X = -2Ø   Y = -12   Z = 19

Do you have another cross product? (Y/N) N
```

1.8 LOG TO ANY BASE

The logarithm of any number is the power to which the base of the logarithm must be raised to produce the number. This is a classic definition of logarithms. Generally, only two bases are used to calculate logarithms: the base 10 and the base e (where e = 2.71828). The latter is often referred to as the natural logarithm.

Sometimes to make the calculations easier, it is desirable to calculate logarithms to bases other than those noted above. This program makes it possible for your computer to calculate the log of any number to any base. For example, the log of 64 to the base 4 equals 3 and the ln e equals 1. This can be helpful in computer applications for calculating what power of 2 a particular number is. For example, if the number 65536 is entered when the program is run and the base is 2, we discover that 65536 is equal to 2 to the 16th power.

Calculations are based on the definition of a logarithm stated above. From this definition, the following formula is derived:

$$\text{base}^{\log} = \text{number}$$

where the number is produced by raising the base to the log power. If the logarithm of both sides of this equation is taken, the result is:

$$\log(\text{base}^{\log}) = \log(\text{number}) \text{ or } \log B^L = \log n$$

The left side of this last equation is, however, equal to L∗log(B), so that the final result is:

$$L * \log B = \log n$$

If both sides of the equation are divided by log B, the formula used by the program to evaluate the log of any number to any base results. This formula is

```
(*==========================================*)
(*                                          *)
(* Title:   Log to Any Base                 *)
(* Program Summary: Program computes        *)
(*          logarithms to any base.  The    *)
(*          logarithm of any number is      *)
(*          the power to which the base     *)
(*          of the logarithm must be        *)
(*          raised to produce the number.   *)
(*                                          *)
(*==========================================*)


Program Logarithms;

 Uses  Transcend;


  var
    answer,
    cont               : Char;
    out                : Text;
    logar,
    number,
    base               : real;

  Procedure Outputs;
    begin
       page(output);
        repeat
          gotoxy(Ø,5);
          Write('S(creen or P(rinter ? ');
          Readln(answer);
        until answer in ['s','S','p','P'];
    end; { outputs }

  Procedure Title;
    begin
       page(output);
        gotoxy(13,5);
        Writeln('Log to any Base');
        Writeln;Writeln;
        Writeln('This program calculates the logarithm');
        Writeln('of any number to any base');
    end; { title }

  Procedure logs;
    begin
      logar := Log( number) / Log ( base) ;
    end; { logs }
```

```
  Procedure Input;
    begin
       Write('Enter the number:  ');
       Readln(number);
       Write('Enter the base:    ');
       Readln(base);
    end; { input }


  Procedure Printout;
    begin
      case answer of
      's','S' : begin
                  Writeln('The Log of ',Number,' to the base ',base);
                  Writeln;
                  Writeln('is equal to ',Logar);
                end;
      'p','P' : begin
                  Rewrite(out,'printer:');
                  Writeln(out,'The Log of ',Number,' to the base ',base);
                  Writeln('The Log of ',Number,' to the base ',base);
                  Writeln(out);Writeln;
                  Writeln(out,'is equal to ',Logar);
                  Writeln('is equal to ',Logar);
                  Close(out);
                end;
      end; { case of }
    end; { printout }

begin  { main }
   Repeat
      Outputs;
        Title;
          Writeln;Writeln;
        Input;
          Writeln;Writeln;
        Logs;
        Printout;
          Writeln;Writeln;
          Write('Do you have another logarithm?(Y/N)');
          Readln(cont);
    until cont in ['n','N'];
 end.
```

L = (log n)/(log B)

Log to any Base

This program calculates the logarithm
of any number to any base.

Enter the number: 65536
Enter the base: 2

The Log of 6.55360E4 to the base 2.00000
is equal to 1.60000E1

Do you have another logarithm? (Y/N) N

1.9 NEW COORDINATES

This program is helpful when it is necessary to convert data from one frame of reference to another. The operation is frequently necessary in the study of physics. If it is desired to translate the coordinate axis of a particular set of data, subtract the

coordinates of the new origin from the old data point. If the axis is to be rotated as well, it must be accounted for by multiplying by the sine and the cosine of the angle involved.

The actual calculation of the new point is done in *Procedure New*.

A point of interest here is that angle arguments of Pascal accept only radians. The input accepts degrees and, in the main body of the program, degrees are converted to radians after input and before the do loop.

```
(*=========================================*)
(*                                         *)
(* Title:   New Coordinates                *)
(* Program Summary: Program computes       *)
(*          coordinates of data points     *)
(*          in a rectangular coordinate    *)
(*          system that has been           *)
(*          rotated about its origin       *)
(*          or translated to a new         *)
(*          origin.                        *)
(*                                         *)
(*=========================================*)


Program New_Coord;

 Uses  Transcend;


  var
    answer,
    cont               : Char;
    out                : Text;
    Data,
    step               : Integer;
    NewX,
    NewY,
    Deg                : Real;
    Xaxis,
    Yaxis,
    NewXaxis,
    NewYaxis           : Array [Ø..255] of Real;

  Procedure Outputs;
    begin
       page(output);
        repeat
          gotoxy(Ø,5);
          Write('S(creen or P(rinter ? ');
          Readln(answer);
        until ans er in ['s','S','p','P'];
    end; { outputs }

  Procedure Title;
    begin
       page(output);
        gotoxy(13,5);
        Writeln('New Coordinates');
        Writeln;Writeln;
        Writeln('This program will calculate the coordinates');
        Writeln('of data points in a rectangular coordinate');
        Writeln('system that has been rotated about its');
        Writeln('origin, or translated to a new origin.');
        Writeln('The new origin is assumed to be at');
        Writeln('point (Ø,Ø).');
        Writeln('Note: Use a space to separate all ');
        Writeln('data entries that call for two entries');
        Writeln('per line ie, (x,y)');
    end; { title }

  Procedure New;
  begin
 NewXaxis[step]:=(Xaxis[step]-NewX)*Cos(Deg)+(Yaxis[step]-NewY)*Sin(Deg);
    NewXaxis[step]:=Round (1ØØ * NewXaxis[step ]+Ø.5)/1ØØ;
```

```
NewYaxis[step]:=-(Xaxis[step]-NewX)*Sin(Deg)+(Yaxis[step]-NewY)*Cos(Deg);
   NewYaxis[step]:=Round (100 * NewYaxis[step]+0.5)/100;
   end; { New }

  Procedure Input;
   begin
      Write('How many data points?  ');
      Readln(Data);
      Write('Enter new origin (X,Y) ');
      Readln(NewX,NewY);
      Write('Enter Degrees of rotation: ');
      Readln(Deg);
   end; { input }

  Procedure Local;
    begin
      Write('Enter Old Location of point (X,Y): ');
      Readln(Xaxis[step],Yaxis[step]);
    end; { local }


  Procedure Printout;
   begin
     case answer of
     's','S' : begin
                 Writeln('Data points on the new axis are:');
                 Writeln;
                 for step := 1 to Data do
                    Writeln('X',step,' = ',Round(NewXaxis[step]),'   Y',step,
                        ' = ',Round(NewYaxis[step]));
                end;
     'p','P' : begin
                 Rewrite(out,'printer:');
                 Writeln(out,'Data points on the new axis are');
                 Writeln('Data points on the new axis are');
                 Writeln(out);Writeln;
                 for step:= 1 to Data do
                  begin
                  Writeln(out,'X',step,' = ',Round(NewXaxis[step]),'   Y',step,
                        ' = ',Round(NewYaxis[step]));
                  Writeln('X',step,' = ',Round(NewXaxis[step]),'   Y',step,
                        ' = ',Round(NewYaxis[step]));
                     end; { for }
                 Close(out);
                end;
     end; { case of }
   end; { printout }

begin  { main }
   Repeat
      Outputs;
        Title;
          Writeln;Writeln;
        Input;
        Deg:= Deg * 3.14159 / 180;
        for step := 1 to Data do
           begin
              Local;
                New;
           end; { for }
          Writeln;Writeln;
        Printout;
          Writeln;Writeln;
          Write('Do you have another conversion?(Y/N)');
          Readln(cont);
   until cont in ['n','N'];
 end.
```


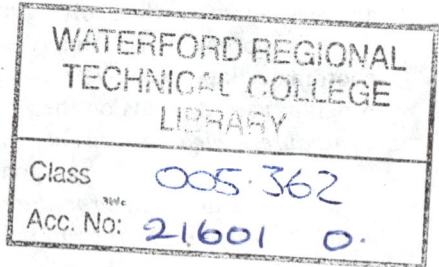

New Coordinates

This program will calculate the coordinates of data points in a rectangular coordinate system that has been rotated about its

```
origin, or translated to a new origin.
The new origin is assumed to be at
point (Ø,Ø).
Note: Use a space to separate all
data entries that call for two entries
per line, i.e., (x,y).

How many data points? 3
Enter new origin (x,y) 3  3
Enter Degrees of rotation: Ø
Enter Old Location of point (x,y): 3  3
Enter Old Location of point (x,y): Ø  Ø
Enter Old Location of point (x,y): 5  2

Data points on the new axis are
X1 = Ø  Y1 = Ø
X2 = -3  Y2 = -3
X3 = 2  Y3 = -1

Do you have another conversion? (Y/N) N
```

1.10 RECTANGULAR-TO-POLAR COORDINATE CONVERSION

Like the previous program, this one is also used to convert data from one frame of reference to another. This time, however, the data is converted from conventional rectangular (or Cartesian) coordinates to polar coordinates. Conversions of this type are very common when solving problems in electrical engineering.

The program is actually two programs in one. It can accept either polar coordinate data and give rectangular coordinates, or vice versa. The segment of program used depends on the answer to the question posed in the input procedure. *Procedure Which* does all the work.

If "R" is entered in response to the input question, the rectangular-to-polar conversion is done. In *Procedure Which*, the angle *theta* is calculated. Where *coord* equals R, *theta* is returned in radians and then converted to degrees.

```
(*=========================================*)
(*                                         *)
(* Title:    Rectangular to Polar          *)
(*           Coordinate Conversion         *)
(* Program Summary: Program converts       *)
(*           points in rectangular         *)
(*           coordinates to polar          *)
(*           coordinates and vice versa.   *)
(*                                         *)
(*=========================================*)


Program Rect_to_Polar;

 Uses  Transcend;

  var
    coord,
```

```
  answer,
  cont              : Char;
  out               : Text;
  Theta,
  Radius,
  X,
  Y                 : Real;

Procedure Outputs;
  begin
     page(output);
      repeat
        gotoxy(0,5);
        Write('S(creen or P(rinter ? ');
        Readln(answer);
      until answer in ['s','S','p','P'];
  end; { outputs }

Procedure Title;
  begin
     page(output);
      gotoxy(5,5);
       Writeln('Rectangular /Polar Coordinate Conversion');
       Writeln;Writeln;
       Writeln('This program converts points in ');
       Writeln('rectangular coordinates to polar');
       Writeln('coordinates and vise versa.  Enter');
       Writeln('the type of data you are supplying.');
       Writeln;
       Writeln('Note: Use a space to separate all ');
       Writeln('data entries that call for two entries');
       Writeln('per line ie, (x,y)');
  end; { title }

Procedure Input;
  begin
    repeat
     Write('Polar or Rectangular (P/R): ');
     Readln(coord);
    until coord in [ 'r','R','p','P'];
  end; { input }

Procedure Input_Theta;
  begin
     Write('Enter Radius, Theta (degrees): ');
     Readln(Radius, Theta);
  end;

Procedure Input_XY;
  begin
     Write('Enter X,Y: ');
     Readln(x,y);
  end;

Procedure Which;
  begin
    case coord of
     'r','R': begin
               Input_XY;
                Theta := ATAN ( Y / X );
                Radius := Y / Sin ( Theta );
                Theta := Theta * 57.2958;
               end;
     'p','P': begin
               Input_Theta;
                Theta := Theta / 57.2958;
                X      := ( Radius * Cos (theta));
                Y      := ( Radius * Sin (theta));
               end;
    end; { case }
  end;

Procedure Printout;
  begin
    case answer of
    's','S' : begin
         case coord of
```

```
                'r','R': begin
                         Write('R := ',round(Radius));
                         Writeln(' Angle := ',round(theta),' degrees');
                         end;
                'p','P': begin
                          Writeln('X := ',round(x),' Y := ',round(y));
                         end;
              end;
                   end;
        'p','P' : begin
                   Rewrite(out,'printer:');
              case coord of
            'r','R':begin
                   Write(out,'R := ',round(Radius));
                   Write('R := ',round(Radius));
                   Writeln(out,' Angle := ',round(theta),' degrees');
                   Writeln(' Angle := ',round(theta),' degrees');
                    end;
            'p','P':begin
                   Writeln(out,'X := ',round(x),' Y := ',round(y));
                   Writeln('X := ',round(x),' Y := ',round(y));
                    end;
            end;
             Close(out);
         end;
      end;
     end;


begin  { main }
   Repeat
      Outputs;
        Title;
          Writeln;Writeln;
        Input;
          Writeln;
        Which;
        Printout;
          Writeln;Writeln;
          Write('Do you have another conversion?(Y/N)');
          Readln(cont);
    until cont in ['n','N'];
 end.
```

Rectangular/Polar Coordinate Conversion

This program converts points in
rectangular coordinates to polar
coordinates and vice versa. Enter
the type of data you are supplying.

Note: Use a space to separate all
data entries that call for two entries
per line, i.e., (x,y).

Polar or Rectangular (P/R): R

Enter X, Y : 3 4

R = 5 Angle = 53 degrees

Do you have another conversion? (Y/N) N

1.11 MAX/MIN LOCATOR

When doing any type of data analysis, it is often necessary to know the largest and smallest numbers of a data set. This need is so widespread, in fact, that the FORTRAN computer language has special max and min functions built in. While no such capability exists in Pascal, it is a simple matter to write a program that will find these numbers.

In this program, variable *Largest* is the largest number and variable *Smallest* is the smallest. Both start out assigned the same value; that is, the first number entered. After all the numbers are entered, the program checks each number in order of entry to find any that are larger than *Largest*. If one is found, it is substituted for *Largest*. A similar substitution occurs for *Smallest*. Substitutions are done in *Procedure Check*.

Since this program is so handy, you may want to use it as a subroutine in a larger program that generates numbers. The subroutine is *Procedure Check*.

```
(*========================================*)
(*                                        *)
(* Title:   Max/Min Locator               *)
(* Program Summary: Program finds         *)
(*         the maximum and minimum        *)
(*         numbers in a group of          *)
(*         numbers.                       *)
(*                                        *)
(*========================================*)


Program MaxMinLocator;

   var
    answer,
    cont                : Char;
    out                 : Text;
    Largest,
    Smallest            : Real;
    Numbers             : Array [Ø..255] of Real;
    Numlist,
    Entries,
    Checks              : Integer;

  Procedure Outputs;
    begin
       page(output);
        repeat
          gotoxy(Ø,5);
          Write('S(creen or P(rinter ? ');
          Readln(answer);
        until answer in ['s','S','p','P'];
    end; { outputs }

  Procedure Title;
    begin
       page(output);
        gotoxy(11,5);
         Writeln('Max / Min Locator');
         Writeln;Writeln;
         Writeln('This program finds the largest');
         Writeln('and smallest numbers in a list');
         Writeln('of numbers');
    end; { title }

  Procedure Input;
    begin
       Write('How many numbers in list? ');
       Readln(Numlist);
    end; { input }
```

```
Procedure Entry;
  begin
     Writeln('Enter numbers: ');
    For entries := 1 to Numlist do
         begin
           Write('A',entries,' = ');
           Readln(numbers [ entries ]);
         end;
  end;

Procedure Check;
  begin
    Smallest:= Numbers [ 1 ];
    Largest := Smallest;
      For checks := 1 to Numlist do
          begin
            if Numbers [ checks ] > Largest
               then Largest := Numbers [ checks ];
            if Numbers [ checks ] - Smallest < Ø
               then Smallest := Numbers [ checks ];
          end;
  end;


Procedure Printout;
  begin
    case answer of
    's','S' : begin
                 Writeln('The Largest number is ',Largest);
                 Writeln;
                 Writeln('The Smallest number is ',smallest);
              end;
    'p','P' : begin
               Rewrite(out,'printer:');
                 Writeln(out,'The Largest number is ',Largest);
                 Writeln('The Largest number is ',Largest);
                 Writeln(out);Writeln;
                 Writeln(out,'The Smallest number is ',Smallest);
                 Writeln('The Smallest number is ',Smallest);
               Close(out);
              end;
    end;
 end;

begin  { main }
   Repeat
      Outputs;
        Title;
          Writeln;Writeln;
        Input;
        Entry;
        Check;
        Printout;
          Writeln;Writeln;
          Write('Do you have another locator?(Y/N)');
          Readln(cont);
   until cont in ['n','N'];
 end.
```

Max/Min Locator

This program finds the largest
and smallest numbers in a list
of numbers.

How many numbers in list? 5

Enter numbers:
A1 = 78
A2 = -9

A3 = Ø
A4 = 3445
A5 = .Ø1

The Largest number is 3.445ØØE3

The Smallest number is -9.ØØØØØ

Do you have another locator? (Y/N) N

1.12 DIMENSION SCALER

Anyone who has made a scaled-down drawing of an object realizes the difficulties involved in converting the dimensions of the object to a scaled-down size that will fit on the paper at hand. This program eliminates tedious work.

After the largest dimension of the object and the largest dimension of the drawing are entered, the program calculates the scale factor. It then requests the units of measure used in the drawing and the units of the object. Next, it asks how many dimensions are going to be converted so that a sufficient number of memory spaces can be set aside. Finally, it asks for the dimensions themselves.

A handy feature is the test in *Procedure Inputdim* that checks the dimensions entered to ensure that none of them are larger than the previously designated largest dimension. If a value is larger, an error message is generated and the user is asked to enter the dimension again.

```
(*==========================================*)
(*                                          *)
(* Title:   Dimension Scaler                *)
(* Program Summary: Program determines      *)
(*          the dimensions to draw          *)
(*          to scale an object from         *)
(*          a different scale.              *)
(*                                          *)
(*==========================================*)


Program Dimension_Scaler;

  var
    answer,
    cont                 : Char;
    out                  : Text;
    step,
    Number               : Integer;
    UnitsDraw,
    Unitsobj             : String;
    DimObject,
    DimDraw,
    Scale                : Real;
    Dim                  : Array [1..254] of Real;

  Procedure Outputs;
    begin
       page(output);
        repeat
          gotoxy(Ø,5);
          Write('S(creen or P(rinter ? ');
          Readln(answer);
        until answer in ['s','S','p','P'];
    end; { outputs }
```

```
  Procedure Title;
    begin
       page(output);
         gotoxy(12,5);
          Writeln('Dimension Scaler');
          Writeln;Writeln;
          Writeln('This program determines the dimensions ');
          Writeln('required to draw to scale given the');
          Writeln('largest dimension of the object and the');
          Writeln('largest dimension that is possible on');
          Writeln('the drawing');
    end; { title }

  Procedure Input;
    begin
      Writeln('What is the largest dimension of');
      Write('object: ');
      Readln(DimObject);
      Writeln;
      Writeln('What is the largest dimension of');
      Write('drawing: ');
      Readln(DimDraw);
      Writeln;
      Writeln('What are your drawing units ');
      Write('(E.G. inches):  ');
      Readln(Unitsdraw);
      Writeln;
      Writeln;
      Write('What are your object units: ');
      Readln(Unitsobj);
      Writeln;
      Writeln('How many dimensions do you want');
      Write('to convert: ');
      Readln(Number);
   end;

 Procedure Inputdim;
   begin
      Writeln;
      Writeln('Enter all dimensions in decimal form');
      Writeln;Writeln;
         for step := 1 to number do
           begin
            Write('Enter dimension ',step,' :');
            Readln(Dim[step]);
            if Dim[step] > DimObject then
              begin
               Writeln(chr(7),'Dimension too large!');
                Write('Reenter dimension ',step,' :');
                Readln(Dim[step]);
              end;

           end;
   end;

Procedure Scaling;
  begin
    scale := DimDraw / Dimobject ;
  end;

Procedure Printout;
  begin
    case answer of
    's','S' : begin
                Writeln;Writeln;Writeln;
                Writeln('Scale factor = ',scale);
                Writeln;Writeln;Writeln;
                 for step := 1 to number do
                  begin
                   Write(Dim[step],' ',unitsobj,' scales to ');
                   Writeln((100 * scale * dim[step])/100,' ',unitsdraw);
                  end;
              end;
    'p','P' : begin
               Rewrite(out,'printer:');
                Writeln(out);Writeln(out);Writeln(out);
                Writeln;Writeln;Writeln;
```

```
                    Writeln(out,'Scale factor = ',scale);
                    Writeln('Scale factor = ',scale);
                    Writeln(out);Writeln(out);Writeln(out);
                    Writeln;Writeln;Writeln;
                     for step := 1 to number do
                       begin
                         Write(out,Dim[step],' ',unitsobj,' scales to ');
                         Write(Dim[step],' ',unitsobj,' scales to ');
                         Writeln(out,(100*scale*dim[step])/100,' ',unitsdraw);
                         Writeln((100 * scale * dim[step] )/100,' ',unitsdraw);
                       end;
                   Close(out);
                  end;
              end; { case }
         end; { printout }

begin  { main }
   Repeat
      Outputs;
        Title;
          Writeln;Writeln;
        Input;
        Inputdim;
        Scaling;
        Printout;
          Writeln;Writeln;
          Write('Do you have another scaling?(Y/N)');
          Readln(cont);
    until cont in ['n','N'];
 end.
```

Dimension Scaler

This program determines the dimensions required to draw to scale given the largest dimension of the object and the largest dimension that is possible on the drawing.

What is the largest dimension of object: 35

What is the largest dimension of drawing: 11

What are your drawing units (E.G. inches): inches

What are your object units: feet

How many dimensions do you want to convert: 4

Enter all dimensions in decimal form

Enter dimension 1 : 35
Enter dimension 2 : 12
Enter dimension 3 : 4Ø36

```
Dimension too large!

Enter dimension 3 : 6
Enter dimension 4 : 23.75

Scale factor = 3.14286E-1

3.50000E1 feet scales to 1.1000E1 inches
1.20000E1 feet scales to 3.77143 inches
6.00000 feet scales to 1.88571 inches
2.37500E1 feet scales to 7.46429 inches

Do you have another scaling? (Y/N) N
```

1.13 HISTOGRAM

This program will plot any function as a histogram. Instead of plotting a point as a single spot, the histogram approach plots as a bar graph everything up to and including the data point so that the data stands out more clearly. The bar in this program is constructed of asterisks. Any other symbol can be used by simply changing the asterisk in *Procedure Printout* to whatever symbol you wish to use.

This program can be confusing if you don't examine it before use. Because Pascal is compiled, you must define the function to be used before compiling and then recompile the program for each function used. The example uses Y = Sin(X*Delta) where delta is the value to be incremented. This is because, unlike other languages such as BASIC, the *For-Do* loops do not have a STEP function to them.

The histogram is plotted with the Y-axis horizontal and the X-axis vertical. The resolution of the plot can be greatly improved by decreasing the value of *Delta X*.

```
(*==========================================*)
(*                                          *)
(* Title:   Histogram                       *)
(* Program Summary: Progam will             *)
(*          plot any function as a          *)
(*          histogram using text            *)
(*          mode graphics.                  *)
(*                                          *)
(*==========================================*)


Program Histogram;

(* This program plots histograms.  You
   enter the function to be ploted under
   Function Y;  The program mustthen be
   compiled to be run.  If any variables
   are entered besides X, you will have
   to enter them and their relationship
   under var and under funct.          *)

 uses transcend;

  var
    answer,
    cont              : Char;
    out               : Text;
    X1,
    X2,
```

```
   X3,
   Max,
   step              : Integer;
   Delta,
   Ymin,
   Ymax,
   Xmin,
   Xmax,
   X,
   Scale             : Real;

 Procedure Outputs;
   begin
      page(output);
       repeat
         gotoxy(Ø,5);
         Write('S(creen or P(rinter ? ');
         Readln(answer);
       until answer in ['s','S','p','P'];
   end; { outputs }

 Procedure Title;
   begin
      page(output);
       gotoxy(15,5);
        Writeln('Histogram');
        Writeln;Writeln;
        Writeln('Note: For all entries that require');
        Writeln('     data entries (x,y); separate ');
        Writeln('     all data with a space.');
   end; { title }


 Procedure Input;
   begin
     Write('Enter Ymin,Ymax : ');
     Readln(Ymin,Ymax);
     Writeln;
     Write('Enter Xmin,Xmax,Delta X : ');
     Readln(Xmin,Xmax,Delta);
     Writeln;
     Writeln;
     Writeln;
   end;

 { Note: In the next function, the variable
         is multiplied by Delta.  This is
         because Pascal does not have a step
         function in it For-Next loop.  If
         this were left out, the Histogram
         would jump more and not have as
         appealing and smooth a look to it }

 Function Y (X    :real): real;
   begin
      { example function is shown }

      Y := Sin ( X * Delta);
   end;

 Procedure Change;
   begin
    X2 := round(Xmin);
    X3 := round(Xmax/delta);
   end;

Procedure Printout;
  begin
    case answer of
    's','S' : begin
                For X1 := X2 to X3 do
                  begin
                   X := X1;
                   Max:= round((39 * (Y(X)- Ymin) / (Ymax -Ymin)));
                     For step := 1 to max do
                          Write('*');
                          Writeln;
```

```
                end;
            end;
  'p','P' : begin
             Rewrite(out,'printer:');
              Writeln(out,'Ymin = ',Ymin,' Ymax = ',Ymax);
              Write(out,'Xmin = ',Xmin,' Xmax = ',Xmax);
              Writeln(out,' Delta X =',Delta);
              Writeln(out);
              For X1 := X2 to X3 do
                begin
                  X := X1;
                  Max:= round((39 * (Y(X)- Ymin) / (Ymax -Ymin)));
               For step := 1 to max do
                  begin
                    Write('*');
                    Write(out,'*');
                  end;
                  Writeln;
                  Writeln(out);
                 end;
             Close(out);
            end;
        end; { case }
   end; { printout }

   begin  { main }
      Repeat
         Outputs;
           Title;
             Writeln;Writeln;
           Input;
            change;
            printout;
             Writeln;Writeln;
             Write('Do you have another Histogram?(Y/N)');
             Readln(cont);
       until cont in ['n','N'];
    end.
```

Histogram

Note: For all entries that require
data entries (x,y); separate
all data with a space.

Enter Ymin,Ymax : -1 1

Enter Xmin,Xmax,Delta X : Ø 4 .2

Do you have another Histogram? (Y/N) N

```
********************
***********************
***************************
*******************************
**********************************
************************************
**************************************
***************************************
***************************************
**************************************
*************************************
***********************************
*********************************
******************************
**************************
**********************
******************
***************
***********
********
*****
```

1.14 CIRCLE FINDER

One of the laws of geometry states that if three points are located in a plane and do not form a straight line, it is possible to draw a circle that will pass through each of the three points. The equations for calculating the radius of that circle and the location of the X and Y coordinates of the center (below) are complex and difficult to handle.

$$Y = \frac{\dfrac{(X3^2 - X1^2) + (Y3^2 - Y1^2)}{2(X3 - X1)} - \dfrac{(X2^2 - X1^2) + (Y2^2 - Y1^2)}{2(X2 - X1)}}{\dfrac{Y3 - Y1}{X3 - X1} - \dfrac{Y2 - Y1}{X2 - X1}}$$

$$X = \frac{(X3^2 - X1^2) + (Y3 - Y1)((Y3 + Y1) - 2Y)}{2(X3 - X1)}$$

$$R = \sqrt{(X3 - X)^2 + (Y3 - Y)^2}$$

This program takes these equations and uses them to calculate the required information. To make it easier to check, the equations have been broken down into four parts located in *Procedure Find.*

The *Procedure Printout* uses a slightly different form of *Writeln*, in that it will give a formatted printout. This is done by defining the second number after ":" as 2, hence *Writeln*(3.00000:5:2) will be written as 3.00.

```
(*=========================================*)
(*                                         *)
(* Title:  Circle Finder                   *)
(* Program Summary: Program will           *)
(*         find the center and radius      *)
(*         of a circle that is formed      *)
(*         by three points not in a        *)
(*         straight line.                  *)
(*                                         *)
(*=========================================*)


Program Cirle_Finder;

 uses transcend;

  var
    answer,
    cont                 : Char;
    out                  : Text;
    X1,X2,X3,
    Y1,Y2,Y3,
    Part1,
    Part2,
    Part3,
    Part4,
    X,Y,
    Radius               : Real;

  Procedure Outputs;
    begin
       page(output);
        repeat
          gotoxy(0,5);
          Write('S(creen or P(rinter ? ');
          Readln(answer);
        until answer in ['s','S','p','P'];
    end; { outputs }
```

```
Procedure Title;
  begin
     page(output);
      gotoxy(14,5);
       Writeln('Circle Finder');
       Writeln;Writeln;
       Writeln('This program will find the center and');
       Writeln('radius of a circle that is determined');
       Writeln('by any three points in a plane that ');
       Writeln('do not form a straight line.');
       Writeln;
       Writeln('Note: For all entries that require');
       Writeln('      data entries (x,y); separate ');
       Writeln('      all data with a space.');
       Writeln;Writeln;
  end; { title }

Procedure Input;
  begin
     Write('Enter first point (X,Y):  ');
     Readln(X1,Y1);
     Write('Enter second point (X,Y): ');
     Readln(X2,Y2);
     Write('Enter third point (X,Y):  ');
     Readln(X3,Y3);
     Writeln;Writeln;Writeln;
  end;

Procedure Find;
  begin
     Part1 := ((X2-X1) * (X2+X1) + (Y2-Y1) * (Y2+Y1)) / (2.Ø * (X2-X1));
     Part2 := ((X3-X1) * (X3+X1) + (Y3-Y1) * (Y3+Y1)) / (2.Ø * (X3-X1));
     Part3 := (Y2-Y1) / (X2-X1);
     Part4 := (Y3-Y1) / (X3-X1);
     Y     := (Part2 - Part1) / (Part4 - Part3);
     X     := Part2 - Part4 * Y ;
     Radius:= SQRT (SQR (X3 - X) + SQR (Y3 - Y));
  end;

Procedure Print1;
   begin
      Writeln('The circle defined by the following');
      Writeln('three points:');
      Writeln;
      Writeln(' X1 = ',X1:7:2,' Y1 = ',Y1:7:2);
      Writeln(' X2 = ',X2:7:2,' Y2 = ',Y2:7:2);
      Writeln(' X3 = ',X3:7:2,' Y3 = ',Y3:7:2);
      Writeln;
      Writeln('has a radius = ',round(radius));
      Writeln;
      Writeln('at a center located at:');
      Writeln;
      Writeln(' X = ',X,'   Y = ',Y);
   end;

Procedure Print2;
   begin
     Rewrite(out,'printer:');
       Writeln('The circle defined by the following');
       Writeln(out,'The circle defined by the following');
       Writeln('three points:');
       Writeln(out,'three points:');
       Writeln;
       Writeln(out);
       Writeln(' X1 = ',X1:7:2,' Y1 = ',Y1:7:2);
       Writeln(out,' X1 = ',X1:7:2,' Y1 = ',Y1:7:2);
       Writeln(' X2 = ',X2:7:2,' Y2 = ',Y2:7:2);
       Writeln(out,' X2 = ',X2:7:2,' Y2 = ',Y2:7:2);
       Writeln(' X3 = ',X3:7:2,' Y3 = ',Y3:7:2);
       Writeln(out,' X3 = ',X3:7:2,' Y3 = ',Y3:7:2);
       Writeln;
       Writeln(out);
       Writeln('has a radius = ',round(radius));
       Writeln(out,'has a radius = ',round(radius));
       Writeln;
       Writeln(out);
       Writeln('at a center located at:');
```

```
          Writeln(out,'at a center located at:');
          Writeln;
          Writeln(out);
          Writeln(' X = ',X,'    Y = ',Y);
          Writeln(out,' X = ',X,'    Y = ',Y);
        Close(out);
     end;

Procedure Printout;
   begin
     case answer of
     's','S' : Print1;
     'p','P' : Print2;
       end; { case }
   end; { printout }

begin  { main }
    Repeat
       Outputs;
         Title;
           Writeln;Writeln;
         Input;
         Find;
         Printout;
           Writeln;Writeln;
           Write('Do you have another 3 points?(Y/N)');
           Readln(cont);
     until cont in ['n','N'];
  end.
```

Circle Finder

This program will find the center and radius of a circle that is determined by any three points in a plane that do not form a straight line.

Note: For all entries that require data entries (x,y); separate all data with a space.

Enter first point (x,y): 3 Ø
Enter second point (x,y): Ø 3
Enter third point (x,y): 6 3

The circle defined by the following three points:

X1 = 3.ØØ Y1 = Ø.ØØ
X2 = Ø.ØØ Y2 = 3.ØØ
X3 = 6.ØØ Y3 = 3.ØØ

has a radius = 3

at a center located at:
X = 3.ØØØØ Y = 3.ØØØØ

Do you have another 3 points? (Y/N) N

1.15 NORMALLY DISTRIBUTED RANDOM NUMBERS[1]

Pascal has the RANDOM function that generates a random number. The properties of these random numbers are not always clear, and unless *Randomize* is used, they will always be the same random numbers. Very often, especially in statistical studies, it is necessary to have a series of random numbers that are normally distributed or, in other words, Gaussian distributed.

Among the properties of normally distributed numbers is that 68.27 percent of them will fall within one standard deviation, 95.45 percent within two standard deviations, and 99.73 percent within three standard deviations. If normally distributed numbers are plotted on an axis (Y-coordinate) versus standard deviation (X-coordinate), a bell-shaped curve will be formed.

This program uses the *Random* function and *Randomize* and performs calculations on the resulting number produced by it, so the resulting new numbers will be normally distributed. The exact characteristics of the distribution (e.g., its standard deviation and average) are entered by the user along with the constant, K. The value of K can vary from one to any high number. The higher its value, the more accurate the results will be and the longer it will take to calculate the desired number. The quantity of random numbers to be generated is requested in *Procedure Input*.

```
(*=========================================*)
(*                                         *)
(* Title:   Random Numbers                 *)
(* Program Summary: Program generates      *)
(*          normally distributed           *)
(*          random numbers.                *)
(*                                         *)
(*=========================================*)


Program Random_Numbers;

 uses transcend,applestuff;

  var
    answer,
    cont               : Char;
    out                : Text;
    Stepping,
    Steps,
    K,
    Nums,
    Stddev,
    RND                : Integer;
    Rnd2,
    Number,
    Mean,
    Temp               : Real;

  Procedure Outputs;
    begin
       page(output);
        repeat
          gotoxy(0,5);
          Write('S(creen or P(rinter ? ');
          Readln(answer);
        until answer in ['s','S','p','P'];
    end; { outputs }
```

[1]Adapted from M. Perlman, "Generate Normally Distributed Random Numbers with BASIC," *Electronic Design*, (Nov. 22, 1970): p. 62.

```
  Procedure Title;
    begin
       page(output);
        gotoxy(3,5);
         Writeln('Normally Distributed Random Numbers');
         Writeln;Writeln;
         Writeln('This program generates normally');
         Writeln('distributed random numbrs.  You ');
         Writeln('enter the desired standard deviaton,');
         Writeln('mean,  and K (Large for accuracy,');
         Writeln('small for speed ).');
         Writeln;Writeln;Writeln;
    end; { title }

  Procedure Input;
    begin
       Write('Enter standard deviation: ');
       Readln(stddev);
       Write('Enter K: ');
       Readln(k);
       Write('Enter the Mean: ');
       Readln(mean);
       Write('How many numbers do you want? ');
       Readln(nums);
       Writeln;Writeln;Writeln;
    end;

  Procedure Find;
    begin
       Randomize;
     Temp := Ø;
       For steps := 1 to K do
         begin
             RND   := random;
             Rnd2  := RND * 3.Ø518E-5 ;
             Temp  := temp + Rnd2 ;
             Number:= Stddev * SQRT (12 / K ) * (Temp - K/2) + Mean;
         end;
    end;


Procedure Printout;
  begin
    case answer of
    's','S' : begin
                For stepping := 1 to nums do
                    begin
                       Find;
                       Writeln(number);
                    end;
              end;

    'p','P' : begin
                 Rewrite(out,'printer:');
                   Writeln(out,'Standard Deviaton = ',stddev);
                   Writeln(out,'K = ',K);
                   Writeln(out,'Mean = ',mean);
                      For stepping := 1 to nums do
                        begin
                          find;
                           Writeln(number);
                           Writeln(out,number);
                        end;
                 Close(out);
              end;

    end; { case }
  end; { printout }

begin  { main }
   Repeat
      Outputs;
        Title;
          Writeln;Writeln;
        Input;
        Printout;
          Writeln;Writeln;
```

```
        Write('Do you want another run?(Y/N)');
        Readln(cont);
   until cont in ['n','N'];
end.
```

Normally Distributed Random Numbers

This program generates normally
distributed random numbers. You
enter the desired standard deviation,
mean, and K (Large for accuracy,
small for speed).

Enter standard deviation: 21

Enter K: 1Ø

Enter the mean: 85

How many numbers do you want? 1Ø

9.8Ø863E1
7.21239E1
6.83448E1
8.84978E1
5.931Ø9E1
8.349Ø8E1
1.1217ØE2
9.78631E1
9.36522E1
4.ØØ165E1

1.16 Nth ROOT OF A NUMBER

Although Pascal does not have the capability to calculate the Nth root of a number, this process can be implemented by using the program below. The program calculates successively finer approximations of the root until a desired level of accuracy is reached. The accuracy can be varied (within the limits of your computer, of course).

```
(*========================================*)
(*                                        *)
(* Title:  Nth Root of a Number           *)
(* Program Summary: Program will find     *)
(*         the Nth root of any number.    *)
(*                                        *)
(*========================================*)


Program Nth_Root;

 uses transcend;

  var
    answer,
    cont               : Char;
```

```
   out              : Text;
   Root,
   Realrt,
   Stepping,
   Steps            : Integer;
   Temp,
   Temp2,
   Numrt,
   Number           : Real;

Procedure Outputs;
  begin
     page(output);
      repeat
        gotoxy(Ø,5);
        Write('S(creen or P(rinter ? ');
        Readln(answer);
      until answer in ['s','S','p','P'];
  end; { outputs }

Procedure Title;
  begin
     page(output);
      gotoxy(1Ø,5);
       Writeln('Nth Root of a Number');
       Writeln;Writeln;
       Writeln('This program will find the Nth ');
       Writeln('root of any number. You must enter');
       Writeln('the number and the root desired.');
       Writeln;Writeln;
  end; { title }

Procedure Input;
  begin
     Write('What is the number: ');
     Readln(number);
     Writeln;
     Write('What root do you want: ');
     Readln(root);
     Writeln;
     If root < Ø then
       begin
            Writeln('The root must be zero or larger');
            Writeln('What is the root: ');
            Readln(root);
          end;
    end;

Procedure Printout;
  begin
    case answer of
    's','S' : begin
                 Writeln('The ',root,' root of ',number,' is ',numrt);
              end;

    'p','P' : begin
                Rewrite(out,'printer:');
                  Writeln(out,'The number = ',number);
                  Writeln(out,'The root   = ',root);
                  Writeln(out);
                 Writeln('The ',root,' root of ',number,' is ',numrt);
                 Writeln(out,'The ',root,' root of ',number,' is ',numrt);
                Close(out);
              end;

    end; { case }
  end; { printout }

Procedure Find;
  begin
     if Root = Ø then Printout;
      Numrt  := Number / Root;
        For steps := 1 to 1ØØ do
          begin
             temp := numrt;
               Realrt := Root - 1 ;
                 temp2:= numrt;
```

```
                numrt := ( number / exp(realrt *Ln(temp2))
                               + realrt * numrt)/ root;
            end;
    end;

begin  { main }
   Repeat
      Outputs;
        Title;
          Writeln;Writeln;
        Input;
        Find;
        Printout;
          Writeln;Writeln;
          Write('Do you want another run?(Y/N)');
          Readln(cont);
     until cont in ['n','N'];
end.
```

Nth Root of a Number

What is the number: 65536

What root do you want: 8

The 8th root of 6.55360E4 is 4.00000

Do you want another run? (Y/N) N

1.17 RATIONAL FRACTIONS[1]

Designers often find a need to express a number as a rational fraction, especially in frequency synthesizer and gear train design applications. This program takes any number (rational or irrational) and produces successively finer rational fraction approximations of it. (Note: Because of the accuracy of Apple Pascal to only six significant digits, the computer will reach a point at which it will call the number "Exact" even though it may be far from exact.)

The output consists of the rough approximation, which is the integer part of the number followed by finer and finer approximations. Since, for an irrational number, these approximations can go on until the accuracy of the computer is reached, the program asks for a limit to the number of iterations desired.

The input of a rational number results in a finite number of iterations followed by the word *Exact*, assuming that the exact fraction is reached in the number of iterations indicated.

```
(*=========================================*)
(*                                         *)
(* Title:   Rational Fractions             *)
(* Program Summary: Program will take      *)
(*          any number and produce         *)
(*          successively finer rational    *)
(*          fraction approximations        *)
(*          of it.                         *)
(*                                         *)
(*=========================================*)
```

[1]Adapted from "BASIC Program Expresses Any Number As a Rational Fraction," *Electronic Design*, (Dec. 7, 1972), p. 102.

```
Program Rational_Fract;

  var
    answer,
    cont              : Char;
    out               : Text;
    Num,
    Steps,
    Times             : Integer;
    Number,
    Debug,
    Temp,
    Part1,
    Part2,
    Bottom,
    Top,
    Temp2             : Real;
    Show1,
    Show2             : Integer[10];

  Procedure Outputs;
    begin
       page(output);
        repeat
          gotoxy(0,5);
          Write('S(creen or P(rinter ? ');
          Readln(answer);
        until answer in ['s','S','p','P'];
    end; { outputs }

  Procedure Title;
    begin
       page(output);
        gotoxy(11,5);
         Writeln('Rational Fractions');
         Writeln;Writeln;
         Writeln('This program will accept any number');
         Writeln('as an input and produce successively');
         Writeln('finer rational fraction approximations');
         Writeln('of it.  If a rational fraction is');
         Writeln('entered , the program will indicate');
         Writeln('that the final answer is exact.');
         Writeln;Writeln;
    end; { title }

  Procedure Input;
    begin
       Write('Enter the number: ');
       Readln(number);
       Writeln;
       Write('How many iterations?: ');
       Readln(times);
       Writeln;
    end;

{ note: due to the way pascal handles
        numbers this procedure is at
        best an approximation.  If you
        were using BASIC you could come
        closer. }

  Procedure Find;
    begin
      Temp   := number;
      Part1  := 0.0;
      Part2  := 1.0;
      Top    := 1.0;
      Bottom := 0.0;
   end;

  Procedure Work;
    begin
     num   := trunc(number);
     temp2 := top;
     top   := num * top + part1;
     part1 := temp2;
     temp2 := bottom;
```

```
      bottom:= num * bottom + part2;
      part2 := temp2;
     end;

Procedure Printout;
  begin
    case answer of
    's','S' : begin
                Writeln;Writeln;Writeln;
                  For steps := 1 to times do
                    begin
                       Work;
                       Writeln;
                       show1 := round(top);
                       show2 := round(bottom);
                       Write(show1,'/',show2);
                        if (number = num) or (temp = top/bottom) then
                          begin
                            Writeln(' Exact');
                             Exit(Printout);
                          end
                           else
                             Writeln;
                               number := 1/ (number- num);
                    end;
              end;

    'p','P' : begin
                Rewrite(out,'printer:');
                Writeln;Writeln;Writeln;
                Writeln(out,'The number = ',number);
                Writeln(out,'The iterations = ',times);
                Writeln(out);Writeln(out);
                  For steps := 1 to times do
                    begin
                       Work;
                       Writeln;
                       Writeln(out);
                        Show1 := round (top);
                        Show2 := round (bottom);
                       Write(show1,'/',show2);
                       Write(out,show1,'/',show2);
                        if (number = num) or (temp = top/bottom) then
                          begin
                            Writeln(' Exact');
                            Writeln(out,' Exact');
                              Close(out);
                             Exit(Printout);
                          end
                           else
                             Writeln;
                             Writeln(out);;
                         number := 1/(number- num);
                       end;
                Close(out);
              end;
    end; { case }
  end; { printout }


begin  { main }
   Repeat
      Outputs;
        Title;
          Writeln;Writeln;
        Input;
        Find;
        Printout;
          Writeln;Writeln;
          Write('Do you have another fraction?(Y/N)');
          Readln(cont);
    until cont in ['n','N'];
 end.
```

Rational Fractions

This program will accept any number as an input and produce successively finer rational fraction approximations of it. If a rational fraction is entered, the program will indicate that the final answer is exact.

Enter the number: 6.12318

How many iterations?: 6
6/1
49/8
398/65
845/138
4623/755
5468/893

Do you have another fraction? (Y/N) N

2 ENGINEERING MATHEMATICS

Some of the major uses of computers today are in science and engineering. This chapter contains six programs that solve frequently encountered problems in the field of engineering. The first solves a set of simultaneous equations by using a process known as Gaussian elimination. Equations of this type crop up in every field from finance to physics.

The next program (2.2) evaluates polynomials by using an efficient algorithm that minimizes the number of computations to be performed. Quadratic equations are solved in 2.3, using a general program that solves for real, repeating, and complex roots. The next program, Newton-Raphson roots, can be used to find one root of an equation. By combining this program with the previous one and factoring out the root it finds, third order equations can be solved. Try it.

The next two programs (2.5 and 2.6) perform functions that are a little more difficult. In order to write a program to do integration or differentiation, the theory underlying these operations must be understood. Some of the necessary details are provided in the text accompanying each of these programs.

2.1 SIMULTANEOUS EQUATIONS

One of the most common problems in science and engineering is finding the solution to a system of linear simultaneous equations. Many methods have been developed to solve such systems. One of the oldest and still most commonly used is known as Gaussian elimination. It replaces certain equations of the system by combinations of other equations. The system is then solved by back substitutions.

The number of equations to be solved is designated by *Number*, but the array definitions can be changed to either increase alloted memory or decrease it. *Procedure Input* takes data from the keyboard, one equation at a time. The variable *Fail* is used to *Flag* whether or not a unique solution to the system can be found. When a unique solution can not be found, *Fail* will be made true. This will cause *Procedure Printout* to be executed and the message "No solution to the system has been found." to be printed out. This occurs, for example, when *Number* = 1 and A[1,1] = 0.

Procedure Work contains the code which takes care of *Number* = 1. The back substitution to get the actual answer takes place in *Procedure Find.* The results are printed out in *Procedure Printout.*

The sample run solves the following equations:

3 X1 + 7 X2 + 32 X3 = 12
9 X1 + 43 X2 + 21 X3 = 31
2 X1 + 8 X2 + 19 X3 = 32

This also shows a run for a system of equations that can not be uniquely solved:

2 X1 + 4 X2 = 6
4 X1 + 8 X2 = 12

```
(*========================================*)
(*                                        *)
(* Title:   Simultaneous Equations        *)
(* Program Summary: Program will solve    *)
(*          simultaneous equations        *)
(*          where the coefficients are    *)
(*          entered.                      *)
(*                                        *)
(*========================================*)


PROGRAM SimultaneousEquations;

 VAR
  Answer,
  Cont             : CHAR;
  Out              : TEXT;
  Number ,
  Steps ,
  Store ,
  Stepping ,
  Counter ,
  Count,
  Step             : INTEGER;
  Factor,
  Flag,
  Sum,
  Temp             : REAL;
  B,
  X                : ARRAY [ 1..50 ] of REAL;
  A                : ARRAY [ 1..50,1..50 ] of REAL;
  Fail             : Boolean;

Procedure Outputs;
  begin
    page(output);
     repeat
       gotoXY(0,5);
        Write('S(creen or P(rinter ? ');
        Readln(answer);
     until Answer in [ 'S','s','p','P' ]
  end;  {outputs}

Procedure Title;
  begin
    page(output);
     gotoXY(9,5);
      Writeln('Simultaneous Equations');
      Writeln;Writeln;
      Writeln('This program solves a set of');
      Writeln('simultaneous equations. You must');
      Writeln('enter the Number of equations involved');
      Writeln('and the value of the coefficients  A. B.');
      Writeln;
  end;  { title }
```

```
Procedure Input;
  begin
    Write('Enter the  Number  of equations: ');
    Readln( Number );
    Writeln('Enter the value for "A" and "B" one');
    Writeln('equation at a time:');
    Writeln;
     for Count := 1 to  Number  do
       begin
         for  Stepping  := 1 to  Number  do
           begin
             Write('A(',Count,',', Stepping ,') = ');
             Readln( A [ Count, Stepping  ] );
           end;
         Write('B(',Count,') = ');
         Readln( B [ Count ] );
       end;
     Writeln
  end;

Procedure Printout;
  begin
   case answer of
   's','S' : begin
              Writeln;Writeln;Writeln;
               if  Fail  = True  then
                 Writeln('No solution to the system has been found')
               else
                 begin
                   Writeln('The solution to the system of equations');
                   Writeln('is as follows:');
                   Writeln;
                   For  Steps  := 1 to  Number  do
                         Writeln('X(', Steps ,') = ',X [  Steps  ]);
                         Writeln
                 end
             end;

   'p','P' : begin
              Rewrite(out,'printer:');
              Writeln;Writeln;Writeln;
              Writeln(out);Writeln(out);Writeln(out);
               if  Fail  = True then
                  begin
                    Writeln(out,'No solution to the system has been found');
                    Writeln('No solution to the system has been found')
                  end
               else
                 begin
                   Writeln('The solution to the system of equations');
                   Writeln(out,'The solution to the system of equations');
                   Writeln('is as follows:');
                   Writeln(out,'is as follows:');
                   Writeln;
                   Writeln(out);
                     For  Steps  := 1 to  Number  do
                         begin
                          Writeln('X(', Steps ,') = ',X [ Steps ]);
                          Writeln(out,'X(',Steps,') = ',X [ Steps ]);
                         end;
                          Writeln;
                          Writeln(out)
                 end;
              Close(out)
             end
       end; { case }
     end;

 Procedure Find;
    begin
      X[  Number  ] := B [  Number  ] / A [  Number  ,  Number  ];
      Steps :=  Number  -1 ;
      Repeat
        Sum := 0;
         Count := Steps +1;
           For  Stepping  := Count to  Number  do
                  Sum := Sum + A [ Steps ,  Stepping  ] * X[  Stepping  ];
           X[ Steps ] := ( B[ Steps ] - Sum ) / A[ Steps, Steps ];
```

```
          Steps :=  Steps -1 ;
       until  Steps = Ø ;
          Printout;
     end;

   Procedure Failure;
      begin
         Fail  := True;
        Printout
     end;

 Procedure Solve;
    begin
      For  Stepping  := Count to  Number  do
          begin
            Factor := A [  Stepping , Steps ] / A [ Steps, Steps ];
               For Step := Count to  Number  do
                      begin
                        A [  Stepping ,Step ] := A [  Stepping , Step ] -
                           Factor * A [ Steps, Step ];
                      end;
               B [  Stepping  ] := B[  Stepping  ] - Factor * B [ Steps ]
          end
    end;

 Procedure Switch;
    begin
      If  Store  =  Steps then
           Solve
      else
         begin
          For  Stepping  := 1 to  Number  do
             begin
               Temp                    := A[  Store ,  Stepping  ];
               A[  Store ,  Stepping  ] := A[ Steps,  Stepping  ];
               A[ Steps,  Stepping  ] :=Temp  ;
             end;
           Temp         := B [  Store  ];
           B [  Store  ] := B [ Steps ];
           B [ Steps ] := Temp;
                Solve;
          end;

    end;

 Procedure Try;
    begin
       Flag := ABS ( A [ Steps, Steps ] );
        Store  := Steps;
         Count := Steps + 1;
          For  Stepping  :=  Count to  Number  do
              if ABS ( A [  Stepping , Steps ] ) >= Flag then
                  begin
                    Flag := ABS ( A [  Stepping , Steps ]) ;
                     Store  :=  Stepping ;
                  end;
              if Flag <> Ø then
                  Switch
              else
                  Failure;
    end;

 Procedure Work;
    begin
        Fail  := False;
         If  Number  > 1 then
           begin
              Counter  :=  Number  -1;
               For Steps := 1 to  Counter  do
                        Try;
           If A [  Number ,  Number  ] <> Ø then
               Find
           else
               Failure;
           end
         else
```

```
          if A[ 1,1 ] = Ø then
              Failure
          else
            begin
             X[1] := B[1] / A[1,1] ;
             Printout
            end
    end;

begin { Main }
  Repeat
    Outputs;
    Title;
    Input;
    Work;
    Write('Do you have more data (Y/N) ');
    Readln(Cont);
   until Cont in [ 'N','n']
  end.
```

Simultaneous Equations

This program solves a set of simultaneous equations. You must enter the number of equations involved and the value of the coefficients A.B.

Enter the number of equations: 3
Enter the value for "A" and "B" one equation at a time:

A(1,1) = 3
A(1,2) = 7
A(1,3) = 32
B(1) = 12
A(2,1) = 9
A(2,2) = 43
A(2,3) = 21
B(2) = 31
A(3,1) = 2
A(3,2) = 8
A(3,3) = 19
B(3) = 32

The solution to the system of equations is as follows:
X(1) = -4.65218E1
X(2) = 9.11913
X(3) = 2.74161

Do you have more data? (Y/N) N

2.2 Polynomial Evaluator

Polynomials pop up in every field from engineering to finance. This program will compute the value of any polynomial. All you have to do is enter the coefficients of each term and the point at which the polynomial is to be evaluated.

The order of the polynomial, the first item asked for by the program, is equal to the highest power appearing in the polynomial. The program overcomes the inefficiency of evaluating each power separately by generating each power from the previous smaller one by a single multiplication. This can be a great time saver if many evaluations of high-power polynomials are done. Thus a polynomial,

$$F(X) = A(1)X^N + A(2)X^{(N-1)} + A(3)X^{(N-2)} + \ldots + A(N+1)$$

can be expressed in a factored form as

$$F(X) = ((\ldots(A(1)\,X + A(2))X + A(3))X \ldots + A(N+1)$$

This form of the equation is extremely easy to program in a FOR-DO loop as is done in *Procedure Work.* These few lines provide the heart of the program, and the remaining lines deal with inputting and outputting the information.

```
(*=========================================*)
(*                                         *)
(* Title:   Polynomial Evaluator           *)
(* Program Summary: Program computes       *)
(*          the value of any polynomial    *)
(*          equation.                      *)
(*                                         *)
(*=========================================*)


PROGRAM PolynomialEvaluator;

 VAR
  Answer,
  Ans,
  Cont             : Char;
  Out              : Text;
  Order,
  Steps            : Integer;
  A                : Array [1..50] of Real;
  Sign             : Array [1..50] of Char;
  X,
  Sum              : Real;

Procedure Outputs;
  begin
    page(output);
     repeat
       gotoXY(0,5);
        Write('S(creen or P(rinter ? ');
        Readln(answer);
     until Answer in [ 'S','s','p','P' ]
  end;  {outputs}

Procedure Title;
  begin
    page(output);
     gotoXY(8,5);
      Writeln('Polynomial Evaluator');
      Writeln;Writeln;
      Writeln('This program evaluates Polynomials');
      Writeln('To use, enter the coefficients');
      Writeln('of the polynomial, the order (highest');
```

```
      Writeln('power) of the polynomial and a value');
      Writeln('for X to be evaluated.');
      Writeln
  end;  { title }

Procedure InputOrder;
  begin
    Write('Enter the order of the polynomial: ');
    Readln( Order );
  end;

Procedure InputPoly;
  begin
    InputOrder;
     if Order < 1 then
       begin
        Writeln(chr(13),'Input Error');
        InputOrder;
       end;
       Order := Order+1 ;
      Writeln('Enter the coefficients:');
       Writeln;
      For  Steps := 1 to Order do
        begin
           Write('A',Steps,' = ');
           Readln( A[ Steps ] );
               Sign[ Steps ] := '-';
                  if ABS( A[ Steps ] ) = A [ Steps ] then
                        Sign[ Steps ] := '+';
          end;
    end;

 Procedure InputX;
   begin
      Writeln;
      Write('What is the value of X: ');
      Readln( X );
   end;

Procedure Work;
   begin
     Sum := A[ 1 ];
           for Steps := 2 to Order do
               Sum := Sum * X + A [ Steps ];
   end;

Procedure Printout;
    forward;

Procedure PrintS;
   begin
    Writeln;Writeln;Writeln;
                Writeln('The polynomial: ');
                Writeln;
                   if Order = 1 then
                     begin
                        Writeln( X,' = ',X);
                        Exit ( PrintS );
                     end;
                   if Order = 2 then
                     begin
                        Writeln( Sign[ Order-1 ],ABS( A[ Order-1 ]),'X ',
                               Sign[ Order ],ABS(A[ Order ]),' = ', Sum );
                        Writeln;
                        Writeln('For X = ', X);
                     end
                   else
                     begin
                        for Steps := 1 to Order-2 do
                           Write( Sign[ Steps ],ABS(A[ Steps ]),'X^',Order-Steps);
                        Writeln( Sign[ Order-1 ],ABS( A[ Order-1 ]),'X ',
                               Sign[ Order ],ABS(A[ Order ]),' = ', Sum );
                        Writeln;
                        Writeln('For X = ', X);
                        Writeln;Writeln;Writeln;
                        Write('Do you want to try another value of X?');
                        Readln( Ans );
```

```
                    if Ans in [ 'Y','y' ] then
                       begin
                         InputX;
                         Work;
                         Printout
                       end
                  end
end;

Procedure Two;
  begin
     if Order = 2 then
       begin
         Writeln( Sign[ Order-1 ],ABS( A[ Order-1 ]),'X ',
                Sign[ Order ],ABS(A[ Order ]),' = ', Sum );
         Writeln(out, Sign[ Order-1 ],ABS( A[ Order-1 ]),'X ',
                Sign[ Order ],ABS(A[ Order ]),' = ', Sum );
         Writeln;
         Writeln(out);
         Writeln('For X = ', X);
         Writeln(out,'For X = ', X);
       end
     else
       begin
         for Steps := 1 to Order-2 do
           begin
            Write(Sign[ Steps ],ABS(A[ Steps ]),'X^',Order-Steps);
            Write(out,Sign[ Steps ],ABS(A[ Steps ]),'X^',Order-Steps);
           end;
       Writeln( Sign[ Order-1 ],ABS( A[ Order-1 ]),'X ',
            Sign[ Order ],ABS(A[ Order ]),' = ', Sum );
       Writeln( out,Sign[ Order-1 ],ABS( A[ Order-1 ]),'X ',
            Sign[ Order ],ABS(A[ Order ]),' = ', Sum );
       Writeln;
       Writeln(out);
       Writeln('For X = ', X);
       Writeln(out,'For X = ', X);
       Writeln;Writeln;Writeln;
       Writeln(out);Writeln(out);Writeln(out);
    end;
  end;

Procedure PrintP;
  begin
   Rewrite(out,'printer:');
           Writeln;Writeln;Writeln;
           Writeln(out);Writeln(out);Writeln(out);
           Writeln('The polynomial: ');
           Writeln(out,'The polynomial: ');
           Writeln;
           Writeln(out);
             if Order = 1 then
               begin
                 Writeln( X,' = ',X);
                 Writeln(out, X,' = ',X);
                 Close(out);
                 Exit ( PrintP );
               end;
             Two;
             Close(out);
                    Write('Do you want to try another value of X?');
                    Readln( Ans );
                      if Ans in [ 'Y','y' ] then
                         begin
                           InputX;
                           Work;
                           Printout
                         end
  end;

Procedure Printout;
  begin
   case answer of
   's','S' : PrintS;

   'p','P' : PrintP;
       end; { case }
  end;
```

```
begin { Main }
  Repeat
    Outputs;
    Title;
    InputPoly;
    InputX;
    Work;
    Printout;
    Write('Do you want another polynomial? (Y/N) ');
    Readln(Cont);
   until Cont in [ 'N','n']
  end.
```

Polynomial Evaluator

This program evaluates polynomials.
To use, enter the coefficients
of the polynomial, the order (highest
power) of the polynomial and a value
for X to be evaluated.

Enter the order of the polynomial: 4
Enter the coefficients:
A1 = 5
A2 = 178
A3 = 3
A4 = Ø
A5 = 9

What is the value of X: 3

The polynomial:
+5.ØØØØØX∧4 + 1.78ØØE2X∧3 + 3.ØØØØØX∧2 + .ØØØØØX + 9.ØØØØØ = 5.247ØØE3 .

For X = 3.ØØØØØ

Do you want to try another value of X? Y

What is the value of X: 5

The polynomial:
+ 5.ØØØØØX∧4 + 1.78ØØØE2X∧3 + 3.ØØØØØX∧2 + Ø.ØØØØØX + 9.ØØØØ = 2.5459ØE4

For X = 5.ØØØØ

Do you want to try another value of X? N

Do you have another polynomial? (Y/N) N

2.3 QUADRATIC EQUATION SOLVER

One of the most familiar equations to engineers and scientists is the quadratic equation. This program solves quadratic equations for all three special cases: real roots, real repeating roots, and complex roots. It solves the equations of the form

$$AX^2 + BX + C = 0$$

and simply requires that the coefficients of the equation, A,B, and C, be entered.

```
(*=========================================*)
(*                                         *)
(* Title:  Quadratic Equation Solver       *)
(* Program Summary: Program solves         *)
(*         quadratic equations for         *)
(*         real, complex, and              *)
(*         repeating roots.                *)
(*                                         *)
(*=========================================*)

PROGRAM QuadraticSolver;

 uses Transcend;

 VAR
  Answer,
  Ans,
  Cont            : Char;
  Out             : Text;
  A,
  B,
  C,
  Z,
  Root1,
  Root2,
  CRoot1,
  CRoot2,
  Image1,
  Image2          : Real;

Procedure Outputs;
  begin
    page(output);
     repeat
       gotoXY(Ø,5);
        Write('S(creen or P(rinter ? ');
        Readln(answer);
     until Answer in [ 'S','s','p','P' ]
  end;  {outputs}

Procedure Title;
  begin
    page(output);
     gotoXY(7,5);
      Writeln('Quadratic Equation Solver');
      Writeln;Writeln;
      Writeln('This program solves quadratic equations');
      Writeln('of the form: ');
      Writeln('               AX^2+BX+C=Ø');
      Writeln;
      Writeln('for real, complex and repeating roots.');
      Writeln;
      Writeln('For multiple entries leave a space');
      Writeln('between entries.');
  end;  { title }

Procedure Input;
  begin
    Write('Enter coefficients A,B,C:  ');
    Readln( A,B,C );
  end;
```

```
Procedure PrintReal;
  begin
   case answer of
   's','S' : begin
               Writeln('The roots to the equation are real');
               Writeln;
               Writeln('Root 1 = ',Root1,'   Root 2 = ',Root2);
             end;
   'p','P' : begin
                Rewrite(out,'printer:');
                    Writeln('The roots to the equation are real');
                    Writeln(out,'The roots to the equation are real');
                    Writeln;
                    Writeln(out);
                    Writeln('Root 1 = ',Root1,'   Root 2 = ',Root2);
                    Writeln(out,'Root 1 = ',Root1,'   Root 2 = ',Root2);
                Close(out)
             end
        end { case }
  end;

 Procedure PrintRRep;
    begin
   case answer of
   's','S' : begin
               Writeln;
               Writeln('The roots to the equation are real and');
               Writeln('repeating.');
               Writeln;
               Writeln('Root 1 = ',Root1,'   Root 2 = ',Root2);
             end;
   'p','P' : begin
               Rewrite(out,'Printer:');
                  Writeln;
                  Writeln(out);
                  Writeln('The roots to the equation are real and');
                  Writeln(out,'The roots to the equation are real and');
                  Writeln('repeating.');
                  Writeln(out,'repeating.');
                  Writeln;
                  Writeln(out);
                  Writeln('Root 1 = ',Root1,'   Root 2 = ',Root2);
                  Writeln(out,'Root 1 = ',Root1,'   Root 2 = ',Root2);
               Close(out);
            end
        end { case }
    end;

Procedure PrintComp;
   begin
    case answer of
    's','S' : begin
                Writeln;
                Writeln('The roots to the equation are complex');
                Writeln;
                Writeln('Root 1 = ',CRoot1,' + i',ABS(Image1));
                Writeln('Root 2 = ',CRoot2,' - i',ABS(Image2));
              end;
    'p','P' : begin
                Rewrite(out,'Printer:');
                  Writeln;
                  Writeln;
                  Writeln('The roots to the equation are complex');
                  Writeln('The roots to the equation are complex');
                  Writeln;
                  Writeln;
                  Writeln('Root 1 = ',CRoot1,' + i',ABS(Image1));
                  Writeln('Root 1 = ',CRoot1,' + i',ABS(Image1));
                  Writeln('Root 2 = ',CRoot2,' - i',ABS(Image2));
                  Writeln('Root 2 = ',CRoot2,' - i',ABS(Image2));
                Close(out);
              end
          end { case }
    end;

 Procedure Complex;
   begin
```

```
      CRoot1 := - B / ( 2 * A );
      CRoot2 := CRoot1;
      Image1 := Sqrt( ABS( Z ) ) / ( 2 * A );
      Image2 := - Image1;
     PrintComp
  end;

Procedure RRepeat;
   begin
      Root1 := - B / ( 2 * A );
      Root2 := Root1;
     PrintRRep
   end;

Procedure Real;
   begin
      Z := Sqr( B ) - ( 4 * A * C );
          if Z < Ø then
             Complex;
          if Z = Ø then
             RRepeat
          else
           begin
             Root1 := (- B + Sqrt( Z )) / ( 2 * A );
             Root2 := (- B - Sqrt( Z )) / ( 2 * A );
           PrintReal
          end
  end;



   begin { Main }
      Repeat
         Outputs;
         Title;
         Input;
         Real;
           Write('Do you want another root? (Y/N) ');
          Readln(Cont);
         until Cont in [ 'N','n']
   end.
```

Quadratic Equation Solver

This program solves quadratic equations
of the form:
AX ∧ 2 + BX + C = Ø
for real, complex and repeating roots.

For multiple entries leave a space
between entries.

Enter coefficients A,B,C:: 7 23 -6

The roots to the equation are real.

Root 1 = 2.42911E-1   Root 2 = -3.52863

Do you want another root? (Y/N) N

2.4 NEWTON-RAPHSON ROOTS

In the solution of equations, knowing one of the equation's roots often makes it much easier to find the others. An approximation technique, known as the Newton-Raphson method, can be used to find the first positive real roots of an equation. The approximation is calculated according to the following formula:

$$X(N + 1) = X(N) - F(X(N)/ F'(X(N))$$

where F(X) is the function, F'(X) is its derivative, and N is its iteration.

By repeating this calculation, it is possible to get successively finer estimates of the root of the equation. The accuracy of the process is determined by the value of *1E-6* in *Procedure Calculate*, which should be small enough for any requirements.

The iteration formula itself is evaluated in *Procedure Calculate*, just above the "if" statement that checks for *1E-6*. The *For-Do* loop has been introduced to limit the number of iterations to 30. It should be understood that this will affect the accuracy in cases where the *1E-6* value is not reached in 30 iterations. The only reason the *For-Do* loop was included was to speed the process. If accuracy is desired, the *For-Do* loop can be eliminated and a *Repeat Until Abs(X - XN) <1E - 6*, can be used.

```
(*========================================*)
(*                                        *)
(* Title:   Newton-Raphson Roots          *)
(* Program Summary: Program calculates    *)
(*          the root of a function,       *)
(*          once knowing one of the       *)
(*          roots.                        *)
(*                                        *)
(*========================================*)


PROGRAM Newton-Raphson  ;

uses Transcend;  (* only if needed *)

 VAR
  Answer,
  Ans,
  Cont              : Char;
  Out               : Text;
  X,
  XN                : Real;
  N                 : Integer;

Procedure Outputs;
  begin
    page(output);
     repeat
       gotoXY(0,5);
        Write('S(creen or P(rinter ? ');
        Read(answer);
     until Answer in [ 'S','s','p','P' ]
  end;  {outputs}

Procedure Title;
  begin
    page(output);
     gotoXY(10,5);
       Writeln('Newton-Raphson Roots');
       Writeln;Writeln;
       Writeln('This program finds the first positive');
       Writeln('real root of an equation which you must');
       Writeln('define in the function block.');
       Writeln('After doing this, compile the program ');
```

```
        Writeln('and run it. An example function');
        Writeln('is shown, any may be substituted');
        Writeln;
   end;  { title }

 Function Fun(X: Real) : Real;
   begin
     Fun := Sqr (X) -25;
   end;

 Function Fun1(X: Real) : Real;
   begin
     Fun1 := 2 * X ;
   end;

 Procedure Calculate;
   begin
     X := Ø.5 ;
       For N := 1 to 3Ø do
         begin
          XN := X - ( Fun( X ) / Fun1( X )) ;
                if Abs( X - XN ) < 1E-6 then
                     Exit ( Calculate );
          X := XN
         end
    end;

  Procedure Print;
     begin
       case answer of
       's','S' : begin
                   Writeln;
                   Writeln('The root := ',X);
                   Writeln;
                 end;
       'p','P' : begin
                   Rewrite(out,'Printer:');
                      Writeln;
                      Writeln(out);
                      Writeln('The root := ',X);
                      Writeln(out,'The root := ',X);
                      Writeln;
                      Writeln(out);;
                   Close(out);
                end
            end { case }
        end;

   begin { Main }
       Outputs;
       Title;
         Calculate;
         Print;
         Writeln('To use another function recompile!');
   end.
```

Newton-Raphson Roots

This program finds the first positive
real roots of an equation which you must
define in the function block.
After doing this, compile the program
and run it. An example function
is shown, any may be substituted.

The root = 5.ØØØØØ

To use another function recompile!

2.5 DERIVATIVE OF A FUNCTION

The derivative (or tangent) of a curve at a particular point can be approximated by moving a small distance along the curve to another point and comparing the change in the Y-coordinate to the change in the X-coordinate. This method of determining the derivative is known as the First Forward Different Method and is very easy to implement on a computer.

The user must enter the required function into the program in *Function Fun*, and then compile it. A sample function Fun = Sqr(X) + 3X – 2 is used to demonstrate how functions are used in Pascal.

When calculating the derivative, the program first evaluates the function at a given value of X (*Procedure Calculate*). It is then evaluated for X equal to the original value of X plus a small difference, or Delta X (Dx).

Finally, the value of the function at X is subtracted from the value of the function at X plus Dx, and the difference is divided by Dx, producing the derivative Dy (*Procedure Calculate*).

```
(*=========================================*)
(*                                        *)
(* Title:   Derivative of a Function      *)
(* Program Summary: Program calculates    *)
(*          the derivative of a           *)
(*          function using the First      *)
(*          Forward Difference Method.    *)
(*                                        *)
(*=========================================*)

PROGRAM Derivative Solver ;

 (* uses Transcend; only if needed *)

 VAR
  Answer,
  Ans,
  Cont              : Char;
  Out               : Text;
  X,
  Y1,
  Y2,
  Dx,
  Dy                : Real;

Procedure Outputs;
  begin
    page(output);
     repeat
       gotoXY(0,5);
         Write('S(creen or P(rinter ? ');
         Readln(answer);
     until Answer in [ 'S','s','p','P' ]
  end;  {outputs}

Procedure Title;
  begin
    page(output);
     gotoXY(8,5);
      Writeln('Derivative of a Function');
      Writeln;Writeln;
      Writeln('This program calculates the derivative');
      Writeln('of a function by using the "First');
      Writeln('forward difference method".  You must');
      Writeln('define the function in the function');
      Writeln('block.  After doing this, compile the');
      Writeln('program and run it. An example function');
      Writeln('is shown, any may be substituted');
```

```
      Writeln;
      Writeln('The program will ask you for the');
      Writeln('point you want evaluated')
  end;   { title }

Function Fun(X: Real) : Real;
  begin
    Fun := Sqr (X) + 3 * X - 2 ;
  end;

Procedure Input;
  begin
    Writeln;
    Write('Enter X to be evaluated:   ');
    Readln( X );
  end;

Procedure Calculate;
  begin
     Dx := 0.0001 ;
     Y1 := Fun( X ) ;
     Y2 := Fun( X + Dx );
     Dy := ( Y2 - Y1 )/ Dx
  end;


 Procedure Print;
    begin
      case answer of
      's','S' : begin
                  Writeln;
                  Writeln('Derivative at X = ',X,' is  ',Round(Dy));
                  Writeln;
                end;
      'p','P' : begin
                  Rewrite(out,'Printer:');
                     Writeln;
                     Writeln(out);
                     Writeln('Derivative at X = ',X,' is  ',Round(Dy));
                     Writeln(out,'Derivative at X = ',X,' is  ',Round(Dy));
                     Writeln;
                     Writeln(out);;
                  Close(out);
               end
          end { case }
      end;

   begin { Main }
     Repeat
       Outputs;
       Title;
       Input;
         Calculate;
         Print;
         Write('Evaluate it at another point? (Y/N) ');
        Readln(Cont);
       until Cont in [ 'N','n']
   end.
```

Derivative of a Function

This program calculates the derivative of a function by using the "First Forward Difference Method." You must define the function in the function block. After doing this, compile the program and run it. An example function is shown, any may be substituted.

The program will ask you for the point you want evaluated.

Enter X to be evaluated: 5

Derivative at X = 5.00000 is 13

Evaluate it at another point? (Y/N) N

2.6 INTEGRATION BY SIMPSON'S RULE

The integral of a function or data is really the area underneath a curve formed by it. This program will calculate that area, or integral, by an approximation technique known as Simpson's Rule, which states:

$$\text{Area} = \frac{\Delta X(Y1 + 4Y2 + 2Y3 + 4Y4 + \ldots 4Y24 + Y25)}{3}$$

The program is actually two programs in one; that is, data integration and function integration are treated in different ways. *Procedure Input* asks which type of integration is to be performed. If data integration is to be performed, the number of data points must be entered, as well as the value for each point. The data points should be taken at equal intervals, and the program will ask for the interval width. The integral is then evaluated in *Procedure WorkData.*

If the function option is chosen, *InputInt* and *WorkInt* are used. Because Pascal is a compiled language, the function must be put in *Function Fun* and then the program must be recompiled. The program automatically asks the limits of integration of the function and then automatically divides the integration range into intervals. The function is then evaluated.

```
(*=========================================*)
(*                                         *)
(* Title:  Integration by Simpson's Rule   *)
(* Program Summary: Program performs       *)
(*         integration according to        *)
(*         Simpson's Rule.  Supply         *)
(*         either the function or          *)
(*         the data.                       *)
(*                                         *)
(*=========================================*)


PROGRAM Integration ;

 (* uses Transcend; only if needed *)

 VAR
  Answer,
  Ans,
  Which,
  Key,
  Cont            : Char;
  Out             : Text;
  Lower,
  Upper,
  Width,
  FirstLast,
  Sum,
  X,
  Integral,
  Sum1,
  Odd,
```

```
  Even          : Real;
  Interval,
  Stepping,
  Stepper,
  Steps,
  Step,
  Cons            : Integer;
  Y             : Array [ Ø.. 255] of Real;

Procedure Outputs;
  begin
    page(output);
     repeat
       gotoXY(Ø,5);
        Write('S(creen or P(rinter ? ');
        Readln(answer);
     until Answer in [ 'S','s','p','P' ]
  end;  {outputs}

Procedure Title;
  begin
    page(output);
     gotoXY(8,1);
      Writeln('Simpson Integration');
      Writeln;Writeln;
      Writeln('This program performs integration');
      Writeln('according to Simpson',chr(39),'s Rule. You');
      Writeln('may integrate either a function or');
      Writeln('data.');
      Writeln;
      Writeln('If you are entering data you must');
      Writeln('enter the number of spaces into which');
      Writeln('the integration interval has been');
      Writeln('divided.  This must be an even number');
      Writeln('You must enter the width of each space.');
      Writeln;
      Writeln('If you are integrating a function');
      Writeln('then you must define it in the ');
      Writeln('Function block.  After this the');
      Writeln('program must be compiled and run.');
      Writeln;
      Write('<<< Hit any key to continue >>>');
      Read(key)
   end;  { title }

Function Fun(X: Real) : Real;
  begin
    Fun := Sqr (X) - 25  ;
  end;

Procedure Input;
  begin
    Repeat
       page(output);
       Writeln;
       Write('Enter Integral (1) or Data(2):  ');
       Readln( Which );
    until Which in [ '1','2' ];
  end;

Procedure InputInt;
  begin
    Writeln;
    Writeln('Place a space between entries!');
    Writeln;
    Write('Enter limits ( Lower, Upper ): ');
    Readln( Lower , Upper )
  end;

Procedure InputData;
  begin
    Writeln;Writeln;
    Write('Enter how many intervals: ');
    Readln( Interval );
    Writeln;
    Write('Enter interval width: ');
    Readln( Width );
```

```
      Writeln;
        For Step := 1 to (Interval+1) do
          begin
            Write('Y(',Step,') = ');
            Readln( Y [ Step ] );
          end
    end;

  Procedure Printout;
     begin
       case answer of
       's','S' : begin
                   Writeln;
                   Writeln;
                   Writeln('The integral is ', Integral );
                   Writeln;
                   Writeln;
                 end;
       'p','P' : begin
                   Rewrite(out,'Printer:');
                       Writeln;
                       Writeln(out);
                       Writeln;
                       Writeln(out);;
                   Writeln('The integral is ', Integral );
                   Writeln(out,'The integral is ', Integral );
                   Writeln;
                   Writeln;
                   Writeln(out);
                   Writeln(out);
                   Close(out);
                 end
           end { case }
       end;

  Procedure WorkData;
   begin
     X    := Ø;
     Sum1     := Ø;
       For  Stepping := 1 to ( Interval Div 2) do
          X := X + Y[ Stepping * 2 ] ;
       For  Stepper := 1 to (( Interval - 1 ) Div 2 ) do
          Sum1  := Sum1  + Y[ Stepper * 2 + 1 ];
      FirstLast := Y[ 1 ] + Y [ Interval + 1 ] ;
      Sum := FirstLast + ( 4 * X ) + ( 2 * Sum1 ) ;
      Integral := Width / 3 * Sum ;
       Printout
    end;

  Procedure WorkInt;
    begin
      X    := Lower ;
      Odd  := Ø;
      Even := Ø ;
      Cons   := 25;
         Width  := ( Upper - Lower ) / ( Cons - 1 );
          For Steps := 1 to Cons do
           begin
                Y [ Steps ] := Fun ( X );
                     X := X + Width ;
           end;
          For Steps := 1 to (( Cons - 2 ) Div 2 ) do
             Odd   := Odd + 2 * ( Y [ Steps * 2 + 1 ] );
          For Steps := 1 to (( Cons - 1) Div 2) do
             Even := Even + 4 * ( Y [ Steps * 2 ] );
       Integral := Width / 3 * ( Y [ 1 ] + Even + Odd + Y [ Cons ] );
         Printout
      end;

    begin { Main }
      Repeat
        Outputs;
        Title;
        Input;
          Case Which of
           '1'      : begin
                        InputInt;
```

```
                WorkInt;
              end;
      '2'     : begin
                WorkData;
                InputData;
                WorkData;
              end;
    end { case };
        Write('Another Integration? (Y/N) ');
           Readln(Cont);
  until Cont in [ 'N','n']
end.
```

Simpson Integration

This program performs integration
according to Simpson's Rule. You
may integrate either a function or
data.

If you are entering data you must
enter the number of spaces into which
the integration interval has been
divided. This must be an even number.
You must enter the width of each space.

If you are integrating a function
then you must define it in the
Function block. After this the
program must be compiled and run.

<<< Hit any key to continue >>>

Enter Integral (1) or Data(2): 2

Enter how many intervals: 4

Enter interval width: 2

Y(1) = Ø
Y(2) = 2
Y(3) = 4
Y(4) = 16
Y(5) = 32

The integral is 7.46667E1

Another Integration? (Y/N) Y

Place a space between entries!

Enter limits (Lower, Upper): Ø 1Ø

The integral is 8.33333E1

3 COMPLEX MATHEMATICS

Complex numbers are frequently found in engineering calculations, and it is convenient to be able to perform calculations with them on the computer. Unfortunately, this is another area in which Pascal is not up to par with FORTRAN. Because of the unique properties of complex numbers, they cannot be handled in the same way as conventional or real numbers. A complex number is really a pair of numbers, A and B, and is expressed as

$$A + iB \text{ or } A + jB$$

where i and j have the property that i^2 or j^2 is equal to −1. The imaginary part of the number is iB and the other part, A, is called the real part.

The programs that follow were designed to overcome the lack of complex capabilities that exist in BASIC and Pascal, and to improve the capability to perform complex computations provided in FORTRAN.

3.1 COMPLEX ADDITION

The addition of complex numbers can be performed by following the general rule:

$$(A + iB) + (C + iD) = (A + C) + i(B + D)$$

The program is very general and lets you add any amount of complex numbers desired. The addition of the real and imaginary parts is done separately in *Procedure Input*. As the numbers are entered, they are automatically added. The last line in *Input* determines whether the sign is positive or negative. *Procedure Printout* prints out the correct formatted answer. *Procedure Polar* converts the complex number to polar notation, where it is expressed as a magnitude and a direction.

```
(*=======================================*)
(*                                      *)
(* Title:  Complex Addition             *)
(* Program Summary: Program performs    *)
(*         complex addition either      *)
(*         in normal notation or        *)
(*         polar notation.              *)
(*                                      *)
(*=======================================*)


 PROGRAM ComplexAddition ;

  uses Transcend;

  VAR
   Resp,
   Answer,
   Ans,
   Sign,
   Cont            : Char;
   Out             : Text;
   Number,
   Steps           : Integer;
   Realtot,
   Imagtot,
   Mag,
   Theta           : Real;
   Reals,
   Imagi           : Array [Ø..5Ø] of Real;

 Procedure Outputs;
   begin
     page(output);
      repeat
        gotoXY(Ø,5);
         Write('S(creen or P(rinter ? ');
         Readln(Answer);
      until Answer in [ 'S','s','p','P' ]
   end;  {outputs}

 Procedure Title;
   begin
     page(output);
      gotoXY(12,1);
       Writeln('Complex Addition');
       Writeln;Writeln;
       Writeln('This program performs Complex Math');
       Writeln('and will add any number of complex');
       Writeln('numbers.');
       Writeln;
       Writeln('For multiple data entries ');
       Writeln('place a space between entries');
       Writeln;
    end;  { title }


 Procedure Input;
   begin
        Writeln;
        Write('How many numbers are you adding? ');
        Readln( Number );
          Realtot := Ø;
          Imagtot := Ø;
            Writeln;
            Writeln('Enter the real and imaginary parts');
            Writeln;
              For Steps := 1 to Number do
                  begin
                         Sign := '-';
                           Write('Number ',Steps,' := ');
                           Readln( Reals [ Steps ], Imagi [ Steps ] );
                             Realtot := Realtot + Reals [ Steps ] ;
                             Imagtot := Imagtot + Imagi [ Steps ] ;
                               If Abs ( Imagtot ) = Imagtot then
                                      Sign := '+'
                   end;
```

```
 end;

Procedure Polar;
  begin
    Mag := Sqrt( Sqr( Realtot ) + Sqr ( Imagtot ));
      if (Realtot = Ø) and (Imagtot < Ø) then
            begin
                Theta := 27Ø ;
                Exit( Polar )
            end;
      if (Realtot = Ø) and (Imagtot > Ø) then
            begin
                Theta :=  9Ø ;
                Exit( Polar )
            end;
      if (Realtot < Ø) and (Imagtot = Ø) then Theta := 18Ø else
           begin
             Theta := Atan ( Imagtot / Realtot ) * 57.2958 ;
               if Realtot < Ø  then Theta := Theta + 18Ø ;
           end
    end;

Procedure Printout;
   begin
     case answer of
     's','S' : begin
                  Writeln;
                  Writeln;
                  Write('The sum is ',Round( Realtot ),' ',Sign);
                  Writeln(' i',Abs(Round( Imagtot )));
                  Writeln;
                  Write('Do you want it in polar form? (Y/N)  ');
                  Readln( Resp );
                      If Resp in ['Y','y'] Then
                         begin
                          Polar;
                          Writeln;
                          Writeln('Magnitude := ',Round(Mag*1ØØ)/1ØØ);
                          Writeln;
                          Writeln('Angle := ',Round(Theta*1ØØ)/1ØØ,' degrees');
                         end;
                  Writeln;
                  Writeln;
                end;
     'p','P' : begin
                  Rewrite(out,'Printer:');
                  Writeln;
                  Writeln;
                  Writeln(out);
                  Writeln(out);
                  Write('The sum is ',Round( Realtot ),' ',Sign);
                  Writeln(' i',Abs(Round( Imagtot )));
                  Write(out,'The sum is ',Round( Realtot ),' ',Sign);
                  Writeln(out,' i',Abs(Round( Imagtot )));
                  Writeln;
                  Write('Do you want it in polar form? (Y/N)  ');
                  Readln( Resp );
                      If Resp in ['Y','y'] Then
                         begin
                          Polar;
                          Writeln;
                          Writeln(out);
                          Writeln('Magnitude ::= ',Mag);
                          Writeln(out,'Magnitude := ',Mag);
                          Writeln(out);
                          Writeln;
                          Writeln('Angle := ',Theta,' degrees');
                          Writeln(out,'Angle := ',Theta,' degrees');
                         end;
                  Writeln;
                  Writeln;
                  Writeln(out);
                  Writeln(out);
                Close(out);
               end
           end { case }
       end;
```

```
begin { Main }
  Repeat
    Outputs;
    Title;
    Input;
    Printout;
          Write('Another Addition? (Y/N) ');
              Readln(Cont);
  until Cont in [ 'N','n']
end.
```

Complex Addition

This program performs complex math
and will add any number of complex
numbers.

For multiple data entries
place a space between entries.

How many numbers are you adding? 4

Enter the real and imaginary parts.

Number 1 = 2 3
Number 2 = Ø -5
Number 3 = 23 12
Number 4 = -3Ø 5

The sum is -5 + i15

Do you want it in polar form? (Y/N) Y

Magnitude = 1.58114E1

Angle = 1.Ø8435E2 degrees

Another Addition? (Y/N) N

3.2 COMPLEX SUBTRACTION

The subtraction of complex numbers is very similar to the addition of complex numbers. The only difference is the signs. The program is performed according to the following rule:

$$(A + iB) - (C + iD) = (A - C) + i(B - D)$$

Operation of this program is identical to the addition program. The polar routine is also included.

```
(*=========================================*)
(*                                         *)
(* Title:  Complex Subtraction             *)
(* Program Summary: Program performs       *)
(*        complex subtraction either       *)
(*        in normal notation or            *)
(*        polar notation.                  *)
(*                                         *)
(*=========================================*)


PROGRAM ComplexSubtraction ;

 uses Transcend;

 VAR
  Resp,
  Answer,
  Ans,
  Sign,
  Cont          : Char;
  Out           : Text;
  Real1,
  Real2,
  Image1,
  Image2,
  RealTot,
  ImageTot,
  Mag,
  Theta         : Real;

Procedure Outputs;
  begin
    page(output);
     repeat
       gotoXY(0,5);
        Write('S(creen or P(rinter ? ');
        Readln(Answer);
     until Answer in [ 'S','s','p','P' ]
  end;  {outputs}

Procedure Title;
  begin
    page(output);
     gotoXY(10,1);
      Writeln('Complex Subtraction');
      Writeln;Writeln;
      Writeln('This program performs Complex Math');
      Writeln('and will subtract complex numbers');
      Writeln;
      Writeln('For multiple data entries ');
      Writeln('place a space between entries');
      Writeln;
   end;  { title }


Procedure Input;
  begin
       Sign := '-';
       Writeln;
       Writeln('What are the numbers you are subtracting?');
       Writeln('Enter the real and imaginary parts');
       Writeln;
       Readln( Real1 , Image1 );
       Readln( Real2 , Image2 );
           RealTot := Real1 - Real2 ;
           ImageTot := Image1 - Image2 ;
               if Abs( ImageTot ) = ImageTot then Sign := '+';
  end;

 Procedure Polar;
   begin
     Mag := Sqrt( Sqr( RealTot ) + Sqr ( ImageTot ));
       if (RealTot = 0) and (ImageTot < 0) then
                begin
                   Theta := 270 ;
                   Exit( Polar )
                end;
```

```
  if (RealTot = Ø) and (ImageTot > Ø) then
         begin
            Theta :=  9Ø ;
            Exit( Polar )
         end;
  if (RealTot < Ø) and (ImageTot = Ø) then Theta := 18Ø else
      begin
        Theta := Atan ( ImageTot / RealTot ) * 57.2958 ;
          if RealTot < Ø  then Theta := Theta + 18Ø ;
      end
end;


Procedure Printout;
   begin
     case answer of
     's','S' : begin
                 Writeln;
                 Writeln;
                 Write('The difference is ', RealTot ,' ',Sign);
                 Writeln(' i',Abs( ImageTot ));
                 Writeln;
                 Write('Do you want it in polar form? (Y/N)  ');
                 Readln( Resp );
                    If Resp in ['Y','y'] Then
                       begin
                        Polar;
                        Writeln;
                        Writeln('Magnitude := ',Round(Mag*1ØØ)/1ØØ);
                        Writeln;
                        Writeln('Angle := ',Round(Theta*1ØØ)/1ØØ,' degrees');
                       end;
                 Writeln;
                 Writeln;
               end;
     'p','P' : begin
                 Rewrite(out,'Printer:');
                 Writeln;
                 Writeln;
                 Writeln(out);
                 Writeln(out);
                 Write('The difference is ', RealTot ,' ',Sign);
                 Writeln(' i',Abs( ImageTot ));
                 Write(out,'The difference is ', RealTot ,' ',Sign);
                 Writeln(out,' i',Abs( ImageTot ));
                 Writeln;
                 Write('Do you want it in polar form? (Y/N)  ');
                 Readln( Resp );
                    If Resp in ['Y','y'] Then
                       begin
                        Polar;
                        Writeln;
                        Writeln(out);
                        Writeln('Magnitude := ',Mag);
                        Writeln(out,'Magnitude := ',Mag);
                        Writeln(out);
                        Writeln;
                        Writeln('Angle := ',Theta,' degrees');
                        Writeln(out,'Angle := ',Theta,' degrees');
                       end;
                 Writeln;
                 Writeln;
                 Writeln(out);
                 Writeln(out);
               Close(out);
             end
         end { case }
      end;


  begin { Main }
    Repeat
      Outputs;
      Title;
      Input;
      Printout;
```

```
          Write('Another Subtraction? (Y/N) ');
              Readln(Cont);
  until Cont in [ 'N','n']
end.
```

Complex Subtraction

This program performs complex math
and will subtract complex numbers.

For multiple data entries
place a space between entries.

What are the numbers you are subtracting?

Enter the real and imaginary parts.

23 -12
14 89

The difference is 9.00000 - i1.01000E2

Do you want it in polar form? (Y/N) Y

Magnitude = 1.01400E2
Angle = -8.4907E1 degrees

Another subtraction? (Y/N) N

3.3 COMPLEX MULTIPLICATION

Multiplication of complex numbers is somewhat different than the multiplication of ordinary numbers. The real part of the final answer is found by subtracting the product of the imaginary parts of the two numbers from the product of the real parts of the two numbers.

The imaginary part of the answer is found by multiplying the real part of the first number by the imaginary part of the second number and adding it to the product of the real part of the second number and the imaginary part of the first number. A more clearly stated form is:

$$(1)\quad (A + iB)(C + iD) = (AC - BD) + i(BC + AD)$$

This result can be obtained by treating the two parts of one complex number as two separate numbers. Multiplying C + iD first by A and then by iB gives

$$(2)\quad A(C + iD) = AC + iAD$$
$$(3)\quad iB(C + iD) = iCB + i^2BD$$

But since $i^2 = -1$, adding equations (2) and (3) give the results in equations (1).

In this program, as in previous ones, the real and imaginary parts of the final answer are evaluated separately (*Procedure Input*). The ability to get the results in polar form is included.

```
(*========================================*)
(*                                        *)
(* Title:  Complex Multiplication         *)
(* Program Summary: Program performs      *)
(*         complex multiplication in      *)
(*         either normal notation or      *)
(*         in polar notation.             *)
(*                                        *)
(*========================================*)


PROGRAM ComplexMultiply ;

 uses Transcend;

 VAR
  Resp,
  Answer,
  Ans,
  Sign,
  Cont           : Char;
  Out            : Text;
  Real1,
  Real2,
  Imag1,
  Imag2,
  RealProd,
  ImagProd,
  Mag,
  Theta          : Real;

Procedure Outputs;
  begin
    page(output);
     repeat
       gotoXY(Ø,5);
        Write('S(creen or P(rinter ? ');
        Readln(Answer);
     until Answer in [ 'S','s','p','P' ]
  end;  {outputs}

Procedure Title;
  begin
    page(output);
     gotoXY(8,1);
      Writeln('Complex Multiplication');
      Writeln;Writeln;
      Writeln('This program performs Multiplication');
      Writeln('of complex numbers.');
      Writeln;
      Writeln('For multiple data entries ');
      Writeln('place a space between entries');
      Writeln;
   end;  { title }


Procedure Input;
  begin
       Sign := '-';
       Writeln;
       Writeln('What are the numbers you are multiplying?');
       Writeln('Enter the real and imaginary parts');
       Writeln;
       Readln( Real1 , Imag1 );
       Readln( Real2 , Imag2 );
           RealProd := Real1 * Real2 - ( Imag1 * Imag2 ) ;
           ImagProd := Imag1 * Real2 + ( Real1 * Imag2 ) ;
               if Abs( ImagProd ) = ImagProd then Sign := '+';
  end;
```

```
Procedure Polar;
  begin
    Mag := Sqrt( Sqr( RealProd ) + Sqr ( ImagProd ));
      if (RealProd = 0) and (ImagProd < 0) then
            begin
                 Theta := 270 ;
                 Exit( Polar )
            end;
      if (RealProd = 0) and (ImagProd > 0) then
            begin
                 Theta :=  90 ;
                 Exit( Polar )
            end;
      if (RealProd < 0) and (ImagProd = 0) then Theta := 180 else
          begin
            Theta := Atan ( ImagProd / RealProd ) * 57.2958 ;
              if RealProd < 0  then Theta := Theta + 180 ;
          end
    end;

Procedure Printout;
   begin
     case answer of
     's','S' : begin
                 Writeln;
                 Writeln;
                 Write('The product is ', RealProd ,' ',Sign);
                 Writeln(' i',Abs( ImagProd ));
                 Writeln;
                 Write('Do you want it in polar form? (Y/N)  ');
                 Readln( Resp );
                     If Resp in ['Y','y'] Then
                        begin
                         Polar;
                         Writeln;
                         Writeln('Magnitude := ',Round(Mag*100)/100);
                         Writeln;
                         Writeln('Angle := ',Round(Theta*100)/100,' degrees');
                        end;
                 Writeln;
                 Writeln;
               end;
     'p','P' : begin
                 Rewrite(out,'Printer:');
                 Writeln;
                 Writeln;
                 Writeln(out);
                 Writeln(out);
                 Write('The product is ', RealProd ,' ',Sign);
                 Writeln(' i',Abs( ImagProd ));
                 Write(out,'The product is ', RealProd ,' ',Sign);
                 Writeln(out,' i',Abs( ImagProd ));
                 Writeln;
                 Write('Do you want it in polar form? (Y/N)  ');
                 Readln( Resp );
                     If Resp in ['Y','y'] Then
                        begin
                         Polar;
                         Writeln;
                         Writeln(out);
                         Writeln('Magnitude := ',Mag);
                         Writeln(out,'Magnitude := ',Mag);
                         Writeln(out);
                         Writeln;
                         Writeln('Angle := ',Theta,' degrees');
                         Writeln(out,'Angle := ',Theta,' degrees');
                        end;
                 Writeln;
                 Writeln;
                 Writeln(out);
                 Writeln(out);
               Close(out);
              end
          end { case }
      end;
```

```
begin { Main }
  Repeat
    Outputs;
    Title;
    Input;
    Printout;
          Write('Another Multiplication? (Y/N) ');
             Readln(Cont);
  until Cont in [ 'N','n']
end.
```

Complex Multiplication

This program performs Multiplication
of complex numbers.

For multiple data entries
place a space between entries.

What are the numbers you are multiplying?

Enter the real and imaginary parts.

15 18
-6 -2

The product is -5.40000E1 - i1.3800E2

Do you want it in polar form? (Y/N) Y

Magnitude = 1.48189E2
Angle = 2.48629E2 degrees

Another Multiplication? (Y/N) N

3.4 COMPLEX DIVISION

Division of complex numbers is performed according to the following rule:

$$\frac{(A + iB)}{(C + iD)} = \frac{(AC + BD)^2 + i(BD - AC)^2}{C^2 + D^2}$$

This is accomplished separately for the real and imaginary parts in *Procedure Input.*

```
(*==========================================*)
(*                                          *)
(* Title:   Complex Division                *)
(* Program Summary: Program performs        *)
(*          complex division either         *)
(*          in normal notation or           *)
(*          in polar notation.              *)
(*                                          *)
(*==========================================*)
```

```
PROGRAM ComplexDivision ;

 uses Transcend;

 VAR
  Resp,
  Answer,
  Ans,
  Sign,
  Cont            : Char;
  Out             : Text;
  Real1,
  Real2,
  Imag1,
  Imag2,
  RealDiv,
  ImagDiv,
  Mag,
  Theta           : Real;

Procedure Outputs;
  begin
    page(output);
     repeat
       gotoXY(Ø,5);
        Write('S(creen or P(rinter ? ');
        Readln(Answer);
     until Answer in [ 'S','s','p','P' ]
  end;  {outputs}

Procedure Title;
  begin
    page(output);
     gotoXY(12,1);
      Writeln('Complex Division');
      Writeln;Writeln;
      Writeln('This program performs Division of');
      Writeln('complex numbers.');
      Writeln;
      Writeln('For multiple data entries ');
      Writeln('place a space between entries');
      Writeln;
   end;  { title }

Procedure Input;
  begin
       Sign := '-';
       Writeln;
       Writeln('What number (first) are you dividing');
       Writeln('by the other (second)');
       Writeln;
       Writeln('Enter the real and imaginary parts');
       Writeln;
       Readln( Real1 , Imag1 );
       Readln( Real2 , Imag2 );
           RealDiv := ( Real1 * Real2 + Imag1 * Imag2 ) /
                             ( Sqr( Real2 ) + Sqr( Imag2 )) ;
           ImagDiv := ( Imag1 * Real2 - Real1 * Imag2 ) /
                             ( Sqr( Real2 ) + Sqr( Imag2 )) ;
               if Abs( ImagDiv ) = ImagDiv then Sign := '+';
  end;

 Procedure Polar;
   begin
     Mag := Sqrt( Sqr( RealDiv ) + Sqr  ( ImagDiv ));
       if (RealDiv = Ø) and (ImagDiv < Ø) then
                    begin
                      Theta := 27Ø ;
                      Exit( Polar )
                    end;
       if (RealDiv = Ø) and (ImagDiv > Ø) then
                    begin
                      Theta :=  9Ø ;
                      Exit( Polar )
                    end;
       if (RealDiv < Ø) and (ImagDiv = Ø) then Theta := 18Ø else
            begin
```

```
                Theta := Atan ( ImagDiv / RealDiv ) * 57.2958 ;
                  if RealDiv < 0  then Theta := Theta + 180 ;
              end
         end;


   Procedure Printout;
      begin
        case answer of
        's','S' : begin
                   Writeln;
                   Writeln;
                   Write('The answer is ', RealDiv ,' ',Sign);
                   Writeln(' i',Abs( ImagDiv ));
                   Writeln;
                   Write('Do you want it in polar form? (Y/N)   ');
                   Readln( Resp );
                      If Resp in ['Y','y'] Then
                         begin
                          Polar;
                          Writeln;
                          Writeln('Magnitude := ',Round(Mag*100)/100);
                          Writeln;
                          Writeln('Angle := ',Round(Theta*100)/100,' degrees');
                         end;
                   Writeln;
                   Writeln;
                  end;
        'p','P' : begin
                   Rewrite(out,'Printer:');
                   Writeln;
                   Writeln;
                   Writeln(out);
                   Writeln(out);
                   Write('The answer is ', RealDiv ,' ',Sign);
                   Writeln(' i',Abs( ImagDiv ));
                   Write(out,'The answer is ', RealDiv ,' ',Sign);
                   Writeln(out,' i',Abs( ImagDiv ));
                   Writeln;
                   Write('Do you want it in polar form? (Y/N)   ');
                   Readln( Resp );
                      If Resp in ['Y','y'] Then
                         begin
                          Polar;
                          Writeln;
                          Writeln(out);
                          Writeln('Magnitude := ',Mag);
                          Writeln(out,'Magnitude := ',Mag);
                          Writeln(out);
                          Writeln;
                          Writeln('Angle := ',Theta,' degrees');
                          Writeln(out,'Angle := ',Theta,' degrees');
                         end;
                   Writeln;
                   Writeln;
                   Writeln(out);
                   Writeln(out);
                 Close(out);
                end
           end { case }
       end;


    begin { Main }
      Repeat
        Outputs;
        Title;
        Input;
        Printout;
              Write('Another Division? (Y/N) ');
                 Readln(Cont);
      until Cont in [ 'N','n']
    end.
```

```
Complex Division

This program performs division of
complex numbers.

For multiple data entries
place a space between entries.

What number (first) are you dividing
by the other number (second)?

Enter the real and imaginary parts.

2 7
3 4

The answer is 4.62400E1 + i6.76000

Do you want it in polar form? (Y/N) Y

Magnitude = 4.67300E1
Angle = 8.32000 degrees

Another Division? (Y/N) N
```

3.5 ABSOLUTE VALUE

Another function provided in the FORTRAN complex math package is the absolute value function. This is essentially the subroutine that does the polar conversion in the previous programs. Unlike the CABS function in FORTRAN, this program will also return the angle as well as the magnitude of the complex number. The magnitude is found by taking the square root of the sum of the squares of the real part and the imaginary part:

$$\text{Magnitude} = (\text{Real}^2 + \text{Imag}^2)^{1/2}$$

The angle is determined by calculating the inverse tangent of the imaginary part divided by the real part:

$$\text{Angle} = \text{Arctan}\ (\text{Imag}/\text{Real})$$

This is a tricky calculation to make, in that the result of the inverse tangent, Arctan, is a value in radians in the range $\pi/2$ to $-\pi/2$. To achieve the final result, it is necessary to convert the answer to degrees and then determine in which quadrant the angle is located. The conversion to degrees is done in *Procedure Polar.* The determination of the quadrant, and thus the true angle is also done in *Procedure Polar* but at the end of the procedure.

```
(*=======================================*)
(*                                       *)
(* Title:  Absolute Value                *)
(* Program Summary: Program calculates   *)
(*          the absolute value, i.e.,    *)
(*          the magnitude and angle      *)
(*          of complex numbers.          *)
(*                                       *)
(*=======================================*)


PROGRAM AbsoluteValue ;

 uses Transcend;

 VAR
  Answer,
  Ans,
  Cont            : Char;
  Out             : Text;
  Realnum,
  Imagnum,
  Mag,
  Theta           : Real;

Procedure Outputs;
  begin
    page(output);
     repeat
       gotoXY(Ø,5);
        Write('S(creen or P(rinter ? ');
        Readln(Answer);
     until Answer in [ 'S','s','p','P' ]
  end;  {outputs}

Procedure Title;
  begin
    page(output);
     gotoXY(14,1);
      Writeln('Absolute Value');
      Writeln;Writeln;
      Writeln('This program calculates the absolute');
      Writeln('value of a complex number and its angle');
      Writeln;
      Writeln('For multiple data entries ');
      Writeln('place a space between entries');
      Writeln;
   end;  { title }


Procedure Input;
  begin
       Writeln;
       Writeln('Enter the real and imaginary parts');
       Writeln;
       Readln( Realnum , Imagnum );
  end;


 Procedure Polar;
   begin
     Mag := Sqrt( Sqr( Realnum ) + Sqr ( Imagnum ));
       if (Realnum = Ø) and (Imagnum < Ø) then
                begin
                  Theta := 27Ø ;
                  Exit( Polar )
                end;
       if (Realnum = Ø) and (Imagnum > Ø) then
                begin
                  Theta :=  9Ø ;
                  Exit( Polar )
               end;
       if (Realnum < Ø) and (Imagnum = Ø) then Theta := 18Ø else
            begin
              Theta := Atan ( Imagnum / Realnum ) * 57.2958 ;
                if Realnum < Ø  then Theta := Theta + 18Ø ;
            end
     end;
```

```
Procedure Printout;
   begin
     case answer of
     's','S' : begin
                 Writeln;
                 Writeln;
                 Writeln('Magnitude := ',Round(Mag*100)/100);
                 Writeln;
                 Writeln('Angle := ',Round(Theta*100)/100,' degrees');
                 Writeln;
                 Writeln;
               end;

    'p','P' : begin
                Rewrite(out,'Printer:');
                Writeln;
                Writeln(out);
                Writeln;
                Writeln(out);
                Writeln('Magnitude := ',Mag);
                Writeln(out,'Magnitude := ',Mag);
                Writeln(out);
                Writeln;
                Writeln('Angle := ',Theta,' degrees');
                Writeln(out,'Angle := ',Theta,' degrees');
                Writeln;
                Writeln;
                Writeln(out);
                Writeln(out);
              Close(out);
             end
        end { case }
     end;


begin { Main }
  Repeat
    Outputs;
    Title;
    Input;
    Polar;

    Printout;
          Write('Another Value? (Y/N) ');
             Readln(Cont);
  until Cont in [ 'N','n']
end.
```

Absolute Value

This program calculates the absolute
value of a complex number and its angle.

For multiple data entries
place a space between entries.

Enter the real and imaginary parts.

3 4

Magnitude = 5.00000
Angle = 5.31301E1 degrees

Another Value? (Y/N) N

3.6 COMPLEX NUMBER TO A REAL POWER

This is one of the programs that provides more capability than the FORTRAN complex functions. With it, Pascal can be used to raise a complex number to a real power. The calculation is based on De Moivre's theorem that states:

$$Z^N = R^N [\ \mathrm{Cos}\ N\ (\Theta) + i\ \mathrm{Sin}\ N\ (\Theta)]$$

where N is the number of roots.

In order to implement this equation, it is necessary to call the polar conversion routine twice, once to calculate the results and a second time to display the results in polar notation.

```
(*=========================================*)
(*                                         *)
(* Title:   Complex Number to a Real       *)
(*          Power                          *)
(* Program Summary: Program calculates     *)
(*          the value of a complex         *)
(*          number raised to a             *)
(*          real power.                    *)
(*                                         *)
(*=========================================*)


PROGRAM ComplexRealPower ;

 uses Transcend;

 VAR
  Resp,
  Answer,
  Ans,
  Sign2,
  Sign1,
  Cont            : Char;
  Out             : Text;
  RealNum,
  ImagNum,
  CompReal,
  CompImag,
  MagPwr,
  Mag,
  Theta           : Real;
  Stepping,
  Expo               : Integer;

Procedure Outputs;
  begin
    page(output);
     repeat
       gotoXY(Ø,5);
        Write('S(creen or P(rinter ? ');
        Readln(Answer);
     until Answer in [ 'S','s','p','P' ]
  end;  (outputs)

Procedure Title;
  begin
    page(output);
     gotoXY(5,1);
      Writeln('Complex Number to a Real Power');
      Writeln;Writeln;
      Writeln('This program will calculate the value');
      Writeln('of a complex number raised to a real');
      Writeln('power.');
      Writeln;
      Writeln('For multiple data entries ');
```

```
      Writeln('place a space between entries');
      Writeln;
   end;  { title }

Procedure Input;
  begin
       Writeln;
       Write('Enter exponent: ');
       Readln( Expo );
       Writeln;
       Writeln('Enter complex number ( Real, Imaginary ) ');
       Writeln;
       Readln( CompReal , CompImag );
       Writeln;
  end;

 Procedure Polar;
   begin
     Mag := Sqrt( Sqr( CompReal ) + Sqr ( CompImag ));
       if (CompReal = Ø) and (CompImag < Ø) then
             begin
                Theta := 27Ø ;
                Exit( Polar )
             end;
       if (CompReal = Ø) and (CompImag > Ø) then
             begin
                Theta :=  9Ø ;
                Exit ( Polar )
             end;
       if (CompReal < Ø) and (CompImag = Ø) then Theta := 18Ø else
            begin
              Theta := Atan ( CompImag / CompReal ) * 57.2958 ;
                if CompReal < Ø  then Theta := Theta + 18Ø ;
            end
     end;

Procedure Work;
  begin
     Polar;
    Theta := Theta / 57.2958 ;
           MagPwr := Mag ;
        for Stepping := 1 to Expo-1  do
            MagPwr := MagPwr * Mag ;  { a replacement for Mag^Expo }
                Mag := MagPwr ;
                   RealNum  := Mag * Cos ( Expo * Theta );
                   ImagNum  := Mag * Sin ( Expo * Theta );
                 Sign1 :='-';
                 Sign2 :='-';
            if Abs ( CompImag ) = CompImag then Sign2 := '+' ;
            if Abs ( ImagNum ) = ImagNum then Sign1 := '+';
  end;

Procedure PPrint1;
     begin
       Polar;
       Writeln;
       Writeln('Magnitude := ',Mag);
       Writeln;
       Writeln('Angle := ',Theta,' degrees');
     end;

Procedure PPrint2;
   begin
    Polar;
    Writeln;
    Writeln(out);
    Writeln('Magnitude := ',Mag);
    Writeln(out,'Magnitude := ',Mag);
    Writeln(out);
    Writeln;
    Writeln('Angle := ',Theta,' degrees');
    Writeln(out,'Angle := ',Theta,' degrees');
   end;
```

```
Procedure Printout;
   begin
     case answer of
     's','S' : begin
                 Writeln;
                 Writeln('The answer is: ');
                 Write('(',CompReal,Sign2,' i',Abs( CompImag ),' )^',Expo);
                 Writeln(' = ',Round(RealNum),Sign1,' i',Abs( ImagNum ));
                   CompReal := RealNum;
                   CompImag := ImagNum;
                 Writeln;
                     Write('Do you want answer in polar form?(Y/N) ');
                     Readln( Resp );
                       if Resp in [ 'Y' , 'y' ] then
                           PPrint1
               end;
     'p','P' : begin
                 Rewrite(out,'Printer:');
                 Writeln;
                 Writeln(out);
                 Writeln('The answer is: ');
                 Writeln(out,'The answer is: ');
                 Write('(',CompReal,Sign2,' i',Abs( CompImag ),' )^',Expo);
                 Write(out,'(',CompReal,Sign2,' i',Abs( CompImag ),' )^',Expo);
                 Writeln(' = ',Round(RealNum),Sign1,' i ',Abs( ImagNum ));
                 Writeln(out,' = ',Round(RealNum),Sign1,' i ',Abs( ImagNum ));
                   CompReal := RealNum;
                   CompImag := ImagNum;
                 Writeln;
                 Write('Do you want it in polar form? (Y/N)  ');
                 Readln( Resp );
                    If Resp in ['Y','y'] Then
                      PPrint2;
               Close(out);
              end
          end { case }
       end;

  begin { Main }
    Repeat
      Outputs;
      Title;
      Input;
       Work;
      Printout;
       Writeln;
            Write('Another Conversion?(Y/N) ');
                Readln(Cont);
    until Cont in [ 'N','n']
  end.
```

Complex Number to a Real Power

This program will calculate the value
of a complex number raised to a real
power.

For multiple data entries
place a space between entries.

Enter exponent: 3
Enter complex number (Real, Imaginary):

Ø 8

The answer is:
(Ø.ØØØØØ + i8.ØØØØØ)^3 = Ø - 5.12ØØØE2

Do you want it in polar form? (Y/N) Y

Magnitude = 5.12ØØØE2
Angle = 2.7ØØØE2 degrees

Another Conversion? (Y/N) N

3.7 COMPLEX ROOTS

By substituting the inverse power to De Moivre's equation and expanding, it becomes possible to calculate the N Nth roots of a complex number. The resulting equation is

$$Z^{(1/N)} = R^{(1/N)} = \left[\mathrm{Cos}\left(\frac{\Theta + 2K}{N}\right) + i\mathrm{Sin}\left(\frac{\Theta + 2K}{N}\right)\right]$$

Where

K = 0,1,2,3,. . . . .,N−1
N = the desired root

To implement this equation, the polar conversion routine is called into action twice. The above equation is located in *Procedure Work.* Since Pascal cannot raise a number to a power, it is accomplished using the following formula:

$$A^{(1/N)} = \exp\left(\frac{\ln A}{N}\right)$$

```
(*=========================================*)
(*                                         *)
(* Title:   Complex Roots                  *)
(* Program Summary: Program calculates     *)
(*          the Nth root of a              *)
(*          complex number.                *)
(*                                         *)
(*=========================================*)


PROGRAM ComplexRoots ;

 uses Transcend;

 VAR
  Resp,
  Answer,
  Ans,
  BS,
  Cont              : Char;
  Out               : Text;
  CompReal,
  Am,
  Mag,
  Temp,
  Theta             : Real;
  Steps,
  Stepping,
```

```
  Root           : Integer;
  RealNum,
  ImagNum        : Array [1..50] of Real;

Procedure Outputs;
  begin
    page(output);
     repeat
       gotoXY(0,5);
        Write('S(creen or P(rinter ? ');
        Readln(Answer);
     until Answer in [ 'S','s','p','P' ]
  end;  {outputs}

Procedure Title;
  begin
    page(output);
     gotoXY(13,1);
      Writeln('Complex Roots');
      Writeln;Writeln;
      Writeln('This program will calculate the "N"');
      Writeln('Nth roots of a complex number.');
      Writeln;
      Writeln('For multiple data entries ');
      Writeln('place a space between entries');
      Writeln;
   end;  { title }

Procedure Input;
  begin
       Writeln;
       Write('Enter root desired: ');
       Readln( Root );
       Writeln;
       Write('Enter Real and Imaginary parts: ');
       Readln( CompReal , Am );
       Writeln;
  end;

 Procedure Polar;
   begin
     Mag := Sqrt( Sqr( CompReal ) + Sqr ( Am ));
       if (CompReal = 0) and (Am < 0) then
             begin
                Theta := 270 ;
                Exit( Polar )
             end;
       if (CompReal = 0) and (Am > 0) then
             begin
                Theta :=  90 ;
                Exit ( Polar )
             end;
       if (CompReal < 0) and (Am = 0) then Theta := 180 else
            begin
              Theta := Atan ( Am / CompReal ) * 57.2958 ;
                if CompReal < 0  then Theta := Theta + 180 ;
            end
     end;

Procedure Work;
  begin
     Polar;
    Theta := Theta / 57.2958 ;
        for Steps := 1 to Root do
           begin
              Temp := Theta + 2 * Steps * 3.14159;
              Temp := Temp / Root ;
              RealNum [ Steps ] := exp( Ln ( Mag ) / Root ) * Cos ( Temp );
              ImagNum [ Steps ] := exp( Ln ( Mag ) / Root ) * Sin ( Temp );
           end;
  end;

Procedure Print1;
     begin
       Writeln;
       Writeln('Magnitude := ',Mag);
       Writeln('Angle := ',Theta,' degrees');
     end;
```

```
      Procedure Print2;
         begin
          Writeln;
          Writeln(out);
          Writeln('Magnitude := ',Mag);
          Writeln(out,'Magnitude := ',Mag);
          Writeln('Angle := ',Theta,' degrees');
          Writeln(out,'Angle := ',Theta,' degrees');
         end;

     Procedure Printout;
      begin
       case answer of
       's','S' : begin
                  Writeln;
                  Writeln('The roots are: ');
                  Writeln;
                  For Stepping := 1 to Root do
                   begin
                     BS := '-';
                      if Abs ( ImagNum [ Stepping ] ) = ImagNum [ Stepping ] then
                                            BS :='+' ;
                      Write(RealNum[ Stepping ],' ',BS);
                      Writeln(' i ',Abs( ImagNum [ Stepping ] ));
                   end;
                      Writeln;
                      Write('Do you want answer in polar form?(Y/N) ');
                        Readln( Resp );
                         if Resp in [ 'Y' , 'y' ] then
                           for Stepping := 1 to Root do
                             begin
                              CompReal := RealNum [ Stepping ] ;
                              Am := ImagNum [ Stepping ] ;
                              Polar;
                              Print1
                            end;
                end;
     'p','P' : begin
                Rewrite(out,'Printer:');
                Writeln;
                Writeln(out);
                Writeln('The roots are: ');
                Writeln(out,'The roots are: ');
                Writeln;
                Writeln(out);
                For Stepping := 1 to Root do
                  begin
                    BS := '-';
                     if Abs ( ImagNum [ Stepping ] ) = ImagNum [ Stepping ] then
                                  BS :='+' ;
                     Write(RealNum[ Stepping ],' ',BS,' i ');
                     Writeln(Abs( ImagNum [ Stepping ] ));
                     Writeln(out);
                     Write(out,RealNum[ Stepping ],' ',BS,' i ');
                     Writeln(out,Abs( ImagNum [ Stepping ] ));
                  end;
                   Writeln;
                   Write('Do you want answer in polar form?(Y/N) ');
                   Readln( Resp );
                     if Resp in [ 'Y' , 'y' ] then
                        for Stepping := 1 to Root do
                          begin
                            CompReal := RealNum [ Stepping ] ;
                            Am := ImagNum [ Stepping ] ;
                            Polar;
                            Print2
                          end;
            Close(out);
           end
       end { case }
    end;

       begin { Main }
         Repeat
           Outputs;
           Title;
```

```
     Input;
      Work;
     Printout;
      Writeln;
           Write('Another Root?(Y/N) ');
              Readln(Cont);
  until Cont in [ 'N','n']
end.
```

Complex Roots

This program will calculate the "N"
Nth root of a complex number.

For multiple data entries
place a space between entries.

Enter root desired: 5
Enter real and imaginary parts: 8 -3

The roots are:
5.78Ø82E-1 + i1.42284
-1.17456 + i9.894772E-1
-1.3Ø4ØØ - i8.11312E-1
3.68642E-1 - i1.49Ø89
1.53184 - i1.1Ø112E-1

Do you want the answer in polar form? (Y/N) Y

Magnitude = 1.53579
Angle = 6.78887E1 degrees

Magnitude = 1.53579
Angle = 1.39889E2 degrees

Magnitude =1.53579
Angle = 2.11889E2 degrees

Magnitude = 1.53579
Angle = -7.61115E1 degrees

Magnitude = 1.53578
Angle = -4.11149 degrees

Another Root? (Y/N) N

3.8 COMPLEX EXPONENTIAL

Calculus often requires the evaluation of the exponential function,

$$\text{Exp}^{Z}$$

where Z = A + iB.

The exponential raised to a complex power can be calculated by using the following formula:

$$\text{Exp}^{Z} = \text{Exp}^{X}(\text{Cos}Y + i\text{Sin}Y)$$

Where Z = X + iY.

Evaluation of this formula is done for separate parts in *Procedure Work*. Once again, polar output capability is provided.

```
(*=======================================*)
(*                                       *)
(* Title:  Complex Exponential           *)
(* Program Summary: Program calculates   *)
(*         the value of the              *)
(*         exponential (e) raised        *)
(*         to a complex power.           *)
(*                                       *)
(*=======================================*)


PROGRAM ComplexExponential ;

 uses Transcend;

 VAR
  Resp,
  Answer,
  Ans,
  Sign2,
  Sign1,
  Cont            : Char;
  Out             : Text;
  Realnum,
  Imagnum,
  RealExp,
  ImagExp,
  Mag,
  Theta           : Real;

Procedure Outputs;
  begin
    page(output);
     repeat
       gotoXY(0,5);
        Write('S(creen or P(rinter ? ');
        Readln(Answer);
     until Answer in [ 'S','s','p','P' ]
  end;  {outputs}

Procedure Title;
  begin
    page(output);
     gotoXY(10,1);
      Writeln('Complex Exponential');
      Writeln;Writeln;
      Writeln('This program calculates the value of');
      Writeln('the Exponential (e) raised to a complex');
      Writeln('power.');
      Writeln;
      Writeln;
```

```
        Writeln('For multiple data entries ');
        Writeln('place a space between entries');
        Writeln;
     end;  { title }

 Procedure Input;
   begin
        Writeln;
        Writeln('What is the complex power to which');
        Writeln('you want the expontential raised?');
        Writeln;
        Write('Enter Real and Imaginary parts: ');
        Readln( Realnum , Imagnum );
        Writeln;
   end;

 Procedure Work;
   begin
      Sign1 := '-';
      Sign2 := '-';
         RealExp :=  Exp ( Realnum ) * Cos ( Imagnum );
         ImagExp :=  Exp ( Realnum ) * Sin ( Imagnum );
      if Abs ( ImagExp ) = ImagExp then Sign1 := '+';
      if Abs ( RealExp ) = RealExp then Sign2 := '+';
   end;

  Procedure Polar;
    begin
      Mag := Sqrt( Sqr( Realnum ) + Sqr  ( Imagnum ));
        if (Realnum = Ø) and (Imagnum < Ø) then
              begin
                 Theta := 27Ø ;
                 Exit( Polar )
              end;
        if (Realnum = Ø) and (Imagnum > Ø) then
              begin
                 Theta :=  9Ø ;
                 Exit  ( Polar )
              end;
        if (Realnum < Ø) and (Imagnum = Ø) then Theta := 18Ø else
             begin
               Theta := Atan  ( Imagnum / Realnum ) * 57.2958 ;
                 if Realnum < Ø  then Theta := Theta + 18Ø ;
             end
      end;


 Procedure PPrint1;
      begin
        Writeln;
        Writeln('Magnitude := ',Mag);
        Writeln;
        Writeln('Angle := ',Theta,' degrees');
      end;

 Procedure PPrint2;
    begin
     Writeln;
     Writeln(out);
     Writeln('Magnitude := ',Mag);
     Writeln(out,'Magnitude := ',Mag);
     Writeln(out);
     Writeln;
     Writeln('Angle := ',Theta,' degrees');
     Writeln(out,'Angle := ',Theta,' degrees');
    end;

  Procedure Printout;
     begin
       case answer of
       's','S' : begin
                   Writeln;
                   Writeln('The answer is: ');
                   Write('Exp^(',Round(Realnum),' ',Sign2,' i');
                   Write(Round(Abs( Imagnum )),') := ', RealExp,' ');
                   Writeln(Sign1,' i',Abs( ImagExp ));
                     Realnum := RealExp;
```

```
                  Imagnum := ImagExp;
                    Write('Do you want answer in polar form?(Y/N) ');
                    Readln( Resp );
                      if Resp in [ 'Y' , 'y' ] then
                          begin
                            Polar;
                            PPrint1
                          end;
              end;
    'p','P' : begin
                Rewrite(out,'Printer:');
                Writeln;
                Writeln(out);
                Writeln('The answer is: ');
                Writeln(out,'The answer is: ');
                Write('Exp^(',Round(Realnum),' ',Sign2,' i');
                Write(Round(Abs( Imagnum )),') := ', RealExp,' ');
                Writeln(Sign1,' i',Abs( ImagExp ));
                Write(out,'Exp^(',Round(Realnum),' ',Sign2,' i');
                Write(out,Round(Abs( Imagnum )),') := ', RealExp,' ');
                Writeln(out,Sign1,' i',Abs( ImagExp ));
                  Realnum := RealExp;
                  Imagnum := ImagExp;
                Write('Do you want it in polar form? (Y/N)  ');
                Readln( Resp );
                   If Resp in ['Y','y'] Then
                       begin
                         Polar;
                         PPrint2;
                       end;
               Close(out);
              end
        end { case }
    end;


begin { Main }
  Repeat
    Outputs;
    Title;
    Input;
     Work;
    Printout;
     Writeln;
          Write('Another Exponential?(Y/N) ');
              Readln(Cont);
  until Cont in [ 'N','n']
end.
```

Complex Exponential

This program calculates the value of the exponential (e) raised to a complex power.

For multiple data entries place a space between entries.

What is the complex power to which you want the exponential raised?

Enter the real and imaginary parts: 2 6

The answer is:
Exp^(2 + i6) = 7.09475 - i2.06462

Do you want it in polar form? (Y/N) Y

Magnitude = 7.38906
Angle = -1.63353E1 degrees

Another Exponential? (Y/N) Y

Complex Exponential

This program calculates the value of the exponential (e) raised to a complex power.

For multiple data entries
place a space between entries.

What is the complex power to which you want the exponential raised?

Enter the real and imaginary parts: 3 Ø

The answer is:
Exp(3 + iØ) = 2.ØØ855E1 + iØ.ØØØØØ

Do you want it in polar form? (Y/N) N

Another Exponential? (Y/N) N

4 MATRIX MATHEMATICS

Most home computer systems do not have the capability to perform any matrix math. However, with the programs in Chap. 4 this disadvantage can be overcome.

The first five programs perform operations on two-dimensional matrices, which are standard in Pascal. The remaining three programs extend the computer's capabilities to a third dimension. When working with three-dimensional matrices, you should be aware that memory space gets used up very quickly.

The output of these programs has been arranged so that the results are printed in columns. This has been done by simulating the BASIC tab function and using the length. This utilizes a minimal number of spaces. Since Pascal outputs numbers in a similar format, and if the space is the same between each number, the numbers will line up in columns, although not necessarily straight ones.

4.1 ADDITION AND SUBTRACTION OF TWO-DIMENSIONAL SQUARE MATRICES

To add or subtract two matrices, simply perform the operation with the corresponding elements in each of the arrays. When this program is run, it asks for the size of the array. It is necessary to answer with only one number since the matrix is assumed to be square.

If you wish to add arrays that are not square, this is simply done by entering zero as elements in those rows or columns not used.

The program asks whether addition or subtraction is to be performed. If any answer is given other than A (for addition) subtraction is assumed. If the answer is "A" or "a" the first part of *Procedure Work* does the job, otherwise the second part goes to work.

```
(*=========================================*)
(*                                         *)
(* Title:  Addition and Subtraction        *)
(*         of Two-Dimensional Square       *)
(*         Matrices                        *)
(* Program Summary: Program adds and       *)
(*         subtracts two dimensional       *)
(*         matrices.                       *)
(*                                         *)
(*=========================================*)


PROGRAM AddSubMatrix ;


 VAR
  Resp,
  Answer,
  Ans,
  Cont            : Char;
  Out             : Text;
  Stepping,
  Stepper,
  Steps,
  Temp2,
  Temp1,
  Size            : Integer;
  A,
  B,
  C               : Array  [1..10,1..10] of Real;
  D               : Array  [1..10,1..10] of Integer;
  Len,
  Space           : String;

Procedure Outputs;
  begin
    page(output);
     repeat
       gotoXY(0,5);
        Write('S(creen or P(rinter ? ');
        Readln(Answer);
     until Answer in [ 'S','s','p','P' ]
  end;  {outputs}

Procedure Title;
  begin
    page(output);
     gotoXY(2,1);
      Writeln('Two-D Matrix Addition and Subtraction');
      Writeln;Writeln;
      Writeln('This program adds and subtracts');
      Writeln('two 2-Dimensional square matrices');
      Writeln;
      Writeln;
   end;  { title }

Procedure Input;
  begin
       Write('Enter size of matrices: ');
       Readln( Size );
       Writeln;
       Write('Addition or Subtraction (A/S): ');
       Readln( Resp );
       Writeln;
       Writeln('Enter the elements of Array "A"');
              for Steps := 1 to Size do
                    begin
                       for Stepping := 1 to Size do
                           begin
                              Write('A (',Steps,',',Stepping,') = ');
                              Readln( A [ Steps, Stepping ] );
                           end;
                       Writeln;
                    end;
       Writeln;
       Writeln('Enter the elements of Array "B"');
              for Steps := 1 to Size do
```

```
                    begin
                       for Stepping := 1 to Size do
                           begin
                              Write('B (',Steps,',',Stepping,') = ');
                              Readln( B [ Steps, Stepping ] );
                              C [ Steps , Stepping ] := Ø;
                           end;
                       Writeln;
                    end;

  end;

Procedure Work;
  begin
   If Resp in [ 'A','a' ]  then
       For Steps := 1 to Size do
        For Stepping := 1 to Size do
           C [ Steps , Stepping ] := C [ Steps , Stepping ] +
                  A [ Steps , Stepping ] + B [ Steps , Stepping ]
   else
       For Steps := 1 to Size do
        For Stepping := 1 to Size do
           C [ Steps , Stepping ] := C [ Steps , Stepping ] +
                     A [ Steps , Stepping ] - B [ Steps , Stepping ]
  end;

   Procedure Tab;
      begin
         D [ Steps , Stepping ] := Round ( C [ Steps , Stepping ] );
         Space :=' ';
         Str ( D [ Steps , Stepping ], Len );
           Temp2 := Length ( Len );
           Temp1  := (8 * Stepping - Temp2 );
             For Stepper := 1 to Temp1 do
                 Space := Concat( Space,' ');
      end;

Procedure Printout;
   begin
     case answer of
     's','S' : begin
                 Writeln;
                 Writeln('The answer is: ');
                  For Steps := 1 to Size do
                    begin
                     for Stepping := 1 to Size do
                       begin
                         Tab;
                         Write(Space,C [ Steps , Stepping ] );
                       end;
                      Writeln;Writeln;
                    end;
                end;
     'p','P' : begin
                 Rewrite(out,'Printer:');
                   Writeln;
                   Writeln('The answer is: ');
                   Writeln(out,'The answer is: ');
                    For Steps := 1 to Size do
                      begin
                       for Stepping := 1 to Size do
                         begin
                          Tab;
                           Write(Space,C [ Steps , Stepping ] );
                           Write(out,Space,C [ Steps , Stepping ] );
                         end;
                        Writeln;Writeln(out);
                        Writeln;Writeln(out);
                     end;
                   Writeln;
               Close(out);
              end
          end { case }
      end;


   begin { Main }
     Repeat
```

```
    Outputs;
    Title;
    Input;
     Work;
    Printout;
     Writeln;
          Write('Another Matrices?(Y/N) ');
              Readln(Cont);
  until Cont in [ 'N','n']
end.
```

2-Dimensional Matrix Addition and Subtraction

This program adds and subtracts
two 2-dimensional square matrices.

Enter size of matrices: 2

Addition or Subtraction (A/S): A

Enter the elements of Array "A"
A(1,1) = 2
A(1,2) = 12

A(2,1) = 18
A(2,2) = 9

Enter the elements of Array "B"
B(1,1) = Ø
B(1,2) = 3

B(2,1) = 4
B(2,2) = 12

The answer is:
 2.ØØØØØ 1.5ØØØØE1

 2.2ØØØØE1 2.1ØØØØE1

Another Matrices? (Y/N) N

4.2 TWO-DIMENSIONAL MATRIX MULTIPLICATION

Matrix multiplication can be performed only when the number of columns of one matrix is equal to the number of rows of the other. In this program, the matrices are assumed to be square. However, as with the former program, nonsquare matrices can be accommodated by entering zero as the value for the unused rows or columns.

The product of two matrices is defined by the following equation:

$$C(i, j) = [A(i, k)][B(k, i)] + C(i, j)$$

```
(*=======================================*)
(*                                       *)
(* Title:  Two-Dimensional Matrix        *)
(*         Multiplication                *)
(* Program Summary: Program multiplies   *)
(*         two square matrices.          *)
(*                                       *)
(*=======================================*)


PROGRAM MultiplyMatrix ;


 VAR
  Answer,
  Ans,
  Cont          : Char;
  Out           : Text;
  Stepping,
  Step,
  Steps,
  Temp1,
  Temp2,
  Size          : Integer;
  A,
  B,
  C             : Array  [1..10,1..10] of Real;
  D             : Array  [1..10,1..10] of Integer;
  Len,
  Space         : String;

 Procedure Outputs;
  begin
    page(output);
     repeat
       gotoXY(0,5);
        Write('S(creen or P(rinter ? ');
        Readln(Answer);
     until Answer in [ 'S','s','p','P' ]
  end;  {outputs}

 Procedure Title;
  begin
    page(output);
     gotoXY(6,1);
      Writeln('Two-D Matrix Multiplication');
      Writeln;Writeln;
      Writeln('This program multiplies two square');
      Writeln('matrices.');
      Writeln;
      Writeln;
   end;  { title }

 Procedure Input;
  begin
       Write('Enter size of matrices: ');
       Readln( Size );
       if Size = 0 then Input;
       Writeln;
       Writeln('Enter the elements of Array "A"');
              for Steps := 1 to Size do
                    begin
                       for Stepping := 1 to Size do
                            begin
                               Write('A (',Steps,',',Stepping,') = ');
                               Readln( A [ Steps, Stepping ] );
                            end;
                       Writeln;
                    end;
       Writeln;
       Writeln('Enter the elements of Array "B"');
              for Steps := 1 to Size do
                    begin
                       for Stepping := 1 to Size do
                            begin
                               Write('B (',Steps,',',Stepping,') = ');
                               Readln( B [ Steps, Stepping ] );
                               C [ Steps , Stepping ] := 0;
```

```
                                end;
                            Writeln;
                         end;

  end;

Procedure Work;
  begin
   For Steps := 1 to Size do
     For Stepping := 1 to Size do
       for Step := 1 to Size do
         C[ Steps , Stepping ] := C [ Steps , Stepping ] +
               A [ Steps , Step ] * B [ Step , Stepping ]
 end;

 Procedure Tab;
    begin
       D [ Steps , Stepping ] := Round ( C [ Steps , Stepping ]);
       Space :=' ';
       Str ( D [ Steps , Stepping ], Len );
         Temp1 := Length ( Len );
         Temp2  := (8 * Stepping - Temp1 );
           For Step := 1 to Temp2 do
               Space := Concat( Space,' ');
    end;

Procedure Printout;
   begin
     case answer of
     's','S' : begin
                 Writeln;
                 Writeln('The answer is: ');
                  For Steps := 1 to Size do
                    begin
                     for Stepping := 1 to Size do
                       begin
                         Tab;
                         Write(Space,C [ Steps , Stepping ] );
                       end;
                      Writeln;Writeln;
                    end;
                end;
     'p','P' : begin
                 Rewrite(out,'Printer:');
                   Writeln;
                   Writeln('The answer is: ');
                   Writeln(out,'The answer is: ');
                    For Steps := 1 to Size do
                      begin
                       for Stepping := 1 to Size do
                         begin
                          Tab;
                           Write(Space,C [ Steps , Stepping ] );
                           Write(out,Space,C [ Steps , Stepping ] );
                         end;
                        Writeln;Writeln(out);
                        Writeln;Writeln(out);
                     end;
                   Writeln;
               Close(out);
              end
        end { case }
     end;


 begin { Main }
   Repeat
     Outputs;
     Title;
     Input;
      Work;
     Printout;
      Writeln;
           Write('Another Matrices?(Y/N) ');
              Readln(Cont);
   until Cont in [ 'N','n']
 end.
```

```
2-Dimensional Matrix Multiplication

This program multiplies two square
matrices.

Enter size of matrices: 3

Enter the elements of Array "A"

A(1,1) = 3.1
A(1,2) = 4
A(1,3) = 1

A(2,1) = 8.3
A(2,2) = 6.6
A(2,3) = Ø

A(3,1) = 3
A(3,2) = 12
A(3,3) = 9

Enter the elements of Array "B"

B(1,1) = Ø
B(1,2) = Ø
B(1,3) = Ø

B(2,1) = 2
B(2,2) = 5
B(2,3) = 8

B(3,1) = 15
B(3,2) = 21
B(3,3) = 6

The answer is:

   2.3ØØØØE1    4.1ØØØØE1    3.8ØØØØE1

  1.32ØØØE1    3.3ØØØØE1    5.28ØØØE1

 1.59ØØØE2   2.49ØØØE2    1.5ØØØØE2

Another Matrices? (Y/N) N
```

4.3 TRANSPOSITION OF TWO-DIMENSIONAL MATRICES

Sometimes it is necessary to exchange the rows and columns of a matrix. This operation is known as transposition.

To perform transposition, the program sets the elements of the original array A(i, j) equal to B(j, i), which is the transposed array. This is done in *Procedure Work*.

```
(*========================================*)
(*                                        *)
(* Title:   Transposition of Two-         *)
(*          Dimensional Matrices          *)
(* Program Summary: Program transposes    *)
(*          rows and columns of a         *)
(*          matrix.                       *)
(*                                        *)
(*========================================*)


PROGRAM Transposition ;


 VAR
  Answer,
  Ans,
  Cont              : Char;
  Out               : Text;
  A,
  B                 : Array [1..20,1..20] of Real;
  Size,
  Steps,
  Stepping          : Integer;
  Space             : String;

Procedure Outputs;
  begin
    page(output);
     repeat
       gotoXY(0,5);
        Write('S(creen or P(rinter ? ');
        Read(Answer);
     until Answer in [ 'S','s','p','P' ]
  end;  {outputs}

Procedure Title;
  begin
    page(output);
     gotoXY(09,1);
       Writeln('2-D Matrix Transposition');
       Writeln;Writeln;
       Writeln('This program transposes a 2-Dimensional');
       Writeln('matrix.');
       Writeln;
       Writeln;
   end;  { title }

Procedure Init;
  begin
     For Steps:= 1 to Size do
      For Stepping := 1 to Size do
        begin
          A [ Steps , Stepping ] := 0 ;
          B [ Steps , Stepping ] := 0
        end;
 end;

Procedure Input;
  begin
       Write('Enter size of matrices: ');
       Readln( Size );
       Init;
       Writeln;
       Writeln('Enter Matrix elements: ');
```

```
        Writeln;
               for Steps := 1 to Size do
                    begin
                       for Stepping := 1 to Size do
                           begin
                              Write('A (',Steps,',',Stepping,') = ');
                              Readln( A [ Steps, Stepping ] );
                           end;
                       Writeln;
                   end;
        Writeln;
      Space :='    ';
   end;

  Procedure Printout;
     begin
       case answer of
       's','S' : begin
                   Writeln;
                   Writeln('The transposed Matrix is: ');
                    For Steps := 1 to Size do
                      begin
                       for Stepping := 1 to Size do
                         begin
                           Write(Space,B [ Steps , Stepping ] );
                         end;
                        Writeln;Writeln;
                     end;
                 end;
       'p','P' : begin
                   Rewrite(out,'Printer:');
                     Writeln;
                   Writeln('The transposed Matrix is: ');
                   Writeln(out,'The transposed Matrix is: ');
                    For Steps := 1 to Size do
                      begin
                       for Stepping := 1 to Size do
                         begin
                           Write(Space,B [ Steps , Stepping ] );
                           Write(out,Space,B [ Steps , Stepping ] );
                         end;
                          Writeln;Writeln(out);
                          Writeln;Writeln(out);
                       end;
                     Writeln;
                 Close(out);
                end
            end { case }
        end;

 Procedure Work;
    begin
      for Steps := 1 to Size do
       begin
         for Stepping := 1 to Size do
             B[ Stepping , Steps ] := A [ Steps , Stepping ]
       end
    end;

    begin { Main }
      Repeat
        Outputs;
        Title;
        Input;
         Work;
        Printout;
         Writeln;
              Write('Another Matrices?(Y/N) ');
                 Readln(Cont);
      until Cont in [ 'N','n']
    end.
```

```
2-Dimensional Matrix Transposition

This program transposes a 2-Dimensional
matrix.

Enter size of matrices: 4

Enter matrix elements:

A(1,1) = 1
A(1,2) = 2.3
A(1,3) = 6
A(1,4) = 3

A(2,1) = 1.1
A(2,2) = 3.3
A(2,3) = 6
A(2,4) = 12

A(3,1) = 3
A(3,2) = 3
A(3,3) = 2
A(3,4) = 1

A(4,1) = Ø
A(4,2) = 6
A(4,3) = 4
A(4,4) = 1

The transposed matrix is:

 1.ØØØØØ  1.1ØØØØ  3.ØØØØØ  Ø.ØØØØØ

 2.3ØØØØ  3.3ØØØØ  3.ØØØØØ  6.ØØØØØ

 6.ØØØØØ  6.ØØØØØ   2.ØØØØØ  4.ØØØØØ

 3.ØØØØØ  1.2ØØØØE1  1.ØØØØØ  1.ØØØØØ

Another Matrices? (Y/N) N
```

4.4 SCALAR MULTIPLICATION OF A TWO-DIMENSIONAL MATRIX

The multiplication of a matrix by an ordinary number (not another matrix) is called scalar multiplication.

Scalar multiplication is accomplished by multiplying each element of the matrix by the scalar number.

```
(*=========================================*)
(*                                         *)
(* Title:   Scalar Multiplication of       *)
(*          a Two-Dimensional Matrix       *)
(* Program Summary: Program multiplies     *)
(*          a matrix by a scalar           *)
(*          value.                         *)
(*                                         *)
(*=========================================*)


PROGRAM ScalarMultiplyMatrix ;


 VAR
  Answer,
  Ans,
  Cont            : Char;
  Out             : Text;
  Stepping,
  Stepper,
  Steps,
  Temp,
  Temp1,
  Size            : Integer;
  A,
  B               : Array  [1..10,1..10] of Real;
  C               : Array  [1..10,1..10] of Integer;
  Len,
  Space           : String;
  Scalar               : Real;
Procedure Outputs;
  begin
    page(output);
     repeat
       gotoXY(0,5);
        Write('S(creen or P(rinter ? ');
        Readln(Answer);
     until Answer in [ 'S','s','p','P' ]
  end;  {outputs}

Procedure Title;
  begin
    page(output);
     gotoXY(5,1);
      Writeln('2-D Matrix Scalar Multiplication');
      Writeln;Writeln;
      Writeln('This program multiplies a matrix');
      Writeln('by a scalar value.');
      Writeln;
      Writeln;
   end;  { title }

Procedure Input;
  begin
       Write('Enter size of matrices: ');
       Readln( Size );
       if Size = 0 then Input;
       Writeln;
       Write('Enter the Scalar value: ');
       Readln( Scalar );
       Writeln('Enter the matrix Elements: ');
              for Steps := 1 to Size do
                    begin
                       for Stepping := 1 to Size do
                           begin
                              Write('A (',Steps,',',Stepping,') = ');
                              Readln( A [ Steps, Stepping ] );
                           end;
                       Writeln;
                    end;
       Writeln;

  end;

Procedure Work;
  begin
```

```
      For Steps := 1 to Size do
        For Stepping := 1 to Size do
           B [ Steps , Stepping ] := A [ Steps , Stepping ] * Scalar ;
 end;

Procedure Tab;
   begin
      C [ Steps , Stepping ] := Round ( B [ Steps , Stepping ]);
      Space :=' ';
      Str ( C [ Steps , Stepping ], Len );
        Temp := Length ( Len );
        Temp1  := (8 * Stepping - Temp );
          For Stepper := 1 to Temp1 do
              Space := Concat( Space,' ');
   end;

Procedure Printout;
   begin
     case answer of
     's','S' : begin
                 Writeln;
                 Writeln('The answer is: ');
                  For Steps := 1 to Size do
                    begin
                     for Stepping := 1 to Size do
                       begin
                         Tab;
                         Write(Space,B [ Steps , Stepping ] );
                       end;
                      Writeln;Writeln;
                    end;
               end;
     'p','P' : begin
                 Rewrite(out,'Printer:');
                   Writeln;
                   Writeln('The answer is: ');
                   Writeln(out,'The answer is: ');
                    For Steps := 1 to Size do
                      begin
                       for Stepping := 1 to Size do
                         begin
                          Tab;
                            Write(Space,B [ Steps , Stepping ] );
                            Write(out,Space,B [ Steps , Stepping ] );
                         end;
                        Writeln;Writeln(out);
                        Writeln;Writeln(out);
                     end;
                   Writeln;
               Close(out);
              end
          end { case }
      end;

  begin { Main }
    Repeat
      Outputs;
      Title;
      Input;
       Work;
      Printout;
       Writeln;
            Write('Another Matrices?(Y/N) ');
               Readln(Cont);
    until Cont in [ 'N','n']
  end.
```

2-Dimensional Matrix Scalar Multiplication

This program multiplies a matrix
by a scalar value.

Enter size of matrices: 3

```
Enter the scalar value: 3

Enter the matrix elements:

A(1,1) = Ø
A(1,2) = 1
A(1,3) = 1

A(2,1) = 3
A(2,2) = 12
A(2,3) = 34

A(3,1) = 7
A(3,2) = 17
A(3,3) = 2

The answer is:

   Ø.ØØØØØ    3.ØØØØØ    3.ØØØØØ

   9.ØØØØØ    3.6ØØØØØE1    1.Ø2ØØØE2

  2.1ØØØØE1    5.1ØØØØØE1    6.ØØØØØ

Another matrices? (Y/N) N
```

4.5 MATRIX INVERSION

Division of matrices is not possible. Instead, another operation, which accomplishes essentially the same thing, is performed. To divide one matrix by another, one of the matrices is first inverted and then multiplied by the other. This is similar to the algebraic operation in which the reciprocal of a number is multiplied by another number to perform the equivalent of division.

Not all matrices can be inverted. Matrices that have a determinant equal to zero cannot be inverted, and are known as singular matrices. If a matrix is multiplied by its inverse, it will result in a unit matrix; that is, a matrix in which all the elements of the diagonal are equal to one, and all other elements are equal to zero.

```
(*========================================*)
(*                                        *)
(* Title:   Matrix Inversion              *)
(* Program Summary: Program inverts       *)
(*          one matrix so it can be       *)
(*          multiplied by another,        *)
(*          which is the same as          *)
(*          division.                     *)
(*                                        *)
(*========================================*)
```

```
PROGRAM InversionMatrix ;


 VAR
  Answer,
  Ans,
  Cont             : Char;
  Out              : Text;
  Stepping,
  Stepper,
  Step,
  Steps,
  Size             : Integer;
  A,
  B                : Array  [1..10,1..10] of Real;
  Len,
  Space            : String;
  Divisor,
  Temp2,
  Temp             : Real;

Procedure Outputs;
  begin
    page(output);
     repeat
       gotoXY(0,5);
        Write('S(creen or P(rinter ? ');
        Readln(Answer);
     until Answer in [ 'S','s','p','P' ]
  end;  {outputs}

Procedure Title;
  begin
    page(output);
     gotoXY(12,1);
      Writeln('Matrix Inversion');
      Writeln;Writeln;
      Writeln('This program inverts a 2-Dimensional');
      Writeln('matrix.');
      Writeln;
      Writeln;
   end;  { title }

Procedure Init;
  begin
     For Steps:= 1 to Size do
      For Stepping := 1 to Size do
        begin
          A [ Steps , Stepping ] := 0 ;
          B [ Steps , Stepping ] := 0
        end;
 end;

Procedure Input;
  begin
       Write('Enter size of matrices: ');
       Readln( Size );
       Init;
       Writeln;
       Writeln('Enter Matrix elements: ');
       Writeln;
              for Steps := 1 to Size do
                    begin
                       for Stepping := 1 to Size do
                           begin
                              Write('A (',Steps,',',Stepping,') = ');
                              Readln( A [ Steps, Stepping ] );
                           end;
                       Writeln;
                       B [ Steps , Steps ] := 1;
                    end;
       Writeln;
     Space :='    ';
  end;

 Procedure Printout;
    begin
```

```
        case answer of
        's','S' : begin
                    Writeln;
                    Writeln('The inverted Matrix is: ');
                     For Steps := 1 to Size do
                       begin
                        for Stepping := 1 to Size do
                          begin
                            Write(Space,B [ Steps , Stepping ] );
                          end;
                         Writeln;Writeln;
                       end;
                  end;
        'p','P' : begin
                    Rewrite(out,'Printer:');
                      Writeln;
                    Writeln('The inverted Matrix is: ');
                    Writeln(out,'The inverted Matrix is: ');
                     For Steps := 1 to Size do
                       begin
                        for Stepping := 1 to Size do
                          begin
                            Write(Space,B [ Steps , Stepping ] );
                            Write(out,Space,B [ Steps , Stepping ] );
                          end;
                           Writeln;Writeln(out);
                           Writeln;Writeln(out);
                        end;
                      Writeln;
                  Close(out);
                 end
             end { case }
         end;

Procedure Work1;
  begin
     if Stepping = Step then Exit (Work1) ;
     Divisor := A [ Step , Stepping ] ;
        for  Stepper := 1 to Size do
            begin
              A [ Step , Stepper ] := A [ Step , Stepper ]
                           - A [ Stepping , Stepper ] * Divisor;
              B [ Step , Stepper ] := B [ Step , Stepper ]
                           - B [ Stepping , Stepper ] * Divisor;
            end;
  end;

Procedure Work;
  begin
    For Step := 1 to Size do
        begin
           Temp := A [ Stepping , Step ] ;
           A [ Stepping , Step ] := A [ Steps , Step ];
           A [ Steps , Step ] := Temp ;
           Temp2 := B [ Stepping , Step ] ;
           B [ Stepping , Step ] := B [ Steps , Step ];
           B [ Steps , Step ] := Temp2
        end;
    Divisor := A [ Stepping , Stepping ] ;
    For Step := 1 to Size do
        begin
          A [ Stepping , Step ] := A [ Stepping , Step ] / Divisor ;
          B [ Stepping , Step ] := B [ Stepping , Step ] / Divisor ;
        end;
    For Step := 1 to Size  do
        Work1
  end;

 Procedure Other;
  begin
    Writeln('The matrix is singular');
    Exit ( Program );
  end;

 Procedure Test;
   begin
       For Steps := Stepping to Size do
```

```
                begin
                 if A [ Steps , Stepping ] <> Ø then
                      Exit ( Test )
                end;
           Other;
     end;

Procedure MainWork;
  begin
       For Stepping := 1 to Size do
         begin
            Test;
            Work
         end;
       Printout;
  end;

   begin { Main }
     Repeat
       Outputs;
       Title;
       Input;
        MainWork;
        Writeln;
             Write('Another Matrices?(Y/N) ');
                Readln(Cont);
     until Cont in [ 'N','n']
   end.
```

Matrix Inversion

This program inverts a 2-Dimensional matrix.

Enter size of matrices: 3

Enter matrix elements:

A(1,1) = Ø
A(1,2) = .2
A(1,3) = -1

A(2,1) = 4
A(2,2) = -.125
A(2,3) = 3

A(3,1) = Ø
A(3,2) = 1Ø
A(3,3) = 2

The inverted matrix is:

7.27163E-1 2.5ØØØØØE-1 -1.14183E-2

1.923Ø8E-1 Ø.ØØØØØ 9.61538E-2

-9.61538E-1 Ø.ØØØØØ 1.923Ø8E-2

Another Matrices? (Y/N) N

4.6 ADDITION AND SUBTRACTION OF THREE-DIMENSIONAL MATRICES

This program offers the capability of adding or subtracting two three-dimensional matrices.

Aside from the fact that an extra *FOR-DO* loop is used, the program is quite similar to the two-dimensional addition/subtraction program.

When working with three-dimensional matrices, be aware that matrix size is much more limited than with two-dimensional matrices, because more memory is needed for a three-dimensional matrix. A 10 by 10 matrix requires only 100 places to be set aside for numbers; a 10 by 10 by 10 matrix needs 1000 places. Therefore, if large three-dimensional matrices are used, a memory problem may be encountered.

```
(*=========================================*)
(*                                         *)
(* Title:  Addition and Subtraction        *)
(*         of Three-Dimensional Matrices   *)
(* Program Summary: Program adds or        *)
(*         subtracts two 3-dimensional     *)
(*         matrices.                       *)
(*                                         *)
(*=========================================*)


PROGRAM ThreeDAddSub ;

 VAR
  Resp,
  Answer,
  Ans,
  Cont           : Char;
  Out            : Text;
  Stepping,
  Step,
  Sign,
  Steps,
  Size           : Integer;
  A,
  B,
  C              : Array  [1..5,1..5,1..5] of Real;
  D              : Array  [1..5,1..5,1..5] of Integer;
  Len,
  Space          : String;

Procedure Outputs;
  begin
    page(output);
     repeat
       gotoXY(0,5);
        Write('S(creen or P(rinter ? ');
        Readln(Answer);
     until Answer in [ 'S','s','p','P' ]
  end;  {outputs}

Procedure Title;
  begin
    page(output);
     gotoXY(6,1);
      Writeln('3D Matrix Addition and Subtraction');
      Writeln;Writeln;
      Writeln('This program adds and subtracts two');
      Writeln('3-Dimensional matrices.');
      Writeln;
      Writeln;
   end;  { title }
```

```
Procedure Input;
  begin
   Write('Enter size of matrices: ');
   Readln( Size );
   Writeln;
   Write('Addition or Subtraction (A/S): ');
   Readln( Resp );
   Writeln('Enter the elements of Array "A"');
   Writeln;
      for Steps := 1 to Size do
         begin
            for Stepping := 1 to Size do
                begin
                   for Step := 1 to Size do
                       begin
                          Write('A (',Steps,',',Stepping,',',Step,') = ');
                          Readln( A [ Steps, Stepping , Step ] );
                       end;
                   Writeln;
                end;
            Writeln;
         end;
      Writeln;
    Writeln('Enter the elements of Array "B"');
      Writeln;
        for Steps := 1 to Size do
              begin
                 for Stepping := 1 to Size do
                     begin
                        for Step := 1 to Size do
                          begin
                             Write('B (',Steps,',',Stepping,',',Step,') = ');
                             Readln( B [ Steps , Stepping , Step ] );
                             C [ Steps , Stepping , Step ] := Ø;
                          end;
                        Writeln;
                     end;
                Writeln;
              end;
      end;

Procedure Work;
  begin
   Sign := 1;
    if Resp in [ 'S','s'] then Sign := -1;
       For Steps := 1 to Size do
        For Stepping := 1 to Size do
         for Step := 1 to Size do
          C[ Steps , Stepping , Step ] := A[ Steps , Stepping , Step ] +
                   Sign * B[ Steps , Stepping , Step ];
     Space :='    ';
  end;

Procedure Printout;
   begin
     case answer of
     's','S' : begin
                 Writeln;
                 Writeln('The answer is: ');
                  For Steps := 1 to Size do
                    begin
                      for Stepping := 1 to Size do
                        begin
                          for Step := 1 to Size do
                           Write(Space,C [ Steps , Stepping ,Step ] );
                           Writeln;
                        end;
                      Writeln;
                    end;
                Writeln;Writeln;
              end;
     'p','P' : begin
                Rewrite(out,'Printer:');
                  Writeln;
                  Writeln('The answer is: ');
                  Writeln(out,'The answer is: ');
```

```
                    For Steps := 1 to Size do
                      begin
                       for Stepping := 1 to Size do
                         begin
                          for Step := 1 to Size do
                             begin
                              Write(Space,C [ Steps , Stepping , Step  ] );
                              Write(out,Space,C [ Steps , Stepping , Step ] );
                             end;
                            Writeln;Writeln(out);
                         end;
                       Writeln;Writeln(out);
                       Writeln;
                    end;
                 Close(out);
                end
            end { case }
        end;

begin { Main }
  Repeat
    Outputs;
    Title;
    Input;
     Work;
    Printout;
     Writeln;
          Write('Another Matrices?(Y/N) ');
             Readln(Cont);
  until Cont in [ 'N','n']
end.
```

3-Dimensional Matrix Addition and Subtraction

This program adds and subtracts two
3-Dimensional matrices.

Enter size of matrices:2

Addition or Subtraction (A/S): S

Enter the elements of Array "A":

A(1,1,1) = 34
A(1,1,2) = 65

A(1,2,1) = 12
A(1,2,2) = 38

A(2,1,1) = 72
A(2,1,2) = 42

A(2,2,1) = 10
A(2,2,2) = 91

Enter the elements of Array "B":

B(1,1,1) = 18
B(1,1,2) = 33

B(1,2,1) = 54
B(1,2,2) = 58

B(2,1,1) = 90
B(2,1,2) = 7

B(2,2,1) = 15
B(2,2,2) = 23

The answer is:

```
 1.60000E1   3.20000E1
-4.20000E1  -2.00000E1

-1.80000E1   3.50000E1
-5.00000    6.80000E1
```

Another Matrices? (Y/N) N

4.7 MULTIPLICATION OF THREE-DIMENSIONAL MATRICES

This program is very similar to its two-dimensional counterpart. The major difference between the two is the addition of an extra *FOR-DO* loop in the input and output statements.

```
(*=========================================*)
(*                                         *)
(* Title:   Multiplication of Three-       *)
(*          Dimensional Matrices           *)
(* Program Summary: Program multiplies     *)
(*          two 3-dimensional matrices.    *)
(*                                         *)
(*=========================================*)

PROGRAM MultiplyThreeD ;


 VAR
  Answer,
  Ans,
  Cont            : Char;
  Out             : Text;
  Step,
  Steps,
  Stepping,
  Size            : Integer;
  A,
  B,
  C               : Array  [1..5,1..5,1..5] of Real;
  Space           : String;

Procedure Outputs;
  begin
    page(output);
     repeat
       gotoXY(0,5);
        Write('S(creen or P(rinter ? ');
        Readln(Answer);
     until Answer in [ 'S','s','p','P' ]
  end;  {outputs}
```

```
Procedure Title;
  begin
    page(output);
     gotoXY(6,1);
      Writeln('3-D Matrix Multiplication');
      Writeln;Writeln;
      Writeln('This program multiplies 2 3-Dimensional');
      Writeln('matrices element by element.');
      Writeln;
      Writeln;
   end;  { title }

Procedure Input;
  begin
    Write('Enter size of matrices: ');
    Readln( Size );
    if Size = Ø then Input;
    Writeln;
    Writeln('Enter the elements of Array "A"');
    Writeln;
       for Stepping := 1 to Size do
             begin
                for Step := 1 to Size do
                    begin
                       for Steps := 1 to Size do
                          begin
                            Write('A (',Stepping,',',Step,',',Steps,') = ');
                            Readln( A [ Stepping , Step , Steps ] );
                          end;
                       writeln;
                      end;
                Writeln;
             end;
    Writeln;
    Writeln('Enter the elements of Array "B"');
    Writeln;
       for Stepping := 1 to Size do
             begin
                for Step := 1 to Size do
                    begin
                       for Steps := 1 to Size do
                           begin
                              Write('B (',Stepping,',',Step,',',Steps,') = ');
                              Readln( B [ Stepping , Step , Steps ] );
                                 C [ Stepping , Step , Steps ] := Ø;
                              end;
                          Writeln;
                       end;
                   Writeln;
              end;

   end;

Procedure Work;
  begin
    For Stepping := 1 to Size do
     For Step := 1 to Size do
      for Steps := 1 to Size do
        C[ Stepping , Step , Steps ] :=
             A[ Stepping , Step , Steps ] * B[ Stepping , Step , Steps ] ;
    Space := '    ';
   end;

Procedure Printout;
 begin
   case answer of
   's','S' : begin
               Writeln;
               Writeln('The answer is: ');
                For Stepping := 1 to Size do
                  begin
                   for Step := 1 to Size do
                     begin
                       for Steps := 1 to Size do
                         begin
                          Write(Space,C [ Stepping , Step , Steps ] );
                         end;
```

```
                    Writeln;
                  end;
                 Writeln;Writeln;
              end;
          end;
 'p','P' : begin
             Rewrite(out,'Printer:');
               Writeln;
               Writeln('The answer is: ');
               Writeln(out,'The answer is: ');
                For Stepping := 1 to Size do
                  begin
                   for Step := 1 to Size do
                     begin
                      for Steps := 1 to Size do
                          begin
                           Write(Space,C [ Stepping , Step , Steps ] );
                           Write(out,Space,C [ Stepping , Step , Steps ] );
                          end;
                      Writeln;Writeln(out);
                     end;
                    Writeln;Writeln(out);
                    Writeln;Writeln(out);
                 end;
               Writeln;
           Close(out);
          end
      end { case }
  end;


begin { Main }
  Repeat
    Outputs;
    Title;
    Input;
     Work;
    Printout;
     Writeln;
          Write('Another Matrices?(Y/N) ');
             Readln(Cont);
  until Cont in [ 'N','n']
end.
```

3-Dimensional Matrix Multiplication

This program multiplies 2 3-Dimensional
matrices element by element.

Enter size of matrices: 3

Enter the elements of Array "A"

A(1,1,1) = 1
A(1,1,2) = 2
A(1,1,3) = 3

A(1,2,1) = 4
A(1,2,2) = 5
A(1,2,3) = 6

A(1,3,1) = 7
A(1,3,2) = 8
A(1,3,3) = 9

```
A(2,1,1) = Ø
A(2,1,2) = 12
A(2,1,3) = 13

A(2,2,1) = 14
A(2,2,2) = 115
A(2,2,3) = 213

A(2,3,1) = 312
A(2,3,2) = 56
A(2,3,3) = 78

A(3,1,1) = 21
A(3,1,2) = 43
A(3,1,3) = 55

A(3,2,1) = 6
A(3,2,2) = 4
A(3,2,3) = 3

A(3,3,1) = 7
A(3,3,2) = Ø
A(3,3,3) = Ø

Enter the elements of Array "B"

B(1,1,1) = Ø
B(1,1,2) = Ø
B(1,1,3) = Ø

B(1,2,1) = 3
B(1,2,2) = 1
B(1,2,3) = 5

B(1,3,1) = 7
B(1,3,2) = 8
B(1,3,3) = 3

B(2,1,1) = 5
B(2,1,2) = 1
B(2,1,3) = 1

B(2,2,1) = 2
B(2,2,2) = 8
B(2,2,3) = Ø

B(2,3,1) = 3
B(2,3,2) = 6
A(2,3,3) = 8
```

B(3,1,1) = 3
B(3,1,2) = 1
B(3,1,3) = 1

B(3,2,1) = Ø
B(3,2,2) = 5
B(3,2,3) = 3

B(3,3,1) = 2
B(3,3,2) = 4
B(3,3,3) = 8

The answer is:

```
 Ø.ØØØØØ   Ø.ØØØØØ   Ø.ØØØØØ
 1.2ØØØØE1   5.ØØØØØ   3.ØØØØØE1
 4.9ØØØØE1   6.4ØØØØE1   2.7ØØØØE1

Ø.ØØØØØ   1.2ØØØØE1   1.3ØØØØE1
 2.8ØØØØE1   9.2ØØØØE2   Ø.ØØØØØ
 9.36ØØØE2   3.36ØØØE2   6.24ØØØE2

6.3ØØØØE1   4.3ØØØØE1   5.5ØØØØE1
 Ø.ØØØØØ   2.ØØØØØE1   9.ØØØØØ
 1.4ØØØØE1   Ø.ØØØØØ   Ø.ØØØØØ
```

Another Matrices? (Y/N) N

4.8 SCALAR MULTIPLICATION OF THREE-DIMENSIONAL MATRICES

With this program, a three-dimensional matrix can be multiplied by a scalar value. This program, as well as other programs performing three-dimensional mathematics, is very similar to its two-dimensional counterpart.

```
(*=========================================*)
(*                                         *)
(* Title:   Scalar Multiplication of       *)
(*          3-Dimensional Matrices         *)
(* Program Summary: Program multiplies a   *)
(*          3-dimensional matrix by        *)
(*          a scalar value.                *)
(*                                         *)
(*=========================================*)

PROGRAM Scalar3DiMatrix ;

  VAR
   Answer,
   Ans,
   Cont           : Char;
```

```
    Out            : Text;
    Scale          : Real;
    Step,
    Stepper,
    Steps,
    Size           : Integer;
    A,
    B              : Array  [1..5,1..5,1..5] of Real;
    Space          : String;

  Procedure Outputs;
    begin
      page(output);
        repeat
          gotoXY(0,5);
            Write('S(creen or P(rinter ? ');
            Readln(Answer);
        until Answer in [ 'S','s','p','P' ]
    end;  {outputs}

  Procedure Title;
    begin
      page(output);
        gotoXY(6,1);
         Writeln('3-D Matrix Scalar Multiplication');
         Writeln;Writeln;
         Writeln('This program multiplies a 3-Dimensional');
         Writeln('matrix by a scalar.');
         Writeln;
         Writeln;
     end;  { title }

  Procedure Input;
    begin
     Write('Enter size of matrices: ');
     Readln( Size );
     if Size = 0 then Input;
     Writeln;
     Write('Enter Scaling Factor: ');
     Readln( Scale );
     Writeln('Enter the elements of the array: ');
     Writeln;
       for Steps := 1 to Size do
         begin
           for Step := 1 to Size do
              begin
                for Stepper := 1 to Size do
                  begin
                   Write('A (',Steps,',',Step,',',Stepper,') = ');
                   Readln( A [ Steps , Step , Stepper ] );
                  end;
               Writeln;
           end;
       Writeln;
     end;
     Writeln;
  end;

  Procedure Work;
    begin
     For Steps := 1 to Size do
       For Step := 1 to Size do
          for Stepper := 1 to Size do
           B[ Steps , Step , Stepper ] := A [ Steps , Step , Stepper ] * Scale;
      Space := '    '
    end;


   Procedure Printout;
      begin
        case answer of
        's','S' : begin
                   Writeln;
                   Writeln('The answer is: ');
                    For Steps := 1 to Size do
                      begin
                        for Step := 1 to Size do
```

```
                    begin
                      for Stepper := 1 to Size do
                        begin
                         Write(Space,B [ Steps , Step , Stepper  ] );
                        end;
                      Writeln;
                    end;
                  Writeln;Writeln;
                end;
            end;
   'p','P' : begin
              Rewrite(out,'Printer:');
                Writeln;
                Writeln('The answer is: ');
                Writeln(out,'The answer is: ');
                 For Steps := 1 to Size do
                   begin
                    for Step := 1 to Size do
                      begin
                        for Stepper := 1 to Size do
                          begin
                            Write(Space,B [ Steps , Step , Stepper ] );
                            Write(out,Space,B [ Steps , Step , Stepper ] );
                          end;
                        Writeln;
                        Writeln(out);
                      end;
                     Writeln;Writeln(out);
                     Writeln;Writeln(out);
                   end;
                Writeln;
            Close(out);
           end
       end { case }
   end;


begin { Main }
  Repeat
    Outputs;
    Title;
    Input;
     Work;
    Printout;
     Writeln;
          Write('Another Matrices?(Y/N) ');
             Readln(Cont);
  until Cont in [ 'N','n']
end.
```

3-Dimensional Matrix Scalar Multiplication

This program multiplies a 3-Dimensional
matrix by a scalar.

Enter the size of matrices: 2

Enter Scaling Factor: 3.3

Enter the elements of the array:

A(1,1,1) = 1Ø
A(1,1,2) = 32

A(1,2,1) = 4
A(1,2,2) = 19

```
A(2,1,1) = 73
A(2,1,2) = Ø

A(2,2,1) = 1
A(2,2,2) = 37


The answer is:

   3.3ØØØØE1    1.Ø56ØØE2
   1.32ØØØE1    6.27ØØØE1

   2.4Ø9ØØE2    Ø.ØØØØØ
   3.3ØØØØ    1.221ØØE2

Another Matrices? (Y/N) N
```

5 DATA ANALYSIS

The programs in this chapter are designed to make the interpretation of experimental data easy. The first three programs require the input of a set of data, which then results in the calculation of either the geometric mean, average, and median, or variance and standard deviation.

The next two programs allow you to enter several data points and interpolate for a value between them. The programs accomplish this by forming the second or third order La Grange polynomial that represents the given data.

The last four programs in this section are curve-fitting programs. From the input the programs try to form the equation that best fits the data. The four programs try to fit four different types of equations. Once the equation is found, it is printed. Then the program permits interpolation of data. In addition to printing the equation, the program also prints the sum of the deviations squared. This value is a measure of how well the equation fits. The smaller its value, the better the fit. A value of zero indicates a good fit but not necessarily one that is perfect.

5.1 GEOMETRIC MEAN

Mathematically, the geometric mean is the Nth root of the product of N numbers. In other words, the geometric mean is the second term of three consecutive terms in a geometric progression. Thus, for the progression 2, 4, 8, the geometric mean is 4 because it is the middle term; and $2 \times 4 \times 8 = 64$, whose third root is 4.

In order to set up a loop to read the input numbers, the user must first specify the quantity of numbers to be entered. The mean is calculated in *Procedure Work.*

```
(*======================================*)
(*                                      *)
(* Title:  Geometric Mean               *)
(* Program Summary: Program calculates  *)
(*         the geometric mean of        *)
(*         any number (N) of data       *)
(*         points by finding the Nth    *)
(*         root of the product of       *)
(*         all the numbers.             *)
(*                                      *)
(*======================================*)


PROGRAM GeometricMean ;

Uses Transcend;

 VAR
  Answer,
  Ans,
  Cont           : Char;
  Out            : Text;
  Number,
  Steps          : Integer;
  P              : Array  [1..50] of Real;
  Mean,
  Temp              : Real;

Procedure Outputs;
  begin
    page(output);
     repeat
       gotoXY(0,5);
        Write('S(creen or P(rinter ? ');
        Readln(Answer);
     until Answer in [ 'S','s','p','P' ]
  end;  (outputs)

Procedure Title;
  begin
    page(output);
     gotoXY(11,1);
      Writeln('Geometric Mean');
      Writeln;Writeln;
      Writeln('This program will compute the geometric');
      Writeln('mean of a set of data points. You must');
      Writeln('enter the number of points and their value.');
      Writeln;
      Writeln;
   end;  { title }

Procedure Input;
  begin
       Write('Enter number of points: ');
       Readln( Number );
       Writeln;
              for Steps := 1 to Number do
                    begin
                      Write('Enter point ',Steps,' : ');
                      Readln( P [ Steps ] );
                    end;
       Writeln;

  end;

Procedure Work;
   begin
       Temp := 1;
     For Steps := 1 to Number do
         Temp := Temp * P [ Steps ] ;
         Mean:= exp ( Ln ( Temp ) / Number ) ;
   end;

 Procedure Printout;
    begin
      case answer of
      's','S' : begin
```

```
                Writeln('The Data are :');
                   for Steps := 1 to Number do
                       Writeln('     ',P [ Steps ] );
                Writeln;
                Writeln;
                Writeln('The geometric mean is ',Mean);
             end;
    'p','P' : begin
               Rewrite(out,'Printer:');
                Writeln('The Data are :');
                Writeln(out,'The Data are :');
                   for Steps := 1 to Number do
                       begin
                          Writeln('     ',P [ Steps ] );
                          Writeln(out,'     ',P [ Steps ] );
                       end;
                Writeln;
                Writeln;
                Writeln(out);
                Writeln(out);
                Writeln('The geometric mean is ',Mean);
                Writeln(out,'The geometric mean is ',Mean);
             Close(out);
            end
        end { case }
    end;

begin { Main }
  Repeat
    Outputs;
    Title;
    Input;
     Work;
     Printout;
     Writeln;
          Write('Another Problem ?(Y/N) ');
             Readln(Cont);
  until Cont in [ 'N','n']
end.
```

Geometric Mean

This program will compute the geometric
mean of a set of data points. You must
enter the number of points and their value.

Enter number of points: 5

Enter point 1 : 75
Enter point 2 : 68
Enter point 3 : 97
Enter point 4 : 55
Enter point 5 : 3Ø

The data are :

7.5ØØØØE1
6.8ØØØØE1
9.7ØØØØE1
5.5ØØØØE1
3.ØØØØØE1

The geometric mean is 6.Ø585ØE1

Another Problem? (Y/N) N

5.2 AVERAGE AND MEDIAN CALCULATION

In the analysis of data, two of the most elementary operations performed are averaging and determination of the median. Averaging is a simple operation. It entails the summing of all data points and the division of this sum by the number of points.

Finding the median is a little more difficult. It is the value of the data point, in an ordered set of data, below and above which there are an equal number of data points. If there is no middle number, which occurs when there are an even number of data points, the median is calculated as the arithmetic mean of the two middle values.

In order to find the median, it is first necessary to order (or sort) the data. The sorting function has a long history, and there are countless subroutines for it. The one used here is fairly fast and memory efficient. The sorting takes place in *Procedure Work*. After the data is sorted, a test is made in the second half of *Procedure Work* to determine if the number of data points is even. If it is, the program calculates the average of the two middle data points as the median. If not, the middle data point is chosen.

An advantage of the program, resulting from the median calculation, is the outputting of data in ordered form, highest value first.

```
(*=========================================*)
(*                                         *)
(* Title:   Average and Median             *)
(*          Calculation                    *)
(* Program Summary: Program calculates     *)
(*          the average and median         *)
(*          of a set of data.              *)
(*                                         *)
(*=========================================*)


PROGRAM AverageMedian;


 VAR
  Answer,
  Ans,
  Cont             : Char;
  Out              : Text;
  Num,
  Inc,
  M,
  Stepping,
  Inthalf,
  Number,
  Steps            : Integer;
  A                : Array  [1..50] of Real;
  Half,
  Med,
  Sum,
  Avg,
  Temp               : Real;

Procedure Outputs;
  begin
    page(output);
     repeat
       gotoXY(0,5);
        Write('S(creen or P(rinter ? ');
        Readln(Answer);
     until Answer in [ 'S','s','p','P' ]
  end;  {outputs}

Procedure Title;
  begin
    page(output);
```

```
      gotoXY(11,1);
       Writeln('Average and Median');
       Writeln;Writeln;
       Writeln('This program calculates the average');
       Writeln('and the median of a set of data.       ');
       Writeln;
       Writeln;
    end;  { title }

Procedure Input;
  begin
       Write('Enter number of points: ');
       Readln( Number );
       Writeln;
       Num := Number -1;
       Writeln('Enter data:');
              for Steps := 1 to Number do
                    begin
                      Write('A',Steps,' = ');
                      Readln( A [ Steps ] );
                    end;
       Writeln;
  end;

Procedure Test;
  begin
        If A [ M ] >=  A [ Stepping ] then
            Exit ( Test );
        M := Stepping
  end;

Procedure Work;
   begin
    for Steps := 1 to Num do
       begin
         Inc := Steps + 1 ;
         M   := Steps ;
          For Stepping := Inc to Number do
             Test ;
          Temp := A [ Steps ] ;
          A [ Steps ] := A [ M ] ;
          A [ M ] := Temp
       end;
    Half := Number / 2 ;
    Inthalf := Round ( Half );
         If Half = Round ( Half ) then
            Med := ( A [ Inthalf ] + A [ Inthalf + 1  ] ) / 2 ;
         If Half <  Round ( Half ) then
            Med := A [ Inthalf ]
        else
            Med := A [ Inthalf +1 ] ;
            Sum := Ø ;
               For Steps := 1 to Number do
                  Sum := Sum + A [ Steps ] ;
            Avg := Sum / Number
   end;

 Procedure Printout;
    begin
      case answer of
      's','S' : begin
                  Writeln;
                  Writeln('Average = ',Avg,'    Median = ',Med);
                  Writeln;
                  Writeln('The Data in descending order are :');
                     for Steps := 1 to Number do
                         Writeln('     ',A [ Steps ] );
                  Writeln;
                  Writeln;
                end;
      'p','P' : begin
                 Rewrite(out,'Printer:');
                  Writeln;
                  Writeln(out);
                  Writeln('Average = ',Avg,'    Median = ',Med);
                  Writeln(out,'Average = ',Avg,'    Median = ',Med);
                  Writeln;
```

```
                Writeln('The Data in descending order are :');
                Writeln(out);
                Writeln(out,'The Data in descending order are :');
                   for Steps := 1 to Number do
                       begin
                          Writeln('     ',A [ Steps ] );
                          Writeln(out,'     ',A [ Steps ] );
                       end;
                Writeln;
                Writeln;
                Writeln(out);
                Writeln(out);
             Close(out);
            end
        end { case }
    end;

begin { Main }
  Repeat
    Outputs;
    Title;
    Input;
     Work;
     Printout;
     Writeln;
          Write('Another Calculation ?(Y/N) ');
             Readln(Cont);
  until Cont in [ 'N','n']
end.
```

Average and Median

This program calculates the average
and the median of a set of data.

Enter number of points: 5

Enter data:

A1 = 75
A2 = 68
A3 = 97
A4 = 55
A5 = 3Ø

Average = 6.5ØØØØE1 Median = 6.8ØØØØE1

The data in descending order are :

9.7ØØØØE1
7.5ØØØØE1
6.8ØØØØE1
5.5ØØØØE1
3.ØØØØØE1

Another calculation? (Y/N) N

5.3 VARIANCE AND STANDARD DEVIATION CALCULATION

The standard deviation is a measure of dispersion of a set of data. Mathematically, it is the square root of the arithmetic mean of the squares of the deviation of each data point from the average of the data set.

The variance of a set of data is simply the standard deviation squared. If all data points are close to the mean, the variance (and standard deviation) will be small. If the values are distributed far from the mean, the variance will be large.

Since the average (mean) must be calculated to find the variance and standard deviation, its value is also printed.

```
(*=========================================*)
(*                                         *)
(* Title:   Variance and Standard          *)
(*          Deviation Calculations         *)
(* Program Summary: Program calculates     *)
(*          the variance and standard      *)
(*          deviation of a set of data.    *)
(*                                         *)
(*=========================================*)


PROGRAM VarianceStdDev;

uses Transcend;

 VAR
  Answer,
  Ans,
  Cont           : Char;
  Out            : Text;
  Stepping,
  Number         : Integer;
  A              : Array [1..10] of Real;
  Sum,
  Avg,
  Square,
  Dev,
  Vari,
  StdDev        : Real;

Procedure Outputs;
  begin
    page(output);
     repeat
       gotoXY(0,5);
        Write('S(creen or P(rinter ? ');
        Readln(Answer);
     until Answer in [ 'S','s','p','P' ]
  end;  {outputs}

Procedure Title;
  begin
    page(output);
     gotoXY(5,1);
      Writeln('Variance and Standard Deviation');
      Writeln;Writeln;
      Writeln('This program calculates the variance');
      Writeln('and the standard deviation of a set of data.');
      Writeln;
      Writeln;
   end;  { title }

Procedure Input;
  begin
       Write('Enter number of points: ');
       Readln( Number );
       Writeln;
       Writeln('Enter data:');
            Sum := 0 ;
```

```
                for Stepping := 1 to Number do
                      begin
                        Write('A',Stepping,' = ');
                        Readln( A [ Stepping ] );
                        Sum := Sum + A [ Stepping ] ;
                      end;
        Writeln;
  end;

Procedure Work;
   begin
     Avg := Sum / Number ;
      Square := 0 ;
    for Stepping := 1 to Number do
        begin
            Dev :=  A [ Stepping ] - Avg ;
           Square := Square + Sqr( Dev )
        end;
    Vari := Square / ( Number - 1 ) ;
    StdDev := Sqrt ( Vari );
   end;

 Procedure Printout;
    begin
      case answer of
      's','S' : begin
                   Writeln;
                   Writeln('Average = ',Avg);
                   Writeln;
                   Writeln('Variance = ',Vari);
                   Writeln;
                   Writeln('Standard Deviation = ',StdDev);
                   Writeln;
                   Writeln;
                end;
      'p','P' : begin
                  Rewrite(out,'Printer:');
                   Writeln;
                   Writeln('Average = ',Avg);
                   Writeln(out);
                   Writeln(out,'Average = ',Avg);
                   Writeln;
                   Writeln('Variance = ',Vari);
                   Writeln(out);
                   Writeln(out,'Variance = ',Vari);
                   Writeln;
                   Writeln('Standard Deviation = ',StdDev);
                   Writeln(out);
                   Writeln(out,'Standard Deviation = ',StdDev);
                   Writeln;
                   Writeln;
                   Writeln(out);
                   Writeln(out);
                Close(out);
               end
           end { case }
        end;

   begin { Main }
     Repeat
       Outputs;
       Title;
       Input;
        Work;
        Printout;
        Writeln;
             Write('Another Calculation ?(Y/N) ');
                Readln(Cont);
     until Cont in [ 'N','n']
   end.
```

Variance and Standard Deviation

This program calculates the variance
and the standard deviation of a set of data.

Enter the number of points: 5

Enter data:

A1 = 75
A2 = 68
A3 = 97
A4 = 55
A5 = 3Ø

Average = 6.5ØØØØE1

Variance = 6.145ØØE2

Standard Deviation = 2.47891E1

Another calculation? (Y/N) N

5.4 THREE-POINT INTERPOLATION

Very often after experimental data is taken, it is necessary to extrapolate to a point not included in the original set. One way of doing this is to use three of the known data points to form a La Grange polynomial. This polynomial represents the second order equation defined by the three points. The technique is also known as quadratic interpolation.

The coefficients of the equation are calculated in *Procedure Work*. The actual interpolation of values takes place in *Procedure Input2* after the value for X is entered.

```
(*========================================*)
(*                                        *)
(* Title:   Three-Point Interpolation     *)
(* Program Summary: Program extrapolates  *)
(*          a point not given in the      *)
(*          data set by using three       *)
(*          other points in the set.      *)
(*                                        *)
(*========================================*)


PROGRAM ThreepointInter ;

 VAR
  Answer,
  Ans,
  Cont              : Char;
  Out               : Text;
  XØ,X1,X2,X,
  YØ,Y1,Y2,Y,
  RØ,R1,R2,R3,
  CØ,C1,C2          : Real;

Procedure Outputs;
  begin
    page(output);
     repeat
       gotoXY(Ø,5);
        Write('S(creen or P(rinter ? ');
        Readln(Answer);
     until Answer in [ 'S','s','p','P' ]
  end;  {outputs}
```

```
Procedure Title;
  begin
    page(output);
     gotoXY(6,1);
      Writeln('Three-point Interpolation');
      Writeln;Writeln;
      Writeln('This program forms a La Grange ');
      Writeln('Polynomial which represents the');
      Writeln('2nd order equation defined by ');
      Writeln('3 points.  It is used to interpolate');
      Writeln('values of Y for given values of X');
      Writeln;
      Writeln('When entering multiple inputs, use');
      Writeln('a space between data entries');
      Writeln;
      Writeln;
   end;  { title }

Procedure Input;
  begin
     Write('Enter First Point (X,Y): ');
     Readln( XØ , YØ );
     Writeln;
     Write('Enter Second Point (X,Y): ');
     Readln( X1 , Y1 );
     Writeln;
     Write('Enter Third Point (X,Y): ');
     Readln( X2 , Y2 );
     Writeln;
     Writeln;
  end;

Procedure Work;
   begin
      RØ := YØ / (( XØ - X1 ) * ( XØ - X2 ));
      R1 := Y1 / (( X1 - XØ ) * ( X1 - X2 ));
      R2 := Y2 / (( X2 - XØ ) * ( X2 - X1 ));
      CØ := RØ * X1 * X2 + R1 * XØ * X2 + R2 * XØ * X1 ;
      C1 := - RØ * ( X1 + X2 ) - R1 * ( XØ + X2 ) - R2 * ( XØ + X1 );
      C2 := RØ + R1 + R2 ;
   end;

Procedure Printout;
  Forward;

Procedure Input2;
  begin
       Write('Enter X :');
       Readln( X );
       Y := CØ + C1 * X + C2 * Sqr ( X );
       Printout;
  end;

 Procedure Printout;
    begin
      case answer of
      's','S' : begin
                  Writeln;
                  Writeln('at X = ',Round(X),'    Y = ',Round(Y));
                  Writeln;
                  Write('Another Point ( Y/N ): ');
                  Readln( Ans );
                  If Ans in [ 'Y','y' ] then
                       Input2;
                end;
      'p','P' : begin
                   Rewrite(out,'Printer:');
                  Writeln;
                  Writeln(out);
                  Writeln('at X = ',Round(X),'    Y = ',Round(Y));
                  Writeln(out,'at X = ',Round(X),'    Y = ',Round(Y));
                  Writeln;
                  Writeln(out);
                  Write('Another Point ( Y/N ): ');
                  Readln( Ans );
                  If Ans in [ 'Y','y' ] then
```

```
                        begin
                          Close(out);
                          Input2
                        end;
                 Close(out);
               end
          end { case }
     end;

begin { Main }
  Repeat
     Outputs;
     Title;
     Input;
      Work;
      Input2;
      Writeln;
            Write('Another New 3 points ?(Y/N) ');
                Readln(Cont);
  until Cont in [ 'N','n']
end.
```

Three-point Interpolation

This program forms a La Grange polynomial which represents the 2nd order equation defined by 3 points. It is used to interpolate values of Y for a given value of X.

When entering multiple inputs, use a space between data entries.

Enter First Point (X,Y): 1 8

Enter Second Point (X,Y): 4 38

Enter Third Point (X,Y): 2 16

Enter X : 3

at X = 3 Y = 26

Another Point? (Y/N) Y

Enter X : 7

at X = 7 Y = 86

Another Point? (Y/N) N

Another New 3 points ?(Y/N) N

5.5 FOUR-POINT INTERPOLATION

While the three-point interpolation technique in the previous program will suffice for most applications, for greater accuracy it is possible to use four points and form a polynomial of the third degree.

The coefficients for the third-degree equation are calculated in *Procedure Work*. Interpolation of values takes place in *Procedure Input2* after the entry of X.

```
(*========================================*)
(*                                        *)
(* Title:  Four-Point Interpolation       *)
(* Program Summary: Program extrapolates  *)
(*         a point not given in the       *)
(*         data set by using four         *)
(*         other points in the set.       *)
(*                                        *)
(*========================================*)


PROGRAM FourpointInter ;

 VAR
  Answer,
  Ans,
  Cont             : Char;
  Out              : Text;
  XØ,X1,X2,X3,X,
  YØ,Y1,Y2,Y3,Y,
  RØ,R1,R2,R3,
  CØ,C1,C2,C3,
  CA,CB            : Real;

Procedure Outputs;
  begin
    page(output);
     repeat
       gotoXY(Ø,5);
        Write('S(creen or P(rinter ? ');
        Readln(Answer);
     until Answer in [ 'S','s','p','P' ]
  end;  {outputs}

Procedure Title;
  begin
    page(output);
     gotoXY(6,1);
      Writeln('Four-point Interpolation');
      Writeln;Writeln;
      Writeln('This program forms a La Grange ');
      Writeln('Polynomial which represents the');
      Writeln('3rd order equation defined by ');
      Writeln('4 points.  It is used to interpolate');
      Writeln('values of Y for given values of X');
      Writeln;
      Writeln('When entering multiple inputs, use');
      Writeln('a space between data entries');
      Writeln;
      Writeln;
   end;  { title }

Procedure Input;
  begin
     Write('Enter First Point (X,Y): ');
     Readln( XØ , YØ );
     Writeln;
     Write('Enter Second Point (X,Y): ');
     Readln( X1 , Y1 );
     Writeln;
     Write('Enter Third Point (X,Y): ');
     Readln( X2 , Y2 );
     Writeln;
```

```
      Write('Enter Fourth Point (X,Y): ');
      Readln( X3 , Y3 );
      Writeln;Writeln;
   end;

 Procedure Work;
    begin
       RØ := YØ / (( XØ - X1 ) * ( XØ - X2 ) * ( XØ - X3 ));
       R1 := Y1 / (( X1 - XØ ) * ( X1 - X2 ) * ( X1 - X3 ));
       R2 := Y2 / (( X2 - XØ ) * ( X2 - X1 ) * ( X2 - X3 ));
       R3 := Y3 / (( X3 - XØ ) * ( X3 - X1 ) * ( X3 - X2 ));
       CØ := - RØ * X1 * X2 * X3 - R1 * XØ * X2 * X3 -
              R2 * XØ * X1 * X3 - R3 * XØ * X1 * X2 ;
       C1 := RØ * ( X1 * X2 + X1 * X3 + X2 * X3 ) +
                R1 * ( XØ * X2 + X2 * X3 + XØ * X3 );
       CA := R2 * ( XØ * X1 + X1 * X3 + XØ * X3 );
       CB := R3 * ( XØ * X1 + X1 * X2 + XØ * X2 );
       C1 := C1 + CA + CB ;
       C2 := - RØ * ( X1 + X2 + X3 ) - R1 * ( XØ + X2 + X3 ) -
                R2 * ( XØ + X1 + X3 ) - R3 * ( XØ + X1 + X2 );
       C3 := RØ + R1 + R2 + R3 ;
    end;

 Procedure Printout;
   Forward;

 Procedure Input2;
   begin
        Write('Enter X :');
        Readln( X );
        Y := CØ + C1 * X + C2 * Sqr ( X ) + C3 * X * Sqr ( X );
        Printout;
   end;

  Procedure Printout;
     begin
       case answer of
       's','S' : begin
                   Writeln;
                   Writeln('at X = ',Round(X),'    Y = ',Round(Y));
                   Writeln;
                   Write('Another Point ( Y/N ): ');
                   Readln( Ans );
                   If Ans in [ 'Y','y' ] then
                       Input2;
                  end;
       'p','P' : begin
                    Rewrite(out,'Printer:');
                   Writeln;
                   Writeln(out);
                   Writeln('at X = ',Round(X),'    Y = ',Round(Y));
                   Writeln(out,'at X = ',Round(X),'    Y = ',Round(Y));
                   Writeln;
                   Writeln(out);
                   Write('Another Point ( Y/N ): ');
                   Readln( Ans );
                   If Ans in [ 'Y','y' ] then
                       begin
                        Close (out);
                        Input2
                      end;
                  Close(out);
                end
           end { case }
       end;

   begin { Main }
     Repeat
       Outputs;
       Title;
       Input;
        Work;
        Input2;
        Writeln;
              Write('Another New 4 points ?(Y/N) ');
                  Readln(Cont);
     until Cont in [ 'N','n']
   end.
```

```
Four-point Interpolation

This program forms a La Grange
polynomial which represents the
3rd order equation defined by
4 points. It is used to interpolate
values of Y for given values of X.

When entering multiple inputs, use
a space between data entries.

Enter First Point (X,Y): 3  47

Enter Second Point (X,Y): Ø  5

Enter Third Point (X,Y): 2  19

Enter Fourth Point (X,Y): 1  7

Enter X : 4
at X = 4 Y = 97

Another Point (Y/N) : Y

Enter X : 1Ø
at X = 1Ø  Y = 1195

Another Point (Y/N) : N

Another New 4 points ? (Y/N) N
```

5.6 LINEAR LEAST SQUARES FIT

After collecting data points from an experiment, it is often desirable to try and find a relationship to define that set of data.

This program uses the least squares technique to fit data points to a linear line whose equation is

$$Y = MX + B$$

Constants M and B are both calculated by the program.

Although the program tries to find an equation to closely approximate the data described, it doesn't always do a good job. An indication of this is a parameter called the "sum of the deviations squared." The smaller this number is, the better the fit is, and vice versa.

If no equation can be found, the program prints out a message to that effect.

```
(*======================================*)
(*                                      *)
(* Title:  Linear Least Squares Fit     *)
(* Program Summary: Program calculates  *)
(*         the least squares fit for    *)
(*         a plot whose abscissa and    *)
(*         ordinate are both linear.    *)
(*                                      *)
(*======================================*)


PROGRAM LeastSquareFit ;

 VAR
  Answer,
  Ans,
  Cont           : Char;
  Out            : Text;
  Number,
  Steps          : Integer;
  X,
  Y              : Array [1..20] of Real;
  X1,
  X2,
  Y1,
  XY,
  J,
  XX,
  YY,
  M,
  B,
  Dev,
  D2             : Real;
  Sign           : Char;

Procedure Outputs;
  begin
    page(output);
      repeat
        gotoXY(0,5);
         Write('S(creen or P(rinter ? ');
         Readln(Answer);
      until Answer in [ 'S','s','p','P' ]
  end;  {outputs}

Procedure Title;
  begin
    page(output);
      gotoXY(6,1);
       Writeln('Least Squares Linear Fit');
       Writeln;Writeln;
       Writeln('This program performs a least-squares');
       Writeln('approximation of a line. It');
       Writeln('accepts the X & Y coordinates of given');
       Writeln('data points and tries to produce a best');
       Writeln('fit line that represents the data.');
       Writeln('The equation of the line is:');
       Writeln;
       Writeln('    Y := MX + B ');
       Writeln;
       Writeln('Where M and B are calculated by the');
       Writeln('program.');
       Writeln;
       Writeln('When entering multiple inputs, use');
       Writeln('a space between data entries');
       Writeln;
       Writeln;
   end;  { title }

Procedure Input;
  begin
     Write('Enter number of data points :  ');
     Readln( Number );
     Writeln;
     Writeln('Enter data: ');
     Writeln;
```

```
            For Steps := 1 to Number do
               begin
                 Write('X(',Steps,') Y(',Steps,') = ');
                 Readln( X [ Steps ] , Y [ Steps ] )
               end;
      end;

    Procedure Interpolate;
       begin
         Write('Do you want to interpolate data (Y/N) ?');
         Readln( Ans );
           if Ans in [ 'Y','y'] then
                Begin
                  Write('Enter X: ');
                  Readln( XX );
                  YY := M * XX + B ;
                  Writeln('Y := ',Round(YY));
                    Writeln;
                    Interpolate
                end
       end;
    Procedure PInterpolate;
       begin
         Write('Do you want to interpolate data (Y/N) ?');
         Readln( Ans );
           if Ans in [ 'Y','y'] then
                Begin
                  Write('Enter X: ');
                  Readln( XX );
                  YY := M * XX + B ;
                  Writeln('Y := ',Round(YY));
                  Writeln(out,'Y := ',Round(YY));
                    Writeln;
                    PInterpolate
                end
       end;

    Procedure Printout;
       begin
         case answer of
         's','S' : begin
                      Writeln;
                      Writeln('The linear Equation that best fits the');
                      Writeln('given data is:');
                      Writeln;
                      Writeln('    Y := ',Round(M),' X ',Sign,Round(Abs( B )));
                      Writeln;
                      Writeln('For this equation, the sum of deviations');
                      Writeln('squared is ',Round(D2));
                      Writeln;
                      Writeln;
                      Interpolate;
                   end;
         'p','P' : begin
                     Rewrite(out,'Printer:');
                      Writeln;
                      Writeln(out);
                      Writeln('The linear Equation that best fits the');
                      Writeln('given data is:');
                      Writeln;
                      Writeln(out,'The linear Equation that best fits the');
                      Writeln(out,'given data is:');
                      Writeln(out);
                      Writeln('    Y := ',Round(M),' X ',Sign,Round(Abs( B )));
                      Writeln;
                      Writeln(out,'    Y := ',Round(M),' X ',Sign,Round(Abs( B )));
                      Writeln(out);
                      Writeln('For this equation, the sum of deviations');
                      Writeln('squared is ',Round(D2));
                      Writeln;
                      Writeln;
                      Writeln(out,'For this equation, the sum of deviations');
                      Writeln(out,'squared is ',Round(D2));
                      Writeln(out);
                      Writeln(out);
                      PInterpolate;
                   Close(out);
```

```
                end
            end { case }
        end;

Procedure Work;
   begin
    Sign := '-';
    X1 := Ø; Y1 := Ø; XY := Ø; X2 := Ø;
      For Steps := 1 to Number do
        begin
          X1 := X1 + X [ Steps ] ;
          Y1 := Y1 + Y [ Steps ] ;
          XY := XY + X [ Steps ] * Y [ Steps ] ;
          X2 := X2 + X [ Steps ] * X [ Steps ]
        end;
     J := Number * X2 - X1 * X1 ;
        if J = Ø then
          begin
            Writeln('No solution found!');
            Exit(Work);
          end;
       M := ( Number * XY - X1 * Y1 ) / J ;
       B := ( Y1 * X2 - X1 * XY ) / J ;
            if Abs ( B ) = B then
                 Sign := '+';
       Dev := Ø;
       D2 := Ø;
          for Steps := 1 to Number do
              begin
                Dev := Dev + Y [ Steps ] - M * X [ Steps ] - B ;
                D2 := D2 + Sqr( Y [ Steps ] - M * X [ Steps ] - B )
              end;
        Printout;
   end;

   begin { Main }
     Repeat
       Outputs;
       Title;
       Input;
        Work;
        Writeln;
             Write('Another data entry?(Y/N) ');
                Readln(Cont);
     until Cont in [ 'N','n']
   end.
```

Least Squares Linear Fit

This program performs a least-squares approximation of a linear line. It accepts the X & Y coordinates of given data points and tries to produce a best fit line that represents the data. The equation of the line is:

$$Y = MX + B$$

Where M and B are calculated by the program.

When Entering multiple inputs, use a space between data entries.

Enter number of data points: 3

```
Enter data:

X(1) Y(1) = Ø 3
X(2) Y(2) = 1Ø 53
X(3) Y(3) = 3 18

The linear Equation that best fits the
given data is:

   Y = 5X + 3

For this equation the sum of deviations
squared is Ø.

Do you want to interpolate data? (Y/N) Y

Enter X: 2
Y = 13

Do you want to interpolate data? (Y/N) N

Another data entry? (Y/N) N
```

5.7 LEAST SQUARES FIT — LOGARITHMIC ORDINATE

Like the previous program, this one tries to fit data to an equation. The equation used, however, is not linear. It is semilogarithmic. The abscissa is linear, but the ordinate is logarithmic. The equation used here is

$$Y = B\,(10^{M})^{X}$$

where once again M and B are calculated by the program.

Once the equation has been found, the program asks if data interpolation is desired and evaluates Y for any value of X if the answer is yes. If no solution is found, the program will tell you.

```
(*=========================================*)
(*                                         *)
(* Title:   Least Squares Fit -            *)
(*          Logarithmic Ordinate           *)
(* Program Summary: Program calculates     *)
(*          the least squares fit for      *)
(*          a semilogarithmic plot         *)
(*          having a linear abscissa       *)
(*          and a logarithmic ordinate.    *)
(*                                         *)
(*=========================================*)
```

```
PROGRAM LogLeastSquareFit ;

uses Transcend;

 VAR
  Answer,
  Ans,
  Cont            : Char;
  Out             : Text;
  Number,
  Steps           : Integer;
  X,
  Y               : Array [1..20] of Real;
  X1,
  X2,
  Y1,
  XY,
  J,
  XX,
  YY,
  MM,
  Ly,
  Lb,
  DTemp,
  M,
  B,
  Dev              : Real;

Procedure Outputs;
  begin
    page(output);
     repeat
       gotoXY(0,5);
        Write('S(creen or P(rinter ? ');
        Readln(Answer);
     until Answer in [ 'S','s','p','P' ]
  end;  {outputs}

Procedure Title;
  begin
    page(output);
     gotoXY(5,1);
      Writeln('Least Squares Fit Log Ordinate');
      Writeln;Writeln;
      Writeln('This program performs a least-squares');
      Writeln('approximation of a semi-logarithmic line');
      Writeln('The abscissa is linear and the ordinate ');
      Writeln('is logarithmic .  This program accepts');
      Writeln('the X & Y coordinates of given data');
      Writeln('points and tries to produce a best');
      Writeln('fit line that represents the data.');
      Writeln('The equation of the line is:');
      Writeln;
      Writeln('   Y := B * ( 10 ^ M ) ^ X  ');
      Writeln;
      Writeln('Where M and B are calculated by the');
      Writeln('program.');
      Writeln;
      Writeln('When entering multiple inputs, use');
      Writeln('a space between data entries');
      Writeln;
      Writeln;
   end;  { title }

Procedure Input;
  begin
     Write('Enter number of data points :  ');
     Readln( Number );
     Writeln;
     Writeln('Enter data: ');
     Writeln;
        For Steps := 1 to Number do
           begin
             Write('X(',Steps,') Y(',Steps,') = ');
             Readln( X [ Steps ] , Y [ Steps ] )
           end;
  end;
```

```
Procedure Interpolate;
   begin
     Write('Do you want to interpolate data (Y/N) ?');
     Readln( Ans );
       if Ans in [ 'Y','y'] then
            Begin
              Write('Enter X: ');
              Readln( XX );
              MM := exp( M * Ln ( 10 )) ;{ 10 ^ M }
              YY := B * exp( XX * Ln ( MM )) ;{ B * MM ^ XX }
              Writeln('Y := ',YY);
                Writeln;
                Interpolate
            end
    end;
Procedure PInterpolate;
   begin
     Write('Do you want to interpolate data (Y/N) ?');
     Readln( Ans );
       if Ans in [ 'Y','y'] then
            Begin
              Write('Enter X: ');
              Readln( XX );
              MM := exp( M * Ln ( 10 )) ;{ 10 ^ M }
              YY := B * exp( XX * Ln ( MM )) ;{ B * MM ^ XX }
              Writeln('Y := ',YY);
              Writeln(out,'Y := ',YY);
                Writeln;
                PInterpolate
            end
    end;

Procedure Printout;
   begin
     case answer of
     's','S' : begin
                 Writeln;
                 Writeln('The equation that best fits the');
                 Writeln('given data is:');
                 Writeln;
                 Writeln(' Y := ',B,'* ( 10 ^ ',M,' ) ^ X');
                 Writeln;
                 Writeln('For this equation the sum of deviations');
                 Writeln('squared is : ', Dev );
                 Writeln;
                 Writeln;
                 Interpolate;
               end;
     'p','P' : begin
                Rewrite(out,'Printer:');
                 Writeln;
                 Writeln(out);
                 Writeln('The equation that best fits the');
                 Writeln('given data is:');
                 Writeln(out,'The equation that best fits the');
                 Writeln(out,'given data is:');
                 Writeln;
                 Writeln(' Y := ',B,'* ( 10 ^ ',M,' ) ^ X');
                 Writeln(out);
                 Writeln(out,' Y := ',B,'* ( 10 ^ ',M,' ) ^ X');
                 Writeln;
                 Writeln('For this equation the sum of deviations');
                 Writeln('squared is : ',Dev);
                 Writeln;
                 Writeln;
                 Writeln(out);
                 Writeln(out,'For this equation the sum of deviations');
                 Writeln(out,'squared is : ', Dev );
                 Writeln(out);
                 Writeln(out);
                 PInterpolate;
               Close(out);
              end
        end { case }
     end;
```

```
Procedure Work;
   begin
    X1 := Ø; Y1 := Ø; XY := Ø; X2 := Ø;
      For Steps := 1 to Number do
        begin
          Ly := Ln( Y[ Steps ] ) / Ln ( 1Ø );
          X1 := X1 + X [ Steps ] ;
          Y1 := Y1 + Ly ;
          XY := XY + X [ Steps ] * Ly ;
          X2 := X2 + X [ Steps ] * X [ Steps ]
        end;
     J := Number * X2 - X1 * X1 ;
        if J = Ø then
          begin
            Writeln('No solution found!');
            Exit(Work);
          end;
       M := ( Number * XY - X1 * Y1 ) / J ;
       Lb := ( Y1 * X2 - X1 * XY ) / J ;
       B := exp ( Lb * Ln( 1Ø )) ;
       Dev := Ø;
          for Steps := 1 to Number do
              begin
                DTemp := Ln( 1Ø ) - M * X [ Steps ] - Lb   ;
                Dev := Dev + Ln ( Y [ Steps ] ) / DTemp * DTemp ;
                 end;
        Printout;
   end;

   begin { Main }
     Repeat
       Outputs;
       Title;
       Input;
        Work;
        Writeln;
             Write('Another data entry?(Y/N) ');
                Readln(Cont);
     until Cont in [ 'N','n']
   end.
```

Least Squares Fit Log Ordinate

This program performs a least squares approximation of a semilogarithmic line. The abscissa is linear and the ordinate is logarithmic. This program accepts the X & Y coordinates of given data points and tries to produce a best fit line that represents that data. The equation of the line is:

Y = B ∗ (1Ø∧M)∧X

Where M and B are calculated by the program.

When entering multiple inputs, use a space between data entries.

Enter number of data points : 3

Enter data:

X(1) Y(1) = 3 95
X(2) Y(2) = Ø 3
X(3) Y(3) = 2 3Ø

The equation that best fits the
given data is:

Y = 2.99941 * (1Ø∧5.ØØ172E-1)∧X

For this equation the sum of deviations
squared is : 9.Ø5369

Do you want to interpolate data? (Y/N) Y

Enter X: 1
Y = 9.48872

Do you want to interpolate data? (Y/N) N

Another data entry? (Y/N) N

5.8 LEAST SQUARES FIT — LOGARITHMIC ABSCISSA

This program also finds the least squares fit for a semilogarithmic plot, but this time the ordinate (Y) is linear and the abscissa (X) is logarithmic. The equation of the curve used here is :

$$Y = M * \ln(X)/\ln(10) + B$$

As in the previous programs, if no fit for the data can be found, the program prints out a message to that effect.

```
(*=========================================*)
(*                                         *)
(* Title:   Least Squares Fit -            *)
(*          Logarithmic Abscissa           *)
(* Program Summary: Program calculates     *)
(*          the least squares fit for      *)
(*          a semilogarithmic plot         *)
(*          having a linear ordinate       *)
(*          and a logarithmic abscissa.    *)
(*                                         *)
(*=========================================*)


PROGRAM LogLeastSquareFitAbsc ;

uses Transcend;

 VAR
  Answer,
  Ans,
  Cont            : Char;
  Out             : Text;
  Number,
  Steps           : Integer;
```

```
   X,
   Y              : Array [1..20] of Real;
   X1,
   X2,
   Y1,
   XY,
   J,
   XX,
   YY,
   MM,
   Lx,
   Lb,
   DTemp,
   M,
   B,
   Dev            : Real;
   Sign           : Char;

 Procedure Outputs;
   begin
     page(output);
      repeat
        gotoXY(0,5);
         Write('S(creen or P(rinter ? ');
         Readln(Answer);
      until Answer in [ 'S','s','p','P' ]
   end;  {outputs}

 Procedure Title;
   begin
     page(output);
      gotoXY(5,1);
       Writeln('Least Squares Fit Log Abscissa');
       Writeln;Writeln;
       Writeln('This program performs a least-squares');
       Writeln('approximation of a semi-logarithmic line');
       Writeln('The ordinate is linear and the abscissa ');
       Writeln('is logarithmic .  This program accepts');
       Writeln('the X & Y coordinates of given data');
       Writeln('points and tries to produce a best');
       Writeln('fit line that represents the data.');
       Writeln('The equation of the line is:');
       Writeln;
       Writeln('Y := M * Ln ( X ) / Ln ( 10 ) + B');
       Writeln;
       Writeln('Where M and B are calculated by the');
       Writeln('program.');
       Writeln;
       Writeln('When entering multiple inputs, use');
       Writeln('a space between data entries');
       Writeln;
       Writeln;
    end;  { title }

 Procedure Input;
   begin
      Write('Enter number of data points :  ');
      Readln( Number );
      Writeln;
      Writeln('Enter data: ');
      Writeln;
         For Steps := 1 to Number do
            begin
              Write('X(',Steps,') Y(',Steps,') = ');
              Readln( X [ Steps ] , Y [ Steps ] )
            end;
   end;

 Procedure Interpolate;
    begin
      Write('Do you want to interpolate data (Y/N) ?');
      Readln( Ans );
        if Ans in [ 'Y','y'] then
             Begin
               Write('Enter X: ');
               Readln( XX );
               YY := M * ( Ln ( XX ) / Ln ( 10 )) + B ;
```

```
                Writeln('Y := ',YY);
                  Writeln;
                  Interpolate
              end
        end;
    Procedure PInterpolate;
       begin
         Write('Do you want to interpolate data (Y/N) ?');
         Readln( Ans );
           if Ans in [ 'Y','y'] then
                Begin
                  Write('Enter X: ');
                  Readln( XX );
                  YY := M * ( Ln ( XX ) / Ln ( 10 )) + B ;
                  Writeln('Y := ',YY);
                  Writeln(out,'Y := ',YY);
                    Writeln;
                    PInterpolate
                end
        end;

    Procedure Printout;
       begin
         case answer of
         's','S' : begin
                     Writeln;
                     Writeln('The equation that best fits the');
                     Writeln('given data is:');
                     Writeln;
                     Writeln(' Y := ',M,'*Ln(X)/Ln(10) ',Sign,Abs(B));
                     Writeln;
                     Writeln('For this equation the sum of deviations');
                     Writeln('squared is : ', Dev );
                     Writeln;
                     Writeln;
                     Interpolate;
                  end;
         'p','P' : begin
                     Rewrite(out,'Printer:');
                     Writeln;
                     Writeln(out);
                     Writeln('The equation that best fits the');
                     Writeln('given data is:');
                     Writeln(out,'The equation that best fits the');
                     Writeln(out,'given data is:');
                     Writeln;
                     Writeln(' Y := ',M,'*Ln(X)/Ln(10) ',Sign,Abs(B));
                     Writeln(out);
                     Writeln(out,' Y := ',M,'*Ln(X)/Ln(10) ',Sign,Abs(B));
                     Writeln;
                     Writeln('For this equation the sum of deviations');
                     Writeln('squared is : ',Dev);
                     Writeln;
                     Writeln;
                     Writeln(out);
                     Writeln(out,'For this equation the sum of deviations');
                     Writeln(out,'squared is : ', Dev );
                     Writeln(out);
                     Writeln(out);
                     PInterpolate;
                  Close(out);
                 end
             end { case }
          end;

    Procedure Work;
       begin
          Sign := '-';
        X1 := 0; Y1 := 0; XY := 0; X2 := 0;
          For Steps := 1 to Number do
           begin
             Lx := Ln( X[ Steps ] ) / Ln ( 10 );
             X1 := X1 + Lx ;
             Y1 := Y1 + Y[ Steps ] ;
             XY := XY + Y [ Steps ] * Lx ;
             X2 := X2 + Lx * Lx
           end;
```

```
  J := Number * X2 - X1 * X1 ;
     if J = Ø then
       begin
         Writeln('No solution found!');
         Exit(Work);
       end;
    M := ( Number * XY - X1 * Y1 ) / J ;
    B := ( Y1 * X2 - X1 * XY ) / J ;
         if Abs( B ) = B then
              Sign := '+';
         Dev := Ø;
           for Steps := 1 to Number do
              begin
                DTemp := Y [ Steps ] - M * Ln ( X [ Steps ] ) / Ln (1Ø)- B
                Dev := Dev + DTemp * DTemp ;
              end;
     Printout;
end;

begin { Main }
  Repeat
    Outputs;
    Title;
    Input;
     Work;
     Writeln;
          Write('Another data entry?(Y/N) ');
             Readln(Cont);
  until Cont in [ 'N','n']
end.
```

Least Squares Fit Log Abscissa

This program performs a least squares approximation of semilogarithmic line. The ordinate is linear and the abscissa is logarithmic. This program accepts the X & Y coordinates of given data points and tries to produce a best fit line that represents that data. The equation of the line is:

Y = M * ln (X)/ln (1Ø) + B

Where M and B are calculated by the program.

When entering multiple inputs, use a space between data entries.

Enter number of data points: 3

Enter data:

X1 Y1 = 2 6.72
X2 Y2 = 1Ø 8.4
X3 Y3 = 75 1Ø.5

The equation that best fits the given data is:

Y = 2.40141 * ln (X) ln (10) + 5.99763

For this equation the sum of deviations
squared is : 1.37501E-6

Do you want to interpolate data? (Y/N) Y

Enter X: 25
Y = 9.35466

Do you want to interpolate data? (Y/N) N

5.9 LEAST SQUARES FIT — FULL LOG

To complete this chapter's set of programs, this program calculates the least squares fit for data in a full log ordinate system (both X and Y coordinates are logarithmic).

In this case, the equation of the curve is :

$$Y = B * X^{M}$$

Again, if no solution is found, the fact is indicated.

```
(*==========================================*)
(*                                          *)
(* Title:   Least Squares Fit -             *)
(*          Full Log                        *)
(* Program Summary: Program calculates      *)
(*          the least squares fit for       *)
(*          a plot whose abscissa and       *)
(*          ordinate are both               *)
(*          logarithmic.                    *)
(*                                          *)
(*==========================================*)


PROGRAM LeastSquareFitLog ;

uses Transcend;

 VAR
  Answer,
  Ans,
  Cont             : Char;
  Out              : Text;
  Number,
  Steps            : Integer;
  Lx,
  Ly,
  X,
  Y                : Array [1..10] of Real;
  X1,
  X2,
  Y1,
  XY,
  J,
  XX,
  YY,
  MM,
  Lb,
  DTemp,
  M,
  B,
  Dev              : Real;
```

```
Procedure Outputs;
  begin
    page(output);
     repeat
       gotoXY(0,5);
        Write('S(creen or P(rinter ? ');
        Readln(Answer);
     until Answer in [ 'S','s','p','P' ]
  end;  {outputs}

Procedure Title;
  begin
    page(output);
     gotoXY(7,1);
      Writeln('Least Squares Fit Full Log');
      Writeln;Writeln;
      Writeln('This program performs a least-squares');
      Writeln('approximation of a logarithmic line');
      Writeln('It accepts the X & Y coordinates of given');
      Writeln('data points and tries to produce a best');
      Writeln('fit line that represents the data.');
      Writeln('The equation of the line is:');
      Writeln;
      Writeln('Y := B * X^ M');
      Writeln;
      Writeln('Where M and B are calculated by the');
      Writeln('program.');
      Writeln;
      Writeln('When entering multiple inputs, use');
      Writeln('a space between data entries');
      Writeln;
      Writeln;
   end;  { title }

Procedure Input;
  begin
     Write('Enter number of data points :  ');
     Readln( Number );
     Writeln;
     Writeln('Enter data: ');
     Writeln;
        For Steps := 1 to Number do
           begin
             Write('X(',Steps,') Y(',Steps,') = ');
             Readln( X [ Steps ] , Y [ Steps ] )
           end;
  end;

Procedure Interpolate;
   begin
     Write('Do you want to interpolate data (Y/N) ?');
     Readln( Ans );
       if Ans in [ 'Y','y'] then
             Begin
               Writeln;
               Write('Enter X: ');
               Readln( XX );
               YY := B * exp( M * Ln ( XX )) ;
               Writeln('Y := ',YY);
                 Writeln;
                 Interpolate
             end
    end;
Procedure PInterpolate;
   begin
     Write('Do you want to interpolate data (Y/N) ?');
     Readln( Ans );
       if Ans in [ 'Y','y'] then
             Begin
               Writeln;
               Write('Enter X: ');
               Readln( XX );
               YY := B * exp( M * Ln ( XX )) ;
               Writeln('Y := ',YY);
               Writeln(out,'Y := ',YY);
                 Writeln;
                 PInterpolate
             end
    end;
```

```
Procedure Printout;
   begin
     case answer of
     's','S' : begin
                 Writeln;
                 Writeln('The equation that best fits the');
                 Writeln('given data is:');
                 Writeln;
                 Writeln(' Y := ',B,' * X^ ', M );
                 Writeln;
                 Writeln('For this equation the sum of deviations');
                 Writeln('squared is : ', Dev );
                 Writeln;
                 Writeln;
                 Interpolate;
               end;
     'p','P' : begin
               Rewrite(out,'Printer:');
                 Writeln;
                 Writeln(out);
                 Writeln('The equation that best fits the');
                 Writeln('given data is:');
                 Writeln(out,'The equation that best fits the');
                 Writeln(out,'given data is:');
                 Writeln;
                 Writeln(' Y := ',B,' * X^ ', M );
                 Writeln(out);
                 Writeln(out,' Y := ',B,' * X^ ', M );
                 Writeln;
                 Writeln('For this equation the sum of deviations');
                 Writeln('squared is : ',Dev);
                 Writeln;
                 Writeln;
                 Writeln(out);
                 Writeln(out,'For this equation the sum of deviations');
                 Writeln(out,'squared is : ', Dev );
                 Writeln(out);
                 Writeln(out);
                 PInterpolate;
               Close(out);
              end
          end { case }
       end;

Procedure Work;
   begin
    X1 := Ø; Y1 := Ø; XY := Ø; X2 := Ø;
     For Steps := 1 to Number do
       begin
         Lx [ Steps ] := Ln( X[ Steps ] ) / Ln ( 1Ø );
         Ly [ Steps ] := Ln( Y[ Steps ] ) / Ln ( 1Ø );
                Y1 := Y1 + Ly [ Steps ] ;
                X1 := X1 + Lx [ Steps ] ;
                XY := XY + Lx [ Steps ] * Ly [ Steps ] ;
                X2 := X2 + Lx [ Steps ] * Lx [ Steps ] ;
       end;
    J := Number * X2 - X1 * X1 ;
       if J = Ø then
         begin
           Writeln('No solution found!');
           Exit(Work);
         end;
      M := ( Number * XY - X1 * Y1 ) / J ;
      Lb := ( Y1 * X2 - X1 * XY ) / J ;
       B := exp( Lb * Ln(1Ø));
           Dev := Ø;
             for Steps := 1 to Number do
                begin
                  DTemp := Ly [ Steps ] - M * Lx [ Steps ] - Lb ;
                        Dev := Dev + ( DTemp * DTemp );
                end;
       Printout;
   end;

   begin { Main }
     Repeat
       Outputs;
```

```
      Title;
      Input;
       Work;
       Writeln;
            Write('Another data entry?(Y/N) ');
                Readln(Cont);
   until Cont in [ 'N','n']
end.
```

Least Squares Fit Full Log

This program performs a least squares approximation of a logarithmic line.
It accepts the X & Y coordinates of given data points and tries to produce a best fit line that represents that data.
The equation of the line is:

Y = B*X∧M

Where M and B are calculated by the program.

When entering multiple inputs, use a space between data entries.

Enter number of data points: 4

Enter data:

X(1) Y(1) = 1 .71
X(2) Y(2) = 2 3.5
X(3) Y(3) = 7 62.5
X(4) Y(4) = 1Ø 142

The equation that best fits the given data is:

Y = 7.1ØØ94E-1 *X∧2.3ØØ99

For this equation the sum of deviations squared is : 1.Ø81Ø6E-8

Do you want to interpolate data? (Y/N) Y

Enter X: 3
Y = 8.89548

Do you want to interpolate data? (Y/N) N

Another data entry? (Y/N) N

6 BASIC ELECTRICITY

For those readers who are new to the field of electricity, the first program in this chapter will help to overcome the problem of interpreting the value of a resistor from its color-coded bands. The information on the value of the resistor and its tolerance are coded in three or four different colored bands. When you enter the color of each band in turn, the computer will calculate the value of the resistor and its tolerance.

The next program (6.2) in this series calculates the value of any number of parallel resistors. Calculating the value of series resistors is easy, just add. For parallel resistors, however, the reciprocals of the values are added together, and this final value is converted to the total resistance by taking its reciprocal. This program can also be used for inductors since their values are calculated the same way. In addition, the program can be used to calculate the value of series capacitors.

The next two programs (6.3 and 6.4) convert resistor circuits from one configuration to another. This comes in handy if the equivalent resistance of one of these configurations is desired.

Programs 6.5 through 6.8 come in handy when doing electrical wiring or designing power transformers. With these programs, the size of wire can be calculated from desired resistance and length or desired current carrying capacity. In addition, given a particular wire size and type material, the resistance of wire is calculated.

The final program (6.8) calculates the turns ratio of primary to secondary coils in a transformer, given the primary and secondary resistances.

6.1 RESISTOR COLOR CODE INTERPRETATION

This program is designed for the novice electronic hobbyist who is not yet familiar with standard resistor color code. The colors of each of the three or four bands are entered into the computer (write out each word in CAPITALS), and the program interprets the colors and calculates the resistance.

When the colors are entered, the string to which it is assigned is then compared with the colors listed in *Procedure Color*. When a match is found in the appropriate *Procedure Input*, the position in the array is used as a multiplier.

The list of colors is displayed with the program, and only these colors should be used. Nonetheless, many people have a tendency to use different forms of the word gray. To compensate, a check is made in the beginning of *Procedure Work* to see if the correct form is used. If not, the incorrect form is converted to the correct form.

The fourth band on the resistor is not always present. The two color bands specify the most significant digits of the value. The third band is the decimal multiplier that is determined by raising 10 to the power of the value of the color. For example, since black = 0 and $10^0 = 1$, black has a multiplier of 1. The color values are as follows:

Black	0
Brown	1
Red	2
Orange	3
Yellow	4
Green	5
Blue	6
Purple	7
Gray	8
White	9

The fourth color band represents the tolerance of the resistor and is either gold for 5 percent, silver for 10 percent or no color for 20 percent.

```
(*=========================================*)
(*                                         *)
(* Title:   Resistor Color Code            *)
(*          Interpretation                 *)
(* Program Summary: Program interprets     *)
(*          the color code of a            *)
(*          resistor and gives its         *)
(*          resistance value and           *)
(*          tolerance.                     *)
(*                                         *)
(*=========================================*)


PROGRAM ColorCode ;

uses Transcend;

 VAR
  Answer,
  Ans,
  Cont          : Char;
  Out           : Text;
  ES            : Array [0..12] of String ;
  Ohms,
  First,
  Second,
  Third,
  Forth         : String;
  Percent,
  Steps         : Integer;
  RR,
  Res,
  Band1,
  Band2,
  Band3         : Real;

Procedure Outputs;
  begin
```

```
      page(output);
       repeat
          gotoXY(Ø,5);
           Write('S(creen or P(rinter ? ');
           Readln(Answer);
       until Answer in [ 'S','s','p','P' ]
   end;  {outputs}

Procedure Color;
  begin
    ES [Ø] := 'BLACK';  ES [1] := 'BROWN';  ES [2] :='RED';
    ES [3] := 'ORANGE'; ES [4] := 'YELLOW'; ES [5] :='GREEN';
    ES [6] :='BLUE';    ES [7] := 'PURPLE'; ES [8] :='GRAY';
    ES [9] :='WHITE';   ES [1Ø] :='NONE';   ES [11] :='GOLD';
    ES [12] :='SILVER'
  end;

Procedure Title;
  begin
    page(output);
     gotoXY(11,1);
       Color;
      Writeln('Resistor Color Code');
      Writeln;Writeln;
      Writeln('This program interprets the resistor');
      Writeln('color code.  You enter the color ');
      Writeln('of each band and the program will tell');
      Writeln('you what the value of the resistor is.');
      Writeln;
      Writeln('The possible colors are :');
        for Steps := 1 to 9 do
           Writeln('    ', ES [ Steps ] );
      Writeln;
      Writeln;
      Writeln('When making entries please make');
      Writeln('all entries in Capitals');
      Writeln;
   end;  { title }

Procedure Input1;
   begin
        Write('Enter the color of the first band : ');
        Readln( First );
         For Steps := Ø to 1Ø do
           if First = ES [ Steps ] then  Exit ( Input1 );
        Writeln('Please check that you use Capitals!!!');
        Writeln('or check your spelling');
        Input1
   end;

Procedure Input2;
   begin
     Writeln;
     Write('Enter the color of the second band : ');
     Readln( Second );
        For Steps := Ø to 1Ø do
          if Second = ES [ Steps ] then  Exit ( Input2 );
     Writeln('Please check that you use Capitals!!!');
     Writeln('or check your spelling');
     Input2
   end;

Procedure Input3;
   begin
     Writeln;
     Write('Enter the color of the third band : ');
     Readln( Third );
      For Steps := Ø to 1Ø do
         if Third = ES [ Steps ] then  Exit ( Input3 );
    Writeln('Please check that you use Capitals!!!');
    Writeln('or check your spelling');
    Input3
   end;

Procedure Input4;
   begin
```

```
        Writeln;
        Writeln('Enter the color of the forth band ');
        Write('(Gold, Silver, or None )  :');
        Readln( Forth );
          For Steps := 10 to 12 do
             if Forth = ES [ Steps ] then  Exit ( Input4 );
        Writeln('Please check that you use Capitals!!!');
        Writeln('or check your spelling');
        Input4
      end;

Procedure Input;
  begin
      Input1;
      Input2;
      Input3;
      Input4;
      Writeln;
  end;

Procedure Work;
   begin
    if First = 'GRAY' then First :='Grey';
    if Second = 'GRAY' then Second :='Grey';
    if Third = 'GRAY' then Third :='Grey';
        for Steps := 1 to 9 do
          begin
            if First = ES [ Steps ] then Band1 := Steps ;
            if Second = ES [ Steps ] then Band2 := Steps ;
            if Third = ES [ Steps ] then Band3 := Steps ;
                if (Third = 'WHITE') or (Third ='GOLD') then Band3 := -1 ;
                if (Third = 'GREY') or (Third ='SILVER') then Band3 := -2 ;
            if Forth ='GOLD' then Percent := 5 ;
            if Forth ='SILVER' then Percent := 10 ;
            if Forth ='NONE' then Percent := 20 ;
          end;
   RR := 10 * Band1 + Band2 ;
    if Band3 < 2 then Ohms :='Ohms';
    If ( Band3 >= 2 ) and ( Band3 < 5 ) then
         begin
            Ohms := 'Kilohms';
            Band3  := Band3 - 3
         end;
    If Band3 >= 5 then
         begin
            Ohms := 'Megohms';
            Band3  := Band3 - 6
         end;
   Res := Round(  RR * exp( Band3 * Ln ( 10 )));
  end;

 Procedure Printout;
    begin
      case answer of
      's','S' : begin
                   Writeln;
                   Writeln;
                   Writeln('Resistor value is ',Round(Res),' ',Ohms);
                   Writeln;
                   Writeln('With a tolerance of ',Percent,' percent');
                   Writeln;
                end;
      'p','P' : begin
                  Rewrite(out,'Printer:');
                   Writeln;
                   Writeln;
                   Writeln(out);
                   Writeln(out);
                   Writeln('Resistor value is ',Round(Res),' ',Ohms);
                   Writeln(out,'Resistor value is ',Round(Res),' ',Ohms);
                   Writeln;
                   Writeln('With a tolerance of ',Percent,' percent');
                   Writeln(out);
                   Writeln(out,'With a tolerance of ',Percent,' percent');
                   Writeln(out);
                   Writeln;
                Close(out);
```

```
            end
        end { case }
    end;


begin { Main }
  Repeat
    Outputs;
    Title;
    Input;
     Work;
     Printout;
     Writeln;
          Write('Another data entry?(Y/N) ');
             Readln(Cont);
  until Cont in [ 'N','n']
end.
```

Resistor Color Code

This program interprets the resistor
color code. You enter the color
of each band and the program will tell
you what the value of the resistor is.

The possible colors are :

BLACK
BROWN
RED
ORANGE
YELLOW
GREEN
BLUE
PURPLE
GRAY
WHITE

When making entries please make
all entries in capitals.

Enter color of the first band : ORANGE

Enter color of the second band : WHITE

Enter color of the third band : Yellow

Please check that you use capitals!!!
or check your spelling.

Enter color of the third band : YELLOW

Enter color of the fourth band

(Gold, Silver, or None) : GOLD

Resistor value is 390 Kilohms
with a tolerance of 5 percent.

Another data entry? (Y/N) N

6.2 PARALLEL RESISTORS

Calculating the equivalent value of several parallel resistors can be time consuming and inconvenient, particularly if the resistors are not of equal value or multiples of each other.

This program takes the tedium out of such calculations and performs them more quickly than can be done manually. The total parallel resistance of any number of resistors is calculated by first converting the resistances to conductances (conductance is the reciprocal of resistance) and adding the conductances together. Once the total conductance is calculated, the program takes the reciprocal, which is the total parallel resistance.

To use the program, it is necessary to enter only the number of parallel resistors involved and the value of each in ohms.

```
(*=========================================*)
(*                                         *)
(* Title:  Parallel Resistors              *)
(* Program Summary: Program calculates     *)
(*         the equivalent resistance       *)
(*         for a group of parallel         *)
(*         resistors.                      *)
(*                                         *)
(*=========================================*)

PROGRAM ParallelResistance ;

 VAR
  Answer,
  Ans,
  Cont             : Char;
  Out              : Text;
  ParRes,
  Res              : Real;
  Num,
  Number           : Integer;

Procedure Outputs;
  begin
    page(output);
     repeat
       gotoXY(0,5);
         Write('S(creen or P(rinter ? ');
         Readln(Answer);
     until Answer in [ 'S','s','p','P' ]
  end;  {outputs}

Procedure Title;
  begin
    page(output);
     gotoXY(11,1);
      Writeln('Parallel Resistors');
      Writeln;Writeln;
      Writeln('This program will calculate the');
      Writeln('resistance of any number of parallel');
      Writeln('resistors.  You must enter the number');
      Writeln('of resistors and their value in ohms.');
      Writeln;
   end;  { title }
```

```
Procedure Input;
  begin
     Write('Enter the number of resistors : ');
     Readln( Number );
     Writeln;
     ParRes := 0;
    for Num := 1 to Number do
       begin
         Write('Enter Res ',Num,' : ');
         Readln( Res );
         Writeln;
           ParRes := ParRes + 1 / Res ;
       end;
     Writeln
  end;

Procedure Work;
  begin
    ParRes := 1 / ParRes ;
  end;

 Procedure Printout;
    begin
      case answer of
      's','S' : begin
                  Writeln;
                  Writeln;
                  Writeln('The equivalent parallel Resistance');
                  Writeln('is : ',Round(ParRes),' ohms');
                  Writeln;
                end;
      'p','P' : begin
                  Rewrite(out,'Printer:');
                  Writeln;
                  Writeln;
                  Writeln(out);
                  Writeln(out);
                  Writeln('The equivalent parallel Resistance');
                  Writeln(out,'The equivalent parallel Resistance');
                  Writeln('is : ',Round(ParRes),' ohms');
                  Writeln(out,'is : ',Round(ParRes),' ohms');
                  Writeln(out);
                  Writeln;
                Close(out);
               end
           end { case }
       end;


   begin { Main }
     Repeat
       Outputs;
       Title;
       Input;
       Work;
        Printout;
        Writeln;
             Write('Another resistance entry?(Y/N) ');
                Readln(Cont);
     until Cont in [ 'N','n']
   end.
```

Parallel Resistors

This program will calculate the
resistance of any number of parallel
resistors. You must enter the number
of resistors and their value in ohms.

Enter number of resistors : 4

```
Enter Res 1 : 1000

Enter Res 2 : 1000

Enter Res 3 : 500

Enter Res 4 : 250

The equivalent parallel resistance
is : 125 ohms

Another resistance entry? (Y/N) N
```

6.3 DELTA TO WYE TRANSFORMATION

Very often in the analysis of electrical circuits, resistors are connected as shown in Fig. 6.3(a). An example of such an instance is a Wheatstone resistance bridge (Fig. 6.3(b)) in which there are actually two such circuits. This form of connection is called a "delta" or, sometimes a "pi" connection.

Calculating the equivalent resistance of a delta circuit can be difficult. To make it easier, the delta circuit is generally converted to an equivalent wye circuit (Fig. 6.3(c)). As far as connections to the terminals are concerned, the two circuits are equivalent.

This program converts the delta resistors Ra, Rb, and Rc to wye resistors R1, R2, and R3. This is done in *Procedure Work*. The subroutine in *Procedure Ohms* selects the correct resistance descriptor string.

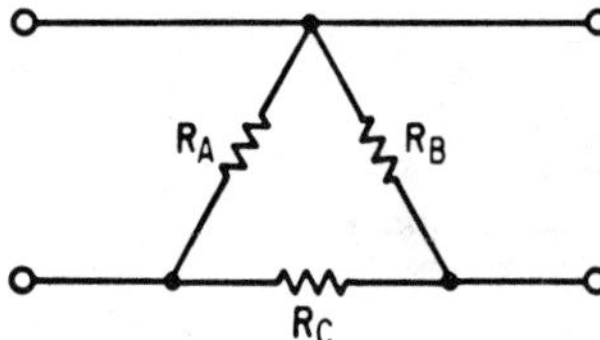

Fig. 6.3 (a) Original delta circuit

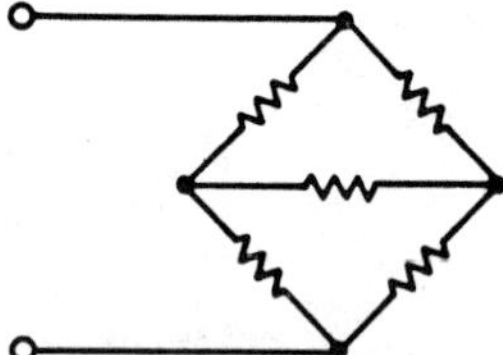

Fig. 6.3 (b) Wheatstone bridge contains delta circuit

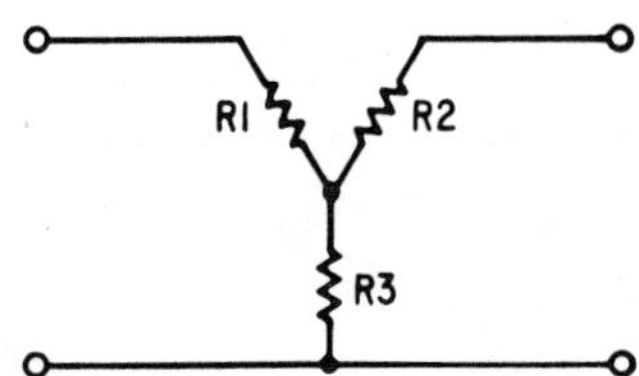

Fig. 6.3 (c) Equivalent wye circuit

```
(*=========================================*)
(*                                         *)
(* Title:  Delta to Wye Transformation     *)
(* Program Summary: Program converts       *)
(*         resistor network from the       *)
(*         Delta configuration to          *)
(*         the Wye configuration.          *)
(*                                         *)
(*=========================================*)


PROGRAM DeltaToWyeTransformation ;

 VAR
  Answer,
  Ans,
  Cont            : Char;
  Out             : Text;
  Ra,
  Rb,
  Rc,
  Temp,
  Rtot,
  R1,
  R2,
  R3              : Real;
  R1ohms,
  ohm,
  R2ohms,
  R3ohms              : String;

Procedure Outputs;
  begin
    page(output);
     repeat
       gotoXY(0,5);
        Write('S(creen or P(rinter ? ');
        Readln(Answer);
     until Answer in [ 'S','s','p','P' ]
  end;  {outputs}

Procedure Title;
  begin
    page(output);
     gotoXY(8,1);
       Writeln('Delta to Wye Conversion');
       Writeln;Writeln;
       Writeln('This program converts resistor networks');
       Writeln('from a Delta configuration to a Wye');
       Writeln('configuration.');
       Writeln;
       Writeln('When making multiple data entries, use');
       Writeln('a space between entries');
       Writeln;
  end;  { title }

Procedure Input;

  begin
    Writeln;
    Write('Enter Ra , Rb, Rc (ohms) : ');
    Readln( Ra , Rb , Rc );
    Writeln;
    Writeln;
  end;

Procedure Ohms;
  begin
   if Temp >= 1E6 then
     begin
       ohm := 'Megohms';
       Temp :=  Temp / 1E6
     end
   else if ( Temp >= 1E3 ) and (Temp < 1E6 ) then
     begin
      ohm := 'Kilohms';
      Temp := Temp / 1E3;
```

```
      end
    else
      ohm := 'Ohms'
   end;

 Procedure Work;
   begin
       Rtot  := Ra + Rb + Rc ;
       R1 := Ra * Rc /  Rtot ;
       Temp := R1 ;
           ohms;
        R1 := Temp ;
        R1ohms := ohm ;
        R2 := Rb * Rc / Rtot ;
        Temp := R2 ;
           ohms;
         R2 := Temp ;
         R2ohms := ohm ;
         R3 := Ra * Rb / Rtot ;
         Temp := R3 ;
           ohms;
          R3 := Temp ;
          R3ohms := ohm ;
    end;

 Procedure Printout;
    begin
      case answer of
      's','S' : begin
                     Writeln;
                     Writeln;
                     Writeln('For a Delta to Wye transformation the');
                     Writeln('new resistor values are: ');
                     Writeln;
                     Writeln('R1 := ',R1,' ',R1ohms);
                     Writeln;
                     Writeln('R2 := ',R2,' ',R2ohms);
                     Writeln;
                     Writeln('R3 := ',R3,' ',R3ohms);
                     Writeln;
                end;

      'p','P' : begin
                  Rewrite(out,'Printer:');
                     Writeln;
                     Writeln;
                     Writeln(out);
                     Writeln(out);
                     Writeln('For a Delta to Wye transformation the');
                     Writeln(out,'For a Delta to Wye transformation the');
                     Writeln('new resistor values are: ');
                     Writeln(out,'new resistor values are: ');
                     Writeln;
                     Writeln(out);
                     Writeln('R1 := ',R1,' ',R1ohms);
                     Writeln(out,'R1 := ',R1,' ',R1ohms);
                     Writeln;
                     Writeln(out);
                     Writeln('R2 := ',R2,' ',R2ohms);
                     Writeln(out,'R2 := ',R2,' ',R2ohms);
                     Writeln;
                     Writeln(out);
                     Writeln('R3 := ',R3,' ',R3ohms);
                     Writeln(out,'R3 := ',R3,' ',R3ohms);
                     Writeln(out);
                     Writeln;
               Close(out);
              end
          end { case }
      end;


   begin { Main }
     Repeat
       Outputs;
       Title;
       Input;
```

```
      Work;
       Printout;
       Writeln;
            Write('Another transformation?(Y/N) ');
               Readln(Cont);
   until Cont in [ 'N','n']
end.
```

Delta to Wye Conversion

This program converts resistor networks
from a Delta configuration to a Wye
configuration.

When making multiple data entries, use
a space between entries.

Enter Ra, Rb, Rc (ohms) : 1ØØ 2ØØ 3ØØ

For a Delta to Wye transformation the
new resistor values are:

R1 = 5.ØØØØØE1 Ohms
R2 = 1.ØØØØØE2 Ohms
R3 = 3.33333E1 Ohms

Another transformation? (Y/N) N

6.4 WYE TO DELTA TRANSFORMATION

It is often desirable to convert wye configurations to delta configurations. Calculations can be done with the wye configuration. The resistors R1, R2, and R3 can be then converted to their delta circuit equivalents: Ra, Rb, and Rc (Fig. 6.4).

The equations for the conversion are located in *Procedure Work.* As in the previous program, the subroutines located in *Procedure Ohms* selects the correct resistance descriptor string.

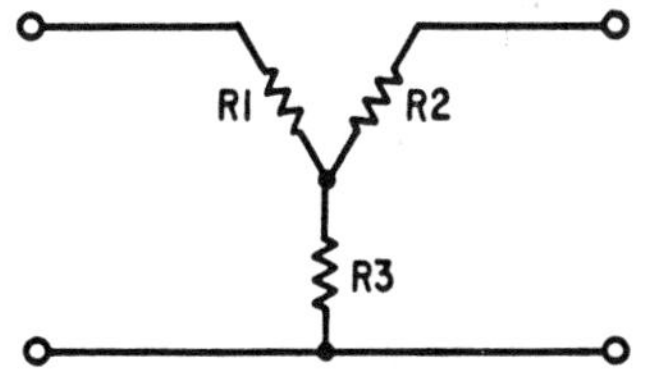

Fig. 6.4 (a) Original wye circuit

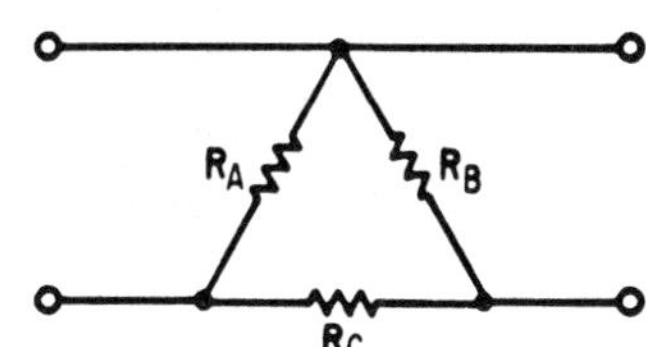

Fig. 6.4 (b) Equivalent delta circuit

```
(*=======================================*)
(*                                       *)
(* Title:  Wye to Delta Transformation   *)
(* Program Summary: Program converts     *)
(*         resistor network from the     *)
(*         Wye configuration to the      *)
(*         Delta configuration.          *)
(*                                       *)
(*=======================================*)


PROGRAM WyeToDeltaTransformation ;

 VAR
  Answer,
  Ans,
  Cont          : Char;
  Out           : Text;
  Ra,
  Rb,
  Rc,
  Temp,
  Rtot,
  R1,
  R2,
  R3            : Real;
  RaOhms,
  Ohm,
  RbOhms,
  RcOhms            : String;

Procedure Outputs;
  begin
    page(output);
     repeat
       gotoXY(Ø,5);
        Write('S(creen or P(rinter ? ');
        Readln(Answer);
     until Answer in [ 'S','s','p','P' ]
  end;  {outputs}

Procedure Title;
  begin
    page(output);
     gotoXY(8,1);
       Writeln('Wye to Delta Conversion');
       Writeln;Writeln;
       Writeln('This program converts resistor networks');
       Writeln('from a Wye configuration to a Delta');
       Writeln('configuration.');
       Writeln;
       Writeln('When making multiple data entries, use');
       Writeln('a space between entries');
       Writeln;
  end;  { title }

Procedure Input;
  begin
    Writeln;
    Write('Enter R1 , R2 , R3 (ohms) : ');
    Readln( R1 , R2 , R3 );
    Writeln;
    Writeln;
  end;

Procedure Ohms;
  begin
   if Temp >= 1E6 then
     begin
       Ohm := 'Megohms';
       Temp :=  Temp / 1E6
     end
   else if ( Temp >= 1E3 ) and (Temp < 1E6 ) then
     begin
      Ohm := 'Kilohms';
      Temp := Temp / 1E3;
```

```
        end
      else
        Ohm := 'Ohms'
    end;

  Procedure Work;
    begin
        Rtot := R1 * R2 + R1 * R3 + R2 * R3   ;
        Ra := Rtot / R2 ;
        Temp := Ra ;
           ohms;
        Ra := Temp ;
        RaOhms := Ohm ;
        Rb := Rtot / R1 ;
        Temp := Rb ;
           ohms;
        Rb := Temp ;
        RbOhms := Ohm ;
        Rc := Rtot / R3 ;
        Temp := Rc;
           ohms;
        Rc := Temp ;
        RcOhms := Ohm
     end;

  Procedure Printout;
     begin
       case answer of
       's','S' : begin
                      Writeln;
                      Writeln;
                      Writeln('For a Wye to Delta transformation the');
                      Writeln('new resistor values are: ');
                      Writeln;
                      Writeln('Ra := ',Ra,' ',RaOhms);
                      Writeln;
                      Writeln('Rb := ',Rb,' ',RbOhms);
                      Writeln;
                      Writeln('Rc := ',Rc,' ',RcOhms);
                      Writeln;
                 end;
       'p','P' : begin
                   Rewrite(out,'Printer:');
                      Writeln;
                      Writeln;
                      Writeln(out);
                      Writeln(out);
                      Writeln('For a Wye to Delta transformation the');
                      Writeln(out,'For a Wye to Delta transformation the');
                      Writeln('new resistor values are: ');
                      Writeln(out,'new resistor values are: ');
                      Writeln;
                      Writeln(out);
                      Writeln('Ra := ',Ra,' ',RaOhms);
                      Writeln(out,'Ra := ',Ra,' ',RaOhms);
                      Writeln;
                      Writeln(out);
                      Writeln('Rb := ',Rb,' ',RbOhms);
                      Writeln(out,'Rb := ',Rb,' ',RbOhms);
                      Writeln;
                      Writeln(out);
                      Writeln('Rc := ',Rc,' ',RcOhms);
                      Writeln(out,'Rc := ',Rc,' ',RcOhms);
                      Writeln(out);
                      Writeln;
                 Close(out);
                end
            end { case }
         end;


    begin { Main }
      Repeat
        Outputs;
        Title;
        Input;
        Work;
```

```
      Printout;
      Writeln;
           Write('Another transformation?(Y/N) ');
              Readln(Cont);
  until Cont in [ 'N','n']
end.
```

Wye to Delta Conversion

This program converts resistor networks
from a Wye configuration to a Delta
configuration.

When making multiple data entries, use
a space between entries.

Enter R1 , R2 , R3 (ohms) : 100 200 300

For a Wye to Delta transformation the
new resistor values are:

Ra = 5.50000E2 Ohms
Rb = 1.10000 Kilohms
Rc = 3.66667E2 Ohms

Another transformation? (Y/N) N

6.5 WIRE SIZE FROM RESISTANCE AND LENGTH

When using this and the following two programs, it is not necessary to use wire gauge tables. Program 6.5 determines the American Wire Gauge (AWG) number to provide a given resistance to a given length of wire.

The program uses the two formulas listed below to perform its calculations. As it stands, the program will handle both copper and aluminum wires. If other materials are desired, *Procedure Input* should be changed and the additional data should be entered in *Procedure Work* by changing the "else" to another *IF* statement with the appropriate Rho and Metal.

$$R = \rho \frac{L}{A}$$

$$AWG = \frac{\ln(A/105532)}{\ln(.79304)}$$

```
(*=======================================*)
(*                                       *)
(* Title:   Wire Size from Resistance    *)
(*          and Length                   *)
(* Program Summary: Program finds        *)
(*          the diameter wire required   *)
(*          to provide a given           *)
(*          resistance over a given      *)
(*          length.                      *)
(*                                       *)
(*=======================================*)
```

```
PROGRAM WireSize ;

uses Transcend;

 VAR
  Answer,
  Ans,
  Cont           : Char;
  Out            : Text;
  Res,
  Temp,
  Len,
  Rho,
  Awg            : Real;
  TypeMetal      : Integer;
  Metal          : String;

Procedure Outputs;
  begin
    page(output);
     repeat
       gotoXY(Ø,5);
        Write('S(creen or P(rinter ? ');
        Readln(Answer);
     until Answer in [ 'S','s','p','P' ]
  end;  {outputs}

Procedure Title;
  begin
    page(output);
     gotoXY(2,1);
       Writeln('Wire Size From Resistance and length');
       Writeln;Writeln;
       Writeln('This program will find the required');
       Writeln('diameter wire to produce a given');
       Writeln('resistance for a given length of the wire.');
       Writeln;
  end;  { title }

Procedure Input;
  begin
     Write('Enter the desired resistance (ohms) : ');
     Readln( Res );
     Writeln;
     Write('Enter the length of wire (ft) : ');
     Readln( Len );
     Writeln;
     Write('Copper (1) or Aluminum (2) : ');
     Readln( TypeMetal );
  end;

Procedure Work;
  begin
     if TypeMetal = 2 then
        begin
          Metal := 'Aluminum';
          Rho := 17.34
        end
     else
        begin
          Metal :='Copper';
          Rho := 1Ø.575
        end;
       Temp := Rho * Len / Res ;
       Awg :=  Log ( Temp  / 1.Ø5532E5 ) / Log (Ø.793Ø4)
  end;

 Procedure Printout;
    begin
      case answer of
      's','S' : begin
                     Writeln;
                     Writeln;
                     Write('A ',Round(AWG),' AWG ',Metal);
                     Writeln(' wire that is ',Round(Len));
                     Writeln('feet long will have a resistance');
                     Writeln('of ',Round(Res),' ohms');
                     Writeln;
```

```
              end;
    'p','P' : begin
                Rewrite(out,'Printer:');
                  Writeln;
                  Writeln;
                  Writeln(out);
                  Writeln(out);
                  Write('A ',Round(AWG),' AWG ',Metal);
                  Writeln(' wire that is ',Round(Len));
                  Write(out,'A ',Round(AWG),' AWG ',Metal);
                  Writeln(out,' wire that is ',Round(Len));
                  Writeln('feet long will have a resistance');
                  Writeln('of ',Round(Res),' ohms');
                  Writeln(out,'feet long will have a resistance');
                  Writeln(out,'of ',Round(Res),' ohms');
                  Writeln(out);
                  Writeln;
              Close(out);
             end
         end { case }
    end;

begin { Main }
  Repeat
    Outputs;
    Title;
    Input;
    Work;
     Printout;
     Writeln;
          Write('Another resistance and length ?(Y/N) ');
             Readln(Cont);
  until Cont in [ 'N','n']
end.
```

Wire Size From Resistance and Length

This program will find the largest
diameter wire that will produce a given
resistance for a given length of wire.

Enter the desired resistance (ohms) : 8

Enter the length of wire (ft) : 100

Copper (1) or Aluminum (2) : 1

A 29 AWG Copper wire that is 100
feet long will have a resistance
of 8 ohms.

Another resistance and length? (Y/N) Y

Enter the desired resistance (ohms) : 8

Enter the length of wire (ft) : 100

Copper (1) or Aluminum (2) : 2

A 27 AWG Aluminum wire that is 100

feet long will have a resistance
of 8 ohms.

Another resistance and length? (Y/N) N

6.6 WIRE SIZE FROM CURRENT CAPACITY

In the design of transformers, chokes, and electrical power circuits, one must determine what size wire can safely handle a particular current. This capacity is also convenient for anyone repairing appliances or adding some additional wiring to his home.

This program will request an input for desired current and ask if the application is for general or for military requirements, where temperature ranges from -55° to 125°C. The military design is slightly more conservative and results in the need for larger wires (smaller AWG number).

```
(*=========================================*)
(*                                         *)
(* Title:   Wire Size from Current         *)
(*          Capacity                       *)
(* Program Summary: Program finds          *)
(*          the smallest diameter wire     *)
(*          that will carry a given        *)
(*          amount of current.             *)
(*                                         *)
(*=========================================*)


PROGRAM WireSize ;

uses Transcend;

 VAR
  Answer,
  Ans,
  Cont           : Char;
  Out            : Text;
  Cma,
  Amps,
  Temp,
  AmpCap,
  Awg            : Real;
  TypeUse       : Integer;
  Use           : String;

Procedure Outputs;
  begin
    page(output);
     repeat
       gotoXY(0,5);
        Write('S(creen or P(rinter ? ');
        Readln(Answer);
     until Answer in [ 'S','s','p','P' ]
  end;  {outputs}

Procedure Title;
  begin
    page(output);
     gotoXY(5,1);
       Writeln('Wire Size From Current Capacity');
       Writeln;Writeln;
       Writeln('This program will find the smallest ');
       Writeln('diameter wire that will carry a desired');
       Writeln('amount of current. `You must indicate');
       Writeln('if it is for general use or Military');
       Writeln('environment (-55 to 125 C)');
```

```
      Writeln;
   end;  { title }

Procedure Input;
  begin
     Write('Enter current capacity (amps) : ');
     Readln( Amps );
     Writeln;
     Write('General use (1) or Military (2) : ');
     Readln( TypeUse );
     Writeln;
  end;

Procedure Work;
  begin
     if TypeUse = 2 then
        begin
          Cma := 700;
          Use  := 'Military'
        end
     else
        begin
          Cma := 400;
          Use  := 'General'
        end;
       AmpCap := Cma * Amps ;
       Temp := exp ( 2 * Ln ( 325 )) / AmpCap ;
       Awg := 10 * ( Log ( Temp )) / Log ( 10 );
    end;

 Procedure Printout;
    begin
      case answer of
   's','S' : begin
               Writeln;
               Writeln;
               Writeln('For ',Use,' applications :');
               Writeln;
               Writeln('a ',Round(AWG),' AWG wire is required to carry');
               Writeln('a current of ',Round(Amps),' amperes');
               Writeln;
            end;
   'p','P' : begin
              Rewrite(out,'Printer:');
               Writeln;
               Writeln;
               Writeln(out);
               Writeln(out);
               Writeln('For ',Use,' applications :');
               Writeln(out,'For ',Use,' applications :');
               Writeln;
               Writeln(out);
               Writeln('a ',Round(Awg),' AWG wire is required to carry');
               Writeln('a current of ',Round(Amps),' amperes');
               Writeln(out,'a ',Round(AWG),' Awg wire is required to carry');
               Writeln(out,'a current of ',Round(Amps),' amperes');
               Writeln(out);
               Writeln;
              Close(out);
            end
          end { case }
      end;


   begin { Main }
     Repeat
       Outputs;
       Title;
       Input;
       Work;
        Printout;
        Writeln;
             Write('Another application ?(Y/N) ');
                Readln(Cont);
     until Cont in [ 'N','n']
   end.
```

Wire Size From Current Capacity

This program will find the largest diameter wire that will carry a desired amount of current. You must indicate if it is for general use or military environment (-55 to 125 C).

Enter current capacity (amps) : 15

General use (1) or military (2): 1

For General applications:

A 12 AWG wire is required to carry a current of 15 amperes.

Another application ? (Y/N) N

6.7 WIRE RESISTANCE CALCULATOR

Long lengths of wire run between two points can often lead to degradation of the signal because the resistance of the wire is too high. With this program, the resistance of a length of wire of a specified size and material can be quickly calculated.

The resistivity constants for aluminum and copper are already in the program, so either of these materials can be specified. If other materials are required, the rho for the material will have to be added, as will a material descriptor string as in *Procedure Work*.

```
(*========================================*)
(*                                        *)
(* Title:   Wire Resistance Calculator    *)
(* Program Summary: Program finds         *)
(*          the resistance of a given     *)
(*          AWG-size wire for a given     *)
(*          length of the wire.           *)
(*                                        *)
(*========================================*)


PROGRAM WireResistance ;

uses Transcend;

 VAR
  Answer,
  Ans,
  Cont            : Char;
  Out             : Text;
  Awg,
  Len,
  TypeMetal,
  Rho,
  Temp,
  Res             : Real;
  Metal           : String;

Procedure Outputs;
  begin
    page(output);
     repeat
       gotoXY(Ø,5);
        Write('S(creen or P(rinter ? ');
        Readln(Answer);
```

```
    until Answer in [ 'S','s','p','P' ]
  end;  {outputs}

Procedure Title;
  begin
    page(output);
     gotoXY(7,1);
       Writeln('Wire Resistance Calculator');
       Writeln;Writeln;
       Writeln('This program will find the resistance');
       Writeln('of a given AWG size wire for a given ');
       Writeln('length of wire.');
       Writeln;
  end;  { title }

Procedure Input;
  begin
     Write('Enter AWG size : ');
     Readln( Awg );
     Writeln;
     Write('Enter the length of wire (ft) : ');
     Readln( Len );
     Writeln;
     Write('Copper (1) or Aluminum (2) : ');
     Readln( TypeMetal );
     Writeln;
  end;

Procedure Work;
  begin
     if TypeMetal = 2 then
        begin
           Metal := 'Aluminum';
           Rho := 17.34
        end
     else
        begin
           Metal := 'Copper';
           Rho := 10.575
        end;
     Temp  := 1.05532E5 * exp( AWG * Ln(0.79304));
     Res  := Rho * Len / Temp ;
    end;

 Procedure Printout;
    begin
      case answer of
   's','S' : begin
                Writeln;
                Writeln;
                Write('A ',Round(AWG),' AWG ',Metal);
                Writeln(' wire that is ',Round(Len));
                Writeln('feet long will have a resistance');
                Writeln('of ',Res,' ohms');
                Writeln;
             end;
   'p','P' : begin
               Rewrite(out,'Printer:');
                Writeln;
                Writeln;
                Writeln(out);
                Writeln(out);
                Write('A ',Round(AWG),' AWG ',Metal);
                Write(out,'A ',Round(AWG),' AWG ',Metal);
                Writeln(' wire that is ',Round(Len));
                Writeln(out,' wire that is ',Round(Len));
                Writeln('feet long will have a resistance');
                Writeln(out,'feet long will have a resistance');
                Writeln('of ',Res,' ohms');
                Writeln(out,'of ',Res,' ohms');
                Writeln(out);
                Writeln;
               Close(out);
             end
          end { case }
       end;
```

```
begin { Main }
  Repeat
    Outputs;
    Title;
    Input;
    Work;
     Printout;
     Writeln;
          Write('Another application ?(Y/N) ');
              Readln(Cont);
  until Cont in [ 'N','n']
end.
```

Wire Resistance Calculator

This program will find the resistance of a given AWG size wire for a given length of the wire.

Enter AWG size : 29

Enter the length of wire (ft) : 100

Copper (1) or Aluminum (2) : 1

A 29 AWG copper wire that is 100 feet long will have a resistance of 8.34332 ohms.

Another application ? (Y/N) N

6.8 TRANSFORMER TURNS RATIO

In addition to being used as voltage-changing devices in power supplies, transformers are also used as impedance-matching devices in audio equipment.

For those who like to "roll their own," this program will calculate the turns ratio required for a transformer given the primary and secondary impedances.

```
(*=========================================*)
(*                                         *)
(* Title:   Transformer Turns Ratio        *)
(* Program Summary: Program calculates     *)
(*          the turns ratio of a           *)
(*          transformer given the          *)
(*          desired input and output       *)
(*          impedance.                     *)
(*                                         *)
(*=========================================*)


PROGRAM TurnsRatio ;

uses Transcend;

 VAR
  Answer,
  Ans,
  Cont           : Char;
  Out            : Text;
  Primary,
  Secondary,
  Turns          : Real;
```

```
Procedure Outputs;
  begin
    page(output);
     repeat
       gotoXY(Ø,5);
        Write('S(creen or P(rinter ? ');
        Readln(Answer);
     until Answer in [ 'S','s','p','P' ]
  end;  {outputs}

Procedure Title;
  begin
    page(output);
     gotoXY(8,1);
       Writeln('Transformer Turns Ratio');
       Writeln;Writeln;
       Writeln('This program calculates the turns');
       Writeln('ratio required for a transformer given');
       Writeln('the desired input and output impedance.');
       Writeln;
  end;  { title }

Procedure Input;
  begin
     Write('Enter primary impedance : ');
     Readln( Primary );
     Writeln;
     Write('Enter secondary impedance : ');
     Readln( Secondary );
     Writeln;
  end;

Procedure Work;
  begin
    Turns := Sqrt ( Primary / Secondary );
  end;

 Procedure Printout;
    begin
      case answer of
   's','S' : begin
               Writeln;
               Writeln;
               Writeln('The turns ratio for a transformer with');
               Writeln;
               Writeln('primary impedance of ',Round(Primary),' ohms');
               Writeln;
               Write('and a secondary impedance of ');
               Writeln(Round(Secondary),' ohms');
               Writeln;
               Writeln('is ',Round(Turns),' / 1');
               Writeln;
            end;
   'p','P' : begin
              Rewrite(out,'Printer:');
               Writeln;
               Writeln;
               Writeln(out);
               Writeln(out);
               Writeln('The turns ratio for a transformer with');
               Writeln(out,'The turns ratio for a transformer with');
               Writeln;
               Writeln(out);
               Writeln('primary impedance of ',Round(Primary),' ohms');
               Writeln(out,'primary impedance of ',Round(Primary),' ohms');
               Writeln;
               Writeln(out);
               Write('and a secondary impedance of ');
               Writeln(Round(Secondary),' ohms');
               Write(out,'and a secondary impedance of ');
               Writeln(out,Round(Secondary),' ohms');
               Writeln;
               Writeln(out);
               Writeln('is ',Round(Turns),' / 1');
               Writeln(out,'is ',Round(Turns),' / 1');
               Writeln(out);
```

```
                Writeln;
               Close(out);
             end
           end { case }
       end;


   begin { Main }
     Repeat
       Outputs;
       Title;
       Input;
       Work;
        Printout;
        Writeln;
             Write('Another application ?(Y/N) ');
                Readln(Cont);
     until Cont in [ 'N','n']
   end.
```

Transformer Turns Ratio

This program calculates the turns
ratio required for a transformer to give
the desired input and output impedance.

Enter primary impedance : 5ØØØ

Enter secondary impedance : 8

The turns ratio for a transformer with
primary impedance of 5ØØØ ohms
and a secondary impedance of 8 ohms
is 25/1.

Another application ? (Y/N) N

7 BASIC ELECTRONICS

In Chap. 6, the programs dealt with some basic calculations in electricity. In this chapter, the programs deal with some basic calculations in electronics.

While electric circuits deal primarily with resistance, electronic circuits deal intimately with inductance and capacitance. The first three programs calculate the inductance of wires. This becomes an important factor when wire lengths are long and high frequencies or very fast logic circuits are used. The fourth program calculates the ac resistance of an inductor, known as the inductive reactance.

Like an inductor, a capacitor has resistance to ac voltage. Program 7.5 calculates this capacitive reactance. In the design of electronic circuits, it is sometimes required to bypass a resistor with a capacitor to provide an alternate path for ac signal. Program 7.6 can be used to calculate the value of the capacitor required.

The final program (7.7) performs resonant circuit calculations. It deals with frequency, capacitance, and inductance. Given any two of the above, the program will calculate the third.

7.1 INDUCTANCE OF A STRAIGHT WIRE

Circuit designs using high speed logic, such as ECL, can be seriously affected by the inductance of interconnecting leads. To overcome such potential problems, it is necessary to determine the inductance of these leads.

For a straight wire, the inductance increases as its length and permeability increases, and decreases as its diameter increases. These relationships are not linear, however, and cannot be determined without calculation.

This program performs these calculations. The user must enter the diameter and length of wire in centimeters, and the permeability.

```
(*========================================*)
(*                                        *)
(* Title:   Inductance of a Straight      *)
(*          Wire                          *)
(* Program Summary: Program calculates    *)
(*          the inductance of a           *)
(*          straight piece of wire.       *)
(*                                        *)
(*========================================*)


PROGRAM Inductance ;

uses Transcend;

 VAR
  Answer,
  Ans,
  Cont             : Char;
  Out              : Text;
  Dia,
  Perm,
  Induct,
  Len               : Real;

Procedure Outputs;
  begin
    page(output);
     repeat
       gotoXY(Ø,5);
        Write('S(creen or P(rinter ? ');
        Readln(Answer);
     until Answer in [ 'S','s','p','P' ]
  end;  {outputs}

Procedure Title;
  begin
    page(output);
     gotoXY(5,1);
       Writeln('Inductance of a Straight Wire');
       Writeln;Writeln;
       Writeln('This program finds the inductance of');
       Writeln('a straight wire.  You must enter the');
       Writeln('diameter of the wire in centimeters,');
       Writeln('and the permeability of the wire (MU)');
       Writeln('The wire length must be entered.');
       Writeln;
  end;  { title }

Procedure Input;
  begin
     Write('Enter the wire diameter (cm) : ');
     Readln( Dia );
     Write('Enter the permeability : ');
     Readln( Perm );
     Write('Enter the wire length (cm) : ');
     Readln( Len );
     Writeln;
  end;

Procedure Work;
  begin
     Induct := Ø.ØØ2 * Len * ( Ln( 4 * Len / Dia ) - 1 + Perm / 4 );
  end;

 Procedure Printout;
    begin
      case answer of
   's','S' : begin
                Writeln;
                Writeln;
                Writeln('For a wire with a :');
                Writeln;
                Writeln;
                Writeln('Diameter := ',Dia,' centimeters');
                Writeln;
```

```
              Writeln('Permeability := ',Perm);
              Writeln;
              Writeln('Length := ',Len,' centimeters');
              Writeln;
              Writeln('Inductance :=',Induct,' microhenries');
              Writeln;
           end;
'p','P' : begin
             Rewrite(out,'Printer:');
              Writeln;
              Writeln;
              Writeln(out);
              Writeln(out);
              Writeln('For a wire with a :');
              Writeln(out,'For a wire with a :');
              Writeln;
              Writeln;
              Writeln(out);
              Writeln(out);
              Writeln('Diameter := ',Dia,' centimeters');
              Writeln(out,'Diameter := ',Dia,' centimeters');
              Writeln;
              Writeln(out);
              Writeln('Permeability := ',Perm);
              Writeln(out,'Permeability := ',Perm);
              Writeln;
              Writeln(out);
              Writeln('Length := ',Len,' centimeters');
              Writeln(out,'Length := ',Len,' centimeters');
              Writeln;
              Writeln(out);
              Writeln('Inductance :=',Induct,' microhenries');
              Writeln(out,'Inductance :=',Induct,' microhenries');
              Writeln(out);
              Writeln;
             Close(out);
           end
        end { case }
    end;

begin { Main }
  Repeat
    Outputs;
    Title;
    Input;
    Work;
     Printout;
     Writeln;
         Write('Another application ?(Y/N) ');
            Readln(Cont);
  until Cont in [ 'N','n']
end.
```

Inductance of a Straight Wire

This program finds the inductance of
a straight wire. You must enter the
diameter of the wire in centimeters,
and the permeability of the wire (MU).
The wire length must be entered.

Enter the wire diameter (cm) : .Ø16

Enter the permeability : 1

Enter the wire length (cm) : 2ØØ

For a wire with a :
Diameter = 1.60000E-2 centimeters
Permeability = 1.00000
Length = 2.00000E2 centimeters

Inductance = 4.02791 microhenries

Another application ? (Y/N) N

7.2 INDUCTANCE OF A PARALLEL WIRE PAIR

Program 7.1 found the inductance of a single straight wire. This program determines the inductance of two parallel wires viewed from one end (Fig 7.2).

As in the previous program, the diameter and length of the wires should be entered in centimeters. In addition, the distance between wires, also in centimeters, must be entered.

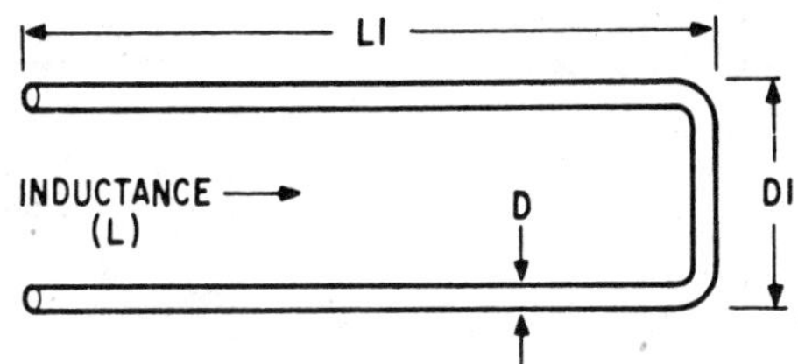

Fig. 7.2 Inductance of a parallel wire pair

```
(*=========================================*)
(*                                         *)
(* Title:   Inductance of a Parallel       *)
(*          Wire Pair                      *)
(* Program Summary: Program calculates     *)
(*          the inductance of two          *)
(*          parallel wires.                *)
(*                                         *)
(*=========================================*)


PROGRAM Inductance ;

uses Transcend;

 VAR
  Answer,
  Ans,
  Cont            : Char;
  Out             : Text;
  Dia,
  Perm,
  Dist,
  Induct,
  Len              : Real;

Procedure Outputs;
  begin
    page(output);
     repeat
       gotoXY(0,5);
        Write('S(creen or P(rinter ? ');
        Readln(Answer);
     until Answer in [ 'S','s','p','P' ]
  end;  {outputs}
```

```
Procedure Title;
  begin
    page(output);
     gotoXY(5,1);
       Writeln('Inductance of a Parallel Pair');
       Writeln;Writeln;
       Writeln('This program finds the inductance of');
       Writeln('parallel wires.  You must enter the');
       Writeln('diameter of the wire in centimeters,');
       Writeln('and the permeability of the wire (MU)');
       Writeln('The wire length must be entered.');
       Writeln;
  end;  { title }

Procedure Input;
  begin
     Write('Enter the wire diameter (cm) : ');
     Readln( Dia );
     Write('Enter the permeability : ');
     Readln( Perm );
     Write('Enter the wire length (cm) : ');
     Readln( Len );
     Write('Enter the distance between wires (cm) : ');
     Readln( Dist );
     Writeln;
  end;

Procedure Work;
  begin
       Induct := 0.004 * Len * ( Ln ( 2 * Dist / Dia ) + Perm / 4 - Dist / Len
  end;

Procedure PrintSc;
  begin
     Writeln;
     Writeln('For a wire with a :');
     Writeln;
     Writeln('Diameter := ',Dia,' centimeters');
     Writeln;
     Writeln('Permeability := ',Perm);
     Writeln;
     Writeln('Length := ',Len,' centimeters');
     Writeln;
     Writeln('Pair distance := ',Dist,' centimeters');
     Writeln;
     Writeln('Inductance :=',Induct,' microhenries');
     Writeln;
  end;

 Procedure PrintPr;
  begin
     Rewrite(out,'Printer:');
      Writeln;
      Writeln(out);
      Writeln('For a wire with a :');
      Writeln(out,'For a wire with a :');
      Writeln;
      Writeln(out);
      Writeln('Diameter := ',Dia,' centimeters');
      Writeln(out,'Diameter := ',Dia,' centimeters');
      Writeln;
      Writeln(out);
      Writeln('Permeability := ',Perm);
      Writeln(out,'Permeability := ',Perm);
      Writeln;
      Writeln(out);
      Writeln('Length := ',Len,' centimeters');
      Writeln(out,'Length := ',Len,' centimeters');
      Writeln;
      Writeln(out);
      Writeln('Pair distance := ',Dist,' centimeters');
      Writeln(out,'Pair distance := ',Dist,' centimeters');
      Writeln;
      Writeln(out);
      Writeln('Inductance :=',Induct,' microhenries');
      Writeln(out,'Inductance :=',Induct,' microhenries');
      Writeln(out);
```

```
      Writeln;
     Close(out);
  end;

Procedure Printout;
   begin
      case answer of
  's','S' : begin
                PrintSc
             end;
  'p','P' : begin
                PrintPr
             end
          end { case }
       end;


  begin { Main }
    Repeat
       Outputs;
       Title;
       Input;
       Work;
        Printout;
        Writeln;
              Write('Another application ?(Y/N) ');
                  Readln(Cont);
    until Cont in [ 'N','n']
  end.
```

Inductance of a Parallel Pair

This program finds the inductance of
parallel wires. You must enter the
diameter of the wire in centimeters,
and the permeability of the wire (MU).
The wire length must be entered.

Enter the wire diameter (cm) : .Ø16

Enter the permeability : 1

Enter the wire length (cm) : 2ØØ

Enter the distance between wires (cm) : 3

For a wire with a:
Diameter = 1.6ØØØØE-2 centimeters
Permeability = 1.ØØØØØ
Length = 2.ØØØØE2 centimeters
Pair distance = 3.ØØØØØ centimeters

Inductance = 4.92954 microhenries

Another application ? (Y/N) N

7.3 INDUCTANCE OF A CIRCULARLY ARRANGED GROUP OF PARALLEL WIRES

Sometimes, to reduce the inductance of wiring, several wires are connected in parallel and arranged in a circle (Fig 7.3). To calculate the inductance of wires with this arrangement, the diameter of the circle must be taken into account.

When entering data for this configuration, the permeability should be entered as one.

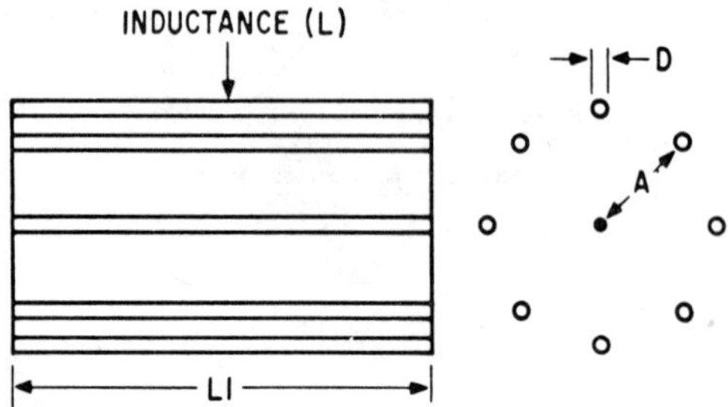

Fig. 7.3 Inductance of a circular group of wires

```
(*=========================================*)
(*                                         *)
(* Title:   Inductance of a Circularly     *)
(*          Arranged Group of Parallel     *)
(*          Wires                          *)
(* Program Summary: Program calculates     *)
(*          the inductance of several      *)
(*          circularly arranged            *)
(*          parallel wires.                *)
(*                                         *)
(*=========================================*)


PROGRAM Inductance ;

uses Transcend;

 VAR
  Answer,
  Ans,
  Cont           : Char;
  Out            : Text;
  Dia,
  Perm,
  Radius,
  Number,
  React,
  Temp,Induct,
  Len             : Real;

Procedure Outputs;
  begin
    page(output);
     repeat
       gotoXY(0,5);
        Write('S(creen or P(rinter ? ');
        Readln(Answer);
     until Answer in [ 'S','s','p','P' ]
  end;  {outputs}

Procedure Title;
  begin
    page(output);
     gotoXY(1,1);
```

```
        Writeln('Inductance of a Group of Parallel Wires');
        Writeln;Writeln;
        Writeln('This program finds the inductance of');
        Writeln('circularly arranged group of parallel');
        Writeln('wires.  To use you must enter the');
        Writeln('diameter of the wire in centimeters ');
        Writeln('and the permeability of the wire (MU)');
        Writeln('The wire length must also be entered.');
        Writeln;
  end;  { title }

Procedure Input;
  begin
     Write('Enter the wire diameter (cm) : ');
     Readln( Dia );
     Write('Enter the permeability : ');
     Readln( Perm );
     Write('Enter the wire length (cm) : ');
     Readln( Len );
     Write('Enter the radius of wire circle (cm) : ');
     Readln( Radius );
     Write('Enter the number of wires : ');
     Readln( Number );
     Writeln;
  end;

Procedure Work;
  begin
      Temp :=   ( Dia / 2 * Number * exp ((Number-1) * Ln ( Radius )));
      React :=exp (( 1 / Number ) * Ln ( Temp ));
      Induct := 0.002 * Len * ( Ln ( 2 * Len / React ) - 1 )
  end;

Procedure PrintSc;
  begin
     Writeln;
     Writeln('For ',Round(Number),' wires with a :');
     Writeln;
     Writeln('Circle radius of ',Radius,' centimeters');
     Writeln;
     Writeln('Wire diameter := ',Dia,' centimeters');
     Writeln;
     Writeln('Permeability := ',Perm);
     Writeln;
     Writeln('Length := ',Len,' centimeters');
     Writeln;
     Writeln('Inductance :=',Induct,' microhenries');
     Writeln;
  end;

Procedure PrintPr;
  begin
     Rewrite(out,'Printer:');
      Writeln;
      Writeln(out);
      Writeln('For ',Round(Number),' wires with a :');
      Writeln(out,'For ',Round(Number),' wires with a :');
      Writeln;
      Writeln(out);
      Writeln('Circle radius of ',Radius,' centimeters');
      Writeln(out,'Circle radius of ',Radius,' centimeters');
      Writeln;
      Writeln(out);
      Writeln('Wire diameter := ',Dia,' centimeters');
      Writeln(out,'Wire diameter := ',Dia,' centimeters');
      Writeln;
      Writeln(out);
      Writeln('Permeability := ',Perm);
      Writeln(out,'Permeability := ',Perm);
      Writeln;
      Writeln(out);
      Writeln('Length := ',Len,' centimeters');
      Writeln(out,'Length := ',Len,' centimeters');
      Writeln;
      Writeln(out);
      Writeln('Inductance :=',Induct,' microhenries');
      Writeln(out,'Inductance :=',Induct,' microhenries');
```

```
      Writeln(out);
      Writeln;
     Close(out);
  end;

Procedure Printout;
   begin
      case answer of
  's','S' : begin
                PrintSc
            end;
  'p','P' : begin
                PrintPr
            end
          end { case }
       end;


  begin { Main }
    Repeat
       Outputs;
       Title;
       Input;
       Work;
        Printout;
        Writeln;
              Write('Another application ?(Y/N) ');
                  Readln(Cont);
    until Cont in [ 'N','n']
  end.
```

Inductance of a Group of Parallel Wires

This program finds the inductance of circularly arranged group of parallel wires. To use, you must enter the diameter of the wire in centimeters and the permeability of the wire (MU). The wire length must also be entered.

Enter the wire diameter (cm) : .Ø16

Enter the permeability : 1

Enter the wire length (cm) : 2ØØ

Enter the radius of wire circle (cm) : 3

Enter the number of wires : 6

For 6 wires with a :
Circle radius of 3.ØØØØØ centimeters
Wire diameter = 1.6ØØØØE-2 centimeters
Permeability = 1.ØØØØØ
Length = 2.ØØØØØE2 centimeters

Inductance = 1.83282 microhenries

Another application ? (Y/N) N

7.4 CAPACITIVE REACTANCE CALCULATOR

The resistance of the flow of ac current exhibited by a capacitor is called "capacitive reactance." The capacitive reactance, denoted by X_C, is inversely proportional to the frequency of the ac voltage.

This program calculates the reactance of a capacitor at a given frequency according to the formula:

$$X_C = \frac{1}{2\pi FC}$$

where $\pi = 3.14159$, F = frequency in hertz, and C equals the capacitance in farads.

When using the program, capacitance is entered in microfarads and not farads. The conversion to farads is done in *Procedure Work*. The reactance returned by the program is in ohms.

```
(*=========================================*)
(*                                         *)
(* Title:    Capacitive Reactance          *)
(*           Calculator                    *)
(* Program Summary: Program calculates     *)
(*           the capacitive reactance      *)
(*           of a capacitor.               *)
(*                                         *)
(*=========================================*)


PROGRAM Capacitance;


 VAR
  Answer,
  Ans,
  Cont           : Char;
  Out            : Text;
  Cap,
  Cap2,
  Freq,
  React          : Real;

Procedure Outputs;
  begin
    page(output);
     repeat
       gotoXY(Ø,5);
        Write('S(creen or P(rinter ? ');
        Readln(Answer);
     until Answer in [ 'S','s','p','P' ]
  end;  {outputs}

Procedure Title;
  begin
    page(output);
     gotoXY(1Ø,1);
       Writeln('Capacitive Reactance');
       Writeln;Writeln;
       Writeln('This program calculates the capacitive');
       Writeln('reactance of a capacitor.  You must');
       Writeln('enter the value of capacitance in');
       Writeln('microfarads and the frequency in hertz.');
       Writeln;
  end;  { title }

Procedure Input;
  begin
     Write('Enter the capacitance (uf) : ');
```

```
      Readln( Cap );
      Writeln;
      Write('Enter the frequency (hz) : ');
      Readln( Freq );
      Writeln;
   end;

 Procedure Work;
   begin
      Cap2 := Cap ;
      Cap  := Cap * 1E-6 ;
      React := 1 / ( 2 * 3.14159 * Freq * Cap )
   end;

  Procedure Printout;
     begin
       case answer of
    's','S' : begin
                 Writeln;
                 Writeln('The reactance of a ',Cap2);
                 Writeln;
                 Writeln('microfarad capacator at ',Freq,' hertz');
                 Writeln;
                 Writeln('is ',React,' ohms');
                 Writeln;
              end;
    'p','P' : begin
                Rewrite(out,'Printer:');
                 Writeln;
                 Writeln(out);
                 Writeln;
                 Writeln(out);
                 Writeln('The reactance of a ',Cap2);
                 Writeln(out,'The reactance of a ',Cap2);
                 Writeln;
                 Writeln(out);
                 Writeln('microfarad capacator at ',Freq,' hertz');
                 Writeln(out,'microfarad capacator at ',Freq,' hertz');
                 Writeln;
                 Writeln(out);
                 Writeln('is ',React,' ohms');
                 Writeln(out,'is ',React,' ohms');
                 Writeln(out);
                 Writeln;
                Close(out);
              end
            end { case }
       end;


   begin { Main }
     Repeat
       Outputs;
       Title;
       Input;
       Work;
        Printout;
        Writeln;
             Write('Another application ?(Y/N) ');
                Readln(Cont);
     until Cont in [ 'N','n']
   end.
```

Capacitive Reactance

This program calculates the capacitive reactance of a capacitor. You must enter the value of capacitance in microfarads and the frequency in hertz.

Enter the capacitance (μf) : .ØØ33

Enter the frequency (Hz) : 1E6

The reactance of a 3.30000E-3
microfarad capacitor at 1.00000E6 hertz
is 4.82288E1 ohms.

Another application ? (Y/N) N

7.5 INDUCTIVE REACTANCE CALCULATOR

Just as a capacitor acts as an ac reactance, so can an inductor. However, unlike the capacitor whose reactance increases with a decrease in frequency, the reactance of an inductor increases with an increase in frequency.

The inductive reactance, denoted by X_L, is calculated according to the following formula:

$$X_L = 2\pi FL$$

where again $\pi = 3.14159$, F is the frequency in hertz, and L is the inductance in henries.

When using the program, the inductance must be entered in henries and the frequency in hertz. The inductive reactance is returned in ohms.

```
(*========================================*)
(*                                        *)
(* Title:   Inductive Reactance           *)
(*          Calculator                    *)
(* Program Summary: Program calculates    *)
(*          the inductive reactance       *)
(*          of an inductor.               *)
(*                                        *)
(*========================================*)

PROGRAM Inductive;

 VAR
  Answer,
  Ans,
  Cont            : Char;
  Out             : Text;
  Induct,
  Freq,
  React           : Real;

Procedure Outputs;
  begin
    page(output);
     repeat
       gotoXY(0,5);
        Write('S(creen or P(rinter ? ');
        Read(Answer);
     until Answer in [ 'S','s','p','P' ]
  end;  {outputs}

Procedure Title;
  begin
    page(output);
     gotoXY(10,1);
       Writeln('Inductive Reactance');
       Writeln;Writeln;
       Writeln('This program calculates the inductive');
       Writeln('reactance of a inductor.  You must');
       Writeln('enter the value of inductance in');
```

```
        Writeln('henries and the frequency in hertz.');
        Writeln;
   end;  { title }

Procedure Input;
   begin
      Write('Enter the inductance (H) : ');
      Readln( Induct );
      Writeln;
      Write('Enter the frequency (hz) : ');
      Readln( Freq );
      Writeln;
   end;

Procedure Work;
   begin
      React :=   2 * 3.14159 * Freq * Induct
   end;

 Procedure Printout;
     begin
       case answer of
    's','S' : begin
                Writeln;
                Writeln('The reactance of a ',Induct);
                Writeln;
                Writeln('henry inductor at ',Freq,' hertz');
                Writeln;
                Writeln('is ',React,' ohms');
                Writeln;
              end;
    'p','P' : begin
               Rewrite(out,'Printer:');
                Writeln;
                Writeln(out);
                Writeln;
                Writeln(out);
                Writeln('The reactance of a ',Induct);
                Writeln(out,'The reactance of a ',Induct);
                Writeln;
                Writeln(out);
                Writeln('henry inductor at ',Freq,' hertz');
                Writeln(out,'henry inductor at ',Freq,' hertz');
                Writeln;
                Writeln(out);
                Writeln('is ',React,' ohms');
                Writeln(out,'is ',React,' ohms');
                Writeln(out);
                Writeln;
               Close(out);
              end
            end { case }
        end;


    begin { Main }
      Repeat
        Outputs;
        Title;
        Input;
        Work;
         Printout;
         Writeln;
              Write('Another application ?(Y/N) ');
                 Read(Cont);
      until Cont in [ 'N','n']
    end.
```

Inductive Reactance

This program calculates the inductive reactance of an inductor. You must enter the value of the inductance in

henries and the frequency in hertz.

Enter the inductance (H) : 3

Enter the frequency (Hz) : 6Ø

The reactance of a 3.ØØØØØ
henry inductor at 6.ØØØØØE1 hertz
is 1.13Ø97E3 ohms

Another application ? (Y/N) N

7.6 BYPASS CAPACITOR CALCULATION

In the design of amplifiers, it is frequently necessary to bypass the resistor in the emitter leg of a circuit. The value of the capacitor can be easily determined by using a variation of the relationship between frequency, resistance, and capacitance shown below (this variation is listed in *Procedure Work* of the program):

$$f = \frac{1}{2\,\pi RC}$$

In this program the frequency (f) is the lowest frequency that is attenuated 3 dB. To calculate the value of the bypass capacitor, the value of the resistor to be bypassed and the frequency must be entered. The answer is given in microfarads.

```
(*=========================================*)
(*                                         *)
(* Title:   Bypass Capacitor Calculator    *)
(* Program Summary: Program calculates     *)
(*          the value of a capacitor       *)
(*          used to bypass an emitter      *)
(*          resistor in an amplifier       *)
(*          circuit.                       *)
(*                                         *)
(*=========================================*)


PROGRAM BypassCapacitance;

Const
 Pi = 3.14159 ;

 VAR
  Answer,
  Ans,
  Cont               : Char;
  Out                : Text;
  Freq,
  Res,
  Cap                : Real;

Procedure Outputs;
  begin
    page(output);
     repeat
       gotoXY(Ø,5);
         Write('S(creen or P(rinter ? ');
         Readln(Answer);
     until Answer in [ 'S','s','p','P' ]
  end;  {outputs}
```

```
Procedure Title;
  begin
    page(output);
     gotoXY(12,1);
       Writeln('Bypass Capacitor');
       Writeln;Writeln;
       Writeln('This program calculates the value of');
       Writeln('a bypass capacitor for a cathode or');
       Writeln('emitter resistor in an amplifier');
       Writeln('circuit.  You must enter the lowest');
       Writeln('frequency to be bypassed and the value');
       Writeln('of the resistor it is bypassing.');
       Writeln;
  end;  { title }

Procedure Input;
  begin
     Write('Enter the bypass frequency (Hz) : ');
     Readln( Freq );
     Writeln;
     Write('Enter the bypass resistor (ohms) : ');
     Readln( Res );
     Writeln;
  end;

Procedure Work;
  begin
     Cap := 1E7 / ( 2 * Pi * Freq * Res )
  end;

 Procedure Printout;
    begin
      case answer of
   's','S' : begin
                Writeln;
                Writeln('Bypass capacitor is ',Cap,' microfarads');
                Writeln;
             end;
   'p','P' : begin
               Rewrite(out,'Printer:');
                Writeln;
                Writeln(out);
                Writeln;
                Writeln(out);
                Writeln('Bypass capacitor is ',Cap,' microfarads');
                Writeln(out,'Bypass capacitor is ',Cap,' microfarads');
                Writeln(out);
                Writeln;
               Close(out);
             end
           end { case }
       end;


   begin { Main }
     Repeat
       Outputs;
       Title;
       Input;
       Work;
        Printout;
        Writeln;
             Write('Another application ?(Y/N) ');
                Readln(Cont);
     until Cont in [ 'N','n']
   end.
```

Bypass Capacitor

This program calculates the value of
a bypass capacitor for a cathode or
emitter resistor in an amplifier
circuit. You must enter the lowest

```
frequency to be bypassed and the value
of the resistor it is bypassing.

Enter the bypass frequency (Hz) : 60

Enter the bypass resistance (ohms) : 200

Bypass capacitor is 1.32629E2 microfarads

Another application ? (Y/N) N
```

7.7 RESONANT CIRCUIT CALCULATIONS

Anyone involved in the design of electronic circuits will sooner or later perform resonant circuit calculations. With this program, it is a simple matter to determine the resonant frequency of an L-C circuit or, given the frequency, to determine the capacitance or inductance when one of the two is known.

When run, the program displays a menu to determine which of the possible calculations will be performed. The determination of the resonant frequency is done in *Procedure FindFreq*. Capacitance is determined in *Procedure FindCap*. Finally, the inductance is determined in *Procedure Findind.*

The formulas used in these calculations are variations of the basic relationship between frequency, inductance, and capacitance for resonant circuits. This relationship is :

$$f = \frac{1}{2\sqrt{LC}}$$

```
(*=========================================*)
(*                                         *)
(* Title:   Resonant Circuit Calculator    *)
(* Program Summary: Program calculates     *)
(*          the resonant frequency of      *)
(*          an L-C circuit.                *)
(*                                         *)
(*=========================================*)


PROGRAM Resonant;

uses Transcend;

Const
 Pi = 3.14159 ;

 VAR
  Answer,
  Ans,
  Cont           : Char;
  Out            : Text;
  Cap,
  Freq,
  Induct                  : Real;
  Choice                  : Integer;

Procedure Outputs;
  begin
    page(output);
```

```
    repeat
      gotoXY(0,5);
       Write('S(creen or P(rinter ? ');
       Readln(Answer);
    until Answer in [ 'S','s','p','P' ]
  end;  {outputs}

Procedure Title;
  begin
    page(output);
     gotoXY(11,1);
       Writeln('Resonant Circuits');
       Writeln;Writeln;
       Writeln('This program calculates the ');
       Writeln('resonant frequency of an L-C circuit.');
       Writeln('If L or C is unknown and the frequency');
       Writeln('is given, it will calculate the ');
       Writeln('unknown');
       Writeln;
  end;  { title }

Procedure FindFreq;
  Begin
     Freq := 1 / (2 * Pi * Sqrt ( Induct * Cap * 1E-6 ));

       case answer of
   's','S' : begin
                Writeln;
                Writeln('F := ',Freq,' hertz');
                Writeln;
             end;
   'p','P' : begin
               Rewrite(out,'Printer:');
                Writeln;
                Writeln(out);
                Writeln;
                Writeln(out);
                Writeln('F := ',Freq,' hertz');
                Writeln(out,'F := ',Freq,' hertz');
                Writeln(out);
                Writeln;
               Close(out);
             end
           end { case }
      end;

Procedure FindCap;
  Begin
    Cap := 1 / ( Induct * ( Freq * 2 * Pi ) * ( Freq * 2 * Pi ));

       case answer of
   's','S' : begin
                Writeln;
                Writeln('C := ',Cap,' microfarads');
                Writeln;
             end;
   'p','P' : begin
               Rewrite(out,'Printer:');
                Writeln;
                Writeln(out);
                Writeln;
                Writeln(out);
                Writeln('C := ',Cap,' microfarads');
                Writeln(out,'C := ',Cap,' microfarads');
                Writeln(out);
                Writeln;
               Close(out);
             end
           end { case }
       end;

Procedure FindInd;
  Begin
     Induct := 1E6 / ( Cap * ( 2 * Pi * Freq ) * ( 2 * Pi * Freq ));

       case answer of
   's','S' : begin
```

```
                Writeln;
                Writeln('L := ',Induct,' henries');
                Writeln;
             end;
'p','P' : begin
              Rewrite(out,'Printer:');
                Writeln;
                Writeln(out);
                Writeln;
                Writeln(out);
                Writeln('L := ',Induct,' henries');
                Writeln(out,'L := ',Induct,' henries');
                Writeln(out);
                Writeln;
              Close(out);
             end
          end { case }
      end;

Procedure InputFreq;
  begin
   Write('Enter Capacitance (uf) : ');
   Readln( Cap );
   Writeln;
   Write('Enter Inductance (henries) : ');
   Readln( Induct );
   FindFreq
  end;

Procedure InputCap;
  begin
   Write('Enter Frequency (Hz) : ');
   Readln( Freq );
   Writeln;
   Write('Ente Inductance (henries) : ');
   Readln( Induct );
   FindCap
  end;

Procedure InputInd;
  begin
   Write('Enter Capacitance (uf) : ');
   Readln( Cap );
   Writeln;
   Write('Enter Frequency (Hz) : ');
   Readln( Freq );
   FindInd
 end;

Procedure Input;
  begin
     Writeln('Enter choice to calculate :');
     Writeln;
     Writeln('1...Frequency');
     Writeln('2...Capacitance');
     Writeln('3...Inductance');
     Writeln;
     Write('Enter choice : ');
     Readln( Choice );
     Writeln;
        if Choice = 1 then
                 InputFreq
        else
        if Choice = 2 then
                 InputCap
        else
        if Choice = 3 then
                 InputInd
        else input;
  end;


   begin { Main }
     Repeat
       Outputs;
       Title;
       Input;
```

```
      Writeln;
            Write('Another application ?(Y/N) ');
                Readln(Cont);
    until Cont in [ 'N','n']
  end.
```

Resonant Circuits

This program calculates the
resonant frequency of an L-C circuit.
If L or C is unknown and the frequency
is given, it will calculate the
unknown.

Enter choice to calculate :

1...Frequency
2...Capacitance
3...Inductance

Enter choice : 3

Enter capacitance (μf) : 2.2

Enter frequency (Hz) : 1ØØØ

L =1.15138E-2 henries

8 COMPUTER-AIDED CIRCUIT DESIGN

One area in which computers have traditionally been used is computer-aided design. Computers are a great design aid because they make it easy to change the design to accommodate a particular parameter. In addition, changes can be made in the original design without the need for building a prototype to find how the change will affect the final result.

The seven programs in this chapter will help you design a particular circuit or tell you how it will react to a particular set of conditions. The first program solves the equations for a class A amplifier and permits the use of either germanium or silicon transistors.

The second program designs a discrete transistor Schmitt trigger, which converts any inputted waveform to a square wave signal. Triggering levels are selectable. The third program in this chapter is not really a design program. It calculates how long a given power supply will continue to provide current to a load after the ac voltage has ceased. It is design related, however, in that various parameters of the power supply can be altered until desired time is reached.

The last four programs (8.4 through 8.7) deal with the design of various types of power supplies. The first of these (8.4) is unusual in that it does not use any power transformer to drop the ac voltage to the required level. Instead, it uses the reactance of a capacitor, saving much weight and expense. This type of power supply, however, is limited to low current and must be used carefully, because it is not isolated from the ac line.

The other three supplies are variations of the same theme and similar in makeup. These are full-capability supplies that should give trouble-free service.

8.1 CLASS A TRANSISTOR AMPLIFIER DESIGN[1]

Computer-aided circuit design is very popular among engineers, because it enables the designer to change the value of one or more components and lets him see what the effect will be on the rest of the design. It does this quickly, without the need for actually building the circuit.

This program solves the circuit equations for a stabilized, self-biased transistor operating as a class A amplifier under small signal conditions. The method used in the program involves simultaneously solving the circuit equation at both extremes. An assumption is made that ICBO = 0 at the low temperature extreme.

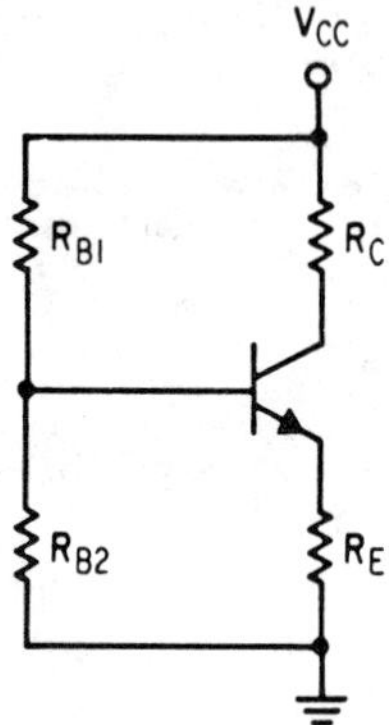

Fig. 8.1 Class A transistor amplifier

```
(*========================================*)
(*                                        *)
(* Title:   Class A Transistor Amplifier  *)
(*          Design                        *)
(* Program Summary: Program solves        *)
(*          the circuit equations for     *)
(*          a class A amplifier.          *)
(*                                        *)
(*========================================*)

PROGRAM TransistorAmplifier;

uses Transcend;

 VAR
  Answer,
  Ans,
  Cont              : Char;
  Out               : Text;
  Rb1,Rb2,Rc,Re,Res,
  Tmin,Tmax,Temp,
  Betamin,Betamax,Beta,Beta2,
  IeMax,IeMin,Icbo,
  Vce,Vs,BiasV,
  Toler,
  Tol               : Real;
  Trans             : Integer;
```

[1] Adapted from C. P. Popenoe, "Basic Program Designs Bias Circuits," *Electronic Design* 13 (June 21, 1970): p. 93.

```
Procedure Outputs;
  begin
    page(output);
     repeat
       gotoXY(0,5);
        Write('S(creen or P(rinter ? ');
        Readln(Answer);
     until Answer in [ 'S','s','p','P' ]
  end;  {outputs}

Procedure Title;
  begin
    page(output);
     gotoXY(09,1);
       Writeln('Class A Transistor');
       Writeln;Writeln;
       Writeln('This program solves the circuit');
       Writeln('equations for a stabilized, self-biased');
       Writeln('transistor operating as a Class A');
       Writeln('amplifier under small signal conditions');
       Writeln;
       Writeln('If multiple data entries are to be made');
       Writeln('per line, separate each with a space.');
  end;  { title }

Procedure Input;
  begin
    Writeln;
    Write('Enter tolerance of resistor (%) : ');
    Readln( Toler );
    Writeln;
    Writeln('Enter Silicon (1) or Germanium (0) ');
    Write('Transistors : ');
    Readln( Trans );
    Writeln;
    Write('Enter Min and Max Temperatures : ');
    Readln( Tmin , Tmax );
    Writeln;
    Write('Enter Min and Max Beta : ');
    Readln( Betamin , Betamax );
    Writeln;
    Write('Enter Supply Voltage : ');
    Readln( Vs );
    Writeln;
    Writeln('Enter a value for the emitter');
    Write('resistor and percent tolerance : ');
    Readln( Re , Tol );
    Writeln;
    Write('Enter V(ce) and I(cbo) in ma : ');
    Readln( Vce , Icbo );
    Writeln;
    Write('Enter I(e) Max,Min in ma : ');
    Readln( IeMax , IeMin );
    Writeln
  end;

Procedure PrintScr;
   begin
     Writeln;
     Writeln('For a stabilized bias circuit :');
     Writeln;
     Writeln;
     Writeln('R(b1) := ',Rb1,' ohms');
     Writeln('R(b2) := ',Rb2,' ohms');
     Writeln('R(c)  := ',Rc,' ohms');
     Writeln('R(e)  := ',Re,' ohms');
     Writeln;
     Writeln('Bias voltage := ',BiasV,' volts');
     Writeln;
  end;

Procedure PrintPrt;
  begin
     Rewrite(out,'Printer:');
      Writeln;
      Writeln(out);
      Writeln;
```

```
      Writeln(out);
      Writeln('For a stabilized bias circuit :');
      Writeln(out,'For a stabilized bias circuit :');
      Writeln;
      Writeln;
      Writeln(out);
      Writeln(out);
      Writeln('R(b1) := ',Rb1,' ohms');
      Writeln(out,'R(b1) := ',Rb1,' ohms');
      Writeln('R(b2) := ',Rb2,' ohms');
      Writeln(out,'R(b2) := ',Rb2,' ohms');
      Writeln('R(c)  := ',Rc,' ohms');
      Writeln(out,'R(c)  := ',Rc,' ohms');
      Writeln('R(e)  := ',Re,' ohms');
      Writeln(out,'R(e)  := ',Re,' ohms');
      Writeln;
      Writeln(out);
      Writeln('Bias voltage := ',BiasV,' volts');
      Writeln(out,'Bias voltage := ',BiasV,' volts');
      Writeln(out);
      Writeln;
    Close(out);
  end;

Procedure Printout;
   begin
     case answer of
          's','S' : PrintScr;
          'p','P' : PrintPrt;
     end { case }
   end;

Procedure Work;
 begin
   IeMax  := IeMax * (1 - 0.03 * Tol );
   IeMin  := IeMin * (1 + 0.03 * Tol );
   Betamin  := Betamin * 0.865 * exp( 0.00575 * Tmin );
   Temp  := Tmax - 25 ;
   Beta  := (0.00895 - 0.00565 * Temp + 0.00048 * Temp * Temp );
   Betamax  := Betamax * ( 0.865 * exp ( 0.00575 * Tmax )
                 - ( Trans - 1 ) * Beta );
       if Trans = 0 then
          Icbo := Icbo * exp ( 0.075 * Temp );
   Rc  := 2000 * ( Vs - Vce) / ( IeMax + IeMin ) - Re ;
   Beta2  := Betamax + 1 ;
   Res  := (( IeMax - IeMin ) * Re + 2.5 * ( Tmax - Tmin ))
            / ( Icbo + IeMin / (Betamin + 1 ) -IeMax / Beta2 );
     if Res <= 0 then
            begin
              Writeln('The range of I(e) is too narrow!');
              Exit( Work );
            end;
   BiasV  := IeMin * 0.001 * ( Res / ( Betamin + 1 ) + Re )
          + 0.2 + 0.5 * Trans - 0.0025 * ( Tmin - 25 );
   Rb1  := Vs * Res / BiasV ;
   Rb2  := Vs * Res / ( Vs - BiasV ) ;
   Printout;
end;

  begin { Main }
    Repeat
      Outputs;
      Title;
      Input;
      Work;
       Writeln;
            Write('Another application ?(Y/N) ');
                Readln(Cont);
    until Cont in [ 'N','n']
  end.
```

Class A Transistor

This program solves the circuit equations for a stabilized, self-biased

```
transistor operating as a Class A
amplifier under small signal conditions.

If multiple data entries are to be made
per line, separate each with a space.

Enter tolerance of resistor (%) : 1Ø

Enter Silicon (1) or Germanium (Ø)
Transistors : Ø


Enter Min and Max Temperatures : -25  75

Enter Min and Max Beta : 5Ø  15Ø

Enter Supply Voltage : 12

Enter value for the emitter
resistor and percent tolerance :  2ØØØ  5

Enter V(ce) and I(cbo) in ma : 5  .ØØ2

Enter I(e) Max,Min in ma : 1.2  .8

For a stabilized bias circuit :

R(b1) = 2.24841E4 ohms
R(b2) = 5.2358ØE3 ohms
R(c) = 5.21649E3 ohms
R(e) = 2.ØØØØØE3 ohms Bias voltage = 2.26659 volts

Another application ? (Y/N) N
```

8.2 SCHMITT TRIGGER CIRCUIT DESIGN[1]

Schmitt Trigger circuits are commonly used in digital designs (Fig. 8.2). They come in handy to convert various shaped signals (sine wave, and so forth) into square wave pulses. The Schmitt trigger has an upper and a lower threshold voltage. The incoming signal is compared to these threshold voltages, and a square wave output is produced with the same frequency as the input wave form.

[1] Adapted from A. C. Caggiano, "Schmitt Trigger Program Uses Standard Resistor Values," *Electronic Design* 12 (June 7, 1970): p. 100.

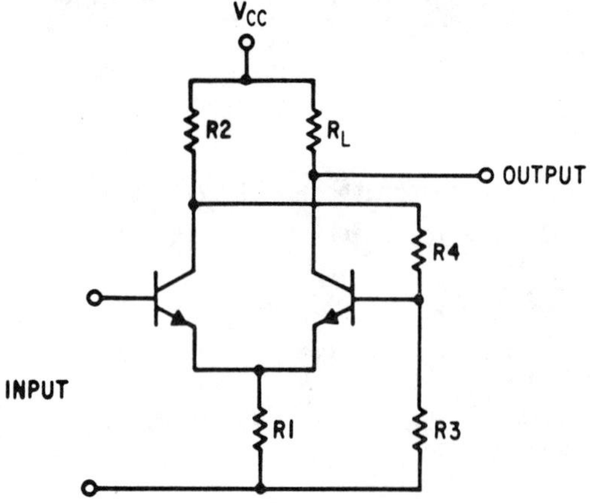

Fig. 8.2 Schmitt trigger

```
(*=========================================*)
(*                                         *)
(* Title:  Schmitt Trigger Design          *)
(* Program Summary: Program solves         *)
(*         the circuit equations for       *)
(*         a Schmitt trigger.              *)
(*                                         *)
(*=========================================*)


PROGRAM SchmittTrigger;

uses Transcend;

 VAR
  Answer,
  Ans,
  Cont           : Char;
  Out            : Text;
  Amps,
  Pwr,
  Upper,
  LoadImp,
  Lower,
  Vu,Vl,
  Vcc,Vbe        : Real;
  Res            : Array [1..5] of Real;
  Steps          : Integer;

Procedure Outputs;
  begin
    page(output);
     repeat
       gotoXY(0,5);
        Write('S(creen or P(rinter ? ');
        Readln(Answer);
     until Answer in [ 'S','s','p','P' ]
  end;  {outputs}

Procedure Title;
  begin
    page(output);
     gotoXY(09,1);
       Writeln('Schmitt Trigger Design');
       Writeln;Writeln;
       Writeln('This program designs a 2-transistor');
       Writeln('Schmitt Trigger circuit.  To use it');
       Writeln('you must enter the upper and lower');
       Writeln('voltage trip points, the value of the');
       Writeln('supply voltage, the base-to-emitter');
```

```
        Writeln('voltage drop of the transistor, and');
        Writeln('the resistance of the load.');
        Writeln;
   end;  { title }

 Procedure Input;
   begin
     Writeln;
     Write('Enter the upper threshold voltage: ');
     Readln( Vu );
     Write('Enter the lower threshold voltage: ');
     Readln( Vl );
     Writeln;
     Write('Enter the supply voltage : ');
     Readln( Vcc );
     Write('Enter the emitter-base voltage :');
     Readln( Vbe );
     Writeln;
     Write('Enter the load impedance (ohms): ');
     Readln( LoadImp );
     Writeln;
   end;

 Procedure PrintScr;
    begin
      Writeln;
      Writeln('For a Schmitt Trigger with an upper');
      Writeln('threshold of ',Vu,' volts and a lower');
      Writeln('threshold of ',Vl,' volts the ');
      Writeln('resistances are :');
      Writeln;
        For Steps := 1 to 4 do
          Writeln('R(',Steps,') := ',Res[ Steps ],' ohms');
      Writeln;
      Writeln('The power dissipated by the load is');
      Writeln(Pwr,' milliwatts.');
    end;

 Procedure PrintPrt;
   begin
     Rewrite(out,'Printer:');
      Writeln;
      Writeln;
      Writeln(out);
      Writeln(out);
      Writeln('For a Schmitt Trigger with an upper');
      Writeln(out,'For a Schmitt Trigger with an upper');
      Writeln('threshold of ',Vu,' volts and a lower');
      Writeln(out,'threshold of ',Vu,' volts and a lower');
      Writeln('threshold of ',Vl,' volts the ');
      Writeln(out,'threshold of ',Vl,' volts the ');
      Writeln('resistances are :');
      Writeln(out,'resistances are :');
      Writeln;
      Writeln(out);
        For Steps := 1 to 4 do
           begin
              Writeln('R(',Steps,') := ',Res[ Steps ],' ohms');
              Writeln(out,'R(',Steps,') := ',Res[ Steps ],' ohms');
           end;
      Writeln;
      Writeln(out);
      Writeln(out,'The power dissipated by the load is');
      Writeln('The power dissipated by the load is');
      Writeln(out,Pwr,' milliwatts.');
      Writeln(Pwr,' milliwatts.');
      Close(out);
    end;

  Procedure Printout;
     begin
       case answer of
            's','S' : PrintScr;
            'p','P' : PrintPrt;
       end { case }
     end;
```

```
Procedure Work;
 begin
   Res[ 1 ] := LoadImp * ( Vu - Vbe ) / ( Vcc - Vu + Vbe );
   Res[ 2 ] := Res[ 1 ] * ( Vcc - Vl + Vbe ) / ( Vl - Vbe );
   Res[ 3 ] := 10 * Res[ 1 ] ;
   Res[ 4 ] := Res[ 3 ] * Vcc / Vu - Res[ 3 ] - Res[ 2 ] ;
   Amps      := ( Vu - Vbe ) / Res[ 1 ];
   Pwr      := 1E3 * Sqr( Amps ) * LoadImp ;
   Upper       := Res[ 3 ] * Vcc / ( Res[ 2 ] + Res[ 3 ] + Res[ 4 ] );
   Lower       := Vbe + Res[ 1 ] * Vcc / ( Res[ 1 ] + Res[ 2 ] );
 end;

  begin { Main }
    Repeat
      Outputs;
      Title;
      Input;
      Work;
      Printout;
       Writeln;
            Write('Another application ?(Y/N) ');
                Readln(Cont);
    until Cont in [ 'N','n']
  end.
```

Schmitt Trigger Design

This program designs a 22-transistor Schmitt trigger circuit. To use it you must enter the upper and lower voltage trip points, the value of the supply voltage, the base-to-emitter voltage drop of the transistor, and the resistance of the load.

Enter the upper threshold voltage: 9

Enter the lower threshold voltage: 5

Enter the supply voltage : 12

Enter the emitter-base voltage : .6

Enter the load impedance (ohms): 300

For a Schmitt trigger with an upper threshold of 9.00000 volts and a lower threshold of 5.00000 volts the resistances are :

R(1) = 7.00000E2 ohms
R(2) = 1.20909E3 ohms
R(3) = 7.00000E3 ohms
R(4) = 1.12424E3 ohms

The power dissipated by the load is 4.32000E1 milliwatts.

Another application ? (Y/N) N

8.3 POWER SUPPLY INTERRUPT TIME

With the increased use of computers in homes and industry, the loss of data due to an interruption of the power supply is of ever increasing importance. In applications where the loss of data is intolerable, backup power supplies are integrated into the system. Unfortunately, there is a certain delay before the backup supply switches in. This program will calculate how long the main power supply will continue to produce a regulated voltage to the load even after the power has been cut. It can then be determined if the period is long enough to cover the time required to switch in the backup. If it is not, the storage capacitor in the main supply can be increased until the desired time is reached.

Another way to increase the time is by increasing the unregulated voltage fed into the voltage regulator. However, the regulator needs to be able to withstand the higher voltage during light loads.

To use this program, values for five variables must be entered: the unregulated voltage, minimum voltage drop across the regulator, the load voltage, the load current in amperes, and the value of the storage capacitor in microfarads.

The program will then calculate in milliseconds the amount of time regulated voltage can be supplied to the load.

```
(*=========================================*)
(*                                         *)
(* Title:   Power Supply Interrupt         *)
(*          Time                           *)
(* Program Summary: Program calculates     *)
(*          how long a power supply        *)
(*          will continue to operate       *)
(*          after its power has been       *)
(*          interrupted.                   *)
(*                                         *)
(*=========================================*)


PROGRAM PowerSupply;

 VAR
  Answer,
  Ans,
  Cont            : Char;
  Out             : Text;
  Vin,
  VoltDrop,
  VLoad,
  ILoad,
  Cap,
  Time            : Real;

Procedure Outputs;
  begin
    page(output);
     repeat
       gotoXY(0,5);
        Write('S(creen or P(rinter ? ');
        Readln(Answer);
     until Answer in [ 'S','s','p','P' ]
  end;  {outputs}

Procedure Title;
  begin
    page(output);
     gotoXY(06,1);
       Writeln('Power Supply Interrupt Time');
       Writeln;Writeln;
       Writeln('This program will calculate the amount');
```

```
      Writeln('of time a regulated power supply will');
      Writeln('continue to provide a regulated output');
      Writeln('after the line voltage has been');
      Writeln('interrupted.');
      Writeln;
  end;  { title }

Procedure Input;
  begin
    Writeln;
    Writeln('Enter: ');
    Write('Unregulated voltage (volts) : ');
    Readln( Vin );
    Write('Minimum regulator drop (volts) : ');
    Readln( VoltDrop );
    Write('Load voltage (volts) : ');
    Readln( VLoad );
    Write('Load current (amperes) : ');
    Readln( ILoad );
    Write('Storage capacitor (microfarads) : ');
    Readln( Cap );
    Writeln
  end;

 Procedure Work;
  begin
    Cap  := Cap * 1E-6 ;
    Time := (( Vin - VoltDrop - VLoad ) * Cap ) / ILoad * 1000
  end;

Procedure PrintScr;
   begin
      Writeln;
      Writeln('This power supply will continue to');
      Writeln('supply the load with regulated voltage');
      Writeln('for ',Round( Time ),' milliseconds after the ');
      Writeln('line voltage has been interrupted.');
      Writeln;
  end;

Procedure PrintPrt;
  begin
     Rewrite(out,'Printer:');
      Writeln;
      Writeln(out);
      Writeln('This power supply will continue to');
      Writeln(out,'This power supply will continue to');
      Writeln('supply the load with regulated voltage');
      Writeln(out,'supply the load with regulated voltage');
      Writeln('for ',Round( Time ),' milliseconds after the ');
      Writeln(out,'for ',Round( Time ),' milliseconds after the ');
      Writeln('line voltage has been interrupted.');
      Writeln(out,'line voltage has been interrupted.');
      Writeln(out);
      Writeln;
     Close(out);
   end;

 Procedure Printout;
    begin
      case answer of
           's','S' : PrintScr;
           'p','P' : PrintPrt;
      end { case }
    end;

   begin { Main }
     Repeat
       Outputs;
       Title;
       Input;
       Work;
       Printout;
        Writeln;
             Write('Another set of data ?(Y/N) ');
                Readln(Cont);
     until Cont in [ 'N','n']
   end.
```

```
Power Supply Interrupt Time

This program will calculate the amount
of time a regulated power supply will
continue to provide a regulated output
after the line voltage has been
interrupted.

Enter:
Unregulated voltage (volts) : 16
Minimum regulator drop (volts) : 3
Load voltage (volts) : 12
Load current (amperes) : 2
Storage capacitor (microfarads) : 5000

This power supply will continue to
supply the load with regulated voltage
for 3 milliseconds after the
line voltage has been interrupted.

Another set of data ? (Y/N) Y

Enter:
Unregulated voltage (volts) : 16
Minimum regulator drop (volts) : 3
Load voltage (volts) : 12
Load current (amperes) : 2
Storage capacitor (microfarads) : 20000

This power supply will continue to
supply the load with regulated voltage
for 10 milliseconds after the
line voltage has been interrupted.

Another set of data ? (Y/N) N
```

8.4 TRANSFORMERLESS POWER SUPPLY

Most low-current dc-operated devices such as radios, tape recorders, and so forth, use small power packs that plug into a wall socket to adapt the device for ac use (Fig. 8.4). Most of the time, these power packs use a transformer to step the line voltage down to a value that is close to the dc voltage value. That is why these supplies are very heavy. In addition to its weight, the transformer is probably the most expensive part of the power supply. If left plugged in for any length of time, it heats up due to losses in the transformer core.

An alternate way to drop the voltage to the desired value without the cost, weight, or heat generation of the transformer is to use a capacitor. While the capacitor

represents an open circuit for dc applications, in ac circuits, it presents a resistance to the flow of electricity. This resistance to current flow varies with the frequency of the ac voltage. It is known as capacitive reactance and is generally denoted by X_C. The formula for capacitive reactance is

$$X_C = \frac{1}{2\pi FC}$$

where $\pi = 3.14159$, F is the frequency in hertz, and C is the capacitance in farads. The formula gives a value for X_C in ohms.

Using this property of the capacitor, it is possible to design a small light-weight power supply. The key design formula is

$$C = \frac{1}{2\pi F(RL)\sqrt{\frac{(VL)^2}{(VO)^2-1}}}$$

where RL is the load resistance, VO is the desired output voltage, and VL is the rms value of the line voltage (most of the time 110 to 117 volts). This equation is implemented in *Procedure Work* with an extra factor of 1E6 to yield an answer in microfarads.

Since the value calculated by the program will most likely be a nonstandard value, the program allows the user to enter the closest standard value (*Procedure InputCapacitor*). The output voltage is then calculated.

Since this is an unregulated supply, the range of output current is rather limited. It can be increased considerably by adding a zener diode to the circuit.

When using a power supply of this type, there are a few important things to remember. First, the output voltage is inversely proportional to the frequency. Therefore, if the frequency goes down (60 to 50 hz), the output voltage will go up. Another important point is that this power supply is not isolated from the line. If the device being powered is not nonconductive or grounded, the possibility of electrical shock exists.

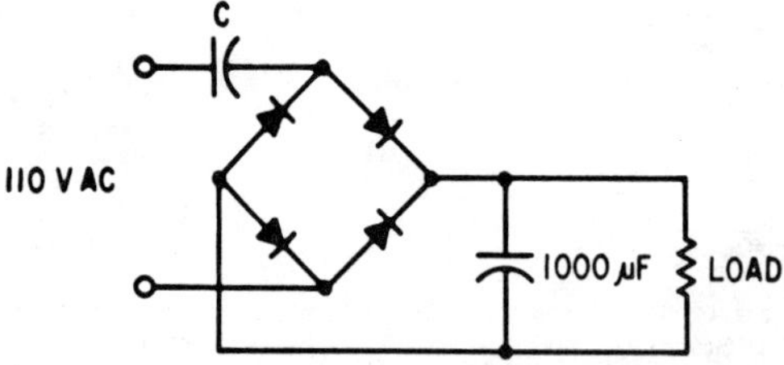

Fig. 8.4 Transformerless power supply

```
(*=======================================*)
(*                                       *)
(* Title:   Transformerless Power Supply  *)
(* Program Summary: Program designs      *)
(*          a low-cost transformerless   *)
(*          power supply to deliver low- *)
(*          current dc voltages.         *)
(*                                       *)
(*=======================================*)
```

```
PROGRAM PowerSupply;

uses Transcend;

 VAR
  Answer,
  Ans,
  Cont            : Char;
  Out             : Text;
  LineVolt,
  Freq,
  Vout,
  RLoad,
  TwoPiF,
  Cap,
  OutputVolts     : Real;

Procedure Outputs;
  begin
    page(output);
     repeat
       gotoXY(0,5);
        Write('S(creen or P(rinter ? ');
        Readln(Answer);
     until Answer in [ 'S','s','p','P' ]
  end;  {outputs}

Procedure Title;
  begin
    page(output);
     gotoXY(06,1);
       Writeln('Transformerless Power supply');
       Writeln;Writeln;
       Writeln('This program designs a low cost');
       Writeln('transformerless dc power supply.  You');
       Writeln('must enter the ac line voltage (110/220)');
       Writeln('and the frequency (50/60 Hz). Also');
       Writeln('required are the load resistance and');
       Writeln('desired voltage');
       Writeln;
  end;  { title }

Procedure Input;
  begin
    Writeln;
    Write('Enter the ac voltage : ');
    Readln( LineVolt );
    Writeln;
    Write('Enter Frequency (Hz) : ');
    Readln( Freq );
    Writeln;
    Write('Enter dc output voltage : ');
    Readln( Vout );
    Writeln;
    Write('Enter Load resistance (ohms) : ');
    Readln( RLoad );
    Writeln;
  end;

Procedure FigureOutput;
  begin
   OutputVolts := LineVolt * RLoad
         / Sqrt( Sqr( RLoad ) + ( 1 / Sqr( TwoPiF * Cap * 1E-6 )));
  end;

Procedure InputCapacitor;
  begin
    Writeln;
    Write('Enter Standard Capacitance (microfarads) : ');
    Readln( Cap );
    FigureOutput;
  end;

 Procedure Work;
  begin
    TwoPiF := 6.281854 * Freq ;
    Cap := 1 / ( TwoPiF * RLoad *
```

```
        Sqrt ( Sqr( LineVolt ) / Sqr ( Vout ) - 1 )) * 1E6 ;
end;

Procedure PrintScr;
   begin
      Writeln;
      Writeln('The value for "C" should be close to');
      Writeln(Cap,' microfarads.  Select the nearest');
      Writeln('standard value.');
      InputCapacitor;
      Writeln;
      Writeln('With a capacitor of ',Cap,' micorfarads');
      Writeln('the output voltage will be ',OutputVolts,' volts');
      Writeln;
      Writeln;
  end;

Procedure PrintPrt;
  begin
     Rewrite(out,'Printer:');
      Writeln;
      Writeln(out);
      Writeln('The value for "C" should be close to');
      Writeln(out,'The value for "C" should be close to');
      Writeln(Cap,' microfarads.  Select the nearest');
      Writeln(out,Cap,' microfarads.');
      Writeln('standard value.');
      InputCapacitor;
      Writeln;
      Writeln(out);
      Writeln('With a capacitor of ',Cap,' micorfarads');
      Writeln(out,'With a capacitor of ',Cap,' micorfarads');
      Writeln('the output voltage will be ',OutputVolts,' volts');
      Writeln(out,'the output voltage will be ',OutputVolts,' volts');
      Writeln(out);
      Writeln;
     Close(out);
   end;

 Procedure Printout;
    begin
      case answer of
           's','S' : PrintScr;
           'p','P' : PrintPrt;
      end { case }
    end;

   begin { Main }
     Repeat
       Outputs;
       Title;
       Input;
       Work;
       Printout;
        Writeln;
             Write('Another set of data ?(Y/N) ');
                Readln(Cont);
     until Cont in [ 'N','n']
   end.
```

Transformerless Power Supply

This program designs a low cost transformerless dc power supply. You must enter the ac line voltage (11Ø/12Ø) and the frequency (5Ø/6Ø Hz). Also required are the load resistance and desired voltage.

Enter the ac voltage : 11Ø

```
Enter Frequency (Hz) : 60

Enter dc output voltage : 6

Enter load resistance (ohms) : 20

The value for "C" should be close to
7.24664 microfarads. Select the nearest
standard value.

Enter standard capacitance (microfarads) : 6.8

With a capacitor of 6.80000 microfarads
the output voltage will be 5.63120 volts.

Another set of data ? (Y/N) N
```

8.5 TWO-DIODE FULL-WAVE REGULATED POWER SUPPLY

A full-wave rectified power supply that uses only two diodes and a center-tapped transformer is a must for low voltage outputs. It has fewer components than the bridge rectifier supply and does not have the additional two-diode voltage drops of the bridge circuit, an important factor in low voltage designs.

This program makes designing full-wave supplies a snap. The design takes into account factors often ignored, such as high and low line voltages. To use the program, the high, low, and nominal line voltages must be entered along with the voltage regulator drop, peak-to-peak ripple voltage, desired output voltage, and load current. All voltages are entered in volts, and the load current is entered in amperes.

The program will indicate what the transformer secondary voltage must be, giving the voltage for each half of the secondary, indicating the secondary current and the volt-amperes of the transformer, and suggesting a value for the filter capacitor.

The generalized circuit for this power supply is shown in Fig 8.5. The regulator, indicated as a block, can be either a discrete regulator or an integrated circuit.

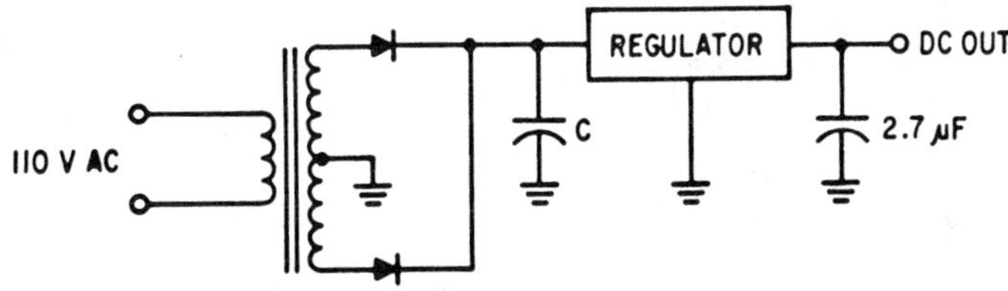

Fig. 8.5 Two-diode full-wave regulated power supply

```
(*=========================================*)
(*                                         *)
(* Title:   Two-Diode Full Wave            *)
(*          Regulated Power Supply         *)
(* Program Summary: Program designs        *)
(*          a low-voltage full-wave        *)
(*          regulated power supply.        *)
(*                                         *)
(*=========================================*)
```

```
PROGRAM PowerSupply;

uses Transcend;

 VAR
  Answer,
  Ans,
  Cont            : Char;
  Out             : Text;
  Vout,
  Vreg,
  Vrip,
  Vnom,
  VLow,
  VHi,
  Amps,
  Ampsec,
  Cap,
  Vsec            : Real;

Procedure Outputs;
  begin
    page(output);
     repeat
       gotoXY(Ø,5);
        Write('S(creen or P(rinter ? ');
        Readln(Answer);
     until Answer in [ 'S','s','p','P' ]
  end;  {outputs}

Procedure Title;
  begin
    page(output);
     gotoXY(Ø4,1);
       Writeln('Full Wave Regulated Power Supply');
       Writeln;Writeln;
       Writeln('This program designs a two-diode full');
       Writeln('wave rectified, regulated power supply');
       Writeln('Enter all voltages in volts');
       Writeln;
  end;  { title }

Procedure Input;
  begin
    Writeln;
    Write('Enter Output voltage : ');
    Readln( Vout );
    Write('Enter Voltage regulator drop : ');
    Readln( Vreg );
    Write('Enter Ripple Voltage (p-p) : ');
    Readln( Vrip );
    Write('Enter nominal line voltage : ');
    Readln( Vnom );
    Write('Enter line voltage low : ');
    Readln( VLow );
    Write('Enter line voltage High : ');
    Readln( VHi );
    Write('Enter required current (amps) : ');
    Readln( Amps );
  end;

 Procedure Work;
  begin
    Vsec  := ( Vout + Vreg + 1.25 + Vrip / 2 ) * Vnom / Ø.92 / VLow / Sqrt (2)
    Ampsec   := 1.2 * Amps ;
    Cap    := (Ampsec / Vrip * 6E-3 ) * 1E6 ;
               if Amps > 1 then Cap := Cap * 3 ;
  end;

Procedure PrintScr;
   begin
      Writeln;
      Writeln('   2-Diode full wave regulated supply');
      Writeln;
      Writeln;
      Writeln('For a ',Vout,' volt, ',Amps,' amp  power supply :');
      Writeln;
      Writeln('Total secondary voltage := ',2 * Vsec,' volts');
      Writeln('Each half of the secondary := ',Vsec,' volts');
```

```
      Writeln('Transformer secondary current  := ',Ampsec,' amps');
      Writeln('Transformer power rating := ',Ampsec * Vsec * 2,' va');
      Writeln('Recommended capacitance := ',Cap,' mfd');
      Writeln;
      Writeln('For light loads and high line voltage');
      Writeln('the voltage drop across the regulator');
      Writeln('will be ',VHi / VLow * Vsec / 2,' volts');
      Writeln;
  end;

  Procedure PrintPr2;
   begin
       Writeln('Transformer power rating := ',Ampsec * Vsec * 2,' va');
       Writeln(out,'Transformer power rating := ',Ampsec * Vsec * 2,' va');
       Writeln('Recommended capacitance := ',Cap,' mfd');
       Writeln(out,'Recommended capacitance := ',Cap,' mfd');
       Writeln;
       Writeln(out);
       Writeln('For light loads and high line voltage');
       Writeln(out,'For light loads and high line voltage');
       Writeln('the voltage drop across the regulator');
       Writeln(out,'the voltage drop across the regulator');
       Writeln('will be ',VHi / VLow * Vsec / 2,' volts');
       Writeln(out,'will be ',VHi / VLow * Vsec / 2,' volts');
       Writeln;
       Writeln(out);
  end;

 Procedure PrintPrt;
   begin
      Rewrite(out,'Printer:');
       Writeln;
       Writeln(out);
       Writeln('   2-Diode full wave regulated supply');
       Writeln(out,'   2-Diode full wave regulated supply');
       Writeln;
       Writeln;
       Writeln(out);
       Writeln(out);
       Writeln('For a ',Vout,' volt, ',Amps,' amp  power supply :');
       Writeln(out);
       Writeln;
       Writeln(out,'For a ',Vout,' volt, ',Amps,' amp  power supply :');
       Writeln(out);
       Writeln;
       Writeln('Total secondary voltage := ',2 * Vsec,' volts');
       Writeln(out,'Total secondary voltage := ',2 * Vsec,' volts');
       Writeln('Each half of the secondary := ',Vsec,' volts');
       Writeln(out,'Each half of the secondary := ',Vsec,' volts');
       Writeln('Transformer secondary current  := ',Ampsec,' amps');
       Writeln(out,'Transformer secondary current  := ',Ampsec,' amps');
       PrintPr2;
      Close(out);
    end;

  Procedure Printout;
     begin
       case answer of
            's','S' : PrintScr;
            'p','P' : PrintPrt;
       end { case }
     end;

    begin { Main }
      Repeat
        Outputs;
        Title;
        Input;
        Work;
        Printout;
         Writeln;
              Write('Another set of data ?(Y/N) ');
                 Readln(Cont);
      until Cont in [ 'N','n']
    end.
```

```
Full-Wave Regulated Power Supply

This program designs a two-diode full-
wave rectified and regulated power supply.

Enter all voltages in volts.

Enter output voltage : 12

Enter voltage regulator drop : 3

Enter ripple voltage (p-p) : 1.2

Enter nominal line voltage : 110

Enter line voltage low : 95

Enter line voltage high : 130

Enter required current (amps) : 2


2-Diode full-wave regulated supply

For a 1.20000E1 volt, 2.00000 amp power supply:

Total secondary voltage = 2.99914E1 volts
Each half of the secondary = 1.49957E1 volts
Transformer secondary current = 2.40000 amps
Transformer power rating = 7.19793E1 va
Recommended capacitance = 3.60000E4 mfd

For light loads and high line voltage
the voltage drop across the regulator
will be 1.02602E1 volts.

Another set of data ? (Y/N) N
```

8.6 BRIDGE-RECTIFIED REGULATED POWER SUPPLY

As mentioned earlier, a bridge-rectified power supply (Fig. 8.6) is not a good choice for low-voltage supplies, because the extra two diodes in the bridge circuit cause an additional voltage drop that may not always be tolerable. For most other applications, however, the bridge rectifier is an excellent choice. Two reasons for this are that center-tapped secondary transformers are not needed, and the voltage of the secondary winding is lower in the bridge circuit. Of course the current will have to be higher to get the same output power.

As in program 8.5, this one also requires high, low, and nominal values of the line voltage to be entered, as well as the minimum voltage regulator drop, load voltage, and current.

The output consists of the transformer secondary voltage, current and volt-ampere rating, and a value for the filter capacitor.

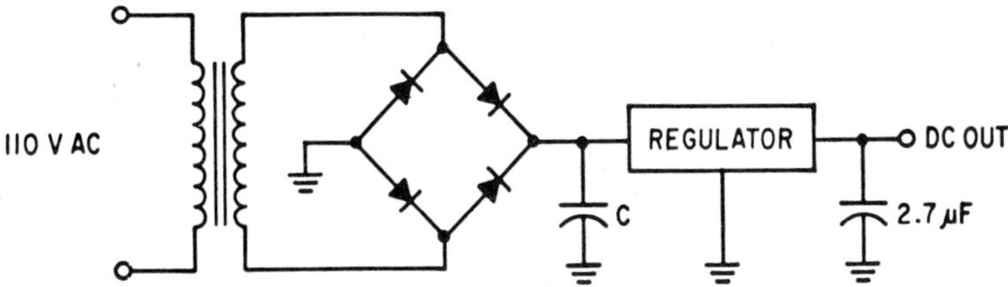

Fig. 8.6 Bridge-rectified regulated power supply

```
(*=========================================*)
(*                                         *)
(* Title:   Bridge-Rectified Regulated     *)
(*          Power Supply                   *)
(* Program Summary: Program designs        *)
(*          regulated power supplies       *)
(*          using a four-diode bridge      *)
(*          rectifier circuit.             *)
(*                                         *)
(*=========================================*)


PROGRAM PowerSupply;

uses Transcend;

 VAR
  Answer,
  Ans,
  Cont           : Char;
  Out            : Text;
  Vo,
  Vreg,
  I,
  Vrip,
  Vnom,
  VLow,
  VHi,
  Vsec,
  Isec,
  Cap            : Real;

Procedure Outputs;
  begin
    page(output);
     repeat
       gotoXY(0,5);
        Write('S(creen or P(rinter ? ');
        Readln(Answer);
     until Answer in [ 'S','s','p','P' ]
  end;  {outputs}

Procedure Title;
  begin
    page(output);
     gotoXY(05,1);
       Writeln('Bridge Regulated Power Supply');
       Writeln;Writeln;
       Writeln('This program designs a bridge full');
       Writeln('wave rectified, regulated power supply');
       Writeln;
  end;  { title }
```

```
Procedure Input;
  begin
    Writeln;
    Write('Enter ouput voltage : ');
    Readln( Vo );
    Writeln;
    Write('Enter voltage regulator drop : ');
    Readln( Vreg );
    Writeln;
    Write('Enter ripple voltage (p-p) : ');
    Readln( Vrip );
    Writeln;
    Write('Enter nominal line voltage : ');
    Readln( Vnom );
    Writeln;
    Write('Enter line voltage low : ');
    Readln( VLow );
    Writeln;
    Write('Enter line voltage high : ');
    Readln( VHi );
    Writeln;
    Write('Enter required current (amps) : ');
    Readln( I );
  end;

 Procedure Work;
  begin
    Vsec  := ( Vo + Vreg + 2.50 + Vrip / 2 ) * Vnom / 0.92 / VLow / Sqrt (2) ;
    Isec   := 1.8 * I ;
    Cap    := (Isec / Vrip * 6E-3 ) * 1E6 ;
              if I > 1 then Cap := Cap * 3 ;
  end;

Procedure PrintScr;
   begin
      Writeln;
      Writeln('    Regulated Power Supply Design');
      Writeln;
      Writeln;
      Writeln('For a ',Vo,' volt, ',I,' amp  power supply :');
      Writeln;
      Writeln('Transformer secondary voltage := ',Vsec,' volts');
      Writeln('Transformer secondary current  := ',Isec,' amps');
      Writeln('Transformer power rating := ',Isec * Vsec,' va');
      Writeln('Recommended capacitance := ',Cap,' microfarad');
      Writeln('For light loads and high line voltage');
      Writeln('the voltage drop across the regulator');
      Writeln('will be ',VHi / VLow * Vsec / 2,' volts');
      Writeln;
  end;

 Procedure PrintPr2;
  begin
      Writeln;
      Writeln('Transformer secondary current  := ',Isec,' amps');
      Writeln(out,'Transformer secondary current  := ',Isec,' amps');
      Writeln('Transformer power rating := ',Isec * Vsec,' va');
      Writeln(out,'Transformer power rating := ',Isec * Vsec,' va');
      Writeln('Recommended capacitance := ',Cap,' microfarad');
      Writeln(out,'Recommended capacitance := ',Cap,' microfarad');
      Writeln('For light loads and high line voltage');
      Writeln(out,'For light loads and high line voltage');
      Writeln('the voltage drop across the regulator');
      Writeln(out,'the voltage drop across the regulator');
      Writeln('will be ',VHi / VLow * Vsec / 2,' volts');
      Writeln(out,'will be ',VHi / VLow * Vsec / 2,' volts');
      Writeln(out);
 end;

Procedure PrintPrt;
  begin
     Rewrite(out,'Printer:');
      Writeln;
      Writeln(out);
      Writeln('    Regulated Power Supply Design');
      Writeln(out,'    Regulated Power Supply Design');
      Writeln;
      Writeln;
```

```
      Writeln(out);
      Writeln(out);
      Writeln('For a ',Vo,' volt, ',I,' amp  power supply :');
      Writeln(out,'For a ',Vo,' volt, ',I,' amp  power supply :');
      Writeln;
      Writeln(out);
      Writeln('Transformer secondary voltage := ',Vsec,' volts');
      Writeln(out,'Transformer secondary voltage := ',Vsec,' volts');
      PrintPr2;
     Close(out);
   end;

 Procedure Printout;
    begin
      case answer of
           's','S' : PrintScr;
           'p','P' : PrintPrt;
      end { case }
    end;

   begin { Main }
     Repeat
       Outputs;
       Title;
       Input;
       Work;
       Printout;
        Writeln;
             Write('Another set of data ?(Y/N) ');
                Readln(Cont);
     until Cont in [ 'N','n']
   end.
```

Bridge Regulated Power Supply

This program designs a bridge full-
wave rectified regulated power supply.

Enter output voltage : 5

Enter voltage regulator drop : 3

Enter ripple voltage (p-p) : 1

Enter nominal line voltage : 110

Enter line voltage low : 95

Enter line voltage high : 130

Enter required current (amps) : 10

Regulated Power Supply Design

For a 5.00000 volt, 1.00000E1 amp power supply:

Transformer secondary voltage = 9.78946 volts
Transformer secondary current = 1.80000E1 amps

Transformer power rating = 1.76210E2 va
Recommended capacitance = 3.24000E5 microfarad

The voltage drop across the regulator
will be 6.69806 volts.

Another set of data ? (Y/N) N

8.7 DUAL-OUTPUT REGULATED POWER SUPPLY

While most digital circuits require only a single supply voltage, a few linear integrated circuits must have symmetrical output voltages that are +V and –V. A power supply to meet these requirements could simply be composed of two supplies in series. But a more effective way of building such a supply is to use a full-wave complementary rectifier design. This circuit combines both the two-diode and bridge full-wave rectifiers into a circuit that will supply both positive and negative voltages.

The circuit in Fig. 8.7 requires a center-tapped secondary transformer. The two voltages of opposite polarity are developed across the bridge and measured with reference to the ground, where the center tap of the transformer secondary is connected.

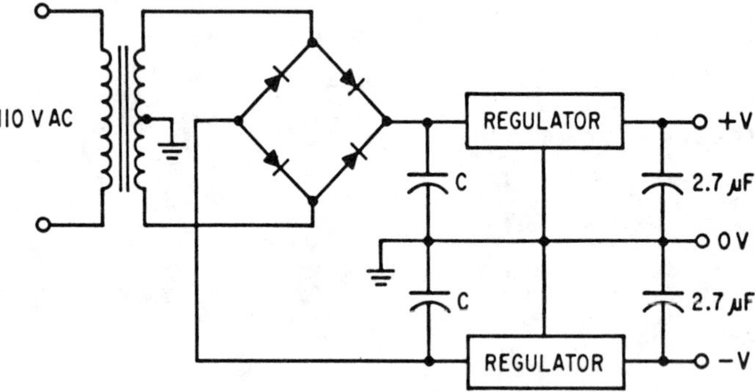

Fig. 8.7 Dual-output regulated power supply

```
(*=========================================*)
(*                                         *)
(* Title:   Dual-Output Regulated          *)
(*          Power Supply                   *)
(* Program Summary: Program designs        *)
(*          a power supply that provides   *)
(*          symmetrical positive and       *)
(*          negative output voltages.      *)
(*                                         *)
(*=========================================*)


PROGRAM PowerSupply;

uses Transcend;

 VAR
  Answer,
  Ans,
  Cont           : Char;
```

```
    Out             : Text;
    Vout,
    Vreg,
    Vrip,
    Vnom,
    VLow,
    VHi,
    Vsec,
    Isec,
    I,
    Cap              : Real;

  Procedure Outputs;
    begin
      page(output);
       repeat
         gotoXY(Ø,5);
          Write('S(creen or P(rinter ? ');
          Readln(Answer);
       until Answer in [ 'S','s','p','P' ]
    end;  {outputs}

  Procedure Title;
    begin
      page(output);
       gotoXY(Ø6,1);
         Writeln('Dual Regulated Power Supply');
         Writeln;Writeln;
         Writeln('This program designs a dual regulated');
         Writeln('power supply');
         Writeln;
    end;  { title }

  Procedure Input;
    begin
      Writeln;
      Write('Enter Symmetrical Output voltage : ');
      Readln( Vout );
      Write('Enter Voltage regulator drop : ');
      Readln( Vreg );
      Write('Enter Ripple Voltage (p-p) : ');
      Readln( Vrip );
      Write('Enter nominal line voltage : ');
      Readln( Vnom );
      Write('Enter line voltage low : ');
      Readln( VLow );
      Write('Enter line voltage High : ');
      Readln( VHi );
      Write('Enter required current (amps) : ');
      Readln( I );
    end;

   Procedure Work;
    begin
      Vsec  := ( Vout + Vreg + 1.25 + Vrip / 2 ) * Vnom / Ø.92 / VLow / Sqrt (2)
      Isec   := 1.8 * I ;
      Cap    := (Isec / Vrip * 6E-3 ) * 1E6 ;
                 if I > 1 then Cap := Cap * 3 ;
    end;

  Procedure PrintScr;
     begin
        Writeln;
        Writeln('  Regulated Power Supply Design');
        Writeln;
        Writeln;
        Writeln('For a ',Vout,' volt, ',I,' amp  power supply :');
        Writeln;
        Writeln('Transformer secondary voltage := ',2 * Vsec,' volts');
        Writeln('Transformer secondary current  := ',Isec,' amps');
        Writeln('Transformer power rating := ',Isec * Vsec * 2,' va');
        Writeln('Recommended capacitance := ',Cap,' mfd');
        Writeln;
        Writeln('For light loads and high line voltage');
        Writeln('the voltage drop across the regulator');
        Writeln('will be ',VHi / VLow * Vsec / 2,' volts');
        Writeln;
    end;
```

```
 Procedure PrintPr2;
  begin
      Writeln('Transformer power rating := ',Isec * Vsec * 2,' va');
      Writeln(out,'Transformer power rating := ',Isec * Vsec * 2,' va');
      Writeln('Recommended capacitance := ',Cap,' mfd');
      Writeln(out,'Recommended capacitance := ',Cap,' mfd');
      Writeln;
      Writeln(out);
      Writeln('For light loads and high line voltage');
      Writeln(out,'For light loads and high line voltage');
      Writeln('the voltage drop across the regulator');
      Writeln(out,'the voltage drop across the regulator');
      Writeln('will be ',VHi / VLow * Vsec / 2,' volts');
      Writeln(out,'will be ',VHi / VLow * Vsec / 2,' volts');
      Writeln;
      Writeln(out);
 end;

Procedure PrintPrt;
  begin
     Rewrite(out,'Printer:');
      Writeln;
      Writeln(out);
      Writeln('  Regulated Power Supply Design');
      Writeln(out,'  Regulated Power Supply Design');
      Writeln;
      Writeln;
      Writeln(out);
      Writeln(out);
      Writeln('For a ',Vout,' volt, ',I,' amp  power supply :');
      Writeln(out);
      Writeln;
      Writeln(out,'For a ',Vout,' volt, ',I,' amp  power supply :');
      Writeln(out);
      Writeln;
      Writeln('Transformer secondary voltage := ',2 * Vsec,' volts');
      Writeln(out,'Transformer secondary voltage := ',2 * Vsec,' volts');
      Writeln('Transformer secondary current  := ',Isec,' amps');
      Writeln(out,'Transformer secondary current  := ',Isec,' amps');
      PrintPr2;
     Close(out);
   end;

 Procedure Printout;
    begin
      case answer of
           's','S' : PrintScr;
           'p','P' : PrintPrt;
      end { case }
    end;

   begin { Main }
     Repeat
       Outputs;
       Title;
       Input;
       Work;
       Printout;
        Writeln;
             Write('Another set of data ?(Y/N) ');
                Readln(Cont);
     until Cont in [ 'N','n']
   end.
```

Dual Regulated Power Supply

This program designs a dual regulated
power supply.

Enter symmetrical output voltage: 18

Enter voltage regulator drop : 3

```
Enter ripple voltage (p-p) : 1.5

Enter nominal line voltage : 11Ø

Enter line voltage low : 95

Enter line voltage high : 13Ø

Enter required current (amps) : 1



Regulated Power Supply Design

For a 1.8ØØØØE1 volt, 1.ØØØØØ amp power supply :

Transformer secondary voltage = 4.Ø9378E1 volts
Transformer secondary current = 1.8ØØØØ amps
Transformer power rating = 7.3688ØE1 va
Recommended capacitance = 7.2ØØØØE3 mfd

For light loads and high line voltage
the voltage drop across the regulator
will be 1.4ØØ5ØE1 volts.

Another set of data ? (Y/N) N
```

9 ACTIVE FILTER DESIGN

In the processing of analog signals, it is invariably necessary to filter the signal to separate desired information from unwanted noise. Active filters are useful for this application because with them it is possible to tailor to any specific set of requirements the filter cutoff characteristics and gain.

The first three programs in this chapter design active bandpass filters. The specific instances when a particular one of these should or should not be used are indicated in the text accompanying the programs.

Programs 9.4 and 9.5 design notch filters, which are generally used to remove a specific noise component, such as 60 Hz hum from audio signals. The final three programs design various active low-pass filters, including an FET input version.

9.1 ACTIVE BANDPASS FILTER[1]

The bandpass filter in Fig. 9.1 is of simple design and can be fabricated with a conventional 741 operational amplifier. As in the program for active low-pass filter No. 2 (see Fig. 9.7), this program contains a data section in the form of *Procedure Equ*, which fills the "C1" array with nine standard values for C1, enabling nine different designs for the same circuit parameters. The program also contains two routines for selecting the proper string descriptors for the capacitors and resistors. *Procedure Caps* selects the proper capacitor descriptors and *Procedure Res* selects the proper resistor descriptors.

When run, the program requests the center frequency, bandwidth, and magnitude of the gain in the bandpass.

[1]Adapted from H. Minuskin, "Active Filter Design Uses BASIC Language," *Electronic Design* 5 (March 1, 1970): p. 83.

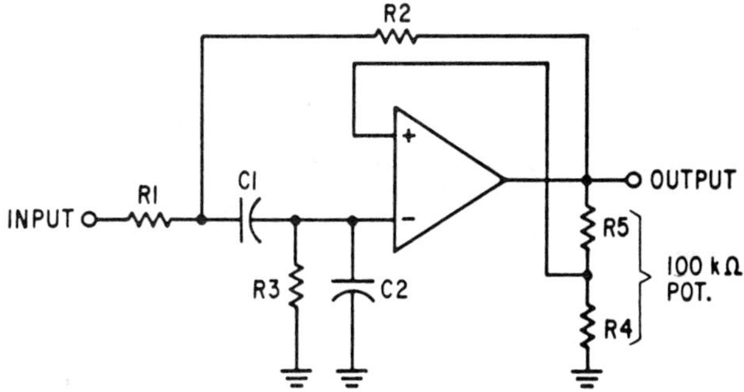

Fig. 9.1 Active bandpass filter

```
(*========================================*)
(*                                        *)
(* Title:   Active Bandpass Filter        *)
(* Program Summary: Program designs       *)
(*          an active bandpass filter.    *)
(*                                        *)
(*========================================*)


PROGRAM ActiveBandpass;

uses Transcend;


 VAR
  Answer,
  Ans,
  Cont             : Char;
  Out              : Text;
  Freq,
  Bw,
  Q,
  K,
  M,
  C2,
  Ca               : Real;
  C1               : Array [1..9] of Real;
  R                : Array [1..5] of Real;
  Rs               : Array [1..5] of String;
  J,
  X,
  Xx               : Integer;
  Ds,
  Es               : String;

Procedure Equ;
  begin
     C1[1] := 1E-9 ; C1[2] :=    1E-8 ; C1[3] := 2.2E-8 ; C1[4] := 4.7E-8 ;
     C1[5] := 1E-7 ; C1[6] := 2.2E-7 ; C1[7] := 4.7E-7 ; C1[8] :=   1E-6 ;
     C1[9] := 1E-5 ;
  end;

Procedure Outputs;
  begin
    page(output);
     repeat
       gotoXY(Ø,5);
        Write('S(creen or P(rinter ? ');
        Readln(Answer);
     until Answer in [ 'S','s','p','P' ]
  end;  {outputs}

Procedure Title;
  begin
    page(output);
```

```
      gotoXY(11,1);
        Writeln('Active Bandpass Filter');
        Writeln;Writeln;
        Writeln('This program designs a bandpass active');
        Writeln('filter.  You must enter the center');
        Writeln('frequency, and the bandwidth.');
        Writeln('The program will supply');
        Writeln('component values for the nine standard');
        Writeln('values of C1.  You can select the most');
        Writeln('convenient set of values to use.');
        Writeln;
  end;  { title }

Procedure Input;
  begin
     Equ;
    Writeln;
    Write('Enter frequency (Hz) : ');
    Readln( Freq );
    Writeln;
    Write('Enter Bandwidth (Hz) : ');
    Readln( Bw );
  end;

 Procedure Res;
    begin
       if R[ J ] >= 1E6 then
            begin
                R[ J ] := R[ J ] / 1E6 ;
                Rs[ J ]:= ' Megahoms';
                Exit( Res )
            end;
       if R[ J ] >= 1E3 then
            begin
                R[ J ] := R[ J ] / 1E3;
                Rs[ J ]:= ' Kilohms';
                Exit( Res )
            end;
         Rs[ J ] := ' Ohms'
    end;

Procedure Work;
  Forward;

Procedure PrintScr;
   begin
   Writeln;
   Writeln('C1 := ',C1[ XX ],Ds);
   Writeln('C2 := ',C2,Es);
   Writeln;
     For J := 1 to 5 do
            begin
              Res;
              Writeln('R',J,' := ',R[ J ],Rs[ J ]);
            end;
   Writeln;
   Write('Another value of C1 (Y/N) : ');
   Readln( Ans );
     if Ans in [ 'N', 'n'] then
        Exit (Work);
     if XX = 9 then Exit (Work)
  end;

 Procedure PrintPrt;
   begin
      Rewrite(out,'Printer:');
       Writeln;
       Writeln(out);
       Writeln('C1 := ',C1[ XX ],Ds);
       Writeln(out,'C1 := ',C1[ XX ],Ds);
       Writeln('C2 := ',C2,Es);
       Writeln(out,'C2 := ',C2,Es);
       Writeln;
           For J := 1 to 5 do
                  begin
                    Res;
```

```
                 Writeln('R',J,' := ',R[ J ],Rs[ J ]);
                 Writeln(out,'R',J,' := ',R[ J ],Rs[ J ]);
               end;
   Writeln;
   Write('Another value of C1 (Y/N) : ');
   Readln( Ans );
     if Ans in [ 'N', 'n'] then
        Exit (Work);
     if XX = 9 then Exit (Work);
      Close(out);
      end;

Procedure Printout;
   begin
     case answer of
          's','S' : PrintScr;
          'p','P' : PrintPrt;
     end { case }
   end;

Procedure Caps;
   begin
         if C1[ XX ] < 1 then Ds := ' Nanofarads';
         if C1[ XX ] < 1 then C1[ XX ]  := Ca  * 1E3 ;
     C2  := Ø.5 * C1[ XX ] ;
         if (C2 >= 1ØØØ) or (Ds = ' Microfarads') then
                        Es := ' Microfarads';
         if (C2 >= 1ØØØ) and (Ds = ' Microfarads') then
                        Exit(Caps);
         if C2 >= 1ØØØ then
                        C2 := ( C2 / 1ØØ ) / 1Ø
   end;

Procedure Mainwork;
  begin
     Ds         := ' Microfarads';
     Es         := ' Nanofarads';
     K          := C1[ XX ] * 6.28 * Freq ;
     M          := Ø.333 * ( 6.5 - ( 1 / Q ));
     Ca         := C1[ XX ] * 1E6 ;
     C1[ XX ]   := Ca ;
          Caps;
            R[ 2 ] := Ø.666 / K ;
            R[ 1 ] := 3 * R[ 2 ] ;
            R[ 3 ] := 4 / K ;
            R[ 5 ] := 1E5 / M ;
            R[ 4 ] := R[ 5 ] * ( M - 1 );
          Printout;
  end;

Procedure Work;
 begin
   Q := Freq / Bw ;
       for XX := 1 to 9 do
            MainWork;
 end;

  begin { Main }
    Repeat
      Outputs;
      Title;
      Input;
      Work;
       Writeln;
            Write('Another set of data ?(Y/N) ');
               Readln(Cont);
    until Cont in [ 'N','n']
  end.
```

Active Bandpass Filter

This program designs a bandpass active filter. You must enter the center

```
frequency and the bandwidth.
The program will supply
component values for the nine standard
values of C1. You can select the
most convenient set of values to use.

Enter frequency (Hz) : 3000

Enter Bandwidth (Hz) : 1000

C1 = 1.00000 Nanofarads
C2 = 5.00000E-1 Nanofarads
R1 = 1.06051E2 Kilohms
R2 = 3.53503E1 Kilohms
R3 = 2.12314E2 Kilohms
R4 = 5.13027E1 Kilohms
R5 = 4.86973E1 Kilohms

Another value of C1? (Y/N) : Y

C1 = 1.00000E1 Nanofarads
C2 = 5.00000 Nanofarads
R1 = 1.06051E1 Kilohms
R2 = 3.53503 Kilohms
R3 = 2.12314E1 Kilohms
R4 = 5.13027E1 Kilohms
R5 = 4.86973E1 Kilohms

Another value of C1? (Y/N) : N

Another set of data? (Y/N) : N
```

9.2 STATE VARIABLE BANDPASS FILTER

This program calculates the component values required for a state variable bandpass filter. This particular filter is built with three operational amplifiers as shown in Fig. 9.2. This active filter has several advantages over other types of filters. It can provide circuit Qs of over 100. It is stable and insensitive to passive component drift.

Components used in the filter should be of high quality. Carbon resistors are suitable, but capacitors designed for bypass and coupling applications are not. Their tolerance and stability are generally poor.

To use the program, the frequency characteristics of the op amp must be known. The popular 741 op amp can easily be used.

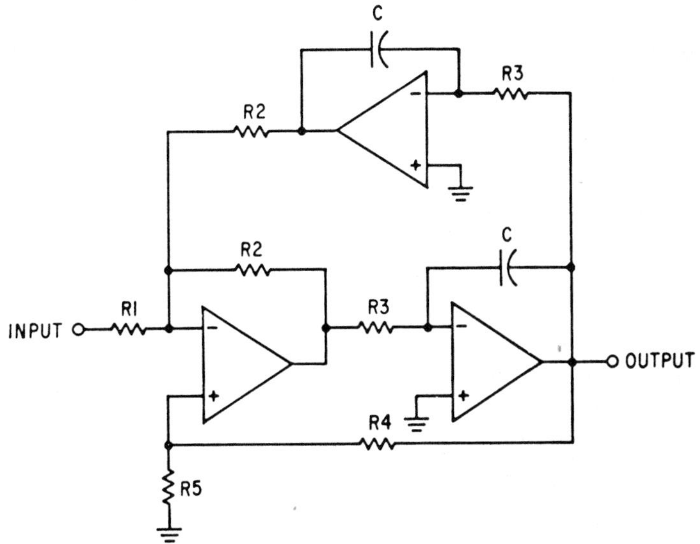

Fig. 9.2 State variable bandpass filter

```
(*=========================================*)
(*                                         *)
(* Title:   State Variable Bandpass        *)
(*          Filter                         *)
(* Program Summary: Program designs        *)
(*          a state variable bandpass      *)
(*          filter.                        *)
(*                                         *)
(*=========================================*)


PROGRAM StateVariBandpass;

uses Transcend;

const
  Pi = 3.1415927 ;

 VAR
  Answer,
  Ans,
  Cont            : Char;
  Out             : Text;
  Ft,
  Gain,
  Q,
  CentFreq,
  C,
  TempCap,
  X,
  Zz,
  V,
  Temp            : Real;
  Steps           : Integer;
  Ohms            : String;
  R               : Array [1..5] of Real;


Procedure Outputs;
  begin
    page(output);
     repeat
       gotoXY(Ø,5);
        Write('S(creen or P(rinter ? ');
```

```
        Readln(Answer);
     until Answer in [ 'S','s','p','P' ]
  end;  {outputs}

Procedure Title;
  begin
    page(output);
     gotoXY(05,1);
       Writeln('State Variable Bandpass Filter');
       Writeln;Writeln;
       Writeln('This program designs a state ');
       Writeln('variable active bandpass filter');
       Writeln('using an operational amplifier');
       Writeln('as the active element.');
       Writeln;
  end;  { title }

Procedure Input;
  begin
    Writeln;
    Write('Enter F(t) of the Op Amp (mHz) : ');
    Readln( Ft );
    Writeln;
    Write('Enter the desired gain (dB) : ');
    Readln( Gain );
    Writeln;
    Write('Enter desired Q : ');
    Readln( Q );
    Writeln;
    Write('Enter the center freq (Hz) : ');
    Readln( CentFreq );
    Writeln;
  end;

Procedure NewC;
  begin
    Write('Enter nearest standard value : ');
    Readln( C );
    TempCap := C;
    C  := C * 1E-9;
    V  := exp(( Gain / 20 ) * Ln ( 10 ));
    R[ 1 ] := R[ 2 ]  * Q / V ;
    R[ 3 ] := ( 1 - ( CentFreq / Ft )) / ( 2 * Pi * CentFreq * C );
    X  :=  2 * ( Q + V );
    ZZ := 1 + ( 2 * X ) / ( Ft / CentFreq );
    R[ 5 ] := R[ 4 ] / ( X / Zz - 1 );
  end;

Procedure Res;
  begin
    Ohms := ' kilohms';
    Temp := R[ Steps ];
    R[ Steps ] := Temp / 1000;
        if R[ Steps ] < 1 then  Ohms := ' ohms';
        if R[ Steps ] < 1 then R[ Steps ] := Temp;
  end;


Procedure PrintScr;
   begin
     Writeln;
     Writeln('The capacitors should have a value of');
     Writeln('about ',Round(C),' Nanofarads');
     Writeln;
     NewC;
     Writeln;
     For Steps := 1 to 5 do
         begin
            Res;
            Writeln('R',Steps,' := ',R[ Steps ],Ohms);
         end;
     For Steps := 1 to 2 do
         Writeln('C',Steps,' := ',TempCap,' Nanofarads');
   end;
```

```
Procedure PrintPrt;
begin
     Rewrite(out,'Printer:');
     Writeln;
     Writeln(out);
     Writeln;
     Writeln('The capacitors should have a value of');
     Writeln(out,'The capacitors should have a value of');
     Writeln('about ',Round(C),' Nanofarads');
     Writeln(out,'about ',Round(C),' Nanofarads');
     Writeln;
     Writeln(out);
     NewC;
     Writeln;
     For Steps := 1 to 5 do
         begin
            Res;
            Writeln('R',Steps,' := ',R[ Steps ],Ohms);
            Writeln(out,'R',Steps,' := ',R[ Steps ],Ohms);
         end;
     Writeln;
     Writeln(out);
     For Steps := 1 to 2 do
        begin
         Writeln('C',Steps,' := ',TempCap,' Nanofarads');
         Writeln(out,'C',Steps,' := ',TempCap,' Nanofarads')
        end;
     Close(out);
    end;

 Procedure Printout;
    begin
      case answer of
           's','S' : PrintScr;
           'p','P' : PrintPrt;
      end { case }
    end;

 Procedure Work;
  begin
     Ft := Ft * 1E6;
     R[ 2 ] := 1E5 ;
     R[ 4 ] := 1E5 ;
     C      := 1 / (6.2832 * CentFreq * 1E-4);
  end;

   begin { Main }
     Repeat
       Outputs;
       Title;
       Input;
       Work;
       Printout;
        Writeln;
             Write('Another set of data ?(Y/N) ');
                Readln(Cont);
     until Cont in [ 'N','n']
   end.
```

State Variable Bandpass Filter

This program designs a state
variable active bandpass filter
using an operational amplifier
as the active element.

Enter F(t) of the op amp (MHz) : 1

Enter the desired gain (dB) : 12

Enter desired Q : 2Ø

Enter the center freq (Hz) : 3ØØØ

The capacitor should have a value of
about 1 Nanofarads.

Enter nearest standard value : 1

R1 = 5.Ø2377E2 kilohms
R2 = 1.ØØØØØE2 kilohms
R3 = 5.28925E1 kilohms
R4 = 1.ØØØØØE2 kilohms
R5 = 2.759Ø6 kilohms

C1 = 1.ØØØØØ Nanofarads
C2 = 1.ØØØØØ Nanofarads

Another set of data ? (Y/N) N

9.3 MULTIPLE FEEDBACK BANDPASS FILTER

Where high circuit Qs are not essential (values of 10 or less), the multiple feedback bandpass filter should be considered. It requires the use of only one operational amplifier and only five other passive components.

Most of the time, it is possible to use the 741 op amp for this circuit. However, if the parameters entered indicate that an amplifier with an F(t) of more than 1 MHz is needed, the program will indicate that a 741 is not suitable and will state the required F(t).

The schematic for the circuit is shown in Fig. 9.3.

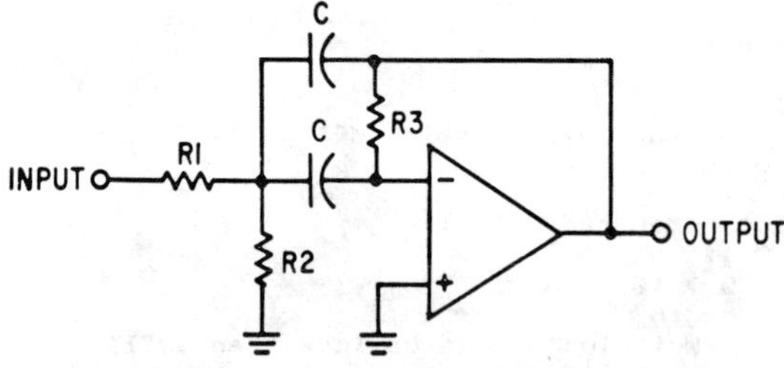

Fig. 9.3 Multiple feedback bandpass filter

```
(*=========================================*)
(*                                         *)
(* Title:    Multiple Feedback Bandpass    *)
(*           Filter                        *)
(* Program Summary: Program designs        *)
(*           a multiple feedback           *)
(*           bandpass filter.              *)
(*                                         *)
(*=========================================*)
```

```
PROGRAM MultiFeedback;

uses Transcend;

const
  Pi = 3.1415927 ;

 VAR
  Answer,
  Ans,
  Cont            : Char;
  Out             : Text;
  Gain,
  Q,
  TempR,
  Freq,
  X,
  Zz,
  Ft,
  C,
  TempC           : Real;
  R               : Array [1..3] of Real;
  ohms,
  Can             : String;
  Steps           : Integer;


Procedure Outputs;
  begin
    page(output);
     repeat
       gotoXY(Ø,5);
        Write('S(creen or P(rinter ? ');
        Readln(Answer);
     until Answer in [ 'S','s','p','P' ]
  end;  {outputs}

Procedure Title;
  begin
    page(output);
     gotoXY(Ø4,1);
       Writeln('Multiple Feedback Bandpass Filter');
       Writeln;Writeln;
       Writeln('This program designs a multiple');
       Writeln('feedback active bandpass filter.');
       Writeln('The "Q" of this filter is limited');
       Writeln('to 1Ø');
       Writeln;
  end;  { title }

Procedure Input;
  begin
    Writeln;
    Write('Enter the desired gain (dB) : ');
    Readln( Gain );
    Writeln;
    Write('Enter desired Q : ');
    Readln( Q );
           if Q > 1Ø then
              begin
                Writeln('Q must be less than 1Ø');
                Write('Reenter Q : ');
                Readln( Q )
              end;
    Writeln;
    Write('Enter the center freq (Hz) : ');
    Readln( Freq );
    Writeln;
  end;

Procedure NewC;
  begin
     Writeln;
     Write('Enter the nearest standard value : ');
     Readln( C );
     TempC := C ;
```

```
      C  := C * 1E-3 ;
      R[ 3 ] := Q / ( Pi * Freq * C * 1E-6 );
      R[ 1 ] := R [ 3 ] / ( 2 * exp(( Gain / 20 ) * Ln( 10 )));
      Zz := Sqr( 2 * Pi * Freq * C * 1E-6 );
      X  := R[ 3 ] - (1 / R[ 1 ] );
      R[ 2 ] := 1 / ( Zz * X )
   end;

 Procedure Res;
   begin
     ohms := ' kilohms';
     TempR := R[ Steps ] ;
     R[ Steps ] := R[ Steps ] / 1000 ;
           if R [ Steps ] < 1 then
                begin
                  ohms := ' ohms';
                  R[ Steps ] := TempR
                end;
   end;

 Procedure PrintScr;
    begin
      Writeln;
      Writeln('For a bandpass frequency of ',Freq,' hertz');
      Writeln('you need an op amp with a F(t)');
      Writeln('of at least ',Ft,' hertz');
      Writeln;
      Writeln('You ',Can,' use a 741 type of amp.');
      Writeln;
      Writeln('The capacitors should have a value of');
      Writeln('about ',Round(C),' nanofarads');
      NewC;
      Writeln;
      Writeln('The components values are: ');
      Writeln;
         for Steps := 1 to 3 do
             begin
               Res;
               Writeln('R',Steps,' := ',R[ Steps ],ohms);
             end;
      Writeln;
          for Steps := 1 to 2 do
               Writeln('C',Steps,' := ',TempC,' nanofarads');
     end;

  Procedure PrintAg;
   begin
         for Steps := 1 to 3 do
             begin
               Res;
               Writeln('R',Steps,' := ',R[ Steps ],ohms);
               Writeln(out,'R',Steps,' := ',R[ Steps ],ohms);
             end;
      Writeln;
      Writeln(out);
          for Steps := 1 to 2 do
               begin
                Writeln('C',Steps,' := ',TempC,' nanofarads');
                Writeln(out,'C',Steps,' := ',TempC,' nanofarads');
               end
  end;

 Procedure PrintPrt;
 begin
      Rewrite(out,'Printer:');
      Writeln;
      Writeln(out);
      Writeln('For a bandpass frequency of ',Freq,' hertz');
      Writeln(out,'For a bandpass frequency of ',Freq,' hertz');
      Writeln('you need an op amp with a F(t)');
      Writeln(out,'you need an op amp with a F(t)');
      Writeln('of at least ',Ft,' hertz');
      Writeln(out,'of at least ',Ft,' hertz');
      Writeln;
      Writeln(out);
      Writeln('You ',Can,' use a 741 type of amp.');
```

```
    Writeln(out,'You ',Can,' use a 741 type of amp.');
    Writeln;
    Writeln(out);
    Writeln('The capacitors should have a value of');
    Writeln('about ',Round(C),' nanofarads');
    Writeln(out,'The capacitors should have a value of');
    Writeln(out,'about ',Round(C),' nanofarads');
    NewC;
    Writeln;
    Writeln(out);
    Writeln('The components values are: ');
    Writeln(out,'The components values are: ');
    Writeln;
    Writeln(out);
    PrintAg;
    Close(out);
  end;

Procedure Printout;
  begin
    case answer of
         's','S' : PrintScr;
         'p','P' : PrintPrt;
    end { case }
  end;

Procedure Work;
 begin
  Ft := Freq * 100 ;
    Can := 'may' ;
     if Ft > 1E6 then
         Can := 'may not';
  C  := 1 / ( 6.2832 * Freq * 1E-5);
 end;

  begin { Main }
    Repeat
      Outputs;
      Title;
      Input;
      Work;
      Printout;
       Writeln;
            Write('Another set of data ?(Y/N) ');
                Readln(Cont);
    until Cont in [ 'N','n']
  end.
```

Multiple Feedback Bandpass Filter

This program designs a multiple
feedback active bandpass filter.
The "Q" of this filter is limited
to 10.

Enter the desired gain (dB) : 10

Enter desired Q : 8

Enter the center freq : 2000

For a bandpass frequency of 2.0000E3 hertz
you need an op amp with a F(t)
of at least 2.00000E5 hertz.

You may use a 741 type op amp.

The capacitors should have a value of
about 8 nanofarads.

Enter the nearest standard value : 8

The component values are:

R1 = 2.51646E1 kilohms
R2 = 6.21699E2 ohms
R3 = 1.59155E2 kilohms

C1 = 8.ØØØØØ nanofarads
C2 = 8.ØØØØØ nanofarads

Another set of data ? (Y/N) N

9.4 WIEN BRIDGE NOTCH FILTER[1]

In applications requiring deep notches and high Qs, a Wien bridge filter is a good choice. Its design is simple. Since it has few frequency-determining components, it is easy to adjust the desired notch frequency.

The Qs in the Wien bridge filter are higher than those for twin tee filters because there is no need for critical balancing between two parallel branch networks. Tuning is simple because there is no interaction between null- and frequency-determining components.

The program calculates component values and allows entry of the nearest standard value.

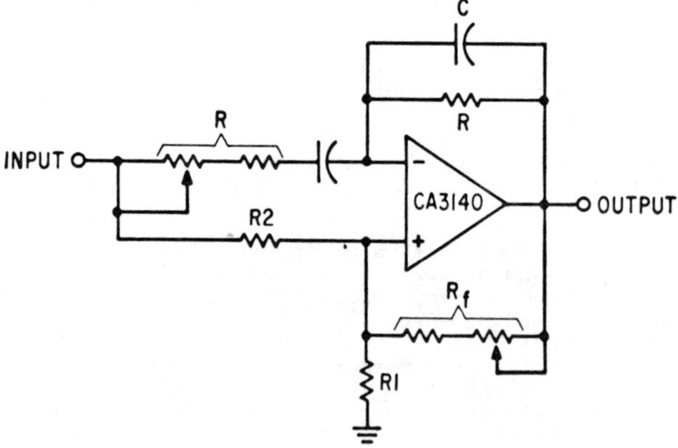

Fig. 9.4 Wien bridge notch filter

[1]G. Darilek and O. Tranbarger, "Try a Wien Bridge Network," *Electronic Design*, (Feb.1, 1978): p. 80.

```
(*======================================*)
(*                                      *)
(* Title:  Wien Bridge Notch Filter     *)
(* Program Summary: Program designs     *)
(*         a Wien bridge notch filter.  *)
(*                                      *)
(*======================================*)


PROGRAM WienBridge;

const
  Pi = 3.1415927 ;

 VAR
  Answer,
  Ans,
  Cont            : Char;
  Out             : Text;
  Freq,
  BandWidth,
  R1,
  R2,
  Rf,
  R,
  Fbw,
  C               : Real;

Procedure Outputs;
  begin
    page(output);
     repeat
       gotoXY(Ø,5);
        Write('S(creen or P(rinter ? ');
        Readln(Answer);
     until Answer in [ 'S','s','p','P' ]
  end;  {outputs}

Procedure Title;
  begin
    page(output);
     gotoXY(Ø4,1);
       Writeln('Wien Bridge Notch Filter Design');
       Writeln;Writeln;
       Writeln;
  end;  { title }

Procedure Input;
  begin
    Writeln;
    Write('Enter center frequency (Hz) : ');
    Readln( Freq );
    Writeln;
    Write('Enter bandwidth (Hz) : ');
    Readln( BandWidth );
    Writeln;
    Writeln;
    Writeln('You must enter a value for R1 in');
    Writeln('the range of 1ØØ kilohms to 1 megohm.');
    Writeln('Your particular application requires');
        Fbw := BandWidth / Freq ;
        if Fbw <= Ø.1 then
            Writeln('a value near the high end of the')
        else
            Writeln('a value near the low end of the');
    Writeln('resistance range.');
    Writeln;
    Write('Enter R1 (Kilohms) : ');
    Readln( R1 );
    Writeln;
  end;

Procedure NearestC;
  begin
     Writeln;
     Write('Enter the nearest standard value (nanofarads): ');
```

```
      Readln( C );
      C  := C * 1E-9 ;
      R2 := Fbw * R1 / ( 3 - Fbw );
      Rf := Fbw * R1 / ( 3 * ( 2 - Fbw ));
      R  := 1 / ( 2 * Pi * Freq * C );
      C  := C * 1E9 ;
      R1 := R1 / 1000 ;
      R2 := R2 / 1000 ;
      Rf := Rf / 1000 ;
      R  := R  / 1000 ;
   end;

 Procedure PrintScr;
    begin
      Writeln;
      Writeln('The value for "C" should be close to');
      Writeln(Round(C),' nanofarads.');
      NearestC;
      Writeln;
      Writeln('For a Wien Bridge Notch Filter with a');
      Writeln('center frequency of ',Freq,' hertz');
      Writeln('and a notch bandwidth of ',BandWidth,' hertz');
      Writeln('component values are as follows: ');
      Writeln;
      Writeln('R1 := ',R1,' kilohms');
      Writeln('R2 := ',R2,' kilohms');
      Writeln('Rf := ',Rf,' kilohms');
      Writeln('R  := ',R ,' kilohms');
      Writeln('C  := ',C ,' nanofarads');
      Writeln;
     end;

  Procedure PrintMore;
   begin
      Writeln('R1 := ',R1,' kilohms');
      Writeln(out,'R1 := ',R1,' kilohms');
      Writeln('R2 := ',R2,' kilohms');
      Writeln(out,'R2 := ',R2,' kilohms');
      Writeln('Rf := ',Rf,' kilohms');
      Writeln(out,'Rf := ',Rf,' kilohms');
      Writeln('R  := ',R ,' kilohms');
      Writeln(out,'R  := ',R ,' kilohms');
      Writeln('C  := ',C ,' nanofarads');
      Writeln(out,'C  := ',C ,' nanofarads')
   end;

 Procedure PrintPrt;
 begin
      Rewrite(out,'Printer:');
      Writeln;
      Writeln(out);
      Writeln('The value for "C" should be close to');
      Writeln(out,'The value for "C" should be close to');
      Writeln(Round(C),' nanofarads.');
      Writeln(out,Round(C),' nanofarads.');
      NearestC;
      Writeln;
      Writeln(out);
      Writeln('For a Wien Bridge Notch Filter with a');
      Writeln(out,'For a Wien Bridge Notch Filter with a');
      Writeln('center frequency of ',Freq,' hertz');
      Writeln(out,'center frequency of ',Freq,' hertz');
      Writeln('and a notch bandwidth of ',BandWidth,' hertz');
      Writeln(out,'and a notch bandwidth of ',BandWidth,' hertz');
      Writeln('component values are as follows: ');
      Writeln(out,'component values are as follows: ');
      Writeln;
      Writeln(out);
      PrintMore;
      Writeln;
      Writeln(out);
      Close(out);
     end;

  Procedure Printout;
     begin
```

```
      case answer of
           's','S' : PrintScr;
           'p','P' : PrintPrt;
      end { case }
    end;

Procedure Work;
 begin
   R1 := R1 * 1ØØØ ;
   C  := 1 / ( 2 * Pi * 1E4 * Freq );
   C  := C * 1E9
 end;

  begin { Main }
    Repeat
      Outputs;
      Title;
      Input;
      Work;
      Printout;
       Writeln;
            Write('Another set of data ?(Y/N) ');
               Readln(Cont);
    until Cont in [ 'N','n']
  end.
```

Wien Bridge Notch Filter Design

Enter center frequency (Hz) : 3ØØØ

Enter bandwidth (Hz) : 5ØØ

You must enter a value for R1 in
the range of 1ØØ kilohms to 1 megohm.
Your particular application requires
a value near the low end of the
resistance range.

Enter R1 (Kilohms): 1ØØ

The value for "C" should be close to
5 nanofarads.

Enter the nearest standard value (nanofarads): 5

For a Wien Bridge Notch Filter with a
center frequency of 3.ØØØØØE3 hertz
component values are as follows:

R1 = 1.ØØØØØE2 kilohms
R2 = 5.8235 kilohms
Rf = 3.Ø3Ø3Ø kilohms
R = 1.Ø61Ø3E1 kilohms
C = 5.ØØØØØ nanofarads

Another set of data ? (Y/N) N

9.5 TWIN TEE NOTCH FILTER

Twin tee filters are generally inconveniently difficult to adjust. Normal variable capacitors have values too low for use below 5 kHz, and fixed capacitors limit adjustment flexibility. The circuit in Fig. 9.5 overcomes these problems by using a capacitance multiplier in which the effective capacitance can be adjusted with a potentiometer.[1]

The circuit below can be adjusted to achieve a rejection of more than 70 dB. The operational amplifier used in this circuit should be of the FET input variety since high output offset voltage would result from input bias current in the feedback network. This in turn would decrease the amplifier's signal swing capability.

At high frequencies, C4 is needed to make the pass-frequency more uniform. Otherwise it can be eliminated.

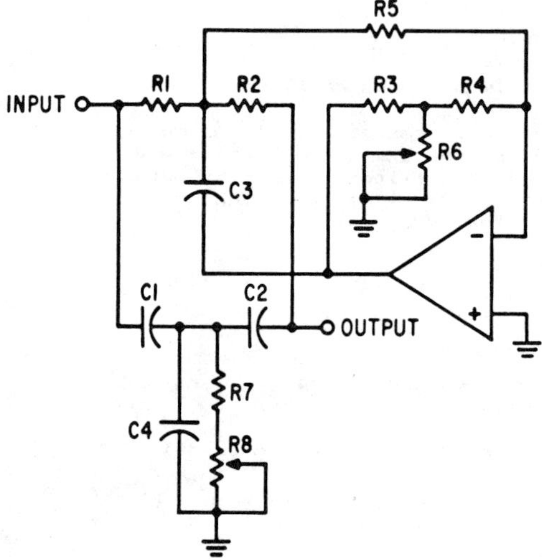

Fig. 9.5 Twin tee notch filter

```
(*=======================================*)
(*                                       *)
(* Title:   Twin Tee Notch Filter        *)
(* Program Summary: Program designs      *)
(*          a twin tee notch filter.     *)
(*                                       *)
(*=======================================*)


PROGRAM TwinTeeNotch ;

uses Transcend;

const
  Pi = 3.1415927 ;

 VAR
  Answer,
  Ans,
```

[1]J. Graeme, "Twin Tee Filter Rejects More Than 70 dB with Capacitance Multiplier Circuit," *Electronic Design*, (Jan. 18, 1978): p. 104.

```
  Cont          : Char;
  Out           : Text;
  Freq,
  C1,
  R1,
  Temp          : Real;
  R             : Array [1..8] of Real;
  C             : Array [1..4] of Real;
  Ohms          : String;
  Steps         : Integer;

Procedure Outputs;
  begin
    page(output);
     repeat
       gotoXY(0,5);
        Write('S(creen or P(rinter ? ');
        Readln(Answer);
     until Answer in [ 'S','s','p','P' ]
  end;  {outputs}

Procedure Title;
  begin
    page(output);
     gotoXY(10,1);
       Writeln('Twin Tee Notch Filter');
       Writeln;Writeln;
       Writeln('This program calculates the values');
       Writeln('of the components for a twin tee');
       Writeln('notch filter.  You must enter the');
       Writeln('frequency of the notch and a value');
       Writeln('for "C".');
       Writeln;
  end;  { title }

Procedure Input;
  begin
    Writeln;
    Write('Enter frequency (Hz): ');
    Readln( Freq );
    Writeln;
    Write('Enter capacitance (microfarads): ');
    Readln( C1 );
    Writeln;
  end;

Procedure Res;
  begin
    Ohms := ' kilohms';
    R1 := 1 / ( Pi * Freq * C1 * Sqrt ( 2 ));
    Temp := R1 ;
    C1 := C1 * 1E6 ;
    R1 := Temp / 1000 ;
       if R1 < 0 then
          begin
            R1 := Temp ;
            Ohms := ' ohms'
          end;
    R[1] := R1 ;
    R[2] := R1 ;
    R[3] :=  10 * R1 ;
    R[4] :=  10 * R1 ;
    R[5] := 100 * R1 ;
    R[6] := 1.5 * R1 ;
    R[7] := 0.2 * R1 ;
    R[8] := 0.1 * R1 ;
    C[1] := C1 ;
    C[2] := C1 ;
    C[3] := 0.7 * C1 ;
    C[4] := C1 / 100 ;
  end;

Procedure PrintScr;
   begin
     Writeln;
     Writeln('For a notch frequency of ',Freq,' hertz');
```

```
  Writeln('and a value "C" of ',C1,' microfarads');
  Writeln('the component values are:');
  Writeln;
        for Steps := 1 to 8 do
           begin
             Write('R',Steps,' := ',R[Steps],Ohms);
             if Steps < 5 then
               Writeln('   C',Steps,' := ',C[Steps],' mfd')
             else
               Writeln
           end;
 Writeln;
 Writeln('*** C4 is only needed for high frequencies');
 Writeln;
end;

Procedure PrintPrt;
begin
     Rewrite(out,'Printer:');
     Writeln;
     Writeln(out);
     Writeln('For a notch frequency of ',Freq,' hertz');
     Writeln(out,'For a notch frequency of ',Freq,' hertz');
     Writeln('and a value "C" of ',C1,' microfarads');
     Writeln(out,'and a value "C" of ',C1,' microfarads');
     Writeln('the component values are:');
     Writeln(out,'the component values are:');
     Writeln;
     Writeln(out);
            for Steps := 1 to 8 do
               begin
                 Write('R',Steps,' := ',R[Steps],Ohms);
                 Write(out,'R',Steps,' := ',R[Steps],Ohms);
                 if Steps < 5 then
                   begin
                     Writeln('  C',Steps,' := ',C[Steps],' mfd');
                     Writeln(out,'   C',Steps,' := ',C[Steps],' mfd')
                   end
                 else
                   begin
                     Writeln(out);
                     Writeln
                   end
               end;
     Writeln;
     Writeln(out);
     Writeln('*** C4 is only needed for high frequencies');
     Writeln(out,'*** C4 is only needed for high frequencies');
     Writeln;
     Writeln(out);
     Close(out);
    end;

 Procedure Printout;
    begin
      case answer of
           's','S' : PrintScr;
           'p','P' : PrintPrt;
      end { case }
    end;

 Procedure Work;
  begin
     C1  := C1 * 1E-6 ;
     Res;
  end;

   begin { Main }
     Repeat
       Outputs;
       Title;
       Input;
       Work;
       Printout;
        Writeln;
            Write('Another set of data ?(Y/N) ');
               Readln(Cont);
     until Cont in [ 'N','n']
   end.
```

```
Twin Tee Notch Filter

This program calculates the values
of the components for a twin-tee
notch filter. You must enter the
frequency of the notch and a value
for "C".

Enter frequency (Hz): 6Ø

Enter capacitance (microfarads): .1

For a notch frequency of 6.ØØØØØE1 hertz
and a value of "C" of 1.ØØØØØE-1 microfarads
the component values are:

R1 = 3.75132E1 kilohms      C1 = 1.ØØØØØE-1 mfd
R2 = 3.75132E1 kilohms      C2 = 1.ØØØØØE-1 mfd
R3 = 3.75132E2 kilohms      C3 = 7.ØØØØØE-2 mfd
R4 = 3.75132E2 kilohms      C4 = 1.ØØØØØE-3 mfd
R5 = 3.75132E3 kilohms
R6 = 5.62698 kilohms
R7 = 7.5Ø264 kilohms
R8 = 3.75132 kilohms

*** C4 is only needed for high frequencies.

Another set of data ? (Y/N) N
```

9.6 ACTIVE LOW-PASS FILTER NO.1

A two-pole active low-pass filter that is built with an FET input operational amplifier (like CA3130 or CA3140) can be designed for upper cutoff frequencies ranging from hundredths of hertz to tens of kilohertz.[1]

The circuit in Fig. 9.6 uses a single amplifier and four other passive components. The program is structured so that the user may enter in component values that are available and close to those desired. The program then uses these component values to calculate the others. As a result, the filter can be designed to operate with the components on hand.

If a value chosen is not of proper value, the program will instruct the user to select another value.

[1] K. Timonthy, "Design Active Low-Pass Filters," *Electronic Design*, (Sept.1, 1977): p. 144.

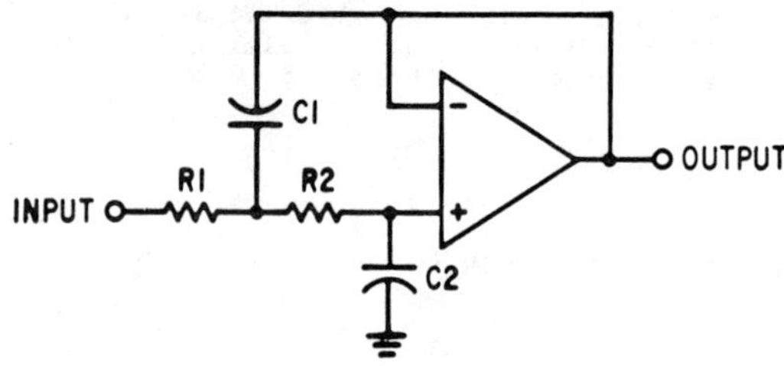

Fig. 9.6 Active low-pass filter No. 1 (FET input)

```
(*========================================*)
(*                                        *)
(* Title:   Active Low-Pass Filter No. 1  *)
(* Program Summary: Program designs       *)
(*          a two-pole active low-pass    *)
(*          filter for an FET input.      *)
(*                                        *)
(*========================================*)


PROGRAM ActiveLowPass;

uses Transcend;

const
  Pi = 3.1415927 ;

 VAR
  Answer,
  Ans,
  Cont            : Char;
  Out             : Text;
  Ic,
  Vc,
  Rmax,
  RR,
  Psi,
  W,
  F,
  Wo              : Real;
  C,
  R               : Array [1..2] of Real;
  I               : Integer;


Procedure Outputs;
  begin
    page(output);
     repeat
       gotoXY(Ø,5);
        Write('S(creen or P(rinter ? ');
        Read(Answer);
     until Answer in [ 'S','s','p','P' ]
  end;  {outputs}

Procedure Title;
  begin
    page(output);
     gotoXY(Ø7,1);
       Writeln('Active Low Pass Filter #1');
       Writeln;Writeln;
       Writeln('This program designs active low-pass');
       Writeln('filters, for best results use a Fet');
       Writeln('input Op Amp (e.g. CA 314Ø ),  You');
       Writeln('must enter the values for the damping ');
       Writeln('factor (Psi) and the ratio of the');
       Writeln('undamped natural frequency to the cutoff');
       Writeln('frequency (wo/wc).');
       Writeln;
```

```
          Writeln('Amplifier input current drift is');
          Writeln('assumed to be  10 na. resulting in a');
          Writeln('1 mV change in voltage.');
     end;  { title }

   Procedure Input;
     begin
       Writeln;
       Write('Do you want to change these values (Y/N)? ');
       Readln( Ans );
         if ans in [ 'N','n'] then
              begin
               Ic := 1E-8 ;
               Vc := 1E-3
              end
         else
          begin
            Writeln;
            Write('Enter current drift (na) : ');
            Readln( Ic );
            Write('Enter voltage drift (mV) : ');
            Readln( Vc );
            Ic    := Ic * 1E-9;
            Vc    := Vc * 1E-3;
         end;
      Writeln;
      Write('Enter Psi : ');
      Readln( Psi );
      Write('Enter Wo/Wc : ');
      Readln( W );
      Write('Enter cutoff frequency : ');
      Readln( F );
   end;

   Procedure Cap1;
     begin
       Write('Enter next highest standard value : ');
       Readln( C[1] ) ;
       Writeln;
       C[1]  := C[1] * 1E-9 ;
       C[2]  := Sqr( Psi ) * C[1] * 1E9
     end;

   Procedure Cap2;
     begin
      Write('Enter next lowest standard value : ');
      Readln( C[2] );
      Writeln;
      C[2]  := C[2] * 1E-9;
      R[1]  := ( Psi + Sqrt( Sqr( Psi ) - C[2] / C[1] )) / ( Wo * C[2] );
      R[2]  := 1 / (Sqr( Wo ) * R[1] * C[1] * C[2] );
      R[1]  := R[1] / 1000;
      R[2]  := R[2] / 1000;
     end;

   Procedure Res;
     begin
        Write('Enter closest standard value : ');
        Readln( R[ I ] );
        Writeln;
        R[I]   := R[I] * 1E3 ;
        C[I]   := C[I] * 1E9 ;
     end;

   Procedure PrintScr;
      begin
        Writeln;
        Writeln('For a ',F,' hertz low pass filter');
        Writeln('the component values are: ');
        Writeln;
        Writeln('R1 := ',R[1],' kilohms');
        Writeln('R2 := ',R[2],' kilohms');
        Writeln;
        Writeln('C1 := ',C[1],' nanofarads');
        Writeln('C2 := ',C[2],' nanofarads');
      end;
```

```
Procedure PrintPrt;
begin
     Rewrite(out,'Printer:');
     Writeln;
     Writeln(out);
     Writeln('For a ',F,' hertz low pass filter');
     Writeln(out,'For a ',F,' hertz low pass filter');
     Writeln('the component values are: ');
     Writeln(out,'the component values are: ');
     Writeln(out);
     Writeln;
     Writeln('R1 := ',R[1],' kilohms');
     Writeln(out,'R1 := ',R[1],' kilohms');
     Writeln('R2 := ',R[2],' kilohms');
     Writeln(out,'R2 := ',R[2],' kilohms');
     Writeln(out);
     Writeln;
     Writeln('C1 := ',C[1],' nanofarads');
     Writeln(out,'C1 := ',C[1],' nanofarads');
     Writeln('C2 := ',C[2],' nanofarads');
     Writeln(out,'C2 := ',C[2],' nanofarads');
     Writeln(out);
     Writeln;
     Close(out);
    end;

 Procedure Printout;
    begin
      case answer of
           's','S' : PrintScr;
           'p','P' : PrintPrt;
      end { case }
    end;

Procedure Compute;
 begin
    Cap1;
    Writeln('C2 should be about ',C[2],' nanofarads');
    Writeln('but not less than 1 nf.');
    Writeln;
    Cap2;
    Writeln;
        For I := 1 to 2 do
          begin
            Writeln('R',I,' should be about ',R[I],' kilohms');
            Res
          end;
   RR  := R[1] + R[2] ;
       if RR > Rmax then
          Begin
            Writeln('C1 is too small, enter next largest');
            Writeln('standard value : ');
            Compute
          end;
 end;

 Procedure Work;
  begin
    Rmax := Vc / Ic ;
    Wo   := 2 * Pi * W * F ;
    C[1] := 2 / ( Psi * Wo * Rmax ) * 1E9;
    Writeln;
    Writeln('C1 should be about ',C[1],' nanofarads');
    Writeln;
    Compute;
      R[1] := R[1] / 1000;
      R[2] := R[2] / 1000;
 end;

   begin { Main }
     Repeat
       Outputs;
       Title;
       Input;
       Work;
       Printout;
```

```
         Writeln;
              Write('Another set of data ?(Y/N) ');
                 Readln(Cont);
      until Cont in [ 'N','n']
    end.
```

Active Low-Pass Filter #1

This program designs active low-pass
filters. For best results use an Fet
input op amp (e.g., CA 314Ø). You
must enter the values for the damping
factor (Psi) and the ratio of the
undamped natural frequency to the cutoff
frequency (wo/wc).

Amplifier input current drift is
assumed to be 1Ø na., resulting in a
1 mv change in voltage.

Do you want to change these values? (Y/N) N

Enter Psi : .866

Enter wo/wc : 1.732

Enter cutoff frequency : 1ØØ

C1 should be about 2.12219E1 nanofarads.

Enter next highest standard value : 27

C2 should be about 2.Ø2488E1 nanofarads
but not less than 1 nf.

Enter next lowest standard value : 2Ø

R1 should be about 4.41993E1 kilohms

Enter closest standard value : 43

R2 should be about 3.53782E1 kilohms

Enter closest standard value : 33

For a 1.ØØØØØE2 hertz low pass filter
the component values are:

R1 = 4.3ØØØØ E1 kilohms
R2 =3.3ØØØØØE1 kilohms

```
C1 = 2.700000E1 nanofarads
C2 = 2.000000E1 nanofarads

Another set of data ? (Y/N) N
```

9.7 ACTIVE LOW-PASS FILTER NO.2[1]

The low-pass filter in Fig. 9.7 is very similar to the previous one with two exceptions. It uses a conventional operational amplifier instead of an FET input and has a feedback network connected to the output. This network is designed to be implemented as a 100-kilohm potentiometer.

The program contains a data section that in *Procedure Cap* fills the array of C1 with nine standard capacitor values. This makes it possible to produce nine different circuit designs for the same parameters so that the most convenient set of components can be chosen.

The program requests a value for the damping ratio. A value of 1.41 will result in an overdamped design to ensure that the phase shift of the circuit never exceeds 170°C. This results in a stable filter design.

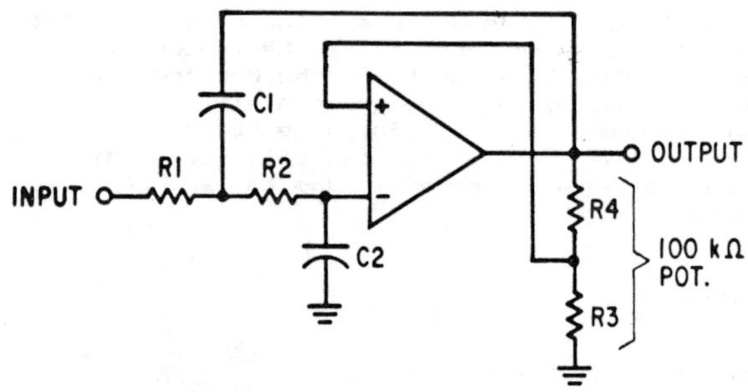

Fig. 9.7 Active low-pass filter No. 2

```
(*=======================================*)
(*                                       *)
(* Title:  Active Low-Pass Filter No. 2  *)
(* Program Summary: Program designs      *)
(*         an active low-pass filter     *)
(*         for an Op Amp input.          *)
(*                                       *)
(*=======================================*)


PROGRAM ActiveLowPass;

 VAR
  Answer,
  Ans,
  Cont          : Char;
  Out           : Text;
  Freq,
  Gain,
  Psi,
  Gain2,
```

[1]Adapted from H. Minuskin, "Active Filter Design Uses BASIC Language," *Electronic Design*, (March 1, 1970): p. 83.

```
    R1,R2,R3,R4,
    TempC,
    C2,
    K              : Real;
    C1             : Array [1..9] of real;
    Steps              : Integer;
    ohms1,
    ohms2,
    ohms34,
    caps1,
    caps2             : String;

  Procedure Work;
     Forward;

  Procedure Outputs;
    begin
      page(output);
       repeat
         gotoXY(0,5);
          Write('S(creen or P(rinter ? ');
          Read(Answer);
       until Answer in [ 'S','s','p','P' ]
    end;  {outputs}

  Procedure Title;
    begin
      page(output);
       gotoXY(10,1);
         Writeln('Active Low Pass Filter #2');
         Writeln;Writeln;
         Writeln('This program designs a low-pass active');
         Writeln('filter. You must enter the frequency');
         Writeln('of the 3 dB point and the desired ');
         Writeln('filter gain.  The program will supply ');
         Writeln('component values for nine standard');
         Writeln('values of C1.  You can select the most');
         Writeln('convenient set of values to use.');
         Writeln;
       end;

  Procedure Input;
    begin
      Writeln;
      Write('Enter frequency (Hz) : ');
      Readln( Freq );
      Writeln;
      Write('Enter Gain : ');
      Readln( Gain );
      Writeln;
      Write('Enter damping ratio ( overdamped := 1.41 ) : ');
      Readln( Psi );
  end;

  Procedure Cap;
    begin
        C1[1] := 1E-9 ; C1[2] :=   1E-8 ; C1[3] := 2.2E-8 ; C1[4] := 4.7E-8;
        C1[5] := 1E-7 ; C1[6] := 2.2E-7 ; C1[7] := 4.7E-7 ; C1[8] :=   1E-6;
        C1[9] := 1E-5 ;
    end;

  Procedure Res;
    begin
     R1  := 2 / ( Psi * K * 1000 );
       if R1 < 1 then
         begin
          ohms1 := ' ohms';
          R1 := 2 / ( Psi * K )
        end;
    R2  := Psi / ( 2 * Gain2 * K * 1000 );
       if R2 < 1 then
         begin
           ohms2 := ' ohms';
           R2 := Psi / ( 2 * Gain2 * K )
         end;
```

```
  R3   := 100 / Gain ;
       if R3 < 1 then
         begin
           ohms34 := ' ohms';
           R3 := 1E5 / Gain
         end;
  R4   :=  R3 * ( Gain - 1 );
end;

Procedure PrintScr;
   begin
     Writeln;
     Writeln('C1 := ',C1[Steps],caps1);
     Writeln('C2 := ',C2,caps2);
     Writeln;
     Writeln('R1 := ',R1,ohms1);
     Writeln('R2 := ',R2,ohms2);
     Writeln('R3 := ',R3,ohms34);
     Writeln('R4 := ',R4,ohms34);
     Writeln;
  end;

Procedure PrintPrt;
begin
     Rewrite(out,'Printer:');
     Writeln;
     Writeln(out);
     Writeln('C1 := ';C1[Steps],caps1);
     Writeln(out,'C1 := ',C1[Steps],caps1);
     Writeln('C2 := ',C2,caps2);
     Writeln(out,'C2 := ',C2,caps2);
     Writeln;
     Writeln(out);
     Writeln('R1 := ',R1,ohms1);
     Writeln(out,'R1 := ',R1,ohms1);
     Writeln('R2 := ',R2,ohms2);
     Writeln(out,'R2 := ',R2,ohms2);
     Writeln('R3 := ',R3,ohms34);
     Writeln(out,'R3 := ',R3,ohms34);
     Writeln('R4 := ',R4,ohms34);
     Writeln(out,'R4 := ',R4,ohms34);
     Writeln;
     Writeln(out);
     Close(out);
    end;

 Procedure Printout;
    begin
      case answer of
           's','S' : PrintScr;
           'p','P' : PrintPrt;
      end { case }
    end;

 Procedure Work;
  begin
     for Steps := 1 to 9 do
       begin
        Cap;
        ohms1  :=' kilohms';
        ohms2  := ohms1;
        ohms34 := ohms1;
        caps1  :=' microfarads';
        caps2  :=' nanofarads';
        K      := C1[Steps] * 6.28 * Freq ;
        Gain2  := Gain;
        TempC  := C1[Steps] * 1E6 ;
        C1[Steps] := TempC ;
             if C1[Steps] < 1 then
               begin
                 caps1 := ' nanofarads';
                 C1[Steps] := TempC * 1E3
               end;
        C2    := Gain2 * C1[Steps] ;
             if (C2 >= 1000) or ( caps1 =' microfarads') then
                     caps2 := ' microfarads';
```

```
          if C2 >= 1000 then C2 := C2 / 1000 ;
          if (C2 >= 1000) and ( caps1 = ' microfarads') then
                   C2 := C2 * 1000 ;
     Res;

     Printout;
          If Steps = 9 then Exit ( Work );
     Write('Another set of values (Y/N) ?');
     Readln( Ans );
          if Ans in [ 'N' , 'n' ] then Exit ( Work );
     Writeln;
    end;
 end;

begin { Main }
  Repeat
    Outputs;
    Title;
    Input;
    Work;
     Writeln;
          Write('Another set of data ?(Y/N) ');
             Readln(Cont);
  until Cont in [ 'N','n']
end.
```

Active Low-Pass Filter #2

This program designs a low-pass active filter. You must enter the frequency of the 3 dB point and the desired filter gain. The program will supply component values for nine standard values of C1. You can select the most convenient set of values to use.

Enter frequency (Hz) : 100

Enter gain : 100

Enter damping ratio (overdamped = 1.41) : 1.41

C1 = 1.00000 nanofarads
C2 = 1.00000E2 nanofarads

R1 = 2.25866E3 kilohms
R2 = 1.12261E1 kilohms
R3 = 1.00000 kilohms
R4 = 9.90000E1 kilohms

Another set of values? (Y/N) Y

C1 = 1.00000E1 nanofarads
C2 = 1.00000 microfarads

R1 = 2.25866E2 kilohms
R2 = 1.12261 kilohms

R3 = 1.00000 kilohms
R4 = 9.90000E1 kilohms

Another set of values? (Y/N) N

Another set of data ? (Y/N) N

9.8 BUTTERWORTH-THOMPSON LOW-PASS FILTER

The active low-pass transitional Butterworth-Thompson filter in Fig. 9.8 actually includes two designs in one. By changing the value for μ, the type of filter designed is changed. For a $\mu = 0$, the filter design is that of a Butterworth filter. For a $\mu = 1$, it is that of a linear-phase Thompson filter. If μ is somewhere in between, the design is a compromise between the flat amplitude response of the Butterworth filter and the flat time delay of the Thompson filter.

The program permits the user to change any or all of the component values to see what effect the change will have on the circuit design.

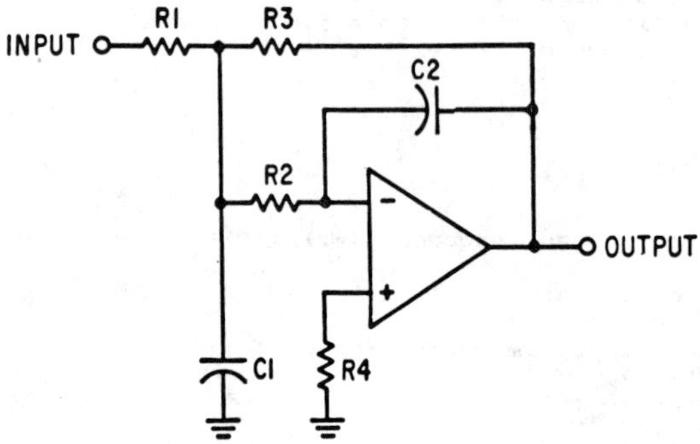

Fig. 9.8 Butterworth-Thompson low-pass filter

```
(*=========================================*)
(*                                         *)
(* Title:   Butterworth-Thompson           *)
(*          Low-Pass Filter                *)
(* Program Summary: Program designs        *)
(*          a Butterworth-Thompson         *)
(*          active low-pass filter.        *)
(*                                         *)
(*=========================================*)


PROGRAM ButterworthThompson;

uses Transcend;

 VAR
  Answer,
  Ans,
  Cont             : Char;
  Out              : Text;
  TempF,
  Freq,
  Gain,
  Mu,
  R1,
  Cap,
  C1,
```

```
  C2            : Real;
  Capsize       : String;
  R             : Array [1..4] of real;
  ResSize       : Array [1..4] of String;
  Steps         : Integer;

Procedure Outputs;
  begin
    page(output);
     repeat
       gotoXY(Ø,5);
        Write('S(creen or P(rinter ? ');
        Read(Answer);
     until Answer in [ 'S','s','p','P' ]
  end;  {outputs}

Procedure Title;
  begin
    page(output);
     gotoXY(Ø2,1);
       Writeln('Butterworth-Thompson low-pass filter');
       Writeln;Writeln;
       Writeln('This program designs a low-pass ');
       Writeln('transitional Butterworth-Thompson ');
       Writeln('filter.  You must enter the corner');
       Writeln('frequency, dc gain, input impedance');
       Writeln('and mu.  Mu varies between Ø and 1.');
       Writeln('When Mu := Ø it corresponds to a ');
       Writeln('Butterworth filter.  When Mu := 1');
       Writeln('it corresponds to a linear phase');
       Writeln('Thompson filter.');
       Writeln;
     end;

Procedure Input;
  begin
    Writeln;
    Write('Enter corner frequency (Hz) : ');
    Readln( Freq );
    Write('Enter dc gain : ');
    Readln( Gain );
    Write('Enter input impedance (ohms) : ');
    Readln( R[1] );
    Write('Enter Mu : ');
    Readln( Mu );
    Writeln;
  end;

Procedure Res;
  begin
     for Steps:= 1 to 4 do
       begin
          ResSize[Steps] := ' ohms';
         if R[Steps] > 1E3 then
           begin
            R[Steps] := R[Steps] / 1E3 ;
            ResSize[Steps] := ' kilohms'
           end;
         if R[Steps] >1E6 then
           begin
            R[Steps] := R[Steps] / 1E6;
            ResSize[Steps] := ' megohms';
           end
      end
 end;

Procedure PrintScr;
   begin
     For Steps := 1 to 4 do
           Writeln('R',Steps,' := ',R[Steps] ,ResSize[Steps]);
     Writeln('C1 := ',C1,' nanofarads');
     Writeln('C2 := ',C2,Capsize);
     Writeln;
     Writeln('The gain is ',Gain);
     Writeln('The actual corner frequency is');
     Writeln(Freq,' hertz');
```

```
      Writeln('The real value of Mu is ',Mu);
      Writeln;
   end;

 Procedure PrintPrt;
 begin
      Rewrite(out,'Printer:');
      Writeln;
      Writeln(out);
      For Steps := 1 to 4 do
              begin
                 Writeln('R',Steps,' := ',R[Steps],ResSize[Steps] );
                 Writeln(out,'R',Steps,' := ',R[Steps],ResSize[Steps] );
              end;
      Writeln('C1 := ',C1,' nanofarads');
      Writeln(out,'C1 := ',C1,' nanofarads');
      Writeln('C2 := ',C2,Capsize);
      Writeln(out,'C2 := ',C2,Capsize);
      Writeln;
      Writeln('The gain is ',Gain);
      Writeln(out,'The gain is ',Gain);
      Writeln('The actual corner frequency is');
      Writeln(out,'The actual corner frequency is');
      Writeln(Freq,' hertz');
      Writeln(out,Freq,' hertz');
      Writeln('The real value of Mu is ',Mu);
      Writeln(out,'The real value of Mu is ',Mu);
      Writeln;
      Writeln;
      Writeln(out);
      Close(out);
     end;

  Procedure Printout;
     begin
       case answer of
            's','S' : PrintScr;
            'p','P' : PrintPrt;
       end { case }
     end;

  Procedure Calc;
    begin
      Mu  := (( R[2] * C2 / R[3] / C1 ) * (Sqr( 1 + Gain + R[3] / R[2]))-2);
      C1 := C1 * 1E9;
      C2 := C2 * 1E9;
      Res;
    end;

 Procedure Different;
   begin
     Write('Do you want to change component values?(Y/N) ');
     Readln( Ans );
     if Ans in [ 'N' , 'n' ] then Exit ( Different );
     Writeln;
     Writeln('Enter resistances in kilohms and');
     Writeln('capacitances in nanofarads');
     Writeln;
         For Steps := 1 to 4  do
             begin
               Write('Enter R',Steps,' := ');
               Readln( R[Steps] );
               R[Steps] := 1E3 * R[Steps]
             end;
    Writeln;
    Write('Enter C1 := ');
    Readln( C1 );
    C1 := C1 * 1E-9 ;
    Write('Enter C2 := ');
    Readln( C2 );
    C2 := C2 * 1E-9;
    Writeln;
    Write('Enter gain := ');
    Readln( Gain );
    Calc;
    Printout;
    Different;
   end;
```

```
Procedure Work;
 begin
    TempF := 6.2832 * Freq ;
    R[3] := Gain * R[1] ;
    R1 := R[1] * R[3] /( R[1] + R[3] );
    Cap  := ( 1 + Gain ) / TempF / R[3] ;
    Writeln('C1 should be larger than ',Cap * 1E9 );
    Write('nanofarads.  Enter C1 : ');
    Readln( C1 );
    Writeln;
    C1 := C1 * 1E-9 ;
    R[2] := R1 / ( TempF * R1 * C1 * Sqrt ( 2 + Mu ) - 1 );
    Capsize :=' nanofarads';
    C2  := 1 / ( R[2] * R[3] * C1 * Sqr( TempF ));
    C2  := C2 * 1E9 ;
    Writeln('C2 should be about ',C2,Capsize);
    Write('Enter nearest standard value : ');
    Readln( C2 );
    C2 := C2 * 1E-9;
    R[4] := R[2] + R1 ;
    Freq  := ( 1 / Sqrt ( R[2] * R[3] * C1 * C2 ) / 6.2832 );
    Calc;
    Printout;
    Different;
   end;

  begin { Main }
    Repeat
      Outputs;
      Title;
      Input;
      Work;
       Writeln;
            Write('Another set of data ?(Y/N) ');
               Readln(Cont);
    until Cont in [ 'N','n']
  end.
```

Butterworth-Thompson Low-Pass Filter

This program designs a low-pass transitional Butterworth-Thompson filter. You must enter the corner frequency, dc gain, input impedance, and Mu. Mu varies between Ø and 1. When Mu = Ø it corresponds to a Butterworth filter. When Mu = 1 it corresponds to a linear phase Thompson filter.

Enter corner frequency (Hz) : 1ØØ

Enter dc gain : 1Ø

Enter input impedance (ohms) : 1ØØØØ

Enter Mu : .5

C1 should be larger than 1.75Ø7ØE2
nanofarads. Enter C1 : 18Ø

```
C2 should be about 9.68499 nanofarads
Enter nearest standard value : 1Ø

R1 = 1.ØØØØØE1 kilohms
R2 = 1.453ØØE1 kilohms
R3 = 1.ØØØØØE2 kilohms
R4 = 2.362Ø9E1 kilohms

C1 = 1.8ØØØØE2 nanofarads
C2 = 1.ØØØØØE1 nanofarads

The gain is 1.ØØØØØE1
The actual corner frequency is
9.84124E1 hertz
The real value of Mu is 5.81313E-1

Another set of data ? (Y/N) N
```

10 COMMUNICATIONS

The programs in this chapter will be helpful in designing communication circuits. In program 10.1, the computer is used to calculate the bandwidth required in a communications channel so that a digital pulse of a specified shape, amplitude, and width can pass.

In programs 10.2 and 10.3, microwave circuit elements, microstrips, and strip-lines are designed. The fourth program in this section greatly simplifies transmission line calculations. Even with the help of the well known Smith chart, the calculations are generally quite cumbersome. With this program, however, they are reduced to trivial exercises. The program calculates the impedances of a line, as well as its normalized impedance. In addition, it calculates the SWR of the line referenced to a particular impedance.

Programs 10.5 and 10.6 calculate the impedance of a line as well as its linear resistance, inductance, and capacitance, given the wire's diameter, length, and frequency at which it will be used. Still in the field of RF, program 10.7 determines the diameter of wire and coil required to produce an inductance of specific value. It also calculates the number of turns required. Programs 10.8 and 10.9 calculate the voltage standing wave ratio and transmission voltage, current, and power ratio to decibels.

10.1 GAUSSIAN PULSE BANDWIDTH CALCULATION

When digital pulses are sent over communications channels, the sharpness of the pulse is determined by the bandwidth of the channel.[1]

For mathematical analysis, it is often possible to assume that the received pulse has a Gaussian wave shape. With this assumption, it is easy to calculate the bandwidth required to pass a pulse with a specified amplitude and pulse width.

[1]F.E. Noel and J.S. Kolodzey, "Nomograph Shows Bandwidth for Specified Pulse Shape," *Electronics*, (April 1, 1976): p. 102.

The bandwidth calculated by this program contains over 95 percent of the pulse power. To use the program, the pulse width is in nanoseconds and the normalized amplitude must be entered. The frequency is rounded off to the nearest megahertz.

```
(*=========================================*)
(*                                         *)
(* Title:   Gaussian Pulse Bandwidth       *)
(*          Calculation                    *)
(* Program Summary: Program determines     *)
(*          the bandwidth needed by        *)
(*          a system to pass a pulse       *)
(*          of a given amplitude and       *)
(*          width.                         *)
(*                                         *)
(*=========================================*)


PROGRAM PulseCalculations;

uses Transcend;

 VAR
  Answer,
  Cont            : Char;
  Out             : Text;
  SlotTime,
  Amplitude,
  Time,
  Freq            : Real;

Procedure Outputs;
  begin
    page(output);
     repeat
       gotoXY(Ø,5);
        Write('S(creen or P(rinter ? ');
        Read(Answer);
     until Answer in [ 'S','s','p','P' ]
  end;  {outputs}

Procedure Title;
  begin
    page(output);
     gotoXY(Ø5,1);
       Writeln('Bandwidth for a Gaussian pulse');
       Writeln;Writeln;
       Writeln('This program determines the bandwidth');
       Writeln('required to pass a Gaussian shaped pulse');
       Writeln('in a system with a given time slot (T)');
       Writeln('and a normalized pulse amplitude (A).');
       Writeln;
       Writeln('You must enter the time slot and ');
       Writeln('the normalized amplitude desired at the');
       Writeln('edge of the slot.');
       Writeln;
     end;

Procedure Input;
  begin
    Writeln;
    Write('Enter time slot (nanoseconds) : ');
    Readln( SlotTime );
    Writeln;
    Write('Enter amplitude (normalized) : ');
    Readln( Amplitude );
    Writeln;
  end;

Procedure PrintScr;
   begin
     Writeln;
     Writeln('The bandwidth required to pass a pulse');
```

```
     Writeln('with a time slot of ',SlotTime,' nanoseconds');
     Writeln('and a normalized amplitude of ',Amplitude);
     Writeln('at the edge of the slot is :');
     Writeln;
     Writeln(Freq,' megahertz');
     Writeln;
  end;

Procedure PrintPrt;
begin
     Rewrite(out,'Printer:');
     Writeln;
     Writeln(out);
     Writeln('The bandwidth required to pass a pulse');
     Writeln(out,'The bandwidth required to pass a pulse');
     Writeln('with a time slot of ',SlotTime,' nanoseconds');
     Writeln(out,'with a time slot of ',SlotTime,' nanoseconds');
     Writeln('and a normalized amplitude of ',Amplitude);
     Writeln(out,'and a normalized amplitude of ',Amplitude);
     Writeln('at the edge of the slot is :');
     Writeln(out,'at the edge of the slot is :');
     Writeln;
     Writeln(out);
     Writeln(Freq,' megahertz');
     Writeln(out,Freq,' megahertz');
     Writeln;
     Writeln(out);
     Close(out);
    end;

 Procedure Printout;
    begin
      case answer of
           's','S' : PrintScr;
           'p','P' : PrintPrt;
      end { case }
    end;

 Procedure Work;
  begin
    Time  := SlotTime * 1E-9;
    Freq  := (0.6366198 / Time ) * Sqrt( 2 * Ln ( 1 / Amplitude ));
    Freq  := Freq / 1E6 ;
    end;

   begin { Main }
     Repeat
       Outputs;
       Title;
       Input;
       Work;
       Printout;
        Writeln;
             Write('Another set of data ?(Y/N) ');
                Readln(Cont);
     until Cont in [ 'N','n']
   end.
```

Bandwidth for a Gaussian Pulse

This program determines the bandwidth required to pass a Gaussian-shaped pulse in a system with a given time slot (T) and a normalized pulse amplitude (A).

You must enter the time slot and the normalized amplitude desired at the edge of the slot.

Enter time slot (nanoseconds) : 10

Enter amplitude (normalized) : .135

The bandwidth required to pass a pulse
with a time slot of 1.00000E1 nanoseconds
and a normalized amplitude of 1.35000E-1
at the edge of the slot is:

1.27403E2 megahertz

Another set of data ? (Y/N) N

10.2 MICROSTRIP DESIGN

A microstrip is a low-loss transmission line used in microwave circuits. It consists of a conductor above the ground plane (Fig. 10.2) and is analogous to a two-wire line in which one of the wires is represented by the ground plane.

When the characteristic impedance of the line, the substrate thickness, and the conductor thickness are entered, along with the dielectric constant of the material separating the conductor, the program will determine the width of the microstrip and the velocity factor of the line.

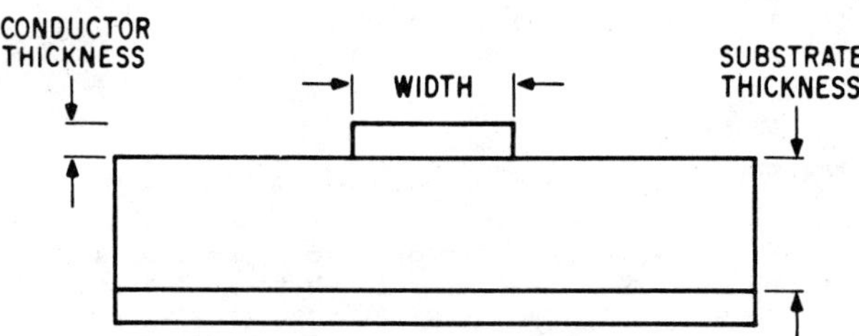

Fig. 10.2 Microstrip cross section

```
(*=========================================*)
(*                                         *)
(* Title:   Microstrip Design              *)
(* Program Summary: Program calculates     *)
(*          the dimensions needed          *)
(*          to produce a microstrip        *)
(*          circuit.                       *)
(*                                         *)
(*=========================================*)


PROGRAM MicroStrip;

uses Transcend;

 VAR
  Answer,
  Cont           : Char;
  Out            : Text;
  Zo,
  Substrate,
  Conductor,
  Dielectric,
```

```
    Temp,
    Width,
    Velocity       : Real;

  Procedure Outputs;
    begin
      page(output);
       repeat
         gotoXY(Ø,5);
          Write('S(creen or P(rinter ? ');
          Read(Answer);
       until Answer in [ 'S','s','p','P' ]
    end;  {outputs}

  Procedure Title;
    begin
      page(output);
       gotoXY(11,1);
         Writeln('Microstrip Design');
         Writeln;Writeln;
         Writeln('This program will calculate the');
         Writeln('dimensions needed to produce a microstrip');
         Writeln('with the desired characteristics.');
         Writeln('You must enter the :');
         Writeln;
       end;

  Procedure Input;
    begin
      Writeln;
      Write('Line impedance (Zo) : ');
      Readln ( Zo );
      Write('Substrate Thickness (inch) : ');
      Readln( Substrate );
      Write('Conductor Thickness (inch) : ');
      Readln( Conductor );
      Write('Dielectric Constant : ');
      Readln( Dielectric );
      Writeln;
    end;

  Procedure PrintScr;
     begin
       Writeln;
       Writeln('For a line impedance of ',Zo,' ohms, a');
       Writeln('substrate thickness of ',Substrate,' inch, a');
       Writeln('conductor thickness of ',Conductor,' inch,');
       Writeln('and a dielectric constant of ',Dielectric);
       Writeln;
       Writeln('a Microstrip with a width ');
       Writeln('of ',Width,' inch is required');
       Writeln;
       Writeln('The velocity factor is ',Velocity);
       Writeln;
    end;

  Procedure PrintPrt;
  begin
       Rewrite(out,'Printer:');
       Writeln;
       Writeln(out);
       Writeln('For a line impedance of ',Zo,' ohms, a');
       Writeln(out,'For a line impedance of ',Zo,' ohms, a');
       Writeln('substrate thickness of ',Substrate,' inch, a');
       Writeln(out,'substrate thickness of ',Substrate,' inch, a');
       Writeln('conductor thickness of ',Conductor,' inch,');
       Writeln(out,'conductor thickness of ',Conductor,' inch,');
       Writeln('and a dielectric constant of ',Dielectric);
       Writeln(out,'and a dielectric constant of ',Dielectric);
       Writeln;
       Writeln(out);
       Writeln('a Microstrip with a width ');
       Writeln(out,'a Microstrip with a width ');
       Writeln('of ',Width,' inch is required');
       Writeln(out,'of ',Width,' inch is required');
       Writeln;
```

```
    Writeln(out);
    Writeln('The velocity factor is ',Velocity);
    Writeln(out,'The velocity factor is ',Velocity);
    Writeln;
    Writeln(out);
    Close(out);
   end;

Procedure Printout;
   begin
     case answer of
          's','S' : PrintScr;
          'p','P' : PrintPrt;
     end { case }
   end;

Procedure Work;
 begin
    Temp       := exp ( Zo * Sqrt( Dielectric + 1.41 ) / 87 );
    Width      := 1.25 * ( 5.98 * Substrate / Temp - Conductor );
    Velocity   := 1 / Sqrt( Ø.475 * Dielectric + Ø.67 );
   end;

  begin { Main }
    Repeat
      Outputs;
      Title;
      Input;
      Work;
      Printout;
       Writeln;
             Write('Another set of data ?(Y/N) ');
                 Readln(Cont);
    until Cont in [ 'N','n']
  end.
```

Microstrip Design

This program will calculate the
dimensions needed to produce a microstrip
with the desired characteristics.

You must enter the :

Line impedance (ZO) : 7Ø
Substrate thickness (inch) : .Ø5
Conductor thickness (inch) : .ØØ1
Dielectric constant : 4.7

For a line impedance of 7.ØØØØØE1 ohms, a
substrate thickness of 5.ØØØØE-2 inch, a
conductor thickness of 1.0000E-3 inch,
and a dielectric constant of 4.7ØØØØ
a Microstrip with a width
of 4.98993E-2 inch is required.

The velocity factor is 5.86967E-1

Another set of data ? (Y/N) N

10.3 STRIP-LINE DESIGN

Like the microstrip, a strip line is also a low-loss transmission line used in microwave circuits. It differs from the microstrip, however, in that it has a second ground plane placed above the conductor (Fig. 10.3).

To use the program, values must be entered for the desired characteristic impedance of the line, the substrate thickness, the conductor thickness, and the dielectric constant of the material separating the conductors. The units used for the thickness are not important as long as they are consistent.

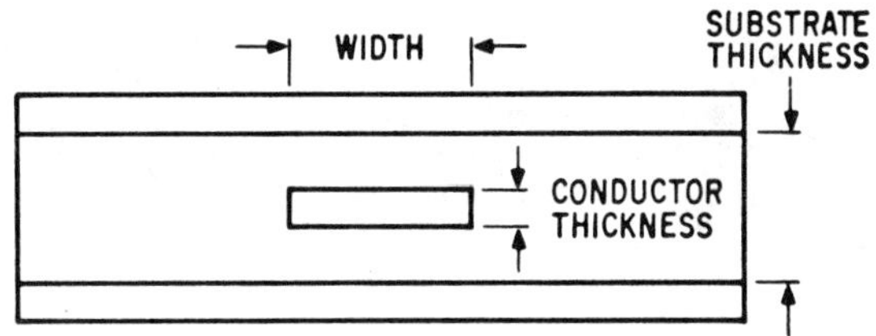

Fig. 10.3 Microstrip cross section

```
(*=========================================*)
(*                                         *)
(* Title:  Strip-Line Design               *)
(* Program Summary: Program calculates     *)
(*         the dimensions needed           *)
(*         to produce a strip-line.        *)
(*                                         *)
(*=========================================*)


PROGRAM StripLine;

uses Transcend;

 VAR
  Answer,
  Cont            : Char;
  Out             : Text;
  Zo,
  Substrate,
  Conductor,
  Dielectric,
  Temp,
  Width,
  Velocity        : Real;

Procedure Outputs;
  begin
    page(output);
     repeat
       gotoXY(Ø,5);
        Write('S(creen or P(rinter ? ');
        Read(Answer);
     until Answer in [ 'S','s','p','P' ]
  end;  {outputs}

Procedure Title;
  begin
    page(output);
     gotoXY(11,1);
       Writeln('Strip-Line Design');
       Writeln;Writeln;
       Writeln('This program will calculate the');
       Writeln('dimensions needed to produce a strip-line');
       Writeln('You must enter the :');
       Writeln;
     end;
```

```
Procedure Input;
  begin
    Writeln;
    Write('Line impedance (Zo) : ');
    Readln ( Zo );
    Write('Substrate Thickness (inch) : ');
    Readln( Substrate );
    Write('Conductor Thickness (inch) : ');
    Readln( Conductor );
    Write('Dielectric Constant : ');
    Readln( Dielectric );
    Writeln;
  end;

Procedure PrintScr;
   begin
     Writeln;
     Writeln('For a line impedance of ',Zo,' ohms, a');
     Writeln('substrate thickness of ',Substrate,' inch, a');
     Writeln('conductor thickness of ',Conductor,' inch,');
     Writeln('and a dielectric constant of ',Dielectric);
     Writeln;
     Writeln('a Strip-line with a width ');
     Writeln('of ',Width,' inch is required');
     Writeln;
     Writeln('The velocity factor is ',Velocity);
     Writeln;
  end;

Procedure PrintPrt;
begin
     Rewrite(out,'Printer:');
     Writeln;
     Writeln(out);
     Writeln('For a line impedance of ',Zo,' ohms, a');
     Writeln(out,'For a line impedance of ',Zo,' ohms, a');
     Writeln('substrate thickness of ',Substrate,' inch, a');
     Writeln(out,'substrate thickness of ',Substrate,' inch, a');
     Writeln('conductor thickness of ',Conductor,' inch,');
     Writeln(out,'conductor thickness of ',Conductor,' inch,');
     Writeln('and a dielectric constant of ',Dielectric);
     Writeln(out,'and a dielectric constant of ',Dielectric);
     Writeln;
     Writeln(out);
     Writeln('a Strip-line with a width ');
     Writeln(out,'a Strip-line with a width ');
     Writeln('of ',Width,' inch is required');
     Writeln(out,'of ',Width,' inch is required');
     Writeln;
     Writeln(out);
     Writeln('The velocity factor is ',Velocity);
     Writeln(out,'The velocity factor is ',Velocity);
     Writeln;
     Writeln(out);
     Close(out);
    end;

 Procedure Printout;
    begin
      case answer of
           's','S' : PrintScr;
           'p','P' : PrintPrt;
      end { case }
    end;

 Procedure Work;
  begin
     Temp      := exp ( Zo * Sqrt( Dielectric ) / 60 );
     Width     := 0.59 * ( 4 * Substrate / Temp - 2.1 * Conductor );
     Velocity  := 1 / Sqrt( Dielectric );
    end;

   begin { Main }
     Repeat
       Outputs;
       Title;
```

```
        Input;
        Work;
        Printout;
         Writeln;
              Write('Another set of data ?(Y/N) ');
                 Readln(Cont);
     until Cont in [ 'N','n']
   end.
```

Strip-Line Design

This program will calculate the
dimensions needed to produce a strip-line.

You must enter the :

Line impedance (ZO) : 7Ø
Substrate thickness (inch) : .Ø5
Conductor thickness (inch) : .ØØ1
Dielectric constant : 4.7

For a line impedance of 7.ØØØØØE1 ohms, a
substrate thickness of 5.ØØØØØE-2 inch, a
conductor thickness of 1.ØØØØØE-3 inch,
and a dielectric constant of 4.7ØØØØ
a strip-line with a width of 8.16759E-3 inch is required

The velocity factor is 4.61266E-1

Another set of data ? (Y/N) N

10.4 LOSSLESS TRANSMISSION LINES

Solving lossless RF transmission line problems with equations and a calculator can be tedious. Even with a Smith chart, you don't always get the accuracy you need. This program combines speed with accuracy to solve these problems.[1]

The program calculates the impedance of the line as well as its normalized impedance. It also yields the SWR referenced to a particular impedance.

To use this program, the characteristic impedance of the line, the reference impedance for the SWR, and the velocity factor must be entered. In addition, the input resistance and reactance, the frequency, and the length of the line must be entered.

When entering the length of the line, a sign must also be given. If the load is at the far end of the line, the sign is negative; if the generator is at the far end of the line, the sign is positive. The length of the line should be entered in meters.

[1] Adapted from R.J. Finger, "Transmission-line Problems Solved Fast with BASIC," *Electronic Design*, (Aug. 17, 1971): p. 76.

```
(*=======================================*)
(*                                       *)
(* Title:  Lossless Transmission Lines   *)
(* Program Summary: Program calculates   *)
(*         the impedance, normalized     *)
(*         impedance,and SWR of an RF    *)
(*         transmission line.            *)
(*                                       *)
(*=======================================*)


PROGRAM LossLess;

uses Transcend;

 VAR
  Answer,
  Cont            : Char;
  Out             : Text;
  LineImp,
  RefImp,
  Vel,
  Len,
  T1,
  A,
  C,
  Freq,
  Tan,
  RealRes,RØ,
  InputRes,
  ImagRes,XØ,
  InputReact,
  RealOut,
  ImagOut,
  RealZo,
  ImagZo,
  SWR,
  D1,D2,
  TempSwr         : Real;
  Sign            : String;

Procedure Outputs;
  begin
    page(output);
     repeat
       gotoXY(Ø,5);
        Write('S(creen or P(rinter? ');
        Read(Answer);
     until Answer in [ 'S','s','p','P' ]
  end;  {outputs}

Procedure Title;
  begin
    page(output);
     gotoXY(Ø6,1);
       Writeln('Lossless Transmission Lines');
       Writeln;Writeln;
       Writeln('This program can be used instead of');
       Writeln('a Smith chart to solve lossless');
       Writeln('transmission line problems.');
       Writeln;
     end;

Procedure Input;
  begin
    Writeln;
    Write('Enter line impedance : ');
    Readln ( LineImp );
    Write('Enter reference impedance : ');
    Readln( RefImp );
    Write('Enter velocity factor : ');
    Readln( Vel );
    Writeln;
    Writeln('Enter Length of line in meters. It is');
    Writeln('negative if load is at far end and');
    Write('positive if generator is there : ');
```

```
      Readln( Len );
      Writeln;
      Write('Enter frequency (mHz) : ');
      Readln( Freq );
      Write('Enter input resistance : ');
      Readln( InputRes );
      Write('Enter input reactance : ');
      Readln( InputReact );
      Writeln;
    end;

  Procedure PrintScr;
     begin
       Writeln;
       Writeln('Frequency (mHz) := ',Freq);
       Writeln;
       Writeln('Output Impedance := ',RealOut,Sign,' i ',ImagOut);
       Writeln;
       Writeln('Normalized Zo := ',RealZo,Sign,' i ',ImagZo);
       Writeln;
       Writeln('Swr := ',SWR);
       Writeln;
    end;

  Procedure PrintPrt;
  begin
       Rewrite(out,'Printer:');
       Writeln;
       Writeln(out);
       Writeln('Frequency (Mhz) := ',Freq);
       Writeln(out,'Frequency (MHz) := ',Freq);
       Writeln;
       Writeln(out);
       Writeln('Output Impedance := ',RealOut,Sign,' i ',ImagOut);
       Writeln(out,'Output Impedance := ',RealOut,Sign,' i ',ImagOut);
       Writeln;
       Writeln(out);
       Writeln('Normalized Zo := ',RealZo,Sign,' i ',ImagZo);
       Writeln(out,'Normalized Zo := ',RealZo,Sign,' i ',ImagZo);
       Writeln;
       Writeln(out);
       Writeln('Swr := ',SWR);
       Writeln(out,'Swr := ',SWR);
       Writeln;
       Writeln(out);
       Close(out);
      end;

   Procedure Printout;
      begin
        case answer of
             's','S' : PrintScr;
             'p','P' : PrintPrt;
        end { case }
      end;

   Procedure Work;
    begin
      T1   := 1.2 * Len * 1.74532925E-2 / Vel ;
      A    := 1E3;
      C    := LineImp / RefImp ;
      Tan    :=  Sin( T1 * Freq )/ Cos ( T1 * Freq );
      RealRes   := InputRes / LineImp ;
      ImagRes   := InputReact / LineImp ;
      D1   := Sqr(1 - ( Tan * ImagRes )) + Sqr( RealRes * Tan );
      RØ   := ( RealRes * ( 1 + Sqr( Tan ))) * C / D1 ;
      XØ   := (( Tan * ( 1 - Sqr( ImagRes ) - ( ImagRes * Tan ) -
                 Sqr( RealRes ))) + ImagRes ) * C / D1 ;
      D2   := Sqr( RØ ) + ( 2 * RØ ) + Sqr( XØ ) + 1 ;
      InputRes   := Sqrt(Sqr( Sqr( RØ ) + Sqr( XØ ) - 1 ) +
                    (4 * Sqr( XØ ))) / D2 ;
      TempSwr    := ( 1 + InputRes ) / ( 1 - InputRes );
      Sign   := ' + ';
      RealOut   := ( RØ * RefImp ) ;
      ImagOut   := Abs ( XØ * RefImp );
```

```
    RealZo  := RØ ;
    ImagZo  := Abs ( XØ );
    SWR := TempSwr
end;

  begin { Main }
    Repeat
      Outputs;
      Title;
      Input;
      Work;
      Printout;
       Writeln;
            Write('Another set of data ?(Y/N) ');
               Readln(Cont);
    until Cont in [ 'N','n']
  end.
```

Lossless Transmission Lines

This program can be used instead of
a Smith chart to solve lossless
transmission line problems.

Enter line impedance : 75

Enter reference impedance : 5Ø

Enter velocity factor : .6

Enter Length of line in meters. It is
negative if load is at far end and
positive if generator is there: -2

Enter frequency (MHz) : 1ØØ

Enter input resistance : 1ØØ

Enter input reactance : 8Ø

Frequency (MHz) = 1.ØØØØØE2

Output impedance = 3.51874E1 + i2.978Ø5E1

Normalized ZO = 7.Ø3749E-1 + i5.9561ØE-1

SWR = 2.16742

Another set of data ? (Y/N) N

10.5 SINGLE-WIRE TRANSMISSION LINE

This program calculates the characteristics of a single-wire transmission line located near the ground.

The diameter of the wire, its distance above the ground, and the frequency at which it is used must be entered. The output is the characteristic impedance and the inductance, capacitance, and resistance per meter.

```
(*==========================================*)
(*                                          *)
(* Title:   Single-Wire Transmission Line   *)
(* Program Summary: Program calculates      *)
(*          the impedance, inductance and   *)
(*          capacitance of an above-ground  *)
(*          transmission line.              *)
(*                                          *)
(*==========================================*)


PROGRAM SingleWire;

uses Transcend;

 VAR
  Answer,
  Cont            : Char;
  Out             : Text;
  Dia,
  Dist,
  Freq,
  Temp,
  Zo,
  L,
  C,
  R               : Real;

Procedure Outputs;
  begin
    page(output);
     repeat
       gotoXY(Ø,5);
        Write('S(creen or P(rinter ? ');
        Read(Answer);
     until Answer in [ 'S','s','p','P' ]
  end;  (outputs}

Procedure Title;
  begin
    page(output);
     gotoXY(Ø5,1);
       Writeln('Single-wire Transmission Line');
       Writeln;Writeln;
       Writeln('This program designs above ground');
       Writeln('single-wire transmission lines.  It ');
       Writeln('calculates the lines characteristic');
       Writeln('inductance, capacitance, and ');
       Writeln('resistance.');
       Writeln;
     end;

Procedure Input;
  begin
    Writeln;
    Write('Enter wire diameter (cm) : ');
    Readln( Dia );
    Write('Enter distance above ground (cm) : ');
    Readln( Dist );
    Write('Enter Frequency (MHz) : ');
    Readln( Freq );
    Writeln;
  end;
```

```
Procedure PrintScr;
   begin
     Writeln;
     Writeln('Line component values are :');
     Writeln;
     Writeln('Zo := ',Zo,' ohms');
     Writeln('L  := ',L,' microhenries/meter');
     Writeln('C  := ',C,' picofarads/meter');
     Writeln('R  := ',R,' ohms/meter');
     Writeln;
  end;

Procedure PrintPrt;
begin
     Rewrite(out,'Printer:');
     Writeln;
     Writeln(out);
     Writeln('Line component values are :');
     Writeln(out,'Line component values are :');
     Writeln;
     Writeln(out);
     Writeln('Zo := ',Zo,' ohms');
     Writeln(out,'Zo := ',Zo,' ohms');
     Writeln('L  := ',L,' microhenries/meter');
     Writeln(out,'L  := ',L,' microhenries/meter');
     Writeln('C  := ',C,' picofarads/meter');
     Writeln(out,'C  := ',C,' picofarads/meter');
     Writeln('R  := ',R,' ohms/meter');
     Writeln(out,'R  := ',R,' ohms/meter');
     Writeln;
     Writeln(out);
     Close(out);
    end;

 Procedure Printout;
    begin
     case answer of
          's','S' : PrintScr;
          'p','P' : PrintPrt;
      end { case }
    end;

   Procedure Work;
  begin
     Temp  := Log ( 4 * Dist / Dia ) ;
     Freq  := Freq * 1E6 ;
     Zo    := 138 * Temp ;
     L     := Ø.46 * Temp ;
     C     := 24.12 / Temp ;
     R     := 8.3 * Sqrt( Freq ) / Dia * 1E-6 ;
    end;

   begin { Main }
     Repeat
       Outputs;
       Title;
       Input;
       Work;
       Printout;
        Writeln;
             Write('Another set of data ?(Y/N) ');
                Readln(Cont);
     until Cont in [ 'N','n']
   end.
```

Single-Wire Transmission Line

This program designs above ground single-wire transmission lines. It calculates the line characteristic inductance, capacitance, and resistance.

Enter wire diameter (cm) : .Ø16

Enter distance above ground (cm) : 2

Enter Frequency (MHz) : 1

Line component values are :

ZO = 3.72458E2 ohms
L = 1.24153 microhenries/meter
C = 8.93674 picofarads/meter
R = 5.18750E-1 ohms/meter

Another set of data ? (Y/N) N

10.6 TWO-WIRE TRANSMISSION LINE

This program calculates the characteristics of a two-wire open transmission line. Details such as the wire diameter, the distance between the wires, and the frequency at which it is to be used are entered. The program then calculates the characteristic impedance of the line and the inductance, capacitance, and resistance per meter.

Although the program specifies that the diameter and distance between wires be entered in centimeters, any consistent units can be used, since the units cancel out in the calculation.

```
(*=======================================*)
(*                                       *)
(* Title:   Two-Wire Transmission Line   *)
(* Program Summary: Program calculates   *)
(*          the impedance, inductance, and *)
(*          capacitance of a two-wire    *)
(*          open transmission line.      *)
(*                                       *)
(*=======================================*)


PROGRAM TwoWire;

uses Transcend;

 VAR
  Answer,
  Cont            : Char;
  Out             : Text;
  Dia,
  Dist,
  Freq,
  Temp,
  Zo,
  L,
  C,
  R               : Real;

Procedure Outputs;
  begin
    page(output);
     repeat
       gotoXY(Ø,5);
        Write('S(creen or P(rinter ? ');
        Read(Answer);
     until Answer in [ 'S','s','p','P' ]
  end;  {outputs}
```

```
Procedure Title;
  begin
    page(output);
     gotoXY(04,1);
       Writeln('Open Two-Wire Transmission Line');
       Writeln;Writeln;
       Writeln('This program designs open two-wire');
       Writeln('transmission lines.  It calculates');
       Writeln('the lines characteristic inductance,');
       Writeln('capacitance, and impedance.');
       Writeln;
     end;

Procedure Input;
  begin
    Writeln;
    Write('Enter wire diameter (cm) : ');
    Readln( Dia );
    Write('Enter distance between wires (cm) : ');
    Readln( Dist );
    Write('Enter Frequency (MHz) : ');
    Readln( Freq );
    Writeln;
  end;

Procedure PrintScr;
   begin
     Writeln;
     Writeln('Line component values are :');
     Writeln;
     Writeln('Zo := ',Zo,' ohms');
     Writeln('L  := ',L,' microhenries/meter');
     Writeln('C  := ',C,' picofarads/meter');
     Writeln('R  := ',R,' ohms/meter');
     Writeln;
  end;

Procedure PrintPrt;
begin
     Rewrite(out,'Printer:');
     Writeln;
     Writeln(out);
     Writeln('Line component values are :');
     Writeln(out,'Line component values are :');
     Writeln;
     Writeln(out);
     Writeln('Zo := ',Zo,' ohms');
     Writeln(out,'Zo := ',Zo,' ohms');
     Writeln('L  := ',L,' microhenries/meter');
     Writeln(out,'L  := ',L,' microhenries/meter');
     Writeln('C  := ',C,' picofarads/meter');
     Writeln(out,'C  := ',C,' picofarads/meter');
     Writeln('R  := ',R,' ohms/meter');
     Writeln(out,'R  := ',R,' ohms/meter');
     Writeln;
     Writeln(out);
     Close(out);
    end;

 Procedure Printout;
    begin
     case answer of
           's','S' : PrintScr;
           'p','P' : PrintPrt;
      end { case }
    end;

   Procedure Work;
  begin
     Temp  := Log ( 2 * Dist / Dia ) ;
     Freq  := Freq * 1E6 ;
     Zo    := 276 * Temp ;
     L   := 0.921 * Temp ;
     C   := 12.06 / Temp ;
     R   := 8.3 * Sqrt( Freq ) / (Dia / 2) * 1E-6 ;
    end;
```

```
begin { Main }
  Repeat
    Outputs;
    Title;
    Input;
    Work;
    Printout;
     Writeln;
          Write('Another set of data ?(Y/N) ');
             Readln(Cont);
  until Cont in [ 'N','n']
end.
```

Open Two-Wire Transmission Line

This program designs open two-wire
transmission lines. It calculates
the line characteristic inductance,
capacitance, and impedance.

Enter wire diameter (cm) : .Ø16

Enter distance between wires (cm) : 3

Enter Frequency (MHz) : 1ØØ

Line component values are :

ZO = 7.1Ø433E2 ohms
L = 2.37Ø68 microhenries/meter
C = 4.68526 picofarads/meter
R = 1.Ø375ØE1 ohms/meter

Another set of data ? (Y/N) N

10.7 RF AIR-CORE INDUCTOR DESIGN

Amateur radio operators and communication engineers find a lot of use for air-core inductors, especially at VHF frequencies. This program will design such inductors, given the diameter of the coil form, the diameter of the wire (both in inches), and the desired inductance (in microhenries).

The program is based on the formula shown below, where N is the number of turns, R is the radius of the coil in inches, and D is the diameter of the wire in inches. The equation is then solved for N. This form of equation is used in the program (*Procedure Work).*

$$N = \frac{10\ DL + \sqrt{100D^2L^2 + 4R^2L(9R + 100)}}{2R^2}$$

```
(*=======================================*)
(*                                       *)
(* Title:  RF Air-Core Inductor Design   *)
(* Program Summary: Program designs      *)
(*         air-core inductors for        *)
(*         RF circuitry.                 *)
(*                                       *)
(*=======================================*)


PROGRAM AirCore;

uses Transcend;

 VAR
  Answer,
  Cont          : Char;
  Out           : Text;
  WireDia,
  CoilDia,
  Induct,
  R,
  A,
  B,
  C,
  Turns         : Real;

Procedure Outputs;
  begin
    page(output);
     repeat
       gotoXY(Ø,5);
        Write('S(creen or P(rinter ? ');
        Read(Answer);
     until Answer in [ 'S','s','p','P' ]
  end;  {outputs}

Procedure Title;
  begin
    page(output);
     gotoXY(Ø6,1);
       Writeln('RF Air-Core Inductor Design');
       Writeln;Writeln;
       Writeln('This program designs air-core inductors');
       Writeln('for RF circuitry.  You must enter the');
       Writeln('diameter of the wire and the diameter');
       Writeln('of the core form, both in inches,');
       Writeln('and the desired inductance in ');
       Writeln('microhenries.');
       Writeln;
     end;

Procedure Input;
  begin
    Writeln;
    Write('Enter wire diameter (cm) : ');
    Readln( WireDia );
    Write('Enter coil form diameter (inches) : ');
    Readln( CoilDia );
    Write('Enter inductance (microhenries) : ');
    Readln( Induct );
    Writeln;
  end;

Procedure PrintScr;
   begin
     Writeln;
     Writeln('Close-wound coil design');
     Writeln;
     Writeln('Wire diameter := ',WireDia,' mils');
     Writeln;
     Writeln('Coil diameter := ',CoilDia,' inches');
     Writeln;
     Writeln('Number of turns required := ',Turns);
     Writeln;
  end;
```

```
Procedure PrintPrt;
begin
     Rewrite(out,'Printer:');
     Writeln;
     Writeln(out);
     Writeln('Close-wound coil design');
     Writeln(out,'Close-wound coil design');
     Writeln;
     Writeln(out);
     Writeln('Wire diameter := ',WireDia,' mils');
     Writeln(out,'Wire diameter := ',WireDia,' mils');
     Writeln;
     Writeln(out);
     Writeln('Coil diameter := ',CoilDia,' inches');
     Writeln(out,'Coil diameter := ',CoilDia,' inches');
     Writeln;
     Writeln(out);
     Writeln('Number of turns required := ',Turns);
     Writeln(out,'Number of turns required := ',Turns);
     Writeln;
     Writeln(out);
     Close(out);
    end;

 Procedure Printout;
    begin
     case answer of
           's','S' : PrintScr;
           'p','P' : PrintPrt;
      end { case }
    end;

   Procedure Work;
  begin
    R  := CoilDia / 2 ;
    A  := 100 * Sqr ( WireDia ) * Sqr ( Induct ) ;
    B  := 4 * Sqr( R ) * Induct * ( 9 * R + 10 * WireDia ) ;
    C  := Sqrt ( A + B ) ;
    Turns  := ( 10 * WireDia * Induct + C ) / ( 2 * Sqr( R )) ;
    WireDia  := WireDia * 1E3 ;
    end;

  begin { Main }
    Repeat
      Outputs;
      Title;
      Input;
      Work;
      Printout;
       Writeln;
             Write('Another set of data ?(Y/N) ');
                Readln(Cont);
    until Cont in [ 'N','n']
  end.
```

RF Air-Core Inductor Design

This program designs air-core inductors for RF circuitry. You must enter the diameter of the wire, the diameter of the coil form, both in inches, and the desired inductance in microhenries.

Enter wire diameter (inches) : .0403

Enter coil form diameter (inches) : .25

Enter inductance (microhenries) : .4

Close-wound coil design

Wire diameter = 4.03000E1 mils

Coil diameter = 2.50000E-1 inches

Number of turns required = 1.32656E1

Another set of data ? (Y/N) N

10.8 VOLTAGE STANDING WAVE RATIO CALCULATION

Ham and CB operators as well as communication engineers are always interested in knowing the VSWR of their antenna feed lines. The VSWR is the ratio of the maximum voltage to the minimum voltage along the line. It is a measure of the mismatch between the line and its load. When the line and the load are perfectly matched, the VSWR is equal to 1.

In mismatch conditions, the VSWR increases in value. As it does, the power loss in the line increases, limiting the amount of power radiated by the transmitter.

In order to calculate the VSWR, the incident power and the reflected power of the line must be entered.

```
(*========================================*)
(*                                        *)
(* Title:   Voltage Standing Wave Ratio   *)
(*          Calculation                   *)
(* Program Summary: Program calculates    *)
(*          the VSWR of antenna feed      *)
(*          lines.                        *)
(*                                        *)
(*========================================*)


PROGRAM SWR;

uses Transcend;

 VAR
  Answer,
  Cont            : Char;
  Out             : Text;
  ReflectPwr,
  IncidentPwr,
  A,
  B,
  VSWR            : Real;

Procedure Outputs;
  begin
    page(output);
     repeat
       gotoXY(0,5);
        Write('S(creen or P(rinter ? ');
        Read(Answer);
     until Answer in [ 'S','s','p','P' ]
  end;  {outputs}

Procedure Title;
  begin
```

```
    page(output);
     gotoXY(12,1);
       Writeln('VSWR Calculation');
       Writeln;Writeln;
       Writeln('This program calculates the voltage');
       Writeln('standing wave ratio when the powers of');
       Writeln('the reflected and incident waves are');
       Writeln('known.');
       Writeln;
     end;

Procedure Input;
  begin
    Writeln;
    Write('Enter the reflected wave power : ');
    Readln( ReflectPwr );
    Write('Enter the incident wave power : ');
    Readln( IncidentPwr );
    Writeln;
  end;

Procedure PrintScr;
   begin
     Writeln;
     Writeln('VSWR := ',VSWR );
     Writeln;
  end;

Procedure PrintPrt;
begin
     Rewrite(out,'Printer:');
     Writeln;
     Writeln(out);
     Writeln('VSWR := ',VSWR );
     Writeln(out,'VSWR := ',VSWR );
     Writeln;
     Writeln(out);
     Close(out);
    end;

 Procedure Printout;
    begin
     case answer of
          's','S' : PrintScr;
          'p','P' : PrintPrt;
      end { case }
    end;

   Procedure Work;
  begin
    A   := 1 + Sqrt ( ReflectPwr / IncidentPwr );
    B   := 1 - Sqrt ( ReflectPwr / IncidentPwr );
    VSWR:= A / B ;
    end;

   begin { Main }
     Repeat
       Outputs;
       Title;
       Input;
       Work;
       Printout;
        Writeln;
             Write('Another set of data ?(Y/N) ');
                Read(Cont);
     until Cont in [ 'N','n']
   end.
```

VSWR Calculation

This program calculates the voltage
standing wave ratio when powers of

```
the reflected and incident waves are
known.

Enter the reflected wave power : 2.5

Enter the incident wave power : 18Ø

VSWR = 1.26719

Another set of data ? (Y/N) N
```

10.9 DECIBEL CONVERSION

The conversion of voltage, current, and power ratios to decibels (dB) is performed by this program.

When run, the program presents a menu. The desired conversion must be selected by entering a number from 1 to 3. Based on this selection, the program, through *Procedure Choose*, then runs either *Procedure Voltage*, *Procedure Current*, or *Procedure Power*.

The program automatically inverts current values to calculate the correct sign.

```
(*=========================================*)
(*                                         *)
(* Title:  Decibel Conversion              *)
(* Program Summary: Program converts       *)
(*         voltage, current, and power     *)
(*         ratios to decibels.             *)
(*                                         *)
(*=========================================*)


PROGRAM Decibel;

uses Transcend;

 VAR
  Answer,
  Cont            : Char;
  Out             : Text;
  Choice          : Integer;
  Common1,
  Common2,
  db              : Real;
  Kind            : String;

Procedure Outputs;
  begin
    page(output);
     repeat
       gotoXY(Ø,5);
        Write('S(creen or P(rinter ? ');
        Read(Answer);
     until Answer in [ 'S','s','p','P' ]
  end;  {outputs}

Procedure Title;
  begin
    page(output);
     gotoXY(13,1);
       Writeln('dB Conversion');
       Writeln;Writeln;
```

```
          Writeln('This program converts voltage, current');
          Writeln('or power ratios to dB.');
          Writeln;
          Writeln('Multiple line entries require data ');
          Writeln('to be separated by a space.');
          Writeln;
        end;

    Procedure Voltage;
      begin
          Write('Enter V2,V1 : ');
          Readln( Common2 , Common1 );
          db   := 20 * Log ( Common2 / Common1 );
          Kind  := 'voltage'
      end;

    Procedure Current;
      begin
          Write('Enter I2,I1 : ');
          Readln( Common1 , Common2 );
          db   := 20 * Log ( Common2 / Common1 );
          Kind  := 'current'
      end;

    Procedure Power;
      begin
          Write('Enter P2,P1 : ');
          Readln( Common2,Common1 );
          db   := 10 * Log ( Common2 / Common1 );
          Kind  := 'power'
     end;

    Procedure Choose;
      begin
       case Choice of

          1  : Voltage ;
          2  : Current ;
          3  : Power ;

       end { case }
     end;

  Procedure Input;
    begin
      Writeln;
      Writeln('Select mode : ');
      Writeln;
      Writeln('1...Voltage');
      Writeln('2...Current');
      Writeln('3...Power');
      Writeln;
      Repeat;
      Write('Enter choice : ');
      Read( Choice );
      Until Choice in [ 1 , 2 , 3 ] ;
      Writeln;
    end;

  Procedure PrintScr;
     begin
       Writeln;
       Writeln('The ',Kind,' ratio ',Common2,' / ',Common1);
       Writeln;
       Writeln('is equal to ',dB,' dB.');
       Writeln;
    end;

  Procedure PrintPrt;
  begin
       Rewrite(out,'Printer:');
       Writeln;
       Writeln(out);
       Writeln('The ',Kind,' ratio ',Common2,' / ',Common1);
       Writeln(out,'The ',Kind,' ratio ',Common2,' / ',Common1);
       Writeln;
```

```
    Writeln(out);
    Writeln('is equal to ',dB,' dB.');
    Writeln(out,'is equal to ',dB,' dB.');
    Writeln;
    Writeln(out);
    Close(out);
  end;

Procedure Printout;
  begin
    case answer of
          's','S' : PrintScr;
          'p','P' : PrintPrt;
     end { case }
  end;

  begin { Main }
    Repeat
      Outputs;
      Title;
      Input;
      Choose;
      Printout;
       Writeln;
             Write('Another set of data ?(Y/N) ');
                Read(Cont);
     until Cont in [ 'N','n']
  end.
```

dB Conversion

This program converts voltage, current,
or power ratios to dB.

Multiple line entries require data
to be separated by a space.

Select mode :

1...Voltage
2...Current
3...Power

Enter choice : 2

Enter I2,I1 : 3 1Ø

The current ratio 1.ØØØØØE1/3.ØØØØØ
is equal to 1.Ø4576E1 dB.

Another set of data ? (Y/N) N

11 PASSIVE FILTERS

For noncritical filtering applications, passive filters are frequently used. The advantage of passive filters is greater reliability. They do not use active devices that can burn out or otherwise be damaged by incorrectly applied voltages or transients.

The filters designed by the programs in this chapter are passive filters of various configurations all of which are built from inductors and capacitors. Given the desired frequency, the program calculates the values of the capacitors and inductors indicated in the accompanying circuit diagrams.

The filters include six types of low-pass filters, four different high-pass filters, and two each of bandpass and band-elimination filters.

11.1 LOW-PASS CONSTANT-K π FILTER

The filter in Fig. 11.1 is composed of two capacitors with a coil connected between them in a π configuration. The filter has a passband frequency range from dc to the cutoff frequency. The program calculates the values of the inductance and the two capacitors, which are equal in value.

To use the program, it is necessary only to enter the terminating resistance of the filter in ohms and the cutoff frequency in hertz.

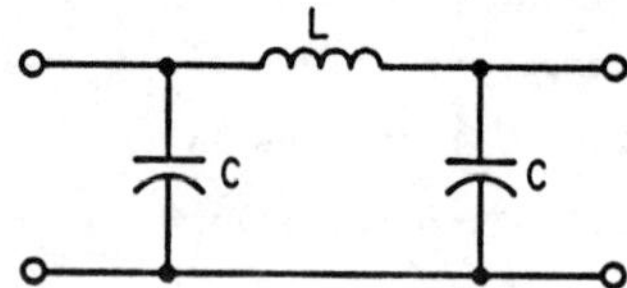

Fig. 11.1 Low-pass constant-K π filter

```
(*=========================================*)
(*                                         *)
(* Title:  Low-Pass  Constant-K Pi         *)
(*         Filter                          *)
(* Program Summary: Program calculates     *)
(*         the inductance and              *)
(*         capacitance of the elements     *)
(*         of a low-pass constant-K        *)
(*         Pi filter.                      *)
(*                                         *)
(*=========================================*)


PROGRAM ConstKPi;

const
 Pi = 3.14159;

 VAR
  Answer,
  Cont          : Char;
  Out           : Text;
  R,
  F,
  Ltemp,
  C,
  L            : Real;

Procedure Outputs;
  begin
    page(output);
     repeat
       gotoXY(Ø,5);
        Write('S(creen or P(rinter ? ');
        Read(Answer);
     until Answer in [ 'S','s','p','P' ]
  end;  {outputs}

Procedure Title;
  begin
    page(output);
     gotoXY(Ø4,1);
       Writeln('Low-Pass Constant-K "Pi" Filters');
       Writeln;Writeln;
       Writeln('This program designs low-pass "Pi" ');
       Writeln('configuration, constant-K filters.');
       Writeln('You must enter the cutoff frequency,');
       Writeln('and the terminating resistance in ohms');
       Writeln;
       Writeln;
     end;

Procedure Input;
  begin
    Writeln;
    Write('Enter the terminating resistance (ohms) : ');
    Readln( R );
    Write('Enter cutoff Frequency (Hz) : ');
    Readln( F );
    Writeln;
  end;

Procedure PrintScr;
   begin
```

```
      Writeln;
      Writeln('Filter component values are : ');
      Writeln;
      Writeln('L := ',L,' millihenries');
      Writeln;
      Writeln('C := ',C,' microfarads');
      Writeln;
   end;

Procedure PrintPrt;
    begin
     Rewrite(out,'Printer:');
     Writeln;
     Writeln(out);
     Writeln('Filter component values are : ');
     Writeln(out,'Filter component values are : ');
     Writeln;
     Writeln(out);
     Writeln('L := ',L,' millihenries');
     Writeln(out,'L := ',L,' millihenries');
     Writeln;
     Writeln(out);
     Writeln('C := ',C,' microfarads');
     Writeln(out,'C := ',C,' microfarads');
     Writeln;
     Writeln(out);
     Close(out);
    end;

 Procedure Printout;
    begin
     case answer of
           's','S' : PrintScr;
           'p','P' : PrintPrt;
       end { case }
    end;

Procedure Work;
  begin
    Ltemp  := R / ( Pi * F );
    C  := 1E6 / ( Pi * R * F * 2 );
    L := Ltemp * 1E3 ;
  end;

   begin { Main }
     Repeat
       Outputs;
       Title;
       Input;
       Work;
       Printout;
        Writeln;
             Write('Another set of data ?(Y/N) ');
                Read(Cont);
     until Cont in [ 'N','n']
   end.
```

Low-Pass Constant-K "Pi" Filters

This program designs low-pass "Pi"
configuration constant-K filters.
You must enter the cutoff frequency
and the terminating resistance in ohms.

Enter the terminating resistance (ohms) : 75
Enter cutoff frequency (Hz) : 3ØØØ

Filter component values are :

L = 7.95775 millihenries
C = 7.Ø7356E-1 microfarads

Another set of data ? (Y/N) N

11.2 LOW-PASS CONSTANT-K T FILTER

The filter in Fig. 11.2 is very similar to the previous one except that it is composed of two inductors and one capacitor arranged in a T configuration. As with the previous filter, the passband is from dc to the cutoff frequency, after which a sharp attenuation occurs. The values of the inductor are identical.

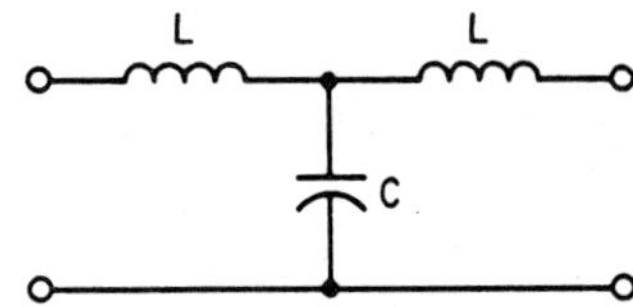

Fig. 11.2 Low-pass constant-K T filter

```
(*========================================*)
(*                                        *)
(* Title:   Low-Pass   Constant-K         *)
(*          T Filter                      *)
(* Program Summary: Program calculates    *)
(*          the inductance and            *)
(*          capacitance of the elements   *)
(*          of a low-pass constant-K      *)
(*          T filter.                     *)
(*                                        *)
(*========================================*)


PROGRAM ConstKT ;

const
 Pi = 3.14159;

 VAR
  Answer,
  Cont            : Char;
  Out             : Text;
  R,
  Freq,
  L,
  C,
  Ltemp              : Real;

Procedure Outputs;
  begin
    page(output);
     repeat
       gotoXY(Ø,5);
        Write('S(creen or P(rinter ? ');
        Read(Answer);
     until Answer in [ 'S','s','p','P' ]
  end;  {outputs}

Procedure Title;
  begin
    page(output);
     gotoXY(Ø4,1);
```

```
         Writeln('Low-Pass Constant-K "T" Filters');
         Writeln;Writeln;
         Writeln('This program designs low-pass "T" ');
         Writeln('configuration, constant-K filters.');
         Writeln('You must enter the cutoff frequency,');
         Writeln('and the terminating resistance in ohms');
         Writeln;
         Writeln;
       end;

Procedure Input;
  begin
    Writeln;
    Write('Enter the terminating resistance (ohms) : ');
    Readln( R );
    Write('Enter cutoff Frequency (Hz) : ');
    Readln( Freq );
    Writeln;
  end;

Procedure PrintScr;
   begin
     Writeln;
     Writeln('Filter component values are : ');
     Writeln;
     Writeln('L := ',L,' millihenries');
     Writeln;
     Writeln('C := ',C,' microfarads');
     Writeln;
  end;

Procedure PrintPrt;
    begin
     Rewrite(out,'Printer:');
     Writeln;
     Writeln(out);
     Writeln('Filter component values are : ');
     Writeln(out,'Filter component values are : ');
     Writeln;
     Writeln(out);
     Writeln('L := ',L,' millihenries');
     Writeln(out,'L := ',L,' millihenries');
     Writeln;
     Writeln(out);
     Writeln('C := ',C,' microfarads');
     Writeln(out,'C := ',C,' microfarads');
     Writeln;
     Writeln(out);
     Close(out);
    end;

 Procedure Printout;
    begin
     case answer of
          's','S' : PrintScr;
          'p','P' : PrintPrt;
      end { case }
    end;

Procedure Work;
  begin
    Ltemp  := R / ( Pi * Freq ) / 2 ;
    C  := 1E6 / ( Pi * R * Freq );
    L := Ltemp * 1E3 ;
  end;

   begin { Main }
     Repeat
       Outputs;
       Title;
       Input;
       Work;
       Printout;
        Writeln;
             Write('Another set of data ?(Y/N) ');
                Read(Cont);
     until Cont in [ 'N','n']
   end.
```

```
Low-Pass Constant-K "T" Filters

This program designs low-pass "T"
configuration  constant-K filters.
You must enter the cutoff frequency
and the terminating resistance in ohms.

Enter the terminating resistance (ohms) : 1000

Enter the cutoff frequency (Hz) : 100

Filter component values are :

L = 1.59155E3 millihenries
C = 3.18310 microfarads

Another set of data ? (Y/N) N
```

11.3 LOW-PASS M-DERIVED T FILTER

Like the other low-pass filters previously described, this one has a passband from dc to the cutoff frequency. What differentiates it from the constant-K type of filter, among other things, is that the cutoff frequency and the frequency at which maximum attenuation is desired are specified when the filter is designed.

As can be seen in Fig 11.3, the T filter is composed of three inductors and a single capacitor. The two series inductors of the input and output lines are identical in value.

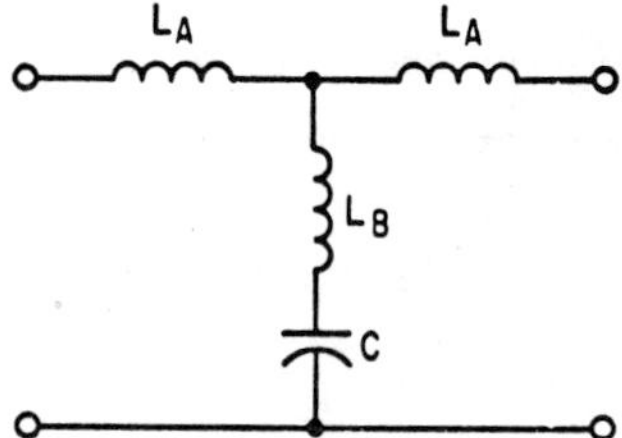

Fig. 11.3 Low-pass M-derived T filter

```
(*=========================================*)
(*                                         *)
(* Title:   Low-Pass  M-Derived T Filter   *)
(* Program Summary: Program calculates     *)
(*          the inductance and             *)
(*          capacitance of the elements    *)
(*          in  a low-pass, M-derived      *)
(*          T filter.                      *)
(*                                         *)
(*=========================================*)


PROGRAM MderivedT;

uses Transcend;

const
  Pi = 3.14159;
```

```
 VAR
  Answer,
  Cont          : Char;
  Out           : Text;
  M,
  Res,
  Freq,
  MaxAttn,
  C,
  La,
  Lb,
  Ltemp         : Real;

Procedure Outputs;
  begin
    page(output);
     repeat
       gotoXY(Ø,5);
        Write('S(creen or P(rinter ? ');
        Read(Answer);
     until Answer in [ 'S','s','p','P' ]
  end;  {outputs}

Procedure Title;
  begin
    page(output);
     gotoXY(Ø5,1);
       Writeln('Low-Pass M-Derived "T" Filters');
       Writeln;
       Writeln;
       Writeln('This program designs low-pass "T" ');
       Writeln('configuration, M-derived filters.');
       Writeln('You must enter the cutoff frequency,');
       Writeln('the frequency of maximum attenuation (Hz)');
       Writeln('and the terminating resistance in ohms');
       Writeln;
       Writeln;
     end;

Procedure Input;
  begin
    Writeln;
    Write('Enter the terminating resistance (ohms) : ');
    Readln( Res );
    Write('Enter cutoff Frequency (Hz) : ');
    Readln( Freq );
    Write('Enter maximum attenuation freq. : ');
    Readln( MaxAttn );
    Writeln;
  end;

Procedure PrintScr;
   begin
     Writeln;
     Writeln('Filter component values are : ');
     Writeln;
     Writeln('La := ',La,' millihenries');
     Writeln;
     Writeln('Lb := ',Lb,' millihenries');
     Writeln;
     Writeln('C := ',C,' microfarads');
     Writeln;
  end;

Procedure PrintPrt;
    begin
     Rewrite(out,'Printer:');
     Writeln;
     Writeln(out);
     Writeln('Filter component values are : ');
     Writeln(out,'Filter component values are : ');
     Writeln;
     Writeln(out);
     Writeln('La := ',La,' millihenries');
     Writeln(out,'La := ',La,' millihenries');
     Writeln;
```

```
      Writeln(out);
      Writeln('Lb := ',Lb,' millihenries');
      Writeln(out,'Lb := ',Lb,' millihenries');
      Writeln;
      Writeln(out);
      Writeln('C := ',C,' microfarads');
      Writeln(out,'C := ',C,' microfarads');
      Writeln;
      Writeln(out);
      Close(out);
     end;

 Procedure Printout;
     begin
      case answer of
            's','S' : PrintScr;
            'p','P' : PrintPrt;
       end { case }
     end;

Procedure Work;
   begin
     M  := Sqrt ( 1 - Sqr ( Freq / MaxAttn )) ;
     Ltemp  := Res / ( Pi * Freq );
     C  := M * 1E6 / ( Pi * Res * Freq );
     La := M * Ltemp * 1E3 / 2 ;
     Lb := ( 1 - Sqr( M )) * Ltemp / 4 / M * 1E3 ;
   end;

    begin { Main }
      Repeat
        Outputs;
        Title;
        Input;
        Work;
        Printout;
         Writeln;
              Write('Another set of data ?(Y/N) ');
                 Read(Cont);
      until Cont in [ 'N','n']
    end.
```

Low-Pass M-Derived "T" Filters

This program designs low-pass "T"
configuration M-derived filters.
You must enter the cutoff frequency,
the frequency of maximum attenuation (Hz)
and the terminating resistance in ohms.

Enter the terminating resistance (ohms) : 100

Enter cutoff frequency (Hz) : 3000

Enter maximum attenuation frequency : 3500

Filter components values are :

La = 2.73258 millihenries
Lb = 3.78357 millihenries
C =5.46516E-1 microfarads

Another set of data ? (Y/N) N

11.4 LOW-PASS M-DERIVED L FILTER

The L configuration filter is used as an end element in a compound filter design. Thus, when several filter sections are stacked to produce a sharper cutoff, the terminating element in the chain is often an L filter.

The L filter in Fig. 11.4 is used as a terminating element when T filter elements are stacked. As with the T filter, the cutoff frequency and the frequency of maximum attenuation must be entered when the program is run.

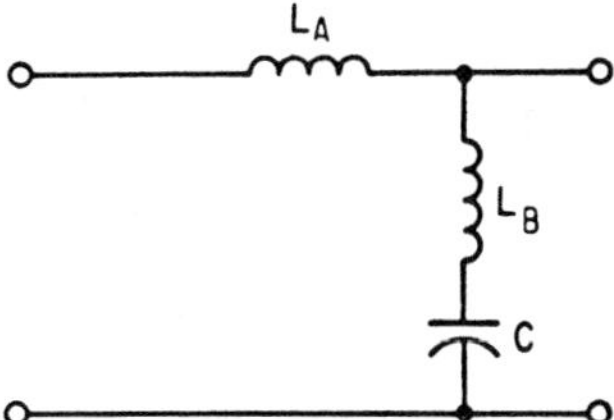

Fig. 11.4 Low-pass M-derived L filter

```
(*=========================================*)
(*                                         *)
(* Title:   Low-Pass  M-Derived L Filter   *)
(* Program Summary: Program calculates     *)
(*          the inductance and             *)
(*          capacitance of the elements    *)
(*          in a low-pass  M-derived       *)
(*          L filter.                      *)
(*                                         *)
(*=========================================*)


PROGRAM MDerivedL;

uses Transcend;

const
 Pi = 3.14159;

 VAR
  Answer,
  Cont              : Char;
  Out               : Text;
  Res,
  Freq,
  MaxAttn,
  C,
  M,
  La,
  Lb,
  L                 : Real;

Procedure Outputs;
  begin
    page(output);
     repeat
       gotoXY(Ø,5);
         Write('S(creen or P(rinter ? ');
         Read(Answer);
     until Answer in [ 'S','s','p','P' ]
  end;  {outputs}

Procedure Title;
  begin
    page(output);
```

```
    gotoXY(Ø5,1);
      Writeln('Low-Pass M-Derived "L" Filters');
      Writeln;Writeln;
      Writeln('This program designs low-pass "L" ');
      Writeln('configuration, M-Derived filters.');
      Writeln('You must enter the cutoff frequency,');
      Writeln('the frequency of maximum attenuation (Hz)');
      Writeln('and the terminating resistance in ohms');
      Writeln;
      Writeln;
    end;

Procedure Input;
  begin
    Writeln;
    Write('Enter the terminating resistance (ohms) : ');
    Readln( Res );
    Write('Enter cutoff Frequency (Hz) : ');
    Readln( Freq );
    Write('Enter maximum attenuation freq. : ');
    Readln( MaxAttn );
    Writeln;
  end;

Procedure PrintScr;
   begin
     Writeln;
     Writeln('Filter component values are : ');
     Writeln;
     Writeln('La := ',La,' millihenries');
     Writeln;
     Writeln('Lb := ',Lb,' millihenries');
     Writeln;
     Writeln('C := ',C,' microfarads');
     Writeln;
  end;

Procedure PrintPrt;
    begin
     Rewrite(out,'Printer:');
     Writeln;
     Writeln(out);
     Writeln('Filter component values are : ');
     Writeln(out,'Filter component values are : ');
     Writeln;
     Writeln(out);
     Writeln('La := ',La,' millihenries');
     Writeln(out,'La := ',La,' millihenries');
     Writeln;
     Writeln(out);
     Writeln('Lb := ',Lb,' millihenries');
     Writeln(out,'Lb := ',Lb,' millihenries');
     Writeln;
     Writeln(out);
     Writeln('C := ',C,' microfarads');
     Writeln(out,'C := ',C,' microfarads');
     Writeln;
     Writeln(out);
     Close(out);
    end;

 Procedure Printout;
    begin
     case answer of
           's','S' : PrintScr;
           'p','P' : PrintPrt;
      end { case }
    end;

Procedure Work;
  begin
    M  := Sqrt( 1 - Sqr ( Freq / MaxAttn ));
    L  := Res / ( Pi * Freq );
    C  := M * 1E6 / ( Pi * Res * Freq * 2 );
    La := M * L * 1E3 / 2 ;
    Lb := ( 1 - Sqr( M )) * L / 2 / M * 1E3 ;
  end;
```

```
begin { Main }
  Repeat
    Outputs;
    Title;
    Input;
    Work;
    Printout;
     Writeln;
          Write('Another set of data ?(Y/N) ');
             Read(Cont);
  until Cont in [ 'N','n']
end.
```

Low-Pass M-Derived "L" Filters

This program designs low-pass "L"
configuration M-Derived filters.
You must enter the cutoff frequency,
the frequency of maximum attenuation (Hz)
and the terminating resistance in ohms.

Enter the terminating resistance (ohms) : 75

Enter cutoff frequency (Hz) : 1000

Enter maximum attenuation freq. : 1100

Filter component values are :

La = 4.97277 millihenries
Lb = 2.36799E1 millihenries
C = 8.84049E-1 microfarads

Another set of data ? (Y/N) N

11.5 LOW-PASS SHUNT M-DERIVED L FILTER

Just as the M-derived L filter was used with the T filter, the shunt M-derived L filter is used with the shunt M-derived π filter.

As indicated in Fig. 11.5, the filter consists of only two capacitors and a single inductor.

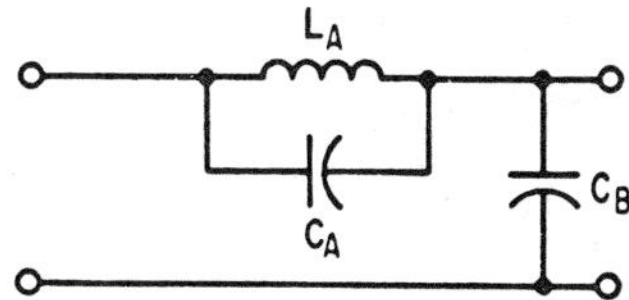

Fig. 11.5 Low-pass shunt M-derived L filter

```
(*=======================================*)
(*                                      *)
(* Title:  Low-Pass  Shunt  M-Derived   *)
(*         L Filter                     *)
(* Program Summary: Program calculates  *)
(*         the inductance and           *)
(*         capacitance of the elements  *)
(*         in a low-pass  shunt         *)
(*         M-derived L Filter.          *)
(*                                      *)
(*=======================================*)

PROGRAM ShuntMDerivedL;

uses Transcend;

const
 Pi = 3.14159;

 VAR
  Answer,
  Cont            : Char;
  Out             : Text;
  Res,
  Freq,
  MaxAttn,
  M,
  Ca,
  Cb,
  C,
  La,
  Ltemp           : Real;

Procedure Outputs;
  begin
    page(output);
     repeat
       gotoXY(Ø,5);
        Write('S(creen or P(rinter ? ');
        Read(Answer);
     until Answer in [ 'S','s','p','P' ]
  end;  {outputs}

Procedure Title;
  begin
    page(output);
     gotoXY(Ø3,1);
       Writeln('Low-Pass Shunt M-Derived "L" Filters');
       Writeln;Writeln;
       Writeln('This program designs low-pass "L" ');
       Writeln('configuration, shunt M-derived filters.');
       Writeln('You must enter the cutoff frequency,');
       Writeln('frequency of maximum attenuation, ');
       Writeln('and the terminating resistance in ohms');
       Writeln;
       Writeln;
     end;

Procedure Input;
  begin
    Writeln;
    Write('Enter the terminating resistance (ohms) : ');
    Readln( Res );
    Write('Enter cutoff Frequency (Hz) : ');
    Readln( Freq );
    Write('Enter maximum attenuation freq. : ');
    Readln( MaxAttn );
    Writeln;
  end;

Procedure PrintScr;
   begin
     Writeln;
     Writeln('Filter component values are : ');
     Writeln;
```

```
      Writeln('La := ',La,' millihenries');
      Writeln;
      Writeln('Ca := ',Ca,' microfarads');
      Writeln;
      Writeln('Cb := ',Cb,' microfarads');
      Writeln;
   end;

Procedure PrintPrt;
    begin
     Rewrite(out,'Printer:');
     Writeln;
     Writeln(out);
     Writeln('Filter component values are : ');
     Writeln(out,'Filter component values are : ');
     Writeln;
     Writeln(out);
     Writeln('La := ',La,' millihenries');
     Writeln(out,'La := ',La,' millihenries');
     Writeln;
     Writeln(out);
     Writeln('Ca := ',Ca,' microfarads');
     Writeln(out,'Ca := ',Ca,' microfarads');
     Writeln;
     Writeln(out);
     Writeln('Cb := ',Cb,' microfarads');
     Writeln(out,'Cb := ',Cb,' microfarads');
     Writeln;
     Writeln(out);
     Close(out);
    end;

 Procedure Printout;
    begin
     case answer of
          's','S' : PrintScr;
          'p','P' : PrintPrt;
      end { case }
    end;

Procedure Work;
  begin
    M  := Sqrt( 1 - Sqr( Freq / MaxAttn ));
    Ltemp  := Res / ( Pi * Freq );
    C  := 1E6 / ( Pi * Res * Freq );
    Cb := M * C / 2 ;
    La := M * Ltemp * 1E3 / 2 ;
    Ca := ( 1 - Sqr( M )) * C / ( 4 * M ) * 2 ;
  end;

   begin { Main }
     Repeat
       Outputs;
       Title;
       Input;
       Work;
       Printout;
        Writeln;
             Write('Another set of data ?(Y/N) ');
                Read(Cont);
     until Cont in [ 'N','n']
   end.
```

Low-Pass Shunt M-Derived "L" Filters

This program designs low-pass "L"
configuration shunt M-derived filters.
You must enter the cutoff frequency,
frequency of maximum attenuation,
and the terminating resistance in ohms.

Enter the terminating resistance (ohms) : 9Ø

Enter cutoff frequency (Hz) : 2ØØØ

Enter maximum attenuation freq. : 21ØØ

Filter component values are :

La = 2.18376 millihenries
Ca = 2.63Ø25 microfarads
Cb = 2.696ØØE-1 microfarads

Another set of data ? (Y/N) N

11.6 LOW-PASS SHUNT M-DERIVED π FILTER

When the use of capacitors is perferred to the use of inductors in a filter circuit, the shunt, M-derived π filter can be used instead of the T configuration. As seen in Fig. 11.6, this filter contains one inductor and three capacitors.

The input and output impedances are identical, as would be assumed by the symmetrical configuration of the filter. This makes it ideal for a middle element in a chain of filter elements.

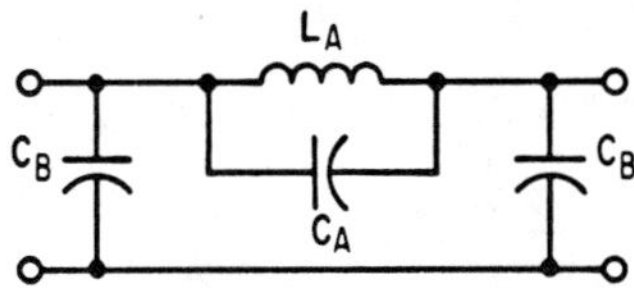

Fig. 11.6 Low-pass shunt M-derived π filter

```
(*=========================================*)
(*                                         *)
(* Title:  Low-Pass  Shunt  M-Derived      *)
(*         Pi Filter                       *)
(* Program Summary: Program calculates     *)
(*         the inductance and              *)
(*         capacitance of the elements     *)
(*         in  a low-pass  shunt           *)
(*         M-derived Pi filter.            *)
(*                                         *)
(*=========================================*)


PROGRAM ShuntMDerPi;

uses Transcend;

const
 Pi = 3.14159;

 VAR
  Answer,
  Cont             : Char;
  Out              : Text;
  Res,
  Freq,
  MaxAttn,
```

```
    M,
    Ca,
    Cb,
    La,
    C,
    Ltemp            : Real;

  Procedure Outputs;
    begin
      page(output);
        repeat
          gotoXY(Ø,5);
            Write('S(creen or P(rinter ? ');
            Read(Answer);
        until Answer in [ 'S','s','p','P' ]
    end;  {outputs}

  Procedure Title;
    begin
    page(output);
     gotoXY(Ø3,1);
       Writeln('Low-Pass Shunt M-Derived Pi Filters');
       Writeln;Writeln;
       Writeln('This program designs low-pass "Pi" ');
       Writeln('configuration, shunt M-Derived filters.');
       Writeln('You must enter the cutoff frequency,');
       Writeln('frequency of maximum attenuation,');
       Writeln('and the terminating resistance in ohms');
       Writeln;
       Writeln;
     end;

 Procedure Input;
   begin
     Writeln;
     Write('Enter the terminating resistance (ohms) : ');
     Readln( Res );
     Write('Enter cutoff Frequency (Hz) : ');
     Readln( Freq );
     Write('Enter maximum attenuation freq. : ');
     Readln( MaxAttn );
     Writeln;
   end;

 Procedure PrintScr;
    begin
      Writeln;
      Writeln('Filter component values are : ');
      Writeln;
      Writeln('La := ',La,' millihenries');
      Writeln;
      Writeln('Ca := ',Ca,' microfarads');
      Writeln;
      Writeln('Cb := ',Cb,' microfarads');
      Writeln;
   end;

 Procedure PrintPrt;
     begin
      Rewrite(out,'Printer:');
      Writeln;
      Writeln(out);
      Writeln('Filter component values are : ');
      Writeln(out,'Filter component values are : ');
      Writeln;
      Writeln(out);
      Writeln('La := ',La,' millihenries');
      Writeln(out,'La := ',La,' millihenries');
      Writeln;
      Writeln(out);
      Writeln('Ca := ',Ca,' microfarads');
      Writeln(out,'Ca := ',Ca,' microfarads');
      Writeln;
      Writeln(out);
      Writeln('Cb := ',Cb,' microfarads');
      Writeln(out,'Cb := ',Cb,' microfarads');
```

```
      Writeln;
      Writeln(out);
      Close(out);
     end;

  Procedure Printout;
     begin
      case answer of
           's','S' : PrintScr;
           'p','P' : PrintPrt;
       end { case }
     end;

Procedure Work;
   begin
     M  := Sqrt( 1 - Sqr( Freq / MaxAttn ));
     Ltemp  := Res / ( Pi * Freq );
     C  := 1E6 / ( Pi * Res * Freq );
     Cb := M * C / 2 ;
     La := M * Ltemp * 1E3 ;
     Ca := ( 1 - Sqr( M )) * C / ( 4 * M );
   end;

    begin { Main }
      Repeat
        Outputs;
        Title;
        Input;
        Work;
        Printout;
         Writeln;
              Write('Another set of data ?(Y/N) ');
                  Read(Cont);
      until Cont in [ 'N','n']
    end.
```

Low-Pass Shunt M-Derived "Pi" Filters

This program designs low-pass "Pi"
configuration shunt M-Derived filters.
You must enter the cutoff frequency,
frequency of maximum attenuation,
and the terminating resistance in ohms.

Enter the terminating resistance (ohms) : 1000

Enter cutoff frequency (Hz) : 5000

Enter maximum attenuation freq. : 5200

Filter component values are :

La = 1.74861E1 millihenries
Ca = 5.35725E-2 microfarads
Cb = 8.74304E-3 microfarads

Another set of data ? (Y/N) N

11.7 HIGH-PASS CONSTANT-K π FILTER

High-pass filters pass all frequencies above the cutoff frequency, but attenuate all frequencies from dc to the cutoff.

This particular filter consists of two inductors, one each across the input and output of the filter, with a capacitor connected between them (Fig. 11.7). This is the π configuration. Since both inductors have the same value, the terminating impedance on the input and output are identical.

To use the program, the termination resistance and cutoff frequency must be entered.

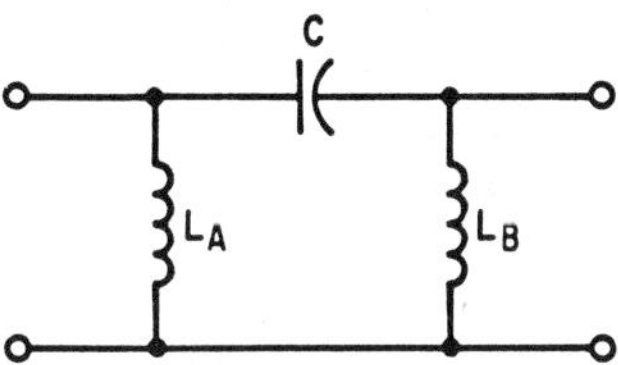

Fig. 11.7 High-pass constant-K π filter

```
(*=========================================*)
(*                                         *)
(* Title:   High-Pass  Constant-K Pi       *)
(*          Filter                         *)
(* Program Summary: Program calculates     *)
(*          the inductance and             *)
(*          capacitance of the elements    *)
(*          in a high-pass  constant-K     *)
(*          Pi filter.                     *)
(*                                         *)
(*=========================================*)


PROGRAM HiConstK;

const
 Pi = 3.14159;

 VAR
  Answer,
  Cont            : Char;
  Out             : Text;
  R,
  F,
  La,
  Lb,
  C,
  L               : Real;

Procedure Outputs;
  begin
    page(output);
     repeat
       gotoXY(Ø,5);
        Write('S(creen or P(rinter ? ');
        Read(Answer);
     until Answer in [ 'S','s','p','P' ]
  end;  {outputs}

Procedure Title;
  begin
    page(output);
     gotoXY(Ø4,1);
       Writeln('High-Pass Constant-K Pi Filters');
```

```
        Writeln;Writeln;
        Writeln('This program designs High-pass Pi');
        Writeln('configuration, constant-K filters.');
        Writeln('You must enter the cutoff frequency,');
        Writeln('and the terminating resistance in ohms');
        Writeln;
        Writeln;
      end;

Procedure Input;
  begin
    Writeln;
    Write('Enter the terminating resistance (ohms) : ');
    Readln( R );
    Write('Enter cutoff Frequency (Hz) : ');
    Readln( F );
    Writeln;
  end;

Procedure PrintScr;
   begin
     Writeln;
     Writeln('Filter component values are : ');
     Writeln;
     Writeln('La := ',La,' millihenries');
     Writeln;
     Writeln('Lb := ',Lb,' millihenries');
     Writeln;
     Writeln('C := ',C,' microfarads');
     Writeln;
  end;

Procedure PrintPrt;
    begin
     Rewrite(out,'Printer:');
     Writeln;
     Writeln(out);
     Writeln('Filter component values are : ');
     Writeln(out,'Filter component values are : ');
     Writeln;
     Writeln(out);
     Writeln('La := ',La,' millihenries');
     Writeln(out,'La := ',La,' millihenries');
     Writeln;
     Writeln(out);
     Writeln('Lb := ',Lb,' millihenries');
     Writeln(out,'Lb := ',Lb,' millihenries');
     Writeln;
     Writeln(out);
     Writeln('C := ',C,' microfarads');
     Writeln(out,'C := ',C,' microfarads');
     Writeln;
     Writeln(out);
     Close(out);
    end;

 Procedure Printout;
    begin
     case answer of
           's','S' : PrintScr;
           'p','P' : PrintPrt;
       end { case }
    end;

 Procedure Work;
   begin
     L  := R / ( 4 * Pi * F ) * 1E3;
     C  := 1E6 / ( Pi * R * F * 4 );
     La := 2 * L ;
     Lb := 2 * L ;
   end;

   begin { Main }
     Repeat
       Outputs;
       Title;
```

```
      Input;
      Work;
      Printout;
       Writeln;
             Write('Another set of data ?(Y/N) ');
                   Read(Cont);
  until Cont in [ 'N','n']
end.
```

High-Pass Constant-K "Pi" Filters

This program designs high-pass "Pi"
configuration constant-K filters.
You must enter the cutoff frequency,
and the terminating resistance in ohms.

Enter the terminating resistance (ohms) : 90

Enter cutoff frequency (Hz) : 1000

Filter component values are :

La = 1.43240E1 millihenries
Lb = 1.43240E1 millihenries
C = 8.84195E-1 microfarads

Another set of data ? (Y/N) N

11.8 HIGH-PASS CONSTANT-K T FILTER

The filter in Fig. 11.8 has a similar frequency response to that of the previous filter. It has series capacitors in the input and output arms with a parallel-connected inductor between them. The two capacitors and the input and output impedances have identical values.

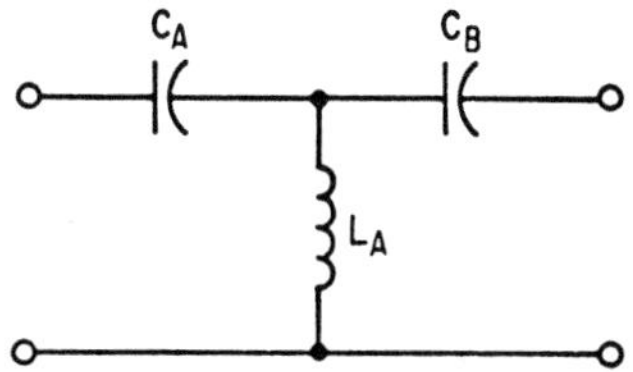

Fig. 11.8 High-pass constant-K T filter

```
(*=======================================*)
(*                                       *)
(* Title:   High-Pass  Constant-K        *)
(*          T Filter                     *)
(* Program Summary: Program calculates   *)
(*          the inductance and           *)
(*          capacitance of the elements  *)
(*          in a high-pass  constant-K   *)
(*          T filter.                    *)
(*                                       *)
(*=======================================*)
```

```
PROGRAM ConstKT;

const
 Pi = 3.14159;

 VAR
  Answer,
  Cont            : Char;
  Out             : Text;
  Res,
  Freq,
  La,
  C,
  Ca,
  Cb,
  Ltemp           : Real;

Procedure Outputs;
  begin
    page(output);
     repeat
       gotoXY(Ø,5);
        Write('S(creen or P(rinter ? ');
        Read(Answer);
     until Answer in [ 'S','s','p','P' ]
  end;  {outputs}

Procedure Title;
  begin
    page(output);
     gotoXY(Ø4,1);
       Writeln('High-Pass Constant-K "T" Filters');
       Writeln;Writeln;
       Writeln('This program designs high-pass "T" ');
       Writeln('configuration, constant-K filters.');
       Writeln('You must enter the cutoff frequency,');
       Writeln('and the terminating resistance in ohms');
       Writeln;
       Writeln;
     end;

Procedure Input;
  begin
    Writeln;
    Write('Enter the terminating resistance (ohms) : ');
    Readln( Res );
    Write('Enter cutoff Frequency (Hz) : ');
    Readln( Freq );
    Writeln;
  end;

Procedure PrintScr;
   begin
     Writeln;
     Writeln('Filter component values are : ');
     Writeln;
     Writeln('La := ',La,' millihenries');
     Writeln;
     Writeln('Ca := ',Ca,' microfarads');
     Writeln;
     Writeln('Cb := ',Cb,' microfarads');
     Writeln;
  end;

Procedure PrintPrt;
    begin
     Rewrite(out,'Printer:');
     Writeln;
     Writeln(out);
     Writeln('Filter component values are : ');
     Writeln(out,'Filter component values are : ');
     Writeln;
     Writeln(out);
     Writeln('La := ',La,' millihenries');
     Writeln(out,'La := ',La,' millihenries');
     Writeln;
```

```
      Writeln(out);
      Writeln('Ca := ',Ca,' microfarads');
      Writeln(out,'Ca := ',Ca,' microfarads');
      Writeln;
      Writeln(out);
      Writeln('Cb := ',Cb,' microfarads');
      Writeln(out,'Cb := ',Cb,' microfarads');
      Writeln;
      Writeln(out);
      Close(out);
     end;

 Procedure Printout;
    begin
     case answer of
           's','S' : PrintScr;
           'p','P' : PrintPrt;
      end { case }
    end;

Procedure Work;
  begin
    Ltemp  := Res / ( 4 * Pi * Freq ) ;
    C  := 1E6 / ( Pi * Res * Freq * 2 );
    Ca := C;
    Cb := C;
    La := Ltemp * 1E3 ;
  end;

   begin { Main }
     Repeat
       Outputs;
       Title;
       Input;
       Work;
       Printout;
        Writeln;
             Write('Another set of data ?(Y/N) ');
                Read(Cont);
     until Cont in [ 'N','n']
   end.
```

High-Pass Constant-K "T" Filters

This program designs high-pass "T"
configuration constant-K filters.
You must enter the cutoff frequency,
and the terminating resistance in ohms.

Enter the terminating resistance (ohms) : 1000

Enter cutoff frequency (Hz) : 60

Filter component values are :

La = 1.32629E3 millihenries
Ca = 2.65258 microfarads
Cb = 2.65258 microfarads

Another set of data ? (Y/N) N

11.9 HIGH-PASS, SERIES, M-DERIVED T FILTER

This filter is used to pass only those frequencies above a certain specified frequency. As with the low-pass M-derived filters, the maximum attenuation frequency must be entered. But unlike the low-pass filters, in which the maximum attenuation frequency was higher than the cutoff frequency, the maximum attenuation for high-pass filters is lower than the cutoff frequency.

The high-pass, series, M-derived T filter is relatively simple. It is built from three capacitors and one inductor (Fig. 11.9).

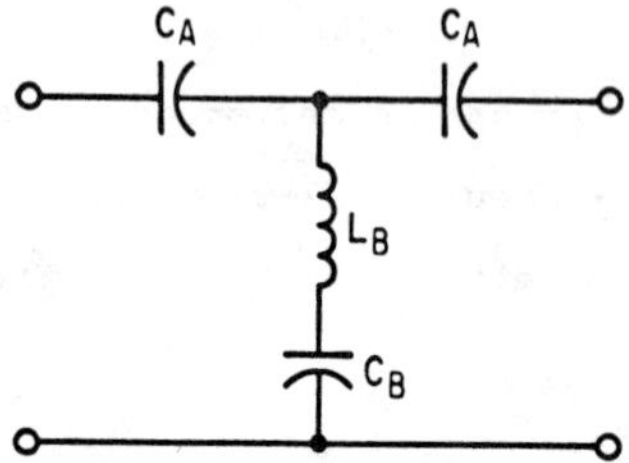

Fig. 11.9 High-pass, series, M-derived T filter

```
(*============================================*)
(*                                            *)
(* Title:   High-Pass   Series   M-Derived    *)
(*          T Filter                          *)
(* Program Summary: Program calculates        *)
(*          the inductance and                *)
(*          capacitance of the elements       *)
(*          in a high-pass  series            *)
(*          M-derived T filter.               *)
(*                                            *)
(*============================================*)


PROGRAM MDerT;

uses Transcend;

const
 Pi = 3.14159;

 VAR
  Answer,
  Cont               : Char;
  Out                : Text;
  Res,
  M,
  Freq,
  MaxAttn,
  Lb,
  Ctemp,
  Ca,
  Cb,
  Ltemp              : Real;

Procedure Outputs;
  begin
    page(output);
     repeat
       gotoXY(Ø,5);
         Write('S(creen or P(rinter ? ');
         Read(Answer);
     until Answer in [ 'S','s','p','P' ]
  end;   {outputs}
```

```
Procedure Title;
  begin
    page(output);
     gotoXY(Ø1,1);
       Writeln('High-Pass Series M-Derived "T" Filters');
       Writeln;Writeln;
       Writeln('This program designs high-pass "T" ');
       Writeln('configuration, series M-derived filters.');
       Writeln('You must enter the cutoff frequency,');
       Writeln('the frequency of maximum attenuation,');
       Writeln('and the terminating resistance in ohms');
       Writeln;
       Writeln;
     end;

Procedure Input;
  begin
    Writeln;
    Write('Enter the terminating resistance (ohms) : ');
    Readln( Res );
    Write('Enter cutoff Frequency (Hz) : ');
    Readln( Freq );
    Write('Enter maximum attenuation freq. : ');;
    Readln( MaxAttn );
    Writeln;
  end;

Procedure PrintScr;
   begin
     Writeln;
     Writeln('Filter component values are : ');
     Writeln;
     Writeln('Lb := ',Lb,' millihenries');
     Writeln;
     Writeln('Ca := ',Ca,' microfarads');
     Writeln;
     Writeln('Cb := ',Cb,' microfarads');
     Writeln;
  end;

Procedure PrintPrt;
    begin
     Rewrite(out,'Printer:');
     Writeln;
     Writeln(out);
     Writeln('Filter component values are : ');
     Writeln(out,'Filter component values are : ');
     Writeln;
     Writeln(out);
     Writeln('Lb := ',Lb,' millihenries');
     Writeln(out,'Lb := ',Lb,' millihenries');
     Writeln;
     Writeln(out);
     Writeln('Ca := ',Ca,' microfarads');
     Writeln(out,'Ca := ',Ca,' microfarads');
     Writeln;
     Writeln(out);
     Writeln('Cb := ',Cb,' microfarads');
     Writeln(out,'Cb := ',Cb,' microfarads');
     Writeln;
     Writeln(out);
     Close(out);
    end;

 Procedure Printout;
    begin
     case answer of
          's','S' : PrintScr;
          'p','P' : PrintPrt;
      end { case }
    end;

Procedure Work;
  begin
    M  := Sqrt( 1 - Sqr( MaxAttn / Freq ));
    Ltemp  := Res / ( Pi * Freq * 4 );
```

```
    Lb := 1E3 * Ltemp / M ;
    Ctemp  := 1E6 / ( Pi * Res * Freq * 4 );
    Ca := Ctemp / M * 2 ;
    Cb := 4 * M / ( 1 - Sqr( M )) * Ctemp ;
  end;

   begin { Main }
     Repeat
       Outputs;
       Title;
       Input;
       Work;
       Printout;
        Writeln;
             Write('Another set of data ?(Y/N) ');
                Read(Cont);
     until Cont in [ 'N','n']
   end.
```

High-Pass, Series, M-Derived "T" Filters

This program designs high-pass "T"
configuration, series, M-derived filters.
You must enter the cutoff frequency,
the frequency of maximum attenuation,
and the terminating resistance in ohms.

Enter the terminating resistance (ohms) : 300

Enter cutoff frequency (Hz) : 1000

Enter maximum attenuating freq. : 950

Filter component values are :

Lb = 7.64556E1 millihenries
Ca = 1.69901 microfarads
Cb = 3.67100E-1 microfarads

Another set of data ? (Y/N) N

11.10 HIGH-PASS SHUNT M-DERIVED π FILTER

The shunt, M-derived π filter is made of three inductors and one capacitor. Two inductors are connected in parallel across the input and output. The other inductor is connected in parallel with a capacitor and is located between the two termination inductors (Fig. 11.10).

To use the program, the termination resistance, cutoff frequency, and maximum attenuation frequency must be entered.

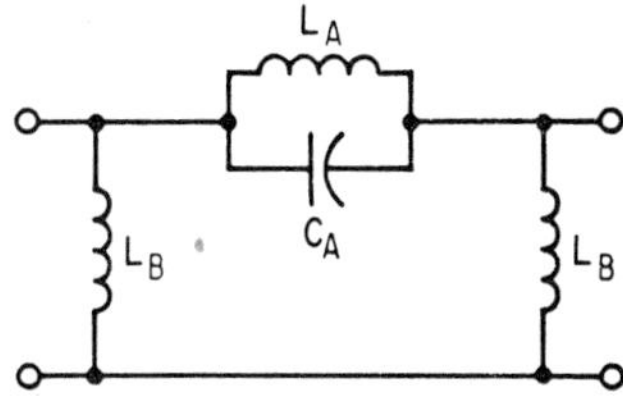

Fig. 11.10 High-pass shunt M-derived π filter

```
(*=========================================*)
(*                                         *)
(* Title:   High-Pass- Shunt- M-Derived    *)
(*          Pi Filter                      *)
(* Program Summary: Program calculates     *)
(*          the inductance and             *)
(*          capacitance of the elements    *)
(*          in a high-pass  shunt          *)
(*          M-derived Pi filter.           *)
(*                                         *)
(*=========================================*)


PROGRAM ShuntMDerPi;

uses Transcend;

const
 Pi = 3.14159;

 VAR
  Answer,
  Cont            : Char;
  Out             : Text;
  Res,
  Freq,
  MaxAttn,
  M,
  La,
  Lb,
  Ca,
  Ctemp,
  Ltemp           : Real;

Procedure Outputs;
  begin
    page(output);
     repeat
       gotoXY(0,5);
        Write('S(creen or P(rinter ? ');
        Read(Answer);
     until Answer in [ 'S','s','p','P' ]
  end;  {outputs}

Procedure Title;
  begin
    page(output);
     gotoXY(01,1);
       Writeln('High-Pass Shunt M-Derived "Pi" Filters');
       Writeln;Writeln;
       Writeln('This program designs high-pass "Pi" ');
       Writeln('configuration, shunted M-derived filters.');
       Writeln('You must enter the cutoff frequency,');
       Writeln('the frequency of maximum attenuation');
       Writeln('and the terminating resistance in ohms');
       Writeln;
       Writeln;
     end;

Procedure Input;
  begin
```

```
    Writeln;
    Write('Enter the terminating resistance (ohms) : ');
    Readln( Res );
    Write('Enter cutoff Frequency (Hz) : ');
    Readln( Freq );
    Write('Enter Maximum Attenuation Freq. : ');
    Readln( MaxAttn );
    Writeln;
  end;

Procedure PrintScr;
   begin
     Writeln;
     Writeln('Filter component values are : ');
     Writeln;
     Writeln('La := ',La,' millihenries');
     Writeln;
     Writeln('Lb := ',Lb,' millihenries');
     Writeln;
     Writeln('Ca := ',Ca,' microfarads');
     Writeln;
  end;

Procedure PrintPrt;
    begin
     Rewrite(out,'Printer:');
     Writeln;
     Writeln(out);
     Writeln('Filter component values are : ');
     Writeln(out,'Filter component values are : ');
     Writeln;
     Writeln(out);
     Writeln('La := ',La,' millihenries');
     Writeln(out,'La := ',La,' millihenries');
     Writeln;
     Writeln(out);
     Writeln('Lb := ',Lb,' millihenries');
     Writeln(out,'Lb := ',Lb,' millihenries');
     Writeln;
     Writeln(out);
     Writeln('Ca := ',Ca,' microfarads');
     Writeln(out,'Ca := ',Ca,' microfarads');
     Writeln;
     Writeln(out);
     Close(out);
    end;

 Procedure Printout;
    begin
     case answer of
          's','S' : PrintScr;
          'p','P' : PrintPrt;
      end { case }
    end;

Procedure Work;
  begin
    M  := Sqrt( 1 - Sqr( MaxAttn / Freq ));
    Ltemp  := Res / ( Pi * Freq * 4 );
    Lb := 2 * Ltemp * 1E3 / M ;
    Ctemp  := 1E6 / ( Pi * Res * Freq * 4 );
    Ca := Ctemp / M ;
    La := 4 * M * Ltemp * 1E3 / ( 1 - Sqr( M ));
  end;

   begin { Main }
     Repeat
       Outputs;
       Title;
       Input;
       Work;
       Printout;
        Writeln;
             Write('Another set of data ?(Y/N) ');
                 Read(Cont);
     until Cont in [ 'N','n']
   end.
```

```
High-Pass Shunt M-Derived "Pi" Filters

This program designs high-pass "Pi"
configuration  shunted M-derived filters.
You must enter the cutoff frequency,
the frequency of maximum attenuation
and the terminating resistance in ohms.

Enter the terminating resistance (ohms) : 9Ø

Enter cutoff frequency (Hz) : 6Ø

Enter maximum attenuation freq. : 55

Filter components values are :

La = 2.27Ø92E2 millihenries
Lb = 5.9735ØE2 millihenries
Ca = 3.68735E1 microfarads

Another set of data ? (Y/N) N
```

11.11 BANDPASS CONSTANT-K π FILTER

The bandpass filter in Fig. 11.11 is a combination of both a low-pass filter and a high-pass filter, in which the cutoff frequency of the low-pass filter is higher than the cutoff frequency of the high-pass filter. Bandpass filters permit a specified band of frequencies to be passed. Above and below this band of frequencies, the signal is significantly attenuated.

This particular filter has two parallel arms in the input and output portions of the circuit, with a series arm connected between them. To use the program, you must enter the upper and lower limits of the passband and the terminating resistance.

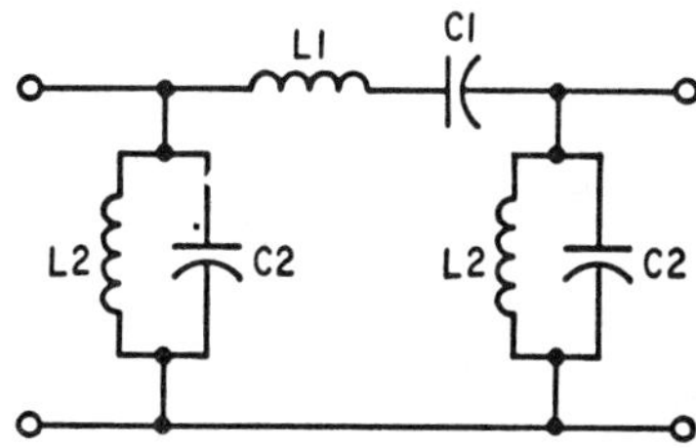

Fig. 11.11 Bandpass constant-K π filter

```
(*=======================================*)
(*                                       *)
(* Title:   Bandpass  Constant-K Pi      *)
(*          Filter                       *)
(* Program Summary: Program calculates   *)
(*          the inductance and           *)
(*          capacitance of the elements  *)
(*          in a bandpass constant-K     *)
(*          Pi filter.                   *)
(*                                       *)
(*=======================================*)


PROGRAM BandKPi;

const
 Pi = 3.14159;

 VAR
  Answer,
  Cont            : Char;
  Out             : Text;
  Res,
  F1,
  F2,
  L1,
  L2,
  C1,
  C2              : Real;

Procedure Outputs;
  begin
    page(output);
     repeat
       gotoXY(Ø,5);
        Write('S(creen or P(rinter ? ');
        Read(Answer);
     until Answer in [ 'S','s','p','P' ]
  end;  {outputs}

Procedure Title;
  begin
    page(output);
     gotoXY(Ø4,1);
       Writeln('Band-Pass Constant-K "Pi" Filters');
       Writeln;Writeln;
       Writeln('This program designs band-pass "Pi" ');
       Writeln('configuration, constant-K filters.');
       Writeln('You must enter the terminating resistance');
       Writeln('and upper and lower bandpass frequencies');
       Writeln;
       Writeln;
     end;

Procedure Input;
  begin
    Writeln;
    Write('Enter the terminating resistance (ohms) : ');
    Readln( Res );
    Write('Enter low frequency (Hz) : ');
    Readln( F1 );
    Write('Enter high frequency (Hz) : ');
    Readln( F2 );
    Writeln;
  end;

Procedure PrintScr;
   begin
     Writeln;
     Writeln('Filter component values are : ');
     Writeln;
     Writeln('L1 := ',L1,' millihenries');
     Writeln;
     Writeln('L2 := ',L2,' millihenries');
     Writeln;
     Writeln('C1 := ',C1,' microfarads');
```

```
      Writeln;
      Writeln('C2 := ',C2,' microfarads');
      Writeln;
   end;

 Procedure PrintPrt;
     begin
      Rewrite(out,'Printer:');
      Writeln;
      Writeln(out);
      Writeln('Filter component values are : ');
      Writeln(out,'Filter component values are : ');
      Writeln;
      Writeln(out);
      Writeln('L1 := ',L1,' millihenries');
      Writeln(out,'L1 := ',L1,' millihenries');
      Writeln;
      Writeln(out);
      Writeln('L2 := ',L2,' millihenries');
      Writeln(out,'L2 := ',L2,' millihenries');
      Writeln;
      Writeln(out);
      Writeln('C1 := ',C1,' microfarads');
      Writeln(out,'C1 := ',C1,' microfarads');
      Writeln;
      Writeln(out);
      Writeln('C2 := ',C2,' microfarads');
      Writeln(out,'C2 := ',C2,' microfarads');
      Writeln;
      Writeln(out);
      Close(out);
     end;

  Procedure Printout;
     begin
      case answer of
           's','S' : PrintScr;
           'p','P' : PrintPrt;
       end { case }
     end;

 Procedure Work;
   begin
     L1  := Res / ( Pi * ( F2 - F1 )) * 1E3 ;
     L2  := ( F2 - F1 ) * Res / ( 2 * Pi * F1 * F2 ) * 1E3 ;
     C1  := ( F2 - F1 ) *1E6 / ( 4 * Pi * F1 * F2 * Res ) ;
     C2  := 1E6 / ( Pi * ( F2 - F1 ) * Res ) / 2 ;
   end;

    begin { Main }
      Repeat
        Outputs;
        Title;
        Input;
        Work;
        Printout;
         Writeln;
              Write('Another set of data ?(Y/N) ');
                 Read(Cont);
      until Cont in [ 'N','n']
    end.
```

Band-Pass Constant-K "Pi" Filters

This program designs band-pass "Pi"
configuration constant-K filters.
You must enter the terminating resistance
and upper and lower bandpass frequencies.

Enter the terminating resistance (ohms) : 75

Enter low frequency (Hz) : 2500

Enter high frequency (Hz) : 3500

Filter component values are :

L1 = 2.38733E1 millihenries
L2 = 1.36419 millihenries
C1 = 1.21261E-1 microfarads
C2 = 2.12207 microfarads

Another set of data ? (Y/N) N

11.12 BANDPASS CONSTANT-K T FILTER

The filter in Fig. 11.12 differs from the previous one in that it has series instead of parallel input and output elements. The frequency response curve of this filter is similar to the earlier bandpass filter.

As with the other program, frequency must be entered in hertz and the terminating resistance in ohms. The component values are given in microfarads and millihenries.

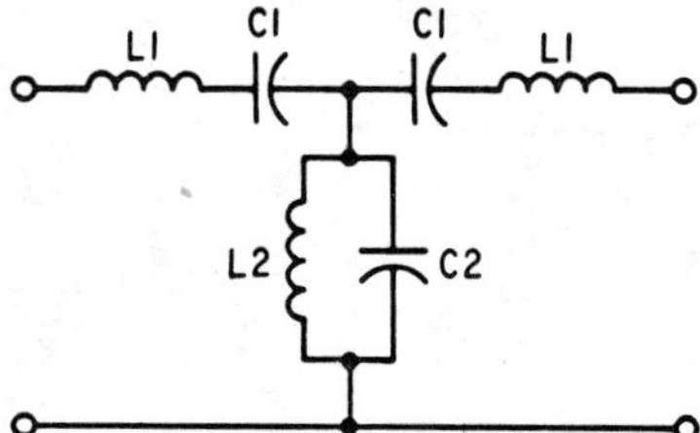

Fig. 11.12 Bandpass constant-K T filter

```
(*=========================================*)
(*                                         *)
(* Title:    Bandpass  Constant-K T Filter *)
(* Program Summary: Program calculates     *)
(*          the inductance and             *)
(*          capacitance of the elements    *)
(*          in a bandpass constant-K       *)
(*          T filter.                      *)
(*                                         *)
(*=========================================*)


PROGRAM ConstKT;

const
 Pi = 3.14159;

 VAR
  Answer,
  Cont             : Char;
  Out              : Text;
  Res,
  LFreq,
  HFreq,
  L1,
```

```
    L2,
    C1,
    C2              : Real;

  Procedure Outputs;
    begin
      page(output);
       repeat
         gotoXY(Ø,5);
          Write('S(creen or P(rinter ? ');
          Read(Answer);
       until Answer in [ 'S','s','p','P' ]
    end;  {outputs}

  Procedure Title;
    begin
      page(output);
       gotoXY(Ø4,1);
         Writeln('Band-Pass Constant-K "T" Filters');
         Writeln;Writeln;
         Writeln('This program designs band-pass');
         Writeln('constant-K "T" filters.');
         Writeln('You must enter the terminating');
         Writeln('resistance and upper and lower');
         Writeln('bandpass frequencies');
         Writeln;
         Writeln;
       end;

  Procedure Input;
    begin
      Writeln;
      Write('Enter the terminating resistance (ohms) : ');
      Readln( Res );
      Write('Enter low Frequency (Hz) : ');
      Readln( LFreq );
      Writeln;
      Write('Enter high Frequency (Hz) : ');
      Readln( HFreq );
      Writeln;
    end;

  Procedure PrintScr;
     begin
       Writeln;
       Writeln('Filter component values are : ');
       Writeln;
       Writeln('L1 := ',L1,' millihenries');
       Writeln;
       Writeln('L2 := ',L2,' millihenries');
       Writeln;
       Writeln('C1 := ',C1,' microfarads');
       Writeln;
       Writeln('C2 := ',C2,' microfarads');
       Writeln;
    end;

  Procedure PrintPrt;
      begin
       Rewrite(out,'Printer:');
       Writeln;
       Writeln(out);
       Writeln('Filter component values are : ');
       Writeln(out,'Filter component values are : ');
       Writeln;
       Writeln(out);
       Writeln('L1 := ',L1,' millihenries');
       Writeln(out,'L1 := ',L1,' millihenries');
       Writeln;
       Writeln(out);
       Writeln('L2 := ',L2,' millihenries');
       Writeln(out,'L2 := ',L2,' millihenries');
       Writeln;
       Writeln(out);
       Writeln('C1 := ',C1,' microfarads');
       Writeln(out,'C1 := ',C1,' microfarads');
```

```
      Writeln;
      Writeln(out);
      Writeln('C2 := ',C2,' microfarads');
      Writeln(out,'C2 := ',C2,' microfarads');
      Writeln;
      Writeln(out);
      Close(out);
     end;

 Procedure Printout;
     begin
      case answer of
            's','S' : PrintScr;
            'p','P' : PrintPrt;
       end { case }
     end;

Procedure Work;
  begin
     L1  := Res * 1E3 / ( Pi * ( HFreq - LFreq ) * 2 );
     L2  := ( HFreq - LFreq ) * Res * 1E3 / ( 4 * Pi * LFreq * HFreq );
     C1  := ( HFreq - LFreq ) * 1E6 / ( 2 * Pi * LFreq * HFreq * Res );
     C2  := 1E6 / ( Pi * ( HFreq - LFreq ) * Res );
  end;

   begin { Main }
     Repeat
       Outputs;
       Title;
       Input;
       Work;
       Printout;
        Writeln;
              Write('Another set of data ?(Y/N) ');
                 Read(Cont);
     until Cont in [ 'N','n']
   end.
```

Band-Pass Constant-K "T" Filters

This program designs band-pass
constant-K "T" filters.
You must enter the terminating
resistance and upper and lower
bandpass frequencies.

Enter the terminating resistances (ohms) : 75

Enter low frequency (Hz) : 9ØØ

Enter high frequency (Hz) : 11ØØ

Filter component values are :

L1 = 5.96832E1 millihenries
L2 = 1.2Ø57 millihenries
C1 = 4.287ØØE-1 microfarads
C2 = 2.122Ø7E1 microfarads

Another set of data ? (Y/N) N

11.13 BAND-ELIMINATION CONSTANT-K π FILTER

Like the bandpass filter, the band-elimination filter can be thought of as a combination of a low-pass filter and a high-pass filter. In this case, however, the cutoff frequency of the low-pass filter would be lower than the cutoff frequency of the high-pass filter, leaving a frequency gap between the two where the frequency is attenuated. Thus, the band-elimination filter is able to attenuate a specific frequency selectively.

As can be seen in Fig. 11.13, this filter has a π configuration. The input and output impedance are identical.

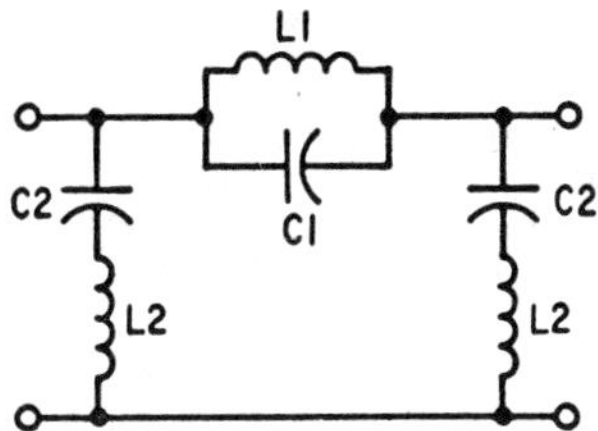

Fig. 11.13 Band-elimination constant-K π filter

```
(*=========================================*)
(*                                         *)
(* Title:   Band-Elimination   Constant-K  *)
(*          Pi Filter                      *)
(* Program Summary: Program calculates     *)
(*          the inductance and             *)
(*          capacitance of the elements    *)
(*          in a band-elimination          *)
(*          constant-K Pi filter.          *)
(*                                         *)
(*=========================================*)


PROGRAM ElimConstKPi;

const
 Pi = 3.14159;

 VAR
  Answer,
  Cont            : Char;
  Out             : Text;
  Res,
  LFreq,
  HFreq,
  L1,
  L2,
  C1,
  C2              : Real;

Procedure Outputs;
  begin
    page(output);
     repeat
       gotoXY(Ø,5);
        Write('S(creen or P(rinter ? ');
        Read(Answer);
     until Answer in [ 'S','s','p','P' ]
  end;  {outputs}

Procedure Title;
  begin
    page(output);
     gotoXY(Ø1,1);
```

```
        Writeln('Band-Elimination Constant-K "Pi" Filters');
        Writeln;Writeln;
        Writeln('This program designs band-elimination');
        Writeln('constant-K "Pi" filters.');
        Writeln('You must enter the terminating');
        Writeln('resistance and upper and lower');
        Writeln('bandpass frequencies');
        Writeln;
        Writeln;
      end;

Procedure Input;
  begin
    Writeln;
    Write('Enter the terminating resistance (ohms) : ');
    Readln( Res );
    Write('Enter low Frequency (Hz) : ');
    Readln( LFreq );
    Writeln;
    Write('Enter high Frequency (Hz) : ');
    Readln( HFreq );
    Writeln;
  end;

Procedure PrintScr;
   begin
     Writeln;
     Writeln('Filter component values are : ');
     Writeln;
     Writeln('L1 := ',L1,' millihenries');
     Writeln;
     Writeln('L2 := ',L2,' millihenries');
     Writeln;
     Writeln('C1 := ',C1,' microfarads');
     Writeln;
     Writeln('C2 := ',C2,' microfarads');
     Writeln;
  end;

Procedure PrintPrt;
    begin
     Rewrite(out,'Printer:');
     Writeln;
     Writeln(out);
     Writeln('Filter component values are : ');
     Writeln(out,'Filter component values are : ');
     Writeln;
     Writeln(out);
     Writeln('L1 := ',L1,' millihenries');
     Writeln(out,'L1 := ',L1,' millihenries');
     Writeln;
     Writeln(out);
     Writeln('L2 := ',L2,' millihenries');
     Writeln(out,'L2 := ',L2,' millihenries');
     Writeln;
     Writeln(out);
     Writeln('C1 := ',C1,' microfarads');
     Writeln(out,'C1 := ',C1,' microfarads');
     Writeln;
     Writeln(out);
     Writeln('C2 := ',C2,' microfarads');
     Writeln(out,'C2 := ',C2,' microfarads');
     Writeln;
     Writeln(out);
     Close(out);
    end;

 Procedure Printout;
    begin
     case answer of
          's','S' : PrintScr;
          'p','P' : PrintPrt;
      end { case }
    end;
```

```
Procedure Work;
  begin
    L2  := Res * 1E3 / ( Pi * ( HFreq - LFreq )) / 2 ;
    L1  := ( HFreq - LFreq ) * Res * 1E3 / ( Pi * LFreq * HFreq );
    C2  := ( HFreq - LFreq ) * 1E6 / ( 2 * Pi * LFreq * HFreq * Res );
    C1  := 1E6 / ( Pi * ( HFreq - LFreq ) * Res * 4 );
  end;

   begin { Main }
     Repeat
       Outputs;
       Title;
       Input;
       Work;
       Printout;
        Writeln;
             Write('Another set of data ?(Y/N) ');
                Read(Cont);
     until Cont in [ 'N','n']
   end.
```

```
Band-Elimination Constant-K "Pi" Filters


This program designs band-elimination
constant-K "Pi" filters.
You must enter the terminating
resistance and upper and lower
bandpass frequencies.

Enter the terminating resistance (ohms) : 75

Enter low frequency (Hz) : 45

Enter high frequency (Hz) : 55

Filter component values are :

L1 = 9.64576E1 millihenries
L2 = 1.19366E3 millihenries
C1 = 1.Ø61Ø3E2 microfarads
C2 = 8.574Ø1 microfarads

Another set of data ? (Y/N) N
```

11.14 BAND-ELIMINATION CONSTANT-K T FILTER

This T filter is similar to the π filter previously described, except for its physical layout. Like the π filter, this one will permit all frequencies to pass except for a specific band designated by the low and high frequencies entered by the user.

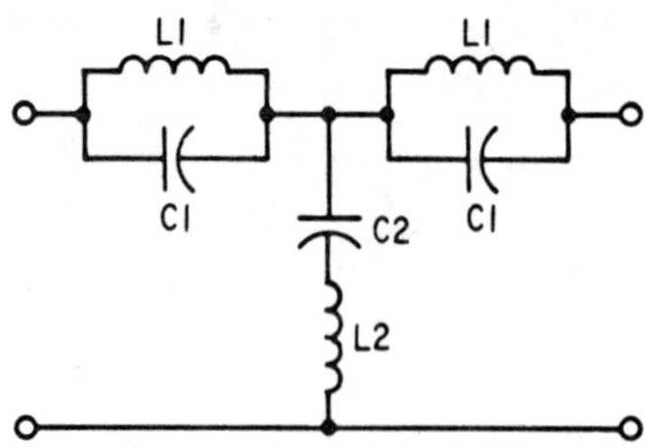

Fig. 11.14 Band-elimination constant-K T filter

```
(*=========================================*)
(*                                         *)
(* Title:   Band-Elimination  Constant-K   *)
(*          T Filter                       *)
(* Program Summary: Program calculates     *)
(*          the inductance and             *)
(*          capacitance of the elements    *)
(*          in a band-elimination          *)
(*          constant-K T filter.           *)
(*                                         *)
(*=========================================*)

PROGRAM ElimConstKT ;

const
 Pi = 3.14159;

 VAR
  Answer,
  Cont           : Char;
  Out            : Text;
  Res,
  LFreq,
  HFreq,
  L1,
  L2,
  C1,
  C2             : Real;

Procedure Outputs;
  begin
    page(output);
     repeat
       gotoXY(Ø,5);
        Write('S(creen or P(rinter ? ');
        Read(Answer);
     until Answer in [ 'S','s','p','P' ]
  end;  {outputs}

Procedure Title;
  begin
    page(output);
     gotoXY(Ø1,1);
       Writeln('Band-Elimination Constant-K "T" Filters');
       Writeln;Writeln;
       Writeln('This program designs band-elimination');
       Writeln('constant-K "T" filters.');
       Writeln('You must enter the terminating');
       Writeln('resistance and upper and lower');
       Writeln('bandpass frequencies');
       Writeln;
       Writeln;
     end;

Procedure Input;
  begin
    Writeln;
```

```
      Write('Enter the terminating resistance (ohms) : ');
      Readln( Res );
      Write('Enter low Frequency (Hz) : ');
      Readln( LFreq );
      Write('Enter high Frequency (Hz) : ');
      Readln( HFreq );
      Writeln;
    end;

  Procedure PrintScr;
     begin
       Writeln;
       Writeln('Filter component values are : ');
       Writeln;
       Writeln('L1 := ',L1,' millihenries');
       Writeln;
       Writeln('L2 := ',L2,' millihenries');
       Writeln;
       Writeln('C1 := ',C1,' microfarads');
       Writeln;
       Writeln('C2 := ',C2,' microfarads');
       Writeln;
    end;

  Procedure PrintPrt;
      begin
       Rewrite(out,'Printer:');
       Writeln;
       Writeln(out);
       Writeln('Filter component values are : ');
       Writeln(out,'Filter component values are : ');
       Writeln;
       Writeln(out);
       Writeln('L1 := ',L1,' millihenries');
       Writeln(out,'L1 := ',L1,' millihenries');
       Writeln;
       Writeln(out);
       Writeln('L2 := ',L2,' millihenries');
       Writeln(out,'L2 := ',L2,' millihenries');
       Writeln;
       Writeln(out);
       Writeln('C1 := ',C1,' microfarads');
       Writeln(out,'C1 := ',C1,' microfarads');
       Writeln;
       Writeln(out);
       Writeln('C2 := ',C2,' microfarads');
       Writeln(out,'C2 := ',C2,' microfarads');
       Writeln;
       Writeln(out);
       Close(out);
      end;

   Procedure Printout;
      begin
       case answer of
             's','S' : PrintScr;
             'p','P' : PrintPrt;
        end { case }
      end;

  Procedure Work;
    begin
      L2  := Res * 1E3 / ( Pi * ( HFreq - LFreq )) / 4 ;
      L1  := ( HFreq - LFreq ) * Res * 1E3 / ( Pi * LFreq * HFreq ) / 2 ;
      C2  := ( HFreq - LFreq ) * 1E6 / ( Pi * LFreq * HFreq * Res );
      C1  := 1E6 / ( Pi * ( HFreq - LFreq ) * Res ) / 2 ;
    end;

     begin { Main }
       Repeat
         Outputs;
         Title;
         Input;
         Work;
         Printout;
          Writeln;
```

```
        Write('Another set of data ?(Y/N)  ');
           Read(Cont);
  until Cont in [ 'N','n']
end.
```

Band-Elimination Constant-K "T" Filters

This program designs band-elimination
constant-K "T" filters.
You must enter the terminating
resistance and upper and lower
bandpass frequencies.

Enter the terminating resistance (ohms) : 5Ø

Enter low frequency (Hz) : 55

Enter high frequency (Hz) : 65

Filter component values are :

L1 = 2.22595E1 millihenries
L2 = 3.97888E2 millihenries
C1 = 3.1831ØE2 microfarads
C2 = 1.78Ø76E1 microfarads

Another set of data ? (Y/N) N

12 ATTENUATOR PADS

When feeding analog signals from one circuit to another, a situation sometimes arises when the applied signal is too strong and its application to a particular circuit causes the circuit to overload and malfunction. Situations such as this commonly arise in RF, audio, and video applications. For example, if you live near a television transmitter, it is possible that the gain produced by your antenna will result in a signal overload, preventing proper reception.

Another application arises when interfacing various sources of audio signals to an amplifier. Too strong a signal on the input will cause distortion in the output of the amplifier.

To overcome these overloading problems, resistor networks, known as attenuator pads, are used. The eleven programs in this chapter design different types of attenuators. Some are for balanced (ungrounded) lines, and some are for unsymmetrical inputs and outputs. Still others are for combinations of these. In addition, the last two programs (12.10 and 12.11) design attenuators that will introduce a minimum of loss when connected between the two systems.

12.1 BALANCED-BRIDGED H-PAD ATTENUATOR

For attenuation of signals on balanced transmission lines with symmetrical input and output impedances, the bridged H attenuator is applicable (Fig. 12.1).

The attenuator is composed of seven resistors. To use the program, it is necessary to enter only the desired symmetrical impedance and the loss introduced by the attenuator.

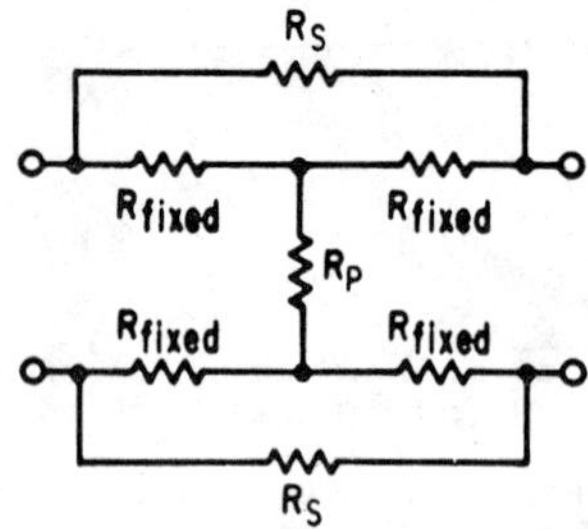

Fig. 12.1 Balanced-bridged H attenuator

```
(*=========================================*)
(*                                         *)
(* Title:   Balanced-Bridged H-Pad         *)
(*          Attenuator                     *)
(* Program Summary: Program calculates     *)
(*          the resistance values for      *)
(*          all resistor components of     *)
(*          a balanced-bridged H-pad       *)
(*          attenuator                     *)
(*                                         *)
(*=========================================*)


PROGRAM BalBridgeHAttnPd;

uses Transcend;

 VAR
  Answer,
  Cont            : Char;
  Out             : Text;
  Imped ,
  dBLoss,
  TempImp,
  K,
  FixRes,
  SerRes,
  ShuntRes                : Real;

Procedure Outputs;
  begin
    page(output);
     repeat
       gotoXY(Ø,5);
        Write('S(creen or P(rinter ? ');
        Read(Answer);
     until Answer in [ 'S','s','p','P' ]
  end;  {outputs}

Procedure Title;
  begin
    page(output);
     gotoXY(Ø9,1);
       Writeln('Balanced Bridged "H" Attenuator');
       Writeln;Writeln;
       Writeln('This program designs balanced');
       Writeln('bridged "H" attenuator pads.');
       Writeln('You must enter the desired impedance');
       Writeln('in ohms and the required voltage');
       Writeln('or current loss in decibels.');
       Writeln;
       Writeln;
     end;

Procedure Input;
  begin
    Writeln;
    Write('Enter desired impedance : ');
```

```
    Readln(  Imped  );
    Write('Enter desired loss in dB : ');
    Readln( dBLoss );
    Writeln;
  end;

Procedure PrintScr;
   begin
     Writeln;
     Writeln('For balanced symmetrical pads of ',Round(TempImp));
     Writeln('ohms and ',Round(dBLoss),' dB loss ');
     Writeln;
     Writeln('Bridged "H" series resistors "S" are');
     Writeln(Round(SerRes),' ohms');
     Writeln;
     Writeln('The shunt resistor "P" is ',Round(ShuntRes),' ohms.');
     Writeln;
     Writeln('This pad uses ',Round(FixRes),' ohm fixed resistors');
     Writeln('for all values of loss.');
     Writeln;
     Writeln('Multiply series resistor values by 2');
     Writeln('for the unbalanced pad values.');
     Writeln;
  end;

Procedure PrintPrt;
    begin
     Rewrite(out,'Printer:');
     Writeln;
     Writeln(out);
     Writeln('For balanced symmetrical pads of ',Round(TempImp));
     Writeln(out,'For balanced symmetrical pads of ',Round(TempImp));
     Writeln('ohms and ',Round(dBLoss),' dB loss ');
     Writeln(out,'ohms and ',Round(dbLoss),' dB loss ');
     Writeln;
     Writeln(out);
     Writeln('Bridged "H" series resistors "S" are');
     Writeln(out,'Bridged "H" series resistors "S" are');
     Writeln(Round(SerRes),' ohms');
     Writeln(out,Round(SerRes),' ohms');
     Writeln;
     Writeln(out);
     Writeln('The shunt resistor "P" is ',Round(ShuntRes),' ohms.');
     Writeln(out,'The shunt resistor "P" is ',Round(ShuntRes),' ohms.');
     Writeln;
     Writeln(out);
     Writeln('This pad uses ',Round(FixRes),' ohm fixed resistors');
     Writeln(out,'This pad uses ',Round(FixRes),' ohm fixed resistors');
     Writeln('for all values of loss.');
     Writeln(out,'for all values of loss.');
     Writeln;
     Writeln(out);
     Writeln('Multiply series resistor values by 2');
     Writeln(out,'Multiply series resistor values by 2');
     Writeln('for the unbalanced pad values.');
     Writeln(out,'for the unbalanced pad values.');
     Writeln;
     Writeln(out);
     Close(out);
    end;

 Procedure Printout;
    begin
     case answer of
           's','S' : PrintScr;
           'p','P' : PrintPrt;
      end { case }
    end;

Procedure Work;
  begin
    TempImp :=  Imped  ;
    K       := exp ( dBLoss * Ln ( 10 ) / 20 );
    FixRes  := TempImp / 2 ;
    SerRes  := TempImp * ( K - 1 ) / 2 ;
    ShuntRes:= TempImp / ( K - 1 ) ;
  end;
```

```
begin { Main }
  Repeat
    Outputs;
    Title;
    Input;
    Work;
    Printout;
     Writeln;
          Write('Another set of data ?(Y/N) ');
            Read(Cont);
  until Cont in [ 'N','n']
end.
```

Balanced Bridged H-Pad Attenuator

This program designs balanced-
bridged H-pad attenuators.
You must enter the desired impedance
in ohms and the required voltage
or current loss in decibels.

Enter desired impedance : 600

Enter desired loss in dB : 6

For balanced symmetrical pads of 600
ohms and 6 dB loss

Bridge "H" series resistors "S" are
299 ohms.

The shunt resistor "P" is 603 ohms.

This pad uses 300 ohm fixed resistors
of all values of loss.

Multiply series resistor values by 2
for the unbalanced pad values.

Another set of data? (Y/N) N

12.2 SYMMETRICAL UNBALANCED-BRIDGED T-PAD ATTENUATOR

The bridged T-pad attenuator (Fig. 12.2) is designed to be used on unbalanced transmission lines (where one side is grounded). The input and output impedances are symmetrical, and the attenuator itself consists of only four resistors.

To use the program, enter the termination impedance and the desired signal loss in dB.

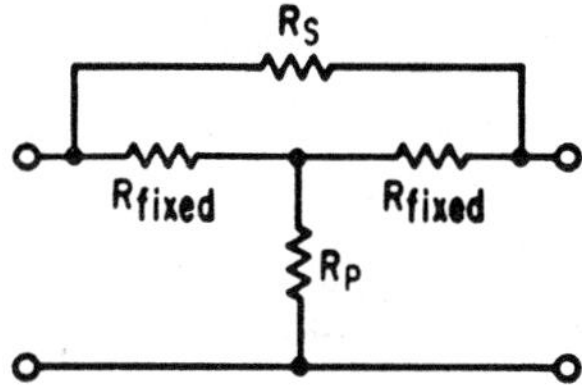

Fig. 12.2 Symmetrical unbalanced-bridged T attenuator

```
(*=======================================*)
(*                                       *)
(* Title:   Symmetrical Unbalanced-      *)
(*          Bridged T-Pad Attenuator     *)
(* Program Summary: Program calculates   *)
(*          the resistance values for    *)
(*          all resistor components of   *)
(*          a symmetrical unbalanced-    *)
(*          bridged T-pad attenuator.    *)
(*                                       *)
(*=======================================*)


PROGRAM SymUnbalBrT;

uses Transcend;

 VAR
  Answer,
  Cont             : Char;
  Out              : Text;
  Imped,
  dBLoss,
  Rfixed,
  K,
  Rs,
  Rp                : Real;

Procedure Outputs;
  begin
    page(output);
     repeat
       gotoXY(Ø,5);
        Write('S(creen or P(rinter ? ');
        Read(Answer);
     until Answer in [ 'S','s','p','P' ]
  end;  {outputs}

Procedure Title;
  begin
    page(output);
     gotoXY(Ø4,1);
       Writeln('Symmetrical Bridged "T" Attenuator');
       Writeln;Writeln;
       Writeln('This program will design unbalanced');
       Writeln('symmetrical bridged "T" attenuators.');
       Writeln('You must enter the desired impedance');
       Writeln('in ohms and the required voltage');
       Writeln('or current loss in decibels.');
       Writeln;
       Writeln;
     end;

Procedure Input;
  begin
    Writeln;
    Write('Enter desired impedance : ');
    Readln(  Imped  );
    Write('Enter desired loss in dB : ');
    Readln( dbLoss );
    Writeln;
  end;
```

```
Procedure PrintScr;
   begin
     Writeln;
     Writeln('For symmetrical pads of ',Round( Rfixed ));
     Writeln('ohms and ',Round( dBLoss ),' dB loss : ');
     Writeln;
     Writeln('Bridged "T" series resistor "Rs" is');
     Writeln(Round( Rs ),' ohms');
     Writeln;
     Writeln('Shunt resistor "Rp" is ',Round( Rp ),' ohms');
     Writeln;
     Writeln('Fixed resistors R1 and R2 are ',Round( Rfixed ),' ohms');
     Writeln;
  end;

Procedure PrintPrt;
    begin
     Rewrite(out,'Printer:');
     Writeln;
     Writeln(out);
     Writeln('For symmetrical pads of ',Round( Rfixed ));
     Writeln(out,'For symmetrical pads of ',Round( Rfixed ));
     Writeln('ohms and ',Round( dBLoss ),' dB loss : ');
     Writeln(out,'ohms and ',Round( dbLoss ),' dB loss : ');
     Writeln;
     Writeln(out);
     Writeln('Bridged "T" series resistor "Rs" is');
     Writeln(out,'Bridged "T" series resistor "Rs" is');
     Writeln(Round( Rs ),' ohms');
     Writeln(out,Round( Rs ),' ohms');
     Writeln;
     Writeln(out);
     Writeln('Shunt resistor "Rp" is ',Round( Rp ),' ohms');
     Writeln(out,'Shunt resistor "Rp" is ',Round( Rp ),' ohms');
     Writeln;
     Writeln(out);
     Writeln('Fixed resistors R1 and R2 are ',Round( Rfixed ),' ohms');
     Writeln(out,'Fixed resistors R1 and R2 are ',Round( Rfixed ),' ohms');
     Writeln;
     Writeln(out);
     Close(out);
    end;

 Procedure Printout;
    begin
     case answer of
           's','S' : PrintScr;
           'p','P' : PrintPrt;
      end { case }
    end;

Procedure Work;
  begin
    Rfixed :=  Imped  ;
    K := exp ( dBLoss * Ln ( 1Ø ) / 2Ø );
    Rs := Rfixed * ( K - 1 );
    Rp := Rfixed / ( K - 1 );
  end;

   begin { Main }
     Repeat
       Outputs;
       Title;
       Input;
       Work;
       Printout;
        Writeln;
             Write('Another set of data ?(Y/N) ');
                Read(Cont);
     until Cont in [ 'N','n']
   end.
```

Symmetrical Bridged T-Pad Attenuator

This program will design symmetrical unbalanced-

```
bridged T-pad attenuators.
You must enter the desired impedance
in ohms and the required voltage
or current loss in decibels.

Enter desired impedance: 300

Enter desired loss in dB : 10

For symmetrical pads of 300
ohms and 10 dB loss :

Bridged "T" series resistor "Rs" is
649 ohms.

Shunt resistor "Rp" is 139 ohms.

Fixed resistors R1 and R2 are 300 ohms.

Another set of data ? (Y/N) N
```

12.3 UNSYMMETRICAL H-PAD ATTENUATOR

The H-pad attenuator in Fig. 12.3 is composed of five resistors. For unsymmetrical attenuators, the input series resistors have a different value than the output series.

The H-pad attenuator designed by the program is a balanced attenuator. For unbalanced attenuators, it is necessary to multiply the series resistor by two.

Input and output impedance values should be entered in ohms. One of the first things the program does is calculate the minimum impedance possible, given the input and output impedances. If this value is larger than the desired loss in decibels, the program terminates execution and prints out a message indicating the minimum loss. If attenuator design is still desired, the value for loss in dB must be made equal to or greater than the minimum loss.

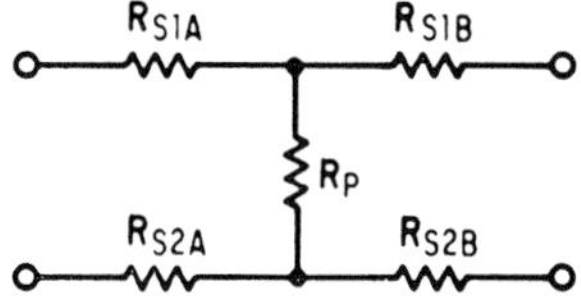

Fig. 12.3 Unsymmetrical H-pad attenuator

```
(*=========================================*)
(*                                         *)
(* Title:    Unsymmetrical H-Pad           *)
(*           Attenuator                    *)
(* Program Summary: Program calculates     *)
(*           the values for all            *)
(*           resistors in an               *)
(*           unsymmetrical H-pad           *)
(*           attenuator.                   *)
(*                                         *)
(*=========================================*)
```

```
PROGRAM UnsymHpad;

uses Transcend;

 VAR
  Answer,
  Cont            : Char;
  Out             : Text;
  InputImp,
  OutputImp,
  dBLoss,
  C,
  D,
  Z1,
  Z2,
  K5,
  L5,
  K,
  N,
  ShuntRes              : Real;
  S               : Array [1..2] of real;

Procedure Outputs;
  begin
    page(output);
     repeat
       gotoXY(Ø,5);
        Write('S(creen or P(rinter ? ');
        Read(Answer);
     until Answer in [ 'S','s','p','P' ]
  end;  {outputs}

Procedure Title;
  begin
    page(output);
     gotoXY(Ø5,1);
       Writeln('Unsymmetrical H Pad Attenuator');
       Writeln;Writeln;
       Writeln('This program will design unsymmetrical');
       Writeln('"H" attenuator Pads.');
       Writeln('You must enter the desired impedance');
       Writeln('in ohms and the required voltage');
       Writeln('or current loss in decibels.');
       Writeln;
       Writeln;
     end;

Procedure Input;
  begin
    Writeln;
    Write('Enter desired input impedance : ');
    Readln(  InputImp  );
    Write('Enter desired output impedance : ');
    Readln( OutputImp );
    Write('Enter desired loss in dB : ');
    Readln( dBLoss );
    Writeln;
  end;

Procedure Scr2;
  begin
     Writeln('Minimum loss for these impedances is');
     Writeln(L5,' dB.');
     Writeln;
     Writeln('For unbalanced pad values multiply');
     Writeln('series resistor values by 2.');
  end;

Procedure PrintScr;
   begin
     Writeln;
     Writeln('For unsymmetrical balanced pads where');
     Writeln('Z1 := ',InputImp,' ohms, Z2 := ',OutputImp,' ohms');
     Writeln('and Loss := ',dBLoss,' dB, the component');
     Writeln('values for an "H" attenuator are :');
     Writeln;
     Writeln('The input series resistors S1A and S1B');
```

```
        Writeln('are ',Round( S[Round(3 - D)] / 2 ),' ohms');
        Writeln;
        Writeln('The output series resistors S2A and S2B');
        Writeln('are ',Round( S[Round(3 - C)] / 2 ),' ohms');
        Writeln;
        Writeln('The shunt resistor P is ',Round(ShuntRes),' ohms');
        Writeln;
        Scr2;
        Writeln;
     end;

   Procedure Prnt2;
     begin
        Writeln('The shunt resistor P is ',Round(ShuntRes),' ohms');
        Writeln(out,'The shunt resistor P is ',Round(ShuntRes),' ohms');
        Writeln;
        Writeln(out);
        Writeln('Minimum loss for these impedances is');
        Writeln(out,'Minimum loss for these impedances is');
        Writeln(L5,' dB.');
        Writeln(out,L5,' dB.');
        Writeln;
        Writeln(out);
        Writeln('For unbalanced pad values multiply');
        Writeln('series resistor values by 2.');
        Writeln(out,'For unbalanced pad values multiply');
        Writeln(out,'series resistor values by 2.');
     end;

   Procedure PrintPrt;
       begin
        Rewrite(out,'Printer:');
        Writeln;
        Writeln(out);
        Writeln('For unsymmetrical balanced pads where');
        Writeln(out,'For unsymmetrical balanced pads where');
        Writeln('Z1 := ',InputImp,' ohms, Z2 := ',OutputImp,' ohms');
        Writeln(out,'Z1 := ',InputImp,' ohms, Z2 := ',OutputImp,' ohms');
        Writeln('and Loss := ',dBLoss,' dB, the component');
        Writeln(out,'and Loss := ',dBLoss,' dB, the component');
        Writeln('values for an "H" attenuator are :');
        Writeln(out,'values for an "H" attenuator are :');
        Writeln;
        Writeln(out);
        Writeln('The input series resistors S1A and S1B');
        Writeln(out,'The input series resistors S1A and S1B');
        Writeln('are ',Round( S[Round(3 - D)] / 2 ),' ohms');
        Writeln(out,'are ',Round( S[Round(3 - D)] / 2 ),' ohms');
        Writeln;
        Writeln(out);
        Writeln('The output series resistors S2A and S2B');
        Writeln(out,'The output series resistors S2A and S2B');
        Writeln('are ',Round( S[Round(3 - C)] / 2 ),' ohms');
        Writeln(out,'are ',Round( S[Round(3 - C)] / 2 ),' ohms');
        Writeln;
        Writeln(out);
        Prnt2;
        Writeln;
        Writeln(out);
        Close(out);
       end;

    Procedure Printout;
       begin
        case answer of
              's','S' : PrintScr;
              'p','P' : PrintPrt;
          end { case }
       end;

   Procedure Work;
     begin
       if OutputImp <= InputImp then
         begin
           C := 1 ;
           D := 2 ;
```

```
      Z1:= InputImp ;
      Z2:= OutputImp
    end
  else
    begin
      C := 2;
      D := 1;
      Z1:= OutputImp;
      Z2:= InputImp
    end;
  K5  := ( Z1 + Sqrt(Sqr( Z1 ) - ( Z1 * Z2 ))) / Z2 ;
  L5  := Ln( K5 / Sqrt( Z1 / Z2 )) / ( Ln (10)/ 20 );
   if dBLoss < L5 then
     begin
      Scr2;
      Exit( Work );
     end;
  K  := exp( dBLoss * ( Ln( 10 ) / 20 )) * Sqrt( Z1 / Z2 );
  N  := exp( dBLoss * ( Ln( 10 ) / 10 ));
  S[1] := Z1 * (( K * Z2 ) * ( K - 2 ) + Z1 ) / (( Sqr(K) * Z2) - Z1);
  S[2] := Z2 * ((Sqr(K) * Z2)- Z1 * ( 2 * K - 1 )) /
          ((Sqr(K) * Z2 ) - Z1 );
  ShuntRes := ( 2 * Sqrt( N * Z1 * Z2 )) / ( N - 1 ) ;
  Printout;
end;

 begin { Main }
   Repeat
     Outputs;
     Title;
     Input;
     Work;
      Writeln;
          Write('Another set of data ?(Y/N) ');
             Read(Cont);
   until Cont in [ 'N','n']
 end.
```

Unsymmetrical H-Pad Attenuator

This program will design unsymmetrical
H-Pad attenuators.
You must enter the desired impedance
in ohms and the required voltage
or current loss in decibels.

Enter desired input impedance : 300

Enter desired output impedance : 150

Enter desired loss in dB : 20

For unsymmetrical balanced pads where
Z1 = 3.00000E2 ohms, Z2 = 1.500000E2 ohms
and Loss = 2.0000E1 dB, the component
values for an "H" attenuator are:

The input series resistor S1A and S1B
are 132 ohms.

The output series resistors S2A and S2B
are 55 ohms.

The shunt resistor P is 43 ohms.

Minimum loss for these impedances is
7.6551 dB.

For unbalanced pad values multiply
series resistor values by 2.

Another set of data ? (Y/N) N

12.4 UNSYMMETRICAL SQUARE-PAD ATTENUATOR

The square-pad attenuator in Fig. 12.4 is composed of four resistors. For unsymmetrical attenuators, the input parallel resistor has a different value than the output parallel resistor. The two series resistors are identical in value.

The square-pad attenuator designed in this program is a balanced attenuator. For unbalanced attenuators, it is necessary to multiply the series resistor by two.

Input and output impedance values should be entered in ohms. One of the first things the program does, given the input and output impedances, is calculate the minimum impedance possible. If this value is larger than the desired loss in decibels, the program terminates execution and prints out a message indicating the minimum loss. If attenuator design is still desired, the value for loss in dB must be made equal to or greater than the minimum loss.

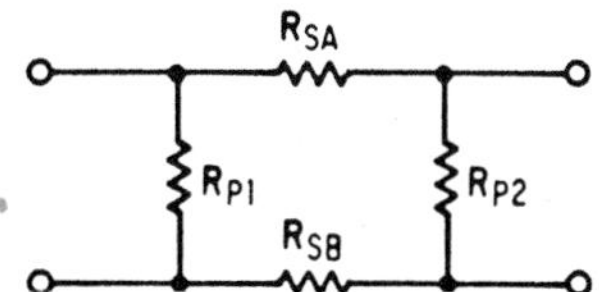

Fig. 12.4 Unsymmetrical square-pad attenuator

```
(*=========================================*)
(*                                         *)
(* Title:   Unsymmetrical Square-Pad       *)
(*          Attenuator                     *)
(* Program Summary: Program calculates     *)
(*          the values for all             *)
(*          resistors in an                *)
(*          unsymmetrical square-pad       *)
(*          attenuator.                    *)
(*                                         *)
(*=========================================*)


PROGRAM UnsymSquare;

uses Transcend;

 VAR
  Answer,
  Cont              : Char;
  Out               : Text;
  InputImp,
  OutputImp,
  dBLoss,
  C,
```

```
  D,
  Z1,
  Z2,
  K5,
  L5,
  K,
  N,
  SeriesRes      : Real;
  P              : Array [1..2] of real;

Procedure Outputs;
  begin
    page(output);
     repeat
       gotoXY(Ø,5);
        Write('S(creen or P(rinter ? ');
        Read(Answer);
     until Answer in [ 'S','s','p','P' ]
  end;  {outputs}

Procedure Title;
  begin
    page(output);
     gotoXY(Ø1,1);
       Writeln('Unsymmetrical Square-Pad Attenuator');
       Writeln;Writeln;
       Writeln('This program will design unsymmetrical');
       Writeln('square attenuator Pads. To use');
       Writeln('you must enter the desired impedance');
       Writeln('in ohms and the required voltage');
       Writeln('or current loss in decibels.');
       Writeln;
       Writeln;
     end;

Procedure Input;
  begin
    Writeln;
    Write('Enter desired input impedance : ');
    Readln(  InputImp  );
    Write('Enter desired output impedance : ');
    Readln( OutputImp );
    Write('Enter desired loss in dB : ');
    Readln( dBLoss );
    Writeln;
  end;

Procedure Scr2;
  begin
     Writeln('Minimum loss for these impedances is');
     Writeln(L5,' dB.');
     Writeln;
     Writeln('For unbalanced pad values multiply');
     Writeln('series resistor values by 2.');
  end;

Procedure PrintScr;
   begin
     Writeln;
     Writeln('For unsymmetrical balanced pads where');
     Writeln('Z1 := ',InputImp,' ohms, Z2 := ',OutputImp,' ohms');
     Writeln('and Loss := ',dBLoss,' dB, the component');
     Writeln('values for a square attenuator are :');
     Writeln;
     Writeln('The input shunt resistor Rp1 is');
     Writeln(Round( P[Round(3 - D)] ),' ohms');
     Writeln;
     Writeln('The output shunt resistor Rp2 is');
     Writeln(Round( P[Round(3 - C)] ),' ohms');
     Writeln;
     Writeln('The series resistors Rsa and Rsb are '
                   ,Round(SeriesRes/2),' ohms');
     Writeln;
     Scr2;
     Writeln;
  end;
```

```
Procedure Prnt2;
     begin
     Writeln;
     Writeln(out);
     Writeln('The series resistors Rsa and Rsb are ',
                                  Round(SeriesRes/2),' ohms');
     Writeln(out,'The series resistors Rsa and Rsb are ',
                                  Round(SeriesRes/2),' ohms');
     Writeln;
     Writeln(out);
     Writeln('Minimum loss for these impedances is');
     Writeln(out,'Minimum loss for these impedances is');
     Writeln(L5,' dB.');
     Writeln(out,L5,' dB.');
     Writeln;
     Writeln(out);
     Writeln('For unbalanced pad values multiply');
     Writeln('series resistor values by 2.');
     Writeln(out,'For unbalanced pad values multiply');
     Writeln(out,'series resistor values by 2.');
  end;

Procedure PrintPrt;
    begin
     Rewrite(out,'Printer:');
     Writeln;
     Writeln(out);
     Writeln('For unsymmetrical balanced pads where');
     Writeln(out,'For unsymmetrical balanced pads where');
     Writeln('Z1 := ',InputImp,' ohms, Z2 := ',OutputImp,' ohms');
     Writeln(out,'Z1 := ',InputImp,' ohms, Z2 := ',OutputImp,' ohms');
     Writeln('and Loss := ',dBLoss,' dB, the component');
     Writeln(out,'and Loss := ',dBLoss,' dB, the component');
     Writeln('values for an square attenuator are :');
     Writeln(out,'values for an square attenuator are :');
     Writeln;
     Writeln(out);
     Writeln('The input shunt resistor Rp1 is');
     Writeln(out,'The input shunt resistor Rp1 is');
     Writeln(Round( P[Round(3 - D)] ),' ohms');
     Writeln(out,Round( P[Round(3 - D)] ),' ohms');
     Writeln;
     Writeln(out);
     Writeln('The output shunt resistor Rp2 is');
     Writeln(out,'The output shunt resistor Rp2 is');
     Writeln(Round( P[Round(3 - C)] ),' ohms');
     Writeln(out,Round( P[Round(3 - C)] ),' ohms');
     Prnt2;
     Writeln;
     Writeln(out);
     Close(out);
    end;

 Procedure Printout;
    begin
     case answer of
          's','S' : PrintScr;
          'p','P' : PrintPrt;
       end { case }
    end;

Procedure Work;
  begin
    if OutputImp <= InputImp then
      begin
        C := 1;
        D := 2;
        Z1:= InputImp;
        Z2:= OutputImp
      end
    else
      begin
        C := 2;
        D := 1;
        Z1:= OutputImp;
        Z2:= InputImp
      end;
```

```
  K5  := ( Z1 + Sqrt(Sqr( Z1 ) - ( Z1 * Z2 ))) / Z2 ;
  L5  := Ln( K5 / Sqrt( Z1 / Z2 )) / ( Ln (10)/ 20 );
   if dbLoss < L5 then
     begin
       Scr2;
       Exit( Work );
     end;
  K  := exp( dBLoss * ( Ln( 10 ) / 20 )) * Sqrt( Z1 / Z2 );
  N  := exp( dBLoss * ( Ln( 10 ) / 20 ));
  P[1] := ( Z1 * ((Sqr(K) * Z2) - Z1)) /
                    ((Sqr(K) * Z2 ) - ( 2 * K * Z1 ) + Z1) ;
  P[2] := (( Sqr(K) * Sqr(Z2)) - ( Z1 * Z2 )) /
                    ( K * Z2 * ( K - 2 ) + Z1 ) ;
  SeriesRes := (P[2] * Z2 * ( K - 1)) / ( P[2] + Z2  ) ;
  Printout;
end;

 begin { Main }
   Repeat
     Outputs;
     Title;
     Input;
     Work;
      Writeln;
           Write('Another set of data ?(Y/N) ');
              Read(Cont);
   until Cont in [ 'N','n']
 end.
```

Unsymmetrical Square-Pad Attenuator

This program will design unsymmetrical
square attenuator Pads. To use,
you must enter the desired impedance
in ohms and the required voltage
or current loss in decibels.

Enter desired input impedance : 300

Enter desired output impedance : 75

Enter desired loss in dB : 15

For unsymmetrical balanced pads where
Z1 = 3.00000E2 ohms, Z2 = 7.50000E1 ohms
and Loss = 1.50000E1 dB, the component
values for a square attenuator are:

The input shunt resistor Rp1 is
907 ohms.

The output shunt resistor Rp2 is
85 ohms

The series resistor Rsa and Rsb are 204 ohms.

Minimum loss of these impedances is
1.14390E1 dB.

For unbalanced pad values multiply
series resistor values by 2.

Another set of data ? (Y/N) N

12.5 SYMMETRICAL SQUARE-PAD ATTENUATOR

The attenuator design in this program (Fig. 12.5) is similar to the one in the previous program. In a symmetrical attenuator, the input and output resistances are identical.

Like the former design, this one is also used for a balanced attenuator. For an unbalanced design, multiply the series resistor values by two.

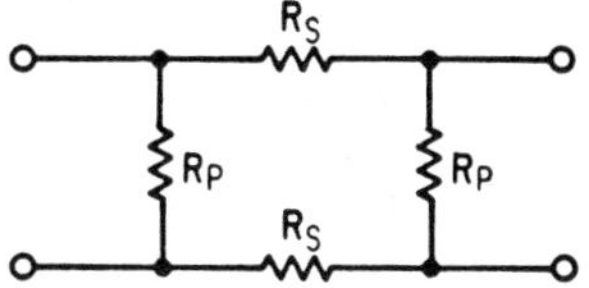

Fig. 12.5 Symmetrical square-pad attenuator

```
(*=========================================*)
(*                                         *)
(* Title:   Symmetrical Square-Pad         *)
(*          Attenuator                     *)
(* Program Summary: Program calculates     *)
(*          the values for all             *)
(*          resistors in a symmetrical     *)
(*          square-pad attenuator.         *)
(*                                         *)
(*=========================================*)


PROGRAM SymSquare;

uses Transcend;

 VAR
  Answer,
  Cont            : Char;
  Out             : Text;
  Imped,
  dBLoss,
  TempImp,
  K,
  SeriesRes,
  ShuntRes                : Real;


Procedure Outputs;
  begin
    page(output);
     repeat
       gotoXY(Ø,5);
        Write('S(creen or P(rinter ? ');
        Read(Answer);
     until Answer in [ 'S','s','p','P' ]
  end;   {outputs}

Procedure Title;
  begin
    page(output);
     gotoXY(Ø4,1);
       Writeln('Symmetrical Square-Pad Attenuator');
```

```
        Writeln;Writeln;
        Writeln('This program will design balanced');
        Writeln('symmetrical square attenuator Pads.');
        Writeln('You must enter the desired impedance');
        Writeln('in ohms and the required voltage');
        Writeln('or current loss in decibels.');
        Writeln;
        Writeln;
      end;

Procedure Input;
  begin
    Writeln;
    Write('Enter desired impedance : ');
    Readln( Imped );
    Write('Enter desired loss in dB : ');
    Readln( dbLoss );
    Writeln;
  end;

Procedure PrintScr;
   begin
     Writeln;
     Writeln('For balanced symmetrical pads of ',Round(TempImp));
     Writeln('ohms and ',Round(dBLoss),' dB loss:');
     Writeln;
     Writeln('Series resistors Rs := ',Round(SeriesRes),' ohms');
     Writeln;
     Writeln('Shunt Resistors Rp := ',Round(ShuntRes),' ohms');
     Writeln;
     Writeln('Multiply series resistor values by 2');
     Writeln('for the unbalenced pad values');
     Writeln;
  end;

Procedure PrintPrt;
     begin
      Rewrite(out,'Printer:');
      Writeln;
      Writeln(out);
      Writeln('For balanced symmetrical pads of ',Round(TempImp));
      Writeln(out,'For balanced symmetrical pads of ',Round(TempImp));
      Writeln('ohms and ',Round(dBLoss),' dB loss:');
      Writeln(out,'ohms and ',Round(dbLoss),' dB loss:');
      Writeln;
      Writeln(out);
      Writeln('Series resistors Rs := ',Round(SeriesRes),' ohms');
      Writeln(out,'Series resistors Rs := ',Round(SeriesRes),' ohms');
      Writeln;
      Writeln(out);
      Writeln('Shunt Resistors Rp := ',Round(ShuntRes),' ohms');
      Writeln(out,'Shunt Resistors Rp := ',Round(ShuntRes),' ohms');
      Writeln;
      Writeln(out);
      Writeln('Multiply series resistor values by 2');
      Writeln('for the unbalanced pad values');
      Writeln(out,'Multiply series resistor values by 2');
      Writeln(out,'for the unbalanced pad values');
      Writeln;
      Writeln(out);
      Close(out);
     end;

  Procedure Printout;
     begin
      case answer of
            's','S' : PrintScr;
            'p','P' : PrintPrt;
       end { case }
     end;

 Procedure Work;
   begin
      TempImp  := Imped ;
      K  := exp( dBLoss * Ln(10) / 20 );
      SeriesRes  := TempImp * (Sqr(K) - 1 ) / ( 4 * K );
```

```
    ShuntRes  := TempImp * ( K + 1 ) / ( K - 1 );
end;

 begin { Main }
    Repeat
       Outputs;
       Title;
       Input;
       Work;
       Printout;
        Writeln;
              Write('Another set of data ?(Y/N) ');
                  Read(Cont);
    until Cont in [ 'N','n']
 end.
```

Symmetrical Square-Pad Attenuator

This program will design balanced
symmetrical square-pad attenuators.
You must enter the desired impedance
in ohms and the required voltage.

Enter desired impedance : 300
Enter desired loss in dB : 6

For balanced symmetrical pads of 300
ohms and 6 dB loss :

Series resistors Rs = 112 ohms.

Shunt resistors Rp = 903 ohms.

Multiply series resistor values by 2
for the unbalanced pad values.

Another set of data ? (Y/N) N

12.6 SYMMETRICAL H-PAD ATTENUATOR

The H-pad attenuator design in this program (Fig. 12.6) is similar to the one in Fig. 12.3. The only difference is that this is a symmetrical attenuator. In a symmetrical attenuator, the input and output resistances are identical.

Like the former design, this one is also used for a balanced attenuator. For an unbalanced design, multiply the series resistor by two.

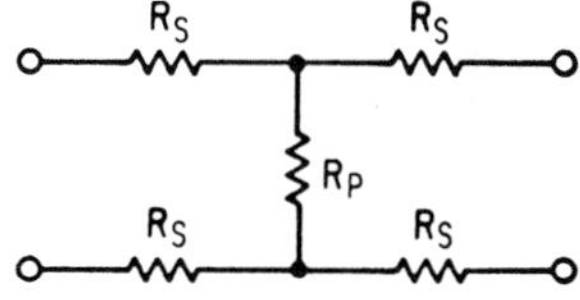

Fig. 12.6 Symmetrical H-pad attenuator

```
(*========================================*)
(*                                        *)
(* Title:  Symmetrical H-Pad Attenuator   *)
(* Program Summary: Program calculates    *)
(*         the values for all             *)
(*         resistors in a symmetrical     *)
(*         H-pad attenuator.              *)
(*                                        *)
(*========================================*)


PROGRAM SymHPad;

uses Transcend;

 VAR
  Answer,
  Cont            : Char;
  Out             : Text;
  Imped,
  dBLoss,
  TempImp,
  K,
  SeriesRes,
  ShuntRes               : Real;


Procedure Outputs;
  begin
    page(output);
     repeat
       gotoXY(Ø,5);
        Write('S(creen or P(rinter ? ');
        Read(Answer);
     until Answer in [ 'S','s','p','P' ]
  end;  {outputs}

Procedure Title;
  begin
    page(output);
     gotoXY(Ø5,1);
       Writeln('Symmetrical H-Pad Attenuator');
       Writeln;Writeln;
       Writeln('This program will design balanced');
       Writeln('symmetrical "H" attenuator Pads.');
       Writeln('You must enter the desired impedance');
       Writeln('in ohms and the required voltage');
       Writeln('or current loss in decibels.');
       Writeln;
       Writeln;
     end;

 Procedure Input;
   begin
     Writeln;
     Write('Enter desired impedance : ');
     Readln( Imped );
     Write('Enter desired loss in dB : ');
     Readln( dBLoss );
     Writeln;
   end;

 Procedure PrintScr;
    begin
      Writeln;
      Writeln('For balanced symmetrical pads of ',Round(TempImp));
      Writeln('ohms and ',Round(dBLoss),' dB loss:');
      Writeln;
      Writeln('Series resistors Rs := ',Round(SeriesRes),' ohms');
      Writeln;
      Writeln('Shunt Resistors Rp := ',Round(ShuntRes),' ohms');
      Writeln;
      Writeln('Multiply series resistor values by 2');
      Writeln('for the unbalenced pad values');
      Writeln;
   end;
```

```
Procedure PrintPrt;
    begin
     Rewrite(out,'Printer:');
     Writeln;
     Writeln(out);
     Writeln('For balanced symmetrical pads of ',Round(TempImp));
     Writeln(out,'For balanced symmetrical pads of ',Round(TempImp));
     Writeln('ohms and ',Round(dBLoss),' dB loss:');
     Writeln(out,'ohms and ',Round(dbLoss),' dB loss:');
     Writeln;
     Writeln(out);
     Writeln('Series resistors Rs := ',Round(SeriesRes),' ohms');
     Writeln(out,'Series resistors Rs := ',Round(SeriesRes),' ohms');
     Writeln;
     Writeln(out);
     Writeln('Shunt Resistors Rp := ',Round(ShuntRes),' ohms');
     Writeln(out,'Shunt Resistors Rp := ',Round(ShuntRes),' ohms');
     Writeln;
     Writeln(out);
     Writeln('Multiply series resistor values by 2');
     Writeln('for the unbalanced pad values');
     Writeln(out,'Multiply series resistor values by 2');
     Writeln(out,'for the unbalanced pad values');
     Writeln;
     Writeln(out);
     Close(out);
    end;

 Procedure Printout;
    begin
     case answer of
           's','S' : PrintScr;
           'p','P' : PrintPrt;
      end { case }
    end;

Procedure Work;
  begin
     TempImp  := Imped ;
     K  := exp( dBLoss * Ln(10) / 20 );
     SeriesRes  := ( 0.5 * TempImp ) * ( K - 1 ) / ( K + 1 );
     ShuntRes  := 2 * TempImp * K / ( Sqr( K ) - 1 );
  end;

   begin { Main }
     Repeat
       Outputs;
       Title;
       Input;
       Work;
       Printout;
        Writeln;
             Write('Another set of data ?(Y/N) ');
                Read(Cont);
     until Cont in [ 'N','n']
   end.
```

Symmetrical H-Pad Attenuator

This program will design balanced
symmetrical H-pad attenuators.
You must enter the desired impedance
in ohms and the required voltage
or current loss in decibels.

Enter desired impedance: 600

Enter desired loss in dB : 3

For balanced symmetrical pads of 600
ohms and 3 dB loss:

Series resistors Rs = 51 ohms.

Shunt resistors Rp = 1703 ohms.

Multiply series resistor values by 2
for the unbalanced pad values.

Another set of data ? (Y/N) N

12.7 SYMMETRICAL LATTICE-PAD ATTENUATOR

This symmetrical lattice-pad attenuator (Fig. 12.7) is a balanced attenuator with identical input and output resistors. In addition, the position of the X and Y resistors may be interchanged.

This attenuator is for balanced lines only. An unbalanced version cannot be formed by multiplying the series resistors by two.

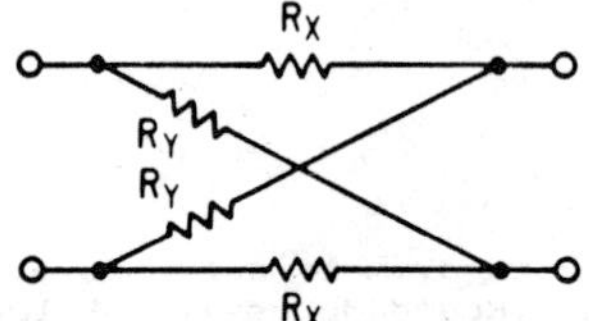

Fig. 12.7 Symmetrical lattice-pad attenuator

```
(*=========================================*)
(*                                         *)
(* Title:    Symmetrical Lattice-Pad       *)
(*           Attenuator                    *)
(* Program Summary: Program calculates     *)
(*           the values for all            *)
(*           resistors in a symmetrical    *)
(*           lattice-pad attenuator.       *)
(*                                         *)
(*=========================================*)


PROGRAM SymLattice;

uses Transcend;

 VAR
  Answer,
  Cont            : Char;
  Out             : Text;
  Imped,
  dBLoss,
  TempImp,
  K,
  ResY,
  TempR,
  ResX            : Real;
```

```
Procedure Outputs;
  begin
    page(output);
     repeat
       gotoXY(0,5);
        Write('S(creen or P(rinter ? ');
        Read(Answer);
     until Answer in [ 'S','s','p','P' ]
  end;  {outputs}

Procedure Title;
  begin
    page(output);
     gotoXY(04,1);
       Writeln('Symmetrical Lattice-Pad Attenuator');
       Writeln;Writeln;
       Writeln('This program will design balanced');
       Writeln('symmetrical lattice attenuator Pads.');
       Writeln('You must enter the desired impedance');
       Writeln('in ohms and the required voltage');
       Writeln('or current loss in decibels.');
       Writeln;
       Writeln;
     end;

Procedure Input;
  begin
    Writeln;
    Write('Enter desired impedance : ');
    Readln( Imped );
    Write('Enter desired loss in dB : ');
    Readln( dBLoss );
    Writeln;
  end;

Procedure PrintScr;
   begin
     Writeln;
     Writeln('For balanced symmetrical pads of ',Round(TempImp));
     Writeln('ohms and ',Round(dBLoss),' dB loss:');
     Writeln;
     Writeln('Lattice resistors Rx := ',Round(ResX),' ohms');
     Writeln;
     Writeln('Lattice Resistors Ry := ',Round(ResY),' ohms');
     Writeln;
  end;

Procedure PrintPrt;
    begin
     Rewrite(out,'Printer:');
     Writeln;
     Writeln(out);
     Writeln('For balanced symmetrical pads of ',Round(TempImp));
     Writeln(out,'For balanced symmetrical pads of ',Round(TempImp));
     Writeln('ohms and ',Round(dBLoss),' dB loss:');
     Writeln(out,'ohms and ',Round(dBLoss),' dB loss:');
     Writeln;
     Writeln(out);
     Writeln('Lattice resistors Rx := ',Round(ResX),' ohms');
     Writeln(out,'Lattice resistors Rx := ',Round(ResX),' ohms');
     Writeln;
     Writeln(out);
     Writeln('Lattice Resistors Ry := ',Round(ResY),' ohms');
     Writeln(out,'Lattice Resistors Ry := ',Round(ResY),' ohms');
     Writeln;
     Writeln(out);
     Close(out);
    end;

 Procedure Printout;
    begin
     case answer of
           's','S' : PrintScr;
           'p','P' : PrintPrt;
      end { case }
    end;
```

```
Procedure Work;
  begin
     TempImp  := Imped ;
     K  := exp( dBLoss * Ln(10) / 20 );
     ResX  := TempImp * ( K + 1 ) / ( K - 1 );
     TempR  := Round(( 0.5 * TempImp ) * ( K - 1 ) / ( K + 1 ));
     ResY  := ( 2 * TempR );
  end;

   begin { Main }
     Repeat
       Outputs;
       Title;
       Input;
       Work;
       Printout;
        Writeln;
              Write('Another set of data ?(Y/N) ');
                 Read(Cont);
     until Cont in [ 'N','n']
   end.
```

Symmetrical Lattice-Pad Attenuator

This program will design balanced
symmetrical lattice-pad attenuators.
You must enter the desired impedance
in ohms and the required voltage
or current loss in decibels.

Enter desired impedance : 300

Enter desired loss in dB : 3

For balanced symmetrical pads of 300
ohms and 3 dB loss:

Lattice resistors Rx = 1754 ohms.

Lattice resistors Ry = 52 ohms.

Another set of data ? (Y/N) N

12.8 SYMMETRICAL π-PAD ATTENUATOR

The symmetrical π-pad attenuator (Fig. 12.8) is for unbalanced lines. It has identical input and output impedances.

The balanced equivalent of the symmetrical π-pad attenuator is the symmetrical square-pad attenuator.

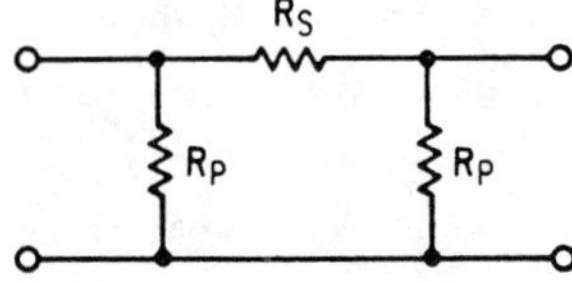

Fig. 12.8 Symmetrical π-pad attenuator

```
(*=========================================*)
(*                                         *)
(* Title:   Symmetrical Pi-Pad             *)
(*          Attenuator                     *)
(* Program Summary: Program calculates     *)
(*          the values for all             *)
(*          resistors in a symmetrical     *)
(*          Pi-pad attenuator.             *)
(*                                         *)
(*=========================================*)


PROGRAM SymPi;

uses Transcend;

 VAR
  Answer,
  Cont          : Char;
  Out           : Text;
  Imped,
  dBLoss,
  TempImp,
  K,
  SeriesRes,
  ShuntRes      : Real;

Procedure Outputs;
  begin
    page(output);
     repeat
       gotoXY(Ø,5);
        Write('S(creen or P(rinter ? ');
        Read(Answer);
     until Answer in [ 'S','s','p','P' ]
  end;  {outputs}

Procedure Title;
  begin
    page(output);
     gotoXY(Ø7,1);
       Writeln('Symmetrical Pi Attenuator');
       Writeln;Writeln;
       Writeln('This program will design balanced');
       Writeln('symmetrical Pi attenuator Pads.');
       Writeln('You must enter the desired impedance');
       Writeln('in ohms and the required voltage');
       Writeln('or current loss in decibels.');
       Writeln;
       Writeln;
     end;

Procedure Input;
  begin
    Writeln;
    Write('Enter desired impedance : ');
    Readln( Imped );
    Write('Enter desired loss in dB : ');
    Readln( dBLoss );
    Writeln;
  end;

Procedure PrintScr;
   begin
     Writeln;
     Writeln('For symmetrical pads of ',Round(TempImp));
     Writeln('ohms and ',Round(dBLoss),' dB loss:');
     Writeln;
     Writeln('Series resistors Rs := ',Round(SeriesRes),' ohms');
     Writeln;
     Writeln('Shunt Resistors Rp := ',Round(ShuntRes),' ohms');
     Writeln;
  end;

Procedure PrintPrt;
    begin
```

```
      Rewrite(out,'Printer:');
      Writeln;
      Writeln(out);
      Writeln('For symmetrical pads of ',Round(TempImp));
      Writeln(out,'For symmetrical pads of ',Round(TempImp));
      Writeln('ohms and ',Round(dBLoss),' dB loss:');
      Writeln(out,'ohms and ',Round(dBLoss),' dB loss:');
      Writeln;
      Writeln(out);
      Writeln('Series resistors Rs := ',Round(SeriesRes),' ohms');
      Writeln(out,'Series resistors Rs := ',Round(SeriesRes),' ohms');
      Writeln;
      Writeln(out);
      Writeln('Shunt Resistors Rp := ',Round(ShuntRes),' ohms');
      Writeln(out,'Shunt Resistors Rp := ',Round(ShuntRes),' ohms');
      Writeln;
      Writeln(out);
      Close(out);
     end;

 Procedure Printout;
    begin
     case answer of
           's','S' : PrintScr;
           'p','P' : PrintPrt;
      end { case }
    end;

Procedure Work;
  begin
     TempImp  := Imped ;
     K  := exp( dBLoss * Ln(10) / 20 );
     SeriesRes  := TempImp * (Sqr(K) - 1 ) / ( 2 * K );
     ShuntRes   := TempImp * ( K + 1 ) / ( K - 1 );
  end;

   begin { Main }
     Repeat
       Outputs;
       Title;
       Input;
       Work;
       Printout;
        Writeln;
             Write('Another set of data ?(Y/N) ');
                Read(Cont);
     until Cont in [ 'N','n']
   end.
```

Symmetrical Pi-Pad Attenuator

This program will design balanced
symmetrical Pi-pad attenuators.
You must enter the desired impedance
in ohms and the required voltage
or current loss in decibels.

Enter desired impedance : 300

Enter desired loss in dB : 12

For symmetrical pads of 300
ohms and 12 dB loss:

Series resistor Rs = 559 ohms.

Shunt resistor Rp = 5Ø1 ohms.

Another set of data ? (Y/N) N

12.9 SYMMETRICAL T-PAD ATTENUATOR

Another attenuator designed for unbalanced lines is the symmetrical T-pad attenuator (Fig. 12.9). It is composed of only three resistors, two series and one parallel.

The balanced equivalent of the symmetrical T-pad attenuator is the symmetrical H-pad attenuator.

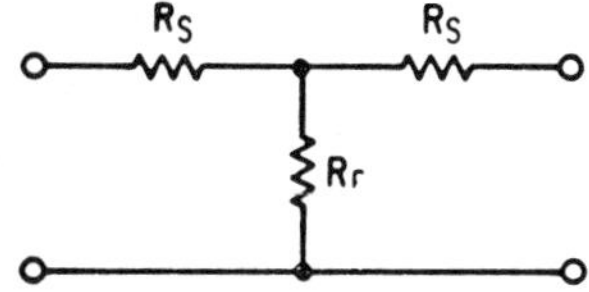

Fig. 12.9 Symmetrical T-pad attenuator

```
(*======================================*)
(*                                      *)
(* Title:  Symmetrical T-Pad Attenuator *)
(* Program Summary: Program calcuates   *)
(*         the values of all            *)
(*         resistors in a symmetrical   *)
(*         T-pad attenuator.            *)
(*                                      *)
(*======================================*)


PROGRAM SymT;

uses Transcend;

 VAR
  Answer,
  Cont            : Char;
  Out             : Text;
  Imped,
  dBLoss,
  TempImp,
  K,
  SeriesRes,
  ShuntRes        : Real;

Procedure Outputs;
  begin
    page(output);
     repeat
       gotoXY(Ø,5);
        Write('S(creen or P(rinter ? ');
        Read(Answer);
     until Answer in [ 'S','s','p','P' ]
  end;  {outputs}

Procedure Title;
  begin
    page(output);
     gotoXY(Ø7,1);
       Writeln('Symmetrical "T" Attenuator');
       Writeln;Writeln;
       Writeln('This program will design balanced');
       Writeln('symmetrical "T" attenuator Pads.');
       Writeln('You must enter the desired impedance');
       Writeln('in ohms and the required voltage');
       Writeln('or current loss in decibels.');
```

```
      Writeln;
      Writeln;
    end;

Procedure Input;
  begin
    Writeln;
    Write('Enter desired impedance : ');
    Readln( Imped );
    Write('Enter desired loss in dB : ');
    Readln( dBLoss );
    Writeln;
  end;

Procedure PrintScr;
   begin
     Writeln;
     Writeln('For symmetrical pads of ',Round(TempImp));
     Writeln('ohms and ',Round(dBLoss),' dB loss:');
     Writeln;
     Writeln('Series resistors Rs := ',Round(SeriesRes),' ohms');
     Writeln;
     Writeln('Shunt Resistors Rp := ',Round(ShuntRes),' ohms');
     Writeln;
     Writeln;
  end;

Procedure PrintPrt;
    begin
     Rewrite(out,'Printer:');
     Writeln;
     Writeln(out);
     Writeln('For symmetrical pads of ',Round(TempImp));
     Writeln(out,'For symmetrical pads of ',Round(TempImp));
     Writeln('ohms and ',Round(dBLoss),' dB loss:');
     Writeln(out,'ohms and ',Round(dBLoss),' dB loss:');
     Writeln;
     Writeln(out);
     Writeln('Series resistors Rs := ',Round(SeriesRes),' ohms');
     Writeln(out,'Series resistors Rs := ',Round(SeriesRes),' ohms');
     Writeln;
     Writeln(out);
     Writeln('Shunt Resistors Rp := ',Round(ShuntRes),' ohms');
     Writeln(out,'Shunt Resistors Rp := ',Round(ShuntRes),' ohms');
     Writeln;
     Writeln(out);
     Close(out);
    end;

 Procedure Printout;
    begin
     case answer of
           's','S' : PrintScr;
           'p','P' : PrintPrt;
      end { case }
    end;

Procedure Work;
  begin
     TempImp  := Imped ;
     K  := exp( dBLoss * Ln(10) / 20 );
     SeriesRes  := TempImp * ( K - 1 ) / ( K + 1 );
     ShuntRes  := 2 * TempImp * K / (Sqr(K) - 1 );
  end;

   begin { Main }
     Repeat
       Outputs;
       Title;
       Input;
       Work;
       Printout;
        Writeln;
             Write('Another set of data ?(Y/N) ');
                Read(Cont);
     until Cont in [ 'N','n']
   end.
```

```
Symmetrical T-Pad Attenuator

This program will design balanced
symmetrical T-pad attenuators.
You must enter the desired impedance
in ohms and the required voltage
or current loss in decibels.

Enter desired impedance : 600

Enter desired loss in dB : 6

For symmetrical pads of 600
ohms and 6 dB loss:

Series resistors Rs = 199 ohms.

Shunt resistors Rp = 803 ohms.

Another set of data ? (Y/N) N
```

12.10 UNBALANCED MINIMUM-LOSS ATTENUATOR PAD

Very often it is desirable to connect two circuits of different impedances with a minimum amount of loss. This can be done with the help of a minimum-loss attenuator pad.

This pad is designed for unbalanced lines and is composed of only two resistors (Fig. 12.10). It is also known as an L-pad.

In addition to selecting the resistor values, the program will indicate the minimum loss in dB.

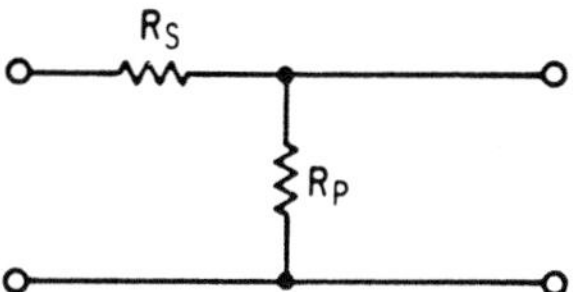

Fig. 12.10 Unbalanced minimum-loss attenuator pad

```
(*=======================================*)
(*                                       *)
(* Title:   Unbalanced Minimum-Loss      *)
(*          Attenuator Pad               *)
(* Program Summary: Program calculates   *)
(*          the values of all            *)
(*          resistors in an unbalanced   *)
(*          minimum-loss attenuator pad. *)
(*                                       *)
(*=======================================*)
```

```
PROGRAM UnBalMinLoss;

uses Transcend;

 VAR
  Answer,
  Cont            : Char;
  Out             : Text;
  InputImp,
  OutputImp,
  K,
  SeriesRes,
  ShuntRes,
  dBLoss              : Real;

Procedure Outputs;
  begin
    page(output);
     repeat
       gotoXY(Ø,5);
        Write('S(creen or P(rinter ? ');
        Read(Answer);
     until Answer in [ 'S','s','p','P' ]
  end;  {outputs}

Procedure Title;
  begin
    page(output);
     gotoXY(Ø5,1);
       Writeln('Unbalanced Minimum Loss Pad');
       Writeln;Writeln;
       Writeln('This program will design a minimum loss');
       Writeln('unbalanced matching Pad.');
       Writeln('You must enter the desired input and ');
       Writeln('output impedance.');
       Writeln;
       Writeln;
     end;

Procedure Input;
  begin
    Writeln;
    Write('Enter desired input impedance : ');
    Readln( InputImp );
    Write('Enter desired output impedance : ');
    Readln( OutputImp );
    Writeln;
  end;

Procedure PrintScr;
   begin
     Writeln;
     Writeln('For unbalanced minimum loss matching');
     Writeln('pads with a ',Round(InputImp),' ohms input resistance');
     Writeln('and a ',Round(OutputImp),' ohms output resistance');
     Writeln;
     Writeln('Series resistor Rs := ',Round(SeriesRes),' ohms');
     Writeln;
     Writeln('Shunt resistor Rp := ',Round(ShuntRes),' ohms');
     Writeln;
     Writeln('The minimum loss is ',dBLoss,' dB');
     Writeln;
  end;

Procedure PrintPrt;
    begin
     Rewrite(out,'Printer:');
     Writeln;
     Writeln(out);
     Writeln('For unbalanced minimum loss matching');
     Writeln(out,'For unbalanced minimum loss matching');
     Writeln('pads with a ',Round(InputImp),' ohms input resistance');
     Writeln(out,'pads with a ',Round(InputImp),' ohms input resistance');
     Writeln('and a ',Round(OutputImp),' ohms output resistance');
     Writeln(out,'and a ',Round(OutputImp),' ohms output resistance');
     Writeln;
```

```
      Writeln(out);
      Writeln('Series resistor Rs := ',Round(SeriesRes),' ohms');
      Writeln(out,'Series resistor Rs := ',Round(SeriesRes),' ohms');
      Writeln;
      Writeln(out);
      Writeln('Shunt resistor Rp := ',Round(ShuntRes),' ohms');
      Writeln(out,'Shunt resistor Rp := ',Round(ShuntRes),' ohms');
      Writeln;
      Writeln(out);
      Writeln('The minimum loss is ',dBLoss,' dB');
      Writeln(out,'The minimum loss is ',dBLoss,' dB');
      Writeln;
      Writeln(out);
      Close(out);
     end;

 Procedure Printout;
    begin
     case answer of
           's','S' : PrintScr;
           'p','P' : PrintPrt;
      end { case }
    end;

Procedure Work;
  begin
     If InputImp < OutputImp then
         Begin
          Writeln('Input Impedance must be larger than');
          Writeln('output impedance');
          Input;
         end;
    K  := ( InputImp + Sqrt(Sqr(InputImp) - ( InputImp * OutputImp ))) /
                             OutputImp ;
    SeriesRes  := InputImp * Sqrt( 1 - ( OutputImp / InputImp ));
    ShuntRes   := OutputImp / Sqrt( 1 - ( OutputImp / InputImp ));
    dBLoss   := Ln( K / Sqrt( InputImp / OutputImp )) / ( Ln(10) / 20 );
  end;

   begin { Main }
     Repeat
       Outputs;
       Title;
       Input;
       Work;
       Printout;
        Writeln;
             Write('Another set of data ?(Y/N) ');
                Read(Cont);
     until Cont in [ 'N','n']
   end.
```

Unbalanced Minimum-Loss Attenuator Pad

This program will design a minimum-loss
unbalanced attenuator pad.
You must enter the desired input and
output impedance.

Enter desired input impedance : 300

Enter desired output impedance : 75

For unbalanced minimum loss matching
pads with a 300-ohm input resistance
and a 75-ohm output resistance

Series resistor Rs = 26Ø ohms.

Shunt resistor Rp = 87 ohms.

The minimum loss is 1.1439ØE1 dB.

Another set of data ? (Y/N) N

12.11 BALANCED MINIMUM-LOSS ATTENUATOR PAD

It is often desirable to connect, with a minimum amount of loss, two circuits of different impedances. This can be done with the help of a minimum-loss attenuator pad.

This pad is designed for balanced lines and is composed of only three resistors (Fig. 12.11).

In addition to selecting the resistor values, the program will indicate the minimum loss in dB.

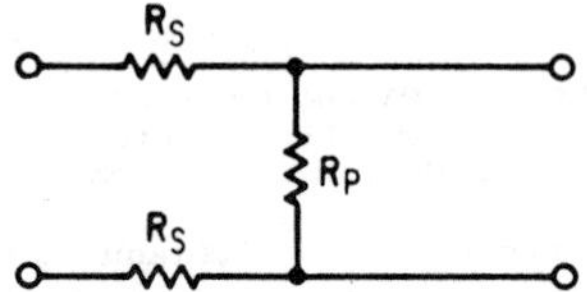

Fig. 12.11 Balanced minimum-loss attenuator pad

```
(*=========================================*)
(*                                         *)
(* Title:   Balanced Minimum-Loss          *)
(*          Attenuator Pad                 *)
(* Program Summary: Program calculates     *)
(*          the values of all              *)
(*          resistors in a balanced        *)
(*          minimum-loss attenuator pad.   *)
(*                                         *)
(*=========================================*)


PROGRAM BalMinLoss;

uses Transcend;

 VAR
  Answer,
  Cont            : Char;
  Out             : Text;
  InputImp,
  OutputImp,
  K,
  SeriesRes,
  ShuntRes,
  dBLoss          : Real;

Procedure Outputs;
  begin
    page(output);
     repeat
```

```
        gotoXY(0,5);
         Write('S(creen or P(rinter ? ');
         Read(Answer);
      until Answer in [ 'S','s','p','P' ]
   end;  {outputs}

 Procedure Title;
   begin
     page(output);
      gotoXY(06,1);
        Writeln('Balanced Minimum Loss Pad');
        Writeln;Writeln;
        Writeln('This program will design a minimum loss');
        Writeln('balanced matching Pad.');
        Writeln('You must enter the desired input and ');
        Writeln('output impedance.');
        Writeln;
        Writeln;
      end;

 Procedure Input;
   begin
     Writeln;
     Write('Enter desired input impedance : ');
     Readln( InputImp );
     Write('Enter desired output impedance : ');
     Readln( OutputImp );
     Writeln;
   end;

 Procedure PrintScr;
    begin
      Writeln;
      Writeln('For balanced minimum loss matching');
      Writeln('pads with a ',Round(InputImp),' ohms input resistance');
      Writeln('and a ',Round(OutputImp),' ohms output resistance');
      Writeln;
      Writeln('Series resistor Rs := ',Round(SeriesRes),' ohms');
      Writeln;
      Writeln('Shunt resistor Rp := ',Round(ShuntRes),' ohms');
      Writeln;
      Writeln('The minimum loss is ',dBLoss,' dB');
      Writeln;
   end;

 Procedure PrintPrt;
     begin
      Rewrite(out,'Printer:');
      Writeln;
      Writeln(out);
      Writeln('For balanced minimum loss matching');
      Writeln(out,'For balanced minimum loss matching');
      Writeln('pads with a ',Round(InputImp),' ohms input resistance');
      Writeln(out,'pads with a ',Round(InputImp),' ohms input resistance');
      Writeln('and a ',Round(OutputImp),' ohms output resistance');
      Writeln(out,'and a ',Round(OutputImp),' ohms output resistance');
      Writeln;
      Writeln(out);
      Writeln('Series resistor Rs := ',Round(SeriesRes),' ohms');
      Writeln(out,'Series resistor Rs := ',Round(SeriesRes),' ohms');
      Writeln;
      Writeln(out);
      Writeln('Shunt resistor Rp := ',Round(ShuntRes),' ohms');
      Writeln(out,'Shunt resistor Rp := ',Round(ShuntRes),' ohms');
      Writeln;
      Writeln(out);
      Writeln('The minimum loss is ',dBLoss,' dB');
      Writeln(out,'The minimum loss is ',dBLoss,' dB');
      Writeln;
      Writeln(out);
      Close(out);
    end;

 Procedure Printout;
    begin
     case answer of
```

```
          's','S' : PrintScr;
          'p','P' : PrintPrt;
      end { case }
    end;

Procedure Work;
  begin
    If InputImp < OutputImp then
        Begin
         Writeln('Input Impedance must be larger than');
         Writeln('output impedance');
         Input;
        end;
    K  := ( InputImp + Sqrt(Sqr(InputImp) - ( InputImp * OutputImp ))) /
                                         OutputImp ;
    SeriesRes  := Ø.5 * InputImp * Sqrt( 1 - ( OutputImp / InputImp ));
    ShuntRes  := OutputImp / Sqrt( 1 - ( OutputImp / InputImp ));
    dBLoss  := Ln( K / Sqrt( InputImp / OutputImp )) / ( Ln(1Ø) / 2Ø );
  end;

   begin { Main }
     Repeat
       Outputs;
       Title;
       Input;
       Work;
       Printout;
        Writeln;
            Write('Another set of data ?(Y/N) ');
               Read(Cont);
     until Cont in [ 'N','n']
   end.
```

Balanced Minimum-Loss Attenuator Pad

This program will design a minimum-loss
balanced attenuator Pad.
You must enter the desired input and
output impedance.

Enter desired input impedance : 75

Enter desired output impedance : 5Ø

For balanced minimum-loss attenuator
pads with a 75-ohm input resistance
and a 5Ø-ohm output resistance

Series resistor Rs = 22 ohms.

Shunt resistor Rp = 87 ohms.

The minimum loss is 5.71947 dB.

Another set of data ? (Y/N) N